W9-BTS-485

A SHEARWATER BOOK

HUMAN NATURES

To Peter
with best
wishes
Paul [illegible]

To Pete and Helen Bing, John and Sue Boething, Larry Condon,
John and Mary Louise Gifford, Max and Isabell Herzstein,
Stanley and Marion Herzstein, Julius Heuscher, Susan Koret,
Walter and Karen Loewenstern, Pete Myers, and
Tim and Wren Wirth—friends indeed

and

for Anne—with love.

HUMAN NATURES

Genes, Cultures, and the Human Prospect

Paul R. Ehrlich

Island Press / SHEARWATER BOOKS
Washington, D.C. • Covelo, California

A Shearwater Book
published by Island Press

Illustrations on pages 28 and 54 by Anne H. Ehrlich.

Illustrations on pages 31, 50, 51, 70, 75, 80, 83, 93, 94, 95, 97, 99, 116, 117, 133, 134, 152, 172, 241, and 292 by John and Judy Waller.

Library of Congress Cataloging-in-Publication Data

Ehrlich, Paul R.
Human natures : genes, cultures, and the human prospect / Paul R. Ehrlich.
p. cm.
Includes bibliographical references and index.
ISBN 1-55963-779-X (acid-free)
1. Human evolution. 2. Social evolution. 3. Human behavior. 4. Human beings—Animal nature. I. Title.
GN281.4 .E374 2000
599.93'8—dc21
00-010436

Printed on recycled, acid-free paper

Manufactured in the United States of America
10 9 8 7 6 5 4 3 2 1

CONTENTS

PREFACE

What is human nature? For thousands of years, philosophers have discussed and debated that question. Underlying almost all those debates, however, has been a key shared assumption: that human nature is a unitary, unchanging thing. This is the "nature" that, along with nurture, is thought to make us who we are.

As the new millennium dawns, that assumption of a single, enduring nature remains widespread, but in my view it has become a major roadblock to understanding ourselves. "Human nature" as a singular concept embodies the erroneous notion that people possess a common set of rigid, genetically specified behavioral predilections that are unlikely to be altered by circumstances. "After all," we're often told, "you can't change human nature." The notion that there is one such nature to change allows us to be painted in the popular mind as instinctively aggressive, greedy, selfish, duplicitous, sex-crazed, cruel, and generally brutish creatures with a veneer of social responsibility.[1] Our better selves are seen to be in constant battle with a universal set of unchanging, primitive "drives," which frequently break through the veneer and create many of the most serious ills that afflict humanity. It is a view as dismal as it is wrong, considering what is actually known about our behavior.

In recent decades, biological and social scientists have made great strides in developing a different view of where we come from and who we are. They have unraveled much of human evolution, those gradual alterations in our genes and cultures that have transformed us into planet-dominating animals. They have illuminated the behavioral flexibility we all possess—such as the capacity to learn one or more of thousands of different languages. And they have documented the inordinate diversity of individuals and societies in areas as different as sexual preferences and political systems. In light of this scientific progress, I want to highlight human *natures:* the diverse and evolving behaviors, beliefs, and attitudes of *Homo sapiens* and the evolved physical structures that govern, support, and participate in our unique mental functioning. Even though our bodies and behaviors share many common attributes, it's far more fruitful to consider not one human nature but many. The universals that bind people together at any point in our evolution are covered in the word *human.* The word *natures* emphasizes the differences that give us our individuality, our cultural variety, and our potential for future genetic and—especially—cultural evolution.

If we really want to know who we are and how we can solve the problems humanity faces, we must try to understand not just human natures as they are today but also their origins. Just as one can have only a partial understanding of World War II without knowing about World War I and the various interacting forces that led to it, to understand our human natures in any depth requires knowing something of our prehistory and the mechanisms of both biological and cultural evolution.

Evolutionary processes created not only us but also the rich array of plants, animals, and microbes that surround and support us. Those other life-forms were responsible for generating many of the most important features of the environments that in turn have shaped our own evolution. In this book, I consider many questions about that shaping. Why did we diverge from the other great apes? How did we acquire our upright posture, powerful brains, complex language, and extraordinary ingenuity? Why and how did our ancestors invent religion, art, agriculture, writing, and political states? In short, where did we come from, and how did we get to where we are? Science now has at least partial answers to these questions about the evolution of our natures. I believe we all need to learn our history, our evolution. Doing so will enhance our well-being, enabling us, for example, to develop more effective strategies for dealing with a broad array of social and medical problems. Furthermore, as the dominant animal, we are now modifying Earth in ways that will profoundly influence the evolutionary future of almost all other living beings—and especially the cultural evolution of future generations of our own species. If we care about our descendants and the world they will inherit, we need to develop a firm grasp of the evolutionary process.

My broad purposes in writing this book are to tell simply but accurately what scientists have learned about the answers to those "why and how" questions and to explore the implications of that knowledge for our present and future practices. Throughout, I want to show how deep are our biological and cultural roots and how an understanding of them can inform our decisions about the future. Evolution is the explanatory principle that connects all biological phenomena, including cultures, into a seamless whole; as the great geneticist Theodosius Dobzhansky put it, "Nothing in biology makes sense except in the light of evolution."[2] And human natures are, certainly, "in biology."

But I have more specific purposes than just to provide a concise, and I hope stimulating, overview of current knowledge of the evolution of human natures. First, I want to give an evolutionist's antidote to the extreme hereditary determinism that infests much of the current discussion of human behavior—the idea that we are somehow simply captives of tiny, self-copying entities called genes.[3] Second, I want to emphasize that much of our biology makes sense only when considered in a context of culture, and our culture is changing through an evolutionary process that is generally thought of as history. A third, related purpose is to call more attention to what a potent force the changes in our store of non-genetic information—our cultural evolution—have been in shaping our past and present. We need to learn how to direct that cultural process in ways more beneficial for the human future. A fourth specific purpose of the book is to explore the puzzles and problems created by differences in rates of evolution. Why did the skulls of our ancestors change at a different rate from the rest of their skeletons? What difference does it make that microbes can evolve much more rapidly than human beings? How can we speed our understanding of ways to organize a more just and sustainable society in order to catch up with our escalating technological capacity to harm one another and our life-supporting environment? Psychologist Daniel Kahneman succinctly stated the basic problem a quarter century ago: "The increase in man's power over his environment has not been accompanied by a concomitant improvement of his ability to make rational use of that power."[4]

Finally, and perhaps most important, I want to show how a greater familiarity with evolution might contribute to our resolving a particular suite of problems that has become known as the human predicament. Human activities are now undermining society's life-support systems—the systems, for instance, that maintain the quality of the atmosphere and, by controlling the cycling of critical gases and nutrients, make it possible for people to grow crops. The human predicament is causing extreme concern among scientists,[5] and an appreciation of its evolutionary roots can only increase our chances of creating a sustainable society. An understanding of the evolution of our perceptual systems makes clear one part of our difficulty in coming to grips with environmental issues—we simply didn't evolve senses capable of detecting some of the most serious problems unaided. Knowledge of that, in turn, suggests directions in which solutions might be found.

I've tried to reach these admittedly ambitious goals while also striving for brevity. For those who would like to investigate the issues in greater depth, I've provided in the endnotes, in addition to full documentation of statements made, suggestions for further reading and amplification of some topics.

The famous cynic H. L. Mencken wrote: "Once apparently the chief concern and masterpiece of the gods, the human race now begins to bear the aspect of an accidental by-product of their vast, inscrutable and probably nonsensical operations."[6] With that statement, he apparently despaired of science illuminating the evolution of human nature. That project of illumination has been a lifelong fascination of mine, and despite my admiration of Mencken, I don't share his despair. We may never reach a full understanding of our natures and the way they have evolved. But it is a goal worth striving toward; there will be more fascinating revelations along the way; and scientists already know enough to enable us to suggest practical applications in some arenas.

Attempts to characterize human natures tend to be very controversial, but the controversy is just one measure of the great interest in understanding them and the importance of our doing so. We need not shrink from debate, but we should strive to keep it civil and directed at solving human problems. As eighteenth-century Scottish philosopher David Hume said, referring to academics: "If any of the learned be inclined, from their natural temper, to haughtiness and obstinacy, a small tincture of pyrrhonism [radical skepticism][7] might abate their pride, by showing them that the few advantages which they may have obtained over their fellows are but inconsiderable if compared with the universal perplexity and confusion which is inherent in human nature. In general, there is a degree of doubt, and caution, and modesty, which, in all kinds of scrutiny and decision, ought forever to accompany a just reasoner."[8]

Chapter 1

Evolution and Us

"Among scientific theories, the theory of evolution has a special status, not only because some of its aspects are difficult to test directly and remain open to several interpretations, but also because it provides an account of the history and present state of the living world."

—François Jacob[1]

" . . . when we regard every production of nature as one which has had a history; when we contemplate every complex structure and instinct as the summing up of many contrivances, each useful to the possessor . . . how far more interesting, I speak from experience, will the study of natural history become!"

—Charles Darwin[2]

Columbine High School, Littleton, Colorado, April 20, 1999. Two young men approached their schoolmates and whipped semi-automatic weapons from under their black trench coats. The slaughter began. According to some accounts, one student was asked whether she believed in God. When she replied, "Yes," she was shot dead. An athlete was gunned down because he was black. All told, thirteen people, most of them students, were killed and twenty-three were wounded before the two gunmen killed themselves. This was only one of a dozen or so incidents of senseless mass shootings within a year in the United States. Why did it happen? Why do some kids behave so differently from others? Did the two gunmen have abusive parents? Were they cursed with "killer genes"? Had television, movie, and video game violence warped their minds?

Such carnage is not a special product of human natures in the United States. Similar murder sprees occur around the world. In 1996 in Dunblane, Scotland, a "peculiar" man used four handguns to slaughter sixteen kindergarten students and their teacher. Shortly thereafter at Port Arthur, Tasmania, a gunman killed thirty-five innocent people, ranging in age from three to seventy-two. Many people also seem ready and willing to participate in much vaster, more organized schemes of murder—to commit genocide, be it in Nazi Germany or Rwanda. And clean-cut American boys, flying bombers, incinerated hundreds of thousands of Japanese and German men, women, and children in World War II. Why do people do such things? Is it human nature? Do we share with chimps a genetically programmed propensity for violence and simply have better weapons and organization than they do? But what, then, of the majority of human beings who *don't* do such things?

Of course, there is a brighter side to our natures. Many human beings risk their lives for others, and some die in the attempt to help. At the Yad Vashem Holocaust memorial outside Jerusalem, there is an "Avenue of the Righteous Among the Nations," along which more than 6,000 trees have been planted, each to memorialize a non-Jew who helped a Jew without expectation of reward.[3] What moved those people and others like them to stand up to one of history's most horrific regimes?[4] Why did they risk their lives to save individuals whom their neighbors were willing to see carted away to their doom? Expression of genes for altruism? Simple conviction, learned in childhood, that it was the right thing to do as one human being to another?

President Jimmy Carter was appalled at having had adulterous thoughts; Bill Clinton and François Mitterrand seem to have had fewer qualms. Is Clinton just a "product of his time," the sexually liberating 1960s, and are the French just innately different? Why do some men apparently seek more sexual variety than others?

Pedro, a middle school student from a family of poor Mexican immigrants, desperately wants to go to college even though no one in his family has ever done so. He studies hard, but his school counselor discourages him. Tests indicate that his intelligence quotient (IQ) is only 98. Grace, on the other hand, a student from a well-to-do Anglo family, scored 125 on the IQ test. She seems bound for Yale University, the college her father attended. She hardly studies at all, yet she gets terrific grades. Did Grace win out in the "smart gene"[5] lottery by being born an Anglo, whereas Pedro had the bad luck to be born into the "wrong" ethnic group? Or is Pedro loaded with smart genes but deprived of an opportunity to develop his potential and held back by a meaningless score on a biased test?

As I write this, my back is aching. That's because I, like the rest of my species, stand on my "hind legs." Why isn't it our nature to run around on all fours like a proper mammal? Why do we end up with such weird posture—a posture that forced me to undergo back surgery and causes virtually all adult *Homo sapiens* to experience back pain?

Where did we get our capacity for conscious awareness and our ability to build long-term plans in our minds and then talk about those plans with other human beings? And with all that capacity for memory and foresight, why do I still eat rich chocolate desserts whenever I get the chance? I remember the articles I've read that tell me how bad saturated fat is for my circulatory system, and I can foresee a coronary in my future unless I'm careful (or lucky!). But still I have a lust for hot-fudge sundaes. A recent magazine headline suggests I'm far from alone: "Fifty Secrets to Fight Fat—No Dieting Required! Plus, Outsmart Your Family Fat Gene!"[6] Why do so many of us have irresistible cravings, and different ones at that, including addictions to much more life-threatening substances than chocolate?

I had been struggling to learn Spanish for almost a decade, starting in my fifties, when my granddaughter Jessica entered a Spanish immersion school at the age of five. A few days after she started, I tried a little Spanish on her, asking her whether she wanted some chocolate. Her response was an immediate correction: "No, Grandpa, not *chah-kah*-lah-tay—it's *choh-koh*-lah-tay." Why are children natural linguists, whereas adults generally aren't? Why is it that only we human beings, of all creatures, talk, and write to one another?

How Can We Explain Human Behavior?

When we think about our behavior as individuals, "Why?" is a question almost always on the tips of our tongues. Sometimes that question is about perceived similarities: why is almost everyone religious; why do we all seem to crave love; why do most of us like to eat meat? But our differences often seem equally or more fascinating: why did Sally get married although her sister Sue did not, why did they win and we lose, why is their nation poor and ours rich? What were the fates of our childhood friends? What kinds of careers did they have; did they marry; how many children did they have? Our everyday lives are filled with why's about differences and similarities in behavior, often unspoken, but always there. Why did one of my closest colleagues drink himself to death, whereas I, who love wine much more than he did, am managing to keep my liver in pretty good shape? Why, of two very bright applicants admitted to our department at Stanford University for graduate work, does one turn out pedestrian science and another have a spectacular career doing innovative research? Why are our natures often so different, and why are they so frequently the same?

The background needed to begin to answer all these *why*s lies within the domain of human biological and cultural evolution, in the gradual alterations in genetic and cultural information possessed by humanity. It's easy to think that evolution is just a process that sometime in the distant past produced the physical characteristics of our species but is now pretty much a matter of purely academic, and local school board, interest. Yet evolution is a powerful, ongoing force that not only has shaped the attributes and behaviors shared by all human beings but also has given every single individual a different nature.

A study of evolution does much more than show how we are connected to our roots or explain why people rule Earth—it explains why it would be wise to limit our intake of beef Wellington, stop judging people by their skin color, concern ourselves about global warming, and reconsider giving our children antibiotics at the first sign of a sore throat. Evolution also provides a framework for answering some of the most interesting questions about ourselves and our behavior.

When someone mentions evolution and behavior in the same breath, most people think immediately of the power of genes, parts of spiral-shaped molecules of a chemical called DNA. Small wonder, considering the marvelous advances in molecular genetics in recent decades. New subdisciplines such as evolutionary medicine[7] and evolutionary psychology[8] have arisen as scientists have come to recognize the importance of evolution in explaining contemporary human beings, the network of life that supports us, and our possible fates. And the mass media have been loaded with stories about real or imagined links between every conceivable sort of behavior and our genes.

Biological evolution—evolution that causes changes in our genetic endowment—has unquestionably helped shape human natures, including human behaviors, in many ways. But numerous commentators expect our genetic endowment to accomplish feats of which it is incapable. People don't have enough genes to program all the behaviors some evolutionary psychologists, for example, believe that genes control.[9] Human beings have something on the order of 100,000 genes,[10] and human brains have more than 1 *trillion* nerve cells, with about 100–1,000 trillion connections (synapses) between them.[11] That's at least 1 *billion* synapses per gene, even if each and every gene did nothing but control the production of synapses (and it doesn't). Given that ratio, it would be quite a trick for genes typically to control more than the most general aspects of human behavior.[12] Statements such as "Understanding the genetic roots of personality will help you 'find yourself' and relate better to others"[13] are, at today's level of knowledge, frankly nonsensical.

The notion that we are slaves to our genes is often combined with reliance on the idea that all problems can be solved by dissecting them into ever smaller components—the sort of reductionist approach that has been successful in much of science but is sometimes totally unscientific.[14] It's like the idea that knowing the color of every microscopic dot that makes up a picture of your mother can explain why you love her. Scientific problems have to be approached at the appropriate level of organization if there is to be a hope of solving them.

That combination of assumptions—that genes are destiny at a micro level and that reductionism leads to full understanding—is now yielding distorted views of human behavior. People think that coded into our DNA are "instructions" that control the details of individual and group behavior: that genetics dominates, heredity makes us what we are, and what we are is changeable only over many generations as the genetic endowment of human populations evolves. Such assertions presume, as I've just suggested, that evolution has produced a level of genetic control of human behavior that is against virtually all available evidence. For instance, ground squirrels have evolved a form of "altruistic" behavior—they often give an alarm call to warn a relative of approaching danger. Evidence does indicate that this behavior is rooted in their genes; indeed, it probably evolved because relatives have more identical genes than do unrelated individuals. But some would trace the "altruistic" behavior of a business executive sending a check to an agency helping famine victims in Africa, or of a devout German Lutheran aiding Jews during the Holocaust, to a genetic tendency as well. In this view, we act either to help relatives or in the expectation of reciprocity—in either case promoting the replication of "our" genes. But experimental evidence indicates that not all human altruistic behavior is self-seeking—that human beings, unlike squirrels, are not hereditarily programmed only to be selfish.[15]

Another false assumption of hereditary programming lies behind the belief that evolution has resulted in human groups of different quality.[16] Many people still claim (or secretly believe), for example, that blacks are less intelligent than whites and women less "logical" than men, even though those claims are groundless. Belief in genetic determinism has even led some observers to suggest a return to the bad old days of eugenics, of manipulating evolution to produce ostensibly more skilled people. Advocating programs for the biological "improvement of humanity"[17]—which in the past has meant encouraging the breeding of supposedly naturally superior individuals—takes us back at least to the days of Plato, more than two millennia ago, and it involves a grasp of genetics little more sophisticated than his.

Uniquely in our species, changes in culture have been fully as important in producing our natures as have changes in the hereditary information passed on by our ancestors. Culture is the nongenetic information (socially transmitted behaviors, beliefs, institutions, arts, and so on) shared and exchanged among us. Indeed, our evolution since the invention of agriculture, about 10,000 years ago, has been overwhelmingly cultural because, as we shall see, cultural evolution can be much more rapid than genetic evolution. There is an unhappy predilection, especially in the United States, not only to overrate the effect of genetic evolution on our current behavior but also to underrate that of cultural evolution. The power of culture to shape human activities can be seen immediately in the diversity of languages around the world. Although, clearly, the ability to speak languages is a result of a great deal of genetic evolution, the specific languages we speak are just as clearly products of cultural evolution. Furthermore, genetic evolution and cultural evolution are not independent. There are important "coevolutionary" interactions between them. To take just one example, our farming practices (an aspect of our culture) change our physical environment in ways that alter the evolution of our blood cells, as we shall see in the next chapter.

Not only is the evolution of our collective nongenetic information critical to creating our natures, but also the rate of that evolution varies greatly among different aspects of human culture. That, in turn, has profound consequences for our behavior and our environments. A major contemporary human problem, for instance, is that the rate of cultural evolution in science and technology has been extraordinarily high in contrast with the snail's pace of change in the social attitudes and political institutions that might channel the uses of technology in more beneficial directions.[18] No one knows exactly what sorts of societal effort might be required to substantially redress that imbalance in evolutionary rates, but it is clear to me that such an effort, if successful, could greatly brighten the human prospect.

Science has already given us pretty good clues about the reasons for the evolution of some aspects of our natures; many other aspects remain mysterious despite a small army of very bright people seeking reasons. Still other aspects (such as why I ordered duck in the restaurant last night rather than lamb) may remain unanswerable—for, as I will argue in a later chapter, human beings have a form of free will. But even to *think* reasonably about our natures and our prospects, some background in basic evolutionary theory is essential. If Grace is smarter than Pedro because of her genes, why did evolution provide her with "better" genes? If Pedro is actually smarter than Grace but has been incorrectly evaluated by an intelligence test designed for people of another culture, how did those cultural differences evolve? If I was able to choose the duck for dinner because I have free will, what exactly does that mean? How did I and other human beings evolve that capacity to make choices without being complete captives of our histories? Could I have exercised my free will to eat a cockroach curry had we been in a restaurant that served it (as some in Southeast Asia do)? Almost certainly not—the very idea nauseates me, probably because of an interaction between biological and cultural evolution.

Every attribute of *every* organism is, of course, the product of an interaction between its genetic endowment and its environment. Yes, the number of heads an individual human being possesses is specified in the genes and is the same in a vast diversity of environments.[19] And the language or languages a child speaks (but not her capacity to acquire language) is determined by her environment. But without the appropriate internal environment in the mother's body for fetal development, there would be no head (or infant) at all; and without genetically programmed physical structures in the larynx and in the developing brain, there would be no capacity to acquire and speak language. Beyond enabling us to make such statements in certain cases, however, the relative contributions of heredity and environment to various human attributes are difficult to specify. They clearly vary from attribute to attribute. So although it is informative to state that human nature is the product of genes interacting with environments (both internal and external), we usually can say little with precision about the processes that lead to interesting behaviors in adult human beings. We can't partition the responsibility for aggression, altruism, or charisma between DNA and upbringing. In many such cases, trying to separate the contributions of nature and nurture to an attribute is rather like trying to separate the contributions of length and width to the area of a rectangle, which at first glance also seems easy. When you think about it carefully, though, it proves impossible.[20]

Diverse notions of inherited superiority or inferiority and of characteristic innate group behaviors have long pervaded human societies: beliefs about the divine right of kings; "natural" attributes that made some people good material for slaves or slave masters; innate superiority of light-skinned people over dark-skinned people; genetic tendencies of Jews to be moneylenders, of Christians to be sexually inhibited, and of Asians to be more hardworking than Hispanics; and so on. Consider the following quote from a recent book titled *Living with Our Genes*, which indicates the tone even among many scientists: "The emerging science of molecular biology has made startling discoveries that show beyond a doubt that genes are the single most important factor that distinguishes one person from another. We come in large part ready-made from the factory. We accept that we *look* like our parents and other blood relatives; we have a harder time with the idea we *act* like them."[21]

In fact, the failure of many people to recognize the fundamental error in such statements (and those in other articles and books based on genetic determinism, such as Richard J. Herrnstein and Charles Murray's famous *The Bell Curve*)[22] is itself an environmental phenomenon—a product of the cultural milieu in which many of us have grown up. Genes do not shout commands to us about our behavior. At the very most, they whisper suggestions, and the nature of those whispers is shaped by our internal environments (those within and between our cells) during early development and later, and usually also by the external environments in which we mature and find ourselves as adults.

How do scientists know that we are not simply genetically programmed automata? First, biological evolution has produced what is arguably the most astonishingly adaptable device that has ever existed—the human nervous system. It's a system that can use one organ, the brain, to plan a marriage or a murder, command muscles to control the flight of a thrown rock or a space shuttle, detect the difference between a 1945 Mouton and a 1961 Latour, learn Swahili or Spanish, and interpret a pattern of colored light on a flat television screen as a three-dimensional world containing real people. It tries to do whatever task the environment seems to demand, and it usually succeeds—and because many of those demands are novel, there is no way that the brain could be preprogrammed to deal with them, even if there were genes enough to do the programming. It would be incomprehensible for evolution to program such a system with a vast number of inherited rules that would reduce its flexibility, constraining it so that it could not deal with novel environments. It would seem equally inexplicable if evolution made some subgroups of humanity less able than others to react appropriately to changing circumstances. Men and people with white skin have just as much need of being smart and flexible as do women and people with brown skin, and there is every reason to believe that evolution has made white-skinned males fully as capable as brown-skinned women.

A second type of evidence that we're not controlled by innate programs is that normal infants taken from one society and reared in another inevitably acquire the behaviors (including language) and competences of the society in which they are reared. If different behaviors in different societies were largely genetically programmed, that could not happen. That culture dominates in creating intergroup differences is also indicated by the distribution of genetic differences among human beings. The vast majority (an estimated 85 percent) is not between "races" or ethnic groups but *between individuals within groups.*[23] Human natures, again, are products of similar (but not identical) inherited endowments interacting with different physical and cultural environments.

Thus, the genetic "make-brain" program that interacts with the internal and external environments of a developing person doesn't produce a brain that can call forth only one type of, say, mating behavior—it produces a brain that can engage in any of a bewildering variety of behaviors, depending on circumstances. We see the same principle elsewhere in our development; for instance, human legs are not genetically programmed to move only at a certain speed. The inherited "make-legs" program normally produces legs that, fortunately, can operate at a wide range of speeds, depending on circumstances. Variation among individuals in the genes they received from their parents produces some differences in that range (in any normal terrestrial environment, I never could have been a four-minute miler—on the moon, maybe). Environmental variation produces some differences, too (walking a lot every day and years of acclimatization enable me to climb relatively high mountains that are beyond the range of some younger people who are less acclimatized). But no amount of training will permit any human being to leap tall buildings in a single bound, or even in two.

Similarly, inherited differences among individuals can influence the range of mental abilities we possess. Struggle as I might, my math skills will never approach those of many professional mathematicians, and I suspect that part of my incapacity can be traced to my genes. But environmental variation can shape those abilities as well. I'm also lousy at learning languages (that may be related to my math incompetence). Yet when I found myself in a professional environment in which it would have been helpful to converse in Spanish, persistent study allowed me to speak and comprehend a fair amount of the language. That was possible even long after I had passed the years during which a new language is easiest to acquire—even if I couldn't teach my tongue to pronounce *choh-koh-lah-tay* properly. But there are no genetic instructions or environmental circumstances that will allow the development of a human brain that can do a million mathematical calculations in a second. That is a talent reserved for computers, which were, of course, designed by human minds.

Are there any behavioral instructions we can be sure are engraved in human DNA? If there are, at least one should be the urge to have as many children as possible. We should have a powerful hereditary tendency to maximize our genetic contributions to future generations, for, as we'll see in the next chapter, that's the tendency that makes evolution work. Yet almost no human beings strictly obey this genetic "imperative"; environmental factors, especially cultural factors, have largely overridden it. Most people choose to make smaller genetic contributions to the future—that is, have fewer children—than they could, thus figuratively thwarting the supposed maximum reproduction "ambitions" of their genes. If genes run us as machines for reproducing themselves, how come they let us practice contraception? We are the only animals that deliberately and with planning enjoy sex while avoiding reproduction.[24] We can and do "outwit" our genes—which are, of course, witless. In this respect, our hereditary endowment made a big mistake by "choosing" to encourage human reproduction not through a desire for lots of children but through a desire for lots of sexual pleasure.

There are environments (sociocultural environments in this case) in which near-maximal human reproduction has apparently occurred. For example, the Hutterites, members of a Mennonite sect living on the plains of western North America, are famous for their high rate of population growth. Around 1950, Hutterite women over the age of forty-five had borne an average of ten children, and Hutterite population growth rates exceeded 4 percent per year.[25] Interestingly, however, when social conditions changed, the growth rate dropped from an estimated 4.12 percent per year to 2.91 percent.[26] Cultural evolution won out against those selfish little genes.

Against this background of how human beings can overwhelm genetic evolution with cultural evolution, it becomes evident that great care must be taken in extrapolating the behavior of other animals to that of human beings. One cannot assume, for example, that because marauding chimpanzees of one group sometimes kill members of another group, selection has programmed warfare into the genes of human beings (or, for that matter, of chimps). And although both chimp and human genetic endowments clearly can interact with certain environments to produce individuals capable of mayhem, they just as clearly can interact with other environments to produce individuals who are not aggressive. Observing the behavior of nonhuman mammals—their mating habits, modes of communication, intergroup conflicts, and so on—can reveal patterns we display in common with them, but those patterns certainly will not tell us which complex behaviors are "programmed" inalterably into our genes. Genetic instructions are of great importance to our natures, but they are not destiny.

Chang and Eng—Nurture and Nature

There are obviously limits to how much the environment ordinarily can affect individual characteristics. No known environment, for example, could have allowed me to mature with normal color vision: like about 8 percent of males, I'm color-blind—the result of a gene inherited from my mother. But the influence on many human attributes of even small environmental differences should not be underestimated. Consider the classic story of the "Siamese twins" Chang and Eng. Born in Siam (now Thailand) on May 11, 1811, these identical twins were joined at the base of their chests by an arm-like tube that in adulthood was five or six inches long and about eight inches in circumference.[27] They eventually ended up in the United States, became prosperous as a sideshow attraction, and married sisters. Chang and Eng farmed for a time, owned slaves before the Civil War, and produced both many children and vast speculation about the circumstances of their copulations. They were examined many times by surgeons who, working before the age of X-rays, concluded that it would be dangerous to try to separate them.

From our perspective, the most interesting thing about the twins was their different natures. Chang was slightly shorter than Eng, but he dominated his brother and was quick-tempered. Eng, in contrast, was agreeable and usually submissive. Although the two were very similar in many respects, in childhood their differences once flared into a fistfight, and as adults on one occasion they disagreed enough politically to vote for opposing candidates. More seriously, Chang drank to excess and Eng did not. Partly as a result of Chang's drinking, they developed considerable ill will that made it difficult for them to live together—they were constantly quarreling. In old age, Chang became hard of hearing in both ears, but Eng became deaf only in the ear closer to Chang. In the summer of 1870, Chang suffered a stroke, which left Eng unaffected directly but bound him physically to an invalid. On January 17, 1874, Chang died in the night. When Eng discovered his twin's death, he (although perfectly healthy) became terrified, lapsed into a stupor, and died two hours later, before a scheduled surgical attempt was to have been made to separate the two. An autopsy showed that the surgeons had been correct—the twins probably would not have survived an attempt to separate them.

Chang and Eng demonstrated conclusively that genetic identity does not necessarily produce identical natures, even when combined with substantially identical environments—in this case only inches apart, with no sign that their mother or others treated them differently as they grew up. Quite subtle environmental differences, perhaps initiated by different positions in the womb, can sometimes produce substantially different behavioral outcomes in twins. In this case, in which the dominant feature of each twin's environment clearly was the other twin, the slightest original difference could have led to an escalating reinforcement of differences.

The nature–nurture dichotomy, which has dominated discussions of behavior for decades, is largely a false one—all characteristics of all organisms are truly a result of the simultaneous influences of both.[28] Genes do not dictate destiny in most cases (exceptions include those serious genetic defects that at present cannot be remedied), but they often define a range of possibilities in a given environment.[29] The genetic endowment of a chimpanzee, even if raised as the child of a Harvard professor, would prevent it from learning to discuss philosophy or solve differential equations. Similarly, environments define a range of developmental possibilities for a given set of genes. There is no genetic endowment that a child could get from Mom and Pop that would permit the youngster to grow into an Einstein (or a Mozart or a García Marquez—or even a Hitler) as a member of an isolated rain-forest tribe without a written language.

Attempts to dichotomize nature and nurture almost always end in failure.

Although I've written about how the expression of genes depends on the environment in which the genes are expressed, another way of looking at the development of a person's nature would have been to examine the contributions of three factors: genes, environment, *and* gene–environment interactions.[30] It is very difficult to tease out these contributions, however. Even under experimental conditions, where it is possible to say something mathematically about the comparative contributions of heredity and environment, it can't be done completely because there is an "interaction term." That term cannot be decomposed into nature or nurture because the effect of each depends on the contribution of the other.

To construct an artificial example, suppose there were a gene combination that controlled the level of a hormone that tended to make boys aggressive. Further, suppose that watching television also tended to make boys aggressive.[31] Changing an individual's complement of genes so that the hormone level was doubled and also doubling the television-watching time might, then, quadruple some measure of aggressiveness. Or, instead, the two factors might interact synergistically and cause the aggression level to increase fivefold (perhaps television is an especially potent factor when the viewer has a high hormone level). Or the interaction might go the other way—television time might increase aggression only in those with a relatively low hormone level, and doubling both the hormone level and the television time might result in only a doubling of aggression. Or perhaps changing the average *content* of television programming might actually reduce the level of aggressiveness so that even with hormone level and television time doubled, aggressiveness would decline. Finally, suppose that, in addition, these relationships depended in part on whether or not a boy had attentive and loving parents who provided alternative interpretations of what was seen on television. In such situations, there is no way to make a precise statement about the contributions of "the environment" (television, in this case) to aggressiveness. This example reflects the complexity of relationships that has been demonstrated in detailed studies of the ways in which hormones such as testosterone interact with environmental factors to produce aggressive behavior.[32]

The best one can ordinarily do in measuring what genes contribute to attributes (such as aggressiveness, height, or IQ test score) is calculate a statistical measure known as heritability. That statistic tells how much, on average, offspring resemble their parents in a particular attribute *in a particular set of environments.*[33] Heritability, however, is a measure that is difficult to make and difficult to interpret. That is especially true in determining heritability of human traits, where it would be unethical or impossible to create the conditions required to estimate it, such as random mating within a population.[34]

Despite these difficulties, geneticists are gradually sorting out some of the ways genes and environments can interact in experimental environments[35] and how different parts of the hereditary endowment interact in making their contribution to the development of the individual. One of the key things they are learning, as will become clear in the next chapter when we go to the Animal House, is that it is often very difficult for genetic evolution to change just one characteristic. That's worth thinking about the next time someone tells you that human beings have been programmed by natural selection to be violent, greedy, altruistic, or promiscuous, to prefer certain facial features, or to show male (or white) dominance. At best, such programming is difficult; often, as we will see, it is impossible.

The Nature of Human Nature

Today's debates about human nature—about such things as the origins of ethics; the meanings of consciousness, self, and reality; whether we're driven by emotion or reason; the relationship between thought and language; whether men are naturally aggressive and women peaceful; and the role of sex in society—trace far back in Western thought. They have engaged thinkers from the pre-Socratic philosophers, Plato, and Aristotle to René Descartes, John Locke, Georg Wilhelm Friedrich Hegel, Charles Sanders Peirce, and Ludwig Wittgenstein, just to mention a tiny handful of those in the Western tradition alone.[36]

What exactly *is* this human nature we hear so much about? The prevailing notion, as I mentioned in the preface, is that it is a single, fixed, inherited attribute—a common property of all members of our species. That notion is implicit in the universal use of the term in singular form.[37] And I think that singular usage leads us astray. To give a rough analogy, *human nature* is to *human natures* as *canyon* is to *canyons*. We would never discuss the "characteristics of canyon." Although all canyons share certain attributes, we always use the plural form of the word when talking about them in general. That's because even though all canyons have more characteristics in common with one another than any canyon has with a painting or a snowflake, we automatically recognize the vast diversity subsumed within the category *canyons*. As with *canyon*, at times there is reason to speak of human nature in the singular, as I sometimes do in the pages that follow when referring to what we all share—for example, the ability to communicate in language, the possession of a rich culture, and the capacity to develop complex ethical systems. After all, there are at least *near*-universal aspects of our natures and our genomes (genetic endowments), and the variation within them is small in relation to the differences between, say, human and chimpanzee natures or human and chimpanzee genomes.

In the pages that follow I argue, contrary to the prevailing notion (and as previewed in the preface), that human nature is not the same from society to society or from individual to individual, nor is it a permanent attribute of *Homo sapiens*.[38] Human nature*s* are the behaviors, beliefs, and attitudes of *Homo sapiens* and the changing physical structures that govern, support, and participate in our unique mental functioning. I will emphasize that there are many such natures, a diversity generated especially by the overwhelming power of cultural evolution—the super-rapid kind of evolution in which our species excels. The human nature of a Chinese man living in Beijing is somewhat different from the human nature of a Parisian woman; the nature of a great musician is not identical with that of a fine soccer player; the nature of an inner-city gang member is different from the nature of a child being raised in an affluent suburb; the nature of someone who habitually votes Republican is different from that of her identical twin who is a Democrat; and my human nature, despite many shared features, is different from yours.

The differences among individuals and groups of human beings are, as already noted, of a magnitude that dwarfs the differences within any other non-domesticated animal species.[39] Using the plural, *human natures*, puts a needed emphasis on that critical diversity, which, after all, is very often what we want to understand. We want to know why two genetically identical individuals would have different political views; why Jeff is so loud and Barbara is so quiet; why people in the same society have different sexual habits and different ethical standards; why some past civilizations flourished for many centuries and others perished; why Germany was a combatant in two horrendous twentieth-century wars and Switzerland was not; why Julia is concerned about global warming and Juliette doesn't know what it is. There is no single human nature, any more than there is a single human genome, although there are features common to all human natures and all human genomes.

But if we are trying to understand anything about human society, past or present, or about individual actions, we must go to a finer level of analysis and consider human nature*s* as actually formed in the world. It is intellectually lazy and incorrect to "explain" the relatively poor school performance of blacks in the United States, or the persistence of warfare, or marital discord, by claiming that nonwhites are "naturally" inferior, that all people are "naturally" aggressive, or that men are "naturally" promiscuous. Intellectual performance, aggression, and promiscuity, aside from being difficult to define and measure, all vary from individual to individual and often from culture to culture. Ignoring that variance simply hides the causative factors—cultural, genetic, or both—that we would like to understand.

Permanence is often viewed as human nature's key feature; after all, remember, "you can't change human nature." But, of course, we *can*—and we do, all the time. The natures of Americans today are very different from their natures in 1940. Indeed, today's human natures everywhere are diverse products of change, of long genetic and, especially, cultural evolutionary processes. A million years ago, as paleoanthropologists, archaeologists, and other scientists have shown, human nature was a radically different, and presumably much more uniform, attribute. People then had less nimble brains, they didn't have a language with fully developed syntax, they had not developed formal strata in societies, and they hadn't yet learned to attach worked stones to wooden shafts to make hammers and arrows.

Human natures a million years in the future will also be unimaginably different from human natures today. The processes that changed those early people into modern human beings will continue as long as there are people. Indeed, with the rate of cultural evolution showing seemingly continuous acceleration, it would be amazing if the broadly shared aspects of human natures were not quite different even a million *hours* (about a hundred years) in the future. For example, think of how Internet commerce has changed in the past million or so minutes (roughly two years).

As evolving mental–physical packages, human natures have brought not only planetary dominance to our species[40] but also great triumphs in areas such as art, music, literature, philosophy, science, and technology. Unhappily, though, those same packages—human behavioral patterns and their physical foundations—are also the source of our most serious current problems. War, genocide, commerce in drugs, racial and religious prejudice, extreme economic inequality, and destruction of society's life-support systems are all products of today's human natures, too. As Pogo so accurately said, "We have met the enemy, and they is us."[41] But nowhere is it written that those problems have to be products of tomorrow's human natures. It is theoretically possible to make peace with ourselves and with our environment, overcome racial and religious prejudice, reduce large-scale cruelty, and increase economic equality. What's needed is a widespread understanding of the evolutionary processes that have produced our natures, open discourse on what is desirable about them, and conscious collective efforts to steer the cultural evolution of the more troublesome features of our natures in ways almost everyone would find desirable. A utopian notion? Maybe. But considering the progress that already has been made in areas such as democratic governance and individual freedom, race relations, religious tolerance, women's and gay rights, and avoidance of global conflict, it's worth a try.

Chapter 2

TALES FROM THE ANIMAL HOUSE

> "Can we doubt (remembering that many more individuals are born than can possibly survive) that individuals having any advantage, however slight, over others, would have the best chance of surviving and of procreating their kind? On the other hand, we may feel sure that any variation in the least degree injurious would be rigidly destroyed. This preservation of favourable variations and the rejection of injurious variations, I call Natural Selection."
>
> —Charles Darwin[1]

The Grand Canyon is an awe-inspiring place. Once you've been an insignificant speck lost in its incredible depths and felt the pressure of the Colorado River on your oars, they say that you're "forever above Lava"—one of the many gut-wrenching rapids of the river. It took government geologist John Wesley Powell and his party of explorers three months in 1869 to make the first trip down the river in four tiny wooden rowboats; some friends and I did it in rubber rafts in less than two weeks and in comparative comfort. Still, that gave us a lot of time to contemplate the canyon's great beauty and vast scale—ranging from 4 to 18 miles wide, 280 miles long, and sometimes more than a mile deep.[2] To me, the canyon's most mind-boggling quality is not its wealth of ever-changing colors—buffs, greens, violets, grays, pinks, chocolate browns—and spectacularly sculpted forms but the fact that the canyon's complexity and grandeur was produced, counterintuitively, by the action of tender water on tough rock. The Grand Canyon is a product of geological evolution taking place over an unimaginable stretch of time; the Colorado River's simple and persistent erosive process took about 20 million years, 20,000 millennia, to carve that wonder.

The secret of biological evolution also is vast amounts of time—time for a simple and persistent process to produce results that often *seem* miraculous to an organism that evolved a life span of some seventy years. As complex as the Grand Canyon seems and as counterintuitive its production, it pales in comparison with our own creation. Arizona's majestic ditch was produced in *less than one one-hundredth* of the 4 million millennia it took biological and cultural evolution to produce our natures.[3] And the basic reason why this generation of complexity from simplicity seems counterintuitive to us is, ironically, part of our evolved human natures. There was no reason for evolution to endow us with an intuitive grasp of time on scales of tens of millions or billions of years, and it hasn't.

The basic explanation of evolution, our own and that of every other organism, traces to one of the most influential books ever written, Charles Darwin's *On the Origin of Species*,[4] published in 1859. The English biologist actually did not originate the idea that living beings were modified descendants of prior beings, but he proposed a mechanism, natural selection, to explain how evolution could occur. Perhaps most important, he supported his hypothesis with a massive body of evidence. Darwin's ideas about evolution were so powerful that a century and a half later, the entire field of evolutionary biology—a discipline engaging thousands of biologists—is still referred to as Darwinism. Today, the central mechanism of evolutionary theory remains Darwin's natural selection.[5] It is with that mechanism that we begin our journey to understanding how the first primitive, extraordinarily simple organisms that evolved in the oceans of our planet[6] could, in some 4 billion years, have been transformed into wonderfully complicated creatures with brains capable of understanding their own origins and with a stunning diversity of natures.

Selection: Natural and Unnatural

My own hands-on introduction to selection came in the Animal House, a ramshackle house that had been converted into a makeshift laboratory. It was located on the campus of the University of Kansas, where I had gone in the fall of 1953 to earn my doctorate under Charles Michener, an outstanding evolutionist. The Animal House stank and sported cockroaches the size of fox terriers—my fellow graduate students said that if they charged, you should try to shoot them between the eyes. The smell came from a colony of beetles that the zoology department of the university kept adjacent to the Animal House. The job of the beetles was to clean the flesh off the bones of animals destined to be skeletal displays for the zoology museum. At the time, they were rumored to have been fed a dead hippopotamus fresh from a zoo and had not proved quite up to the job.

It was strange enough for a young easterner whose primary interest was in women to be transported to Kansas, where the social scene was rather different from that in Philadelphia, where I'd done my undergraduate work. It was stranger still, for someone whose main interest aside from the opposite sex was butterflies, to sit at a microscope at midnight determining the sex of fruit flies that had been knocked unconscious with ether. I couldn't even smoke to mask the smell of rotting flesh; because ether is highly explosive, a spark could have been lethal. But eventually I grew accustomed to the stench. I learned to mix the two kinds of substances (media, as they are known in science) we needed to culture fruit flies—one on which the adult flies would lay their eggs and one to grow the yeasts that the flies' larvae (maggots) ate. I carefully recorded what happened to the flies we raised in pint milk bottles and myriad glass vials and analyzed data using what now are antique mechanical adding machines. Despite the tedium that science often involves, the excitement that laboratory research can generate—even working on small insects with the latinized (or "scientific") name of *Drosophila melanogaster*—began to sneak into my brain alongside my late-evening plans. Fruit flies, it turns out, have their charms, which may be why they became the organism whose doings in microcosms of milk bottles and vials provided much of the basic foundation of the science of genetics.

Michener had obtained a graduate assistantship for me with Robert Sokal, now one of the world's most distinguished senior scientists but then a Young Turk and assistant professor of entomology (the study of insects). Sokal had a grant from the Office of Naval Research to study the evolution of resistance to the pesticide dichlorodiphenyltrichloroethane (DDT) in fruit flies. In the early 1950s, insecticide resistance had just begun to be recognized as a big problem, and fruit flies, then as now, were important tools for understanding evolutionary processes. *Drosophila* are easily raised in the laboratory, and they grow rapidly—a mere dozen days or so are required for them to complete a generation—which makes them ideal for study because evolution operates not on clock time but on generation time.[7] As a powerful insect poison, DDT had saved countless human lives from the scourge of typhus at the end of World War II (when dusted under the clothes of troops and refugees, DDT killed the lice that transmitted the disease) and had effectively controlled populations of other noxious insects. But DDT was already showing signs of losing its miraculous insect-killing power. Many bugs could by then survive and reproduce in its presence, and our lab was one of several trying to expand knowledge of how and why.

We raised the fly maggots on a gooey medium containing enough DDT to kill most of them. We used as parents in each generation the few flies who as maggots had survived the exposure to DDT, formed a pupa (the resting stage in which the maggot is transformed into a winged adult fly—analogous to the chrysalis of a butterfly or the stage of a moth inside the cocoon), and produced an adult capable of breeding.[8] For succeeding generations, we mixed up media with higher and higher concentrations of the insecticide. After ten generations or so, flies had evolved to be so resistant to the poison that they could almost use the DDT solution as an aperitif. They thrived in the presence of levels of poison far higher than those that would wipe out the flies that we simply raised generation after generation on the medium without DDT—our "control" strain. Unnatural selection, so to speak, was at work.[9] We had stood in for nature and used selection to cause the evolution of a strain of DDT-resistant fruit flies to take place right before our eyes, in a matter of months. "Unnatural" our engineering may have been, but the microcosmic process we were observing was simply the laboratory equivalent of natural processes that have changed the characteristics of checkerspot butterflies, daisy-like plants, bacteria that live in the human gut, and mountain lions—indeed, of all living organisms, including human beings.

Selection, unnatural or natural, is the differential contribution of offspring to the next generation by individuals that have different genetic endowments.[10] In our laboratory, fruit flies that happened to have some innate resistance to the poison had been more likely to live and reproduce than had those lacking such a built-in advantage. Because it's tough to breed when you're dead, that difference produced the requisite differential in rates of reproduction we call selection. In any sexually reproducing population, some individuals will be genetically different from others. Some pigs will have genes that make them a little fatter than other pigs that lack those genes. Some poppies will have genes that predispose them to grow taller than other poppies. Some people will have genes that give them blue eyes; others will have genes for brown eyes. Often, some genetic types will have more offspring than others. If that happens—if some genetic kinds of pigs, poppies, people, or what-have-you outreproduce other kinds—natural selection has occurred, and the genes carried by those types will be more prevalent in the following generation.[11] The key point here is that some of the fruit flies in our Animal House experiments had genes that made them slightly more resistant to the DDT, and their offspring survived to become adult flies that were able to reproduce. They reproduced more, on average, than did those flies whose genes made them more susceptible to DDT. Selection took place.

In the language of evolutionary biology, those DDT-resistant flies that lived to lay eggs and yield a new generation are said to have been more fit (or to have had higher fitness) than those that succumbed to the DDT. Fitness in evolutionists' jargon, then, does not necessarily refer to any physical characteristic, such as keen sight in a fruit fly or great flight ability. It is purely a measure of relative reproductive contribution.[12] An ugly ninety-pound human weakling who fathers a dozen children is much more fit in the evolutionary sense than a tall, handsome, muscular man who is childless.[13]

Another critical aspect of the evolutionary process was vividly demonstrated in Sokal's Animal House lab. It has long been clear to biologists that selection operating on one trait normally will also change others—a fact that too frequently is lost on nongeneticists writing about human evolution. Fruit flies normally scatter their pupae at random over the surface of the gooey medium on which the maggots grow. The Animal House fruit fly populations that were selected for DDT resistance, however, behaved differently. As in the normal strains, the maggots crawled all over the surface of the nutritional goo that we had poured into the vials. But increasingly, they tended to form their pupae just around the edges of the medium. We had selected just for DDT resistance and got a different behavioral pattern in the bargain.

Something more striking occurred if we started with normal flies and created no selection pressure for DDT resistance (that is, did not mix DDT into the medium) but instead selected those individuals that pupated near the edge of the goo to be the parents of each generation. In this case, we got edge-pupating strains, as one might expect, but those strains, when tested, also proved to be resistant to DDT.[14] In other words, when we selected for DDT resistance, we got strains of flies that also pupated at the edges of the medium; when we selected for edge-pupation behavior, we got strains that also were DDT resistant.[15]

What we found in fruit flies has been found over and over again in selection experiments using many different organisms: *it is usually very difficult to select for just one characteristic.* That suggests, among other things, just how unlikely it is that many human behavioral traits claimed to be the direct result of natural selection actually are. Suppose, for example, that a gene occurred in a population of our ancestors that produced a tendency to prefer mates with small noses. Selection might never increase the frequency of such a gene if bigger noses would allow better detection of both food and enemies and smaller noses would thus expose individuals to risks that greatly outweighed the sexual benefits of small noses. New genes most often don't produce more evolutionarily fit individuals, and selection frequently does not operate to make a characteristic better serve an organism's interests because of genetic (or other) constraints. Our bad backs and the weak, easily broken legs of horses selected to run fast in races are monuments to the difficulties of selecting just for strong backs and high speeds.

There are two reasons why selection may affect characteristics that are not direct causes of differences in reproduction. First, the same genes frequently (perhaps almost always—we can't be sure) influence more than one trait (e.g., an enzyme that helps a fruit fly break down DDT may also influence its choice of pupation site).[16] Second, genes that are physically close together on chromosomes—the filaments in the cell nucleus that carry the genes—tend to be transmitted together (that is, they show linkage); thus, selection that favors a trait associated with one gene will often carry along the trait or traits associated with another gene.

Behind the explanations just given are the fine-scale workings of the genes themselves. Genes, the physical embodiments of genetic information that ultimately help to generate traits such as DDT resistance and choice of pupation site, are segments of molecules of deoxyribonucleic acid (DNA). The gene-bearing DNA molecules in the chromosomes of animals and plants[17] have the famed "double helix" structure: they are like twisted chemical ladders with identical chemical "uprights" connected by four different kinds of "rungs." It is the sequence of those rungs (chemically described as base pairs)[18] that encodes the information. A gene consists of a DNA segment—a series of rungs—that determines the sequence of the amino acid subunits that make up a single protein.[19] Earlier, when I discussed natural selection in fruit flies, I was referring to differentials in reproduction among individual flies. Those were actually traceable to differences in the sequence of DNA rungs that code for the production of different proteins in the flies' cells. Some of these proteins may have been enzymes that destroyed DDT faster than other enzymes. In our own distant ancestors, natural selection for upright posture or for eyes that form detailed images occurred in a process basically similar to what took place in my selection experiments with fruit flies, but it went on for many, many more generations.

Shortly, we'll want to look at what an understanding of genetics can tell us about our own behavior, and there is a first thing to remember about differences in the ways individual human beings (and other organisms) act and appear. Ordinarily, such differences do not correspond directly to differences in the genetic endowment of those individuals. To deal with this lack of correspondence, biologists distinguish between phenotypes and genotypes. Phenotypes are the observable characteristics of individuals. Fruit flies that are resistant to DDT and fruit flies that are susceptible to DDT are said to have different phenotypes. So do flies that pupate near the periphery of the medium and flies that pupate far from the periphery, people with blue eyes and people with brown eyes, and scientists who are mediocre and scientists who are brilliant. Many phenotypic differences are produced entirely by environmental factors and do not reflect any underlying genetic variation—an example would be a difference in skin color in identical twins (individuals with the same genetic endowment) when one has been sunning in Savannah and the other freezing in Fargo.

In contrast to phenotypic differences are differences in the unseen complement of genes of two individuals—differences in their genotypes.[20] It turns out that some such differences in DNA are not reflected by differences in phenotypes. That is, two distinct genotypes may produce indistinguishable phenotypes—for example, two different genotypes produce people with brown eyes. That's why some brown-eyed couples (those in which both individuals carry genes for blue eyes) can have blue-eyed children and other brown-eyed couples (those in which at most one individual has a gene for blue eyes) can't—it's because of differences in the parental genotypes that are not expressed in their phenotypes.[21] Similarly, the same genotype, given different environments of development, can lead to quite different expressions—like the tanned Savannah and pale Fargo twins. Aggressive and submissive are phenotypes; tall and short are phenotypes; smoker and nonsmoker are phenotypes.[22] The genotypes of those phenotypes are any parts of the genetic endowment, or genome, responsible for interacting with the environment to produce the differences. But I emphasize that many phenotypic differences will not reflect genotypic differences.

Genes and Environments

For individuals to have different genetic endowments so that selection can operate, there must be variation in genotypes from individual to individual—which means that there must be different kinds of genes. The ultimate source of variation in the DNA—that is, the creation of different kinds of genes—is mutation: the accidental alteration of DNA that changes genes.[23] Accidents happen—an X-ray knocks out a rung or a cell's delicate micromachinery inserts the wrong partner across a rung when DNA is duplicated in the process of reproduction. Mutant genes often interact with the environment to cause changes in observable qualities of individuals, although many mutations have no phenotypic effect.[24] In short, genetic variation has its basic source in mutation. And normally, there is plenty of such variation.[25]

A key axiom of modern evolutionary theory is that mutations do not occur in response to the needs of the organism. If aggressive males make more money in the stock market than do submissive males, that does not mean that in the latter, "genes for submissiveness" will tend to mutate into "genes for aggression." Put another way, mutations are not called forth by environmental factors; if a mutation improves the functioning of the individual, it is simply by chance. When smoking was added to the environment of Europeans, for example, their DNA didn't mutate to produce genes that would make lungs resistant to hot smoke and cancer-causing chemicals from the burning of tobacco. Related to the first axiom, that environmental need does not direct mutation, is a second axiom of modern evolutionary theory: that "acquired characters"—those, like suntans, developed through experience and not inheritance—are not themselves inherited. The idea that evolution occurred through the inheritance of characteristics that individuals acquired during their lives was once popular. It is generally referred to as Lamarckism, named after eighteenth-century evolutionist Jean-Baptiste de Lamarck. Lamarck's ideas were pioneering for his time, and he was more visionary than he is now often given credit for.[26] Unhappily for Lamarck and happily or unhappily for us, life experiences cannot be passed on genetically to our children, reasonable as the idea once seemed. Building up your muscles by vigorous exercise won't make the child you have later any more brawny.

Because mutations are random relative to need and because organisms generally fit well into their environments, mutations normally are either neutral or harmful; only very rarely are they helpful—just as a random change made by poking a screwdriver into the guts of your computer will rarely improve its performance.[27] A proto-giraffe stretching its neck to reach leaves high on an acacia tree would not have caused genetic mutations that produced longer necks when passed on to its offspring.[28] But in a population of short giraffes living on a savanna with tall trees, an occasional mutation that led to slightly longer necks or legs apparently gave those longer-necked or longer-legged phenotypes a slight reproductive advantage. They presumably got a little more of the foliage to eat than did their shorter cohorts and, on average, had slightly more young. Over hundreds of thousands of generations, an accumulation of such altered DNA produced those wonderful, graceful beasts we can still see wandering the African plains.

In addition to mutation, a second cause of genetic differences among individuals is recombination, a reshuffling of genes that occurs during the process of sexual reproduction.[29] In organisms that reproduce sexually, most cells have two sets of chromosomes, one received from the father and one from the mother. But during the specialized type of cell division that precedes fertilization—as cells in males divide to produce sperm and cells in females divide to make eggs—the chromosome number is reduced by half: each sperm and each egg contains only a single chromosome set. If that didn't happen, when the sperm fertilized the egg, adding the male's genes to those of the female, the number of chromosomes would double—and a doubling of chromosomes in each generation would soon produce so many chromosomes that the cell (indeed, the entire world) could not contain them.[30] In this process of cell division that halves the chromosome number, the paternal and maternal chromosomes carried by both males and females are mixed so that the resulting sperm and eggs contain chromosomes from both parents of the male and chromosomes from both parents of the female. As the chromosome number is halved, DNA is transferred physically from chromosome to chromosome, increasing genetic mixing by making new combinations of genes on the chromosomes of the sperm and egg. That mixing process is recombination.[31]

Mutation and recombination guarantee that virtually every individual will have a unique genotype; even cells within the same individual (and those of identical twins) will often differ slightly because of new mutations. The influence of those genetic differences is often reflected in our different natures. Moreover, we all develop in different environments, which also shape our natures. This, too, often is ignored in popular accounts that emphasize differences only in genes, not environments—as one can see in the frequent news stories of the "scientists discover gene for cancer" or "homosexuality is in the genes" genre. But that view of genetics is too simple to provide a basis for clear thinking about evolutionary processes. The genes that make up genotypes interact with environments (during development[32] from fertilized egg to adult) to produce the phenotypes of adult organisms. The environment is fully as important as the DNA sequence: as I said earlier, exactly the same genotype can lead to quite different phenotypes in different environments.[33] Raise one member of a pair of identical twins in an environment in which there is barely enough food to stay alive and give her no education, and raise the other in a food-rich environment and send her to Stanford University, and you'll get individuals with very different phenotypes. If one of a pair of identical twins is blinded very early in life and his sight is later restored, he will never learn to see properly like his brother.

But does the environment actually count for much? Aren't the genes really in charge? Genes are sometimes described as "selfish," self-replicating (self-copying) elements that, although reshuffling every generation and occasionally mutating, have the "goal" of maximizing the production of more copies of themselves. In this view, we are, so to speak, just a way DNA has found to make more DNA. But of course genes are utterly devoid of goals, selfishness, or any other such psychological attributes we might ascribe to people or other animals—the title of Richard Dawkins's well-written book *The Selfish Gene* notwithstanding.[34] Genes, remember, are just strings of chemical subunits in a long molecule, capable of determining the sequence of amino acids that a complex biochemical apparatus will put into a protein but quite incapable of any emotion. Genes contain the information to do that sequencing job, but they are no more selfish than this information-containing sentence. Furthermore, genes are self-replicating only in a *very* limited sense—in the same sense that this page would be "self-replicating" if it were parked in a copying machine loaded with the same kind of paper. Genes are "designed" to be copied, but in order for them to be replicated, they must be supported by living cells. There is no replication of genes or use of their information unless they are embedded in a complex cellular mechanism. And if those cells are to survive, the genes must operate in concert with one another and with other components of the cells to allow the right combinations of proteins to be produced. So it would make as much or as little sense to call genes cooperative as to describe them as selfish, and it is much less misleading to avoid such analogies altogether.

There is an additional reason why the description of a dichotomy between genes and environment, or, as it is often put, between nature and nurture, is so problematic. Not only does the environment influence (or prevent) the expression of genes, but also genetic differences, in their influence on the actions of individuals, can change the environments of those individuals. A combination of selection and historical accident has given each species, including fruit flies and human beings, a unique evolutionary history. Both the genes of a species and the environments in which it lives change as a result of the interactions between populations and those environments. Others of the same species ordinarily are important elements of the environment of each individual, especially in highly social animals that, like human beings, stay with and rear their offspring. For example, the chances of children with various genetic defects being able to mature and reproduce often depends on the social environment. Children born with the metabolic disorder phenylketonuria, caused by a deleterious genetic mutation, once were doomed. Today, they can survive to reproduce in a society that understands the defect and can supply the affected individual with a diet in which the amino acid phenylalanine is tightly regulated.[35] Thus, people can create environments that will change the proportions of genes[36] (in this case, increasing the proportion of the phenylketonuria mutant gene) and therefore change the biological evolution of human populations.

In fact, *Homo sapiens* has profoundly altered its own total environment and has also altered that of virtually every other organism on Earth. Our species, for instance, has changed the gaseous composition of the atmosphere by adding carbon dioxide and ozone-destroying chemicals to it, altered the character of much of the planet's surface (by cutting down forests, by farming, by building roads and cities, etc.), and created novel toxic chemicals that are now found everywhere. Those environmental changes in turn cause changes in selection pressures in human beings and other organisms. Some changes, such as increased levels of radioactive substances in the environment, even increase the rates at which genes mutate.

The Forces of Change

In response to environmental conditions, then, natural selection gradually changes the proportions of various kinds of genes within populations and thus gradually alters the proportions of organisms with corresponding traits within those populations. In the Animal House, the original fruit fly populations had a few individuals with combinations of genes that made them a little bit resistant to DDT. Those resistant individuals were more likely to live to reproduce than to die barren, and thus they outreproduced the less resistant individuals. More resistant flies, breeding with one another, brought together new combinations of genes that further favored resistance. Those new genotypes influenced the development of phenotypes that were even more able to tolerate exposure to DDT. After relatively few generations, this selection had resulted in populations in which most individuals had genotypes that made them very resistant to DDT, and the fruit fly strain had adapted to its new, DDT-polluted environment. In nature, many kinds of insect pests rapidly adapted in the same way when DDT was first introduced into their environments.[37]

Selection is the only known process that, given adequate time, can produce the complex features of organisms that adapt them to their ever-changing environments. It can result in the evolution of structures or behaviors that appear to be designed to meet the needs of the creatures that possess them. Although mutations do not occur in response to need, any that enhance reproduction will be promoted, which, of course, is what we describe as selection. DDT resistance in flies living in a DDT-polluted environment, the thorns of cacti threatened with attack by hungry desert animals, the flowers of daisies that must attract butterflies and other insects to pollinate them, the wings that bats need to sweep through the night gobbling up mosquitoes, and the brains of people that let them maneuver in complex societies—all were "designed" by natural selection. A century and a half of intensive search and debate by scientists since Darwin has turned up nothing else that can explain the vast majority of adaptive features, features that improve the reproductive capabilities of the organism.[38]

Selection may be the only factor that can explain most adaptations,[39] but it isn't the only factor that can change the proportions—frequencies—of genes in natural populations, changes that are fundamental to the process of evolution. Mutation can change those proportions by transforming some genes into others, as we've seen. But migrant individuals entering or leaving a population can also change gene frequencies if they have different genetic characteristics from those of the population as a whole. If a thousand blue-eyed people suddenly land on an island already populated by a thousand brown-eyed individuals, the proportion of genes coding for blue eyes in that island population will change dramatically. Such migration is sometimes referred to as gene flow because it simply adds or subtracts different genes from the gene pool of the population.

Another factor that can change a population's genetic composition is random chance, whether in the process of reproduction or in the happenstance of which individuals get to reproduce. For example, if a hot, wind-whipped wildfire sweeps suddenly through a forest, it may wipe out half of the forest's population of a species of land snail. There are numerous genetic differences among individual snails, but snails with certain genotypes are not likely to be more vulnerable to being cremated than are snails with other genotypes. The genetic composition of the snail population will doubtless change, but in this case in what is likely to be a random way, rather than in relation to the survival abilities of snails with different genotypes. Such random change is known as genetic drift, and it is especially important in small populations.[40]

In sum, mutation and recombination produce the genetic variants; selection, gene flow, and drift determine their fate.[41] Because selection is the creative evolutionary process, responsible for most of the characteristics that make organisms fit into their environments, I'll return to it frequently in the pages that follow. To understand selection is to understand the guts of the evolutionary process.[42]

Selection in Action

Armed with an understanding of natural selection, evolutionists have been able to construct reasonable explanations for all sorts of seemingly weird adaptations. One is the adaptation that allows bats to operate in the pitch black of night. Bats navigate and hunt in the dark by sending out high-pitched squeaks and detecting the echoes that return from objects in their world. It's a system similar to the sonar (*so*und *na*vigation *r*anging) used by naval surface ships to detect the presence and position of submarines. Another is an even stranger, related example. A certain type of mite has adapted to make its home in just one ear of a moth, even though moths have a pair of ears, one on each side of the base of the abdomen. The mites have evolved a path-following behavior such that when a mite climbs onto the head of a moth that is visiting a flower for nectar, it follows the trail of other mites and ends up in the same ear as its predecessors. With mites in one ear, the moth can still hear the sonar "pinging" and evade the bats that prey on it, but if mites infected both ears, the moth would be deaf. Mites that traveled their own routes would tend to end up infecting both ears of the moth and leaving fewer offspring than those that followed the leader—because ear mites on a moth that is devoured by a bat don't have a bright reproductive future.[43]

Consider as well an adaptation much more challenging to explain evolutionarily—part of the life cycle of the human botfly.[44] These tropical flies, which look a lot like giant common houseflies, capture mosquitoes[45] as they approach human beings and other animals and lay their eggs on the mosquitoes, usually on their mouthparts. When the mosquito bites a person (or another mammal), the body heat of the host causes the eggs to hatch, and the maggots then crawl down the mosquito's mouthparts and onto the victim's skin. There they bore in and develop, breathing through openings in their hind ends, which they poke out of their burrow in a developing boil. When they've matured, the maggots crawl out, drop off, form pupae in the soil, and emerge as adult flies to restart the cycle. Friends of mine who have picked up these charming guests during their fieldwork say that if you get one behind the ear, you can hear it chewing!

This could be seen as an example of special creation—a sort of joke by a capricious deity. But maybe not. Relatives of the botfly, like some other insects, distribute their eggs more or less at random, which provides a clue to a possible evolutionary route. Occasionally, eggs are attached to leaves and hatch when a warm-blooded animal brushes past. A small change in behavior, caused by a lucky mutation or recombination, could start a trend among the botflies away from laying eggs at random on various motionless objects and toward laying them on biting insects. Flies that followed this trend would be more likely to have their maggots end up on suitable mammalian hosts. The result: those with the new genotype would reproduce more successfully than those with the old. That differential reproduction is selection. Even those who recognize that selection (as an after-the-fact description) is mindless are likely to say, in shorthand, that selection "favored" the flies' laying eggs on mosquitoes.

Natural selection, as you might suppose, lies at the root of why people are different from chimpanzees (or butterflies or athlete's foot fungi). Human beings and chimpanzees[46] have undergone on the order of 250,000 generations of natural selection since our ancestors took different evolutionary paths. For the more than 5 million years since that split, individuals in each line with certain genotypes outreproduced those with other genotypes. That differential reproduction of genotypes is a basic part of the explanation of why we can write books and chimps can't.

In the process of selection—whether in human beings, chimpanzees, or fruit flies—which genes in any particular generation are the most reproductively valuable depends almost entirely on the environments in which the individuals of that generation mature. This is a crucial point. Genes in fruit flies that are of great value in an environment containing DDT—those genes that confer resistance to DDT—may well handicap an individual carrying those genes in a DDT-free environment (perhaps wasting energy in producing an unnecessary enzyme for defense). When DDT-resistant fruit fly strains are returned to a DDT-free environment, natural selection often leads to the evolution of susceptibility once again.

Selection in Nature

Is the picture of selection that I'm painting based purely on a few Animal House experiments, some theoretical considerations, and a lot of speculation about various adaptations? Not at all. Shortly after World War II, DDT was overused in many places in the midwestern United States in order to control houseflies—for instance, in Topeka, Kansas, not too far from the Animal House. At first, this tactic was very successful, and fly populations plummeted. As a result, people stopped engaging in standard fly-control practices that denied flies places to breed, such as carefully covering garbage cans and disposing of lawn clippings. When the inevitable evolution of DDT resistance took place, the houseflies, freed from previous constraints, rebounded to even higher population sizes.

There is abundant other evidence that bears on the efficacy of natural selection in nonhuman organisms,[47] ranging from the evolution of resistance to antibiotics by disease-causing bacteria all the way to the evolution of color patterns in guppies.[48] We know a great deal about selection, even though it can be difficult to document in nature. Slight differences in reproductive success are difficult or impossible to discriminate from chance variation except in laboratory experiments. Over thousands of generations, however, it is just such slight differences that transform the genetics of a population. Despite the problems of detecting low levels of selection in short periods of time, biologists are continually finding new and informative examples of evolution occurring in nature—sometimes occurring fast enough to be observed in action in human lifetimes.

Changes in the frequencies of color forms in a British moth known as the peppered moth represent the classic case of selection detected in nature.[49] It was an example of a phenomenon known as melanism (black or almost black coloration), and the development of that melanism illustrated how rapidly evolution can proceed if there is a sudden environmental change.[50] Among biologists, this example was a great instance of catching evolution in action, a visually compelling demonstration of Darwin's mechanism actually at work outside the laboratory.

The peppered moth exists in two different forms—a "speckled" (white with black speckles) form, which is camouflaged against lichen-covered tree trunks, and a melanic (black or nearly black) form, which is camouflaged against sooty, lichen-free trunks (see the figure). In 1848, the speckled form made up more than 99 percent of the peppered moth populations in the area around Manchester, England. Fifty years later, however, more than 99 percent of the Manchester-area moths were of the melanic form.[51] Genetic studies showed that the difference was caused primarily by the substitution of one form of the same gene for another.

The spread of the gene associated with the melanic phenotype was highly correlated with soot pollution caused by the industrialization of the area—hence, the darkening is referred to as industrial melanism. Apparently, the tree trunks became sooty, the lichens were killed, and the previously camouflaged speckled forms became conspicuous and were frequently eaten by birds and other predators that hunt by sight. Thus, the genes that produced the speckled pattern became less frequent in the population. The speckled phenotype of the peppered moth became rare, and melanic moths took over the populations.

Speckled and melanic (dark) forms of the peppered moth against natural, lichen-covered tree trunks (above) and sooty trunks on which air pollution has killed the lichens (below). Moths in unpolluted and polluted environments have evolved appropriate color patterns that enhance their survival.

In unpolluted areas, the melanic individuals still tend to be eaten disproportionately by the birds, as they always had been, because they are easily seen against the speckled background of lichens growing on tree trunks. The genes producing the speckled phenotype confer an advantage on their owners in clean woods, and the speckled moths dominate in populations occupying such habitats. This explanation is supported by the geographic coincidence of pollution and melanism and by studies of bird predation on the moths. And it recently received strong support when abatement of air pollution led to reestablishment of lichens and a subsequent decline in the frequency of melanic individuals.[52]

Were it not for the human-induced changes in the environment, the strong selection favoring the "speckled genotype" would have kept the melanic form as a rare variant, produced by an occasional mutation of the key gene. This is a crucial point in understanding the role of environment in evolution. Atmospheric pollution favored genes for melanic phenotypes in the peppered moth, and the frequent evolution in other organisms of resistance to pesticides and antibiotics follows the same pattern. Human beings "pollute" the environment of pests and bacteria by purposely exposing them to poisons. That favors genes that produce phenotypes capable of detoxifying or excreting the poison, as we are discovering to our distress in an increasing number of instances. Because constant exposure to poisons has the same effect in agricultural fields as it did in fruit fly vials in the Animal House and as it does in other laboratory selection experiments,[53] the number of insect and mite pests now resistant to one or more pesticides is pushing 500.[54]

It is clear from both laboratory experiments and field observations that we will be plagued by resistant insects as long as we continue to depend on routine use of pesticides.[55] A very conservative estimate of the costs of pesticide resistance worldwide is about $1 billion annually.[56] A similar and rising cost is being incurred because of antibiotic resistance among deadly pathogens such as the tubercle bacillus, which is responsible for tuberculosis.[57] Sadly for us, disease-causing bacteria worldwide have, in a few decades, become adjusted to living in environments polluted (from the bacterium's perspective) with one or more of a large number of antibiotics.[58] The persistent habit of using antibiotics even when it is not known whether a susceptible bacterium is responsible for an infection encourages the evolution of resistance. So does the widespread use of antibiotics as growth promoters in animal feeds. The short generation times of bacteria and their ability to share genes easily[59] has thus already provided us with widespread examples of rapid evolution. Antibiotic-resistant bacteria and pesticide-resistant insects, mites, and rats are a grim testimonial both to the efficacy of natural selection and to the folly of ignoring what is known about evolution.

As the cases of the peppered moth and resistance show, large-scale, rapid changes produced by human activities have caused selection in nature to change populations rapidly enough for scientists to detect the ongoing evolution easily.

Strong selection in nature is not brought about only by human influence, however.[60] For instance, unusual circumstances have permitted natural selection to be observed in the field in two kinds of European land snail that show a variety of banding and color patterns on their shells (see the figure).[61] A characteristic behavior of several kinds of British thrush, birds related to the robin, allowed two biologists, Arthur Cain and Phillip Sheppard, to show that selection by birds could play a major role in determining the snails' shell phenotype.[62]

Frequencies of different kinds of shells turned out to be correlated with the microhabitats in which the snails were located—such as beech, oak, and mixed deciduous woods, hedgerows, and open areas with long, coarse herbage. The thrushes, which find their prey by sight, were acting as selective agents. These birds use favorite stones, "thrush anvils," to crack open snails they pick up in various habitats. The researchers ascertained the frequencies of shell colors and patterns in living snails by collecting very carefully in the various microhabitats. They then compared those frequencies with the frequencies of types in the piles of shells around thrush anvils, which showed what the thrushes had collected. In early spring, for instance, when there is exposed earth and brown leaf litter on the woodland floor, thrushes are more likely to spot and eat snails with yellow (rather than brown or pink) shells. As the background greens up in late spring, it becomes advantageous for a snail to be yellow.

Although there is often strong visual selection favoring one type of shell, several shell types frequently are found in a single population living against a single background. Why didn't selection lead to just one type of shell? The reason why such a mixture of types[63] persists appears to be, among other possibilities, that there is countervailing nonvisual selection. Climate apparently plays a selective role, promoting the reproduction of different genes under different conditions of temperature and humidity. Those genes influence at least egg production and development time[64] and possibly shell color and banding patterns.[65] Thus, the simplicity of the visual selection situation found in the British research does not provide a universal key to the variety of shell colors in land snails—indeed, it appears that a multitude of selective forces may be operating in these populations. As we'll see, many "simple" evolutionary explanations of physical and behavioral characteristics of human beings—such as that our upright posture evolved to free our hands to carry tools and weapons or that homosexual behavior is caused by "gay genes"—also are not so simple under careful scrutiny.

Islands, starting with Charles Darwin's famous visit to the Galápagos Islands, have often served as splendid outdoor laboratories for study of the evolution of everything from bird wing lengths to the creation of biodiversity (I'll discuss the latter in chapter 3).[66] The reason is circumscription. Island-dwelling organisms typically can't disperse easily; they're stuck in their environments and must adjust to the conditions there—sometimes very rapidly. In this respect, islands can serve as larger analogs of the fruit fly vials. But there is a curious

Land snails of the genus *Cepaea* showing variability in shell patterning. Above, banded and unbanded snails on coarse grass (snails are about three-fourths of an inch across); below, thrush anvil (rock) with broken shells. Drawings based on photographs supplied by the late W. H. Dowdeswell.

aspect to many island organisms: they commonly have less physical ability to move about than do similar species that live on continents. Evolutionists interpret these differences as the result of selection against dispersal ability. For example, flightless rails (birds in the coot family) once inhabited many Pacific islands—until invading people, and the rats that hitched rides around the world with people, exterminated most of them. Rails that live on continents, by contrast, fly well. One can easily construct an evolutionary explanation for this. In the absence of terrestrial predators, individual rails with smaller wings spent less energy developing those useless organs (the pectoral muscles that power the wings account for a disproportionately large fraction of a flying bird's metabolism) and presumably were less likely to be blown off the island by storms and perish at sea. Thus, rails with smaller wings would have had more surviving offspring than those with wings of normal size. Similar examples of flightlessness evolving on islands can be seen in dodos, kiwis, and other birds as well as in a variety of insects.[67]

Recently, biologists studying annual plants of the sunflower family on about 200 small islands near Vancouver Island have shown that selection to reduce dispersal ability can operate in an unexpectedly short time.[68] Populations frequently went extinct, and then islands were recolonized from the readily dispersed mainland populations. The seeds of these plants are blown over the water attached to fluffy "parachutes" like those of dandelions. Within a decade after plants recolonized the islands, the parachutes of two species were shown to have decreased in size in subsequent generations and the seeds of one species had increased in weight—changes that in both cases decreased dispersal ability. In essence, biologists actually observed a sequence similar to that hypothesized for the loss of rail wings, in this case doubtless largely a result of the loss of seeds with big parachutes in the surrounding sea.[69] In all these examples, remember that the basic evolutionary process consists of individuals with certain combinations of genes outreproducing those with other combinations, thereby raising the frequency of those particular combinations of genes in the population—and in the process modifying the kinds of individuals that occur in that population.

Rates of Evolution

Despite the special benefits of studying natural selection in the field, in most cases laboratory populations of rapidly reproducing organisms remain the best systems in which to study the details of the process of evolution. That's because speed is of the essence—short generation times can allow us to see the sorts of processes in weeks or months that would take tens or hundreds of thousands of years to observe in a human population (which could not be experimented on in any case). Bacteria, fruit flies, and flour beetles[70] are commonly used for such experiments. Some of the most important recent work has been done with bacteria in which a generation spans periods of minutes rather than of months, years, or decades. For instance, twelve populations of a gut bacterium[71] have now been tracked by Richard Lenski and his colleagues for more than 20,000 generations—more than ten years—in an experimental laboratory system. Not only were the bacteria not in their natural environment, the human intestine, but also they were exposed to a changed laboratory environment in which their sugar source, glucose, was in limited supply.[72] All populations introduced into the experimental environment[73] were started with a single cell from the same genetic strain, so all subsequent hereditary changes must have had their roots in mutation.[74] The size of the individual bacteria and their ability to compete could be measured. Competitive ability was measured against that of the bacteria's own ancestors "resurrected" from freezers (where the bacteria remained alive but could not reproduce and evolve). The system is analogous to a series of fossil beds from which both the fossils and their environments can be recreated at will because samples of the original strain and, periodically, of the evolving strains were "fossilized" in a low-temperature freezer.

The results of the first 10,000 generations of the experiment have now been analyzed. Both the size of the bacterial cells and the competitive ability of the populations evolved rapidly for about 2,000 generations in all twelve populations, indicating that natural selection was adjusting the bacteria to the experimental environment. Interestingly, though, progress then rapidly slowed during the next 3,000 generations, and little change occurred over the final 5,000 generations. One important aspect of this result is that the isolated populations in the same novel environment all showed the same pattern of evolution of fitness,[75] as a result of selection operating on genetic variation produced by mutation. All the populations evolved so that individuals of each population at the end of the experiment had a competitive advantage over the original kinds of individuals (that is, they became more fit in glucose-limited media). Such a parallel increase in fitness shows that to a degree, evolution by natural selection is repeatable.[76] On the other hand, different lines of bacteria evolved different cell sizes and different fitnesses, apparently as a result of their different histories and of chance events (the particular suite of mutations that occurred and were thus available to be subsequently acted on by selection).

Results of long-term experiments with fruit flies show some similarities to those seen in the bacteria. In one experiment, six fruit fly populations were subjected to selection for a higher number of abdominal bristles—hair-like projections on the fly's belly. All lines showed roughly similar increases, but there were many small differences, such as in the rate at which the lines changed, differences that provided the investigators with insight into the genetic details of the process.[77] The bacterial experiment continues and doubtless will yield further interesting results.[78] The evolution of humanity since the dawn of our species has occurred over roughly 20,000 generations (400,000 years), so having a laboratory system in which similar numbers of generations of a rapidly reproducing organism can be studied could prove very enlightening.

Stephen J. Gould, perhaps the best essayist among evolutionists, sees a mismatch between rates of evolution observed in field and laboratory populations and those inferred from the fossil record. He claims that the former rates of evolutionary change are too fast to explain the slow changes (those taking place over millions of years) seen in the organisms of the past.[79] As he points out, rates measured in studies of change in natural populations today (often referred to as investigations of microevolution) are often 10,000 or more times faster than those commonly measured by paleontologists.

The fossil record lacks fine-scale resolution, however, and selection pressures often fluctuate as, say, droughts alternate with periods of normal precipitation.[80] Thus, rapid changes may have been occurring all the time, say from larger to smaller individuals and back again every few decades, even though the average size seems constant in samples of fossils from a series of strata each representing, say, 2,000 years. And, of course, examples of selection measured in nature are a biased sample because they're largely limited to situations in which selection is especially powerful and they focus on those traits undergoing rapid evolution. The average levels of selection implied by the fossil record are almost impossible to detect in what is called "ecological time" (tens to hundreds of generations). After all, a 1 percent differential in the reproductive performance of genotypes is quite strong selection—to measure it, we'd need to be able to show that, say, 100 flies with one genotype produced 100 offspring that bred in the next generation while 100 flies with another genotype produced 101. In nature, as indicated earlier, detecting such a difference and showing that it was not a result of chance would usually be impossible: it would involve tracking the fates of hundreds of thousands of flies of known genotypes with enormous precision over numerous generations. Yet a 1 percent differential probably would be enough to account for all of evolution between the origin of life (at least 3.5 billion years ago) and the appearance of upper-class Englishmen (assumed by upper-class Englishmen of Darwin's day to be the pinnacle of evolution). Thus, much evolution, probably most of that now occurring in nature, is undetectable by scientists.

Nevertheless, evolutionary processes can be rapid enough to effect important biological changes in ecological time—months, years, decades, or centuries during which Earth's ecosystems do not change dramatically—as experience with antibiotic and pesticide resistance shows.[81]

Evolutionary Hangovers

As indicated earlier, natural selection has no foresight. It cannot purposely produce behaviors that will be adaptive in all future circumstances. And it can change the genetic characteristics of populations of long-lived organisms such as human beings only very slowly. This has been dramatically demonstrated over the past few millennia. Clever human beings, exploding on the world scene in an evolutionary eye-blink, have changed the environment in ways that alter the adaptive value of many traits in numerous organisms, including *Homo sapiens* itself. Rapid environmental changes can create what I call evolutionary hangovers—structures or behaviors that once were adaptive but whose positive influence on reproductive performance has declined or disappeared. Genetic evolution ordinarily can cure such hangovers only at what, when the hangovers are ours, is often a distressingly slow pace.

Distraction displays, in which birds pretend to have been injured in order to draw predators away from their nests, may provide an example of an evolutionary hangover in a nonhuman organism. These maneuvers have become less adaptive since human beings became common predators, given that the display simply alerts a person to the presence of a nest. To the degree that the benefits of such displays (fooling stupid predators) outweigh their costs (attracting smart ones), selection should lead to their retention. But we smart predators are rapidly becoming more common, and distraction displays may soon be an evolutionary hangover. Then, selection will extinguish the displays—or, if the birds can't evolve rapidly enough, they themselves will go extinct.

The latter was the fate of the passenger pigeon, whose main defense against predators was to form gigantic nesting colonies at different places each year, "swamping" the local predators. In the early nineteenth century, the size of some colonies may have exceeded a billion birds. The colonies were so large that the local hawks, squirrels, and other enemies of the birds and their young could eat only a limited proportion before breeding was complete and the passenger pigeons moved on. The predators could not build large populations of their own on the basis of such an ephemeral resource, so the strategy worked well for the pigeons. It worked well, that is, before people with guns and railroads penetrated the upper Midwest after the Civil War. Then, railroads transported slaughtered pigeons to big eastern markets, and hunting to satisfy those markets made the previously adaptive behavior an evolutionary hangover that led to extermination of the birds. In 1878, some 3 million birds were shipped to the East by a single hunter in Michigan; by 1889, passenger pigeons were extinct in that state.[82] The last survivor of the species, a female named Martha, died in the Cincinnati Zoo in 1914 and now reposes, stuffed, in the National Museum of Natural History in Washington, D.C.

Of course, we, too, have suffered negative (as well as positive) effects of the industrialization that produced the trains and thus was instrumental in the extinction of the passenger pigeon. We have evolutionary hangovers when cultural evolution makes deleterious changes in our environment more rapidly than genetic evolution can respond to them. Some of the rapid environmental changes brought about by industrialization seem to have caused an increased incidence of cancer in human populations. As have all other organisms, we human beings have evolved mechanisms that protect us from some of the harmful influences of the environment—mechanisms that repair damage to DNA are a classic example of defense against carcinogens (cancer-causing substances).[83] Millions of years of eating plants have led hominids to evolve detoxification mechanisms for many of the nasty chemicals that plants have evolved to poison herbivores. But human cultural evolution has introduced numerous new environmental threats, including novel synthetic carcinogens, to which there has not yet been time for us to evolve responses. Our perceptual and defensive systems evolved in a world lacking those threats; therefore, we have difficulty recognizing and dealing with them, and our "normal" behavior becomes an evolutionary hangover. Indeed, the tendency of *Homo sapiens* to continue with "business as usual" in the face of a rapidly deteriorating environment is producing what may be the most gigantic hangover of all time.

Because in most cases the selection pressures created by evolutionary hangovers are rather weak, biological evolution to cure hangovers can be very slow. The first scientific recognition of an environmental health threat was in 1775, when London surgeon Percivall Pott noted that severe and chronic contact with soot was responsible for scrotal cancer in men who had been chimney sweeps since boyhood.[84] Our ancestors, such as *Homo erectus*, didn't have chimneys. Thus, soot appearing in the environment influenced more than the color of peppered moths. But cultural evolution, which created the hangover, also cured it (by eliminating the need for chimney sweeps), long before genetic evolution could produce men with soot-proof scrotums.

On the other hand, the long use of fire by *Homo sapiens*, often in caves or shelters with poor ventilation, may have accidentally "preadapted" our lungs somewhat for life in a smoggy world. Preadaptation is roughly the opposite of an evolutionary hangover; it's simply having had properties evolve in response to one environmental situation that later by chance prove functional in a new environment. Describing something as a preadaptation is a restrospective judgment. Thus, a genetically evolved response improving the lungs' ability to remove contaminants from simple wood smoke would later prove functional in an environment in which lungs are assaulted with a more complex mix of pollutants.

In any case, selection could not preadapt our species to withstand much of the chemical assault to which it is now subjected because that assault is totally unprecedented. Earth is now awash in millions of tons of novel synthetic chemicals; tens of thousands of different ones are continually released into the environment. Many of them are toxic, carcinogenic, or mutagenic (damaging to genetic material) or cause developmental problems by mimicking hormones, chemical messengers that help to control many of the life processes of organisms. Defenders of this dangerous chemical inundation often claim that concern is misplaced[85] because there are many natural poisons and carcinogens out there, especially in the foods we eat. What is missing from their analysis is, among other things, an evolutionary perspective. There have been at least millions of generations of primate evolution during which resistance to the dangerous (to us) plant defensive chemicals could evolve, but there have been only two generations or so to evolve responses to most synthetic compounds.

Although there do not appear to be strong selection pressures from natural carcinogenic hazards (cancer, like cardiovascular disease, largely kills people after their reproductive years), a long time would have been available for weak selection pressures to produce resistance. Of course, many defensive chemicals of plants tend to be immediately distasteful or poisonous. It does a plant little good if the animal that chews it up dies ten years later of a tumor. And people do appear to have evolved responses that minimize exposure to residual plant toxins (those that haven't been removed from crops by selection). One scientist proposed that morning sickness is a response to spicy (well-defended) vegetable foods.[86] Presumably, developing fetuses are much more sensitive to toxins that reach them through the mother's bloodstream than adults are to the same toxins in their environment. One can imagine that vulnerability to toxins also explains children's dislike of brussel sprouts!

Cultural evolution may have led people inadvertently to attack themselves, using a technique similar to the hormonal strategies that plants have evolved genetically to defend themselves against bugs. Some of the natural plant defensive chemicals are similar to the hormones that control insect development and, as evolutionist Douglas Futuyma has put it, "can severely affect growth, development, and survival of insects."[87] A wide variety of synthetic chemicals similar to mammalian hormones, as well as hormones of other organisms, is now entering the environment in pesticides, plastics, cleaning compounds, and the like—and these chemicals are often much more difficult for the body to break down than the natural hormone mimics found in plants. Our perceptual systems, which emphasize sight and minimize the ability to detect small concentrations of dangerous chemicals, may prove to be a most serious hangover. That's because today, one possible selection pressure tracing to those hard-to-detect hormone-mimicking chemicals could be severe. There is increasing evidence that human sperm counts are declining, perhaps in response to the now-widespread exposure to synthetic hormone mimics.[88] If men become less fertile, we must hope that enough genetic variation will remain to allow humanity to evolve resistance to the hormone mimics quickly. Otherwise, the population explosion may end disastrously, not in ecological collapse helped along by the adverse effects of synthetic chemicals on the organisms that run our life-support systems but in a loss of people's ability to reproduce themselves.

Even changes in cultural evolution much more positive than the invention of dangerous synthetic chemicals can have unfortunate consequences that in turn could influence future genetic evolution. For instance, there has been a reduction in the number of children borne per woman, especially in industrial countries, and a simultaneous decline in breast-feeding. Because lactation suppresses menstruation, this has created evolutionarily unusual patterns of hormonal activity by putting women through almost continuous menstrual cycling. That could be a factor in recently observed rising rates of uterine and breast cancer.[89] The prevalence of those cancers in turn may be modifying selection pressures to favor women who are not prone to induction of cancers by those hormonal patterns. But once again, that selection would be weak because most of the mortality is fairly late in life. Another problem apparently related to cultural evolutionary change is an upsurge of asthma and other respiratory allergies in recent years, which presumably are related to environmental alterations.

Unfortunately, we are quite ignorant of the evolutionary significance of the part of the immune system that seems to have no function (at least in affluent countries) except to cause allergies.[90] Knowledge of which function that part of the system evolved for (it presumably protects us from something) might very well show us how to ameliorate its unfortunate effects.[91]

Evolution and Behavior: The Case of Altruism

Sometimes it is very hard to understand how natural selection can actually produce certain characteristics. One of the toughest challenges has been to understand how apparently selfless behavior can be tied to those putatively selfish genes. If evolution by natural selection works to perpetuate the characteristics of individuals that reproduce more than others, how can we explain the evolution of altruism?[92] After all, sacrificing one's own good for that of others doesn't often make a person reproductively successful. During World War II in Europe, for example, getting caught by the Nazis while helping a Jew was not a way to increase one's fitness. How could a person's "selfish" genes permit this? The issue of the origin of altruism is a key one that very frequently comes up in discussions of the behavior of social organisms, especially human beings. And it is illustrative of the complexities and differences of opinion that attend attempts to figure out how natural selection works.

The basic outline of an answer for how natural selection could lead to altruistic behavior was provided in the 1960s by a fine evolutionary theoretician, William Hamilton, who developed the notion of inclusive fitness.[93] The idea is simple. Selection operates when carriers of some genes outreproduce carriers of other genes. If altruistic behavior were inherited, it would be more likely to spread if the altruism were directed at close relatives, because relatives share genes. For example, on average, one-half of the genes of brothers and sisters (and of parents and offspring) are identical. Between uncles or aunts and their nieces and nephews, on average one-quarter of the genes are identical.[94] Suppose a brother remains single and works to help support a sister and brother-in-law who otherwise would be too poor to have a family. If, as a result, the sister bears and raises five children, the brother's effort will have added more copies of his genes to the next generation than would have his fathering two children of his own. The fifth child will have added about an extra 25 percent more of his genes to the next generation than would have been passed on if the altruistic brother had personally had two offspring but had not helped his sister. Selection on such acts of altruism depends on the relationship of the actors, the costs to the altruist, and the benefits to the recipient.[95]

The more closely related the aided individuals are to the altruists, the more readily the evolution of genetically based altruism can be explained. Laying down one's life for a brother (letting him live to reproduce) increases the representation of genes identical to the helper's in the next generation more than would laying down one's life for a cousin. Differential reproduction of genes caused by the influence of individuals on the success of their relatives (who carry many of the same genes) is often referred to as kin selection. Kin selection is used to explain the phenomenon of "helpers at the nest," in which young animals aid their parents in raising brothers and sisters rather than going out and starting families of their own. This behavior, which can be considered a form of altruism, has been especially well documented in birds.[96]

Inclusive fitness is not the only possible "biological" explanation for the evolution of altruistic behavior. Another involves various forms of reciprocity,[97] in which the donor suffers a small cost, the recipient makes a large gain, and the donor has the expectation of a future benefit that will more than compensate for the cost of the altruistic behavior. One of the few reported cases of such reciprocity in nonhuman animals was observed in the highly social vampire bat.[98] In a well-designed study, it was shown that vampire bats that have an unsuccessful night of foraging and do not find an animal to bite for a blood meal are in danger of starving to death. (By the way, the bats don't suck blood, as many believe, but slice open an animal's flesh with their sharp teeth and lap up the blood.) The research showed that successful members of a colony fed unsuccessful foragers by regurgitation and that the recipients would reciprocate on future nights.[99]

Kin selection is often evoked to explain patterns of human social behavior, especially in connection with various forms of altruism, but the degree of its involvement is not known. A basic problem is that although assumptions are made about how genes for altruism can be favored in people, there are essentially no data that allow selection of such genes to be demonstrated or measured. Certainly, much human altruism is determined primarily by cultural influences,[100] which, as we'll see, are almost certainly why many gentiles tried to aid Jews during the Nazi era. But the universality and persistence of systems for keeping track of relatives and determining behavior toward them (kinship systems) in human societies can be seen as an argument for a general influence of kin selection on the genome of *Homo sapiens.*[101] After all, what other reason could there be for human societies to have developed complex systems of terminology to allow individuals to keep track of their genetic relationships to one another?[102] Why do women (and in most cases men) in all societies recognize and care for their children?[103]

Group selection—the differential reproduction of entire populations—is also sometimes proposed to account for the evolution of traits such as altruism that may benefit the entire population. Under a group selection hypothesis, for instance, it would be assumed that golden-mantled ground squirrels giving alarm calls would save other members of their population from predators while risking their own lives by calling the predators' attention to themselves. Thus, populations including individuals that give alarm calls are more likely to avoid extinction than those lacking that behavior. Eventually, all surviving populations of golden-mantled ground squirrels would consist of individuals that give the calls, and group selection would have modified the species from one that did not warn against approaching predators into one that did. The problem with such group selection explanations for altruistic behavior is that individual selection operating in the opposite direction will usually overwhelm it. Individuals with a free-rider (nonaltruistic) genotype would benefit from the behavior of the altruists but would not reciprocate. Ground squirrels that remained silent, didn't call attention to themselves, and just dived into their holes when a hawk appeared would be less likely to be eaten than their altruistic cohorts who gave a shout. These free riders would get a benefit but would pay no cost. Assuming that the free riders were not recognized and cut off from benefits, the free-rider genotype would be the most fit, outreproducing the carriers of altruistic genes, and silent ground squirrels would take over the population.

In some circumstances, however, group selection may work to encourage traits unlikely to be favored by individual selection, such as altruism. Those circumstances obtain if the groups in question are small; if rates of turnover (extinction and reestablishment of groups) are high, limiting the time in which selection can operate among individuals within the group; if there is a fluctuating environment; or if a combination of these conditions exists.[104] Clearly, selection can also favor any trait through differential reproduction of groups. Those groups can range in scope from various subdivisions of a population to species or even more inclusive taxonomic categories such as genera (our genus, and that of some fossil people, is *Homo*) and orders (*Homo*, along with all the genera of apes, monkeys, lemurs, tarsiers, and so on, is in the order Primates). Grizzly bear populations went extinct in Colorado and California but survive in Wyoming and Alaska—an example of differential reproduction of populations.[105] Trilobites and dinosaurs died out, but birds and mammals thrived; these are examples of differential reproduction of higher taxonomic categories. But how important selection is above the level of differential reproduction of individuals in producing altruism or other characteristics remains a complicated and controversial question.[106] My guess, based on the immensely larger numbers and rapid turnover of individuals compared with groups, is that the vast majority of what we would consider the creative results of evolution—production of the sharp eyes of hawks, the streamlined bodies of porpoises, the clever brains of human beings—occur because of differences in the genotypes of individuals: individual selection. And it is likely that much of human altruism has not been programmed into our genomes by selection but is a result of characteristics of the human mind that evolved for other reasons. But I'll defer that discussion until we look at evolutionary ethics, near the end of the book.

Selection in Human Beings

The process of evolution seems relatively easy to observe in the world of bacteria, fruit flies, peppered moths, land snails, and the like—but how easily can it be observed in people? Direct evidence for natural selection is more difficult to find in human populations than in those of many other organisms. First, evolution apparently does not often change human populations very much over ecological time, largely because human beings have such long generation times, and there is a limit to the number of generations that can be observed over a scientific career. Second, there are ethical constraints on experimenting on human populations. Nonetheless, evidence of natural selection in human populations is present in various forms.

Rapid change over a handful of generations usually requires very strong selection pressures—levels of differential reproduction (say, some genotypes having a tenfold or hundredfold advantage over others) that are virtually unknown in human populations except perhaps when assailed by lethal epidemic diseases.[107] When there are relatively few survivors of epidemics, as when European diseases were carried to the New World[108] or when more than 95 percent of the people of northwestern China died in a plague in A.D. 310–312,[109] populations can rapidly evolve immunity.[110] Today, there are already signs of natural selection for human genotypes that are better able to delay death from acquired immune deficiency syndrome (AIDS).[111] Discernible changes in most human traits that are not subject to such dramatic selection pressures (say, in ear size or in the tendency to be generous or conciliatory) will take place only over thousands or hundreds of thousands of years, if genetic variation exists for those traits.

As an example of one of the most direct forms of differential reproduction of human genotypes, let's look at variation in our red blood cells. Some Africans and people of African descent living outside Africa have an inherited condition that causes their red blood cells to distort ("sickle"), changing from disks into forms that often look like crescent moons. Those individuals suffer severe, often fatal sickle-cell anemia. It turns out that in each cell, they have two copies of a gene that codes for a special form of the hemoglobin molecule (hemoglobin makes red blood cells red and plays the crucial role of transporting oxygen to our cells). People who produce only that special kind of hemoglobin are at grave risk of dying from anemia. Those who have just one copy of that unusual gene paired with a normal gene produce two kinds of hemoglobin, one of which is the normally functioning hemoglobin that most of us produce. These people's red blood cells don't sickle as readily, and they usually are asymptomatic (they are said to have the sickle-cell trait).

People with both kinds of hemoglobin are relatively resistant to the most lethal species of malarial parasite *(Plasmodium falciparum)*, which, when it enters the human body, spends part of its life cycle in the red blood cells.[112] People in Africa have been exposed to deadly *P. falciparum* malaria for thousands of generations, and individuals who carry both versions of the gene (the two different alleles, in the language of genetics) have been at a selective advantage.[113] Those people suffer less from malaria and are not anemic, and therefore they outreproduce those who carry identical alleles and thus produce only one kind of hemoglobin, either the normal one or the one that causes sickling.[114] Sickle-cell trait is still found in people in the United States despite the absence of malaria, but it will gradually disappear. Selection operating against the genes that produce sickling hemoglobin can be seen whenever people move into areas where malarial parasites do not occur. The frequency of the sickling allele is much lower in populations of people of African origin now living in Georgia and the malaria-free Caribbean island of Curaçao than, for example, in those living in nearby, malarial Surinam.[115] In the relatively few generations since the slave trade was finally stopped, the disadvantages of sickle-cell anemia have led to a slow decline in the frequency of the formerly protective genetic trait.[116]

The sickling situation clearly illuminates the difficulty selection has in influencing just a single characteristic. It is also an example of how genetic and cultural evolution can interact—after all, the slave trade and malaria control are cultural phenomena. Furthermore, the form of agriculture practiced by a society influences the risks of getting malaria and the selection imposed by it. For instance, is flooded rice cultivation, which provides breeding places for malarial mosquitoes, used? Has the culture dispensed with domestic animals that provide preferred blood sources for the insects?[117] Has cultural evolution led people to build houses with window screens, learn the pertinent habits of the mosquitoes in order to improve control techniques,[118] or provide chemical prophylaxis against the malarial parasites? All are factors in determining people's exposure to malaria and thus the proportion of sickling and nonsickling hemoglobin genes they possess.[119]

We can infer indirectly that strong predispositions for the development of many of human beings' most characteristic features, ones much more obvious than the shape of blood cells, are also programmed into our genes by natural selection and are not often significantly modified by the environment in normal individuals under normal circumstances. Failure of an organism to produce major phenotypic characteristics (e.g., head, arms, eyes, lungs, flowers, leaves) or production of them only in greatly modified form virtually always decreases the organism's reproductive potential in any environment. Selection pressure is strong against any genes that might produce wingless butterflies, plants without chlorophyll, or human beings who are brainless or whose blood will not clot. There is much less reason to believe, as we will see, that many of the most interesting aspects of our natures—such as our differing sexual orientations, religions, languages, aggressiveness, intelligences, senses of humor, and so on—are products of natural selection.

So far, we've seen how natural selection can gradually modify populations of bacteria and fruit flies, snakes and moths, flowers and people. We've taken a rather reductionist approach. To return to the canyon analogy, we've seen the equivalent of how a stream of water from our garden hose, turned on too hard, can erode soil from around the roots of our begonias and change the configuration of our flower bed. But how does evolution lead to that next level of organizational complexity, to communities consisting of different kinds of organisms? How did evolution produce the necessary diversity—bacteria and fruit flies, snakes and moths, flowers and chimpanzees, dinosaurs, human beings with diverse natures, and all the rest? How has the microprocess of natural selection produced the complexity and diversity of the biological "Grand Canyon" revealed by the fossil record and by the living world in which we are embedded? That topic, the generation of the biodiversity that is absolutely critical to humanity and of which we are a part, will entertain us next.

Chapter 3

Our Natures and Theirs

"A man consists of some seven octillion (7×10^{27}) atoms, grouped in about ten trillion (10^{13}) cells. This agglomeration of cells and atoms has some astounding properties; it is alive, feels joy and suffering, discriminates between beauty and ugliness, and distinguishes good from evil. There are many other living agglomerations of atoms, belonging to at least two million, possibly twice that many, biological species. . . . How has this come about?"

—Theodosius Dobzhansky[1]

Evolution is often presented as occurring in more or less of a vacuum and pictured in our minds as a succession of fruit flies ever more eager to sip DDT while dining, or perhaps a sequence of solitary figures marching across the page: ape → ape-man → shambling "Neanderthal" → fully upright human being. In such renderings, it is easy to focus so much on lineage that we lose sight of a crucial fact about our past and our present: without the existence of many other kinds of organisms and the environments they are instrumental in creating, there would be neither fruit flies nor people. Our human natures are totally dependent on the "natures" of other species. We coevolve with many of them, influencing one another's evolutionary paths. Those mental sequences also allow us to think of our natures and theirs as strictly a product of genetic change and to lose sight of the crucial roles of the environment and (especially in human history) of cultural evolution.

Clever as we are, we human beings still require for our survival a diversity of other life-forms, some proportion of today's 10 million species or more with which we share Earth. Our dependencies today on these communities of life are somewhat different from those of our hunter-gatherer forebears or of early primates snatching up insects in bushes during the Paleocene epoch, 60 million years ago. But the dependencies are no less complete. Like all other organisms, we must exchange materials and energy with our environments, and thus we ourselves are elements in ecosystems—communities of species and the physical environments with which they interact.

We breathe the oxygen that was added to Earth's atmosphere by the photosynthetic activities of a multitude of plants and microorganisms, past and present. We drink water circulated and purified by a cycle that depends on plants to draw water from beneath Earth's surface and release it into the atmosphere, and to anchor soils that mediate surface and underground water flows. We live in climates stabilized by such natural ecosystems, a stability that is essential to the success of agriculture. We eat foods derived directly from numerous species in nature or produced by a subset of those species that have been domesticated—all of which themselves depend on other species for their existence. That's why the diversity of other life-forms—often called biodiversity[2]—is critical to the survival and the future of humanity and why its current decline on a global scale is of such concern.[3]

We know why biodiversity is disappearing—the primary reason is that *Homo sapiens* is destroying natural habitats, and our capability of so doing, as we'll see, is largely due to our cultural evolution. Knowing how a vast array of species, including our own, evolved and how these species shaped one another and their environments may help us to stanch the flow of extinction and even to regenerate some of our lost biological heritage. The process of diversification has occurred over billions of years, and an understanding of it is integral to our understanding the generation of our modern human natures. Many aspects of our natures evolved both biologically and culturally to deal with elements of the diversity that surrounds us and with which we have long interacted as food, predators, parasites, shelter, and so on.

It was the world's diversity of life-forms and their intricate patterns of distribution and interaction with their environments that first attracted the attention of Charles Darwin to evolution. Some of his most powerful arguments for the evolutionary origin of biodiversity, and against the prevailing idea that each species was independently created, were drawn from his observations of geographic patterns of species diversity. For example, organisms living in very similar habitats in different parts of the world were often very different from one another, which wouldn't make sense if a master designer had made each species equipped for life in a particular habitat. Moreover, as Darwin noted, species found on an island were more often "related to those of the nearest continent, or of other near islands"[4] than to species on similar islands in different parts of the world. Furthermore, the degree of similarity of island species to mainland species was often related to the depth of the channels separating the islands from the mainland, a sign of how recently the land areas must have been connected. This was "an inexplicable relation[ship] on the view of independent acts of creation,"[5] Darwin argued; if different species of organisms on each island were specially created, why would a creator be limited to forms similar to those on adjacent continents?

Why Not Green Slime?

So where *do* different species come from? Why isn't Earth covered with a rather uniform layer of green slime, a mob of similar descendants of the first organism to evolve from a nonliving mixture of chemicals?

The very same processes described in chapter 2 that cause evolution within populations (microevolution) also cause populations to differentiate and form different kinds of organisms. The creation of new species is, along with evolutionary changes within species, the mechanism that generates major evolutionary patterns—such as the diversification of vertebrates into fishes, amphibia, reptiles, birds, mammals, and so on.

The awesome process of generating the whole diversity of life, going back to the first appearance of living proto-organisms in Earth's early oceans—the grand diversification that is often called macroevolution—all started with variation in environments. No two environments were identical then. Tides, salt concentrations, bottom composition, depths, currents, and other factors varied from place to place. Nor are any two environments identical today. And no environment was constant then or is constant today. Seasonal weather cycles, changes in climate, changes in topography through such mechanisms as erosion, seabed spreading, mountain building, glaciation, and alteration of the courses of streams—these are just a few of the physical processes that promote the evolution of different species. In turn, changes in each species population usually alter the environment of other species (predators, competitors, etc.) that interact with them in the same locality.

For example, DDT—my old friend from the Animal House—is especially deadly to those insects that normally eat mites. DDT was banned in the United States in 1973, but before that, it caused a great deal of environmental harm. When applied to a field, it greatly reduced the populations of mite predators and caused mites to undergo population explosions, in turn altering the environments of many other populations of plants and animals. Before World War II, spider mites[6] were a group of minor pests. After widespread use of synthetic organic insecticides (of which DDT is just one) had destroyed many of their natural enemies, spider mites became, in the words of one group of scientists, "the most serious arthropod pests affecting agriculture world-wide."[7] Similarly, the extermination of passenger pigeons nearly a century ago, ending with Martha's lonely demise, may curiously have been an important factor in encouraging outbreaks of Lyme disease. The pigeons occurred in flocks of billions and fed on acorns and beechnuts (mast); with the birds' extinction, more mast was available to deer mice,[8] and populations of the rodents flourished. That made the environment more favorable for mouse ticks,[9] which transmit the spirochaetes[10] that cause Lyme disease. Commercial hunters, making the environment more favorable for the mice, ticks, and spirochaetes by driving the passenger pigeon to extinction, made it less favorable for *Homo sapiens.* Martha's revenge?

Such changes in the living environment can be as important to the diversification of organisms, to the process of creating new species, as changes in the physical environment. All populations, human and nonhuman, are exposed to different combinations of selection pressures, depending on the species around them. In addition, mutation, migration, and genetic drift will also, by chance, ordinarily affect separate populations differently, generating more geographic variation within species. Different populations of sexually reproducing species—that is, groups of individuals of the same species occupying different places—are always at least slightly different genetically. It is this geographic variation that forms the basis for the most common process by which new species evolve.

Geography and the Origins of Species

The great evolutionist Ernst Mayr first clearly synthesized what has become the standard view of how most new species arise—that is, of the source of Earth's astonishing species diversity.[11] Because this view is intimately related to the geographic variation in organisms observed by Charles Darwin and many others before and since, the diversification process is sometimes called geographic speciation. A starting point in understanding Mayr's model is the accepted view of taxonomists and evolutionists that when two kinds of organisms live together in the same region (that is, are sympatric) without interbreeding, they belong to different species.

Swainson's thrushes and hermit thrushes around the Rocky Mountain Biological Laboratory (RMBL) in Colorado fit that description. Although these two kinds of thrushes don't interbreed, they closely resemble each other (they are most easily distinguished by the different songs of the males) and are considered members of the same genus—the taxonomic category that groups together similar species.[12] How could these two species have evolved? They presumably arose from a common thrush ancestor, which we might call the "swermit" thrush, but how? Under Mayr's model of geographic speciation, at some point in the past, two populations of the single ancestral species became spatially isolated from each other. Perhaps a period of climatic drying produced a grassland barrier to swermit thrush movement in what was once a continuous stretch of suitable woodland. Or perhaps a chance group of swermit thrush colonists was blown across a mountain barrier in a storm and then occupied previously vacant habitat. Whatever happened, the two ancestral populations became allopatric—isolated from each other—and could follow different evolutionary paths. Their environments were different (remember, no two are identical), so their selection pressures would have been different. Over time, chance mutations and the random effects of genetic drift would have influenced the two populations differently.

As a result, the two swermit thrush populations began to diverge, each on a separate evolutionary trajectory. Thousands (or millions) of years later, the populations again expanded in range and came to occupy the same area; that is, they became sympatric. Perhaps the climate turned moist again and woodlands recolonized the former grassland. Or perhaps another group of colonists was blown across the mountains. If sufficient divergence had occurred, the reunited groups would have been unable to breed with each other and thus would have remained distinct. Perhaps they had already evolved the different songs of the males, and sing as they might, those of one group could not attract females from the other group. Perhaps there was some intergroup mating but the genetic endowments of the two groups had become so different that the resulting hybrids were sterile. Whatever the cause, there were now two noninterbreeding kinds of thrushes where previously there had been only the swermit thrush. Speciation had occurred, and today my wife, Anne, and I have both Swainson's thrushes and hermit thrushes around our mountain cabin at RMBL.

My description of the Mayr model is necessarily simplified, but a key thing to keep in mind is that it predicts the existence of very diverse patterns of geographic variation in organisms at any given time.[13] Those patterns would be related to the mobility of organisms, the composition of communities, the strength of environmental differences (and thus selection pressures) across geographic gradients, and the degree and length of time of isolation. The "snapshot" of organic diversity that one gets today fits that picture; it shows every conceivable level of diversification. At one end of the spectrum are a multitude of geographically isolated but extremely similar populations that are difficult even to distinguish statistically. At the other end are many clearly distinct species living together without interbreeding, in which every individual can be indisputably assigned to one species or the other (such as wolves and coyotes, mallard and pintail ducks, or tiger swallowtail and black swallowtail butterflies).

The Meaning of Species Revisited

Because Swainson's thrushes and hermit thrushes remain distinct even though they live together and because mating of the two types of thrush does not ordinarily result in hybrid individuals, they are what Mayr originally called biological species. Such failure to interbreed is fundamental to his biological species concept, which emphasizes the reproductive isolation that keeps biodiversity from disappearing through interbreeding and leads taxonomists to try to classify together in the same species organisms that are actually interbreeding or that can *potentially* interbreed.[14]

Unfortunately for taxonomists, there are numerous situations in between those same-species/different-species extremes. For example, there is often hybridization between members of populations that are sufficiently different for taxonomists to have designated them as different species.[15] Mules are hybrids between male donkeys and female horses, both distinct species. Hybrids between indigo buntings and lazuli buntings (two sparrow-like birds) are common in the Great Plains, as are hybrids between other bird species.[16] Taxonomists usually consider the parental types to be different species in these instances, but the species have not yet evolved large enough differences between them to prevent an exchange of genes when their populations intermingle.[17]

All this would be fine except that human beings like to categorize things neatly, and taxonomists are human beings. They would like to have some easy rule for determining whether two individuals (or two populations) belong to the same species. The interbreeding criterion works pretty well for organisms that are living together (where one can observe actual interbreeding),[18] but it has long been recognized as hopeless when it comes to allopatric organisms.[19] There, one is faced with guessing whether the two populations are sufficiently distinct that if they reunited naturally they would not interbreed often enough[20] to meld the two populations. Laboratory tests of the capacity to hybridize can help us make better guesses but cannot definitively address the issue of whether or not hybridization would occur under natural conditions (where, for instance, behavioral mating displays may be better expressed) or what the viability of offspring would be under those conditions. Mallards and black ducks rarely interbreed in the wild, even when living in the same ponds. But they readily interbreed and produce fertile hybrids when a male of one species is isolated with a female of the other in a zoo. Similarly, polar bears and brown bears do not interbreed when they come into contact in nature, but they can produce fertile hybrids in zoos.[21]

Human "Racial" Variation

Is the obvious geographic variation in modern *Homo sapiens* a sign of incipient speciation? Not in the least. The most important thing about human variation is that patterns in one characteristic (say, skin color) are not congruent with those in another (say, nose width).[22] If you map variation in those characteristics, the patterns of change in one do not correspond with those in others, as shown on the accompanying maps. Geographic variations in various characteristics in our species are not strongly correlated with one another,[23] nor are geographic variations of different attributes within other organisms that have been studied thoroughly.[24] That's exactly why Harvard University evolutionist Edward O. Wilson (of sociobiology fame)[25] and the late Cornell University taxonomist William L. Brown long ago pointed out that subspecies, which are taxonomic subdivisions of a species based on the geographic pattern of a few characteristics, often are not biological entities.[26] Widespread species thus can be divided into any number of different sets of "subspecies" simply by selecting different characteristics on which to base them. Races in people are arbitrarily defined entities; individuals from different geographic regions are capable of mating and producing fully fertile offspring. Indeed, they are often eager to interbreed, as has been seen throughout history in the reproductive consequences of armies occupying new territories. Attempts to treat divisions of humanity based primarily on skin color as natural evolutionary units have always

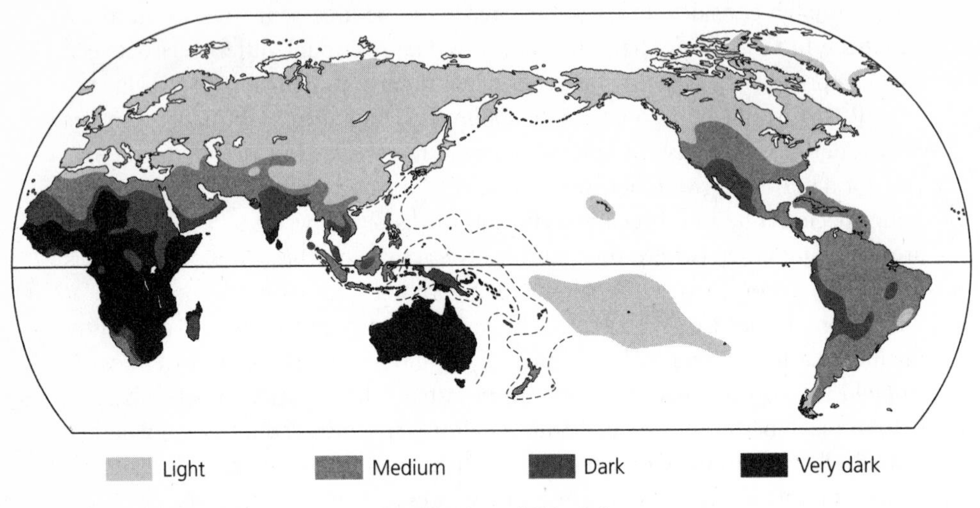

a. Distribution of Skin Color

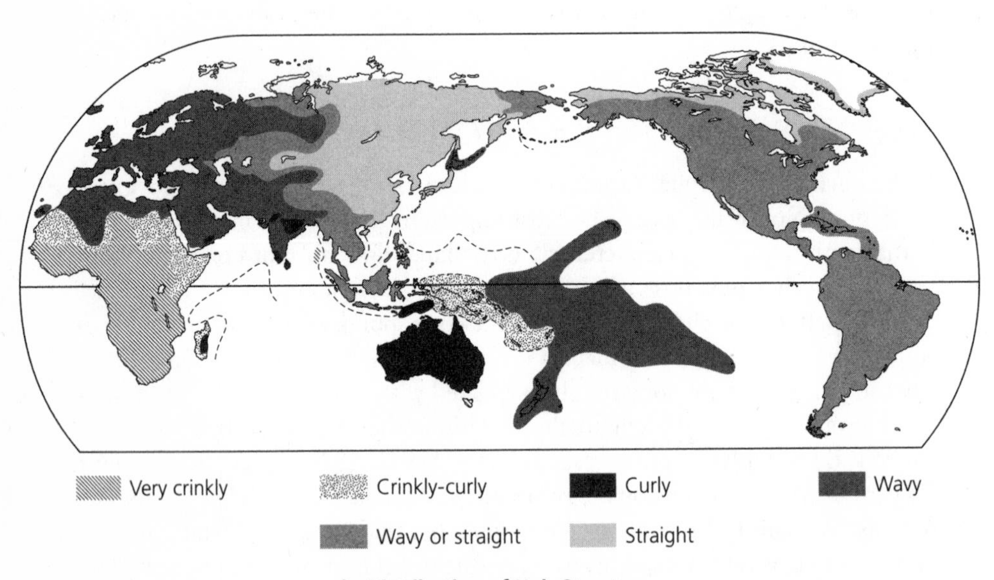

b. Distribution of Hair Structure

Discordant geographic variation in human characteristics. These maps indicate that there is no way to divide humanity into a series of distinct races because different characteristics do not vary in the same patterns. Data from Biasutti 1967.

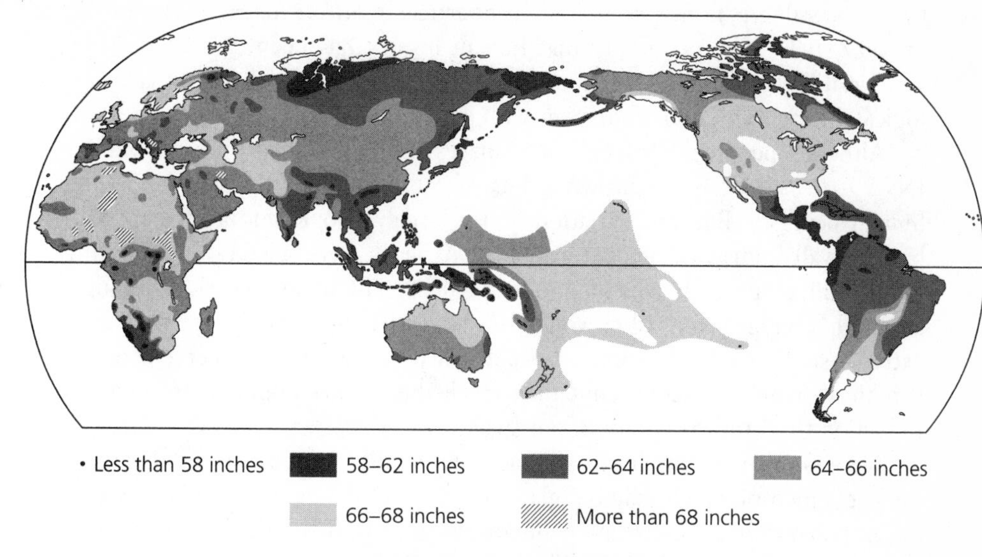

c. Distribution of Average Height

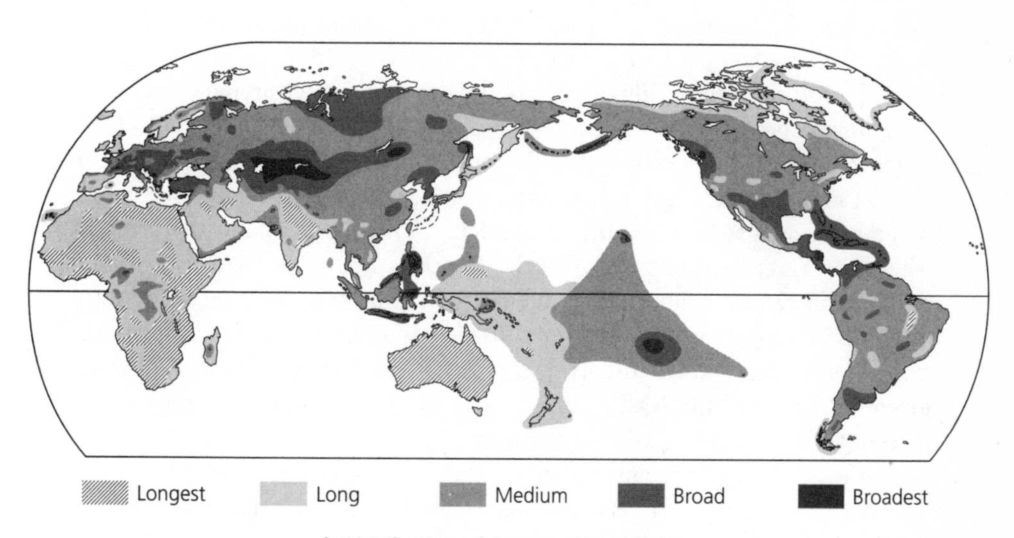

d. Distribution of Average Head Shape

been, and still are, nonsensical.[27] A recent article in *Science* magazine summed it up very well: "The possibility that human history has been characterized by genetically relatively homogeneous groups ('races'), distinguished by major biological differences, is not consistent with genetic evidence."[28]

Although people differ from chimpanzees only in less than 2 percent of their DNA base pairs,[29] that still leaves lots of room for genetic variation among human beings.[30] But this variation is not largely geographically (or racially) based. If all human beings except native Africans were wiped out, humanity would still retain somewhat more than 90 percent of its genetic variability, although average skin color would be darker. If only the Xhosa people of extreme southern Africa survived, we'd still have lost only 20 percent of the genetic variation currently contained in all the world's populations—a trivial reduction from the viewpoint of our species.[31]

Geographic variation in the genetic background of characteristics that catch the eye, much of which may be relatively trivial in the evolution of *Homo sapiens*, has obviously been extremely important socially, being the basis of notions of "race." Color variation, as we'll see, has some adaptive significance. But the erroneous notion that it is possible to "color-code" human beings for quality (as one racist scientist put it) has, of course, been a source of endless trouble, causing violence and discord within states that have diverse populations and contributing to the causes and conduct of various wars.[32]

What is the selective importance of the geographically varying characteristic in human beings to which the most attention has been paid—skin color?[33] It is generally agreed that skin color is in some way related to history of exposure to solar radiation because the dark brown or black pigment melanin serves as a natural screen against damaging ultraviolet (UV) radiation.[34] But that's almost where the agreement ends. It is likely that the value of melanin as a barrier against skin cancer, sunburn, and the destruction by UV radiation of a photosensitive B-complex vitamin known as folate[35] is an important factor for selection favoring dark skin in the Tropics.[36] But it can hardly be the whole story, given that some of the least sunny areas in the world are in tall tropical rain forests, which also house very dark-skinned peoples.[37]

Although its normal onset after the main reproductive years might make skin cancer appear to be a relatively weak selective agent on skin color, there is evidence from the typically early deaths of albinos in tropical areas that other negative effects of a relative absence of melanin appear early in reproductive life.[38] Lighter skin in populations living in the polar regions has been assumed to be related to the production of vitamin D by the action of ultraviolet radiation on the vitamin's precursors in the deeper layers of the skin.[39] For example, Swedes and other northern farming peoples were thought to have light skins so that they could produce vitamin D in this way from precursors obtained from their vegetable food. The Inuit people, Lapps, and other darker-skinned northern peoples get their vitamin D from fish and other dietary sources that are rich in it. At one time, it was erroneously thought that even dark-skinned people can produce adequate vitamin D in polar regions.[40]

What factor other than solar radiation is involved in the origins of skin-color differences? Anthropologist Marvin Harris believes that natural selection for dark skin was reinforced by culturally induced mate choices. In sun-baked areas, darker-skinned infants were preferred because social knowledge told people that heavily pigmented children were more likely to grow up free of disfiguring malignancies. Darker-skinned mates were considered superior because they survived longer and reproduced more. Harris points out that on original contact, African peoples considered Europeans repulsive because they associated white skin with the devil and death.[41] The jury is still out, but this is at least a plausible theory.

In any case, though it is quite common for a species as widespread as *Homo sapiens* to display geographic variation, there is no sign of substantial genetic differences among human groups, certainly none that would presage speciation. Indeed, it is likely, as we will see, that all modern human populations are derivatives of a small group of African populations that lived a mere few thousand generations ago. There are also signs of substantial gene flow having occurred among human populations ever since, and there are high levels now. So it is hardly surprising that we show only superficial interpopulation variation and no signs of splitting into several human species.

Ernst Mayr, Jack Dempsey, and Adaptive Radiation

One generally accepted process of speciation, as noted earlier, involves isolated populations evolving independently until they become so differentiated that they do not interbreed when they come into contact again.[42] Such differentiation in isolation appears to be the dominant mechanism of speciation in sexually reproducing organisms,[43] but there are still controversies about how ubiquitous that mechanism is.[44] Other models of speciation have been proposed to supplement Mayr's. In those models, populations may be able to differentiate even when they meet at the edges of their ranges but do not overlap (parapatric speciation) or even when they are not isolated geographically (sympatric speciation).[45] Furthermore, we know from the experimental populations of bacteria tracked for many thousands of generations in Richard Lenski's laboratory (see chapter 2) that neither different selection pressures nor small population size leading to divergence through drift is always necessary for diversification. Those populations of bacteria did not evolve to exactly the same points, even though they apparently were subjected to the same pressures. This suggests that diversification, at least in the relatively short run, can occur simply because of historical accidents such as the order in which mutations appear in a population.[46] But these are all details, and I suspect that Ernst Mayr's model will remain the principal explanation of the generation of diversity in sexually reproducing animals, including our ancestors.[47]

There are, however, patterns in some groups of animals that strongly suggest that the Mayr model needs the sort of supplementation just mentioned. One of the truly amazing of those, and a pattern that also provides insights into present-day conservation problems created by human planetary dominance, is the existence of "species flocks" or "species swarms" of cichlid fishes[48] in the three largest East African lakes—Lake Malawi, Lake Tanganyika, and Lake Victoria.[49] Cichlids are familiar to tropical fish fanciers—the family Cichlidae includes species such as angel fishes and Jack Dempseys (so named for their belligerence, a typical cichlid attribute). They are favorites of the aquarium trade—fish that millions of people, me included, have enjoyed breeding in aquariums tens of thousands of miles from the species' native habitats. The cichlid species in the African lakes range in length from a couple of inches to more than two feet, and they are among the best examples of the way in which evolution can provide great aesthetic benefits to human beings. They supply it not only in the form of great physical beauty (the males of some species are as gorgeously colored as many birds and butterflies) but also in their beautifully intricate behavior.

In Lake Victoria, there were until recently at least 500 cichlid species that once were all included in the same genus.[50] That group of cichlids dominates all the East African lakes, where they have time and again produced an astonishing array of forms, differing mostly in the ways they eat and the equipment they eat with. There are bottom, midwater, and surface feeders; herbivores that scrape algae from rocks, ones that scrape algae from aquatic plants, and ones that eat the plants themselves. There are all manner of carnivores, some truly bizarre.

African cichlids from Lake Victoria. Above, surface feeder; middle, midwater feeder; bottom, bottom feeder. These are just a few of the species in this famous adaptive radiation.

Fishes that eat scales they snatch off other fishes have evolved independently in each of the three largest lakes.[51] Some of the cichlids look and act like herring, eating tiny floating organisms[52] from midwater. Others eat snails, some crushing the shells with their jaws and others crushing them with a second set of jaws in their throats.[53] Many eat other fishes; some of these cichlids look and act like bass, mackerel, or barracuda. One Lake Malawi species lies on its side looking like rotting meat, soon to snap up unwary scavengers that chance upon the "free" meal.

Some of the cichlids' breeding habits are also peculiarly oral.[54] A majority of African lake cichlids use mouth-brooding—protecting eggs and young by sheltering them in the mouth—which is carried out by both sexes or by either sex alone. In most mouth-brooders, the female gobbles up the eggs as soon as they're laid, before the male fertilizes them. Then she attempts to snatch up bright spots ("egg dummies") that adorn the anal fin or pelvic fins of the male. While she's doing that, the male releases sperm, which are sucked into the female's mouth, thereby fertilizing the eggs. The eggs are subsequently carried in the female's mouth, where the young hatch and grow for days or weeks, until the mother releases them in an auspicious habitat. Mouth-brooders tend to have only a few, quite large eggs. Oral brooding must provide better protection from predation than the more widespread cichlid habit of laying eggs and defending the young in a nest. But the system is far from foolproof. The African lakes that are full of animated incubators are also haunted by other cichlids called paedophages—infant-eaters—specialized to smash or suck the babies out of the mouths of the brooding parents. Several species don't even wait that long, instead rushing between a spawning pair to gobble up each egg as it is laid. The phenomenal behavioral diversity of the cichlids also creates selection pressures on noncichlid fish species that occupy the lakes. For example, in Lake Tanganyika, a species of catfish has evolved an extraordinary strategy. Its females search for mating cichlids. When one finds a spawning pair, she swims between its members, eats one of their eggs, and lays an egg of her own. The male catfish follows and fertilizes the egg, and he in turn is followed by the female cichlid, which circles around frantically, filling her mouth with catfish eggs. The female cichlid then fosters the young catfish until the mouthful of spiny young becomes too much to bear!

Speciation in some of these great lakes' cichlids apparently occurred rapidly. The members of the cichlid species flock in Lake Victoria are all quite closely related to one another, and apparently all the cichlid groups found there were derived from a common ancestor only a few hundred thousand years ago.[55] Populations became fragmented into isolated units in diverse ways: in satellite lakes periodically rejoined with Lake Victoria;[56] perhaps in fragments of the main lake split off during prolonged dry spells (as seen today in Lake Kyoga);[57] on underwater rock reef "islands" isolated from one another by mud bottom;[58] and so on.

Lake Victoria apparently was dry or nearly dry during the most recent glaciation, so the more than 500 cichlid species that are found nowhere else must have evolved there in the past 12,400 years[59]—though not, as originally thought, from one common ancestor but from a relatively few surviving species from an earlier diversification. That is the most extreme documented example of rapid vertebrate speciation.[60] It has been suggested, however, that cichlid species restricted to rocky outcrops in Lake Malawi may have evolved in even less time—in a mere 200 years![61] All of this strongly suggests that differentiation without geographic isolation (sympatric speciation) was at least somehow involved,[62] as it may have been in generating some fish species diversity in recently formed northern Canadian lakes.[63]

The basic explanation for the generation of cichlid species flocks in Lake Victoria involves three kinds of selection pressure acting in various combinations. First, some differentiation was due to geographic differences among isolates around the margin of the lake, especially during times when the lake level was much lower.[64] However, this is probably the least important of the contributing factors. Second, and potentially the major force for creating new species, is the reproductive (but not geographic) isolation of color forms that apparently traces to sexual selection for spectacularly (and differently) colored males; the females prefer colorful mates.[65] This can be a form of sympatric speciation. Finally, there was also some disruptive selection—selection in which two or more different phenotypes were favored over intermediates between them.[66] Response to disruptive selection, primarily based on different diets, is hypothesized to have been relatively rapid because of the cichlids' evolutionarily "flexible" jaw apparatus.[67] This would allow coexistence of the reproductively isolated units because individuals with intermediate jaw structures resulting from hybridization between those units are less successful than nonhybrid individuals. Thus, sexual selection has promoted speciation, and disruptive selection has produced a wonderful example of what is called adaptive radiation.[68] Adaptive radiation is the relatively rapid diversification of a single evolutionary line into an array of species adapted to use different resources or habitats.[69]

Careful studies in laboratory and field suggest that Lake Victoria is again losing cichlid diversity because of declining visibility as the lake is increasingly overfertilized[70] by runoff from regional agricultural activities.[71] This is compounded by elimination of algae-eating cichlids caused by the introduction by people of exotic predators, which caused a buildup of floating algae in the water and further reduced visibility.[72] As a result, there is less color diversity among rock cichlids in murky water than among those in clear water not far away.[73] Mate choice apparently becomes more difficult, thus rendering sexual selection less potent and lowering barriers between species as the water gets cloudier and visibility is reduced.[74]

It is sobering to think that human activities now operate on a scale commensurate with that of the natural forces that create and destroy entire biological communities. At least half of the cichlid species of Lake Victoria disappeared between 1982 and 1986, a period during which human intervention sent the lake through convulsive changes in the water quality and the biological community that echo to this day.[75] Indeed, the diversity of all of the East African great lakes is now threatened by human activities.[76] The wonderful scale-eaters, baby-suckers, herring cichlids, and other fishes of Lake Victoria now survive chiefly in a handful of tiny marginal lakes[77] or, in a few cases, only in a captive breeding program launched by African, American, and European scientists as a communal act of last resort.[78]

Adaptive radiations sometimes occur when a new resource becomes available—as when the diversification of flowering plants some 100 million years ago provided a wide variety of new food resources for plant-eating insects.[79] They also have often occurred when colonizing plants or animals have reached habitats previously unoccupied by their sort of organism. That was the situation that presumably greeted the first cichlid invaders of the African lakes. Classic cases are the radiations of Darwin's finches on the Galápagos Islands[80] and of honeycreepers on the Hawaiian Islands[81] (see the figure). In each case, the descendants of a single flock of invaders apparently spread gradually from island to island of the archipelago, where they speciated in the relative isolation of individual islands. After they had passed the point where they could no longer interbreed, occasional migrants presumably recolonized their previous home islands and reestablished sympatry with the ancestral populations. Unfortunately, although these models seem eminently reasonable, it has not as yet been possible to test them. But they follow the classic Mayr model of differentiation in isolation; there is no necessity for sympatric speciation to be involved in those adaptive radiations. Most, in fact, can probably be explained without invoking it.

Many adaptive radiations also show up in the fossil record, a classic example being that of mammals after the extinction of the dinosaurs some 65 million years ago. Many organisms disappeared from the fossil record at the Cretaceous-Tertiary (K-T) boundary, which is why there is that boundary—the limits of geological time periods are defined largely by sharp changes in the composition of the fossil floras and faunas. That extinction event, almost certainly caused by the collision of an extraterrestrial body with Earth, was responsible for wiping out the previously dominant dinosaurs and emptying many ecological niches, into which the mammals then evolved.[82] The realization that such a collision could wipe out human civilization (and perhaps generate an eventual adaptive radiation that would replace the mammals) has spawned a great deal of worry and a number of profitable movies.

We hear much about extinction now, and for good reason, but it was rarely spoken of before Darwin brought focus to the issue.[83] He viewed extinction as the converse of speciation: "On the theory of natural selection the extinction of old forms and the production of new and improved forms are intimately connected together."[84] He even analogized rarity as a prelude to extinction as being like sickness as a prelude to death.[85] Before Darwin, most people believed that

A
B
C
D
E
F
G
H
I
J
K
L
M
N
O
H. Douglas Pratt
1977

Earth's plants and animals had been produced in a single miraculous episode and represented a carefully designed "great chain of being" from which no links would disappear.[86] The concept of extinction was as contrary to this belief as was speciation.

The Fossil Record and Evolutionary Patterns

The fossil record[87] forces us to look at the many constraints that history puts on what natural selection can accomplish in the generation and persistence of species. For instance, were it not for that cosmic event destroying the dinosaurs, it is likely that human beings never would have evolved. The ecological niches for our ancestors might never have opened up, and Earth today might be dominated by intelligent, social reptilian descendants of the dinosaurs, with a few scruffy and pestiferous egg-eating mammals living in their shadows. This is but one of many examples of the way in which historical contingency influences the course of evolution.[88]

Another is that the fossil record often shows how the evolutionary process unexpectedly co-opts organs that evolved in response to one set of selection pressures to fill another function. Classic examples of such co-option are the evolution of the protective scales of reptiles into the flight feathers of birds and the evolution of bones that were originally parts of the jaws of reptiles into the tiny bones of our inner ears. That natural selection has had to work with the variation available has sometimes led to designs that are clearly less than optimal. This is why so many people suffer from lower back pain by the time they've passed the age of sixty.[89]

One of the most interesting questions generated by the fossil record has to do with the *rates* of evolution not only within lineages (as discussed in the previous chapter) but also in the formation of new species. Many scientists have thought of evolutionary history as a story of gradual and more or less constant change. But some evidence in the record has been interpreted as showing that most periods of geological time have been periods of evolutionary stasis or equilibrium, when not much appeared to be going on—the same old fossil species is found in layer after layer of rocks, with each species appearing much the same as specimens from younger and younger strata are examined. Such periods of

---

Opposite: Hawaiian honeycreeper adaptive radiation. All of these birds are descendants of a common ancestral species that diversified in the islands. A, Mamo, *Drepanis pacifica;* B, Iiwi, *Vestiaria coccinea;* C, Crested Honeycreeper, *Palmeria dolei;* D, Ula-ai-hawana, *Ciridops anna;* E, Apapane, *Himatione sanguinea;* F, Akiapolaau, *Hemignathus wilsoni;* G, Kauai Akialoa, *Hemignathus procerus;* H, Akepa, *Loxops c. coccinea;* I, Amakihi, *Loxops v. virens;* J, Creeper, *Paroreomyza maculata bairdi;* K, Maui Parrotbill, *Pseudonestor xanthophrys;* L, Ou, *Psittirostra psittacea;* M, Grosbeak Finch, *Psittirostra kona;* N, Nihoa Finch, *Psittirostra cantans ultima* (female); O, *Melamprosops phaeosoma.* From painting by H. Douglas Pratt, in Raikow 1977.

evolutionary equilibrium were "punctuated" by periods of simultaneous rapid change within species and a proliferation of new species.[90]

There is, of course, no reason why evolution should always proceed at the same pace, and there are reasons why speciation might be especially common at the same time that changes within species are accelerating. Rapid environmental change—the opening of myriad new niches after a mass extinction event or by the ending of an ice age, human-caused transformation of land surfaces, or the toxification of Earth by synthetic organic chemicals—may dramatically change selection pressures on populations in general, enhance geographic differences, and spatially fragment previously continuous populations. It could thus increase rates of both microevolution—changes within species—and speciation. Whether or not the punctuated equilibrium model, as opposed to a model in which evolution largely proceeds gradually (although with varying rates), accurately depicts the history of evolution has been controversial ever since Niles Eldredge and Stephen J. Gould first proposed it in 1972.[91] Unhappily, the fossil record normally represents a biased sample of past life,[92] and it may be inadequate to provide answers. There may well be a mix of gradual change and punctuation, with one or the other occurring at different times and in different places and with various degrees of both represented. Indeed, the Lenski group's long-term laboratory studies of bacterial populations showed periods of stasis and rapid evolution that may be akin to the punctuated equilibria claimed for the fossil record.[93] Other studies suggest that at least some patterns of what appear to be punctuation are, in essence, statistical artifacts generated by microevolutionary processes.[94] In any case, punctuated equilibrium is sometimes cited in the anthropological literature as an "explanation" for the relatively abrupt appearance of language or other human traits,[95] circumstances in which it in fact has no explanatory power whatsoever (and such power was not claimed by Eldredge and Gould).[96] The theory of punctuated equilibrium simply states that evolutionary change happens in bursts; an explanation of why this change occurs still relies on the microevolutionary processes described in chapter 2. The theory does not suggest new evolutionary mechanisms, nor does it cast light, as one linguist has suggested, on how so complex a characteristic as a grammatical language could appear suddenly.

Ever since people began to wonder about evolution, there has been an interest in whether the process has some kind of goal. Indeed, one of the most persistent questions about evolution has been "Is it progressive?" It turns out to be a complex issue, at the center of which is a semantic problem: what exactly is meant by *progressive*? In the sense of there being a great chain of life advancing from the primordial slime and single-celled organisms to mammals and those upper-class Englishmen, the answer is a clear no. There is no sign that evolution has "progressed" from, say, bacteria to people. Indeed, one might claim that our beautifully designed, highly efficient, bacterial distant relatives are actually the pinnacle of evolution. They do a great job of rapidly reproducing DNA; if their genes were capable of being selfish and feeling joy, those genes would be very happy. The total weight—that is, the biomass—of bacteria may be the greatest of any group of organisms.[97]

Stephen J. Gould is correctly skeptical about the general progressiveness of evolution—that it moves toward goals or in some way improves organisms[98]—but his evolutionist sparring partner Richard Dawkins points out some limited ways in which evolution might be considered progressive.[99] Within an evolutionary line, the "fit" of an organism into its ever-changing environment inevitably changes (progresses) under the lash of natural selection—or the line dies out. This is especially obvious when major selective forces operating on an organism are generated by other species in a biological community. But this is a narrow view of progress and not what is usually meant by the term. Ernst Mayr expressed the consensus in a recent book: "There simply is no indication in the history of life of any universal trend to, or capacity for, evolutionary progress. Where seeming progress is found, it is simply a byproduct of changes effected by natural selection."[100]

Evolution of Communities

Evolution shapes the characteristics of species, but in so doing, it also shapes the characteristics of communities. Remember, selection pressures are generated not only by physical alterations of the environment but also by other organisms in it. When two species are ecologically intimate, closely influencing each other's lives as do predators and prey or hosts and parasites, each normally becomes a major source of selection operating on the other; in such situations, coevolution occurs. As a species, human beings are ecologically intimate with lots of organisms, from cows and crop pests to mackerel and malarial mosquitoes, and coevolution affects us in many ways.

The classic case of coevolution is the interaction between plants and the organisms that feed on them. Butterflies and the plants their caterpillars eat was the case study with which plant evolutionist Peter Raven and I first explored coevolution in the early 1960s. Because plants cannot run away from their predators, they have been selected for static defenses, the most obvious example being the spines that cacti grow. Less obvious, but overall more important, are the chemical defenses that plants have evolved to poison, disorient, intoxicate, starve, or trap (in goo such as pine resin) the creatures that try to eat them.[101] A zebra eats parts of a plant with a hallucinogenic defensive chemical, goes off and tries to mate with a lion, and never molests the plant again. Human beings adopt those chemical defenses to their own purposes, as spices, medicines, pesticides, and recreational drugs.[102]

Coevolution is, by definition, a two-way street. While the plants were evolving chemical defenses, the organisms that eat plants (herbivores)—from viruses to mammals—were evolving ways to avoid or detoxify the chemicals.[103] In these sorts of coevolutionary races,[104] each player's evolution must be progressive in Dawkins's sense or it will go extinct.

We live with a partial legacy of the coevolutionary races between plants and herbivores; the long evolutionary experience of herbivores with plant poisons almost certainly makes it easier for "pests" among them to evolve resistance to the poisons we invent to kill them. The herbivorous pests often will already have evolved enzyme systems that detoxify the defensive chemicals of certain plants, systems that with some evolutionary modification also will detoxify synthetic poisons. The carnivores that normally eat herbivores, on the other hand, have less experience with poisons and are normally more susceptible to pesticides. Presumably, they are typically less adept at evolving detoxifying systems of their own. Our ignoring the lessons of coevolution is one more reason why the use of synthetic organic pesticides has been a relatively unsuccessful approach to controlling the pests that attack crops. The poisons tend to have much more severe effects on the predators that attack the pests than on the pests themselves. The results can be plagues such as those of spider mites mentioned earlier.

Understanding coevolution—of plants and herbivores, parasites and hosts, predators and prey, competitors, and mutualists (species that benefit from association)—has been the focus of a major effort among biologists in recent decades,[105] and very substantial progress has been made. For instance, new methods of reconstructing evolutionary history have allowed examination of coevolutionary interactions over geological time. As a result, there is growing evidence that, as Peter Raven and I first speculated, plant–herbivore coevolution was a major factor in the diversification of both flowering plants and insects.[106] Moreover, experiments using bacteria and the viruses that attack them (bacteriophages) show that, like evolution in response to physical and chemical changes in environments, coevolution can occur very rapidly.[107]

The definition of coevolution Peter Raven and I adopted has subsequently been stretched to include interactions between any two evolving systems, an extension that sometimes seems quite useful.[108] One such extension that is important to the themes of this book is the coevolutionary interaction between Earth's biota (plants, animals, and microorganisms) and Earth's climate.[109] Organisms influence the climate by changing the reflectivity of the planet's surface—where forests grow, for instance, more solar energy is absorbed than in deserts, where more of the energy is reflected away. They influence it by adding greenhouse gases to and subtracting them from the atmosphere, and they influence it by altering the hydrologic (water) cycle, for instance by plants holding soil in place with their roots and slowing water's flow.[110] Climate changes, in turn, alter selection pressures on virtually all organisms; as we'll see, a change in climate may well have played a key role in the splitting of our lineage from that of chimpanzees.

Cultural Evolution

Human beings, like our close relatives the chimpanzees and all other organisms, are products of Darwinian processes—processes by which the inherited (genetic) information carried by organisms is modified and reorganized generation after generation. But those processes, continually operating to maintain or increase genetic fitness in the face of environmental challenges, have given human beings a special ability, the capacity to invent, modify, store, and transmit a large body of culture. Culture is nongenetic (extragenetic) information embodied in stories, songs, tools, customs, morals, art objects, oral histories, books, television shows, computer databases, satellite images, electron micrographs, and so on.[111] Although other animals, from birds to chimps, possess culture in the sense of showing their offspring traditional ways of doing things, no other creature exhibits culture remotely on the scale that humanity does.

The evolution of that body of extragenetic information—cultural evolution—has been centrally important in making us the unique beasts we are.[112] Cultural evolution rests on a foundation of genetic (or biological) evolution—especially that of our brains and tongues—but can proceed at what by comparison is a lightning pace. It allows cultural (nongenetic) differentiation of human populations with a speed no Mayrian process could match.

There are several important differences between genetic and cultural evolution besides the most obvious one, that cultural evolution can vastly outpace genetic evolution because it's not constrained by generation time. Our genes are passed only from one generation to relatives in succeeding generations. In contrast, the units of culture—ideas, basically—are passed among both relatives and nonrelatives not only between generations (in both directions) but also within generations.[113] Children, for example, can exchange ideas among themselves and also influence the thinking of their parents, grandparents, and teachers. In addition, cultural evolution can involve transmission over many generations to nonrelatives in totally different cultures. More than a century after Darwin's death, his ideas continue to influence people on every continent speaking dozens of languages.[114]

Cultural evolution, unlike genetic evolution, also *can* involve the inheritance of acquired characters. Jean-Baptiste de Lamarck, the eighteenth-century evolutionist who thought such transmission possible, was correct with respect to cultural evolution but not, of course, with respect to biological evolution. Suppose, for instance, you invent a new drug that helps cure colds. That drug (or its formula) could be passed on to your children, and it could make them, and future generations, less likely to suffer bad colds. But no matter how much of it you take, you won't change the genes you pass on to your children one whit—and unless they follow your example by taking the drug (cultural transmission), they won't be a bit more resistant to colds.

Unhappily, the details of the process of cultural evolution are not well understood. Cultural evolution does not ordinarily involve natural selection among individuals, but much more often than biological evolution it may involve differential survival of groups. For example, it is possible that certain religious beliefs may spread because they increase the stability of societies that acquire them.[115] But some aspects of cultural evolution's astonishing complexity are beginning to be worked out. For example, researchers are beginning to understand why some innovations spread rapidly and some do not spread at all. The extent of an innovation's diffusion and the consequences of its adoption depend on a wide range of cultural factors, from the method of the innovation's implementation to the degree of stratification within the society in which diffusion is occurring.[116] The existence of what have been termed social traps, in which short-term local reinforcement of individual behavior counters the long-term best interests of the individual and society, is another factor shaping cultural evolution.[117] The term *social trap*, which originally came from an analogy with funnel-shaped fish traps, refers to situations in which societies get themselves headed in a direction that later may prove unpleasant or lethal but that they see no way in the present to avoid or back out of. A simple example is the many individual short-term rewards of fossil-fuel use, which have led to organized opposition to attempts to reduce that use, despite its potentially catastrophic long-term effects through its contribution to global warming.[118]

Of great significance in the context of human natures is the coevolution of genes and culture.[119] There are many ways in which culture can alter selection pressures, a classic case being sickle-cell anemia, in which, remember, farming and building practices influence selection by changing exposure to malaria-carrying mosquitoes.[120] Another well-known example involves selection on genes affecting lactose tolerance. Many human adults (termed malabsorbers) do not produce enough of the digestive enzyme lactase to break down lactose, an energy-rich sugar in milk, and drinking milk or eating milk products gives them gut trouble. Other adults (absorbers) don't face diarrhea, flatulence, and nausea as a penalty for eating ice cream. The proportions of malabsorbers and absorbers in human populations are related to the cultural trait of dairy farming, which varies in its history from culture to culture.[121] It also seems likely that as selection led to more individuals in dairying cultures being absorbers (there are, after all, nutritional advantages of being able to absorb lactose), the idea of continuing to drink the milk of nonhuman animals after weaning would itself have spread, as it evidently did.[122]

The influence of culture on the genetic composition of the human population has also been studied in connection with cultural biases in the sex ratio among offspring—a critical issue today, when it is often possible to "choose" the sex of children. Whether or not an individual is genetically male or female depends on genes transmitted via the father's sperm.[123] With respect to sex determination, there are two kinds of sperm, which can be called X and Y, and one kind of egg, X. If an egg is fertilized by an X sperm, the child is genetically female (XX); if by a Y sperm, it is male (XY). Preference for sons in places such as China, India, and Pakistan is often expressed in female infanticide (or, now, by differential abortion of female fetuses) or neglect of female children, which increases the male bias of population sex ratios.[124] If there is a fitness loss (reduction in representation of a couple's genes in the next generation by the couple's not replacing the unwanted female offspring), favoring of male offspring will result in selection to alter the genetically controlled primary sex ratio (that is, ratio at birth) toward producing more males, normally 105 or 106 male births per 100 female births.[125]

It is relatively easy to find situations like those in which change in the cultural environment will alter selection pressures on human populations. It has been more difficult to produce examples in which biological evolutionary changes in *Homo sapiens* have influenced cultural evolution—except, of course, that the evolution of big brains made cultural evolution possible. But other influences in the past also have been substantial—for instance, the genetic evolution of erect posture and highly agile tongues and fingers, to which we'll turn in the next few chapters, has had profound effects on cultural evolution.

There is no doubt that the large human brain is a complex organ "designed" by natural selection. It is highly adapted to serve the needs of the most complicated social animal ever to walk the planet. But when big brains interact with the extensive culture they made possible, the results may be very maladaptive in terms of the fitness either of individuals or of groups. Therefore, we cannot assume that because a behavior exists, we can explain its presence by saying it must increase genetic fitness. There are too many behaviors, and they can change much too rapidly (consider popular music fads, use of the World Wide Web, etc.), for them to be explained in terms of genetic evolution—genetic evolution just can't keep up the pace.[126] The development of a celibate clergy may be a good example of a culturally induced behavior that reduces genetic fitness. In some cases, behavior that reduces the fitness of individuals and is also maladaptive for the group as a whole[127] can spread if it is sufficiently often adopted by imitation (especially by individuals' learning from other individuals who are perceived as successful).[128] Addiction to alcohol, tobacco, and other drugs is an instance in which a genetically determined physiological response interacts with a substance whose use is transmitted culturally and causes that use to spread even though its effects on fitness are generally negative. Other examples of behavior that reduces fitness of individuals (and sometimes groups) include eating of the dead (which spreads diseases such as kuru), genital mutilation, and various behaviors (late marriage, use of condoms) that reduce birthrates.[129]

More generally, major problems in analyzing the roots of human behavior trace to the possible complex interactions between genetic evolution and cultural evolution,[130] our primitive state of knowledge of cultural evolution itself,[131] and a lack of understanding of the genetic factors that could influence various behaviors, as well as a fundamental misframing of questions in a nature-versus-nurture context. To complicate things further, processes similar to mutation, migration, and drift are present in cultural evolution.[132] The appearance of charismatic leaders or individuals with novel ideas may be a random phenomenon in human society. Thus, one could consider the advent of an Aristotle, a Buddha, an Elizabeth I, a Newton, a Tecumseh, or a Hitler to be rough analogs of mutations. People with these individuals' proclivities are doubtless present from time to time in all cultures, but as with gene mutations, an appropriate environment is necessary for them to gain prominence.[133] As a possible example of "cultural drift," the long-isolated Tasmanians, living on their small island off the southeastern corner of Australia, had the least complex material culture of any group of modern human beings. This was partly because they were isolated about 10,000 years ago by the rise of sea level as the Pleistocene glaciers retreated. A more modern stone-tool culture penetrated from Asia to Australia about 5,000 years later, but it could not reach Tasmania across the water barrier of Bass Strait (a cultural analog of differential migration). Both bone tools and fishing had disappeared from the Tasmanians' cultural repertoire by 3,500 years ago, for no apparent reason, quite possibly by accident, just as gene variants disappear from populations by accident as a result of genetic drift. Once they were gone, there was no way for them to become reestablished in that isolated population except by reinvention, which did not occur.[134] The diversification of human languages also appears to occur by a cultural process of random changes and spread of neologisms more or less analogous to a combination of migration and drift.

Throughout this book, as we sort through the complexities of how genetic and cultural evolution produced our natures and have gradually made them a threat to the integrity of the ecosystems that support us and the rest of biodiversity, there's a particularly important point to keep in mind. When it comes to social policy, our ignorance and our ethics necessarily limit us largely to using environmental tools to modify behavior.[135] Despite a huge speculative literature, for instance, scientists know little about the influence of genes on human behavior.[136] One main reason for this is that many of the experiments, involving such things as planned matings and rearing of children in strictly controlled environments, that would be required to increase knowledge would be both impractical and unethical. Even if we did know a great deal about the relationship of genetics and behavior, many of us would be ethically opposed to trying to modify most forms of human behavior by genetic engineering. But almost everyone favors engineering the environments in which genes that can influence behavior are expressed, as attested by the existence of institutions such as Stanford University and San Quentin State Prison.

The mechanisms of genetic evolution that have helped to create human natures and the natures of our living companions on Earth are not a closed book. Scientists are continually reexamining evolution in general, selection as the creative mechanism driving it, and speciation as the major engine of diversification.[137] That examination is becoming ever more intensive as molecular biology provides tools that permit the asking and answering of increasingly detailed and sophisticated questions,[138] including questions about such things as the patterns of differentiation among our ancestors. Perhaps the biggest problem of all is that most of the routes from DNA to phenotype—how the genetic code interacts with cellular environments to produce legs or brains—are yet to be mapped. The field of developmental biology remains one of the least understood in all of biology. Nevertheless, genetic evolution, with natural selection as a major driving mechanism, uniquely provides a coherent framework for understanding the diversity of Earth's life-forms and the position of humanity within it. Genetic evolution combined with cultural evolution and the primate history we'll start to examine in the next chapter tells us much about who we are, where we came from, and even where we might end up.

Chapter 4
Standing Up for Ourselves

"In the beginning there was the foot."

—Marvin Harris[1]

One Victorian lady was reported to have told a friend, "Mr. Darwin says we are descended from monkeys." The friend replied, "My dear, I pray that it is not true, but if it is true, I pray it will not become widely known."[2] That prayer has been partially answered, for even most otherwise well educated individuals know precious little about evolution. It's even still common for "animals" and "people" to be seen as two mutually exclusive categories. Not only are we human beings very much mammals and, more specifically, primates, but also there are aspects of the physiology and behavior of living monkeys that, as I shall try to show in the pages that follow, can help us in thinking about our own natures. The way our nearest living relatives comport themselves is not just "monkey business"—it's our business as well.

That we have important similarities with those relatives has been recognized by scientists for a long time. But even scientists have tended either to hastily draw parallels between the behavior of apes and the behavior of people, declaring similarities to be the result of the same genetic heritage, or to overemphasize differences and thus miss important continuities. As Darwin's great champion Thomas Henry Huxley noted in 1863, "It would be no less wrong than absurd to deny this chasm [between apes and people], but it is at least equally wrong and absurd to exaggerate its magnitude."[3]

Our Place among the Apes

The almost century and a half of inquiry since Huxley made his comment has seen the once-wide gap in evolutionary evidence between us and our closest living relatives largely filled in by the discovery of numerous fossils. The fossil record, greatly enhanced particularly in the past half century, shows a remarkable series of creatures with characteristics of various degrees of intermediacy between those of human beings and those of other great apes. It is a continuity that many people find fascinating. Recently, I purchased a series of beautiful skull replicas[4] and lined them up on the file cabinets in my office: gorilla and chimpanzee at one end, modern human being at the other, and five fossil representatives of the human family—hominids[5]—in a graded sequence in between. I use them in teaching, but they inevitably attract great interest among a wide array of visitors. People seem almost as interested in the genealogy of their species as in their personal genealogies.

In addition to the continuity shown in skulls, similar continuity has been illustrated in detailed comparisons of our other physical structures, our genes, and our behavior with those of living great apes, monkeys, and other primates. Who were those distant ancestors, and how have natural selection and speciation entered into our own history? In short, what do we know of the physical foundations of human nature? This chapter tries to answer those questions; the next explores how an upright but still small-brained human being became the culture-wielding genius of today, with all our capacities and shortcomings.

We need only trace back our family tree some 30 million years (just about halfway to the age of the dinosaurs) to find the common ancestor we share with Old World monkeys. If we consider chimps rather than true monkeys, the distance to a common ancestor is only about one-sixth of that. Just a few decades ago, scientists believed that human beings belonged to a different group of primates, separate from chimpanzees and bonobos (the latter formerly called pygmy chimps),[6] gorillas, and orangutans. But in recent years, through the use of modern techniques of molecular biology, researchers have discovered that in terms of branching of the evolutionary tree, we're positioned right in the midst of the great apes.[7] It was only about 5 million years ago that our ancestors split off from the evolutionary line leading to modern chimps and bonobos,[8] whereas the ancestors of gorillas[9] left the chimp–human line about 7 million years ago (see the figure).[10] Genealogically, we are more closely related to chimps and bonobos than chimps and bonobos are to gorillas.[11]

Our physical resemblance to the other great apes is apparent to any unbiased observer, as it was to the father of taxonomy, Carolus Linnaeus, who in 1758—almost exactly a century before publication of Darwin's *On the Origin of Species*—classified us as primates and placed us in the same genus, *Homo*,[12] with the other great apes (which at that time were barely known to Europeans). Among these, our closest living relatives in terms of physical and behavioral attributes are probably the bonobos, not the common chimps.[13] Bonobos are more slender than chimps and more frequently stand upright; they have shorter arms compared with their legs, smaller skulls, and smaller teeth; and there is less size difference between the sexes.[14] Physically, they may be the best living primate model for comparisons with *Homo sapiens*.[15] If nothing else, they resemble us in being continually interested in sex. Our fundamental perceptual abilities are very chimp-like,[16] and probably bonobo-like as well (the latter have been

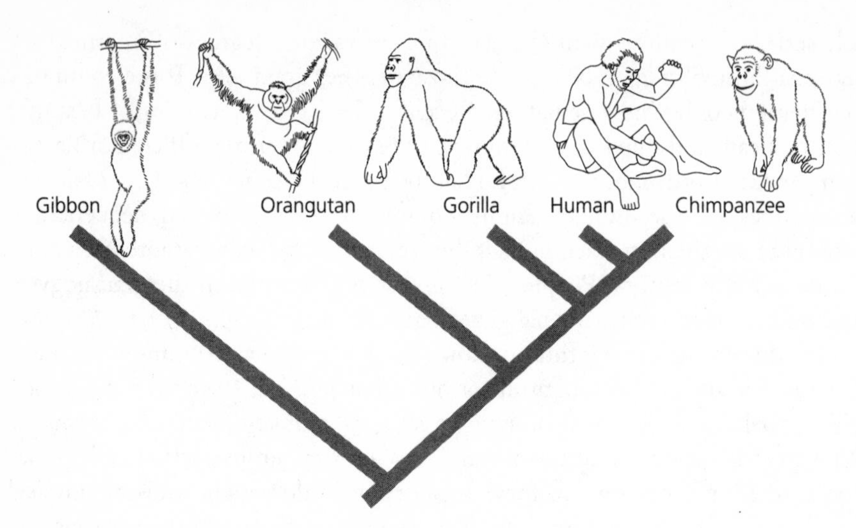

Evolutionary relationships of great apes.

much less thoroughly studied). There are interesting similarities in the parts of human brains involved in language and the homologous parts of chimp brains,[17] and our behavioral resemblance is often striking as well.

Among chimps and occasionally among the other apes (and elsewhere in the animal kingdom) behaviors can be found that once were thought to be characteristic only of human beings. (Technically, I could have said "chimps and bonobos" again, but for convenience I'll call them both chimps from here on and differentiate these two close relatives only when it matters.) Tool use is a classic example.[18] Chimpanzees use a variety of tools and weapons, crudely "manufacturing" some of them[19] and sometimes using two or more types of tool in sequence to accomplish a goal.[20] Like people, they raid neighbors' territories and sometimes deliberately kill trapped members of rival "tribes"—behavior that can be interpreted as a primitive form of warfare.[21] Chimps have also been shown to be capable of quite complex causal reasoning.[22] For instance, presented with a food-containing box tied shut with string, they can be taught to smash stones in order to make sharp pieces to use in cutting the string. And they are, like us, skillful at forming alliances and at making up after quarrels.[23]

Chimps not only use tools but also, as noted earlier, have culture[24]—a body of nongenetic information that is passed from individual to individual and from generation to generation. Chimp culture varies from community to community in the ways in which tools are used and in patterns of grooming and of gaining attention.[25] Chimp natures, like human natures, can differ from group to group.

Cultural transmission among chimps is best exemplified by individuals learning to use tools by observing other individuals,[26] by the appearance of actual teaching sessions,[27] and by local variation in the "tool kits" chimps develop, which indicates that each community has its own cultural tradition.[28] Chimps at Bossou in Guinea, for example, but not those in other communities, use stones as hammers and anvils for cracking oil-palm nuts[29] and sometimes use an additional stone as a shim to level an anvil.[30] This additional stone is what's termed a metatool, a tool used to improve the functioning of another tool. Chimps also appear to become attached to some of these objects: they carry the stones around with them—conceivably an early form of the concept of possessions and of planning ahead. Hand preference, usually for the right hand, we know is a well-developed[31] (but perhaps exaggerated)[32] characteristic of *Homo sapiens.*[33] Handedness can also be found in chimps for some tasks. Especially when using tools, chimps often preferentially use one hand, though some individuals favor the right hand and others the left.[34]

Other "human" behaviors are also characteristic of the societies of our close relatives. Chimps live in patriarchal societies; so do people. Male gorillas often kill offspring of their mates that they did not father, a behavior rare but not unknown in human societies, in which boyfriends and stepfathers are more likely to abuse children than are natural fathers.[35] And, like males of most mammals, including chimps, male human beings often try to inseminate more than one mate.

Of course, these similarities don't mean that people are "just the same" as other apes. There are important differences in both degree and kind between us and our nearest living relatives. Much is often made of our more than 98 percent genetic resemblance to chimps, but that is primarily useful in indicating how recently the evolutionary lines leading to us and them diverged. A great many critical phenotypic differences can be traced to that less than 2 percent difference.[36] Some of those phenotypic differences are, of course, striking and physical. Human beings are the only living apes with fully upright posture. We are also virtually hairless compared with chimps, have much smaller incisors and canines, and have fully opposable thumbs. Female human beings are different from female chimps, indeed from all our other relatives, in having large, fatty breasts before bearing children; and human males are approximately tied with bonobos for having proportionately the largest penises (but not testicles).[37] There are differences related to lifestyles as well. For example, chimps can orient themselves more rapidly than humans to three-dimensional worlds such as forest canopies.[38]

Another major difference in kind associated with our human natures lies in the realm of sexual behavior. We are the only organisms that are known to recognize consciously the connection between copulation and reproduction and, as a result, use contraception and other tactics to avoid pregnancy.[39] People deliberately choose to limit their reproduction at levels that do not maximize an individual's genetic contribution to future generations—and we've done it for thousands, maybe tens of thousands, of years.[40] We are also the only ape species that normally copulates in private and in which older relatives often arrange matings among the younger generation.

The main reason why people can manage these enterprises, as well as split atoms in addition to nuts,[41] is that the brains of both men and women are roughly three to four times as large as those of chimpanzees.[42] Indeed, the biggest difference in kind between human beings and other apes is in the capabilities of our brains—the organ that provides us with many of the unique aspects of our nonphysical human nature, human capacities that are universal or virtually universal.[43] The most obvious universal aspect of human nature is our use of language, a system for communicating thoughts and feelings precisely to other individuals. Language is unique in its use of syntax—the set of rules for how elements of the language work together to produce symbols for objects and actions.[44] Also unique to our species is the oral expression of language—the facility of speech—among all normal individuals. There is little evidence that our close relatives use vocal communication significantly more than do other mammals—for giving alarm calls, sounding territorial warnings, and the like. If they do communicate extensively with one another, it is not primarily through sounds.[45]

We human beings use our unique capabilities extensively in connection with a brain that also allows us to improvise behavior appropriate to almost any situation, basing the behavior on clear understandings of complex causes and effects. No other organism has remotely similar abilities in this area (this is not to say that others cannot associate causes and effects), and we are the only species known to develop belief systems. It is also quite possible that we are the only living animals that have cognitive empathy[46]—the ability to picture ourselves in the position of another individual. This is an invaluable survival tool for a very intelligent organism that lives in a highly social environment.

Our forebears, more than any other primates, specialized in living by their wits—a specialization that has produced another difference in kind. They evolved the strategy of dealing with both prey and enemy through planning, carefully manufactured tools, and a complex, evolving culture; as anthropologists John Tooby and Irven DeVore put it, people evolved into the cognitive niche[47]—a niche that hominids themselves created. We can't be certain how our ancestors got onto that route of specialization, but it certainly has made our brains very different from those of even our nearest relatives. Our brains are fine machines that invented substitutes for fangs to tear meat, for grinding molars to chew up grass, for long tongues to fish delicacies out of anthills, for tough hides or sharp horns to protect us from big cats, and so on. Because of the characteristics of our brains, we no longer need to depend on the stately pace of natural selection to solve the problems of survival—natural selection instead has given us an apparatus for generating quick solutions.

The difference of degree in cultural endowment between ourselves and other animals is so great that it has generated many other differences in kind. The reason is that with language and the problem-solving and information-storage abilities of the human brain, the knowledge contained in human culture can easily and rapidly be expanded and shared. Each of us has access, at least through spoken language and imitation but also in most cases through writing and electronic media, to a vast store of knowledge that human beings have created. Enhanced communication allows our culture to evolve enormous complexity, resulting in the development of behaviors (e.g., my writing this book by means of a computer) that no lone individual or isolated small group could invent. Cultural evolution not only has given us the capability of creating complex artifacts ranging from finely honed stone points to heart–lung machines but also now allows us to trade these and many other commodities around the world—and it has given us the potential to exterminate millions of other human beings in a fraction of a second. Cultural evolution is rarely seen at all outside of the primates, but it may occur at a modest level in chimpanzees.[48] There's a crucial difference, though: chimps' cultural information can be passed on only by imitation—because speech (and writing) are limited today to *Homo sapiens.* And, of course, it is precisely our human capacities to store and share culture that have led to the evolution of the arts, religion, humanities, and science that are such a characteristic part of human nature.

Observing humanity's close relatives in the wild enormously increased my own interest in the roots of our amazing nature. I did research work on butterflies at Gombe Stream Reserve in Tanzania more than a quarter of a century ago. But it wasn't the butterflies that intensified my curiosity—it was a group of chimpanzees that occupied our study site and were themselves being studied by Jane Goodall and her colleagues. I had been looking forward to being in the field with the chimps and was determined to observe them without anthropomorphizing their behavior. Within minutes of my first close-up look, a mother chimp gathered a distressed infant into her arms and comforted it by patting it on the head—and my resolve went out the window. That observation and other incidents at Gombe gave me a very different view of chimps and their behavior. Before my experiences at Gombe, I had known that we human beings, as the dominant animal on the planet, had been steadily pushing chimpanzees toward extinction in the wild,[49] and I hated that on general principles. But in those days, I did not know how closely we are related to the chimps genetically, even though their physical and (as I learned at Gombe) behavioral resemblance is unmistakable. Watching them in action brought home to me that we human beings were losing cousins from whom we might learn a great deal about ourselves—something that had occurred to Jane Goodall and her colleagues long before.

How did chimps and people get to be so similar, yet so different? Why did we end up threatening them with extermination, and not vice versa?[50] The answers to questions about our relationships with our nearest relatives and our progenitors have been sought largely from a scattering of mineral-impregnated bones of our ancestors and their relatives and chipped stone tools laboriously excavated from sites around the world, but especially on the continent of Africa. Unhappily, this fossil record is not as complete as one would like. Whether or not an animal ends up in the fossil record depends in part on the sort of habitat in which it lives; denizens of shallow seas are more likely to end up as recognizable fossils in rocks than are those of tropical forests. Whether animal bones are preserved is determined largely by whether deposits were accumulating (as opposed to eroding) and whether soils were alkaline (as opposed to acidic). And, of course, pure chance enters into the picture as well. Nonetheless, for more than two centuries, scientists have been gradually uncovering a great wealth of fossils and, by careful analysis, putting together an impressive record of the course of evolutionary history. Such remains, combined with the behavior of other living primates and, recently, with genetic analysis, are the lens through which scientists try to peer at the human past.

The Monkeys' Descent

When *On the Origin of Species* was first published,[51] despite the massive evidence that Darwin presented, the notion of evolution was much closer to a "theory" in the popular sense of a speculative idea. The word *theory* has several meanings, and today many nonscientists still interpret its meaning in the phrase "the theory of evolution" as simply a conjecture about the history of life. In science, however, a theory is not just a conjecture, and its meaning is connected to nature as the ultimate arbiter of its usefulness. Scientific hypotheses are, in one way or another, tested against nature—the "real world" that all scientists conventionally agree is "out there."[52] Only when hypotheses are sufficiently tested and bind together information from relatively diverse areas that previously had not been connected do they properly become theories. But that is the opposite of the popular understanding of the term; its scientific meaning is much closer to that of the word *fact* in common parlance.[53] Theories embody the highest level of certainty for comprehensive ideas in science. Thus, when someone claims that evolution is "only a theory," it's roughly equivalent to saying that the proposition that Earth circles the sun rather than vice versa is "only a theory." Evolution is, in fact, a very useful theory.

An early argument against Darwin's ideas was that there were many "missing links" in the fossil record between kinds of organisms in the evolutionary sequence proposed—links that according to Darwin's hypothesis must have existed.[54] The gap between birds and other vertebrates seemed especially large. But a mere two years after the publication of *Origin*, the first, and to this day perhaps the most spectacular, "missing link" was discovered. It was a beautifully preserved fossil of *Archaeopteryx lithographica* (see the figure) from a quarry that mined slabs of Jurassic limestone for use in lithography. *Archaeopteryx* had the head and tail of a reptile but was covered with feathers.[55] Subsequent digs around the world have uncovered a wealth of organisms showing both avian and reptilian characteristics, and now it appears that *Archaeopteryx* was very likely not a direct ancestor of modern birds but an evolutionary offshoot. Nonetheless, it served brilliantly as a "link," showing as it did a mix of bird and reptilian features. The issue is no longer whether there are missing links demonstrating an evolutionary relationship between reptiles and birds but what the details are of the origin of birds and, especially, whether birds are really surviving dinosaurs.[56]

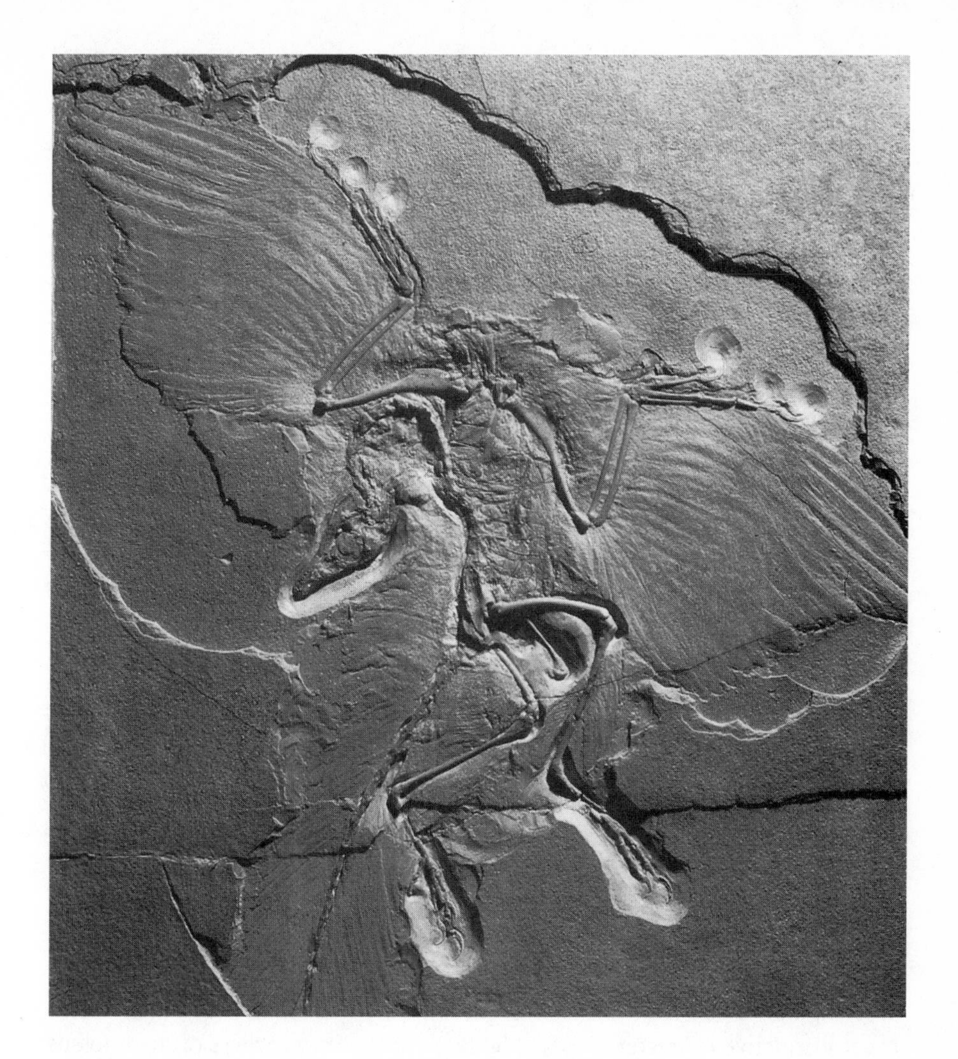

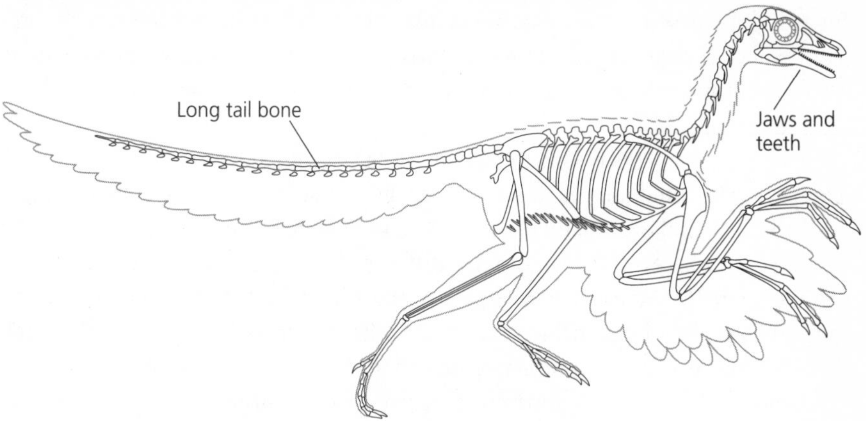

Archaeopteryx. Above, actual fossil; below, reconstruction showing the combination of reptilian and avian characteristics. Photo courtesy of the Humboldt Museum für Naturkunde, Berlin. Drawing based on Chatterjee 1997.

Archaeopteryx was not the only missing link that turned up in the decades after *Origin* was published. *Homo erectus*, a near-perfect intermediate between apes and human beings discovered in the 1890s by Eugène Dubois, a Dutch physician and anatomist turned anthropologist, was another, and subsequently many other links between people and distant ape-like ancestors have been uncovered in the fossil record. It's to that record that we now turn to see how run-of-the-mill mammals evolved into our marvelous selves.[57]

On the road to dominance of Earth, the first step hypothesized to have been taken by our mammalian ancestors was upward. We will probably never know why, but perhaps 80 million years ago in a population of shrew-like[58] organisms, individuals that tended to climb into shrubs and trees apparently started to produce more offspring than did the members of their cohorts that were more terrestrial. Part of the reason for this almost certainly was that during the Cretaceous period, 140–65 million years ago (mya), flowering plants became diverse and abundant, opening new ways of life for tree-living creatures that ate the berries, nuts, fruits, soft leaves, and nectar those plants produced in profusion. Many of those creatures were insects, themselves coevolving with the flowering plants[59] and in the process helping to provide a rich source of insect flesh to nourish our forebears.

Those shrew-like forebears were mammals that, like us, employ placentas to nourish their young during gestation. This group of early placental mammals had become distinct from marsupials (which lack placentas and hold their developing young in a pouch) some 120 mya. Those ancient placentals, which have been placed in the order Insectivora (their teeth suggest a diet of small invertebrates), are represented now by animals such as moles and genuine shrews. They gave rise to the primates, a group of placentals most of which have nearly globular brain cases, flattened faces with forward-facing eye sockets, short jaws, grasping hands and feet, flat nails in place of the usual mammalian claws, and other distinctive characteristics.[60] The teeth in the very earliest primate fossils suggest that these creatures ate more plant materials than did the insectivores from which they diverged. Thus, scientists think that gradually, starting some 80 mya, our primate ancestors became arboreal; that is, they spent most of their time in trees.[61]

The exact patterns of divergence of different primate groups, and just which groups represented in the fossil record gave rise to the monkeys and apes (anthropoids) and eventually to us, are still in dispute.[62] But in any case, the first primates took an evolutionary course different from that taken by the ancestors of that other frequently arboreal group, the squirrels.[63] Rather than gathering nuts and avoiding bugs, primates appear to have increasingly evolved the habit of gathering insects at night, snatching them from the vegetation with dexterous fingers. Primate eyes apparently evolved to a position toward the front of the head to improve the quality of the optical image and to allow both eyes to focus on near objects being manipulated.[64] Large, forward-oriented eyes are also characteristic of other nocturnal predators such as owls and cats. The form of the teeth of early primates supports this view of arboreal hunting. They were similar to those of small modern-day insect-eating primates such as tarsiers. So, like tarsiers, our earliest primate ancestors probably climbed stealthily through bushes and trees at night, searching for insects or other small animals to devour.[65]

Then, starting about 50 mya, archaic primates began to diversify and give rise to early monkeys and apes. Even though, as mentioned earlier, forests are not ideal places for the preservation of fossil material, there is a pretty good fossil record of our arboreal primate lineage in Africa, where much of the action was taking place, in the early and middle Miocene epochs (between approximately 22 and 12 mya).[66] A broad scattering of fossil apes from this period[67] appears to have connections to, or may actually be, ancestors of ours. These links are creatures with names such as *Dendropithecus*, *Proconsul*, and *Kenyapithecus*.[68] But it is likely that there has been so much convergent and parallel evolution—evolution in which similar changes occurred in different lines[69]—that it has not been possible to determine clearly the evolutionary relationships of these apes to the australopithecines, who we'll meet later in this chapter. The latter are the first known links between us and the "monkeys" from which we descended.[70] There actually seems to have been more potential ancestors than we would have needed prior to 12 mya—an embarrassment of fossil riches.[71]

Our ancestors' moving into the trees was very important: we owe our binocular vision and adept hands to their arboreal, bug-catching habits. But their departure from the trees was also critical (think how nondominant squirrel and monkey civilizations are today). Farming can't be done in the treetops, minerals can't be mined there, and cities can't be built. If you want an animal armed with a high-powered rifle or able to build a computer, it can't evolve in an exclusively arboreal setting.

That emergence from the trees occurred as Africa's climate dried out and savannas formed between 10 and 5 mya (in geological terms, the late part of the Miocene epoch). In East Africa, thick forests had become scarce by about 2.6 mya (mid-Pliocene epoch). A shift in global climatic patterns had produced the drier climates at lower latitudes,[72] but the replacement of forests with savannas (or open woodlands) seems to have been more directly a response to rapid geological uplift along what is now the Uganda-Congo border of a north–south mountain range, the "roof of Africa." That range reached a height of almost 17,000 feet (5,100 meters)[73] and created a rain shadow over much of the area to the east.[74] All this is indicated by a growing occurrence of antelopes and other savanna animals in the fossil record and by physical changes visible in the fossilized skulls of some species as their feeding apparatus adapted to a changing vegetation.[75] This climate change was perhaps the most important of many climatic events that helped to shape hominid history.[76] While gorilla ancestors presumably stayed in the forest, the ancestors of chimps and hominids began to invade the more open country. Those are creatures we know very little about because in contrast to the richness of the ape fossil record from 22 to 12 mya, that of the critical time between 10 and 5 mya is, unfortunately, poor.[77]

It is even possible that the east–west climatic division in East Africa was the basis of the eventual evolutionary split between the lines leading to chimpanzees and to human beings. The chimps may have stayed west of the rift valleys and adapted evolutionarily to the more heavily wooded environments there. The hominids could have remained in the drier east, adapting to savannas and open woodlands.[78]

While our closest cousins, the great apes, remained in the forest, representatives of one other group of Old World monkeys, the baboons, moved out into savannas along with our ancestors. Some scientists think that baboons, because of their ecological niche, make a more satisfactory model for the early stages of human evolution than do the chimpanzees.[79] Therefore, it would be a good idea to keep the savanna-dwelling baboons in mind as we take a look at the fossil record of our forebears.

The Bones of Our Ancestors

How much do we need to know about the evolutionary sequence that led from monkeys to us in order to develop a reasonably good picture of the origin of our natures? One might argue that simply knowing there are fossils showing clearly that we are descended from "monkeys" is sufficient. In the early 1960s, when I first started writing about human evolution, the physical sequence in the fossil record that linked us to our ape-like ancestors seemed quite simple and adequate to explain our physical roots.[80] However, the fossil record now shows much more than that. We know that our roots are substantially more complex than previously thought, and even so, the record still contains mysteries about how and when some of humanity's most important attributes evolved—such as why we became fully upright and other apes didn't. Knowing about such things enriches our understanding of human nature. It does so in the same way that knowledge about the founding fathers enriches a U.S. resident's sense of what it means to be a citizen of that country, or familiarity with Benito Juárez, Pancho Villa, Emiliano Zapata, and Francisco Madero enhances Mexicans' understanding of their roots.

What, actually, is now known about our mob of fascinating ancestors and their relatives? To begin, the first fossil hominid recognized as such was a Neanderthal—a long-ago inhabitant of the Neander Valley, near Düsseldorf, Germany.[81] The most important specimen was discovered in 1856, three years before Darwin published *On the Origin of Species.* Although he was aware of the fossil, Darwin chose not to mention it; he realized that his book was sufficiently controversial without explicit reference to human evolution.[82] His was probably the wise course, for, as we'll see in the next chapter, Neanderthals are even today a center of some controversy.

Later in the nineteenth century, the remains of some of our earlier ancestors were exhumed from their ancient graves. Eugène Dubois started the process when he went to the Dutch East Indies (now Indonesia) in search of a missing link between people and apes. He picked Java because as a Dutchman, he could get there easily, and he knew it was home to some cousins of ours, orangutans, one of the three[83] types of great ape.[84] In 1891, he found what he had been looking for. It was the most famous of missing links, the first true "ape-man," which was the meaning of the Latin generic name, *Pithecanthropus*, that he gave to it.[85] Dubois was lucky that, in addition to a partial skull, his first discovery included a femur (thigh bone), which he thought showed that *Pithecanthropus* walked upright. Dubois thus came up with the specific name *Pithecanthropus erectus* for creatures that were, after further discoveries near Peking (now Beijing), China, often referred to as Java and Peking "men." Interestingly, the femur was probably a modern one—although the upright posture of *Pithecanthropus* has been clearly demonstrated by subsequent discoveries in China and East Africa.[86]

The luck of anthropologists held out into the twentieth century, for the next find also was a wonderful missing link, one that had lived millions of years before the Dubois finds. In 1924, a baboon-sized fossil skull of a juvenile primate was dynamited out of a quarry near Taung, north of Johannesburg, and fell into the hands of Raymond Dart, a South African anatomist. Dart gained international fame by immediately recognizing its great importance. The skull was of no young baboon—its braincase was too capacious, it had too small a face, and its canine teeth were too small. Dart named the fossil *Australopithecus africanus* ("the southern ape from Africa"), but the specimen itself became known colloquially as the Taung child. Once, while in South Africa, I was privileged to examine the skull. It's quite a thrill to hold in one's hand the mortal remains of a likely human ancestor from millions of years in the past. And it seemed wholly appropriate that I was instructed to keep my hand and the precious fossil skull over the cushioned top of the examination table.

Dart correctly declared *A. africanus* a link between apes and human beings. The apparent position of the foramen magnum (the hole in the base of the skull through which the spinal cord passes) and the scars showing where muscles supporting the head had attached indicated that australopithecines had walked upright. Just as in the later *Homo erectus*, their heads were balanced at the top of their spines (see the figure). Although Dart's australopithecine discovery and later ones by paleontologist and physician Robert Broom[87] were not immediately accepted by the scientific community, by 1939 studies of the dentition of these fossils had justified Dart's and Broom's claims of their position in the human family tree.[88]

About the same time as Dart was getting lucky, a treasure trove of fossil human beings was found at Zhoukoudian, in China,[89] and named *Sinanthropus pekinensis* (Latin for "Chinese man" but known as "Peking man"). Peking man was soon reclassified as a member of the same species as *Pithecanthropus erectus*.

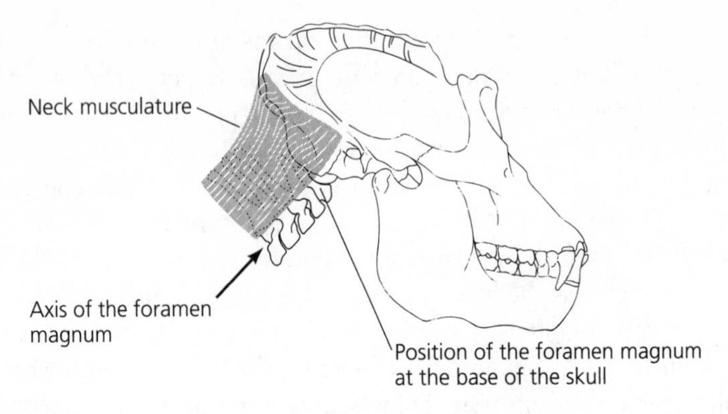

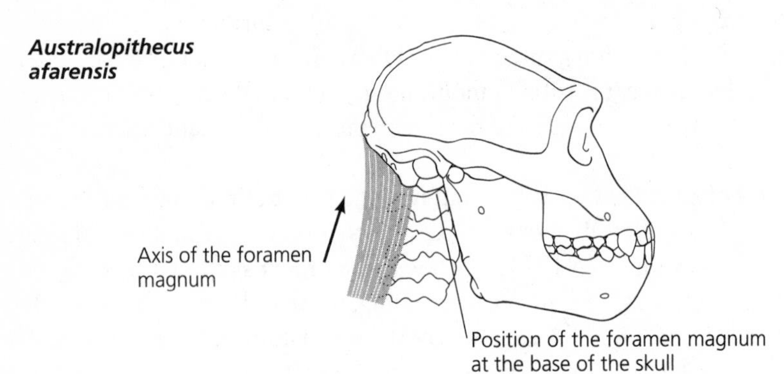

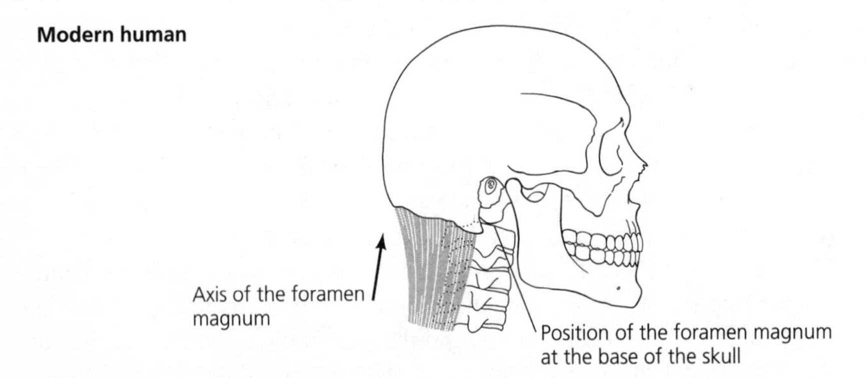

Skulls of gorilla (above), *Australopithecus afarensis* (middle), and modern human being (below), showing position of spinal column in relation to the head. Note that in the gorilla, the big hole (foramen magnum) through which the spinal cord communicates with the brain is located relatively far back on the base of the skull and its orientation is more backward and less downward than in human beings. In normal posture, the gorilla's head is in front of the spinal column, whereas in people, the head is balanced on top of the column. The position and orientation of the foramen magnum are fundamentally human-like in *A. afarensis*, and together with the anatomy of the leg, this shows that *A. afarensis*, like the later *A. africanus*, habitually stood upright in basically the same way that we do. Note further, however, that *A. afarensis* resembled an ape in its forward-projecting face and its relatively large canine teeth. The fossil record could not provide better evidence that we evolved from creatures we would include with the other apes had they survived to the present. Modified from Klein 1999.

The Javan fossils are now dated to sometime around 800,000 years ago (although the dating is quite uncertain).[90] The Chinese sample was perhaps 350,000 years younger but clearly the same beast.[91] More fossil remains of *P. erectus* were subsequently discovered in Java,[92] and *P. erectus* was later transferred to our own genus, *Homo*.[93] By the early 1960s, a few other non-Asiatic human fragments that were also assigned to *H. erectus*, a relative wealth of fossils of our nearest extinct relatives, the Neanderthals, and the discoveries[94] of more robust australopithecines in South and East Africa pretty much made up the rest of the human fossil history known at the time.

A lot has been learned in the succeeding decades, and the basic story has been improved both in detail and in continuity. Modern dating methods and more fossil discoveries now place *Australopithecus (A. africanus)* at about 3.2 to 2.5 mya, *Homo erectus* at 1.5 to 0.4 mya (or even more recent),[95] and *H. sapiens* at about 0.5 mya to the present.[96] Estimates of brain volume have remained pretty stable since the earliest discoveries, although it now appears that in the brainiest *A. africanus*, the volume of the inside of the skull (the endocranial capacity, a good estimate of brain volume) was only about 515 cc[97] instead of 600 cc, as previously thought.[98] That compares with about 400 cc for a chimpanzee and an average of 1,350 cc for modern people[99] and confirms the view that upright posture preceded brilliance by a substantial period.[100]

Lucy and Her Kin

Variations such as the readjustment in estimated brain size of *A. africanus* are hardly significant in the big picture of evolution. If vertebrate paleontologists were trying to reconstruct the ancestry of any species besides *Homo sapiens*, they would long since have been satisfied. *H. erectus* alone would be considered a sufficient "missing link" between people and the other living great apes, much as the feathered but toothy *Archaeopteryx* served to link birds and reptiles. But it is our own ancestors and our own evolution that paleoanthropologists and many of the rest of us seek to understand. Indeed, interest in ancestors and even worship of them in various forms is widespread in our species.[101] So the search continues and has even accelerated in recent decades as new techniques of uncovering our physical past have been deployed and a wealth of new detail has been added to the sketch of human evolutionary history embodied in the sequence fossil ape → *Australopithecus* → *Homo erectus* → *Homo sapiens*.

That detail has shown that our phylogenetic tree (the tree-like diagram representing our genealogy—see the figure) has a lot more branches than previously thought. Hominids, rather than evolving as more or less a single lineage, appear to have undergone something of an adaptive radiation.[102] A look at the discoveries of the past four decades brings this out. Of all the fossil finds of our very early ancestors, arguably the most important was one in 1974 at Hadar, Ethiopia, of a well-preserved *Australopithecus* fossil with a relatively complete postcranial skeleton (all the bones minus the skull and jawbone). The individual, discovered by a team led by Donald Johanson, was given the nickname Lucy because the famous Beatles song "Lucy in the Sky with Diamonds" was playing in the paleontologists' camp on the day of the discovery. Lucy represented a species that lived some 3.5 million years ago and was quite likely ancestral to *Australopithecus africanus*. Lucy and related individuals from Hadar and from Laetoli, Tanzania (almost 1,000 miles from Hadar), were formally named *Australopithecus afarensis*.[103]

A few years after Johanson's discovery, footprints of Lucy's species were found fossilized in volcanic ash at Laetoli, where a nearly complete lower jaw of the new species had been recovered.[104] Anatomical remains and the footprints showed unambiguously that these small-brained ancestors of ours stood upright even before *Australopithecus africanus*—although their skeletons suggest that they could have been quite agile tree-climbers.[105] Subsequent fossil finds in Kenya and Ethiopia of even earlier australopithecines[106] pushed the occurrence of bipedalism back to some 4.4 million years ago (see the figure on page 83).[107]

It has also become clear in recent decades, especially after the 1959 discovery by now-legendary anthropologists Mary Leakey and Louis S. B. Leakey of a skull in the Olduvai Gorge of Tanzania, that there was more than one line of australopithecine evolution. The East African skull the Leakeys found[108] had a more heavily built cranium[109] than that of *Australopithecus africanus*. The skull was highly crested to allow for the insertion of large jaw muscles—muscles matched by large premolars and molars that seemed overdeveloped in relation to its small incisors and canines. *Paranthropus boisei*, as the famous Olduvai skull became known,[110] and its close South African relative, *Paranthropus robustus*, which Robert Broom found more than two decades earlier, are generally called robust australopithecines to distinguish them from the gracile (graceful, slender) line that culminated in *A. africanus*. But *Paranthropus* is more robust than *Australopithecus* only in its skull; in the postcranial skeleton, *A. africanus* and *Paranthropus* were gracile.[111]

It is now assumed that both robust and gracile australopithecines evolved from *A. afarensis*. One line, the robusts, adapted to the new grassland environments by developing the chewing apparatus to feed on roots, tubers, and other coarse plant substances.[112] The graciles remained more generalist in their feeding habits, eventually substituted brains and tools for powerful jaws and big teeth, and gave rise to us.[113] The robust strategy may have been tried at least twice; at least one early offshoot of *A. afarensis* was almost certainly the large-toothed *Australopithecus aethiopicus*. *A. aethiopicus*, like the *Paranthropus* (whose diets were presumably similar), was probably an evolutionary dead end.[114]

A recent australopithecine discovery has further complicated the picture of human origins. At present, we can't tell whether the find, christened *Australopithecus garhi* (see the figure on page 83), may have been a relatively direct ancestor of *Homo* or another dead end like *Paranthropus*, adding to the growing "bushiness" of our phylogenetic tree.[115] But the bush was gradually trimmed. The gracile australopithecines disappeared from the fossil record, presumably

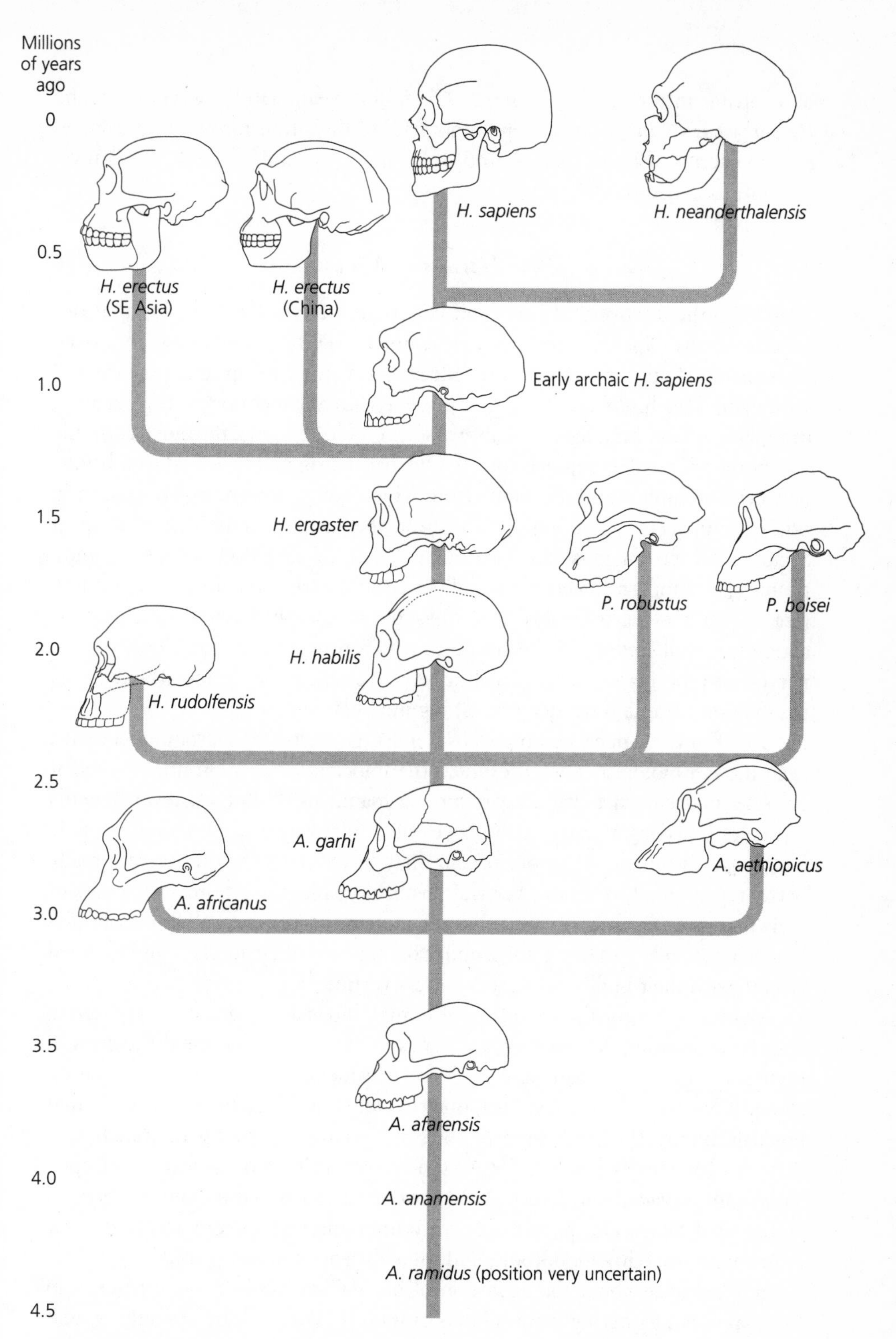

The human family tree. The recent finds of early upright australopithecines, *Australopithecus anamensis* and *Ardipithecus ramidus*,[107] are not illustrated because fossil material is too sparse to allow reconstruction of the skulls. Modified and redrawn from Klein 1999.

by evolving into early *Homo*, about 2.5 million years ago.[116] In contrast, the *Paranthropus* soldiered on as contemporaries of their transformed brethren, an evolving series of *Homo* species, until less than 1.4 mya,[117] possibly just 1 mya, when it, too, died out.[118]

The Human Ascent

Thanks to the discovery of Lucy and her friends, the task of showing that our ancestors were bipedal before they developed large brains was easy compared with that of figuring out why our ancestors evolved an upright posture and walked on their hind legs in the first place. As already mentioned, the penalties of walking on our hind legs have always seemed severe to me; the move certainly was not a triumph of engineering.[119] Our backs originally evolved as a bridge between forelimbs and hind limbs from which our guts were slung, as seen in other mammals today. Going upright threw the back into a difficult-to-support S-curve and let our guts slop into our pelvises, making back pain a common problem for modern *Homo sapiens* and perhaps for earlier versions of our design. Cradled in the pelvis, males' intestines try to escape through holes in the abdominal wall leading to the scrotum, which provide exits for blood vessels, nerves, and spermatic cords. That frequently results in men being afflicted with painful (and, at one time, dangerous) inguinal hernias.

To offset these penalties, bipedalism must have brought substantial benefits, even though biologists once presumed that bipedal locomotion was inefficient because its occurrence was a rarity among mammals.[120] But whatever benefits accrue to bipedalism may actually have come relatively easily because, possibly by pure chance, the ancestors of our first terrestrial forebears were already partly on the road to a more vertical posture. Unlike the arboreal ancestors of early baboons, which scampered along the tops of branches on all fours, our distant arboreal relatives often hung under branches by their forelegs and climbed around with their bodies held more or less vertical.[121]

Many reasons for the evolution of human bipedalism have been suggested, and the continuing debate over this key issue is indicative of the difficulties of understanding the human past from fossil evidence and current observation alone.[122] Perhaps the earliest idea, one that still has a certain attraction, is that bipedalism freed the hands to carry objects,[123] to make better use of branches,[124] or to hurl stones as weapons.[125] As the climate dried and savannas and open woodlands replaced rain forests, food for our ancestors became more dispersed and less palatable. This presumably put a premium on an ability to forage over larger areas, carrying food, tools, and the kids from place to place.[126]

Related ideas about the evolution of bipedalism focus on the efficiency of forearm use in gathering various kinds of food.[127] We can learn something here from the behavior of our savanna-dwelling relatives the baboons. Baboons today occupy savannas and deserts over much of Africa.[128] There are also two forest-dwelling species,[129] the drill and the mandrill,[130] and one additional species,[131] the gelada, whose range is restricted to the cold, dry highland areas of central Ethiopia. Most baboons are opportunistic omnivores, eating everything from crustaceans and other marine life along shorelines to snakes, lizards, small mammals (including hares, which are chased, and young gazelles when they are serendipitously discovered), fledgling birds, and insects to a wide variety of plant materials, of which ninety-four species are known to be eaten by South African baboons alone.[132]

The gelada took an unusual dietary course, feeding largely on grasses, which make up 80 percent or more of their diet.[133] They are manual grazers, dealing efficiently with seeds and other fibrous foods and able to dig for roots and tubers with their stubby fingers. They have a specialized chewing apparatus and evolved the ability to carry out what is called hindgut fermentation[134]—like zebras, they have a symbiotic microbial fauna in their guts that permits them to digest more than half of the crude fiber in their diet.[135]

Another hypothesis about the roots of bipedalism focuses on the development in our ancestors of a precise hand grip like the grip gelada baboons use today in feeding on grass and other seeds. Feeding like geladas would have resulted in much time spent sitting upright or shuffling for short distances on two feet (as geladas now do) while gathering food with both hands. These activities would have preadapted our ancestors to fully bipedal locomotion.[136] A related idea sees the gelada's occasional bipedal shuffle[137] as a device for avoiding the energetic costs of constantly raising and lowering the trunk while feeding. Bipedalism might have developed directly, of course, without our semiterrestrial ancestors going through the shuffle stage, if they did much of their feeding by reaching into fruiting bushes, especially thorny ones, that were difficult to climb. The individuals best able to move around upright might well have outbred those less able to walk bipedally.[138] In either case, being predominantly bipedal would have originated as a feeding strategy and later would have been perfected for locomotion and other purposes.[139]

Another suggestion is that walking upright helped keep our ancestors from overheating in relatively shadeless savanna habitats both by lessening the surface area exposed to the midday sun and by raising the upper body into cooling breezes.[140] This may have been a factor, but recent analyses have cast doubt on whether the advantage is large enough to be evolutionarily significant.[141] Indeed, the "savannas" inhabited by our earliest ancestors might have been open woodlands, where lack of shade probably would not have been a major problem.[142]

An intriguing hypothesis was recently put forward by anthropologists Nina Jablonski and George Chaplin. It ties together the origin of bipedalism with a reduction in the often violent aggression of chimpanzees that may also have characterized the behavior of our early ancestors as they moved into savannas, where their foods were scarcer and more patchily distributed.[143] Jablonski and Chaplin suggest that bipedal threat and charging displays and the appeasement responses these displays elicited (such as the head-down deferral or "bowing" used by chimps and gorillas today) played a major role in suppressing intra- and intergroup aggression as competition over scarce resources escalated with the transition to a savanna environment.[144] Such suppression would have been essential to the survival of our hominid ancestors, who could not have borne the level of injury that the increased competition, and increased frequency of contact as groups foraged more widely, would otherwise have caused. Upright displays might have been especially effective in partly open country, where they could have been observed from a considerable (and safe) distance.[145] For the moment, however, we'll have to withhold judgment on this hypothesis because it is not even certain that bipedalism in our human ancestors arose originally in savanna habitats.

Along with the development of bipedalism, in the evolution of the hominid line there has been a reduction in the degree to which males are bigger than females, a characteristic called sexual size dimorphism. Old World primates—including our nearest relatives—show much greater sexual dimorphism on average than do Western Hemisphere monkeys or prosimians (tarsiers, lemurs, and lorises). Although it is difficult to judge from fossil materials, it appears that most early hominids, like the living great apes, were substantially more dimorphic than modern *Homo sapiens*,[146] a factor that, as we shall see, can be connected to some important aspects of our natures. The reduction in sexual dimorphism observed between our earliest hominid ancestors and us is also seen between the forebears of forest-dwelling bonobos and their modern counterparts. Bonobos, perhaps not coincidentally, are also both more bipedal and more peaceful than chimps.[147]

The work of anthropologist Owen Lovejoy[148] has linked upright posture with human social structure and sexual behavior. He suggested that with the lengthening of the helpless period of infancy and early childhood that accompanied brain expansion, it became necessary for male hominids to carry food and help provision the mother and child. This implies a nonpromiscuous sexual system because there would be strong selection against one male promoting the propagation of another male's genes—which would happen if males provisioned females in a promiscuous species, in which the offspring often would be sired by rivals. Lovejoy's hypothesis also implies that the advantages of bipedalism started in only one sex (which would have weakened the selection pressure), and it implies an importance of provisioning far beyond that seen in our living primate relatives.[149]

What are we to make of all this? Much of the literature, perhaps too much, is focused on finding *the* reason for the evolution of two-legged walking and assumes it to be a uniquely human trait. My own guess is that numerous selection pressures were involved, which operated simultaneously or sequentially. They were made effective by a combination of environmental changes and the peculiarities of our ancestors. One might assume, for example, that if some bonobos and chimps began today to invest energy in making relatively complex stone tools that increased their reproductive advantage, bipedalism might spread more rapidly in bonobos, which are better designed for carrying tools around than are chimps. That, in turn, could favor a trend toward a more upright carrying gait, which would allow the bonobos to make more impressive displays, to snatch food from trees without climbing them, and so on. Of course, "could favor a trend" here is shorthand for the following: "Individuals who happen to have a genetic constitution that produces a more upright stance would leave more offspring than their cohorts who tend to be more quadrupedal. Gradually, the frequency of 'upright genes' in the population would increase, and the population as a whole, over time, would become more upright and bipedal." A similar trend in the past could account for our own bipedalism.

We have yet to put together a reasonably assured and agreed-on scenario for the evolution of bipedalism—except that apes were preadapted to a bipedal terrestrial gait; that the ecological shift from forest to savanna very likely played a major role in the evolution of *human* bipedalism; and that the abilities to carry children, tools, and weapons, to see farther without effort, and to brandish weapons at distant enemies may well have been important. We may never know the full details, but research is beginning to narrow the possibilities. Some uncertainty is a hallmark of science, if not always of popular science writing.

Chapter 5

BARE BONES AND A FEW STONES

"Brains exist because the distribution of resources necessary for survival and the hazards that threaten survival vary in space and time. There would be little need for a nervous system in an immobile organism or an organism that lived in regular and predictable surroundings . . . brains are buffers against environmental variability."

—John Allman[1]

If a carefully groomed and dressed *Australopithecus africanus* were to show up at your dinner party, you quite likely would think Mr. Australopithecus was a curiously slender, upright ape that another guest had brought along from the local zoo. In contrast, a similarly cleaned-up *Homo erectus* seated at the dinner table, with a brain volume of about 1,100 cubic centimeters (cc), would be judged a very strange-looking individual but unmistakably human. Even though your modern adult guests would have brains that averaged some 20 percent larger, Mr. Erectus would not look strikingly pinheaded.[2]

There may be uncertainty about why human beings evolved an upright posture, but there is no question about the fact that they did—or about the significance of that for our situation today. Picture what it would be like to try to build an industrial society if that meant running around on all fours. In our evolutionary story, we now enter the era of big brains, where not just our evolution itself but the *interpretation* of our past begins to have some important implications for the present. It is here that we encounter the debate over whether different groups of early human beings took divergent paths that led to the evolution of discrete races, possibly with discrete capabilities. And here, we must wonder what happened to our ancestors' big-brained Neanderthal contemporaries. Did our forebears breed with them, outcompete them, or kill and eat them, or did the Neanderthals just fade away? Whatever did happen, does it tell us anything about our human natures in the twenty-first century? What clues do fossil bones and stone tools provide?

One thing they reveal is that *Paranthropus* (the "robust" australopithecine) stood on its hind legs and roamed the African landscape for about a million years. But *Paranthropus* really was quite unlike us. If Mr. Paranthropus came to your dinner party, he'd seem like a version of Mr. Australopithecus with bigger molars and more powerful jaw muscles. Although upright posture is one of our outstanding characteristics, it appears that our predecessors achieved it while still having brains little bigger than those of chimpanzees. And important as an erect posture has been to the development of our natures, our brains and the minds they generate virtually *are* our natures. There is not the slightest sign that *Paranthropus* individuals had anything like modern human natures as they grubbed for roots, grabbed an occasional mouthful of meat from the lucky find of a newborn antelope, or copulated in a grassy knoll, always remaining alert for a marauding leopard. They were like the other australopithecines, including whichever ones were our direct ancestors: interesting, but not us.

The focus of our story now shifts from the evolving foundations of human nature to its essence—the development of bigger brains, how our ancestors put them to good use in evolving the general kinds of natures we have today, and the spread of our forebears from Africa to carry those natures over all of Earth.

From Homo habilis to the Naked Ape

After researchers unearthed the australopithecines, the next major "missing link" to be found was perhaps the oldest representative of our own genus, *Homo*. This creature was also fully upright and probably was able to employ its freed forelimbs to carry and use tools and weapons. Indeed, it linked upright, smaller-brained *Australopithecus africanus* to upright, much larger-brained *Homo erectus*—it was a contemporary of the persistent *Paranthropus* that we met in the previous chapter and just saw again. The fossil material of this first-discovered *Homo*, *H. habilis*, was another great find by Louis and Mary Leakey at Olduvai Gorge in Tanzania. Soon after the discovery of *P. boisei*, bits and pieces of another gracile hominid began to turn up, this one with a big braincase and small teeth instead of the reverse. The local robust-toothed australopithecine, it seems quite certain, had coexisted with what was probably one of our direct ancestors.

Evidence was also found at the site indicating that *H. habilis* had made and used stone tools. It is usually assumed that they were the first of our forebears to do so. Indeed, paleoanthropologists are now nearly unanimous in the view that these early human beings, not, as was once thought, the robust australopithecines,[3] were the original makers of stone tools. There is still some uncertainty, however, about whether this conclusion is correct.[4] Considering chimpanzees' extensive use of stones as tools and their capability of learning to make them,[5] it is possible that the australopithecines were responsible for some of the stone artifacts at Olduvai.[6] But it seems likely, given the archaeological evidence, that australopithecines did not systematically flake stone.[7] It was in light of the association of the *Homo* remains with tools that Louis Leakey and colleagues christened the new gracile creature *Homo habilis* ("dexterous man"),[8] and for convenience in this book I'm going to assume that *H. habilis* was responsible for developing the stone tools found at Olduvai Gorge.

As is so often the case with the human fossil record, uncertainties remain. The name *H. habilis* has been given to what actually may turn out to be two roughly contemporaneous East African species (and possibly even a third South African one). Differences in brain and tooth size, among other things, are found in the first specimens of *Homo* and the somewhat larger-brained (about 750 cc versus about 610 cc),[9] larger-toothed individuals that have been placed in a separate species, *Homo rudolfensis.*[10] But the sampling is too sparse to allow us to draw clear conclusions about how many species of *Homo* there were so long ago, intriguing as the question is. In addition, there is some evidence that *H. habilis*, although upright, retained some skeletal characteristics of the australopithecines that presumably made them excellent climbers and able to suspend themselves easily beneath tree limbs. *H. habilis* apparently had not completed the full transition to today's kind of human bipedalism and may well have spent considerable time in trees, foraging, sleeping, and avoiding predators. *H. habilis* and its close relatives were, however, the first of our ancestors to have modern opposable thumbs[11] and to show physical traces (as seen on the insides of their fossilized skulls) of the sort of asymmetrical brain development[12] that has been well studied in modern people. That asymmetry goes along with the right-handedness that is reflected in the way early stone tools were shaped[13] and with specialization of functions in the two halves of the brain.[14]

Homo habilis/rudolfensis lived from roughly 2.5 million years ago (mya) to about 1.7 mya. If *Homo rudolfensis* was a separate species, it may be the oldest member of our genus, given that it is that type which shows up first in the fossil record. We know roughly when these early people lived, and from the artifacts they left, we can make some guesses about how they lived. The tools associated with these early *Homo* fossils are a complex of stones from which flakes had been struck with other stones, and the flakes themselves, which often had been further modified. The remains of the original stones, called cores, are thought to have been used for coarse work, whereas the flakes were crafted for more delicate jobs. The earliest stone tools tend to be small (two to two and a half inches in their longest dimension), and they don't fall into discrete categories that allow researchers to easily establish their uses. At first sight, these artifacts are rather unimpressive, but they can be identified as having been purposely shaped because natural processes do not ordinarily flake stone in the same way and the artifacts are often made of types of stone that only people could have introduced into the deposits in which they are found.[15] Although it is possible for high-energy environments (e.g., fast-running streams) occasionally to flake stone, nature does not concentrate flaked stones at geological sites such as those at Olduvai and elsewhere in East Africa. As a group, these early tools were christened the Oldowan technology, named after Olduvai Gorge, where some of the best samples have been found.[16] Simple and crude though they are, they clearly had been shaped for cutting, scraping, and other functions involved in butchering animals, for which tasks modern experiments show they can be very effective.[17]

Sharp stone tools are not easy to make. It takes considerable practice even for modern *Homo sapiens* to produce good ones consistently. Two anthropologists from Indiana University, Kathy Schick and Nicholas Toth, have done extensive experiments in fabricating and using stone tools. As they put it, just detaching flakes from cores requires "strong intuitive knowledge of three-dimensional geometry as well as sophisticated motor skills"—skills that can take hours for modern people even to begin to master.[18]

Could nonhuman primates possibly have made these tools? Some time ago, a captive orangutan was taught to make sharp stone flakes to cut the cord on a food container, but the core from which the animal made the flakes had to be preshaped and strapped down before the ape was able to strike the core so as to produce a flake.[19] In recent experiments, a captive bonobo named Kanzi, famous for studies of his communication abilities, learned hard-hammer percussion (the making of flake tools by hitting one rock against another). Kanzi independently developed the technique of throwing stones on a hard tile floor to produce sharp flakes. To get a food reward, he used these to cut a cord that held a box shut or to slice through a plastic "skin" stretched over the top of a drum (a task thought to be analogous to the splitting of hides in carcass processing). At that stage, he discovered the importance of using great force and transferred that technique to his percussion efforts. But even with a great deal of practice, Kanzi never learned to make a stone tool as sophisticated as those in the Oldowan technology.[20] He did not grasp the importance of (or lacked the coordination or hand structure to produce) the accurate, forceful blows required to consistently make usable tools.[21]

That our distant hominid ancestors were able to produce such tools demonstrates that they had not only manual dexterity but also a capacity for planning (of how tools were to be used), substantial foresight (because stones, often partially processed cores, were carried around in anticipation of need, as evidenced by their frequent discovery far from sources of the parent rock), and appreciation of the characteristics of various kinds of materials.[22]

The consensus now, based, among other things, on careful studies of the effects of carnivore teeth and Oldowan tools on present-day carcasses in the African veldt, is that *Homo habilis* obtained the meat in its primarily vegetarian diet mostly as a scavenger.[23] Like many modern hunting-gathering people, however, it probably opportunistically hunted small game.[24] But the Oldowan technology of *H. habilis* was no flash in the pan. It was *the* human stone technology for more than 800,000 years. Recent work, however, suggests that the technology differed from place to place more than previously thought, apparently in response to both environmental conditions and the skills of the hominids making the tools.[25] Human natures may have been differentiated geographically more than 2 million years ago, just as chimp natures are today.

The Turkana Boy and His Relatives

The next known phase of human evolution is represented by an important recently recognized African link between *Homo habilis* and *H. erectus.* That link was *Homo ergaster,* who lived from about 1.8 or 1.7 mya until perhaps 1 mya, when it was replaced by *Homo erectus* and other forms. *H. ergaster* has been represented first and foremost by a famous, nearly complete adolescent male skeleton known as the Turkana boy, found at Nariokotome, Kenya, in 1984[26] (see the figure). The lad (gender can be told by the characteristics of the pelvis),[27] who met an untimely end some 1.6 mya, had an endocranial capacity of about 880 cc, which might have topped 900 cc had he lived to become an adult.[28] That is an endocranial capacity some 30 percent or so greater than that of *H. habilis/rudolfensis* and well more than half that of the average modern *Homo sapiens* (which is variable from population to population but has an overall average in the vicinity of 1,350 cc and a spread of about 1,200–1,700 cc).[29] The remarkable Turkana find suggests that *H. ergaster* could have been ancestral to *Homo erectus, Homo neanderthalensis,* and *Homo sapiens.* The Turkana boy showed no signs of having lived in trees. He represented a species that may have averaged about 5 feet, 7 inches in height and in which males may have reached 6 feet, 1 inch,[30] and he had more pronounced, modern human-like brain asymmetry than *H. habilis/rudolfensis.* That *H. ergaster* was relatively tall and had long arms and legs may well have been due to selection for proportionately large body surface to allow ease of heat dissipation, both directly and by evaporation, in its sunny tropical habitat.[31] It may have been the first truly "naked ape"—the first of our ancestors to have lost most of its furry covering[32]—and therefore it may also have been the first to evolve relatively deeply pigmented skin.[33]

Although scientists are now fairly sure that bipedalism did not evolve in order to reduce the area of body surface exposed to the sun and prevent overheating,[34] about the time of *Homo habilis* our ancestors did start to evolve a relatively large brain that required cooling. That and a need to forage widely in savannas[35] apparently created selection pressures for relatively hairless skin and an abundance of heat-dissipating sweat glands, which, in combination, efficiently cool us.[36] One would expect, by the way, that the selection pressure to lose hair for cooling would have had to be quite strong. Hair has important advantages: it not only provides insulation and protection from cold rains but also limits damage from blows and bites and makes it more difficult for insects (often disease vectors such as mosquitoes or tsetse flies) to pierce the skin.[37]

H. ergaster also was associated with much more refined stone tools than its predecessors were, including well-made hand axes (see the figure on page 94). These axes are roughly eight- to ten-inch-long stone core tools that were flaked around the periphery and over both surfaces to produce teardrop-shaped arti-

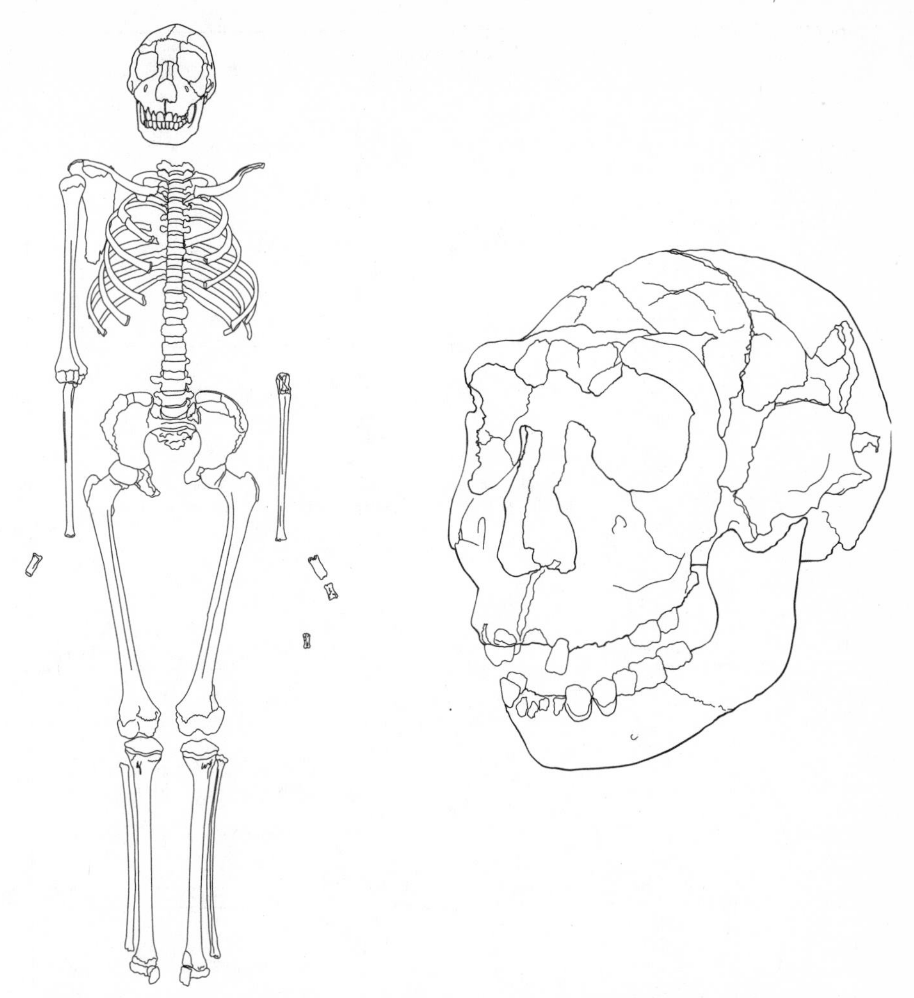

The Turkana boy *(Homo ergaster);* skeleton left, skull right. This individual died as an immature young man, already more than five feet tall. Note that the skull is similar to that of *H. erectus* shown in the next skull figure (p. 95), with a somewhat smaller braincase. Based on Johanson and Edgar 1996.

facts, sometimes with a sharp point. *H. ergaster* also made cleavers, two-faced tools about the same size as the axes but with a rather straight, guillotine-like edge.[38] This advance over the less specialized and smaller Oldowan tools is known as Acheulean technology.[39] The new technology began to appear in Africa with *H. ergaster* 1.7–1.6 mya but was later carried to western Asia and Europe[40] by *H. ergaster* or its descendants. Oddly, the Acheulean tradition itself never seems to have reached the Far East (the signature tool, a hand ax with a teardrop shape, has not been found there), but recent discoveries show that 800,000 years ago in China *H. erectus* made tools fully as sophisticated as those being manufactured in Africa.[41] Still, the Acheulean technology represents humanity's longest-running and most widespread technology; its disappearance

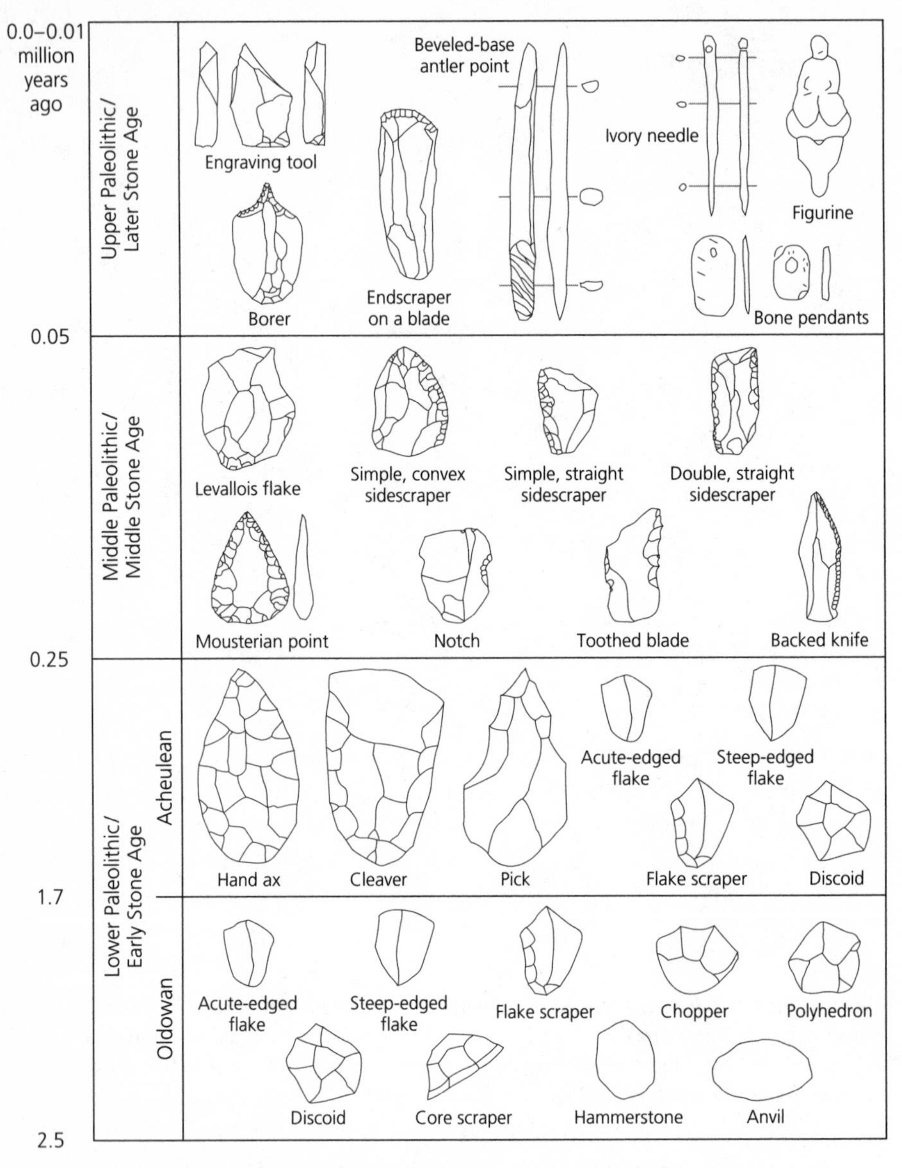

Tools of prehistoric peoples. Note that the divisions of the time scale are not of equal length. Modified and redrawn from Klein 1999.

marked the end of the Early Stone Age or Lower Paleolithic only a few hundred thousand years ago, after well more than a million years of near stasis in stone-tool design.

Not only was *H. ergaster* the first hominid to occupy very seasonal and arid habitats in Africa, but also, about a million years ago and possibly considerably earlier,[42] it dispersed from Africa, perhaps in a series of movements triggered by climate changes.[43] *H. ergaster* was thus the actor in humanity's first "out-

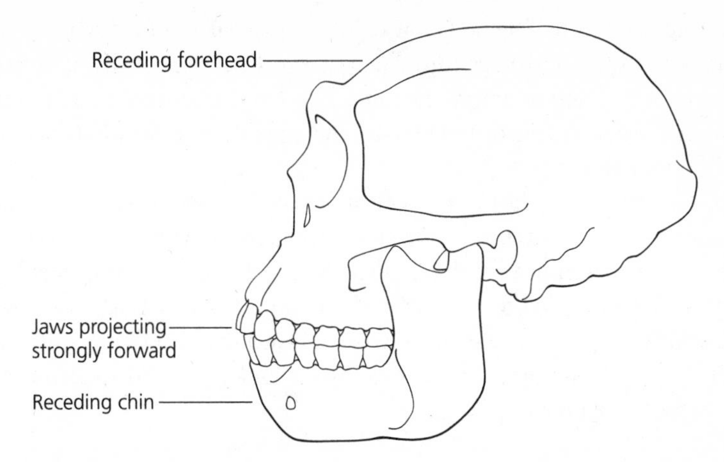

Skull of *Homo erectus* from Java. Compare with those of *H. neanderthalensis* and *H. sapiens* in the next figure and note the somewhat smaller braincase, more sloping forehead, and projecting jaw combined with receding chin of *H. erectus*. Redrawn from Klein 1999.

of-Africa" event—I'll call it "out-of-Africa 1," to distinguish it from "out-of-Africa 2,"[44] the presumed spread of *Homo sapiens* from its African homeland, a topic to which we'll turn later in this chapter.

H. ergaster was ancestral to the first inhabitants of Eurasia, including *H. erectus* in the Far East, and it was ancestral to the common progenitor of *H. sapiens* and *H. neanderthalensis* in Africa. That common ancestor was arguably the first human species to colonize Europe where it gave rise to the classic Neanderthals. Sometime around 900,000 years ago, *H. ergaster* may have reached Java and northern China, where its descendants hung around to be discovered later as the Java and Peking "men." How early they arrived in Europe remains controversial. In any case, continuous habitation there is clear only for the last half million years or so.[45]

If *H. ergaster* were the first naked apes, as they invaded cooler climes they may also have been the first of our ancestors to wrap themselves in the skins of other animals. Although the first archaeological evidence of hominids having trapped animals and sewed their skins together dates from slightly less than 20,000 years ago,[46] it seems certain that they used crude clothing much, much earlier than that to protect themselves from adverse weather. Because cloth and fur rot away in relatively short times except under extraordinary circumstances, we have no fossil remains of these materials. But lacking traces of such coverings themselves, we might still have a way to determine when our predecessors started wrapping themselves in animal skins. Clothing has been around long enough for one human louse, a species exclusive to *Homo sapiens* today, to have differentiated into two forms.[47] One, the head louse, is adapted to the head hair, lays its eggs on the hairs, and clings to its host (from which it sucks blood) through thick and thin. The body louse, in contrast, sucks blood but lays its eggs in the host's clothes and flees into them at the first sign of disturbance. Biochemical genetic analysis might establish the length of time that has elapsed since the two louse forms differentiated and thus determine when people first covered their shame.

The discovery of African *H. ergaster*, especially the Turkana boy, and the conclusion that those hominids were closely related to the later *Homo erectus* but distinct from them appear to have filled in another interesting detail of our ancestry.[48] *H. ergaster* led to *H. erectus*, the famous missing link, which was our first ancestor to occupy Eurasia from what is now northern China in the east to southern Great Britain and Spain in the west and to spread over the African continent to its southern tip.

Neanderthals and the Emergence of Modern Human Beings

In the 1960s, some paleoanthropologists held the view that different populations of *Homo erectus* in various parts of the world gave rise to numerous populations of *H. sapiens*, including those then classified as *H. sapiens neanderthalensis*.[49] Those scientists assumed that *H. sapiens neanderthalensis* disappeared as the result of a combination of causes, including having been outcompeted by and having interbred with *H. sapiens sapiens*.[50] More recent evidence indicates a different geographic pattern and a more interesting story, but one not free of controversy. Some of the most telling evidence comes from the developing field of molecular genetics, which is beginning to have an important influence on paleoanthropology. Molecular biologists are now able to determine the sequences of molecular building blocks in mitochondrial DNA.[51] Mitochondrial DNA (mtDNA) is passed from mothers to offspring in the mitochondria—energy-producing organelles (organs within cells) that are present in the egg but not in the part of the sperm that penetrates the egg.[52] Analysis of mtDNA can provide important evidence about when two different populations of people last had a common ancestor.

As a result of such analysis, most scientists now agree on two things. First, the earliest version of *Homo sapiens* (archaic *H. sapiens*),[53] one with characteristics that could make it close to a common ancestor for Neanderthals and modern people, evolved in Africa from *Homo ergaster* at least 600,000 years ago.[54] Second, rather than being considered a variant of early *Homo sapiens*, Neanderthals are now thought to represent a species in their own right, one that was contemporaneous with evolving *H. sapiens* for hundreds of thousands of years.[55] Evidence obtained by the astonishing technique of sequencing a small sample of DNA retrieved from a Neanderthal fossil[56] suggests that the line leading to *H. neanderthalensis* may well have split off from its common ancestor with modern *Homo sapiens* sometime between 690,000 and 550,000 years ago, with the spread of the estimate indicating the problems of establishing a precise date.

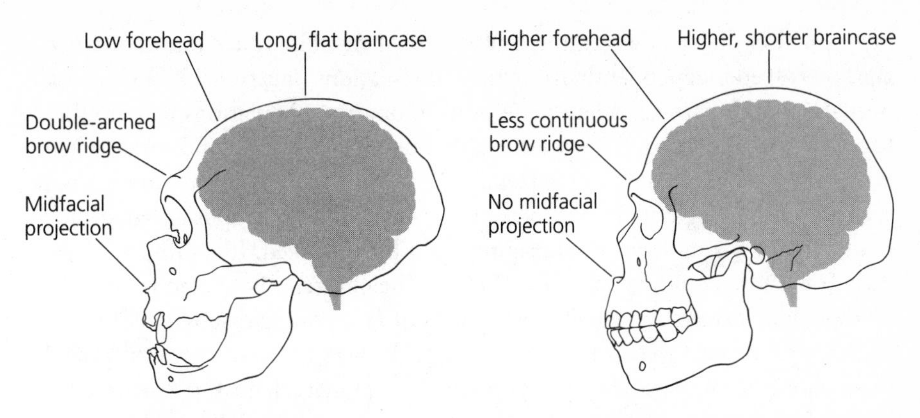

Skull of *Homo neanderthalensis* from La Chapelle-aux-Saints, France, about 50,000 years old (left), and of *H. sapiens* from Skhul cave, Israel, about 90,000 years old (right). Some of the major differences between the skulls of the two species are labeled. The Chapelle-aux-Saints individual was crippled by arthritis and had various ailments and injuries that led early paleoanthropologists to misinterpret Neanderthal anatomy. Based on Johanson and Edgar 1996 and Stringer and Gamble 1993.

Neanderthals were primarily a European and, later, a western Asian people who, it now appears, first evolved somewhere in Europe from archaic *Homo sapiens.*[57] Despite the molecular estimate, the oldest known skeletal remains that can be described with assurance as those of classic Neanderthals date from only some 130,000 years ago,[58] though there are earlier fossils anticipating the classic Neanderthals that date to at least 300,000 years ago. Establishing that the Neanderthals evolved in Europe and trace back at least 300,000 years is one of the major paleoanthropological achievements of the 1990s. The Neanderthals almost made it to the present—they appear to have died out only 40,000–28,000 years ago, perhaps even more recently.[59]

Compared with *H. sapiens, H. neanderthalensis* was barrel-chested with a relatively large but low skull and a projecting face[60] (see the figure). The range of Neanderthal brain volume broadly overlapped with that of both early and modern *H. sapiens* (and averaged higher than that of today's people), but the shape of the cranium was different, with the critical frontal cortex (associated with "higher thought," including long-term planning, deferral of gratification, and other uniquely human attributes) somewhat restricted.[61] Neither *H. neanderthalensis* nor its contemporary mid-Pleistocene *H. sapiens* was any more sexually dimorphic than modern people.[62]

Out-of-Africa 2

If the modern human beings who replaced the Neanderthals are not their evolutionary descendants, where did they come from? There are two basic hypotheses about the emergence of the latest version of *Homo sapiens.*[63] The one mentioned earlier, known as the multiregional model, argues that modern *H. sapiens* evolved independently from many populations of earlier *Homo* species all over the world, with sufficient gene flow among the populations to keep them together on the road to modernity.[64] This view, which once was widely accepted,[65] now seems almost certainly wrong.[66] The genetic data increasingly seem to militate against it, and no one has been able to propose a satisfactory model for how a multiregional origin would have worked. It is difficult to picture how there could actually have been sufficient gene flow among scattered (and possibly extremely sparse) populations of *H. erectus* and subsequently early archaic *H. sapiens* to keep them evolving in lockstep toward modernity while those populations simultaneously retained regional characteristics, such as browridges in Australians and shovel-shaped incisors in East Asians, for about a million years.[67] Members of our species from different geographic areas are remarkably similar genetically to one another (remember the Xhosa people of southern Africa, who alone retain 90 percent of human genetic variability).[68]

Evolution in lockstep in different places around the Eastern Hemisphere is especially difficult to embrace when one considers the diverse selection pressures to which widely separated local or regional populations almost certainly would have been subjected for tens of thousands of generations. Much shorter periods of selection—hundreds of generations or less—have resulted in geographic genetic variation in a wide suite of human characteristics, from blood type to skin color to lactose tolerance. It seems highly unlikely that today's genetically very similar human populations in Eurasia and Africa could have evolved largely in place from *Homo erectus* populations inhabiting the same regions hundreds of thousands or millions of years ago. It is easier to envision them having spread rapidly and recently from a common ancestral population and only then evolving differences in superficial characteristics such as skin color, hair form, tooth shape, and brow ridging under local selection pressures and the influences of gene flow and genetic drift.

Such an alternative spread-and-replacement scenario is provided by the out-of-Africa 2 model. It assumes that Africa was the source of modern *H. sapiens* and that our ancestors (by this time already evolved enough to be considered anatomically modern members of our species) spread a second time from that continent within the past 100,000 years, starting perhaps 60,000–45,000 years ago, or even more recently,[69] and replacing premodern populations[70] (see the figure). There are several variants of the hypothesis, depending on how much interbreeding might or might not have occurred between *H. sapiens* coming out of Africa and premodern populations such as European Neanderthals[71] and *H. erectus* in China and Java.[72] Although the evidence for rapid replacement is quite strong in Europe, more digging will be required in the Far East to give us a secure picture of events there.[73]

The out-of-Africa 2 model receives support from studies in molecular genetics. Some of the original genetic evidence presented came from analysis of

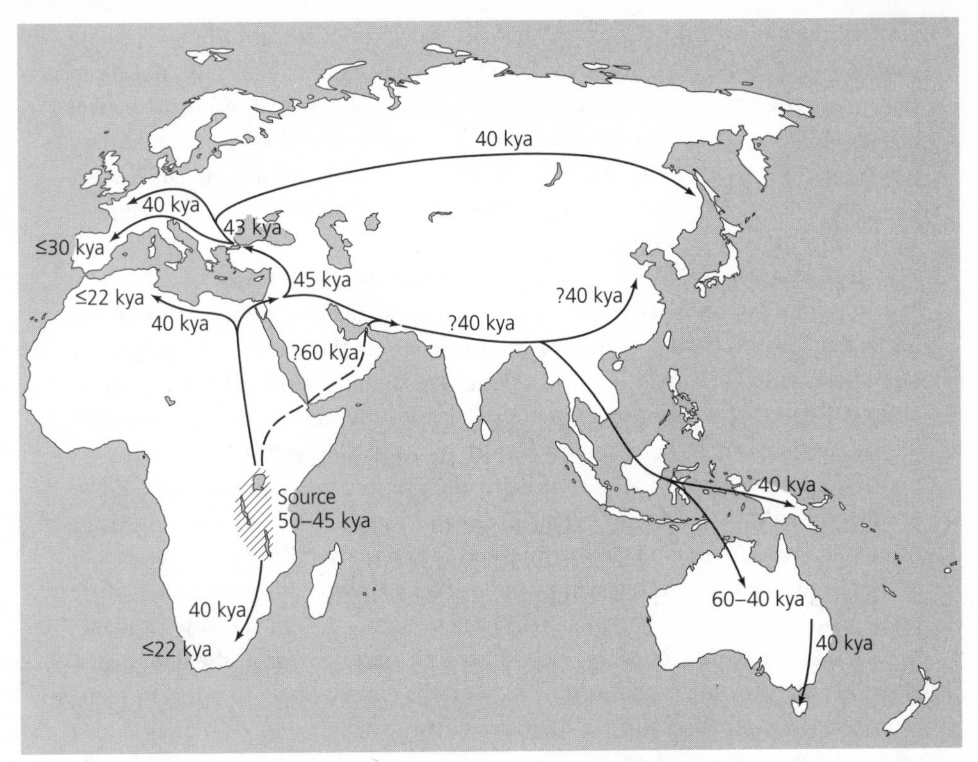

Out-of-Africa 2. Hypothesized dispersal routes from the East African source area are shown by solid lines, with approximate dates in thousands of years (kya = thousands of years ago). Dashed line indicates a possible older (60 kya) dispersion across the mouth of the Red Sea that could explain a postulated arrival in Australia about that time. Adapted from Klein 2000b.

the mitochondrial DNA of different living human populations. That research indicated that all modern human beings trace their ancestry to a single African population that lived 200,000 to 100,000 years ago.[74] This gave rise to the concept of a "mitochondrial Eve"—a single woman from whom we are all descended. But even if our mtDNA all traces back to one individual woman, the rest of our genetic endowment does not. All modern human beings do *not* share just a single female ancestor.

In order to see how both of these statements could be true, we can analogize the transmission of mtDNA through the female line with that of family names through the male line, assuming that in the first generation each man has a unique surname. If a man in that generation has two children, the odds are approximately 0.25 that both will be daughters, and in that case his surname will be lost. After 10 generations, more than 99 percent of all surnames will be gone from the population, and after 10,000 generations, only one surname will be left. But that does not mean that the descendants of the man who originally had that surname are disproportionately represented in the population 10,000 generations later or that the people in that descendant population had only a single male ancestor. Similarly, the survival of one sequence of mtDNA does not mean that it was the only one in the original population. There are serious technical questions, however, about the rate at which mutation and other evolutionary forces change mtDNA and about the way the results of the first mtDNA work were analyzed.[75] As a result, nothing definitive can be inferred about out-of-Africa 2 from these early mtDNA studies alone.

More recent mtDNA results, however, also suggest an African Eve.[76] These are supported by studies of nuclear genes (those carried not in the mitochondria but rather on the chromosomes of the nucleus), so the combined evidence is quite persuasive.[77] Some of the most definitive studies examined the variability and the pattern of association of genes (alleles at different loci) that were found in different regions.[78] They show a pattern of decreasing variability and greater nonrandom association (alleles associated more or less frequently than would be predicted purely on the basis of chance)[79] as one moves from sub-Saharan Africa to northeastern Africa and on to Eurasia. It's just the sort of pattern we would expect to find if a relatively small group of *Homo sapiens* colonists had left their homeland in Africa and begun a trek that eventually spread their descendants throughout the world, replacing populations of *Homo erectus* in the east and of *Homo neanderthalensis* in the west. The patterns clearly point to a recent common African origin for all non-African human groups.[80] Out-of-Africa 2 also recently gained more genetic support from detailed studies of some newly observed geographic variation in the Y chromosome (the one carried only by men). Analysis of these patterns also indicates that anatomically modern *H. sapiens* arose in Africa and that some groups of them gradually spread out of Africa between 200,000 and 100,000 years ago and eventually occupied the entire world.[81]

All of this new genetic evidence builds on previous paleoanthropological research that had been pointing in the same direction. Perhaps the most interesting evidence arose when skulls of near-modern *Homo sapiens* from sites at Qafzeh and Skhul in Israel were redated and found to be roughly 100,000 years old.[82] That is substantially older—perhaps by 40,000–30,000 years—than Neanderthal remains from nearby Israeli caves at Kebara and Amud.[83] If the dates are correct, this, combined with data on the relative rapidity of replacement of Neanderthals by modern people with whom they apparently coexisted with no signs of interbreeding,[84] pretty much puts to rest the notion that Neanderthals were either progenitors of *H. sapiens* or members of the same species. The same conclusion, that Neanderthals were a separate branch of the hominid family tree, is supported by statistical analyses of head shapes of fossil human beings. They show late archaic *Homo sapiens* from Africa to be slightly more similar to modern people than archaic East Asians, and Neanderthals to be much less similar.[85] All of these results are fully congruent with the out-of-Africa 2 hypothesis.[86]

The Neanderthal Mystery

Although there is growing evidence for the out-of-Africa 2 scenario, many intriguing questions remain. For instance, if the Neanderthals didn't evolve into *Homo sapiens*, why did they disappear? It is interesting to consider the possible adaptive differences between our ancestors and the Neanderthals, differences that may help explain the demise of the latter.[87] Some of the robustness of the Neanderthal physique may trace to better adaptation to cold conditions[88]—heftier people, for example, have less body surface per unit volume and thus are better able to conserve body heat. The large brain size of Neanderthals also may be related to their greater metabolic efficiency in cold climates—it's no accident that Inuit people have about the largest brains of all living people. The Neanderthals' large front teeth may have been an Inuit-like feature as well. It could be that the Neanderthals employed their teeth extensively as tools, as do the Inuit[89] (see the figure). Other features of the Neanderthal physique that differentiate Neanderthals from modern *Homo sapiens*, such as the proportions of their limbs, may be traced to genetic evolutionary responses to differences in their hunting technology and foraging techniques; for instance, they may have evolved arms that were better adapted to close-contact hunting with spears and other short-range weapons.[90] That would be an example of gene–culture coevolution: the adoption of that way of hunting could have changed selection pressures on Neanderthal populations to favor the safest and most effective physiques for that activity.

One reason why the Neanderthals disappeared may be that they were not a competitive match for modern *Homo sapiens*, perhaps less able to find food in times of shortage or even not as clever in interspecies combat. Their artifacts were fine stone tools, but these were not of the quality or variety of those of the culturally modern *H. sapiens* who eventually replaced them.[91] Several hundred thousand years ago, the Acheulean technology began to be replaced by a more refined technology, referred to as Middle Paleolithic or Mousterian[92] and frequently associated with the Neanderthals, though human beings who were *anatomically* (but not culturally) modern shared this technology with Neanderthals for many thousands of years.[93] The new technology centered on flakes, often struck from cores that had been modified to predetermine flake size and shape.[94] Hand axes are usually absent among artifacts from this period, and where they do occur, they are much smaller than most Acheulean examples. No one knows why Middle Paleolithic people stopped making hand axes, but one plausible explanation is that they had learned how to mount flakes on wooden handles or shafts and the resulting tools and weapons were more efficient than hand axes for hunting, butchering, woodworking, and other routine tasks. The Middle Paleolithic marked the end of the long Acheulean stasis and the start of another roughly 200,000-year period of technological stagnation. That Mousterian stasis, in which the Neanderthals were embedded, was, like earlier tech-

Inuit woman using her teeth as tools, much as Neanderthals are thought to have done. She is chewing a sealskin boot, or *komik*, to keep it soft, while watching a soccer game. Photograph by P. R. Ehrlich, Coral Harbour, Northwest Territories, Canada, summer 1952.

nologies, characterized by a lack of the worked bones and art objects that are often found among the artifacts of prehistoric culturally modern *Homo sapiens*. At one point, as we'll soon see, our ancestors leaped ahead of their Neanderthal contemporaries in the type of cultural evidence they left behind—possibly indicative of the Neanderthals then being at a significant technological disadvantage.

The planning and coordination skills of the Neanderthals may have been less well developed than those of modern *Homo sapiens*, but of course differences in such skills are difficult to surmise from scanty archaeological evidence. Despite the Neanderthals' physical adaptations to cold, they may have had less ability to adapt technologically—with better clothes, shelters, tools, and weapons—to the extreme environments (and rapid changes) of ice ages. And their robustness did not seem to serve them as well as technological virtuosity served *H. sapiens*. For instance, skeletal remains show that Neanderthals were much more injury-prone and subject to arthritis than their fully modern successors, often referred to as Cro-Magnons.[95] These are possibilities, but there has probably been too much emphasis on the overall "inferiority" of the Neanderthals in recent decades,[96] perhaps a form of speciesism tracing to the sorts of attitudes that can lead to racism. One need only consider how Spanish culture overwhelmed and destroyed the quite advanced Inca and Aztec empires because of what might be viewed as relatively small technological advantages and a lot of luck. The former included the domestication of horses and possession of relatively recently acquired firearms; the latter, such random factors as being carriers of potent Old World diseases to which the Spaniards were immune but against which people of the New World had little resistance.[97] Modeling of birthrates and death rates by anthropologist Ezra Zubrow suggests that the disadvantage of having a slightly higher death rate than that of contemporary *Homo sapiens* could have meant that Neanderthal populations dropped below replacement reproduction levels (death rates became higher than birthrates) and dwindled to extinction while *H. sapiens* populations grew and spread. Zubrow wrote: "This superiority may be as paltry as a one percent difference in mortality [between *H. sapiens* and *H. neanderthalensis*], and the extinction may be as rapid as 30 generations. In other words, Neanderthals could have become extinct in a single millennium."[98]

When I lived with the Inuit in the Canadian Arctic in 1952, I found them a brilliant and admirable people, still in possession of much of their highly adapted arctic culture but rapidly being overwhelmed by the technological, political, economic, and religious power of the industrialized West. The Inuit, using their own skills and technologies, had survived when the Vikings perished in Greenland in the fifteenth century. They watched representatives of the most powerful European empire of the nineteenth century die as members of Sir John Franklin's expedition to find the Northwest Passage succumbed to a harsh environment for which they were ill prepared. In the process, they inadvertently supplied further powerful evidence that cultural factors, geographic happenstance, and luck, not innate cleverness, explains the current dominance of European cultures.[99] Nonetheless, the Inuit have proven largely helpless in the face of societies that purely by happenstance were a few thousand years (an eye-blink in geological time) more technologically advanced. Could the Neanderthals have disappeared because of a similarly small technological disadvantage?

In summary, we don't know what happened to the Neanderthals. The fossil record simply isn't adequate to determine exactly the fate of the only other European human beings that lived into the era of modern *Homo sapiens*.[100] With low Neanderthal population densities, we should not underrate the possible significance of small cultural differences and chance events that, in the long run, might have proven fatal. Because there is no indication that the Neanderthals either evolved into modern *H. sapiens* or disappeared because of hybridization with them,[101] it is logical to assume that they were displaced—outcompeted and perhaps even physically attacked.

It is difficult not to notice some similarity between the rapid disappearance of Neanderthals that coincided with the appearance of modern people and the fate of the native inhabitants of the Western Hemisphere that followed the invasion by Europeans. *Homo sapiens* has not been notable for a tolerance of differences or a drive toward coexistence with differing cultures—to say nothing of competing species. But, unlike the Western Hemisphere invasions 500 years ago, the prehistoric situation would not have been one of invasion by a people with a vastly superior technology backed by a huge population base. Populations of both *Homo sapiens* and *H. neanderthalensis* were sparse,[102] and their technologies were rather similar (compared with that of, say, the horse-riding, gun-toting Europeans who invaded a New World that lacked both horses and guns), although various aspects of their behavior, such as their social organization, might have been rather different. Thus, the fate of the Neanderthals is likely to remain a rich source of controversy and metaphor for some time.[103]

Whatever their fate, the Neanderthals hung around until the time of the greatest cultural transformation in the history of our species. In most places, the Neanderthals disappeared suddenly, sometimes in what appears to be a matter of decades. In one area of western France, there are signs that Neanderthals adopted some of the techniques of the revolutionary Cro-Magnon people *(Homo sapiens)* who eventually replaced them.[104] There, the Mousterian technology of the Neanderthals changed. It was replaced by a tool kit[105] that combined the Mousterian flake tools with longer, blade-like tools characteristic of the Upper Paleolithic Aurignacian[106] stone technology, the earliest stone-tool kit of the Cro-Magnons. The change to that Upper Paleolithic technology, which appeared first in the Middle East about 50,000–40,000 years ago,[107] was the start of the most rapid and radical cultural change ever recorded in the hominid line—what ecologist Jared Diamond has recently popularized as the Great Leap Forward.[108] It is a leap into new technologies, art, and population growth—perhaps even into a new mode of speaking—that is the cause of endless speculation,[109] and we'll return to it frequently when we switch from discussing the bony parts of our natures to discussing the soft ones.

Out-of-Africa Revisited

Other questions aside from the disappearance of *Homo neanderthalensis* arise from the out-of-Africa 2 hypothesis,[110] and they illustrate the complexities of sorting out the routes by which evolution has shaped human natures. For example, why did the climb to sapience happen to occur in Africa? And was the out-migration of early *Homo sapiens* from Africa a single event, a steady trickle, or a series of waves?[111] The genetic evidence with respect to the latter question seems to point toward a single primary event, but other interpretations are possible—for instance, that a small source population generated emigrants over time. It is also unclear how much (if any) hybridization occurred between *Homo sapiens* and *Homo erectus* in Asia. Some recent evidence indicates that relatively isolated groups of *H. erectus* may have persisted there to within 53,000–27,000 years of the present,[112] and there are signs of some possible interbreeding and re-migration that brought Asian premodern genes back to Africa.[113]

But the exact status and fate of Far Eastern *H. erectus* remains an area of controversy.[114] There even is some evidence suggesting that *H. erectus* was capable of crossing straits of a dozen miles or so, island-hopping as it colonized the archipelago east of Java.[115] Palaeoanthropologist Mike Morewood summarized his view of the situation thus: "*Homo erectus* was not just a glorified chimp; they hunted animals with well-crafted spears. We now believe that they made sea crossings to reach Flores and other Indonesian islands. The evidence suggests that the cognitive capabilities of *H. erectus* may be due for a reappraisal."[116] Other paleoanthropologists are more reserved in their judgments, suspecting either that the straits were actually much narrower than now estimated or that an unsuspected land bridge existed.[117]

If *H. erectus* could make such voyages, that would be impressive indeed, given that there is no evidence that even early *Homo sapiens* had sailed across the Strait of Gibraltar from Africa to Europe. With lower glacial sea levels, the Gibraltar crossing would have been only five miles or so. But the situation is confusing because other groups of spreading *Homo sapiens* were able to achieve sea crossings of considerable distance in reaching the fused New Guinea–Australia landmass.[118] One possible explanation of these apparent inconsistencies is that cultural evolution was as spotty in both *H. erectus* and prehistoric *H. sapiens* as it is today among us. Not all coastal peoples in historical times had equivalent attitudes toward the sea or skills as mariners. Seagoing cultures could have thrived in Southeast Asia, for example, while Africans remained landlubbers. Human natures may have been as variable tens of thousands of years ago as they are today.

There is also substantial uncertainty about when Australia and New Guinea were colonized by *H. sapiens*—perhaps as long as 60,000 years ago,[119] but almost certainly by 40,000 years ago.[120] The trip to Australia required the crossing of a substantial stretch (about sixty miles) of ocean even when sea levels were quite low, and this indicates that our ancestors were capable of building seagoing craft some 40,000 years ago.[121] Similar questions about timing of human occupancies arise for other regions of the globe. There is, for example, considerable debate about the date (or dates) and sequencing of the peopling of the Western Hemisphere[122] and about whether or not people resembling Europeans were involved in the original colonization more than 10,000 years ago,[123] perhaps more than 22,000 years ago.[124]

So what is one to make of the mass of sometimes contradictory evidence about the origin or origins of modern *Homo sapiens*? Gradually, the weight of scientific opinion has shifted toward some version of the out-of-Africa 2 scenario,[125] which has become the dominant view among human geneticists.[126] But the multiregional hypothesis still lurks in the background, supported by some fossil evidence that could be interpreted as showing regional parallel evolution from *H. erectus* to *H. sapiens*, in the event that the genetic evidence proves to have been grossly misinterpreted.[127]

The scenario that makes the most sense to me, on the basis of the evidence now available, is that modern human beings evolved in Africa and spread to other continents relatively recently, possibly in several waves, first over most of the Eastern Hemisphere and later to the Americas. The last out-of-Africa wave, as we shall see, may have consisted of people who were little different physically from their predecessors but had reorganized brains and the first fully modern speech. The Africans replaced (and, in the process, to some degree may have interbred with) resident archaic populations. Such an out-of-Africa 2 scenario would explain the genetic differences among present populations by means of a mix of selection, exchange of genes with archaic populations, migration (including re-migration from Eurasia to Africa), and genetic drift operating over the past 100,000–40,000 years.

Darwin had information only about the first Neanderthal finds when he started considering how our natures evolved. Now, however, the history of our natures as revealed by bones and stones is extraordinarily rich. If another organism had a fossil record as fine as that of humanity, it would be considered complete. Instead of a search for a missing link between the apes and *Homo sapiens*, we now have an embarrassment of riches, even though the fossil sampling of East African Pliocene hominids in the critical period of 3 to 2 mya is still quite sparse.[128] It looks as if, as I suggested in the previous chapter, hominids went through an adaptive radiation, and the tree of hominid evolutionary history, rather than being a largely unbranched sequence from *Australopithecus* to us, is more of a phylogenetic bush. Our physical-cultural history now shows more clearly that rates of physical and cultural evolution have long been only partially coupled: early cultures remained amazingly constant during a period of substantial growth in brain *size*, although size alone does not tell us all we need to know about evolving brains (if it did, an elephant or a blue whale might be writing this book). But then, cultural changes took place at astonishing speeds with no significant change in the physical appearance of people or in the characteristics of their brains that can be divined from fossil skulls. And if determined exploration can add more detail to the fossil record, the bony remains of our ancestors seem poised to provide more complete answers to some of the most puzzling why's and how's of human evolution—such as why we became bipedal, why brain evolution lagged behind the evolution of upright posture, and how modern people populated Eurasia.

Questions about the details of our physical evolutionary history have considerable pertinence to policy issues that show up on today's evening news, where hate crimes and affirmative action often make top stories. The out-of-Africa 2 hypothesis reinforces the view that human beings are physically much the same and have the same mental abilities over the entire planet. There is no reason to believe that the substrata on which our diverse natures have been built are fundamentally different among different groups. In short, the out-of-Africa 2 hypothesis adds to the weight of evidence that racism can find no support in science. We know all this in large part because of the successful efforts made by a small army of smart and dedicated paleoanthropologists. Their discovery and analysis of secrets of our past has been a triumph of nonexperimental science.[129]

Nonetheless, big and interesting questions about our past persist. The job of paleoanthropology in helping us understand the roots of our human natures is far from finished. Many of these questions are about the evolution of human behavior and culture, topics that are difficult to trace from the fossil-archaeological record: Why did the first stone-tool culture last for so very long? Why did it then undergo a couple of hundred thousand years of gradual transition and then suddenly blossom into a culture that rapidly, in a few tens of millennia, led to the creation of nuclear weapons and supercomputers? In other words, did the physical evolution of our ancestors' brains cause the Great Leap Forward—or did only the "software" of culture change, not the "wetware" of brain structure? And what, if any, role was played by the evolution of language in that cultural Great Leap?[130] By combining knowledge of the behavior of modern people and of living relatives of *Homo sapiens* with what we *do* know from analyses of the bones, artifacts, home sites, and other traces of our prehistoric ancestors, we can begin to answer those questions about the origins of our unique natures.

Chapter 6

EVOLVING BRAINS, EVOLVING MINDS

> "[O]ur minds would not be the way they are if it were not for the interplay of body and brain during evolution, during individual development, and at the current moment."
>
> —Antonio Damasio[1]

I'm writing this passage at the edge of a tropical rain forest in Costa Rica. It is near the turn of the millennium, yet I can still cause my brain to conjure up the sights and sounds of the arctic tundra where I did research in the summer of 1952. The willow bogs, my first capture of an *Erebia rossii* butterfly, the call of an old squaw duck, and the attempts of a semipalmated plover to distract me from her nest—all these come to mind immediately. So does the sensation of sponging myself off while straddling a blazing Primus stove in my freezing shack and of devouring the box of Barton's Truffles chocolates that my parents kindly sent me. Bits of the local Inuit language still dance through my brain's synapses, as do memories of a very smart young woman whom I'd met in a University of Pennsylvania zoology course just before departing for the far North.

Even in an aging man, my brain still sends out a stream of complex commands that allow my legs to propel me up and down steep rain-forest trails with a surefootedness that no computer-instructed machine could match. In the process, my brain subconsciously solves extraordinarily difficult problems in mathematics and physics that I could never solve with conscious effort. It can so coordinate my movements that I can still (sometimes) hit a pitched ball with a bat or (always, so far) dodge a speeding car as I walk along a Costa Rican highway at night. As I hike through the forest, alert for a glimpse of a dusky antthrush, buff-rumped warbler, or other interesting bird whose salient characteristics are stored in my memory, my brain simultaneously processes the ever-changing kaleidoscopic view of trees, slopes, leaves, butterflies, spiderwebs, and leaf-cutter ant trails that greets me, and which the optics of my eyes are incapable of seeing without cerebral interpretation. And, without my being conscious of its activity, my brain also monitors the functioning of my physiological systems as well as the pressure waves traveling through the atmosphere to my ears and the odorant content of the air around me.

Philosophers and scientists have long wondered about the human brain and the myriad mental processes centered in that organ. Those functions—perceiving, feeling, thinking, remembering, and such—are central to our visions of ourselves. Subjectively, we are largely our minds, as I've suggested. The evolution of human natures, and the human place *in* nature, is also largely a matter of the evolution of minds and thus of brains. After all, the brain is the organ that is central to that complex of functions, just as legs are central to walking, running, kicking, and jumping. Other organs are involved in those acts of motion, too, and those motions are only some of the things legs can do. Similarly, the brain does many things besides generate the mind, and other organs, such as various glands, are crucial to the mind's functioning. But when we think of brains, we ordinarily think of minds, just as when people think of legs they think of walking and running. Even more than the legs, however, the brain is thoroughly integrated with the rest of the body, not only in its reliance on the body for physical and physiological support but also through its control functions and, especially, through feelings and emotions. This integration is sometimes described as our conceptual life being "embodied."[2]

Evolution is the key to the mind.[3] The great expansion of hominid brains revealed in the fossil record clearly is the reason why human beings instead of baboons or chimpanzees burst out of Africa, occupied the entire planet, and shaped Earth to their own uses. That enlargement created minds with enough storage capacity for each individual to develop a personal self, one able to reflect on a gradually stored history. And that, in turn, gave every human being a distinctive nature—far more distinctive than the natures found among any troop of monkeys or pack of hounds. Yes, we can distinguish individuals in populations of gorillas, German shepherds, even butterflies—but their natures are as uniform as peas in a pod when compared with even the natures of Chang and Eng, the congenitally joined identical twins discussed in chapter 1. Our brains are not only the source of our individual natures: the evolved ability of those brains to communicate detailed information to one another allows every group of human beings to evolve a group nature as well. Careful investigation is usually needed to distinguish two baboon troops, but no study whatsoever is required to tell a farming community in the midwestern United States from a Machiguenga tribe in the rain forests of Peru or to tell a street gang from a class of divinity students.

Changing Minds

Virtually all scientists agree that our brains are organs whose "design" by natural selection owes much to our primate ancestors' needs first to forage in bushes and trees and then to survive in a savanna habitat. For example, the parts of our brain that help us see are relatively large, whereas the parts that let us perceive the world through smell, as dogs do, are relatively small. We are "sight animals."[4] Evolution has also provided us with a relatively large cerebral cortex, the brain's outer shell, where most "higher" functions occur, so in addition to being sight animals, we are "thinking animals." For our purposes in this book, a particularly important subdivision of the cortex is the neocortex, a new structure that evolved in the mammals and is central to the way human beings perceive and think about the world.[5]

Evolution seems to have built the neocortex, with its capacity for thought, right on top of other structures in the brain. This is not surprising, for as brain scientist John Allman points out, "the brain can never be shut down."[6] All the old control systems must remain in place and functional while new ones, providing new capabilities, evolve. Regulation of limb movement and blood chemistry remains as essential for our proper functioning as it was for early amphibians and for our rodent-sized forebears who ate the eggs of dinosaurs. But the brains of those predecessors were not faced with the sort of planning and memory requirements characteristic of increasingly social and long-lived hominids. Those control capabilities were added on—perhaps through a process of gene duplication that gave natural selection the opportunity to produce new functions while allowing old, essential ones to remain.[7]

Aside from the development of a neocortex larger than those of the other living primates, many major reorganizational changes have taken place in the human brain since the hominid line diverged from that of the chimpanzees—changes related to the development of social skills suitable for dealing with increasingly complex societies and to the development of language, tool manufacture and use, accurate throwing, increased memory capacity, and the like.[8]

The Midget between Our Ears

Along with these other evolutionary developments, our brains evolved the overall ability to keep us, when awake, continuously and acutely conscious of what's going on around us. Much more than did our distant ancestors, we analyze our physical and social surroundings, remember past events, and "talk" to ourselves about those analyses and memories and their meaning for our future. We have a continuous sense of "self"—of a little individual sitting between our ears—and, perhaps equally important, a sense of the threat of death, of the potential for that individual—our self—to cease to exist.

I call all of this sort of awareness "intense consciousness"; it is central to human natures[9] and is perhaps the least understood aspect of those natures. A more poetic way of describing intense consciousness is that of philosopher John Dewey: "Man differs from the lower animals because he preserves his past experiences. What happened in the past is lived again in memory. About what goes on today hangs a cloud of thoughts concerning similar things undergone in bygone days. . . . [Man lives] in a world of signs and symbols. A stone is not something hard, a thing into which one bumps, but is a monument of a deceased ancestor."[10] Dewey might have added that human beings are also the only animals that seem fully aware of the consciousness of other individuals and thus have been able to develop empathy, the capacity to identify emotionally with others. This is a key to understanding some of the most important aspects of our natures, as we will see.

Consciousness itself, a broader concept than that implied by the term *intense consciousness* as I am using it, has many meanings.[11] I prefer to define consciousness simply as the capacity of some animals, including human beings, to have, when awake, mental representations of real-time events that are happening to them or are being perceived by them.[12] Consciousness, ordinary or intense, is, as psychologist Robert Ornstein said, "the 'front page' of the mind."[13] That is a striking metaphor, but it might mislead if taken too far because unlike the situation with a newspaper, what is on the front page of the mind is apparently generated by what is happening on the inside pages—brain activities of which we are unaware.[14] An interesting example comes from American philosopher Daniel Dennett. He points out that if someone knocks over an empty glass on the table in front of you, you do not "automatically" leap to your feet, but if someone knocks over your cup of coffee, you do. Yet the front page of your mind is not aware of the back-page "thinking" that integrates the possible results of each event and that leads in the second case to nearly instant action.[15]

British psychologist Nicholas Humphrey has developed an intriguing view of the evolution of consciousness.[16] He ties consciousness to bodily sensations—the presence of mental representations rich with "feeling," of "something happening here and now to me." Sensations contrast with perceptions, which are bits of news without emotional content about "what's happening out there."[17] The distinction is a fine one: the chemical exuded by a rose, for example, is perceived as having a sweet scent but also gives a person the sensation of being sweetly stimulated.[18] Humphrey's hypothesis is that sensations occur at the boundary between an organism and its environment and, in human beings, generally are registered in the mode of sight, touch, sound, smell, or taste. When direct sensations are absent, mental representations are accompanied by reminders of sensation; for instance, some thoughts are "heard" as quiet voices within the head and would fade without that component of sensation.

Many of Humphrey's claims can be tested only subjectively, and subjective testing may well be an unreliable guide to what actually is occurring. Do you hear some of your thoughts? I *think* I hear many of mine. I also seem to *see* many of them, visualizing them as in my dreams, although language is certainly lurking around the edges of even my most vivid pictorial representations. Do you find it hard to retain an imagined visual image, of, say, a forest fire while looking at a cloudy sky? I do; the lack of sensation of heat and color makes it hard to retain consciousness of the fire. Humphrey's short summary of consciousness is not the famous "Cogito, ergo sum"[19] of René Descartes but rather "I feel, therefore I am."[20] He reinforces that view with a quip from novelist Milan Kundera: "'I think, therefore I am' is the statement of an intellectual who underrates toothaches."[21]

In Humphrey's view, there has been a gradual evolution of consciousness through a shortening of the sensory loops. Originally, in our distant ancestors, the input of sensation from the organism–environment interface triggered a nervous response to the stimulated part of the body surface. Pricking the surface of the organism sent a message to a nerve center, which responded with, say, a nerve signal to pull in the surface at that point. Then, gradually through evolutionary time, that circuit shortened and the target of the responses moved inward, to a surrogate area on the surface of a slowly evolving nerve center, which eventually became a brain. Nerve impulses arriving in the brain created conscious experiences. These experiences, in turn, could result in the generation of more nerve signals, allowing the organism to make a more complex response.[22] It seems a plausible evolutionary story, but it's at most a starting point for thinking about one of the most difficult of all questions, that of how consciousness, awareness of one's own existence, evolved.

To what degree are other animals "conscious" or "intentional" (able to have beliefs and desires directed toward things and *about* things, purposeful, goal directed)? Problems in determining what it's like to be a nonhuman animal are legion, involving many fine philosophical points.[23] Consciousness may well be limited to higher vertebrates, perhaps restricted to *Homo sapiens*.[24] Before we really attempt to decide who is conscious, it would be nice to know how to explain human consciousness itself. But some philosophers, notably Colin McGinn, doubt that we will ever understand how a pattern of electrochemical impulses in our nervous systems is translated into the rich experience of, say, watching an opera or flying an airplane.[25] He believes that our minds did not evolve in such a way as to enable us to answer that question, which may be fated to remain unanswered for a very long time, if not forever. Daniel Dennett takes a more optimistic view, arguing that at the very least we should not take a defeatist position.[26] In *Consciousness Explained*, he takes an interesting cut at the problem, but he does not "explain" consciousness to my satisfaction. The reason is doubtless rooted in Dennett's and my different subjective definitions of what constitutes "understanding" of how nerve impulses and synaptic connections (the electrochemical links between nerve cells) translate into consciousness. That problem is somewhat analogous to the question of whether knowing the mathematical equations that describe energy transformations allow us to really "understand" energy.

Although there are many such difficulties in understanding the nature of consciousness, Humphrey's evolutionary view implies some sort of continuum, from animal "automata" (amoebas, earthworms, mites) through conscious nonhuman animals capable of intentional behavior (maybe dogs, surely chimps) to us.[27] Under such a view, patterns of apparently more and more thoughtful behavior tend to appear in animals as their evolutionary relationships to *Homo sapiens* grow closer. Right or wrong, assuming various degrees of animal consciousness spares us the hubris of seeing ourselves as the sole possessors of consciousness, dodges ancient and unresolved (probably unresolvable) solipsistic debates, and opens human thought processes and the characteristics of our minds to evolutionary investigation. In any case, for the purposes of our discussion, I will assume that a lesser degree of conscious awareness than you and I possesses is present in many other animals. Even vervet monkeys, as primatologists Dorothy Cheney and Robert Seyfarth put it, "are adept at understanding each other's behavior and relationships," whether or not they think about one another's thoughts.[28] But intense consciousness, as I've defined it, appears unique to *Homo sapiens* among modern organisms and is central to human nature.

What could have been the selective advantage that led to the evolution of intense consciousness? This type of consciousness helps us to maneuver in a complicated society of other individuals, each of whom is also intensely conscious. Intense consciousness also allows us to plan without acting out the plans and to consider that other individuals probably also are planning. I thus agree substantially with Humphrey that "the higher intellectual faculties of primates have evolved as an adaption to the complexities of social life" and that our "styles of thinking which are primarily suited to social problem-solving colour the behavior of man and other primates even towards the inanimate world" (think of such things as physicists' use of the term *charm* to describe quarks, inanimate elementary particles).[29] That the size of the primate neocortex appears to be correlated with the complexity of social structure, with the most social creatures having the largest neocortexes, adds credence to this view.[30] I suspect that in the evolving hominid line, mind and consciousness eventually were further honed by pressures generated by increasing group size and, especially, by the complexity of patterns of social contact within the group and with other groups of hominids.[31] Resource shortages may also have played important roles.[32] My favorite conjecture, to put this in other words, is that human natures have been strongly shaped by multilevel human societies—families, bands, tribes,[33] trade networks, and so on—and by population pressures, all interacting in a positive feedback system. The more adept people became at dealing with social dilemmas and the more individuals there were, the more difficult were the social dilemmas that human beings as a group were able to create and the more that brain evolution was stimulated.

Brains within Brains?

If we wish to understand consciousness, social behavior, and the way in which both of these important aspects of our nature evolved—who we are and where we came from—it is obviously essential to know something about the brain. Behavior is the public display of our human natures; the brain and the mind are its primary generators.

What today is an increasingly prominent scientific view of the properties of our thinking-feeling-remembering-controlling organ traces to, among others, linguist Noam Chomsky[34] and, most explicitly, to psychologist Jerry Fodor, both of the Massachusetts Institute of Technology. It is a modern version of the idea that different physical parts of the brain control different mental "faculties." That view dates at least to Franz Gall (1758–1828), a famous German neuroanatomist and the founder of phrenology, the practice of evaluating mental abilities and character by examining the shape of the skull. His assignments of positions in the brain to particular mental faculties were pure guesswork, but he did stimulate others to experiment on animal brains, and these researchers quickly demonstrated that various brain functions were indeed localized. Scientists since then have gradually unraveled some of the brain's mysteries. Their research has been greatly aided in recent decades by new ways of studying the brain in action, such as positron-emission tomography (PET)[35] and functional magnetic resonance imaging (fMRI),[36] both of which can show the distribution of activity in the brain at a given time. Our understanding of the brain has also benefited greatly from rapid advances in molecular biology that have revealed much about such important issues as how the nerve cells (neurons) of the brain communicate with one another across the tiny gaps (synapses) between them.[37] Long before the development of PET scans and the techniques of molecular biology, however, scientists were able to make significant progress in understanding the geography of the brain by studying people whose brains had been damaged.

Perhaps the most celebrated brain-damage case of all time was that of railway worker Phineas Gage. On September 13, 1848, Gage made a serious error when tamping powder in place to blast some rock during construction of the Rutland and Burlington Railroad in Vermont. Momentarily distracted, he forgot to wait for the powder to be covered by sand and tamped it directly, striking a spark and setting off the charge. The 1.25-inch-thick, yard-long tamping iron shot through his head and into the air, landing yards away. It caused severe frontal lobe damage, but Gage miraculously was not even knocked unconscious and successfully fought off the infection that followed before the wound healed (see the figure). Even more amazingly, most of his brain functions survived: he could walk, talk, and learn new things, and in a general sense, he seemed to retain his intelligence. But his personality changed drastically, so much so that his co-workers said he was "no longer Gage."[38] After the accident, he lost much of his self-restraint and became uncharacteristically vulgar and obstinate. He was unable to carry through with plans, even though skillful execution of plans had once been his hallmark.[39] This remarkable case was the first to show that the ability to make rational decisions is, at least in part, located in structures at the front of each hemisphere of the brain known as the prefrontal cortex (the cortex of the forward-facing portion of each of the two frontal lobes).[40]

Studies of unfortunate individuals who have suffered brain injury from accidents, tumors, strokes, and the like have since proven to be rich sources of information about the localization of the brain's functions. Behavioral deficits result-

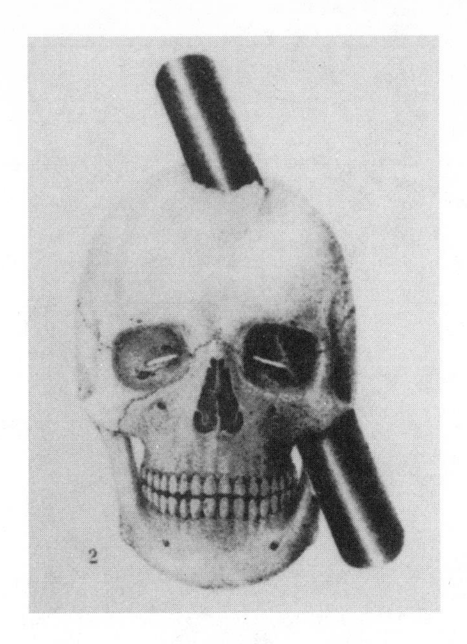

The skull of Phineas Gage, showing where the tamping iron penetrated the head. Courtesy of the National Library of Medicine.

ing from brain injury, for instance, are analyzed and compared with autopsy results and evaluations of damage obtained by other methods.[41] These "natural experiments" allow scientists to construct a picture of which parts of the brain are involved in what functions and how those functions are interrelated. For example, different roles in various types of brain processing are localized in the right and left cerebral hemispheres.[42] For example, studies of most right-handed split-brain patients—those in whom the great bundle of nerves connecting the two hemispheres (the corpus callosum) has been severed—show the function of speech largely localized on the left side of the brain.[43] When split-brain patients are asked to choose among objects by feel only—say, keys or a book—with the left hand (which is connected to the *right* hemisphere; the "wires" cross), they can do it, but they cannot say the name of the object they have chosen. In the split-brain patient, the left hemisphere remains uninformed of what the right hemisphere is doing.[44] Overall, the evidence suggests that the right half of the brain tends to deal with large elements of perception, general outlines of figures in the visual field, gross movements of limbs, and strong emotions. The left, in contrast, generally specializes in fine analysis and precise movements (such as are involved in vocalization) and meanings[45] (see the figure).

Building on a foundation of results from brain-damage studies and results from modern imaging techniques, linguistic studies, and so on, Fodor concluded that perceptual and computational processes are not only localized but also more autonomous—isolated from the brain's background knowledge—than brain researchers had previously thought.[46] Unlike those (especially Swiss psychologist Jean Piaget and his associates[47]) who have argued that neither brain processing nor memory storage is localized[48] and that the brain produces a single, unitary mind,[49] Fodor suggests that those processes go on in semi-isolated

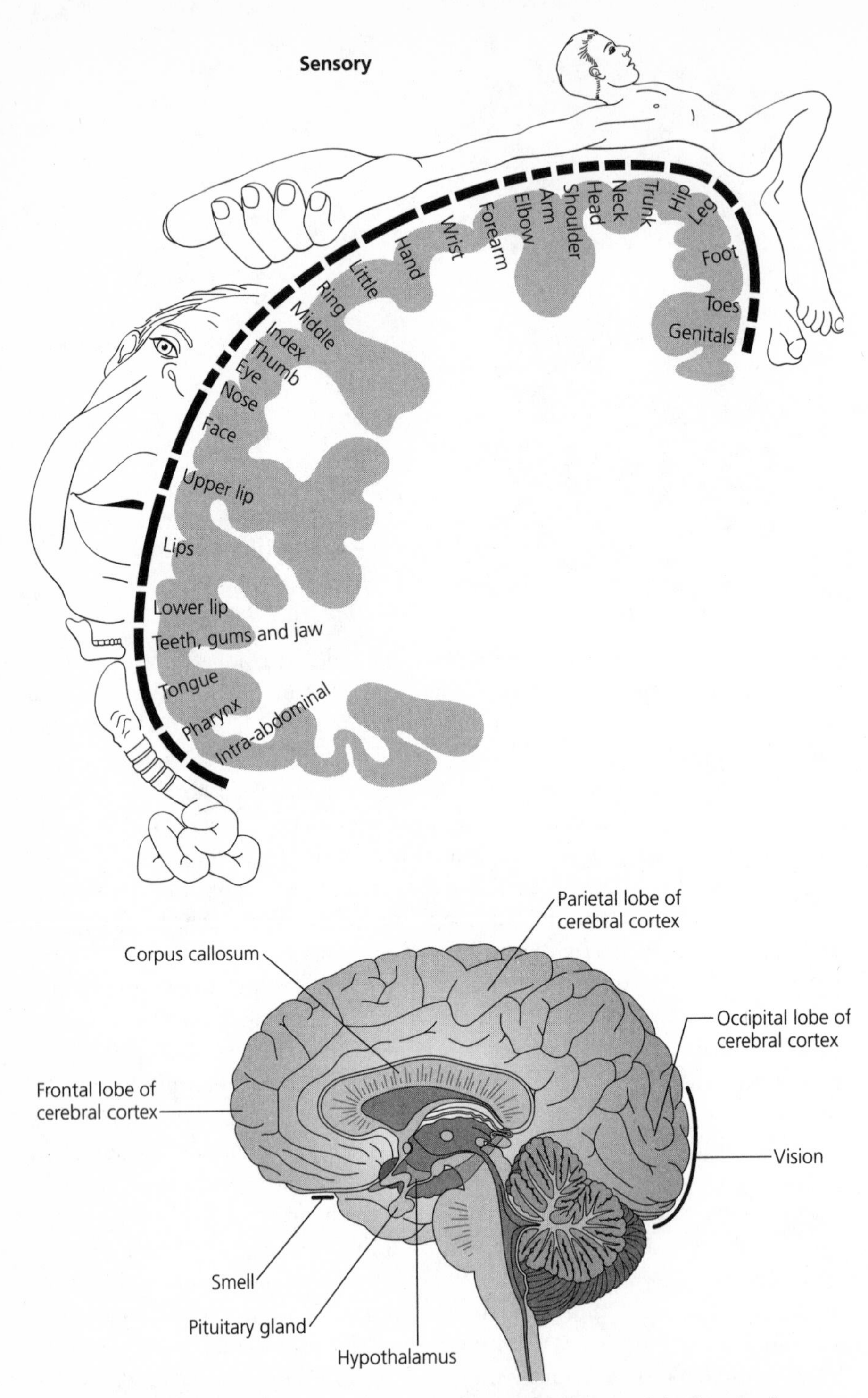

Major areas of the brain shown in cross-section (below) and a map of the functional areas of the cortex (above and opposite). Notice that the parts of the brain that control complex activities such as speech and manipulation of objects have disproportionate amounts of the cortex allocated to them. Redrawn from Raven and Johnson 1999 and Ornstein 1988.

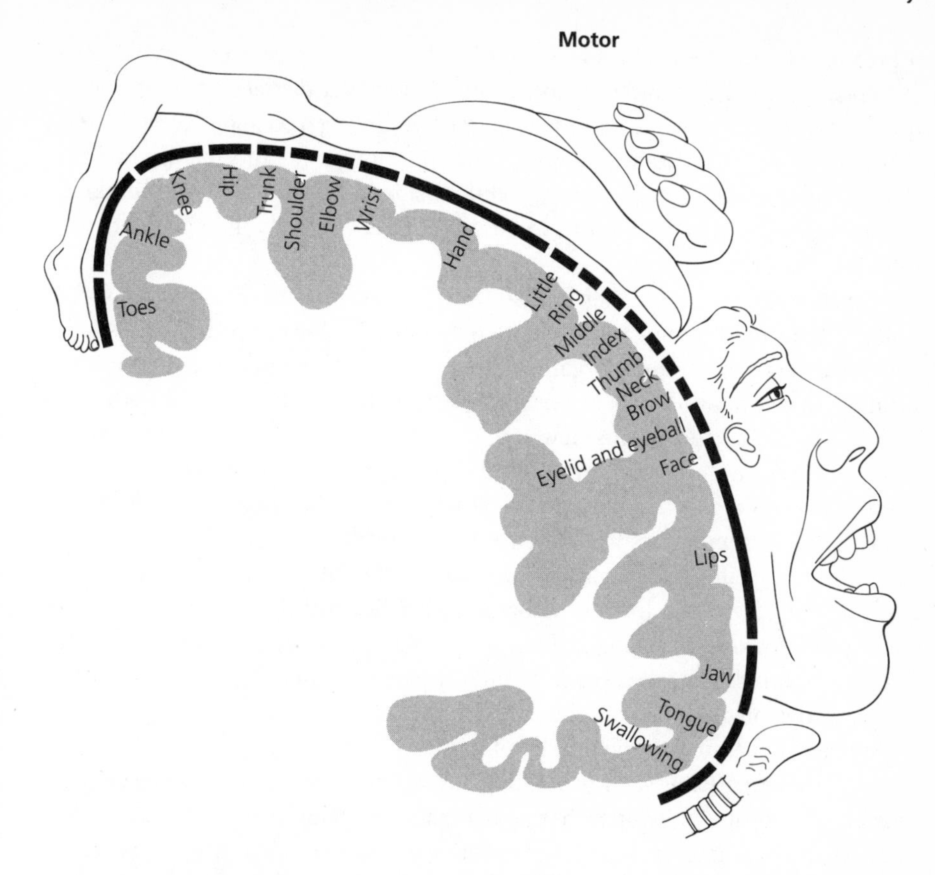

subunits, which he calls modules. These could be analogized to numerous separate "minds" crammed inside our skulls, ones Fodor describes as "domain specific"[50]—that is, specialized for dealing with certain kinds of information and problems. Among other things, evolution, by concentrating functions in local areas, would have helped to minimize the total length of neural "wires" required to make the necessary synaptic connections.[51]

Modules have been proposed by Fodor and others as ways to envision such diverse operations as detecting cause and effect, acquiring language, recognizing faces,[52] maintaining kinship and social bonds (a function doubtless closely tied to face recognition),[53] and determining sexual attraction.[54] One argument Fodor makes for this modularity of the mind is that some modules are sufficiently autonomous that they can't be turned off. When we run our fingers over the top of a table, for instance, we can't avoid feeling a surface. When we look at a scene, it is impossible to turn off the impression of three-dimensional objects against a background. Even when experimental subjects are asked to concentrate on the sound characteristics of spoken words, they cannot help but identify the words.[55] As Fodor wrote, "You can't hear speech as noise even if you would prefer to."[56] This, however, is an argument that could cut both ways. One could argue that if modules were truly autonomous, selection would have made it possible to turn them off independently when so doing favored survival. For example, it might be beneficial to be able to turn off distracting background noise when visual concentration is required for aiming a weapon. Overall, scientific opinion is far from unanimous that the brain is as thoroughly divided into modules as Fodor believes it is,[57] and we should be cautious about thinking of modules as independent entities that might have relatively independent evolutionary histories.

Psycholinguist Steven Pinker, also of the Massachusetts Institute of Technology, describes the mind as a "neural computer."[58] He and many of his colleagues also see the brain as organized into specialized modules[59] that communicate with one another and, at certain points, with other parts of the body and with the world outside. In this view, the brain is seen basically as computer hardware (or "wetware") in which the mind is a program or a connected series of programs. Researchers in this group point out, for instance, that the brain doubtless does some of its work by means of "parallel processing"—using several different chains of neurons to deal with different aspects of a perception (such as movement and form in visual input).[60] But that doesn't mean that the brain can be properly analogized with a parallel-processing computer. Saying that the brain is like a computer is sort of like explaining the operation of a Cadillac to someone by saying it's a device that operates like a horse by changing chemical energy into the energy of motion.

We can learn from such analogies, but we must be cautious about taking them too seriously.[61] Scientists have a tendency to analogize human brains with some aspect of whatever is the latest technology; early in the twentieth century, brains were thought to be organized like telephone switchboards.[62] Computers are much more "stand-alone" devices than brains are; they can compute answers, but they cannot ask intelligent questions and have no capacity for understanding.[63] Furthermore, one should not think of the brain's modules as being single, spatially discrete units, even though that organ has areas specialized for certain functions. Although we experience reality as coherent in space and time, it is represented in parts of the brain that may be widely separated.[64]

Cognitive scientists now often write computer programs to test ways in which they think modules manage the many incredibly difficult tasks that our brains accomplish with such ridiculous ease. The brain's modules, they believe, must be both complex and tightly organized. There are some 10 billion neurons[65] in the "thinking" part of the brain, the cerebral cortex. That cortex is a thin, highly folded layer of "gray matter" that makes up much of the outer surface of the brain.[66] In addition, brain cells are much more energy efficient than silicon chips, and brains can compute more rapidly than most digital computers.[67] Nonetheless, the cortex does not have the capacity, for example, to remember all possible meaningful sentences in a language by assigning a neuron or group of neurons to each. The possible sentences vastly outnumber the available neurons. The cortex is capable of amazing feats of processing, but flexibility is clearly the key to its virtuosity. The modules must contain the information necessary to interpret various inputs and convert them into appropriate responses without permanently assigning each neuron to a highly specific task or tasks. A neuron that is involved in changing *tengo* to *tiene* while one is speaking Spanish may well also participate in remembering the rule for when to use *which* instead of *that*, and perhaps also in deciding whether to eat sushi or steak, and making many other decisions.

If much of the brain is organized into more or less discrete modules, even if elements of these modules are not geographically all of a piece in the brain, this raises important questions as to how coordination among the modules is achieved and how more general abilities of the brain are produced. For example:

- Precisely what neural processes produce consciousness?[68]
- How are short- and long-term memory organized, and how is retrieval accomplished?[69]
- Where and how is higher-order "understanding" achieved?
- How does the brain accomplish seemingly miraculous tasks of computation with blinding speed?[70]
- How do emotion and cognition interact?[71]

Although many bright people are working on these and related problems, many puzzles remain. Even the conclusions of cognitive scientists such as Fodor, Ornstein, and Pinker that much brain activity is modular remain controversial. Scientists are just beginning to understand the mechanisms by which brains store and process information, but what they have learned has already led to a great deal of interesting and ingenious speculation.

What Our Brains Know about Our Brains

Despite the uncertainties, several general points with particular evolutionary relevance about the mechanisms of the human brain seem quite well established. In summary, the key points are as follows:

1. The brain is an organ that, like other organs, has evolved a structure that serves its various functions.
2. The brain can compensate for partial damage and, often, keep thinking.
3. The brain has many "programs"—connected sets of neurons—that have been built in over hundreds of millions of years by natural selection.
4. More recently evolved programs in the brain enable us to solve problems of relationships and causation that are difficult or impossible for other animals to solve.
5. Although selection has led to these capabilities by creation of appropriate genotypes, appropriate environments (both internal and external) are essential to produce the behavioral characteristics we observe.
6. Natural selection has designed the brain's programs to bias certain perceptions and behaviors.

7. Nonetheless, the genetic code does not build specific instructions into the brain's structure for dealing with every conceivable behavioral situation or even large numbers of them.
8. Natural selection has trouble doing just one thing at a time with respect to the brain, just as with other organs.[72] It is unlikely, for example, that selection could produce a brain program that predisposes females to desire males with curly hair without changing other programs of the brain or, perhaps, other aspects of the human phenotype.

Let's look at these points in more detail.

Point 1: *The brain is an organ* that evolved by the same processes as did fins, lungs, hearts, red blood cells, stripes on a land snail, DDT resistance in a fruit fly, or the parachute on a dandelion seed. The evolution of the brain is not fully understood (nor, for that matter, is that of a snail shell, the dandelion seed's parachute, or the genetic mechanism by which fruit flies become DDT resistant), but many reasonable deductions can be made about its function and capabilities.[73] Consciousness, thinking, reasoning—activities of the "mind"—are all centered in the brain and are phenomena as natural (and material) as the legs' actions of standing, walking, running, and kicking. Seventeenth-century French philosopher René Descartes thought there was a divide between the material body and brain and the immaterial, nonphysical mind.[74] His dualism, that of seeing the mind as the "ghost in the machine" of the body,[75] is so contrary to the view of most researchers today that neurologist Antonio Damasio titled his fine book on the brain *Descartes' Error.*[76] Scientists now assume that there cannot be thought without a brain and that thoughts are physical events with (often) physical consequences beyond the firing of nerve cells. For example, mental stress, such as that caused by the loss of a loved one, can depress the functioning of the immune system.[77] There was a lot of truth in the old high school joke-question: "What's the lightest thing in the world?" The answer was an erection, "because it takes only a thought to raise it." If thoughts didn't have a physical basis, the required vasodilation couldn't be triggered.[78]

The physical nature of the mind is, of course, also demonstrated by the effects of accidental brain lesions (as in the case of Phineas Gage), surgical destruction of parts of the brain to treat psychiatric disorders,[79] and mind-altering hormones and therapeutic and recreational drugs.[80] Many of the latter, remember, have been evolved by plants as defensive chemicals to alter the behavior of herbivores, specifically in response to the physical natures of the herbivores' minds.[81]

Point 2: *The brain can compensate* for partial damage and keep thinking, just as a heart with clogged coronary arteries can partly compensate by developing collateral circulation. With the brain, the degree of possible compensation depends not only on the type of damage but also on the age at which the damage occurs.

Early in development, before the basic modular structure has been created (largely by a genetic program interacting with cellular environments), an environmental change may cause damage that can be completely compensated. In young children, for instance, damage to the brain's left hemisphere, where much of language processing is typically concentrated,[82] can be compensated for by migration of functions to the right hemisphere.[83] Indeed, there is a rare brain disorder, Rasmussen's encephalitis, that manifests as seizures arising from one hemisphere of the brain. The most effective treatment for this illness is removal of the entire hemisphere. If a child with the disease is under twelve years old, the loss of half of the brain does not cause serious problems.[84] Similarly, in those who are born deaf, brain areas that are normally devoted to hearing are commandeered for visual processing;[85] conversely, some evidence suggests that individuals who are blind from an early age develop better hearing than sighted individuals.[86]

Later in life, a lesion that severely damages or destroys a section of the brain is more likely to cause a lasting deficit—as in the case of Phineas Gage. Even in adulthood, however, the brain continues to show remarkable plasticity, as is often seen in older persons' recoveries from strokes.[87] This plasticity and resilience of the brain supports the general view that many of the brain's most crucial characteristics are not permanently fixed by genetic instructions.[88]

Point 3: *Many "programs"* have been built into the brain by natural selection over hundreds of millions of years of evolution. Those programs (actually patterns of connection among neurons) contain many well-tested assumptions about the physical and biological world in which we dwell. The brain assumes such things as the presence of gravity and relatively uniform lighting and the unity of solid objects.[89] These programs process sensory inputs and allow us (and many other animals) to develop perceptions based on what is sometimes remarkably little information.[90] In short, the brain imposes structure on the world our eyes "see."[91] Programs in our brains also allow us, on the basis of surprisingly little experience, to do automatically what are, in theory, extremely difficult tasks such as throwing a stone or recognizing a friend's voice on the telephone.

In human beings, and presumably in at least some other primates, emotions play many roles (for example, informing other people of our moods, motives, and intentions),[92] but a particularly important role for them is to coordinate and assign priorities to those brain programs.[93] Emotions can be thought of as subjectively experienced conscious states of awareness that are focused primarily on the perceived goodness or badness of something.[94] They tend to involve intense activation of brain systems that promote impulsive action. (By contrast, moods are generally considered to be more sustained, lower-intensity activations that do not tend to lead to immediate action.)[95]

Human emotional capacities evolved along with our cognitive capacities.[96] Without the ability to respond to stimuli with appropriate emotions, critical decision making becomes impossible. This has been demonstrated by studies of patients with damage to one portion of the frontal lobes of the brain. Such patients do not respond emotionally to stimuli that produce strong emotional reactions in people with intact brains, and these patients are afflicted with crippling indecision. Not only can they not plan their lives; they also find it virtually impossible even to make such simple decisions as where to dine in the evening. Their brains, by their own accounts, become flooded with information pertinent to coming to a decision, but they can't "make up their minds."[97]

Point 4: *Recently evolved programs* in the human brain enable us to solve problems of relationships and causation that are difficult or impossible for most other animals to solve. Human beings, and quite likely some of our closer relatives, have notions of causality built into the nervous system.[98] We human beings also seem predisposed to interpret the behavior of others in terms of our perceptions of their beliefs and desires (rather than imputing others' actions to external forces).[99] I suspect that predisposition may be present to a degree in other higher primates—a suspicion you might share with me after reading the discussion of chimpanzee politics in chapter 9. But the difference in degree of cognitive capacity between *Homo sapiens* and other living primates is so great as to amount to a difference in kind. There is no sign that chimps can do the complex plotting and analysis of other chimps' behaviors that we can do of other people's behaviors.[100] We will never see chimp psychoanalysts or even guidance counselors. Moreover, within the past few million years and possibly much more recently, natural selection endowed our ancestors with brains able, early in an individual's development, to acquire easily one or more languages. Understanding the pattern of brain–language coevolution (or, indeed, determining whether there was significant coevolution) is clearly central to understanding human natures.

Point 5: *Genes and environments work together* in creating the mind. Some human capabilities, such as the capability of developing intense consciousness, and some widespread similarities among human natures are strongly influenced by genetic factors. But environmental factors are also crucial in developing our individual characteristics.[101] This is not the "behaviorist" view that people are born as essentially identical tabulae rasae—blank slates with only the innate capacity to experience love, fear, and rage.[102] A fundamental idea of behaviorism was well expressed by the father of the field, psychologist John B. Watson, in a famous statement some seventy-five years ago, as his views were becoming accepted by psychologists of the day:[103] "Give me a dozen healthy infants, well-formed, and my own specified world to bring them up in and I'll guarantee to take any one at random and train him to become any type of specialist I might select—doctor, lawyer, artist, merchant-chief and, yes, even beggar-man and thief, regardless of his talents, penchants, tendencies, abilities, vocations, and race of his ancestors."[104]

That behaviorist view, erroneous as it appears to be, seems alive and well today in some social science (and political) circles. So, though, is the equally erroneous view that genes have mostly "hard-wired" our brains. Current science can really support only a "substantial genetic influence at some critical points" position, cowardly as such a position may seem in what are often tough debates over whether genes play an important role in creating behavioral differences among people. The view of anthropologist Clifford Geertz that "there is no such thing as human nature independent of culture"[105] seems pretty close to the mark. But it is obvious that a large amount of genetic evolution was involved in developing the human capacity to have an extensive culture, and that evolution has resulted in a variety of human behavioral predispositions. It therefore seems wise to try to understand how and how much those genetic influences contribute to our natures.

Edward O. Wilson, the father of sociobiology (and a major figure in the battle to preserve Earth's precious biodiversity),[106] made an enormous contribution to the study of human nature by bringing modern evolutionary thinking fully into the picture. Sociobiology is the discipline that investigates the bases of social behavior in biological evolution[107]—and that became controversial through attempts by sociobiologists to find selectional reasons for various aspects of human natures. Before Wilson's work, the social sciences had long been dominated by the tabula rasa view. The behaviorists, for example, believed the characteristics of the human mind and patterns of human behavior to be entirely socially determined.[108] That notion, which had flowered partly in response to the racist eugenics movement of the early twentieth century,[109] swung the pendulum to the opposite extreme. At worst, it recommended oppressive programs of its own[110] based on the incorrect behaviorist assumption that there is a single, completely malleable human nature, an assumption that sociobiology has been instrumental in correcting. The most obvious large-scale example of behaviorism gone berserk is the brutal record of Stalin and his henchmen in a Soviet Union that was violently opposed to the findings of the science of genetics.[111] Indeed, the antibiological views of the Stalinists and many far-left radicals were truly mind-boggling. I well remember a discussion in the 1960s in which I tried to explain to several leftist students the biophysical constraints on the efficiency of photosynthesis. The response was, "That depends on whether you use fascist-capitalist photosynthesis or people's photosynthesis."[112]

One result of the intellectual revolution Wilson began has been scientists' search for genetic evolutionary explanations for behavioral differences. Much research has been stimulated, producing results that are interesting regardless of uncertainties about how to partition the roles of biological and genetic evolution in generating the patterns discovered. And a great deal of novel theory has been explored (gene–culture coevolution is an interesting case in point). Unhappily, some have carried interpretation of the results of the search much further than is justified by the available data.

Point 6: *Biological evolution has given the brain certain predispositions.* Some of the brain's built-in programs interact with widely varying environmental stimuli yet generate very similar perceptions and behaviors. They lead us, for instance, to try to protect our offspring, to have sexual desires, to see causes and effects in events with certain relationships, to avoid the danger of falling from heights, and to fight or flee when confronted with enemies, whether we're high-rise dwellers in New York City or yak herders on the Plateau of Tibet. However, even though we know that natural selection has predisposed us to certain behaviors, we can in most cases only roughly estimate the degree to which a given behavior will be expressed under different environmental conditions, and these estimates often are controversial. Note that I am referring to genetic predispositions or biases here—not gross genetic defects that remove behavior from what might be considered the "normal" range.[113] The distinction may be difficult to make in theory, but it is ordinarily easy enough in practice. Most important, it is clear that environmental influences (especially cultural ones) can overwhelm any of those predispositions. In some cultural environments, people will kill their offspring, voluntarily forgo sexual activity, ignore certain kinds of causation, walk high tightropes, or submit to genocide with neither fight nor flight. A woman drowns her children by rolling the family car into a lake; priests are celibate; addicts smoke cigarettes; people go bungee jumping; victims of a roundup for mass execution go quietly to their deaths. This makes cultural evolution a most critical aspect of evolution to consider when trying to understand differences in the natures of modern *Homo sapiens.*

Point 7: *Gene shortage.* Genes cannot incorporate enough instructions into the brain's structure to program an appropriate reaction to every conceivable behavioral situation or even a very large number of them. Here is an instance in which a little bit of reductionist analysis suggests the hopelessness of seeking a genetic reductionist explanation for most of human behavior. Remember, there are some 100,000 genes in the human genome, whereas there are roughly 100–1,000 trillion connections (synapses) between more than a trillion nerve cells[114] in our brains. As I pointed out in chapter 1, that's at least 1 *billion* synapses per gene, even if every gene in the genome contributed to creating a synapse. Clearly, the characteristics of that neural network can be only partially specified by genetic information; the environment and cultural evolution *must* play a very large, often dominant role in establishing the complex neural networks that modulate human behavior. To give one example, it seems highly unlikely that human nature includes genetic programs for fear of snakes and spiders, as is often claimed.[115] Cultural evolution is a much more likely source of such behaviors—which are far from universal in any case.[116] To put it in shorthand terms that I'll use in later chapters, we could be said to have a "gene shortage," a point that is lost on some writers of popular science in their enthusiasm to find "a gene" to explain certain mental traits or a genetic basis for individual or group differences in various human behavioral characteristics.[117]

That enormous complexity of our brains can also, in a way, explain humanity's famed "free will."[118] No modern computer, even if it started at the beginning of life on Earth (perhaps 4 billion years ago), could calculate fast enough to specify all the possible trees of interactions that can be generated by the nerve impulses that travel among the many billions of nerve cells in your brain in a few seconds. Each nerve impulse would occur with a certain probability, would have a "cause," and would contribute to one of a probabilistic array of effects. It is, then, only in a probabilistic sense that every move we make could be predetermined, but those ultimate "causes" of apparently free choices can never be traced, and we can never be aware of them.[119] Thus, although in the abstract there may be no free will, in practice the brains of human beings evolved so that intentional individuals can make real choices and can make them within a context of ethical alternatives.[120]

In any case, it would not have made sense evolutionarily to prevent development of the human capacity to make real choices by imposing strong genetic control over the "wiring" of every synapse; indeed, it would defeat the purpose of the complexity. That wiring pattern in the hominid line evolved to change throughout life with learning and memory. After all, brains basically evolved as devices for dealing with environmental variation and uncertainty.[121] Whatever are the exact physical bases of memory or thought, wiring the brain to store specific instructions for all prospective situations would be a serious "design mistake" unlikely to be made by natural selection even if the genetic code had the capacity to do the job. The adaptive response to a situation today might well not be the same tomorrow. Your reaction to footsteps on the stairs at night when you are expecting a lover, for example, would probably be very different from your reaction after you've heard that a neighbor had been murdered by an intruder.

Natural selection has endowed us with the capacity to figure out a course of action in virtually any situation, "accepting" the possibility that a chosen course may prove unfortunate. Selection has provided us with the capacity to continuously "update" our mental programs on the basis of experience, on average making our responses ever more adaptive. It has not, in most cases, provided us with a predisposition to a given response because chances are that the environmental information of the moment will be a better guide on which to base action than the evolutionary past. In sum, for long-lived organisms dealing with other such organisms, individuals with only stereotyped responses are unlikely to outreproduce those with more behavioral flexibility. Nonetheless, our brains apparently are designed to permit some stereotypes and prejudices to play important roles in governing our behavior.[122] The individual who stands and mentally debates the intentions of an approaching lion that is licking its chops is unlikely to be a super-reproducer, but flexibility is required in deciding whether to react with a call for help, a hurled rock, or an attempt to flee.

What does tend to be strongly under genetic influence in our species (and in others) is the basic blueprint of brain structure—the pattern of major groups of neurons connecting regions of the brain—which is established before the neurons become functional. This pattern is generated by four biochemical guidance mechanisms of short- and long-range attraction and repulsion.[123] When the neurons are in the proper neighborhoods, the embryo "throws the switch," the electrochemical activity that constitutes nerve impulses begins, and that in turn fine-tunes the connections, making the patterns of synaptic connection more firmly established.[124] Such fine-tuning of the genetically dominated basic structure continues with environmental inputs throughout life in the process of learning and storing memories,[125] and each of us, because we receive different inputs, develops a brain with unique and ever-changing "wiring"—a physical manifestation of our distinct and evolving natures.[126]

Even if genes could precisely and permanently "wire together" the billions of neurons with trillions of synapses, the resultant loss of flexibility would be crippling. All those connections could not, as I've indicated, be enough to signal even the possible permutations of words in any single language. Nonetheless, genes do have an important influence on what we do. Modifications of a single gene may affect the functioning of other genes or modify the products of genes so that the structure behind quite complex behaviors is altered. For instance, one gene may increase the production of a hormone in an individual and elicit a wide variety of behavioral changes. But the changes elicited will vary from one individual to another and from one environment to another. The exact degree to which a genome consisting of some 100,000 genes can effectively be modified to fine-tune human behavior remains an open question, but the possibilities must be very highly constrained in relation to the number of possible behaviors. When someone confidently writes about genetic controls on behavior or refers to "instincts" or innate behavior,[127] the claim in most cases overrates both the power of the genes and our knowledge of them—the information in the human environment is enormously more extensive than that in the human genome, and much of it is created by human actions. It is within that body of information that we must seek most of the answers to questions about the evolution of specific human behaviors.

Indeed, the fine details of the patterns of synapses are themselves known to be largely determined environmentally—by local interactions among neurons. The mechanisms involve chemical signals similar to those that are important in the cellular environments of many nonneural aspects of developing embryos. For example, if "target" cells toward which a nerve fiber is growing are killed, the neuron sending out the fiber is likely to perish. Similarly, proper development of the wiring of the visual system depends on visual experiences in early childhood. Newborn monkeys experimentally deprived of visual inputs for a few months cannot discriminate simple shapes when their vision is restored. Similar results are seen when human beings who are born blind because of cataracts have the cataracts removed between the ages of ten and twenty years. They do not see the world in the same way as do those born with normal vision, and they have special difficulty in properly seeing things that they hadn't touched while blind.[128]

Thus, both the biologically evolved properties of our nervous systems and environmental inputs during our development help to construct the "reality" that we see. This is a key point. In mammals, development of the neurons involved in vision continues after birth, and the connections that are made result from a competition among neurons that is mediated by visual inputs.[129] Children, for instance, must learn depth perception, and their appreciation of spatial relationships changes with maturation. That's one reason why when adults return to their kindergarten classrooms, they are often amazed at how small they seem compared with their childhood recollections.[130] Sensory input, especially sensory input at the appropriate time, is very important to the proper development of the nervous system.[131] As John Allman put it, "the brain is unique among the organs of the body in requiring a great deal of feedback from experience to develop its full capacities."[132] Thus, although the patterns of neuronal development are constrained by genetic instructions, the details of brain wiring—which neurons are linked together and how—depend largely on intercellular environments and, in many cases, inputs from the world outside the organism, which may cause a sort of natural selection among neurons.

This selection process is one element of a view of the way the brain functions that's rather analogous to the Darwinian view of the way microevolution proceeds. Not only do some scientists see the formation and fading of synaptic connections as taking place during learning; they also see, in the very short term, spatial and temporal patterns of activity in groups of nerve cells spreading by copying or fading and shrinking as they compete for "attention."[133] Thus, a sort of super-rapid selection is thought to occur for neuronal participation in a particular brain response to a stimulus. "Neural Darwinism"[134] is a very controversial, and obviously somewhat confusing, idea. At the moment, one can only guess that genes and intercellular environments are responsible for some things being more or less innately established in the brain (e.g., a capacity for ready acquisition of language and, perhaps, for acquisition of ethical beliefs),[135] whereas learning involves influences from the environment outside the organism (e.g., the language or norms to which a child is exposed).[136]

Point 8: *Natural selection has trouble doing just one thing at a time.*[137] Just as selection for DDT resistance has changed pupation behavior in fruit flies, in order for selection to produce, say, "a preference for purchasing turtleneck sweaters," it might inadvertently change for the worse other mental functioning that is much more crucial for our survival and reproduction. Thus, a preference such as this would be unlikely to evolve. Few scientists would dispute this point. But suppose I said that it is unlikely that selection would produce a preference in men to mate with many different women because it would very likely alter other things much more crucial for our survival and reproduction. On this point, a scientific debate might be generated about how much collateral disruption of the genome (from selection favoring more promiscuous male behavior) could be tolerated before fitness would be reduced sufficiently to block a growing tendency to promiscuity (which by itself could increase fitness). But overall, the amount of genetic programming possible would be limited by both gene shortage and the imprecision of selection. Thus, mental predispositions would most likely be restricted to those crucial to maintaining a phenotype endowed with the characteristics most important to successful reproduction. In the brain, those characteristics would be the basic patterns of connection that make subsequent *environmental* fine-tuning possible.

The Brain and the World

That fine-tuning depends on the nature of the inputs to our brain from the environment—including the ones we do not consciously perceive. *Homo sapiens* has evolved to screen possible perceptions in certain ways so that individuals are cognizant of only a small part of the potential stimuli that are "out there." And perceptions—our inputs from the outside world—and our feelings about them clearly are the drivers of a great deal of the behavior, the human natures, that we all want to understand.

According to psychologist Nicholas Humphrey, our consciousness is closely connected to feelings—remember, "I feel, therefore I am." But our feelings are strongly shaped by the evolutionary course taken by our ancestors, especially during that long bush- and tree-clambering, insect-snatching period. The main physiological characteristics of our perceptual system were established tens of millions of years ago. That lifestyle gave our ancestors stereoscopic color vision connected with highly controllable mobile fingers with very sensitive touch pads. We don't orient ourselves primarily by smells,[138] echoes, or electric fields, even though these are abundant in the human environment. We don't use their potential aid to help us put together a functional worldview from the odors of our surroundings, echoes of our own "sonar" pinging, or the distortions of electrical fields generated by our muscles, as do dogs, bats, and electric fishes, respectively. Because our perceptions are an interaction between the external world and the evolved characteristics and constraints of the human nervous system, we miss a great deal that is detectable by other organisms. Of course, dogs, bats, and electric fishes miss a great deal of what we perceive, but we at least are able to more fully inform ourselves about the content of the world through use of extensions to our innate perceptual abilities—clever devices such as microscopes, telescopes, radio receivers, PET scans, magnetometers, and the like.[139]

A simple experiment can show how your brain helps to determine what you see. Turn your head rapidly from side to side, and you will perceive your head moving in a room that is stationary. Try it. Now, if you have a video camera, turn it on and move it around in the same way as you moved your head, and then play back the sequence. On the videotape, the room moves at a dizzying pace, and you may even feel a little seasick watching it. Why the difference? Because your brain receives information from special sensory receptors[140] that inform it about the position and movement of body parts. Your brain knows that your head is moving, and it is programmed to compensate for that movement and keep your mental representation of the environment stationary. If you don't have a video camera, you can do a quick test with this book. Move your head from side to side or hold your head still and sweep your eyes over the page. Note that you can still read the text—the words stay still. Now, shut one eye, and with your finger tap the corner of the other eye. See the words jump? Because the movement in your eye when you tapped it was not caused by your neck or eye muscles, the receptors didn't send a message to your brain that would allow it to compensate.

Our brains are also programmed to let us see what we expect to see. The photograph depicts two people in an "Ames room"—they look like a giant and a midget sharing the same space. But the people are both of normal height; the room is abnormally constructed to create the illusion of size difference when the scene is viewed with one eye, without moving the head (something a standard camera on a tripod simulates perfectly). We *expect* the room to be normally "carpentered," with ninety-degree angles at the corners and with rectangular windows, and it is those assumptions that shape the perceptions of anyone viewing people in an Ames room.[141] Put another way, visual (and other) sensory inputs can be analogized as data sent to the brain, and the brain then forms a hypothesis about the state of the real world from those data. The hypothesis that is inferred from the data is what we call a perception.[142] Those "well-tested assumptions about the physical and biological world in which we dwell" men-

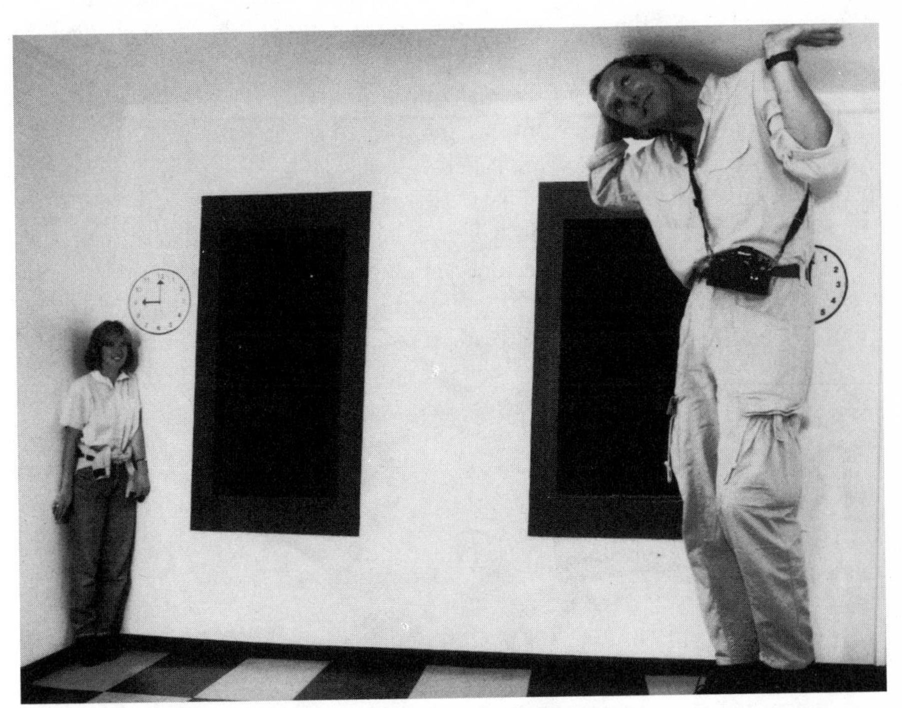

Individuals in an Ames room. The two people are actually the same height. Photograph by S. Schwartzenberg, copyright © 1991 The Exploratorium.

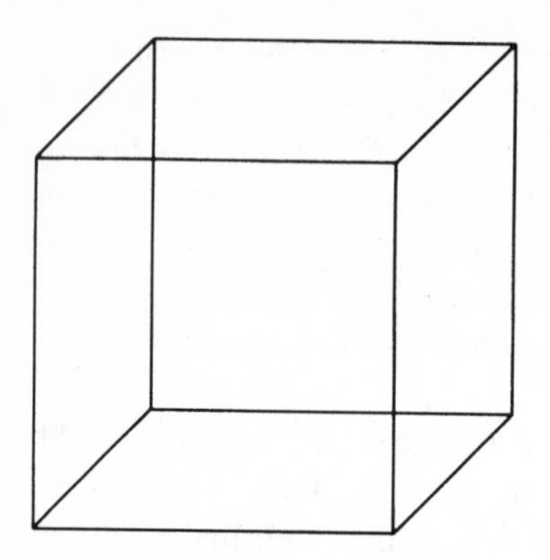

Ambiguous figures. The "cube" (left) can be flipped mentally so that its base and top are visualized from below or above. The figure on the right can be seen as either a vase or two faces.

tioned earlier are part of our hypothesis-manufacturing machinery—but so are our experiences, including experiences with normally carpentered rooms.

Our reactions to ambiguous images such as those in the drawings above attest that a given set of inputs can lead to different hypotheses and that those hypotheses can easily change. In these two images, the visual inputs—that is, the flows of incoming data—remain the same, but our minds construct two hypotheses that we "flip" between as our interpretation of the input changes. In contrast, we can also form a single, solid hypothesis on the basis of a combination of different inputs (data streams) and experience. Thus, all the different configurations in the figure below are interpreted as the letter *R*.

There are a number of ways, called "Gestalt laws,"[143] in which our brains tend to organize incoming visual data. Two examples are shown in the figure on the facing page.[144] These laws probably are largely genetically determined parts of our hypothesis-generating machinery, though ones that are activated by early visual experience. Without such programming, based on the evolutionary experience of hundreds of millions of years, it is difficult to imagine how the arriv-

Different configurations of the letter *R*. Despite their differences, we see all of these as the same letter.

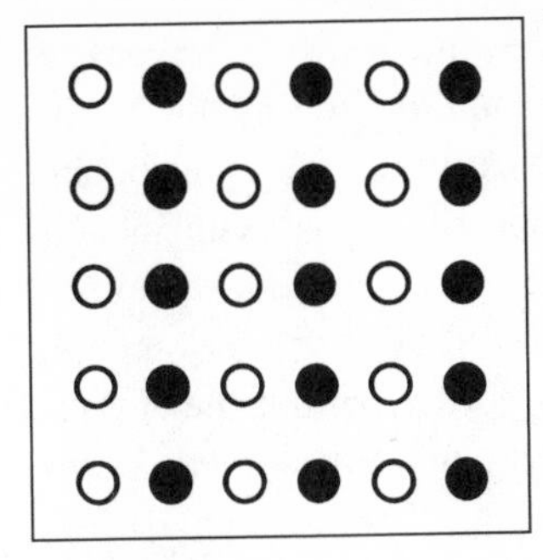

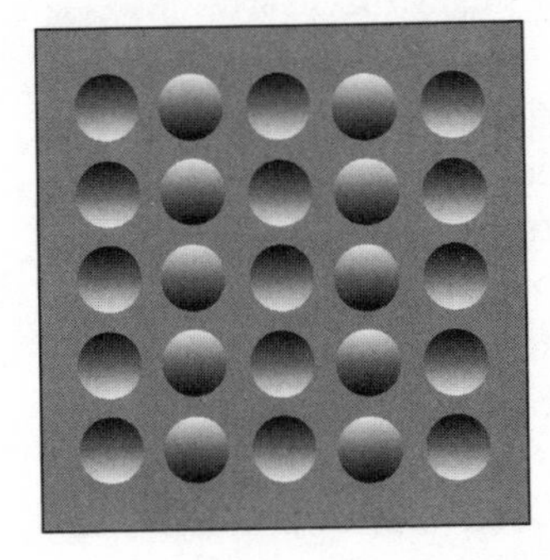

Top: we mentally group the dots in columns because of their shade, even though those in rows are closer together. Middle: we tend to group together figures that are similar in size and shape. The upper two figures illustrate Gestalt laws. Bottom: we also generally interpret figures as if they were illuminated from above, so the outside and middle columns are interpreted as pits and the other two columns as bumps. Note that if you turn the page upside down, the pits are transformed into bumps and vice versa.

ing millions of electrical impulses that physically constitute our visual input could begin to be assembled into a useful subjective view of the world.[145]

Perhaps the most interesting perceptual discovery from our point of view is that a concept of cause and effect seems to be built into our brains early in the course of development. If we see a rock tumble down a slope and hit another rock, which in turn starts off in the same direction, we automatically assume cause and effect. Automatic perception of cause-and-effect relationships was first postulated by eighteenth-century philosopher David Hume[146] and was investigated in the 1940s by psychologist A. E. Michotte,[147] who supported the idea with a series of clever experiments. He asked people to watch filmstrips in which a black square was shown moving toward a stationary red one. When it reached the red one, it stopped and the red one began moving on the same trajectory on which the black one had been, with the red square moving either more slowly or at the same speed. Even people who knew that there was no physical connection between the squares, that the black square was not bumping into the red one and "launching" it, tended to perceive such a collision and launch. On the other hand, the causal illusion disappeared for virtually all subjects if, after the black square made contact with the red square, the red moved off at right angles to the trajectory of the black. If red moved off more rapidly than black had been moving, the illusion was of "triggering," and if they moved off together and eventually stopped together, it was of "entrainment," with black pushing red.

Although there has been some question about universality in interpretation of the specific illusion,[148] recent work makes it seem highly likely that notions related to causation are part of the repertoire of the brain acquired during our long evolutionary history (and doubtless, to a degree, part of the capabilities of many other animals).[149] Michotte's conclusions about built-in perception of cause-and-effect relationships are reinforced by the work of psychologist Alan Leslie and his colleagues.[150] Leslie and Stephanie Keeble, working with twenty-seven-week-old infants, produced evidence suggesting that even these very young infants already perceived a difference between a "launch" event and a similar but noncausal event. The discovery of the "Michotte effect" seems a very important one that, as we will see, may help to explain the nearly universal occurrence of religions.

Another aspect of visual perception that seems tightly tied to our social evolution is that of face recognition. Among the great deal of research that has been done on visual perception, much has focused on recognition of objects.[151] As perception of shapes and images goes, it appears that we are born with a great talent for recognizing faces—a critically important skill for highly social hominids.[152] The capacity for face perception appears to be heavily influenced by genetic factors, occupies specific cells in specific parts of our brains, and seems to be more "holistic" than is the system involved in recognition of inanimate objects (which seems to focus more on arrangements of parts).[153] But it is not clear whether face recognition involves a completely separate neural module or is simply a type of object recognition at which people are highly expert.[154] The evidence for at least a special brain area for face recognition is quite strong, given that brain damage can cause a total loss of ability to recognize faces (to name previously known individuals from their photographs) while sparing the ability to recognize inanimate objects (e.g., a hat or a screwdriver).[155] Brain damage can also cause exactly the opposite—retention of face recognition but loss of the ability to recognize everyday objects. For example, one patient who sustained brain injury in an automobile accident called an asparagus "a rose twig with thorns," a tennis racket "a fencer's mask," a pair of pliers "a clothes peg," and a dart "a feather duster." Like similarly damaged patients, he seemed to try to recognize objects piecemeal rather than holistically. Nonetheless, he had no difficulty in attaching names to faces and could easily name famous people in pictures.[156]

The apparent existence of a face-recognizing module or submodule seems to support the general notion that much of human cognitive power evolved to deal with an ever-escalating social complexity—because what could be more critical to maneuvering in the social world than recognizing other individuals in one's society? Face recognition is a talent that seems to be developed in chimps—they are even able to recognize mother–son pairs among unknown individuals from photographs.[157] Other animals also can identify their cohorts, but none, to my knowledge, to the degree that human beings can.[158]

Cultural Differences in Perception

Although certain aspects of perception were strongly influenced by humanity's common biological evolutionary heritage and are part of every normal person's sensory equipment, the process of perception itself can clearly be modified by experience. At one level, such modification blends with learning—as when trained individuals instantly perceive the significance of a line on a graph or the gestalt of a bird or butterfly species. But a particularly interesting and more pervasive kind of modification produces culture-specific differences in perception. For example, some cultures do not commonly use pictures that are two-dimensional representations of three-dimensional scenes, and people from these cultures do not employ cues of object size, superimposition, and perspective in interpreting pictures, as people from picture-using cultures do as a matter of course. When viewing the image shown in the figure, a person from a non-picture-using culture would say that the elephant is closer to the man than is the antelope—that is, would give a response that is correct if the picture is interpreted two-dimensionally.[159]

In a more general, if anecdotal, context, there is anthropologist Colin Turnbull's report of the response of a Mbuti Pygmy named Kenge to the world out-

Viewing this picture of a hunting scene, an observer unaccustomed to the conventions of three-dimensional interpretation of two-dimensional figures will consider the spear to be aimed at the elephant. Redrawn from Hudson 1960.

side his Ituri Forest home in the Democratic Republic of Congo. Kenge, on accompanying Turnbull to the edge of the forest, for the first time in his life saw a distant vista. The Ruwenzori mountain range was visible in the distance, and Kenge asked Turnbull what the "things" were—"Were they hills? Were they clouds? Just what were they?" Later, when Kenge looked over extensive plains and saw Cape buffalo grazing miles away, he at first took them for insects. As Turnbull recounted: "He asked me what kind of insects they were, and I told him they were buffalo, twice as big as the forest buffalo known to him. He laughed loudly and then told me not to tell such stupid stories." Kenge was first frightened and then puzzled by the way the buffalo "grew" as the car in which he and Turnbull were riding approached them.[160]

Some of the most sophisticated work on intercultural differences in visual perception was done in the 1960s by psychologists Marshall Segall, Donald Campbell, and Melville Herskovits.[161] People in fifteen different societies were exposed to a series of geometric figures that people in Western cultures see as containing optical illusions. The susceptibility of those in non-Western cultures to the illusions depended primarily on whether the environments of those cultures were carpentered (i.e., had dwellings with rectangular rooms, walls that meet in right angles, etc.) and whether the people lived in areas with wide vistas (plains) or restricted vistas (rain forests). People from carpentered environments who are used to right-angle joinings, for example, when looking at the Müller-Lyer illusion illustrated here, see the right vertical line as the longer of the two, though they are actually the same length. Such people apparently interpret the right figure as a back edge (so the line is seen as farther away) and the

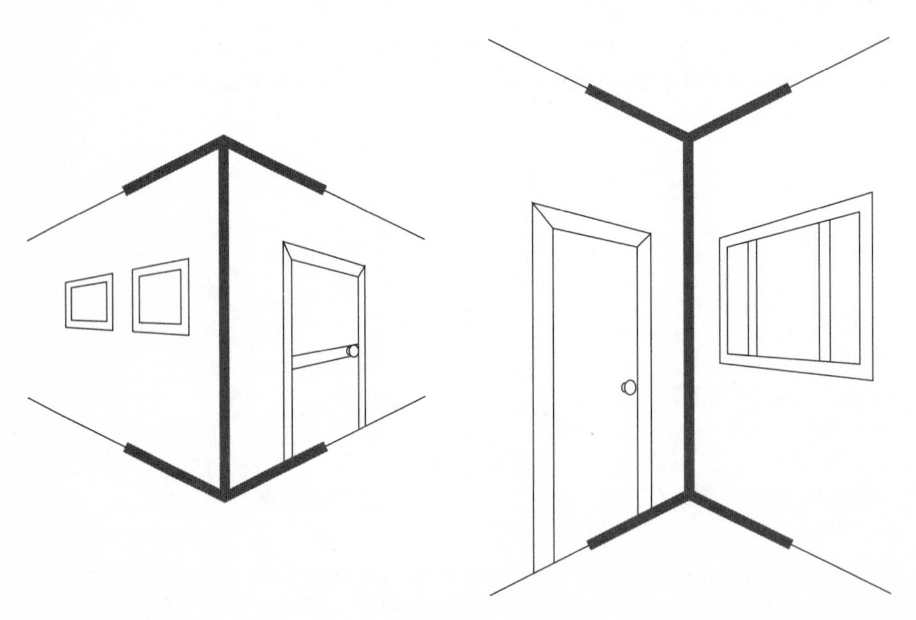

Müller-Lyer illusion (thick lines), shown in the context of a carpentered world.

left figure as a front edge and thus closer. Their automatic compensation for distance creates the illusion, given that in a three-dimensional situation, the farther away of two identically long lines would appear shorter and, if drawn the same length, would be interpreted as longer.[162]

People raised in places where they are accustomed to seeing broad horizons prove to be more often taken in by the horizontal-vertical illusion, seeing the vertical one of two identical-length lines as longer than the horizontal one. Those people, presumably, have become accustomed to interpreting long distances in a horizontal plain in front of them from short vertical retinal images. People who live in forests or on canyon bottoms are less susceptible to this illusion. Once again, we see that perceptions are combinations of the way our neural apparatus has evolved and the environments in which that apparatus develops from infancy and operates subsequently.

Perceiving the Environment

Several aspects of our perceptual system provide some insight into humanity's general failure to come to grips with environmental problems. One, as discussed in the earlier head-movement example, is that perception tends to hold the environmental background constant. Related to this is the phenomenon of habituation, in which a constant stimulus is "tuned out" of consciousness. It is easily demonstrated with reference to sound; one may hear the air conditioner start up but may quickly become unconscious of its continuing hum. Habituation can apply to other kinds of stimuli as well. For instance, when you sat down to read this book, your body went through a series of complex motions monitored by the sensory receptors that inform you of the position and state of the parts of your body. When your behind hit the seat, your brain started getting a series of messages that your behind was compressed, and the messages continue coming. But no doubt you had tuned them out—habituated them—until this passage called them to your attention just now. Our nervous systems contain "filters" and "feature detectors" that make sure that all the possible stimuli in the environment don't reach our consciousness indiscriminately. Thus, the noise of a running air conditioner is filtered out, but a creak of the stairs at midnight is detected.

Habituation keeps one's consciousness from being distracted by, say, the constant sound of a nearby brook when it needs to be paying attention to the possible arrival of important new stimuli—the crackle of a dry leaf as a predator sneaks up or the scent of a potential mate approaching. Movement against a constant background is the way a birder normally first spots a bird; windy days on which trees are waving around are lousy ones on which to try to increase one's "life list" (of birds seen). Keeping the environmental background constant makes it easier to perceive new threats or opportunities as the ecological play proceeds. Our ancestors lived in situations in which that was of paramount importance, and we're still quite good at dealing with sudden changes in our environments, be it a car swerving at us from another lane, a baby's cry of distress, or an unexpected bold proposal by a potential lover. In contrast, there was little or no adaptive value in our ancestors being able to detect background alterations occurring over years or decades. But that situation has changed, and holding the background too constant over long periods appears to be an unfortunate evolutionary hangover.

The difficulty is that the most serious threats now faced by humanity are slow, deleterious changes in the environmental background itself, changes our perceptual systems have evolved to encourage us to ignore.[163] These are changes that take place over decades—population growth, gradual alteration of the climate through global warming, loss of biodiversity, land degradation, accumulation of hormone-mimicking chemicals, dangerous changes in the epidemiological environment, and the like.[164] It is very difficult for people to react to global warming, for instance, and not just because our perceptual systems can't detect the rise in greenhouse gas concentrations in the atmosphere. Even if the gases were visible, we wouldn't detect the change because the increase has been too gradual. The only way to recognize the change is by interpreting graphs made by scientific instruments that were designed to extend our perceptual systems and track the changes. Those graphs condense representations of changes made over many decades into visual differences occurring over the space of a few inches—changes we *can* perceive.

Finally, the evolution of the tendency to habituate apparently allows us to behave in ways that otherwise would be inexplicable, because we have difficulty sensitizing ourselves to continuing trends that develop gradually. Obvious examples are the loss of sensitivity to the effects of horrendous violence that affected many of Hitler's executioners and that today influences children who watch multitudes of murders on television. Here, the constant environment of violence gradually becomes the "normal" environmental background and fails to elicit socially appropriate responses.

Brains, Biases, and Behavior

Our large brains, then, use their great capacities not to store a one-to-one mapping of incoming sensory data but rather to exercise a high degree of selectivity in filtering and interpreting those inputs. They are fed a view of the world that is shaped by our biologically evolved sensory receptors and further biased by some biologically evolved characteristics of the brains themselves. Our nervous systems help us to construct the "reality" we believe is "out there," and the characteristics of those systems inevitably bias our worldviews and thus our natures. There may well be other biases in our neural circuitry about whose origins and characteristics we can only speculate. For instance, one "just so"[165] story would be that our brains develop a bias to dichotomize a more or less continuous world—self or nonself; mind or body; organism or environment; big or small; fast or slow; theory or practice; friend or enemy; nature or nurture; free will or determinism; right or wrong; good or evil; and so on.[166] We see this in the urge to decide whether to fight or flee when threatened or in the urge to decide whether *Homo ergaster* is a separate species from *Homo erectus*. In the first case, one can imagine considerable adaptive value would accrue to a quick and firm decision in lieu of dithering if a warrior of an enemy tribe approaches. The interesting question is whether the advantages of dichotomizing in such life-threatening circumstances influence the way the brain deals with issues less likely to affect fitness, such as whether two segments of a continuous evolutionary lineage should be given different latinized names. Another puzzler is whether a built-in predilection for coming to quick, clean, simple (rather than slow, fuzzy, complex) conclusions on which to base decisions may contribute to the apparent urge to find quick, clean, simple explanations of our behavior—explanations of the "we're just naked apes" sort.

How and why did such biases arise? I believe that some evolved as responses to living in trees and eating bugs, but besides an emphasis on vision (and a de-emphasis on the capacity to detect chemical stimuli) and the ability to manipulate objects cleverly with our forepaws, what is known of our arboreal history tells us little about the origins of the wide range of behaviors characteristic of modern *Homo sapiens*. Unhappily, no fossil record of behavior as such exists. It seems unlikely, for example, that an arboreal past created greater selection pressures to dichotomize than a terrestrial one would have, but there is no way to test that assumption. Attempts to reconstruct the early evolution of human society are limited to extrapolation from the behavior of contemporary primates and hunter-gatherers and to inferences drawn from often spotty archaeological evidence. The question of what conclusions can be drawn about the evolution of human behavior is a complex and controversial one because it is so closely related to that other vexing question I referred to earlier: what role do genes play in shaping human behavior? Is the dichotomizing just mentioned an inherited predisposition, or is it entirely learned? How about a preference for quick, simple answers? Yes, all human characteristics are the products of gene–environment interactions, and that by definition must include behavioral characteristics. Is it parsimonious, then, to conclude that behaviors that are similar in both chimpanzees and human beings are the result of similar genotypes interacting with similar environments? After all, chimps and people share some 98 percent of their genes, so is it really surprising that chimp and human mothers use nearly identical gestures to comfort their offspring?

If not, the resemblance of the gestures seems difficult to explain, because female chimps and female human beings don't live in environments that appear 98 percent similar. The forest at Gombe Stream is a long way from the bedroom of a child who has just awakened from a nightmare in Palo Alto, California. But the Gombe–Palo Alto chasm may not be as wide as it seems. Some aspects of human behavior can be remarkably similar in physical environments as different as those of a high-rise apartment in Singapore, a shack in a community of slash-and-burn agriculturalists in a rain forest on Peru's Río Tambopata, or a shelter in a camp of pygmy hunter-gatherers in an African rain forest.[167] And, of course, the physical environments of the hunter-gatherers don't seem so strikingly different from that of the Gombe Stream chimpanzees.[168]

The reasons behind these puzzling similarities and differences may be traceable to an additional kind of environment in which people in all places reside, as do all chimpanzees and bonobos—a highly organized social environment. When we look at the evolution of the apes (including ourselves), attention must always be paid to the influences of both their physical and social environments—and increasingly to the latter.[169] And two elements that are central in our social environments and absent from those of our closest living relatives are a vast, evolving culture and that which makes such culture possible—our unique facility for language, to which we now turn.

Chapter 7

From Grooming to Gossip?

"*Homo sapiens* . . . can rightfully be called the babbling ape."

—Edward O. Wilson[1]

At the age of twenty, I was fascinated with the notion of learning a non-Indo-European language and communicating with members of a totally different culture. In 1952, during a ten-week stay with the Inuit on Southampton Island in the northern reaches of Hudson Bay, in the Northwest Territories of Canada, I tried to learn the Aivilikmiut[2] Eskimo language and worked with a well-known Inuit, Tommy Bruce (Eenerook), who was eager to improve his English.[3] Tommy and I both made a little progress, and we enjoyed telling each other stories. But the language didn't come all that easily for me, and communication often was frustrating. Once, Tommy asked, "What does it mean, *also*?" Imagine trying to put that idea over with no shared third language! It took me twenty minutes, using every example that came to mind. Although we could communicate on some things and laughed together, I always wondered what he was "really" thinking.

Language is not just our communication system; it is intimately involved in our *thinking* system. Perhaps the biggest difference in kind between human beings and chimpanzees is not our greater problem-solving abilities but our unique linguistic abilities. We're so accustomed to being able to converse with one another that most of us rarely reflect on what a marvelous talent language communication is or on the roles it plays in everything from mating to living in groups to transmitting culture. In addition, it's through language that we generally best express the diversity of our natures. Not only does each culture have a different language or dialect, but also each person has a distinctive mode of speech and a unique talent for using the spoken and, in literate societies, written word.

Language is also an excellent area in which to investigate how heredity and environment work together: all normal human beings are born with the capacity to acquire language, but the language or languages they acquire depends entirely on the environment in which they live. How did this unique communication system evolve, and how is it embedded in our natures?[4] Did it appear gradually before *Homo sapiens* left Africa and spread out over the Old World, perhaps giving our species an advantage over the descendants of *Homo ergaster* already in place? Or did its evolution come later and suddenly, causing the subsequent Great Leap Forward—that amazing surge of cultural evolution marked by the appearance of refined Aurignacian technology, art, signs of religious belief in burials, and the apparent capacity for further revolutions—some 50,000 years ago? To appreciate what is known of the answers to these questions, it helps first to take a brief look at the characteristics of language itself.

What Is a Language?

A language has three main elements: vocabulary, syntax, and meaning. A vocabulary can be simply defined as the words that one must learn in order to speak a language. For this discussion, however, a vocabulary might more usefully be viewed as a set of arbitrary symbols that substitute for objects, actions, and relationships in the real world (or the representations of such in the mind). To these must be added another set of symbols that help connect and organize the first set to allow complex ideas to be communicated. *Apple* and *eat* are examples from the first set—stand-alone symbols of an object and an action. *The* and *didn't* are examples from the second set. The key element to notice is the predominant arbitrariness of symbols in all language systems. Some words, such as *rattle* and *buzz*, resemble the sounds they denote, although different languages still use different symbols: *buzz* in Spanish, for example, is *zumbido*. For the vast majority of vocabulary items, though, there is no logical connection between the item symbolized and the symbol. Neither *apple* nor *manzana* (Spanish for *apple*) nor *Apfel* (German) nor *pomme* (French) bears the slightest resemblance in size, shape, smell, or sound to the fruit these words symbolize.

The symbolic aspect of human language and its vast complexity differentiate it from the communication systems of other animals. Other animals may use arbitrary symbols (badges) for communication; for example, among the bird species known as European great tits, the size of the breast stripe (bib) can symbolize dominance status.[5] Animals as different as pigeons and chimpanzees can be taught the meaning of symbols (peck at the circle and get some pigeon feed; peck at the square and go hungry), but no living nonhuman animal possesses a communication system based on arbitrary symbols and a set of rules about the order in which these symbols *are permitted to be used* (that is, syntax), rules that generate further symbols with additional meanings. Thus, *Tom* and *apple* are arbitrary symbols (in this case, content words—symbols referring to definable objects, actions, or feelings)[6] for, respectively, a person and an edible fruit. *Eat* is an arbitrary symbol for consuming something, and *full* is an arbitrary symbol for the way one feels after eating—both of them also content words. Content words are quite specific to the topic of a sentence. In contrast, *an*, *the*, and *not* are arbitrary syntactic symbols[7] (function words) that specify relationships among content words and are not so much tied to topic. Also in this category are special verbs called auxiliaries (e.g., *does*, *will*), which add elements of speaker attitude or time to sentence symbols. Other syntactic symbols include endings such as *-s* and *-ing*, changes such as *I* → *me* and *they* → *them*, and alterations in pitch of voice that specify number, tense, and case (role of the actor—subject, object, etc.) in sentences. Such changes in form that word symbols undergo are called inflection, and inflection plus syntax (in the restricted sense of word order) equals what we call grammar.

Sentences can be thought of as symbols, too. Thus, "Tom eats an apple," "Tom eats the apple," "If Tom were not full, he would eat the apple," and "Will Tom eat an apple?" are four symbols written in a sort of code that conveys explicit information from person to person, but the symbols bear no conceivable resemblance to the players or acts involved. All of these "apple" sentence symbols conform to the restrictions of syntax, which is why I emphasized "permitted to be used" earlier, given that syntax is not just a question of word order. Syntax exists even in languages in which the functions of words are so well marked by inflection that virtually any word order will convey meaning.

The development of syntax opens up an infinity of possible worlds to hominid communication by allowing the creation of novel symbols that are nonetheless recognizable to those who understand the syntax.[8] This expands possibilities for group planning and individual thought far beyond what is possible for any other animal. The "If Sam were not dead . . ." type of symbol is especially interesting because it conveys the idea of something counterfactual (a logical state in which what goes before is contrary to fact)—a kind of idea that, as far as I know, cannot be transmitted by anything but language. Language really opens up the world of the imagination: "If Og goes behind those bushes and I throw a stone so it lands behind the mammoth, . . ." or, perhaps, even "If I didn't exist, would Og and the mammoth still be there?"

Arbitrary symbols are thus the basis of both vocabulary and syntax, and such symbols of various complexities convey meaning. Both the simple symbol *apple* and the complex symbol "Tom ate the apple that he stole from the blind storekeeper in Pittsburgh last week" are meaningful. But a combination of symbols may be grammatical and yet convey no meaning whatever. A famous example showing that proper syntax (sentence structure) can exist without meaning was invented by the father of modern linguistics, Noam Chomsky: "Colorless green ideas sleep furiously."[9]

An Innate Grammar?

The existence of syntax without meaning has been cited to support the notion that there is a genetic program that permits rapid language acquisition in young children, regardless of the actual language they acquire.[10] According to that notion, people are programmed with a basic set of rules, or constraints, that underlie the grammars of all languages. This idea was pioneered by Chomsky as the hypothesis of a universal grammar.[11] He proposed that in the universal grammar, sentences are not seen as simple linear strings of words; rather, they consist of underlying hierarchical structures of phrases grouped into even bigger phrases. In other words, the rules of grammar are structure-dependent. One line of evidence suggesting that children have a built-in understanding of structure dependence—that they don't divine meaning just from the linear arrangement of words—is that they appear to pick up language faster than they apparently could by simply imitating their parents' speech. In linguistic jargon, language is claimed in this view to be "underdetermined"; that is, there is a poverty of stimuli from which children can learn language. Therefore, some rules of grammar must be innate. Consider the following example of a statement being made into a question by repositioning of a word:

> The car parked outside the store is idling its motor.
> Is the car parked outside the store idling its motor?

So, to change a statement into a question, just move the auxiliary *is* to the front of the sentence. A good rule, right? How about this:

> The car that is parked outside the store is idling its motor.
> Is the car that parked outside the store is idling its motor?

The rule didn't work so well that time, did it? For this sentence, our mental rule depends on our understanding the phrase structure. Thus, the rule for transforming a declarative sentence into a question seems to be not to move the first auxiliary but rather to move the auxiliary that connects a subject phrase with a verb phrase. In the example that follows, the subject phrase is enclosed in braces, {}, and the verb phrase is enclosed in slashes, //:

> {The car that is parked outside the store} is /idling its motor/.
> Is {the car that is parked outside the store} /idling its motor/?

Very young children don't hear a lot of sentences that have a second auxiliary buried within a subject phrase. It's not a feature of the high-pitched, simplified "motherese" that mothers use in talking to their babies, which serves many functions other than improving the children's language skills.[12] The results of a series of careful experiments by linguists Stephen Crain and Mineharu Nakayama of the University of Connecticut suggest that children nevertheless acquire structure-dependent language rules very early in life, perhaps because of genetic predispositions, and thus avoid the type of grammatical mistake shown in the previous paragraph.[13]

Crain and Nakayama gave thirty children ranging in age from three years, two months, to five years, eleven months, an opportunity to ask nine questions of a doll representing the *Star Wars* character Jabba the Hutt. The doll was manipulated to provide yes or no answers to questions about a series of pictures;

when the answer given was correct, the child would pretend to feed Jabba. From the kids' point of view, it was a great game. The first three questions were pretested to be sure the children understood the procedure: for example, "Ask Jabba if the girl is tall." Then the children moved on to the six test questions, which were of the form "Ask Jabba if the boy who is being kissed by his mother is happy."

The youngest fifteen children (average age four years, three months) made grammatical errors 62 percent of the time; the oldest (average age five years, three months), only 20 percent of the time.[14] But even the youngest children did not make the structure-independent error of simply moving the leftmost auxiliary in a linear string of words to the beginning of the sentence. There were no errors of the form "Is the boy who being kissed by his mother is happy?" Instead, errors involved the addition of an auxiliary ("Is the boy who is being kissed by his mother is happy?") or "restarting" ("Is the boy who is being kissed by his mother, is he happy?"). Such results[15] have been claimed to support Chomsky's notion that human infants are born with brain circuitry that functions as a sort of blueprint for universal grammatical rules. This circuitry, it is said, allows very young children to properly employ the grammar of any language to which they are exposed[16] rather than learn all the rules piecemeal by listening to adult speakers.

But, unhappily, there was no experimental control over the actual prior linguistic experience of the children Crain and Nakayama studied. Thus, we cannot be sure that the children had not simply adopted the most common syntactical patterns by imitation or associative learning (learning that takes place by linking experiences with ideas, images, and the like—in this context, by linking certain speech patterns with successful adult communication). Indeed, there is recent evidence that infants as young as eight months of age can start picking up critically important elements in speech such as rhythm, cadence, and meter.[17] It is thus possible that relatively limited periods of listening might account for the acquisition of language, including rules of syntax.[18] On the other hand, the argument that imitation or normal associative learning, which goes on throughout life, is all that is involved goes against the abundant evidence, both anecdotal (remember my granddaughter Jessica and *choh-koh-lah-tay*) and systematic, for the existence of a genetically determined critical period for initial language acquisition early in life.[19]

In addition, observations of pidgin and creole languages support the notion that there is a built-in neural structure that interacts with language learning. Pidgins are very crude languages—strings of content words with little syntax, which the speakers of different languages develop to communicate across linguistic barriers. Tommy Bruce and I started out with a pidgin that mixed the words we had learned from each other's languages: "Where *komatikjuak*?" (Where is the truck?). Pidgins characteristically differ greatly from individual to individual.[20] Creoles are more sophisticated, more grammatical languages that contain function words and rules for word order. Creoles appear to be developed spontaneously by children of pidgin speakers who have no parental models to learn from, and different creoles in disparate places share many characteristics. Furthermore, creoles show similarities to the languages spoken by children three to four years of age.[21] This supports the Chomskyan notion that patterns of language acquisition and appreciation of syntax trace to a genetic predisposition.[22]

Early studies of individuals with various kinds of brain lesions seemed to show that language skills were concentrated in two areas of the cerebral cortex on the left side of the brain's surface, known as Broca's and Wernicke's areas, and this seemed to lend support to the idea of a localized unit that could contain the "wiring" for a supposed universal grammar. However, more recent studies of the brain's functional architecture show that language involves, in a complex fashion, not only these areas but also structures below the cerebral cortex and parts of the prefrontal cortex.[23] Although language (and other) functions are somewhat localized in the brain, they are certainly not contained in isolated "plug-in" units as if they were computer chips. The reality of the brain is much more complex than the reality of the powerful computer on which I'm typing this, but this complexity does not rule out the existence of a universal grammar.

Nonetheless, brilliant as Chomsky's ideas are, a considerable stretch of evolutionary theory is needed to imagine natural selection laboriously constructing a complicated all-purpose grammatical template in the brain. It seems unlikely that *Homo sapiens* has the genes to spare—remember the "gene shortage" discussed in the previous chapter. It is more parsimonious to assume a genetically determined period in which the developing wiring in the brain of a child exposed to any language can make the right connections to instill the grammar of that language. It seems possible, though, that selection has programmed the brain to have the capacity to deal with some syntactical problems, such as association of pronouns with distant nouns and what's called recursive embedding (sentences within sentences, such as "He believed he had tried as hard as he could to do it").[24] Even more difficult to explain by experiential learning alone is the presence in syntax of "empty categories" (subjects and objects that are not overtly expressed) and references to them. Consider two sentences:

Bill wants someone to work for.
Bill wants someone to work for him.

As linguist Derek Bickerton put it, "Absent some innate mechanism, how is the child supposed to learn that the subject of *work for* in the first sentence is *Bill* and in the second, *someone*?"[25]

It seems reasonable that some of human brain structure is designed to help youngsters acquire the ability to deal with such complexities. It also seems reasonable—and the evidence indicates—that genetic predispositions make the patterns in which children acquire adult syntax quite regular, just as they make the patterns in which children acquire adult dentition and genitalia quite regular.[26] There is other evidence for some degree of genetic hard-wiring of systems that help with language acquisition. An important clue can be seen, ironically, in the development of communication systems by deaf children. Even when they lack environmental stimuli in the form of either speech or a sign-language model, there are regularities in their language acquisition.[27] The most common sentence structure (syntax) in their language is two signs signifying actions, joined with pointing to indicate an agent for the action.[28] Thus, although there may be no built-in syntactic template that is good for all languages, deaf children raised in a world rich with moving lips and gestural (not formal sign-language) communication may respond by developing some sort of regular sign-based communication system.[29]

It also seems likely that the brain wiring dedicated to helping in language acquisition is at least partially distinct from that involved in general learning. For example, people with aphasias (difficulties with language) traceable to brain damage may retain other cognitive abilities, and people with other disabilities (e.g., Phineas Gage, the railway worker mentioned in the previous chapter) may retain their language skills.[30] Defects in the wiring, apparently genetic, can also cause children to acquire language late and to develop abnormal speech sounds and grammar, all without associated problems in hearing or thinking or other physical or psychological disorders.[31] And, at the receiving end, infants can detect a wider array of speech sounds than can the adults in their societies, who lose the ability to distinguish sounds not discriminated in their native language (such as *b* and *v* sounds in Spanish or *l* and *r* sounds in some Asian languages).[32] This implies that environmental feedback modifies (introduces constraints on) a neural system designed by natural selection to permit acquisition of any human language.

Interestingly, similar brain injuries result in parallel language deficits among deaf people who communicate by American Sign Language (ASL) and people who are able to speak. In addition, deaf people who use ASL may retain the ability to use facial muscles to produce expressions used in ASL when brain damage prevents them from producing facial expressions otherwise. Apparently, the neural ability to produce nonlinguistic facial expressions is separate from that involved in generating ASL. Special parts of the brain thus appear to be devoted to the production of language. Moreover, just as our nervous systems participate in determining what we "see," it is clear that characteristics of our brains help us to construct the linguistic "reality" that we experience. But the idea of a complex, full-scale universal syntax module in the brain—blueprints for grammatical rules—seems to me to fall into the category of "interesting even if not true," my doubts being based primarily on gene shortage.[33]

Language and Thought

Regardless of the degree to which genetic predispositions are involved in language structure and acquisition, the relationship of language to thought has great evolutionary implications. For example, could it be that language, by giving us the power to "think," caused the Great Leap Forward?

Some psychologists hypothesize that language is essential to thought, but I think the data suggest otherwise. Language is related to thought, but we certainly don't do all our thinking in our familiar language. Consider the number of times you've written or spoken a sentence and then taken it back: "That isn't what I meant to say."[34] We often have trouble expressing our thoughts (as any writer can tell you), so we clearly aren't forming them entirely in language. There are also serious problems with normal language as a medium for thought. One is that although the symbols of language are often ambiguous, our thoughts often are not. The ambiguity arises from the way our thoughts, in all their richness and complexity, are condensed and encoded in arbitrary symbols for communication. How do you interpret the following newspaper headlines?[35]

Squad Helps Dog Bite Victim
Child's Stool Great for Use in Garden
Stud Tires Out

Each of these ambiguities could be resolved by the use of further symbols (such as a hyphen between *dog* and *bite* in the first headline); in these cases, the condensation is too extreme for clarity.

Even when statements are not ambiguous, the thought–symbol mismatch is evident in that the symbols we use often do not communicate all the information embodied in the thoughts we are representing. Consider the following:[36]

Sam is a zebra.
Zebras live in Africa.
Zebras have stripes.

Thus, Sam lives in Africa and has stripes. You, of course, know some additional information from elsewhere that is not transmitted in that particular set of symbols: *the stripes are Sam's own* (that is, there's not a pile of stripes somewhere in Africa that zebras jointly own; if I had written instead "Zebras have herd behavior," Sam would share that attribute), *and Sam shares Africa with other zebras* (he doesn't live in Australia with a bunch of other zebras while yet another herd lives in Africa). As psycholinguist Steven Pinker points out, that's just common sense. But English isn't explicit enough to embody the information the brain needs to transmit that bit of common sense in the first set of symbols. Another set, the words in italics above, would need to be added for the benefit of a Martian who lacked the common knowledge stored in the average human mind.

Here's one more example of thought–symbol mismatch, this one known as the co-reference problem:

> The neatly dressed professor turned from the blackboard and approached the crying student. The *professor* pulled out a handkerchief and *he* handed it gently to her.

The professor and *he* and the *student* and *her* are ways of shortening the communication; the mind thinks of them as exactly the same item, but our system of arbitrary symbols (English) differentiates them.[37]

Thus, there clearly cannot be a one-to-one correspondence between our thinking and the symbols with which we express our everyday thoughts to others. Nonetheless, even if some thinking occurs without language, it is very difficult to imagine how some thoughts—such as "I believe that eventually someone will show that chaos theory is hopelessly inadequate to explain the vagaries of weather"—could be represented in the mind by anything other than language. I certainly can't picture it.

Steven Pinker believes that we do our thinking in a deep, rich, built-in universal language (parallel to Chomsky's universal grammar), which he calls mentalese.[38] Mentalese presumably lacks the ambiguities introduced by the demands for efficiency of communication in the slow "channels" of speech and gesture (as compared with the much faster channels in the neurons of the brain). Does that mean that, contrary to what I said about consciousness earlier, we never "hear" our thoughts in our own "surface" language? Again, we're stuck with subjective evaluations. I think I "hear" many of my thoughts in English but not others. Some I picture (for example, that of Tommy Bruce); others I even feel (silk) or smell (1945 Chateau Mouton). I suspect that other people have the same sort of subjective experiences. For example, people who are learning foreign languages often indicate success by saying something like "I can now think in German." One might even argue (in contradiction to the thought–language ambiguities mentioned earlier) that when one says, "I can't express my thoughts," those thoughts remain unclear and await translation into surface language before becoming clear. Perhaps many important thoughts are actually thought largely in one's own acquired language rather than in mentalese. None of this is to say that there can't be substantial individual differences in the ways people think. Years ago, psychologist Anne Roe suggested that there were two kinds of scientists—those who feel compelled to form mental pictures of problems in order to solve them (that is, who solve problems through the use of visual imagery) and those who are quite comfortable with verbal or other forms of symbolic thinking, such as mathematical formulas or imageless thought.[39] Having discussed this with many colleagues over the years, I believe that Roe was on to something.

It might seem unparsimonious for natural selection to have produced a mechanism of the mind that would require continuous translation of all thoughts, as Pinker's notion of mentalese seems to suggest. But then, it wouldn't be the first unparsimonious product of selection. The answer is simply not known, but Pinker's hypothesis seems untestable and not particularly helpful. My guess (I emphasize *guess*) is that we all use a mix of surface and deep language, depending on topic and speed of use (for example, a warning to "watch out" may be formed entirely in subconscious thought—call it mentalese if you wish—and instantaneously translated into a shout, whereas I may be thinking about this parenthetical expression entirely in English).

Worrying about the connections between thinking and speaking may strike you as similar to proverbial debates over how many angels can dance on the head of a pin. In fact, however, the connection between thought and language (and thus culture) may have had considerable significance in human affairs ever since groups with different languages began to deal with one another, and it may well have present-day application. The issue was alive in eighteenth- and nineteenth-century European philosophy[40] and culminated in the twentieth century with the work of the great Viennese philosopher Ludwig Wittgenstein.[41] Wittgenstein asserted that problems of philosophy were essentially problems of language—which, right or wrong, illuminates the importance of the connection of thought and language. In twentieth-century science, discussion of the connection traces back to Franz Boas, the founding father of North American anthropology; to Edward Sapir, Boas's brilliant linguistics student; and, especially, to Benjamin Lee Whorf, Sapir's student. The basic idea, which Whorf brought to its most prominent form (called the Sapir-Whorf hypothesis[42] or the linguistic relativity principle), was well expressed by Whorf himself. He said that "users of markedly different grammars are pointed by the grammars toward different types of observations and different evaluations of externally similar acts of observation, and hence are not equivalent as observers but must arrive at somewhat different views of the world."[43] Unhappily, this has often been oversimplified as "language completely determines one's worldview," to the detriment of rational debate on what Whorf (who, like many important thinkers, was not always consistent) actually proposed.[44]

Whorf's ideas were based heavily on comparisons of Hopi and English,[45] especially concepts of time in the two languages. His analysis of Hopi has since been challenged,[46] and the Sapir-Whorf hypothesis itself has been severely criticized or completely dismissed by some.[47] Others, such as John Lucy of the University of Pennsylvania, support what they consider a weaker version of the popular "language creates worldview" hypothesis.[48] Lucy compared people who spoke English with speakers of Yucatec Maya, a language of southeastern Mexico.[49] He found a difference in the way the two groups mark their nouns to make them plural and in the way the two languages deal with count nouns (nouns such as *girl* and *mouse*, which can be made plural: *girls* and *mice*) and mass nouns (nouns that cannot be made plural: e.g., *oxygen* and *blood*, but not *oxygens* and *bloods*). On the basis of such differences, he found in nonverbal tests that English speakers paid more attention to the numbers of objects perceived than did Yucatec speakers and that English speakers had a greater preference for classification systems based on shapes, whereas Yucatec speakers preferred those based on materials.[50]

Still other observers have simply been guarded in their judgments, remaining more or less agnostic regarding the Sapir-Whorf hypothesis.[51] It seems to me that enough thought involves language that the different surface structures of languages cannot help but affect the way people view the world, just as experience and environment can alter visual perceptions.[52] This weak version of the hypothesis recently received support from the results of a study of color perception among members of the Berinmo Tribe of Papua New Guinea.[53] People in this tribe categorize colors differently from the way Westerners do, and they *see* them differently. To that extent, at least, language seems to shape worldview.[54]

The Tower of Babel

The existence of some 4,000 to 6,000 languages[55] in the world today reflects the diversity of human cultures that have evolved through random changes, migration, and exchange of ideas over hundreds of thousands of years and, in the process, that have met a great variety of environmental and social challenges. Languages illustrate the rapidity with which cultures can evolve compared with the rates most often observed in genetic evolution. For example, since the collapse of the Roman Empire (fifth century A.D.), various dialects of Latin have evolved into, among others, Catalan, French, Italian, Portuguese, Romanian, Sardinian, and Spanish—speakers of which cannot communicate very effectively with one another without knowing the other's language. In contrast to the perhaps 500 years, or some twenty-five generations, of cultural divergence required to produce these quite distinct languages, closely related North American birds such as blue and Steller's jays and eastern and western bluebirds may have taken millions of generations of genetic divergence from common ancestors to be considered distinct species, and the very closely related timberline and Brewer's sparrows must have required tens of thousands of generations.[56] To return to the African lake cichlids described in chapter 3, even if they evolved at the maximum rates hypothesized, it still took them hundreds to thousands of generations to speciate.

As a measure of the speed of linguistic change—the rate at which a language "erodes"—there is a loss of about 20 percent per millennium from a standard 100–200 item word list; therefore, 6,000 years after splitting, two languages can be expected to share only about 7 percent of words obviously derived from the same root.[57] Thus, if their rates of evolution have remained relatively constant since fully developed languages appeared, we can assume that languages varied from group to group of hunter-gatherers, with the differences increasing with distance. Even if human beings have been speaking for only about 50,000 years, many, many, languages will have evolved and disappeared during that period.[58] The high rate of linguistic drift—random changes in languages—makes it difficult, if not impossible, to reconstruct the distant history of humanity's some 5,000 now-existing languages.[59] Linguists thus have difficulties answering such interesting questions as when (or whether) all extant languages trace back to a single ancestral language. (If they do, that language in theory could have been either the only language ever evolved or the survivor of many independent previous evolutions of language.) There is, at least, very suggestive evidence that all extant languages share a fairly recent common origin, no more than 50,000 years in the past.[60]

There is an important process of feedback from language evolution to genetic evolution. People with different languages are much less likely to mate with each other than are people who share a language. Physical differences often are present between people of different linguistic groups (because their superficial characteristics evolved in different environments), and those differences also influence mate choice. These two factors result in geographic patterns in human populations in which changes in gene frequencies often occur at places where such groups meet rather than at places where there are sharp discontinuities in the physical environment.[61] The correspondence between gene frequency and linguistic frontiers supports the idea that the social environment plays a key role in human genetic evolution, given that in some cases linguistic differences maintain patterns of hereditary differentiation of populations by impeding intermarriage. One wonders whether linguistic differences may have helped to keep *Homo neanderthalensis* and *Homo sapiens* from interbreeding significantly when they lived in the same areas.[62]

The Origin of Language

Having looked at some of the characteristics of languages and their diversity, we can now turn to the puzzling question of when and how language first evolved.[63] This question appears closely related to another problem: why we evolved such large brains. Language is so tightly connected with other unique elements involved in communication in societies of *Homo sapiens*, especially religion and art, that, as mentioned earlier, some scholars believe it evolved only perhaps 50,000 years ago, in time to be responsible for the cultural Great Leap Forward.[64] On the contrary, I suspect (as do many other scientists)[65] that language has been evolving for a very long time, with some of its longest roots possibly tracing far back in the history of nonhuman animals, even to frogs and birds, in which the left hemisphere of the brain is, as in human beings, more heavily involved in vocalization.[66] It seems reasonable, in my view, to assume that essentially a continuum existed from the verbal communications of our great-ape ancestors—perhaps resembling those found even in the relatively small-brained present-day vervet monkey, which is capable of labeling different predators and using distinct alarm calls for raptors, snakes, and leopards[67]—to fully modern language. In between would be a graded series of stages, including what Bickerton called a protolanguage[68] with simple syntax, which possibly existed as far back in our history as did late groups of *Australopithecus africanus*, some 2.5 million years ago.

Symbolic behavior is not restricted to *Homo sapiens.* Kanzi, the captive bonobo described in chapter 5, impoverished as his communication skills were in comparison with human language (he never learned anything resembling human sign language), still was clearly able to learn the meanings of an impressive list of spoken words and visual symbols.[69] He could also perform relatively complex tasks on hearing commands issued by an unseen observer, such as "Go get the tomato that's in the microwave."[70] Many other living species deal with symbols, from dogs that can recognize more than sixty words[71] to monkeys able to count,[72] mongooses that transmit complex information about predators,[73] and chimps that learn bits of sign language.[74] But even if no other present-day nonhuman animal used symbols, that would not preclude the possibility that such behavior occurred in our hominid ancestors hundreds of thousands or even millions of years ago. There are good reasons to believe that brains, intense social interaction, and language coevolved gradually as hominid groups, and competition among them, grew. The notion that human beings first evolved a high level of intelligence and that language then suddenly appeared as an epiphenomenon, or by-product, of our powerful brains[75] seems to defy logic.[76] Let's look at some of the evidence.

Fossil bones, of course, can't tell us anything about language directly, but if the right parts are found, they can tell us something about the capacity for speech as we know it. The position of the larynx in infant *Homo sapiens* is the same as it is in apes and all other mammals. But by the time a child is a year and a half or two years old, the larynx has dropped, creating a larger air space (known technically as the supralaryngeal vocal tract, or SVT) above it, which allows more modification of sounds (see the figure). The development of an SVT testifies to the extreme importance of spoken communication in modern human beings. The arrangement we have is great for talking, but it poses serious perils. In *On the Origin of Species*, Darwin noted "the strange fact that every particle of food and drink we swallow has to pass over the orifice of the trachea, with some risk of falling into the lungs, notwithstanding the beautiful contrivance by which the glottis is closed."[77] Choking to death on food is commonplace (despite the invention of the Heimlich maneuver, which saves some 10,000 lives annually in the United States)[78] and, needless to say, does not increase an individual's reproductive output. In our distant ancestors, this doubtless represented a countervailing selective force, but that was apparently overcome by selection favoring those with genes that helped expand the range of sounds they could produce. It is obviously advantageous for human infants to retain the more choke-proof configuration.[79]

In good fossil material, the position of the larynx can be determined from the flexion (upward arching) of the base of skull. As far as researchers can tell, *Australopithecus* had an ape-like configuration. Unfortunately, fossils of *Homo habilis* are not complete enough to permit determination of the position of the larynx, whereas *H. erectus* appears to have an intermediate position (once again, it is a true missing link). In archaic *H. sapiens* some 300,000 years ago, the position was fully modern.[80] The anatomical status of *H. neanderthalensis* is in dispute, but it almost certainly had some speech capability.[81] Furthermore, on the basis of the diameter of the canal through which the hypoglossal nerve traverses the skull to innervate the tongue, this nerve also had reached modern size more than 300,000 years ago and was fully developed in the Neanderthals.[82] But even if the Neanderthal SVT had been small, restricting the vowel sounds that a

Chimpanzee

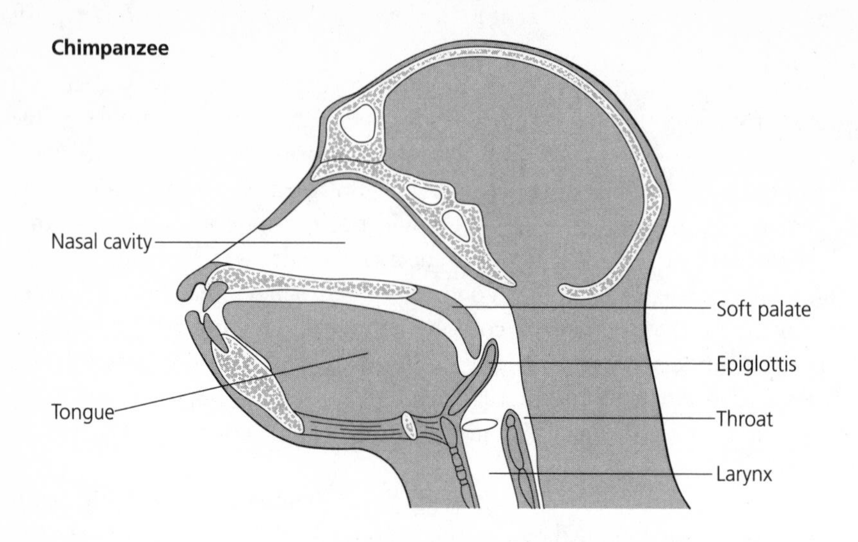

Human Adult

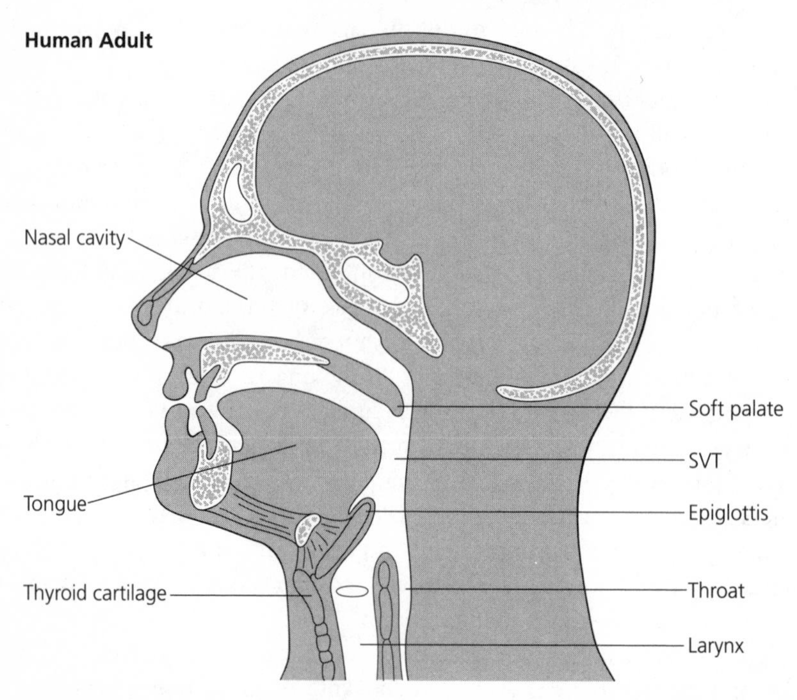

Cross section of the head of a chimpanzee and the head of an adult human being showing the relationship of the larynx (the upper part of the respiratory tract leading down to the lungs) to the tongue and nasal cavity. Note that in the chimp, the tongue is entirely within the mouth, whereas in the human being, the back of the tongue forms the front of the supralaryngeal vocal tract (SVT), giving it part of the flexibility that permits speech. The higher position of the chimp's larynx permits it to move upward and close on the nasal cavity during breathing, thus air can then go unobstructed to the lungs while food passes into the throat on either side of it. The lower position of the adult human pharynx (labeled *SVT*) means that air and food travel a common pathway behind the tongue, increasing the chance of choking. In newborn human beings, the structures are arranged as in the chimp, so babies can safely suckle and breathe simultaneously. Modified from Aiello and Dean 1990.

well-innervated tongue could theoretically produce, that would not have prevented speech. As Steven Pinker noted, "E lengeege weth e smell nember ef vewels cen remeen quete expresseve."[83]

But bones can never give us the full answer about how and when various levels of verbal communication originated. Part of the difficulty in discussing the beginnings of language might be that some anthropologists put too much emphasis on speech rather than on language itself. Remember, our ancestors stood up before they got smart, and in the process they freed their dexterous hands. Our closest relatives, chimpanzees and bonobos, are accomplished gesturers[84] and can learn at least very rudimentary sign language and associate words and symbols.[85] There is even some indication that bonobos may leave symbolic damage in vegetation, indicating directions of group travel (if an understanding of the symbols were substantiated, this would indicate intentional communication at a fairly high level—which might be called writing).[86] Wild chimpanzees in the Taï Forest of Africa's Ivory Coast appear to use symbol-like communication (in the form of drumming and hooting) to transmit information about intended travel direction. The system also includes an iconic[87] element in the positions of the trees on which they choose to drum (i.e., first on a tree along the trail previously traveled and then on one in the proposed direction). This pattern of communication seems to have evolved in response to a low-visibility environment combined with heavy predation pressure from leopards, which might be attracted by actual vocalizations.[88] Perhaps latter-day australopithecines first communicated by means of a symbolic system made up of a combination of gestures and grunts, like the systems used by medieval monks under vows of silence and modern hunters and soldiers.[89] Indeed, it's hard to think of language having originated without a lot of pointing. How else would one get agreement to associate, say, *bolk* with leaf and *crem* with pebble? And perhaps, as is true among modern Aka Pygmies in the tropical forests of central Africa,[90] one of the first uses of a pointing system among early hominids was the instruction of children rather than general communication within a group. Such "vertical transmission" of traits that increase the fitness of offspring (by, for instance, teaching them which plants are poisonous) is a characteristic that would promote the development of language through gene–culture coevolution.[91]

Consider a hypothetical example of how syntax can enter the impromptu sign language in modern *Homo sapiens* when there is a need for silence. Imagine that in Germany's Hürtgen Forest in late 1944, the point man of a U.S. Army patrol stops and signals for quiet by putting an index finger to his lips. Next, he points to the east and holds up four fingers with his left hand and then makes a walking motion with the index and second fingers of his right hand: he has spotted a Wehrmacht foot patrol of four soldiers. The lieutenant in charge points to several members of the patrol and, with a sweeping motion, signals them to take up positions on one side of the trail; then he similarly assigns others to the opposite side. With another series of gestures, including head shaking, shrugging, and pointing first at the others and then at himself while holding up a hand with the index and second fingers extended to simulate a pistol, he conveys the message "Don't shoot unless I do."[92] Planning and counterfactual content are improvised on the spot—and a level of communication is displayed that is far beyond the abilities of Kanzi and his peers.

It is the sort of process well known to anyone who has pulled off a surprise party or played charades. Protolanguage could well have leaned much more heavily for communication on gestures[93] combined with sounds than modern language does. Chimpanzees often use pointing rather than vocalization to indicate a need or desire,[94] children use gestures to transmit information, and deaf children acquire sign language as readily as hearing children acquire spoken language.[95] Protolanguage with a strong gestural component may have been evolutionarily linked with the growing dexterity of tool use, given that the neuromuscular elements involved in both are quite similar.[96] Indeed, a persuasive case has been made that the neural machinery that allows the speedy stringing together of motor acts, such as the rapid transmission of nerve impulses to different muscles required for accurate throwing,[97] is involved in other seemingly disparate sequences, such as the precise hammering necessary for making stone tools and the mouth and tongue gymnastics required to control speech.[98] Similar improved neural control would have allowed the gestural component of protolanguage to become more rapid and effective.

Furthermore, superior communication techniques, even if still heavily gestural, could have played an important role in the cooperation that is often observed in groups of people producing sophisticated tools—as when one individual shapes a stone ax head while another shapes a wooden haft and a third works on a thong to bind the two together.[99] And, of course, gesturing would have greatly facilitated the teaching of toolmaking and hunting strategies. Social behavior, language, and toolmaking appear to be tightly related.

Gradually, as the sets of symbols grew more abundant and control over sounds improved, the shift could have been to nearly exclusive dependence on sounds, which would have had important advantages: an ability to converse without visual contact, while performing other tasks with the hands, and over longer distances.[100] It should be noted that in all respects, the sign languages based on gestures and expression used by deaf people are fully inflected, genuine languages. Thus, the original shift to dependence on sound may not necessarily have involved an increase in the complexity of thoughts that could be transmitted.

The Evolution of Syntax

Could natural selection gradually have brought syntax into an early hominid communication system? The evolution of syntax is central to the great "gradual and early" versus "sudden and late (Great Leap Forward)" debate in evolutionary linguistics. At the "sudden" extreme, Derek Bickerton believes that syntax is an all-or-none type of system, which had to have been achieved in one single, rapid step.[101] His argument is, however, much more plausible than superficially similar ones suggesting that the vertebrate eye had to have arisen in one step because "what good would half an eye be?"[102] Bickerton, who is very knowledgeable about linguistics, and much of whose argument on language evolution I find persuasive, appears to have been somewhat misled here by a misinterpretation of the sometimes muddy concept of punctuated equilibrium (see chapter 3). Punctuated equilibrium, as traditionally visualized, is a phenomenon of geological time scales and speciation. It does not postulate dramatic evolutionary changes within a line in a matter of a relatively few generations—only in a short time compared with the stasis period. The thought that some rather sharp, language-related cognitive divide must account for the Great Leap Forward is widespread among anthropologists and archaeologists, however, and the arguments over what exactly occurred are as abundant as they are sometimes baroque.[103]

At the other extreme, where, as I indicated earlier, my sympathies lie, is the view that syntax and the other features of language evolved rather gradually in the hominid line.[104] On the basis of his own work and that of others as well as the undeniable uniqueness of human language as a communication system among living species, Bickerton asserts that "it is a waste of time to look for antecedents of syntax in ancestral species: syntax could not have come into existence until there was a sizable vocabulary whose units could be organized into complex structures."[105] Bickerton suggests that preadaptations such as the neural machinery that evolved in connection with throwing and the construction of stone tools, as well as a sort of mental filing system for accurate bookkeeping on reciprocity among interacting individuals, provided much of the background for the rapid evolution of syntax.[106]

But I see no reason why a simple "performer, action, object-acted-on" (PAO) syntax predisposition could not have been built in early in human evolution, with the actual sequence determined by the convention of the social group. Given context, "Tom eat dog," "Eat dog Tom," and "Dog eat Tom" are all understandable. So would be selection to remove possible ambiguity in such statements by controlling development of the brain's architecture so that the sequence, be it PAO, AOP, or OAP, had to be standardized in any language environment. It is also possible that cultural evolution achieved the standardization by vertical transmission (from generation to generation) and refinement. It is important to remember that individual development does not necessarily mimic evolutionary history and that linguistic problems for very young children, such as embedded phrases and sentences without actors, could have been present in the language of, say, *Homo ergaster* (if indeed *H. ergaster* had one). The sequence of two full languages giving rise to a pidgin, which then is transformed into a creole and subsequently into another full language, can help us think about the problem, but it cannot supply a definitive answer. It also is not surprising that traces of elements of language, such as Kanzi's ability to learn symbols, are found in our nearest living relatives, even though there is no particular reason why they must be, given that millions of years of evolution separate us. There are, for instance, no traces of flight in many of the bipedal archosaurs (dinosaurs, crocodilians) that were contemporaries of Jurassic flying birds, even though they included the closest living relatives of birds, as crocodiles and alligators are today. I have little doubt, however, that the bonobo Kanzi's ability to use symbols, a component of language, is evolutionarily related to our own development of language.[107] And the very complexity of modern language is one reason to believe that language with substantial communication capacity (perhaps including a primitive syntax) evolved well before the Great Leap. Selection is not known to rapidly produce complex products without obvious precursors on the basis of a fortuitous mutation, and a couple of thousand generations is not much in evolutionary time.

Better Communication through Selection

It is widely thought that our brains (and possibly our language) were honed to fit the situation of a highly social animal that lived by its wits, hunting and gathering in savannas, as I've mentioned. Tasks for which our brains had to equip us include coping with members of our social group and taking maximum advantage of the benefits of group living. A chief way to reap those benefits must have been to gain access to knowledge possessed by other members of the social group. Ordinarily, any social animal benefits by being a member of the group; by learning about food resources, as birds and bees do from other members of their societies; or by being alerted to dangers, as vervet monkeys are by being signaled what type of predator is near. A major selective pressure on brain size would thus have been the advantage of having more storage and processing capacity to deal with inputs from the group and relationships with other members. An individual who could remember whether it was Kug or Glog who cared more about his dominance status, recruit Visti and Stiga to back him in a fight with Kug, and learn from Glog and Visti that hunts should be planned so that deer were approached from downwind would tend to be more reproductively successful than one lacking those capacities. But dealing with knowledge of the group, remembering, arranging, and planning require more than communication—they require *thought*. Those with bigger brains almost certainly would have been better thinkers and probably also would have benefited from a higher level of self-awareness, allowing them, among other things, to better evaluate their own capabilities and consider the intentions of others.

In this context, some fascinating work by psychologist Robin Dunbar shows that the number of individuals in primate groups is related to the ratio of size of neocortex to size of the rest of the brain. (The neocortex is the part of the cortex of the mammalian brain, which, you will recall, is the site with special structures involved in the complex functions we lump into perceiving and thinking.) Dunbar's correlational studies indicate that the more neocortex a social primate has, the greater is its possible group size.[108] Correlation does not necessarily mean causation,[109] but Dunbar and his colleagues have built an interesting series of conjectures on this observation.

Dunbar tied information-processing capacity to grooming, the social glue that bonds primate groups.[110] Grooming—brushing and picking through the fur to remove seeds, parasites, and other materials—is the most common behavior that seems to unite societies of nonhuman primates.[111] Dunbar hypothesized that primates suffer from information overload and group cohesion is lost when the number of individuals with whom each must groom becomes too large for the neocortex to keep track of. The growing size of groups thus created selection pressure for the information-processing capacity to increase—that is, for the neocortex to grow. But there are limits on the amount of time an individual can spend picking through her friends' fur; one must also forage, mate, and watch out for predators, no matter how much one enjoys the tactile pleasures of grooming and nibbling on tasty lice. Thus, grooming can hold together only relatively small groups.

Building on this work, Dunbar and anthropologist Leslie Aiello connected the need to groom, constraints on group size, and the evolution of language.[112] Anatomically modern human beings, their research indicated, have a basic group size of 90–220 individuals,[113] which brackets the number, 148, that was statistically predicted from the relationship between the neocortex–brain volume ratio and group size. By group size is meant the number of people (or other primates) in a subdivision of a population in which each individual knows every other, at least by sight and usually by personal interaction. Within human groups so defined, people generally know how each person is related to the others, and Dunbar and Aiello's statistical prediction conforms quite well with observed group sizes of acquaintances in modern societies (130–150 individuals).[114]

There is a significant positive relationship among nonhuman primates between group size and time spent grooming,[115] and extension of that relationship to groups of early *Homo sapiens* suggests they would have had to spend 30–45 percent of their daytime hours grooming—fun, perhaps, but certainly time-consuming. That's where Aiello and Dunbar see language entering in as a substitute for physical grooming.[116] The much-loved human pastime of gossip, they suggest, is a replacement for combing through one another's fur—especially now that our fur has disappeared.[117]

Aiello and Dunbar hypothesize that the maximum amount of time that could reasonably be occupied in maintaining social cohesion by grooming would be approximately one-quarter of the day. Above this, the pressure to substitute gossip would be severe: one also needs to eat, mate, care for young, and so forth.[118] Using the same sort of analysis involving estimated neocortex volume, they see *Homo habilis/rudolfensis* at about the break point, grooming 23 percent of the day—a rate just higher than that of a group of geladas for which the greatest portion of time spent grooming has been recorded for a living primate.[119] Not only do geladas live in the largest groups of any nonhuman primate; they also use an unusual range of vocalizations, many of them similar to those in human speech, even though they have a neocortex no larger than those of other baboons. They also have some human-like facial features that help in producing the human-like sounds.[120] All of this suggests, again, that some of the forerunners of language could easily have been present in the common ancestor of the great apes.

Aiello and Dunbar conclude that there would have been little pressure for australopithecines to switch from grooming to language but that *Homo habilis/rudolfensis*, with its brain asymmetry and toolmaking capability, might have represented a linguistic watershed. "By the later part of the Middle Pleistocene (about 250,000 years ago)," they suggest, "groups would have become so large that language with a significant social information content would have been essential."[121] That might mean the appearance of a rudimentary syntax—that critical element in modern language. At this point, the grooming hypothesis is a castle built on the foundations of an outhouse, but it certainly is a fascinating castle. Remember also that, as I indicated in the previous chapter, human beings have for a very long time been embedded in complex, multilevel societies in which actual group size may be very difficult to determine, and the rights and obligations of any individual may differ dramatically depending on which group (family, sorority, office staff, church, state, etc.) the person is interacting in at the moment.[122]

But why should hominids have evolved groups of more than 100 individuals? Could it simply have been a result of population growth? Given the lack of evidence about the actual sizes of groups at various times, all scientists can do is speculate.[123] One explanation is that large groups were needed for protection from predators, and there is statistical evidence that group size (and the proportions of males and females in groups) in Old World monkeys is related to predation risk.[124] But it seems a weak hypothesis because other primates have successfully invaded the hominid savanna habitat without developing especially large groups or large brains to defend themselves against savanna predators. Large brains require a great deal of energy to run,[125] and they impose a heavy child-care penalty on parents. A relatively long period of postnatal helplessness results from the necessity for the infant's head to be small enough to pass through the mother's pelvis at birth and the related necessity for much of the baby's brain development to be deferred until after birth. Therefore, the benefits that early hominids derived from having large brains, including group cohesion and communication, must have been particularly great to outweigh those costs. One benefit may have been greater success in scavenging (and possibly in hunting), leading to a richer diet containing more meat. That would help support the brain's need for energy despite the relatively reduced digestive system of hominids (compared with those of other primates).[126] The invention of cooking, which is a way of externalizing part of the digestive process, may also have helped support later brain-size expansion.

As hominid populations grew, competition among groups may well have increased, and that could have added pressure for improved communication within each group[127] to enable it to defend itself against the others. Finally, social relations probably were expanded to include individuals spread over large areas, if only thinly. Formation of regional umbrella groups would have permitted subgroups to have access to concentrated food resources and water holes within the large areas needed by seminomadic or nomadic groups. Of course, all this would be possible only after relatively sophisticated communication systems had evolved.

All of these pressures could well have been significant factors in the evolution of large groups and language, if the two are indeed tightly related. But how was increased language ability spread through groups by individual selection? A "just so" story here might be that individuals who increased the verbal portion of their "grooming" activities had more time for other activities (and perhaps more social success) and thus higher reproductive rates. There may even have been selection among grooming cliques made up of close relatives. Eventually, individuals with more mental storage capacity might have gained more social status (and reproductive access) by retaining more of the available nongenetic information—that is, information about the group's culture—and putting it to good use. As the culture expanded and individuals became more ingenious at manipulating it, the demand for mental storage space would also expand—a partial analogy to the exponential increase of computer memory seen in recent decades as programs have become more complex. In both cases, the expansions stimulated each other in a process of positive feedback. The importance of memory as a storage device in nonliterate cultures cannot be overestimated; the memories of older members of societies are repositories of the group's history and of its techniques for dealing with its problems.

The Great Leap: Two Views

This returns us to the subject of the Great Leap Forward, the dramatic advances in technology that occurred some 50,000 years ago. In my reading of the available information and the debate about it, the people making the leap probably already had quite sophisticated language—as probably did the Neanderthals as well. But if, therefore, a sudden acquisition of language doesn't account for the Great Leap Forward, what does? In one sense, this is a great problem: following the long Acheulean and Mousterian periods of apparent cultural stasis, why should there have been such a flowering of fine toolmaking, with the addition of sculpture, cave painting, manufacture of body ornaments, and other signs of ceremonies such as grave provisioning? Why is there evidence that trading networks suddenly expanded substantially?[128] How did all this take place in just a few thousand years in Africa, western Asia, and Europe, and perhaps also in eastern Asia (which is still archaeologically poorly known)? There are two basic, competing views: cultural change in response to an environmental trigger and sudden, serendipitous, genetically based brain reorganization.

The first view is that the leap was the result of continuing cultural evolution in rapidly changing environments. One can construct (with the help of the paleoanthropological literature) a scenario that doesn't lean on what is, in my view, a genetically unlikely rapid reorganization of hominid brains. The brains of our hunter-gatherer ancestors probably were growing because of the positive feedbacks of an increasingly complex social system in which those most socially adept were outbreeding the less skillful.[129] Plotting, formation of alliances, planning, and thus abstract thinking and clear communication all add up to social adeptness. A developing protolanguage[130]—one with minimal syntax[131]—would be most effective if long lists of words were available. Advantages would accrue to those who could remember more words and use them more effectively in communication, and thus, quite likely, to those with bigger brains.

This scenario (which is far from unanimously admired) could have everyone from *Australopithecus* onward, including many hunter-gatherer groups of *Homo sapiens*,[132] for the most part living a life of relative relaxation and abundance, not often being stressed by factors external to the group. As is the case today for most people in rich countries, individuals' problems would have been primarily with one another and about dominance and access to mates rather than making a living. And in making a living, individuals would have depended more on their wits and their coordination with others in their group than on their teeth and weapons—which would have put a further premium on planning and communication and thus on more sophisticated language. According to this view, our ancestors had indeed entered the cognitive niche.[133]

Some historical hunter-gatherers do appear to have led lives of relaxed abundance; others, however (such as many Inuit groups, the Yanomamö people of the Amazon Basin, and the Berinmo of Papua New Guinea), certainly have not. And the leopard damage recorded on australopithecine skulls suggests that our early ancestors had serious worries beyond social climbing. But because acquiring an abundant life also demands intimate knowledge of one's surroundings, both genetic and cultural evolution probably went not only in the direction of our honing social skills but also toward our gaining superior knowledge about and greater manipulation of the environment outside of the group. Evolving language would have helped here, too, in communicating about the environment.

Why did the first tools appear? In this scenario, arguably because *Homo habilis* found itself in competition with *Paranthropus*, which put its evolutionary stake in big teeth rather than big brains, planning, and communication. Improvement in tools presumably added to *Homo*'s ability to deal with tough foods and thus to compete. As hominid populations grew and group sizes increased, further selection pressure was put on the capacity of brains for information storage and manipulation—and for the ability to communicate. It is not clear when networks of related groups and intergroup trade appeared, but even in the period of the Oldowan technology (2.5–1.7 million years ago), hominids were carrying rocks over distances of several kilometers, from quarries to sites of use.[134] We also cannot be sure how much advantage would have accrued to individuals with slightly superior linguistic-cognitive skills as both competition and networking became more important.

So in this first view, what caused the Great Leap? Necessity could have been the mother of invention. One possibility is that populations of *Homo sapiens* were growing, which might have required further improvement in hunting and gathering techniques and put them in competition, at least in some areas, with the powerful, cold-adapted, and clever Neanderthals. That competition could have made the development of more sophisticated technologies even more important.[135] The leap was also taken at a time of dramatic climatic fluctuations, which in some places could have increased population density (and group size) by concentrating people in remnant pockets of high-quality environment.[136] Relatively high-density conditions could have led to verbal magic (e.g., casting of spells) and body painting to signal status being supplemented by painting, sculpture, and beadwork undertaken to help with the hunt and show who was the cleverest stalker or the wisest leader.[137]

The Great Leap wasn't necessarily any more spectacular a cultural event than the subsequent agricultural and industrial revolutions, which were equally rapid or more so—and which no one argues were the result of a sudden evolutionary restructuring of our brains.[138] After all, while the Arabs were inventing numbers and the Chinese had a highly advanced civilization, the inhabitants of the British Isles were still living a quite primitive existence. Nonetheless, in a few thousand years, the English learned to read, write, do calculus, discover natural selection, and gain political and military advantages over the Arabs and Chinese. Even today, people whose grandparents were hunter-gatherers use computers and fly jet aircraft.

Imagine a scientist from another galaxy coming to Earth 30,000 years from now and looking at the hominid fossil record. Think of the alien scientist trying to figure out what new species had invaded a mere 500 years after our planet had been dominated by a hominid that left largely wood and stone artifacts. That new species had left a layer of fossil automobiles, jet engines, nuclear waste, computers, plastic pipe, television sets, and the like. Sure, the visiting paleontologist would conclude, the fossil bones of the new species of hominid looked like those of the old, but what a rewiring of the thinking apparatus must have occurred between the time of Columbus and the beginning of the twenty-first century.

On the other hand, this first view—that the cultural Great Leap Forward took place without genetically based change in the brain—could be dead wrong. Other scientists prefer a very different scenario. Under this second view, they think there was some basic neurological transformation that allowed, by means of the sort of preadaptations Derek Bickerton hypothesized (or others), the rapid development of spoken language with true syntax. In turn, the resulting greatly improved communication caused the Great Leap without a detectable change in fossil morphology.[139] Changes in brain organization would not necessarily be simultaneous with changes in gross morphology (and internal rewiring would not be detectable as changes in fossil skulls), another example of the way rates of evolution can differ from structure to structure within an organism.[140] And the perfection of spoken language seems as good a candidate for the engine of the Great Leap as any.

This neurological transformation scenario is favored by my Stanford University colleague Richard Klein.[141] He tends to believe that genetic reorganization modified human cognition fundamentally and triggered the Great Leap Forward. New human capabilities then made almost inevitable a continuous sequence of revolutions following those of the leap—revolutions in hunting efficiency, adoption of agriculture, the invention of writing and of printing, the building of industrial societies, development of high technology, and so on. Basically, in this view, the big move in later human evolution was not the appearance of "anatomically modern" *Homo sapiens* in the sense of bone structure but a rapid genetic reorganization of the human brain that allowed our ancestors to become culturally modern and much more innovative, perhaps as a result of a great improvement in their language and planning capabilities.[142]

Klein suggests that such a reorganization occurred just 50,000 years ago or so, probably in equatorial East Africa. Over the ensuing 5,000–10,000 years, those newly talented people spread into Asia and Europe, displacing Neanderthal populations and populations of *Homo sapiens* that had not undergone the reorganization.[143] They left a trail of their characteristic Aurignacian stone technology. The critical neurological transition Klein and others hypothesize could be very difficult to demonstrate because, again, empty fossil skulls say little or nothing about the inner structure of the brain. Our alien scientist 30,000 years in the future would have been right about brain reorganization and its role but wrong about how recently it had occurred.

If the brain-reorganization account is correct, it was an astonishingly important and arguably rather major genetic change that took place in something like 100 generations. At a time of climatic stress, selection pressures could have been high, and a very fortunate mutation or recombination may have come along at just the right time. But it does stretch the imagination of someone who started out in the Animal House, preadaptations or not.[144] Even under intense selection pressures, change progressed rather smoothly in our vials of fruit flies. But the brain is a unique organ, especially in depending so much on environmental input for its development. A single fortuitous mutation or recombination conceivably could have pushed our ancestors' abilities over a threshold. The truth is that we don't yet know exactly what caused the Great Leap, and we may never know. What is indisputable is that human beings were a pretty dull, slowly evolving bunch of critters for millions of years before the leap—which carried us to the status of rulers of Earth and explorers of space in a few tens of millennia.

Why Us and Not Them?

All of this raises another important question. Why haven't other great apes gone down the same evolutionary route, toward big brains and language, as did the hominids? That their ancestors were subject to somewhat different selection pressures goes without saying, but this is hardly an informative response. One possible answer is that our early ancestors were primarily scavengers in a savanna environment, lacking the powerful jaws and claws that would have made them physical competitors with lions, leopards, hyenas, and hunting dogs. Brains and cooperation made up the difference, especially if early successes led to increases in group size. Given this scenario, it would not have been profitable for the forest apes, with a more vegetarian diet, less opportunity to scavenge, and fewer large predators to deal with, to pay the price of larger brains. But what about baboons, which also occupied savanna habitats? Maybe their ancestors were not quite bright enough in outsmarting predators to start making the transition, especially in view of the high costs of producing and running brains and of enduring long periods of postnatal helplessness.

But such speculation does little to clear up the issue. Often, one of the most difficult evolutionary problems to solve is what happened at "choice points" to push one lineage in one direction and another lineage in another direction. Small differences in initial conditions can lead to very different evolutionary trajectories, but those differences may not be easily understood when one is confronted only with their results. Why did some groups of fishes evolve to produce huge numbers of tiny eggs and others to produce much smaller numbers of big eggs (which in some groups hatch within the females' bodies)? Both are highly successful strategies today. Why did some composites (plants of the sunflower family), all descendants of a single ancestral species, evolve into herbs while others became shrubs and still others trees? We are largely stuck with speculating about such questions, just as we are about why big brains and spoken language evolved in people and not in baboons and chimps. Despite enormous progress, scientists are just beginning to scratch the surface of how brain and body interact to produce the human mind and intense consciousness.

We may never know exactly what happened in the evolution of human brains and language, but this does not indicate a cosmic mystery about their origins. There is every reason to assume that the same processes that produced the grace of flying swallows, the marvelously skilled and sensitive trunks of elephants, the stupendous size of sequoia trees, and deadly strains of the gut bacterium *Escherichia coli* also produced the underpinnings of our diverse natures: our amazing brains and our equally amazing linguistic abilities. And they, in turn, are central to our obtaining nourishment and mates, battling our fellows, dealing with the spirit world, and carrying out all the other behaviors human beings have evolved, behaviors to which we can now turn our attention.

Chapter 8

Blood's a Rover

"To date, the hunting way of life has been the most successful and persistent adaptation man has ever achieved."

—Richard B. Lee and Irven DeVore[1]

"[S]ex is not a necessary condition for life. Many organisms have no sexuality and yet look happy enough."

—François Jacob[2]

Before or after the cultural Great Leap Forward that occurred about 50,000 years ago and during which our modern natures first appeared, we can be pretty sure what was on the minds of our ancestors much, if not most, of the time. It's what's on most of our minds day in and day out, even when we're flying airplanes or working at word processors: food and sex. Try to remember the last waking hour when your synapses didn't fire about one or the other of those topics. I didn't think you could. There are some predispositions we can be sure that selection has programmed into our DNA. The one I have already mentioned is the predisposition to maximize our reproduction—as expressed by people's love of sexual activity. Another is to eat. Both are activities so basic that all living creatures have genetic programs that lead them to reproduce and eat (provided we broaden the definition of *eating* to include the activities of organisms that "eat" sunlight or extract energy from what seem, to us, strange chemical reactions).

Acquiring energy and perpetuating one's kind are requirements for being an organism—from tiny germs to giant redwood trees, they all do it. Without energy, organisms could not function; without reproduction, evolution as we know it couldn't take place. Evolution has found lots of ways for organisms to obtain energy, from a rosebush holding its leaves up to the sun to a lion gobbling down a zebra's guts. And reproduction can involve simply dividing in half (amoebas) or squirting eggs and sperm into the sea to meet and produce young that drift out to sea (most coral reef fishes). With the possible exception of occasionally ingesting zebra meat, our early hominid forebears did none of these—those behaviors are not part of our human natures. What our forebears did do, and the evolutionary implications of what they did, is the subject of this chapter, starting with getting food and then moving on to reproduction—which is the natural order, given that we need energy to have sex and not vice versa.

Hunting and Gathering

"Clay lies still, but blood's a rover," wrote English poet A. E. Housman,[3] and in terms of biological evolution, we're rovers. Do our contemporary tastes in food and mates trace to those days when our ancestors wandered over the landscape, using crude tools to extract a precarious living from nature? Was that, as evolutionary psychologists put it, in an infelicitous phrase, our "environment of evolutionary adaptiveness" (EEA), the environment that supplied the selection pressures responsible for many of the more universal aspects of our modern natures?[4] Perhaps so, although evolutionary psychologists tend to overrate both the role of natural selection in the molding of those natures and the constancy of the EEA, which recent studies suggest was highly variable.[5] Intense consciousness and speech evolved while our forebears were living as hunter-gatherers, which they did for the vast majority of human history. It's in that hunting-gathering context that we now must seek answers to questions about the origins of our interesting dietary preferences, sexual habits, and penchants for violence, reconciliation, and religion.

Until the agricultural revolution, some 10,000 years ago, the lifestyle of all hominids was one of hunting and gathering, rather like that of chimpanzees or, indeed, of all omnivorous animals. For most hominids, hunting and gathering simply means subsisting by catching and killing game, scavenging animal remains, and gathering edible wild plants.[6] Even though the conventional label of this mode of subsistence gives pride of place to hunting, as far as is known, most people over most of history gathered more than they hunted and ate more plant materials than meat. Perhaps the name sticks because we tend to be mesmerized by the vision of shaggy cavemen draped in animal skins, waving spears, and circling a trumpeting mammoth and have been less entertained by the image of women patiently gathering nuts and excavating nutritious roots. Similarly, the mammoth-hunting image has made it easy to overlook what was a key hunting-gathering activity for some groups—obtaining food from aquatic environments, such as gathering mussels on rocky shorelines.[7]

If there are lessons to be learned from our biological and cultural past, one would expect this longest epoch of our history to be a principal textbook. Hunting and gathering was the basic hominid way of life for some 5 million years—about 250,000 generations[8]—first on savannas and later in myriad habitats ranging from tropical forests (although perhaps not tropical rain forests)[9] to arctic tundras. During that time, our ancestors not only developed intense consciousness and the capacity for speech but also came a long way technologically and, presumably, socially. Techniques for making stone tools evolved and spread; for more than 100,000 generations—from the time of *Homo habilis*, more than 2 million years ago, almost to the present—our ancestors made extensive use of a tool kit based on chipped rocks, a kit that changed only very gradually.

In contrast to the length of time our ancestors hunted and gathered, the first human lineages to take up agriculture did so only about 400 generations ago, and industrial societies have been around for only a dozen or so generations. In considering what evolutionary changes might have occurred since the agricultural or industrial revolution, remember that genetic evolution operates on generation time, not clock time (that's why bacteria can evolve hundreds of thousands of times faster than human beings). Only very strong selection pressures, involving great disparities in reproductive success (of the sort seen when more than 95 percent of the fruit flies were killed by DDT in each Animal House generation), would produce much change in 400 generations, let alone 12 generations.[10] It is thus reasonable to assume that to whatever degree humanity has been shaped by genetic evolution, it has largely been to adapt to hunting and gathering—to the lifestyles of our pre-agricultural ancestors.

To build a picture of the natures of our hunter-gatherer ancestors, we can draw evidence from several sources. One source is the characteristics of our nearest living primate relatives; another is the behavior of groups of hunter-gatherers in historical times. However, both sets of evidence must be used with caution. Chimpanzees have been evolving on their own path away from our common ancestor as long as we have, and thus today's chimps may not yield an accurate view of our own deep past. And historical hunter-gatherers, as we will see, may be poor representatives of ancient ones.

A third source of information about hunter-gatherer natures is the archaeological record. In reconstructing the evolution of human societies, though, it is important to remember that the prehistoric evidence is relatively sparse, and most of it is preserved in the form of tools. Even there, difficulties abound: the first implements undoubtedly were sticks and unaltered rocks (such as chimps currently use), which cannot now be identified as tools,[11] and it is often not clear how undoubted tools were actually employed.[12] In addition to careful analysis of tools, archaeological study of food debris (especially collections of animal bones) and of occasional "ruins" (fireplaces and the like) adds some information. Finally, there are analyses of presumed adaptations (structural, physiological, psychological) of human beings, some theory, and a lot of debate. They complete the armamentarium of the scientist studying prehistoric hunter-gatherers.

We don't know exactly how much variability there was in the lifestyles of our pre-agricultural forebears over the past several million years (e.g., some diversity, as represented by, say, the differences between African bushmen and Inuit people, might have developed quite recently). Moreover, there are reasons to believe that the behaviors of recent hunter-gatherers in many areas are adaptations to the post-Pleistocene (occurring within the past 10,000 years) shortage of large game animals—making it a parallel development with agriculture rather than a precursor of it and not necessarily representative of hunter-gatherer behavior during times of game abundance.[13] Despite the shortcomings of each type of evidence, it is clear that hunting and gathering was not a unitary phenomenon but a complex and variable mix of subsistence strategies, at least by the time humanity had spread over most of Earth.[14] People following game on the savannas of eastern Africa would have foraged in a very different manner from those dwelling in a seasonal forest in tropical America, which lacked large herds of hoofed animals but had a much greater diversity of plants. In turn, those in a tropical forest who lived on a river's floodplain doubtless had a different lifestyle from those who found themselves on somewhat higher ground. Thus, it is highly unlikely that there was just one "environment of evolutionary adaptiveness."

Life beyond the Forests

No one knows just how the early australopithecines lived when they began to venture out onto the savannas some 5 million years ago. If they used stones as tools, they probably did so very much the way chimpanzees do today—altering them not at all or very little. And, again like chimps, they probably used broken-off branches as tools and weapons, as well as leaves and twigs for appropriate tasks. Raymond Dart, the anatomist who found the Taung child, believed that the australopithecines also used bones, teeth, and horns as tools—an osteodontokeratic culture,[15] he called it. Most scientists now think that the accumulated remains on which Dart based that judgment were actually the result of the activities of hyenas and other scavengers, but it is certainly possible that such materials were sometimes used as tools.

One might suppose that the australopithecines had other habits that were not too different from those of today's chimps. Chimps use twigs to fish for termites and stones to crack nuts. Our ancestors may have begun this way and soon graduated to using bigger sticks, stones, or bones as tools to break open termite mounds in order to add valuable protein and fat to their diets. The same tools could have been used to dig up roots and tubers, giving hominids an advantage over ground-living monkeys such as baboons that dig out roots with their hands.[16] When the australopithecines got lucky, they came across an infant warthog or a carnivore kill that had not been totally stripped by scavengers. In some circumstances, our forebears may have hunted monkeys or other prey, as chimpanzees frequently do.[17] But interestingly, unlike chimps, bonobos (which in some ways, remember, resemble hominids more than chimps do) do not appear to hunt the primates that live in their habitats—their meat-eating seems to be largely restricted to flying squirrels and baby duikers (small antelopes).[18]

Much of what the early hominids ate would have depended on availability, and different australopithecine populations may have had quite different diets.[19]

Primates generally eat animal foods whenever they can get them, but they may not get them often.[20] For example, baboons—the prototypical savanna primate—spend about 98 percent of their feeding time eating vegetable foods.[21] Nonetheless, they do feast on insects, mammals such as infant Thomson's gazelles, and other high-quality sources of protein and fat whenever they get the chance.[22] Access to rich foods is thought to be one of the keys to being able to grow a large, energy-hungry brain. For australopithecines, sticks and unaltered stones would not have been of much use in hunting large hooved animals or even in appropriating the kills of feline predators or hyenas, but they could have been helpful in chasing jackals and vultures away from carcasses.[23] Something has been learned of the hunting behavior of *Homo habilis* by analysis of the fossil remains of antelopes and other prey animals,[24] and it appears most likely that *H. habilis* was primarily an herbivore and occasionally feasted on meat obtained by hunting or scavenging.[25]

My own guess is that the importance of scavenging is overrated in discussions of hominid evolution; on the African savanna, carcasses usually disappear very quickly and are guarded by ferocious predators as long as they still have substantial meat on them. Why fight lions if you can dig up tubers and ambush an occasional young gazelle? Recent discoveries of bones marked by stone tools in conjunction with fossil material of *Australopithecus garhi*[26] (the newly discovered fossil from about 2.5 million years ago) unfortunately add little to our understanding of early hominid diets beyond suggesting that stone-tool use slightly preceded *H. habilis* and that 2.5 million years ago, someone was seeking and processing animal remains far more intensively than chimpanzees ever do.[27]

Considering the apparent scarcity of big game meat even in the diets of hunters with the more advanced Acheulean technology,[28] it is very doubtful that it constituted a major element in the diet of *H. habilis* or its australopithecine contemporaries. The presence of some big game in the diet of these hominids, however, suggests that they engaged in some cooperative hunting (or chasing of predators from kills). They may also have shared the meat. Both behaviors are found in nonhuman hunting animals as different as lions and chimps, and cooperative hunting and sharing of food[29] is a widespread practice among modern human foragers.[30] It is not clear, however, whether *H. habilis* had yet acquired the typical modern human pattern of establishing home bases—campsites to which they would have returned every day to exchange meat, eat, sleep, and socialize.[31] It seems likely that they had not. Although they may have frequently dined in the same cool grove,[32] they probably retained a habit shared with modern chimpanzees: sleeping individually in nests in trees near where the group last fed.[33]

Artifacts, Brains, and Revolutions

Homo ergaster/erectus were another story and, with their bigger brains, a more "human" one.[34] The morphology of these hominids indicates that unlike their predecessors, they did not spend much time in trees. Their technological skills had increased to the point at which, instead of the more or less random flakes and cores that characterized the Oldowan technology of *Homo habilis*, they produced the Acheulean tool kit, in which symmetrical designs, typified by hand axes, appeared. The hand axes were flaked on both sides and often had sharp points. These improved tools should have enhanced our forebears' skills in hunting and rapid butchering and thus added more meat to their diet, presumably an important factor enabling the continuing enlargement of their brains. The absence of any sign that stone heads were attached to spears makes it unlikely that *H. ergaster/erectus* hunted with throwing spears, but the possibility that they thrust sharpened poles into large animals cannot be excluded.[35] From the archaeological evidence and observations of modern hunter-gatherers, it seems unlikely that these successors to *H. habilis* moved the human line into hunting big game on a large scale. As with their ancestors, most of their subsistence probably came from gathering of vegetable foods.[36] Unfortunately, we have little evidence about whether these people supplemented the meat component of their diet with one another—the first solid evidence of cannibalism in hominids was recently turned up at a Neanderthal site some 100,000 years old.[37]

One important missing piece of the puzzle of hominid technological evolution is the date at which our progenitors first gained control of fire. Although Dart originally credited the first use of fire to australopithecines more than 2 million years ago,[38] it is now thought that our ancestors' use of fire might not have begun until about 1.6–1.5 million years ago at the earliest.[39] *H. ergaster/erectus* may have had some control of fire more than a million years ago,[40] which would have helped them make the first out-of-Africa excursion. But full mastery of fire by our ancestors may not actually have come until less than 200,000 years ago.[41]

Fire is one tool used by pre-agricultural human beings that has never been part of the tool kit of other animals; chimps have never employed it to provide warmth to expand their ranges. Fire can do more than provide warmth; it can be used to herd game to places of slaughter, and it can make relatively indigestible grains and other plant materials more easily processed by the gut and, often, less toxic. Cooking has been a critical winning strategy in the age-old coevolutionary races between plants trying to poison their predators and predators trying to neutralize the toxins.[42] Most plant toxins are heat sensitive and can be made harmless by cooking. These heat-sensitive poisons include, for instance, the oxalates found in spinach. Oxalates bind calcium, sodium, and potassium so they cannot be absorbed by the body. The lectins found in legumes are another type of heat-sensitive toxin that, if not destroyed, cause red blood cells to clump and can seriously damage intestinal walls.[43] Cooking and other processing can also play a critical role in converting tubers, which are often well defended chemically and keep their nutrients in relatively inaccessible forms, into sources of nourishment for hominids.[44] In nutrition, as in other areas, human beings started early to use cleverness to expand their resource base.[45]

Interestingly, one of the most active debates in paleoanthropology is whether cooking tubers or eating meat did more to fuel the expansion of the energy-hungry human brain—the expansion so critical to making us what we are today.[46] The tuber school[47] has the support of a large body of evidence that hominids gathered more than they hunted, but it has the difficulty of explaining the absence of evidence for control of fire nearly 2 million years ago, when brain expansion began in earnest. The meat-and-brains school[48] is supported by recent findings that suggest *Australopithecus garhi* was a flesh-eater.[49] My bet is on the meat.[50]

The appearance of *Homo sapiens*, which had a much greater brain volume than its predecessors, is not accompanied by dramatic changes in the archaeological record.[51] The Acheulean technology associated with *H. ergaster/erectus* persisted in early populations of our species in Africa and Europe until 250,000 years ago or so. It was not until about half the history of *Homo sapiens* had passed that the transition to Middle Paleolithic (Mousterian) technology occurred. This technology was based on smaller, lighter tools made primarily from flakes. Many of the tools were denticulates, flakes retouched to give a saw-toothed edge. At that point, the Acheulean hand axes disappeared, presumably because the invention of mounting smaller stones on wood or bone handles was so advantageous that it spread very rapidly.[52] Middle Paleolithic technology lasted until about 50,000 years ago.

Then the Great Leap Forward[53] was signaled by the appearance of new weapons, more elaborate stone tools, and fine stone points, accompanied by tools and weapons made of substances other than stone: bone points, antler harpoons, and ivory needles. String, thread, and sewn clothing were invented.[54] Perhaps more important, starting about 40,000 years ago, art suddenly flourished, yielding magnificent cave paintings, statuettes, and jewelry.[55] Burials began to be accompanied by artifacts to support, defend, and amuse the deceased, and decorations were placed on corpses.[56]

Following the Great Leap, hunter-gatherer groups expanded their ecological niche to include less hospitable habitats, first in temperate Siberia, and then in arctic regions and probably in moist tropical forests. More impressive shelters appeared in cooler regions. One can guess that by this time, social differences within groups were well established, with certain especially skilled hunters providing leadership when cooperative hunting was necessary and shamans interacting with the spirits believed to control the group's destiny. The first signs of social distinctions in the form of jewelry and other personal effects can be imputed from Old World burials going back some 25,000–30,000 years.[57] In the New World, the record of personal effects is more limited, generally dating to only 5,000–4,000 years ago,[58] although analysis of a collection of ancient sandals from a cave in the midwestern United States showed that fashion in footwear (and perhaps some degree of social differentiation) existed some 8,300 years ago among inhabitants of pre-agricultural Missouri.[59] It is generally assumed, however, that status was earned in those societies, which were generally egalitarian, and that true social stratification with hereditary classes did not develop until the origin of political states.

This expansion of human geographic distribution may have been made possible by enhanced hunting skills and organization. A new range of tools and weapons appeared—notably spear-throwers (atlatls), harpoons, and bows and arrows with stone heads—that enabled hunters to kill large animals from a safe distance.[60] Apparently the associated skills eventually developed to the point at which, in combination with unusually rapid climate change, prehistoric human beings caused the Pleistocene overkill.[61] In that episode, along with other extinction events around the world, the invaders of the Americas are thought to have wiped out the so-called Pleistocene megafauna, an extraordinary collection of large animals ranging from ground sloths to mammoths[62] (see the figure). In the process of exterminating large herbivores, the invaders greatly altered the flora of North America.[63] These peoples doubtless used fire in their hunts, which itself would have severely affected plant communities, and the changes became permanent because the surviving herbivores had different dietary preferences from those of some of the more spectacular vanished members of the megafauna.

Whatever role human hunters played in the extinction of some species of their prey, there is no question that well before the agricultural revolution, some human groups had become efficient at killing large game.[64] One widely used technique was to drive herds over cliffs, slaughtering large numbers outright.[65] Use of that method persisted into historical times. The Inuit, for example, used to drive caribou herds into lakes, where the animals could be dispatched at leisure from kayaks.[66] On southern Coats Island in 1952, my Inuit friend Santianna made me a small model of an *inukshuk*, one of the people-sized piles of stones (cairns) the local people used in a similar kind of hunt (see the figure). The cairns were arranged in a V shape, with women and children stationed behind every few *inukshuks* and the hunters concealed behind those at the narrow end of the arrangement. A herd of caribou was then driven toward the cairns, and as they passed by, the women and children shouted and waved. The terrified animals interpreted the unmanned cairns as additional people, and the herd was funneled to its destruction.

Current behavior by hunter-gatherer groups suggests, as does Pleistocene overkill, that as our ancestors intensified their use of resources, their cultural evolution typically did not produce conservation techniques to ensure that their resources could be sustained indefinitely to support a hunter-gatherer subsistence mode.[67] Jared Diamond reported that New Guineans decimated populations of important food animals while still equipped only with stone tools.[68] In addition, it has become clear in recent years that once tropical forest hunter-gatherers are equipped with firearms, they decimate populations of large animals within many miles of their camps. That, in turn, has had substantial consequences for both the structure of the ecosystems in their region and the hunter-gatherers' own subsistence.[69]

For at least tens of thousands of years, modern human beings used their big brains and language capabilities to intensify—to increase the productivity of—

Examples of the now-extinct Pleistocene megafauna of the Americas. Above left, giant ground sloth *(Megatherium)*, about eighteen feet long. Above right, giant beaver *(Castoroides)*, about the size of a black bear. Middle left, sabretooth *(Smilodon)*, about as big as an African lion. Middle right, western camel *(Camelops)*, about the size of a dromedary. Below, mammoth *(Mammuthus primigenius)*, about ten feet high at the shoulder.

Santianna, an Aivilikmiut Inuit, with a model *inukshuk*. Photograph by P. R. Ehrlich, Coats Island, Northwest Territories, Canada, summer 1952.

their hunting and gathering activities. In general, edible plants and animals were widely scattered, and groups of hunter-gatherers are likely to have moved more or less continuously to avoid ever-lengthening foraging trips as local resources were depleted. Some groups may have been able to remain more or less sedentary, however, if they inhabited areas of concentrated resources—coastal areas with rich supplies of fish and shellfish, for example. Until the Great Leap Forward, most human groups were probably still small in size, partly because of their need to exploit relatively scattered resources. Considering what is known about historical hunter-gatherers, anthropologists surmise that those prehistoric local groups might have been on the order of 25–50 individuals in size. Group members traveled together in several bands based on one or more extended families, and these bands formed overnight camps. Groups eventually expanded to perhaps 100–200 individuals (the basic group size in the speculations by Robin Dunbar and Leslie Aiello described in chapter 7) and consisted of closely related families among which individuals of both sexes were exchanged as mates. Beyond that level of social organization, there may have been tribes—groupings of 500 to several thousand people who shared a single language dialect[70]—and often a myth of descent from a common ancestor. These modest levels of social organization indicate that *Homo sapiens*, in terms of any genetic predispositions, must still be basically a small-group animal, accustomed to living in units of at most hundreds or thousands, not millions or billions, of individuals. That may be the most important lesson to learn from our long existence as hunter-gatherers.[71]

Not every place is blessed with the same resources, so people early on started to exchange rocks, shells, hides, and other goods. Heterogeneity of the environment made trade an early part of human prehistory. We can only guess about patterns of exchange among prehistoric hunters and gatherers, but we can be quite sure that trade did occur from the presence in the archaeological record of nonindigenous materials at sites hundreds of miles from their points of origin (for instance, perforated ancient seashells found in central France)[72] and from patterns of trade found in historical groups.[73] Much of the trade was doubtless "relay trade," in which items traveled long distances by being passed along in short segments from one group to the next.[74] This form of trade would have had the advantage of not requiring the exchange of materials between groups with very different languages. As we've seen, there must have been many linguistic barriers between different hunter-gatherer groups, barriers that may have reduced the possibilities of gene exchange among groups.

Sex and Sexual Selection

This brings us at last to sex. The techniques our prehistoric ancestors employed to feed themselves are at least hinted at in the fossil record. But aside from what little can be gleaned from fossil bones, there is no direct way to trace the evolution of human sexual behavior. That's a pity because even eating does not seem to occupy such a central place in the behavioral repertoire of modern societies or seem to be so rich in policy implications. "Sex" in human beings goes far beyond the mechanics of reproduction; it pervades almost every area of our societies, from power relations between men and women to the selling of commercial products. Today, much of human nature is human *sexual* nature, and its basics evolved while our ancestors were hunter-gatherers.

Many aspects of our natures (and those of other animals) have evolved in response to the needs and desires of the opposite sex as well as of our own. Darwin himself recognized that in animals, a major selective force was likely to be mate choice. Sexual selection is simply a form of natural selection in which differential reproduction among genotypes of one sex is caused by competition among members of that sex for opportunities to mate or by the mating preferences of the other sex.

In nonhuman animals, classic traits imputed to sexual selection include the huge, ornate tails of peacocks—which make these male birds more conspicuous but less agile and almost surely more vulnerable to predation. Presumably, the advantage to the males with the most ornate tails in the eyes of peahens more than compensates for their lessened survival value. But why should peahens care about the glories of peacocks' tails? One suggestion has been that the ability of a male to invest great amounts of energy in creating such an ungainly appendage and then to survive with it is an indicator of generally good health (including resistance to parasites), strength, and so on—good genes to help the peahen's offspring to be excellent survivors and reproducers as well.[75]

The evolutionary role that sexual selection has played in determining visually obvious characteristics (such as skin color, hair type, and eye color) in our own species is potentially very interesting but difficult to sort out.[76] That's because selection on such characteristics is changeable as culture changes over time—a charismatic male leader with, say, a lisp might quickly change the reproductive luck of men with lisps.[77] Nevertheless, there is reason to believe that both types of sexual selection, mate choice and competition for mates, have shaped human evolution.[78]

Why Men Rule

Throughout history, there appears to have been a universal male domination of human groups. There are no reliable historical accounts of matriarchal societies. As anthropologist Joan Bamberger put it, "diligent searches into the prehistory of Mediterranean cultures as well as into the present conditions of primitive societies around the world have not uncovered a single undisputed case of matriarchy."[79] Indeed, in many societies, the myth of a previous matriarchy is used to subjugate women; ostensibly, the women had their chance but bungled the running of society, and males had to take over to prevent chaos. "The myth of matriarchy is but the tool used to keep woman bound to her place," Bamberger notes. "To free her, we need to destroy the myth."[80]

Another scholar, Sherry Ortner, commented, "The universality of female subordination, the fact that it exists within every type of social and economic arrangement and in societies of every degree of complexity, indicates to me that we are up against something very profound, very stubborn, something we cannot root out simply by rearranging a few tasks and roles in the social system, or even by reordering the whole economic structure."[81]

I agree with Ortner that those of us who believe in equal treatment of the sexes are up against a very stubborn problem. I also think that it is a problem rooted in genetics that is gradually being solved by cultural evolution. The most important genetically influenced differences between the sexes are differences in size—on average, men are bigger and stronger than women—and in the biologically greater commitment that women make to childbearing and nurturing. Fortunately for those trying to trace the evolution of social behavior, difference in size is a major feature related to our sexual behavior that is often preserved in the fossil record.

Sexual size dimorphism (size difference) in hominids, as I indicated earlier, is very likely a result of sexual selection caused by competition for mates.[82] It is related to the prevalence of polygyny (one male mating with more than one female) in mammals generally, in nonhuman primates more specifically, and (in perhaps a mild form)[83] in *Homo sapiens* itself. Interestingly, the degree of sexual size dimorphism in bonobos is the least among the great apes and is even somewhat less than in modern human beings. Male bonobos are only about 15 percent larger than females, and females are basically codominant with them.[84] But in other apes, the size dimorphism is much greater than in human beings.[85] Male gorillas, for example, are almost twice as big as females, and they also have larger canine teeth. Both of these features are characteristic of species in which males strive against one another to maintain exclusive access to harems.[86]

Substantial male–female size dimorphism is thought to have extended in the hominid line through australopithecines (in which males were about 30 percent larger—about the same as in chimps) and *Homo habilis* (in which males seem to have been as much as 60 percent larger).[87] *H. ergaster/erectus* were the first of our ancestors to have a degree of size difference roughly similar to the modern level (males 15–20 percent larger).[88] This reduction in dimorphism may indicate selection to reduce battles over dominance, perhaps in response to the advantages of cooperative hunting and group defense. Or it may reflect the rising importance of intelligence and skill in determining position in a dominance hierarchy. It could also be a result of increased investment in provisioning females and offspring as opposed to guarding females against copulations with other males—which one would expect to occur when the selective advantage of provisioning became greater than that of warding off cuckoldry.

Male–female size dimorphism in *Homo sapiens* remains substantial enough that most males can assert dominance over females simply by using brute strength. But this dominance is not a matter simply of men against women.[89] Big males dominate both females and small males, and female alliances can turn the tables on individual strong males.

The strength differences between men and women are largely genetically determined—but the consequences of those differences for our natures are subject to environmental modification. In modern Western societies, many a man who has been tempted to sexually harass a female co-worker or to give preferential promotion to a male colleague rather than a talented woman can testify that male–female relationships have been highly modified in this cultural environment. And in some circumstances in modern societies, big human males pick on smaller ones at their own peril, both because of laws and because of the existence of weapons that are effective at a distance and whose efficacy depends not on strength but on skill of use. After all, the Colt .44 pistol was called the "great equalizer" on America's western frontier for a reason! But further cultural modification of attitudes and behavior will be required before, even in the workplace, both females and males are judged without consideration of gender, size, strength, or appearance.

Not only are human males generally bigger and stronger than females; they are also biologically less committed to their offspring. Paternal investment can be limited to a single brief copulation, whereas the mother is burdened by nine months of gestation followed by, in early societies, years of nursing. It thus seems reasonable that the roots of male dominance (and the ubiquity of patriarchy)[90] can be found in a combination of size and sex role. It's tough to be boss when you're smaller, weaker, and tied down with child care. It may even be hard to use one key weapon, the withholding of intercourse, because it can be countered by rape or by changing of partners. Indeed, male dominance pervades human sexual behavior, perhaps reaching its extreme among the Yanomamö, a tribe that lives in the highlands near the headwaters of the Amazon and Orinoco Rivers in Brazil and Venezuela, and among various New Guinea societies. In reprimanding their wives, Yanomamö husbands frequently beat them, shoot arrows into their nonvital parts, chop at them with machetes, or burn them with glowing embers.[91] Among New Guinea males, there is a close intertwining of religion, violence in ritual and warfare, male dominance, and a series of notions about the high value of semen and the dangers to men of both heterosexual intercourse and menstrual blood.[92]

Even if any genetic tendencies toward making one's own decisions (or seeking sexual pleasure) were identical in the two sexes, the inherent male biological advantages of strength and relative freedom from reproductive commitment would probably make men dominant in societies and thus better able to achieve their ends. It is not clear, however, that male and female genetic (or social) programming *is* identical, and there is no reason to believe, a priori, that it should be. Indeed, I suspect that long-term differences in the roles of the sexes have left their evolutionary mark. The group structure of our common ancestor with the chimpanzees probably was similar in many basic characteristics to the group structures of chimps and people today. Our ancestor probably foraged in patriarchal groups of kin structured by dominance hierarchies (which defined individual relationships both between and within sexes). In both chimps and people, fathers, brothers, and sons form the core domestic group; it is ordinarily females that change groups.[93] (In many other primates, such as baboons, however, it is the young males that leave and seek mates in other groups.)[94] The core domestic groups generate lineages, defined as groups that trace their descent to a common ancestor and that take corporate action (such as planning and carrying out hunts).[95]

It also seems reasonable to assume that some collections of prehistoric male-based lineages in which individuals recognized their relationships to one another—collections anthropologists call clans—were originally territorial.[96] Territoriality basically involves the defense of all or part of a home range, the area customarily used by an individual or group in the course of its normal activities. Territories are typically established to protect or monopolize resources, mates, or offspring,[97] and animals may defend territories against a wide variety of potential competitors, against only conspecifics (members of the same species), or against only conspecifics of a certain age and sex. There is evidence that present-day hunter-gatherer groups defend resources primarily when the benefits outweigh the costs of defense, a situation that has been well documented in birds.[98] There is no point, for example, in expending effort to protect either superabundant resources or resources so sparse that huge areas would need to be defended. Thus, prehistoric territoriality presumably would have occurred when valued resources were at an intermediate level of abundance.[99]

Not only did these groups defend their home ranges; on occasion they also raided outside those ranges, killing members of other clans and sometimes destroying those clans and acquiring their territories. Such behavior has been recorded in chimps and has been widely (but not universally) recorded in human hunter-gatherer groups. The capture of women may well have been a goal of combat between males of prehistoric hunter-gatherer clans, just as in some historical groups[100]—one of many connections between violence and sex. Such activities in our long hunter-gatherer history, as well as the need for strength in the hunt,[101] may have provided some selection pressure for maintaining the sexual size dimorphism that still characterizes *Homo sapiens.*

If early clans warred over territories and women, one of the earliest social controls to appear through cultural evolution may have been one to restrain interclan warfare and thereby permit the enlargement of social groups. A modern-day model might be the Xinguano culture of the upper Xingu River basin of central Brazil. There, complex social arrangements focusing on exchange of individuals, goods, and services have been developed that provide benefits of peace recognized by the people of ten single-village tribes whose total population is about 1,200 people.[102] Similarly, most societies try to exert social control over the more directly violent aspects of another apparent cultural universal, rape of women by men.[103] It is easy to imagine that this control is an ancient cultural practice based on the need for cooperation within clans, similar to the controls described earlier for promoting cooperation among clans.

Modern human groups defend territories for the same sorts of reasons as did their ancestors, as well as for reasons (e.g., status) that we can't be sure were present prehistorically. Today, we often draw rather fine distinctions between categories of human individuals who are to be excluded from spaces that an individual or group "owns."[104] Just think of the reaction when a street person enters a university seminar or a woman seeks entry into an all-male club. This human "territoriality" may be strictly a culturally evolved phenomenon related to our unique ability to create specialized social groups, each with its own rules of conventional behavior.[105] Or it could arise from a common genetic background with other territorial mammals. The answer is not clear,[106] although I'm inclined, on the basis of the idea of gene shortage, to lean toward the former. It *is* crystal clear that various forms of human territoriality can be altered by cultural evolution, however, as demonstrated by the decline in racial and sexual segregation in the United States in the second half of the twentieth century.

Sexual Behavior

Maybe high-powered language is the most interesting product of our brains, and maybe not. It could be that our sexual behavior is—after all, sex can be a lot more entertaining than listening to a lecture or reading this book. Sex pervades most people's lives, is (along with marriage) a central concern in all human societies, and plays many roles beyond simple reproduction—as it does in one of our closest living relatives, the bonobo.[107] Few things in life are so important, so pleasurable,[108] so laced with tribulations and taboos, and (partially as a result of the taboos) so little understood. Being human beings, scientists, too, are fascinated by sex. They have studied everything from marriage, divorce, casual sex, positions of intercourse, and the frequency of oral-genital sex to the evolutionary psychology of female attractiveness and the causes of homosexuality. It's a lot more interesting than studying the pollination of plants.

But being sexual creatures, scientists naturally bring great personal prejudice to the study of sex. All of us *really* know about the sexual attitudes and behavior of only one individual (although we may be pretty well informed about our sexual partners), and there is a clear tendency to project our attitudes onto others. As in evaluation of religions and other cultural constructs, there is a tendency to consider one's own experience normal. Moreover, discovering the sex habits of other people is extremely difficult. Survey efforts are haunted by people's propensity to lie about sex and to have difficulty recalling without bias what their activities have been.[109]

Reconstructing the sexual habits of distant ancestors, who cannot be interviewed, is, of course, infinitely more difficult. There is no fossil record of their mating systems—patterns of formation or nonformation of pair bonds, those social associations between males and females in the cause of reproduction. We are forced to use theory, speculation, comparisons with known mating systems of our modern primate relatives, and knowledge of modern human mating systems to deduce the origins of today's patterns of marriage bonds.[110] Those bonds seem to be formed for a variety of reasons—economic (division of labor), social (status), and reproductive (help in child rearing)—with reproductive success being the ultimate goal of all three.[111]

Behavioral scientists break down mating behavior into a series of broad categories, which are sometimes misunderstood, as the category of monogamy illustrates. Monogamy entails a male and female pair-bonded together for at least one reproductive period; but contrary to its popular use, the term *monogamy* technically carries no particular implication of the genetic contributions of the individuals to the young being raised. Monogamous bird species are so classified because of their pair-bonding behavior, notwithstanding the high frequency of multiple parentage among the offspring raised by a bonded pair.[112] Among primates, gibbons and siamangs form lifelong pairs.[113] Some scientists[114] classify human beings as monogamous because of the prominence of the pair bond and because there is normally only one partner in a single ovulatory cycle, even though, as with monogamous birds, children are frequently fathered by males other than the pair-bonded husband.[115]

Polygyny, in which one male is bonded with more than one female—as opposed to polyandry (one female bonding with more than one male) or promiscuity (no bonding)—is often considered the "basic" mating system of mammals. Although most birds are monogamous, presumably because of the need for two adults to care for the vulnerable and rapidly growing young,[116] polygyny is most frequent in mammals because of the relatively small investment required of the male. Males of harem-forming polygynous species (such as gorillas) stay with and guard their females rather than foraging for them and their offspring.[117] Edward O. Wilson quite reasonably described human beings as "moderately polygynous,"[118] and physiologist Roger Short wrote that "we are basically a polygynous primate in which the polygyny usually takes the form of serial monogamy."[119]

Harem size in mammals tends to be correlated with sexual size dimorphism.[120] Male southern elephant seals average forty-eight mates and weigh more than eight times as much as the females. Male gorillas average three to six mates and weigh nearly twice as much as females. Male gibbons, which are monogamous, don't outweigh their consorts. Human males' small weight advantage over females fits the characterization of a low-grade polygyny suggested by Wilson and Short. Whatever the typical system today, given that chimps are promiscuous, there is little reason to believe that our early ancestors were particularly monogamous. Evidence that *Australopithecus africanus* had, like gorillas, marked sexual dimorphism, also suggests that our early ancestors were more polygynous than we are today[121]—although today, cultural factors certainly have much more influence on mating patterns than they did among our distant forebears.

Not only do modern human beings show some sexual size (but not canine tooth) dimorphism, but also men have relatively small testes compared with chimpanzees. The size difference between human males and females suggests that men are equipped to fight over women, though much less so than are, say, male gorillas. The testes size difference is testimonial to our mating system not being basically promiscuous, like that of chimps,[122] among which the ability to deposit large numbers of sperm is important in improving the chances of fertilization because of competition with other males' sperm.[123] In contrast to the size of their testes, human males have relatively large penises.[124] The possibility has been raised that substantial penis length in human beings is related to sperm competition and ejaculation as close as possible to the egg,[125] an explanation that seems unlikely on the ground of variability in both length of penis and position of the partner's cervix forward of where a longer penis would ejaculate.

Another suggestion related to sperm competition was put forward by biologists Robin Baker and Mark Bellis. They hypothesize that the size and shape of the human penis allows it to act as a piston to suck and scrape previously deposited seminal fluid, sperm, and cervical mucus away from the cervix, enhancing the opportunities for passage of the sperm from its ejaculation.[126] It's an interesting idea that implies rapid shifting of partners in our ancestors. Furthermore, human males, like various other mammals,[127] apparently deposit enough seminal fluid in an ejaculation to form an ephemeral plug.[128] The first half of the ejaculate contains most of the sperm, and the last fraction of human seminal fluid deposited is spermicidal.[129]

The question of why women have permanently prominent breasts has similarly provoked considerable debate. Breast size and shape are unrelated to the quantity of milk that can be produced and usually are of erotic interest to males only in societies in which breasts are normally covered, although there are differences in preferred shape and size among different cultures (and, at least as fashions change in modern societies, at different times).[130] One of the most sensible answers to the question is that sexual attraction is one function—once hominids got up on their hind legs, large breasts became conspicuous signals of fertility in hunter-gatherer females, most of whom were pregnant or lactating nearly all the time. Another idea that has been proposed is that breasts made useful cushions for sleeping large-headed infants being carried around by their foraging mothers.[131] This seems *very* unlikely, but it conceivably could be selected for on the basis of willingness of the female to bear more children. Our upright posture may be related to other secondary sexual characteristics in addition to breasts. It could explain the residual hair in our armpits, which serves as a wick to disperse odors (thought to be sexual attractants) from the sweat glands located there. That function was presumably served by glands closer to the genitalia earlier in our evolution, when they were situated at roughly the same height as the nose of a quadrupedal primate.[132]

Not Tonight, Dear

Women's interest in sexual activities clearly fluctuates through time, but not nearly so much as that of many other female mammals, even though all female mammals undergo basically similar reproductive cycles. Most mammals have a clearly defined period of estrus (heat), during which the female is sexually attractive to males and is receptive to copulation—or even actively solicits it—at a time when an egg is available for fertilization. Most of the time (often for most of the year if there is a well-defined breeding season), the majority of nonhuman mammals, male and female, show no sexual activity, and in all nonhuman animals except bonobos, copulation is periodic.

No chimp or bonobo (or human observer of chimps or bonobos) is in much doubt about whether or not a female is ovulating. Female chimps and bonobos go through a cycle of genital swelling that lasts somewhat longer than one month, and they copulate primarily during the ten-day period of maximum swelling. In chimps, the maximum swelling is a very conspicuous bright pink and, at the extreme, has a volume of some 1,400 cubic centimeters, about that of a modern human being's brain (see the figure).[133] With obvious estrus swelling and copulation in public, the sex lives of our closest relatives are relatively easy to study in the field. Not so with people, given that copulation in virtually all human societies takes place in private, and women, like females of

Female chimpanzee at Arnhem Zoo with estrus swelling. Photograph by P. R. Ehrlich.

some monkeys (e.g., vervet and spider monkeys) and other apes (orangutans), largely conceal their time of ovulation not only from males but also from themselves.[134] I say largely because there is evidence that women can detect the timing of other women's menstrual cycles, especially if they are close friends, probably by the dispersal of specific odors. Men also may be able to detect the fertile part of the cycle.[135] The cycles of women who live together tend to become synchronized,[136] but the evolutionary significance of this, if any, is unknown.[137]

Two basic questions that often arise about the sexual cycles of *Homo sapiens* are "Why do human beings (and some other primates) menstruate, whereas most mammals do not?" and "Why is more or less constant sexual activity, rather than a well-defined estrus period, a feature of human natures?"[138] The answer to both questions is that we don't know for sure, but some intriguing hypotheses have been advanced.

The answer to the first question may, in fact, be related to the second, that regarding a lack of estrus. Surprisingly, in early human societies menstruation was a relatively rare event. For most of their adult lives, women were either pregnant or lactating, the latter condition releasing hormones that suppressed the menstrual cycle (creating what is technically called lactational amenorrhea).[139] For instance, women of the Dogon, a sedentary agricultural people of Mali, in western Africa, menstruate about 100 times during their reproductive life, as opposed to roughly 400 menses for many women in the United States.[140] Anthropologist Barbara Harrell summed it up in a rather flowery style: menstruation may have evolved as a signal of a "'coming into heat,' an emotional and genital ripeness for copulation, interjected between the stable, pallid, parasitized state of lactational amenorrhea and the transforming, tumescent, parasitized state of pregnancy. Menstrual flow would thus symbolize the height of female sexual capacity vis-à-vis the male, as well as the female's liminal freedom from biologically or culturally mandated intensive infant parasitism."[141] At a more physiological level, it has been proposed that menstruation functions as a mechanism to flush from the uterus dangerous bacteria that hitchhike in on sperm.[142] Other researchers counter that menstruation is more efficient than maintenance of the uterine lining in the high-energy state necessary for implantation of a fertilized egg.[143] The jury is still out, and any final explanation will need to deal with the question of why not only human females but also females of those other primates menstruate.[144]

The apparent rarity of menstruation among hunters and gatherers may help to explain the widespread custom among the Dogon and other groups of isolating menstruating women from society, often in a special hut. It has often been interpreted simply as an attempt to avoid "pollution" of men by the menstrual flow.[145] However, such social reactions could more often be connected to the relative rarity and special character of menstruation in tribal groups. The taboos among the Dogon appear to be related to great concern about cuckoldry: visits to the huts advertise the reproductive status of the women—everyone is publicly informed about who is or isn't pregnant.[146] Careful hormonal studies by anthropologist Beverly Strassmann indicated that there were negligible signs of misleading signaling among the women (i.e., going to the hut when not menstruating or not going when menstruating)—the taboo was effective.[147]

In some situations, the "pollution" explanation may be close to the truth. Although men and women among the Machiguenga, agriculturalists who live in the rain forests of Peru, get along well, men have great fear that contact with menstrual blood will weaken them, compromising their ability to carry out male tasks, and that threat is a major theme in dreams and folktales.[148] A Machiguenga man who is not a strong hunter is described as one who "likes to be womanly" (with no implication of homosexuality), and a strong woman is characterized as "likes to be manly." Menstruating women make a special exit from the family hut so that there is no chance that if any menstrual blood drips to the floor, a male might step in it and be weakened.[149]

Loss of Estrus

The second question, about the absence of estrus in human beings, has probably been the center of more speculation than any other physical aspect of our sexual natures. The debates, which exemplify the difficulties of untangling the complexities of evolution and behavior, are worth examining in some detail. The classic explanation for the lack of estrus and resultant year-round sexual activity in human beings has focused on the activity's possible role in binding pairs and societies together—it putatively kept men around to help care for helpless infants and led to establishment of the family group as the basic unit of society. But there are difficulties with this explanation. One is the problem of group selection, which is implied if benefits accrue primarily to social groups, as opposed to individuals.[150] Another problem is that other, nonhuman primates are also highly social animals, some with strong pair bonds as well, but show little tendency to extend the time occupied by sexual activity. Although more sex might help maintain social cohesion in some sense, it is an energy-intensive activity that often leads to social conflict, and because it is difficult to keep a sharp lookout while copulating, it could also increase the risk of being killed by a predator. Most serious of all, those who claim that the absence of estrus promotes pair-bonding[151] neglect the fact that pair-bonded (monogamous) animals usually engage in *less* sexual activity than those with more promiscuous mating patterns.

What other possible evolutionary scenarios could account for the disappearance of estrus? Because its loss is universal in human females, we can assume that the lack is traceable to a feature of the human genome that differs from those of chimpanzees—a loss of the capacity both to detect the time of ovulation and to concentrate copulatory activity around that time. In the all-too-common vernacular, it's "in our genes." Lack of estrus wouldn't be there unless it were to the females' reproductive advantage, either directly or because it is genetically coupled to some other advantageous trait (the evolutionary interests of males and females are not necessarily congruent).[152] It wouldn't occur if it interfered with a female's ability to attract males for copulation and to conceive children, unless that was more than counterbalanced by other factors (e.g., increased life expectancy or greater survivability of any young conceived) that enhanced women's chances of passing on their genes. The scenario for the loss of estrus already introduced simply focuses on it as a device to recruit more paternal effort in child rearing.[153] The male must hang around because frequent copulation is required to ensure fertilization of the female, who is concealing her fertile period. If paternal investment in offspring is valuable, then the already homebound male would lose little reproductive opportunity. If he carried out any extra-pair copulations, they would have a low probability of success because of concealed ovulation in his other female consorts. On the other hand, he would gain a great deal (higher survivorship of offspring he almost certainly sired) in the evolving mating system.

One of many other explanatory scenarios for loss of estrus,[154] called the "many-fathers theory" by Jared Diamond,[155] was developed by anthropologist Sarah Blaffer Hrdy.[156] It is based on an assumption drawn from the occurrence of infanticide by male primates.[157] For example, when a new dominant male silverback gorilla takes over a family group, sometimes he will kill offspring of the previous silverback.[158] This, presumably, is to bring the mother back into estrus, to permit the new silverback to father another offspring, and to allow the male to avoid wasting parenting effort on the offspring of his predecessor. One reason for the loss of human estrus, therefore, could be a female strategy to reduce males' knowledge of paternity. If mating is sufficiently promiscuous, males would be unlikely to kill infants because they might have fathered the victims themselves.[159] Here is a clear case in which the interests of males and females are not congruent.

As a test of this hypothesis, animal behaviorist Sandra Andelman was able to monitor ovarian function in a group of vervet monkeys, which do not have a well-defined estrus.[160] She did this by the clever technique of holding up a funnel (leading to a bottle) under female monkeys perching in acacia trees and analyzing the hormone content of the urine collected (it clearly takes great dedication to science to employ such a protocol). Andelman showed that ovulation was indeed largely concealed from the males: copulation began weeks before ovulation and continued during the first half of pregnancy, when females seemed especially receptive even though the males were "wasting" their efforts. One-third of the males actually mated with every female in the group. Except for male transfers into a group, vervet males clearly would risk killing their own young if they were tempted to commit infanticide, and indeed infanticide is rare in vervet monkey populations.

Questions still to be answered include exactly how much male infanticide occurs among most primates[161] and to what degree infanticide is motivated by the reasons suggested by the many-fathers theory. Infanticide has been rec… for example, but the patterns do not ind… s into estrus, and males do not put effo… w of the subject, animal behaviorist Ala… ther *widespread* nor of *general* importan… or sexual behavior."[163] In human bei… t infanticide, though uncommon, is mo… rts who know, despite concealment of … , stepfathers, boyfriends).[164] Fitness is … children. On this, Dixson wrote, "T… nt for the deaths of young children by… hard to escape the conclusion that thi… d by such mechanisms."[165] Overall, th… ather unlikely as a general explanatio… fertile periods. The frequency of inf… its occurrence does have elements re…

… sociobiologist Nancy Burley's "not fo… raightforward. At some point in the evolution of the human brain, women started to figure out the connections between copulation, estrus, and childbearing. The latter is a painful and dangerous experience, and women who were able to recognize their fertile period began to avoid copulation during that time. Those women who were able to detect ovulation thus reproduced less well than those who were not; as a result of this selective differential, the inability to detect ovulation eventually became widespread—though possibly not universal—in human populations.[167] Lack of estrus, under this scenario, evolved to conceal ovulation from the women themselves.[168]

Underlying such explanations of concealed ovulation is the assumption that males cannot guess with any accuracy when ovulation occurs. But, as I indicated earlier, both males and females may be able to detect females' cycles subliminally by means of odor signals (sometimes called pheromones—external hormones).[169] And the "not for me" case is weakened because astronomical levels of endorphins (internally produced opium-like substances) are present during childbirth that make women tend to forget the intense pain of the experience.[170]

A more recent attempt to explain why ovulation is concealed in our species was made by biologists Birgitta Sillén-Tullberg and Anders Møller.[171] It is based on what is understood of the phylogeny (evolutionary history) of *Homo sapiens* and assumptions about the behavior of our ancestors. The researchers found that 47 percent of sixty-eight primate species concealed their ovulation and another 26 percent (including gorillas) lacked prominent ovulatory signals. Eleven of those primate species, including *Homo sapiens*, they considered monogamous; of those, none spectacularly advertises ovulation and all but one fully conceal it. But many nonmonogamous species, such as vervet monkeys, also conceal ovulation. At the other extreme, fourteen prominent advertisers (out of eighteen) are promiscuous species. But among thirty-four kinds of promiscuous primates, only those fourteen leave no doubt about when a female is in heat.

No tight one-to-one correspondence was found between patterns of sexual bonding or nonbonding and the occurrence of concealed ovulation. Therefore, lack of advertising must serve different functions in different mating systems.[172] Different mating systems and degrees of concealment of ovulation, this suggests, are quite readily changed by natural selection, and different combinations must have evolved numerous times during primate evolutionary history. Sillén-Tullberg and Møller concluded that in the evolutionary line leading to *Homo sapiens*, concealed ovulation came first, followed by the evolution of monogamy. Concealed ovulation might have developed during a promiscuous stage, when the sexual system was like that of vervets, possibly as a strategy for females to counter the infanticidal proclivities of males. Later, moderate polygyny (or monogamy) evolved, perhaps under selective pressure for the male to help provision the especially helpless human offspring and thus perpetuate his genes; perhaps it also evolved as a device for helping to ensure paternity.

What is one to make of these varied hypotheses relating to such an important aspect of human sexuality? One thing, in my view, is that although there has been some fascinating speculation, science has not yet produced a persuasive, complete story of the significance of concealed ovulation.[173] It does seem clear, however, that loss of estrus is an evolutionary response to the human sociocultural environment, not to our physical and biological environment. Our ancestors have lived for at least 3,000 generations in a wide variety of habitats (since out-of-Africa 2), and geographic variation in many human characteristics has evolved in response. Yet throughout the worldwide range of the human species, females show the same lack of estrus. Its presence and degree are a species-wide characteristic that does not vary perceptibly with physical, biological, or sociocultural environments. Thus, it is apparently a response to selection pressures that must have acted in ancestral populations of *Homo sapiens* and quite likely much further back in our ancestry. It seems almost certain as well that different patterns of estrus among primate species evolved as a function of different social rather than physical and biological environments. Some chimps, for example, live in environments very similar to those occupied by gorillas, yet chimp females signal estrus very conspicuously and gorillas do not. Some chimps and baboons also live in environments that are assumed to be similar in many ways to those of our African ancestors, but they did not evolve concealed estrus, whereas vervet monkeys, which co-occur widely in the same physical and biological environments as baboons, did.

There are many further twists and turns to the arguments over concealed ovulation that I've not covered here, but it should be clear that trying to develop even a single agreed-on "just so" story to explain the evolution of concealed ovulation is a very difficult task. My guess is that many sociocultural and selection pressures combined in complex ways to produce the human mating system.[174] Anthropologist Donald Symons, a pioneer in thinking about the evolution of human sexual behavior, points out that the function of concealed ovulation is a particularly tough issue to deal with because one is trying to explain the design of something that is absent.[175] In any case, when you see facile evolutionary explanations in the media of why "human nature" has certain characteristics or why people behave in certain ways, you might first remember concealed ovulation, the multitude of explanations for it, and how easy it is to overlook important factors in reasoning about our natures.[176]

Cryptic Copulation

One common human behavior that demands such careful scrutiny is our predilection for concealing our sexual activity from uninvolved members of the community. Not only is human ovulation concealed; ordinarily, so is human copulation. Unlike our primate relatives, human beings seldom mate in public.[177] The origins of cryptic copulation, like those of cryptic ovulation, can only be guessed at. Perhaps as human culture became more complex, group sizes larger, and potential competitors and cooperators more intelligent, cleverer but less dominant "sneaky" males may have become more successful reproductively than those who were simply physically strong and more dominant.[178] Cryptic copulation may have evolved along with the elaboration of deception, becoming especially important when language developed to the point at which verbal communication could be used to arrange hidden trysts. At the same time, gossip about who was mating with whom could supplement grooming,[179] making it easier for philandering to be exposed and, possibly, brutally punished.

We can be sure that there is a very strong genetic component to the concealment of ovulation and reasonably certain that there is a strong genetic influence on the human mating system (mating systems tend to be species-wide in most primates but somewhat labile in *Homo sapiens*). But whether there is much genetic influence on cryptic copulation is not known. Cryptic copulation, it should be said, also occurs in chimpanzees and gorillas.[180] In a fascinating study, primatologist Pascal Gagneux and his colleagues investigated patterns of paternity in a group of chimps long habituated to human observation in the Taï Forest of the Ivory Coast.[181] The researchers climbed trees in order to collect hairs left by known individuals in their precariously sited arboreal nests and retrieved inner cheek cells from chewed, dropped fruit. By extracting DNA from these two sources, they were able to determine which chimps had fathered thirteen different infants. Seven of the infants were not offspring of males from the main group, even though during seventeen years of study, females had never been seen near males of other communities except during group confrontations. Clearly, the females were leaving the group for a significant portion of their copulations, outwitting both the group males and the scientist observers.[182]

Worse yet for cherished ideas about behavior, four of the males in the chimp community were dominant during the study period, but two of them fathered no offspring (within the group) while dominant. So much for the notion that dominance is necessarily closely coupled with high reproductive output, at least in chimps. Apparently, one must always consider the chaps lurking in the bushes. Indeed, the idea that dominant individuals do most of the reproducing in species with dominance hierarchies is fading fast; there's more than one way for a male to increase his reproductive contribution, and sneaking around behind the backs of dominant males is one apparently successful strategy. Nonetheless, the costs of climbing to the top of a dominance hierarchy are not negligible; thus, averaged over evolutionary time, the strategy must have a reproductive payoff.

Cryptic copulation makes sense when pairs are trying to avoid the attention of dominant males. This behavior might have evolved genetically in these primates, but being beaten up by a dominant male can also be a very potent learning experience. To assume that cryptic copulation in human beings is primarily a genetic program shared, at least in part, with those of other primates, seems to me a stretch. Isn't it more reasonable just to assume that smart social organisms quickly learn to conceal behavior that is likely to be severely sanctioned in some way if carried out in public? Concealment might be socially imposed to avoid the disruption that more overt sexual competition might otherwise generate and that might interfere with the smooth running of the group—a form of what might be called *cultural* group selection. Unlike concealed ovulation, it is reversible culturally, as demonstrated by the widespread occurrence in human societies of sex shows and group sex.[183]

The Evolution of Desire

Increasingly, studies of other animals, especially bonobos, have shown that human sexual behavior is not as unusual as once believed. For example, the customary face-to-face,[184] or "missionary," position for intercourse, once thought a human peculiarity, is extensively used by bonobos. Also like human beings, bonobos do not restrict their sexual activity to periods of estrus (which, compared with those of chimpanzees, are prolonged), although copulations peak at that time.[185] But there are limits to the analogy. As psychologist Frans de Waal put it: "Even though the bonobo is a serious contender for the title of sex champion of the primate world, including the prize for the longest penis. . . . [w]ith the average copulation lasting 13 seconds at the San Diego Zoo and 15 seconds in the wild, sexual contact among bonobos is rather quick by human standards. Thus, instead of an endless orgy, we see a social life peppered by extensive sex play and punctuated by brief moments of copulatory activity."[186]

More generally, recent work suggests that human females are just as "sexy" as males. The proximate driving forces of sexual behavior appear to have rather deep evolutionary roots in both sexes. Female monkeys clearly have orgasms, as casual observation[187] and electrophysiological experiments[188] indicate. "Female non-human primates not only like sex, and are driven to it," wrote anthropologist Meredith Small, "but . . . they often mate with many different males and certainly more than is directly necessary for conception."[189] Recent anatomical work, furthermore, has shown that—probably partly as a result of the prudishness of male anatomists—the clitoris has been miscast as a tiny but highly innervated homologue of the penis. The clitoris actually consists of the well-known external tip (or glans) plus an internal mass of erectile tissue that is much larger than previously thought.[190] Women are apparently both psychologically as desirous as males and in possession of considerable ability to have "erections" as well.

This brings us to what is arguably the most central and contentious evolutionary question in the area of human sexual behavior: to what extent are the male–female differences in sexual attitudes among human beings (or other great apes) a matter of genetic programming, as opposed to cultural conditioning? After all, to put it in anthropomorphic terms, why would natural selection leave anything so central to its operation to environmental chance? Surely sexual behavior is "hard-wired"? The current tendency among scientists is to view male–female differences in attitude as controlled by genes found on X or Y chromosomes.[191] And if sexual attitudes are genetically programmed, a standard evolutionary view is that natural selection will have produced a "profligate male–choosy female" strategy in mammals in general and human beings in particular.[192] Meredith Small expressed it well: "Because of basic biological differences at the gamete level, males and females have different patterns of behavior. Males have those expendable gametes and the potential to spread their sperm about. Females have only so many chances to pass on genetic material in the form of infants."[193]

The notion that male mammals (including men) need to put only a small effort into fathering a child, whereas females make a much greater commitment—in terms of both their time and the proportion of their lifetime supply of eggs—is superficially attractive and at some levels certainly correct.[194] Males of nonhuman animals do, for example, try to distribute their genes widely. In a no doubt apocryphal story, the taciturn Calvin Coolidge and his wife were touring a government farm. Mrs. Coolidge was in the lead and saw a rooster busily copulating with a hen. "How many times a day does he do that?" she asked. "Dozens of times," the farmer replied. "Tell that to the president," she said. When the story was relayed to Coolidge, he queried: "Same hen each time?" When told it was a different one each time, Coolidge simply replied, "Tell that to Mrs. Coolidge."[195] That the presence of new partners stimulates copulation in male nonhuman animals is widely documented in the behavioral literature, where it has become known as the "Coolidge effect."[196]

Men certainly are capable of producing many more children if they spread their gametes around than if they are faithful to a single mate. Mulay Ismail the Bloodthirsty, a former emperor of Morocco (1672–1727), is reputed to have fathered almost 900 children—on the order of twenty-five times the number that could be expected of the most fecund woman.[197] Mulay may even beat out some notorious modern-day rock stars and sports heroes in this department.[198] But does that translate into a hereditary tendency for the Coolidge effect in men? Or, to put it slightly differently, do human males seek sexual experiences with different women more avidly than women seek a variety of men? After all, there very well could be a parallel "Mae West effect," an urge among women for sexual variety.[199]

It appears that men indeed seek more variety than women do. One piece of evidence is that observed patterns of promiscuity differ between homosexual men and homosexual women. Where "male attitudes" are found on *both* sides of such relationships, levels of promiscuity become very high. One man, for example, was reported as having been willingly sodomized by forty-eight others in one evening, and one study revealed that more than three-quarters of gay men had more than thirty male partners, but none of the heterosexual "controls" were that promiscuous with women.[200] In contrast to gay men, lesbians much more often form long-term stable relationships.

Men in Western society, at least, seem more ready than women to seek out and accept casual sexual encounters.[201] Is it "in their genes"? That's not at all clear. There is plenty of evidence that women enjoy sexual variety as well,[202] and the vast majority of societies view women as possessing sexual interests fully as strong as men's.[203] Women are the physically weaker sex and are burdened with pregnancy and breast-feeding, so they are in less of a position to maintain control over their own and their partners' activities, however. How would women behave sexually if they were free from male domination, social constraints, and worries about pregnancy? If women are the permanently sexually receptive but passive creatures they long seemed in the minds of male biologists and anthropologists, why would many African societies use female genital mutilation seemingly to control women's sexuality?[204] Why were women's activities and social interests so rigidly restricted in Europe and the United States during the nineteenth century, as well as in many other societies at many other times? Are men genetically programmed to be intolerant of *any* sign of female sexual wandering?

For a long time, male anthropologists and biologists focused on the roles of males in producing hierarchically organized primate societies.[205] As biologist Michael Ghiglieri put it, "It is my contention, and that of other workers,[206] that the reproductive strategies of *males* as shaped by sexual selection to maximize opportunities to sire offspring are the ultimate causes of the social structures of the African hominoids and perhaps orangutans as well."[207] In the background of this conclusion is the obvious factor that it is physically possible for reproductive "success" in males to show greater variance than that in females (remember Mulay and the rock stars). But this view ignores the importance of female mate choice in the evolution and organization of primate societies, especially human societies, in which it often appears to be important.[208]

Some of the most interesting evidence of a major role for female mate choice in hominid evolution is related to the discovery that human females are able to retain sperm differentially from different copulations. Robin Baker and Mark Bellis had the audacity to go where other researchers feared to tread. They enlisted courageous couples to carefully preserve in chemical fixatives condom tips containing ejaculates. The researchers had the volunteers make careful records of the source (e.g., masturbation or copulation) and detailed records of sexual activities in periods preceding the particular climax. They even persuaded a series of women to catch "flowbacks" of seminal fluid.[209] The sample sizes were smaller than might have been hoped, and the participating couples clearly did not constitute a random sample of all couples. Nonetheless, Baker and Bellis gathered some of the most provocative data in the history of research into human sexual behavior.

What did Baker and Bellis discover? There have been three basic theories about the evolutionary significance of female orgasm. One is that it is vestigial, with no adaptive value—the physiological equivalent of nipples in males.[210] A second is the "pole-axe" hypothesis—that orgasm induces lethargy and thus keeps the female prone after intercourse, reducing loss of ejaculate.[211] The third is the "upsuck" hypothesis, which suggests that female orgasm serves to draw seminal fluid deeper into the vagina during intercourse to facilitate conception.[212] The vestigial hypothesis suffers from the impressive nature of female orgasm—a much more spectacular phenomenon than the functionless and relatively insensitive male nipples. Vestigial? Maybe, but I can't see it. At the very least, as a source of great pleasure, it should encourage females to reproduce. The pole-axe hypothesis has always seemed weak in view of the evidence that female quadrupedal mammals can experience orgasm. The experimental work by Baker and Bellis supports the upsuck hypothesis, with more seminal fluid being retained (about 70 percent of sperm) when the female climaxed.[213]

Furthermore, there appears to be a series of physiological mechanisms by which copulatory female orgasms that take place more than one minute before the male ejaculates, copulations without female orgasm, and noncopulatory female orgasms all tend to reduce the uptake of seminal fluid at the next copulation. Thus, by climaxing early, not climaxing during intercourse, or masturbating, females might influence the chances of conception at later copulations. Counterintuitively, by such means, women may *enhance* their chances of pregnancy by limiting their uptake of seminal fluid because too many sperm reaching the egg can prevent the development of a viable embryo.[214] Males also are claimed to adjust the volume of their ejaculate in response to the need to limit the number of sperm attacking an egg, although if this actually happens, the mechanism of adjustment is unknown.[215]

Thus, in monogamous situations, both males and females may act to limit the movement of seminal fluid into the uterus.[216] When there is promiscuity, the issue of sperm competition makes the situation more complex.[217] Women who are copulating with more than one male tend to change their behavior so as to reduce retention of their more closely bonded partner's sperm and increase that from their other partner. This manipulation is accomplished, almost certainly unconsciously, by changing the frequency of orgasms from covert masturbation and by varying the timing of copulatory orgasms. A survey of 3,679 British women showed that those having affairs also unintentionally timed their extra-pair copulations to improve chances of being impregnated by their lovers.[218] Interestingly, contraception was used less frequently during extra-pair copulations than during within-pair copulations.

One might claim that this is a genetic spread-the-risk strategy that results in different offspring being fathered by different men; a woman could thus insure against having chosen a genetic loser as a husband. It could also be interpreted as straightforward, genetically influenced encouragement of sperm competition. It wasn't just any extra-pair copulations that the women apparently timed for pregnancy; those copulations were ones that occurred *within five days* of the last within-pair copulation.[219] Females fertilized by the sperm that win such a competition presumably would increase the fitness of their male offspring were it to turn out that sperm competitiveness is a heritable trait. If further studies bear out these results, which depend on notoriously undependable survey research,[220] they amount to an argument for the female-driven sperm competition hypothesis. But one could also build a psychological explanation for these data in which within-pair copulations simply make women long for their extra-pair partners (a female equivalent of the Coolidge effect?) and the excitement of the affair makes the couples less careful with contraception.

Baker and Bellis, as mentioned earlier, also have produced data suggesting that males have strategies in which, by some unknown mechanism, they unconsciously adjust the quantity of ejaculate they produce. Less is produced if there is no sign of the possibility of sperm competition, more if the presence of rivals is suspected.[221] Apparently the better chances of successful fertilization with fewer sperm can be counterbalanced by swamping out of another male's contribution. Furthermore, in Baker and Bellis's study, when the time between within-pair copulations exceeded four days, masturbation intervened until its frequency reached 100 percent at ten days, with the amount of sperm in the ejaculate lower at first but increasing with time since last orgasm.[222] Despite smaller sperm numbers, females tended to retain more sperm if the male had masturbated shortly before intercourse, probably because of the ability of younger sperm to last longer in the female's reproductive tract. Males seemed to be attempting to optimize the chances of fertilization, not the number of sperm present, which may be negatively correlated with probability of conception. Thus, pronatalist religions (those encouraging high birthrates) that condemn masturbation may be adopting a taboo that runs counter to their goals—and one that may be the least effective of all known taboos. Overall, for an optimal chance of conception, males seem to adjust the number of sperm in their ejaculate by balancing the different requirements for the absence and presence of competition: fewer sperm in its absence and more sperm in its presence.[223] All of this, at the very least, would seem to indicate that the human mating system did not evolve genetically for strict monogamy.

Although all of these adjustments in the behavior and physiology of both sexes appear to be triggered by social situations, it does not seem unreasonable to assume that they have their basis largely in biological rather than cultural evolution. The scientific effort needed now is to check Baker and Bellis's results with different samples. Such spectacular results from one team of investigators cry out for replication by others.

The nearly universal human trait of sexual jealousy being more violent in males than in females[224] may be an evolutionary response related to the absence of strict monogamy and consequent uncertainty about paternity, which has no female equivalent of uncertainty about maternity. A cultural response to an evolutionary concern about who is carrying copies of whose genes apparently explains the ubiquity of kinship systems in human societies. The enduring importance of those systems is suggested by the persistence of nepotism and the development of pseudo-kinship systems (including "old boy" networks and "sisterhoods" of Jewish women) in societies too large for genuine kinship to be an organizing principle.

Although male reproductive strategies are important in the organization of human societies, we've seen that those strategies must necessarily take into account the female strategies (and vice versa). Both must take into account the possible penchant for the other to seek extra-pair copulations. So where do we stand on the two key questions related to this: are women more choosy than men, and if they are, is this largely a biological difference? The answer to both is that we don't yet know. The *average* number of partners of each sex must, obviously, be equal. Maybe men are programmed for promiscuity and women for restraint, and a relatively small corps of prostitutes or otherwise promiscuous women is forced by circumstances in male-dominated societies to take up the slack caused by the reluctance of the rest of the female population. Or maybe men are programmed to be more promiscuous than women but many lack the opportunity—frustrated by women's restraint. Or maybe both sexes are equally choosy (or promiscuous), but men overstate and women understate their number of partners.

If a male–female difference exists in the desire to have multiple mates, it does not necessarily need to have a genetic basis. Even if women enjoy sexual variety as much as or more than men, for thousands of years they have not needed to be rocket scientists to understand that they make a greater potential commitment per copulation than do men. The restraint could simply be learned. The difference in behavior among homosexual men and women (which constitutes some of the strongest evidence indicating a difference grounded partly in genes) could indicate the power of cultural biases about the behavior of the sexes, which may carry male heterosexual attitudes into the behavior of gay men and female heterosexual attitudes into the behavior of lesbians. Support for a cultural explanation of male–female differences was gained during the sexual revolution of the 1960s, in which many women seemed to participate as eagerly as men.[225] The contraceptive pill reduced the male–female difference in commitment even as new societal attitudes reduced cultural constraints. Yet gender differences seem to persist in expressed desires, if not behavior, despite changed attitudes and availability of the pill and abortion.[226]

It seems almost certain that (as in most aspects of human behavior) genes, environment, and gene–environment interactions all play roles in generating gender differences in sexual attitudes. If women really are predisposed to seek sexual variety as much as men, and if they quickly shed ingrained cultural attitudes when contraception improved and the culture changed, it seems a little precious to assume that cultural attitudes could not also be rapidly shed by gay men and lesbians. If women, otherwise unconstrained, prefer variety, lesbian behavior should have converged on that of promiscuous males. There probably is a somewhat greater genetically influenced tendency for women to focus on the quality of relationships and not to seek casual sexual encounters with multiple partners as frequently as do men. But like other human behavior with a genetic component, cultural conditions can clearly overcome that tendency.

The Evolution of Homosexuality

Perhaps the most controversial sexual behavior in the United States today is homosexual behavior. The notion that homosexuality is in some sense "unnatural" and its condemnation for that reason seem a little strange in modern Western societies. After all, people in those societies have discovered many different ways to separate sexual pleasure from reproduction, something that homosexual activity does quite effectively. Homosexual behavior is present in most, and possibly all, human societies. In Western society today, the frequency of some sort of homosexual behavior may be as high as 10 percent, although that of more or less exclusive homosexuality is probably about 2–4 percent, and possibly higher in men than in women.[227] Not only that, but when thinking about its evolution, one should remember that homosexuality plays a variety of roles,[228] and it is widely distributed in the animal kingdom.[229] In that sense, it is certainly "natural."

At first glance, the occurrence of homosexual behavior appears to be counter-evolutionary. After all, if the name of the game is to outreproduce your competitors, homosexuality would seem to be a poor strategy. But that view ignores both what is known about the roles of such behavior in nonhuman animals and the complexity of the human brain and behavior and the sociocultural environments in which human beings operate.

Homosexual behavior is seen in animals as far removed from *Homo sapiens* as seagulls and elephants. When there are too few males in a seagull colony, occasionally two females will bond and jointly raise the young of one. In seagulls, as with many other birds, two cooperating individuals are necessary to successfully fledge young. Human beings also, as is obvious, often form same-sex social bonds between nonrelatives that are unconnected to sex or reproduction—as exemplified by "bands of brothers" and sorority "sisters."

Same-sex relationships are also found in our nearest relatives. Male–male bonding seems to be most intense in chimpanzees, female–female bonding in bonobos.[230] Chimpanzees use a series of nonsexual, rather stereotyped greeting behaviors—vocalizations such as "pant-grunting," kissing, touching, and embracing—to avoid male–male conflict.[231] Males sometimes finger one another's scrotums to resolve tensions, a behavior that also indicates friendly intentions in some human groups. In certain New Guinea tribes, a gentle lifting of the sensitive testicles is a standard male–male greeting without apparent sexual content.[232] Bonobos, on the other hand, use male–male and female–female genital-to-genital contact, in addition to heterosexual intercourse, to resolve tensions.[233]

Male homosexual practices have become institutionalized in some human societies.[234] Thus, it could simply be that homosexual sex can readily become a substitute for heterosexual sex if conditions in the uterus or the social environment provide the proper triggers. The search for social triggers that produce or facilitate homosexual behavior in societies that generally disapprove of homosexuality, however, has not turned up convincing results. Overly protective mothers, contrary to legend, don't seem to "create" homosexual sons.[235]

Part of the solution to the mystery of the widespread occurrence of homosexuality in both human and nonhuman animals may lie in delicate hormonal balances in developing embryos that influence the gender of an individual. Remember, the physical sexual apparatus of an individual depends on genes transmitted via the father's sperm. The genes determining the type of apparatus are carried on a pair of sex chromosomes that come in two varieties, X and Y. There are two types of sperm, one type containing an X and the other a Y chromosome, and one kind of egg, with an X chromosome. If an egg is fertilized by an X-containing sperm, the child is genetically female (XX); if by a Y-containing sperm, it is male (XY).[236] But this genetic "switch" controlling external sexual characteristics, such as whether certain tissues develop into a penis or a clitoris, is only part of the gender story. There appear also to be male–female size differences in at least one structure of the brain necessary for sexual behavior.[237] But because of environmental factors (such as prenatal stress that might change the timing of a pulse of hormones),[238] there is in a sense a range of "genders" in human beings—not a series of sharp male-gay-lesbian-female boundaries.[239] And there is abundant evidence from animal experiments that hormonal shifts early in development, causing changes in the cellular environment of the developing embryo, can have profound effects on gender-related behavior.[240] It has been possible to show in rhesus monkeys, for example, that gender-related behavior can be influenced by hormone levels during several overlapping sensitive periods.[241]

I suspect that some human gender-related behavior is controlled in much the same way. For example, transsexual individuals—those who from childhood are concerned about whether they were born into the "right" sex—may have different brain structures from those of individuals who are not transsexual. A study of six XY (physically male) transsexual individuals showed that at least in this small sample, sex-related brain structures were the same size they normally are in women.[242] Presumably, the mistiming of hormonal pulses in development can alter brain development and result in gender-related behavior not coordinated with the X-Y sex-determining mechanism.

Gender-related behavior, once entrained by such hormonal pulses, can be very difficult to alter. In one famous case, an eight-month-old baby boy, John, suffered a "circumcision accident" in 1967. An electrocautery needle somehow malfunctioned and totally destroyed his penis. A then well known physician and "authority on gender disorders" had concluded that an individual's gender was determined entirely by the way he or she was raised—that the genetic determination of sex could be overridden by culture and surgery.[243] The case of baby John shows how blind environmental determinism can lead to results as horrifying as eugenic interventions. After substantial debate, it was decided to remove the baby's testicles and, later, to surgically fashion a vagina and inject the child with female hormones, converting John to Joan. That was done and the case became a classic success story, written up in textbooks[244] and in *Time* magazine. In part on the basis of this case, which strengthened some physicians' views of gender lability (changeability), gender assignment surgery today is often done when a child is born with genitalia that are damaged or ambiguous.[245] Surgical conversion of genetic males into genital females is more frequent than conversion of genetic females into genital males because it remains very difficult for plastic surgeons to create a satisfactory penis.[246]

But a follow-up examination of the John/Joan case revealed a very different story. Even as a toddler, John (we'll stick with the male pseudonym, even though the family switched to a female one) rejected femininity. He more frequently mimicked his father than his mother. Once, when his mother was putting on makeup and his father was shaving, John put on shaving cream and pretended to shave himself. When he was told to mimic his mother, he responded: "No, I don't want no makeup—I want to shave." He persisted in rejecting girls' clothes and toys and avoided girls' activities even though they were repeatedly offered to him. "Joan" made the mental transition back to being John when he was between nine and eleven years of age and started trying to urinate standing up despite the lack of a penis. At age twelve, when put on an estrogen regime, he rebelled. He frequently did not take his daily dose because he did not want to become a woman, and he adopted a gender-neutral wardrobe of jeans and shirts. His responses on psychological tests were more typical of boys than of girls. When he was fourteen and developing breasts, he told his endocrinologist that he had suspected he was a boy since the second grade.

That led to his being told of the accident and a decision for him to return physically to being John. He underwent a mastectomy and penis reconstruction in his fifteenth and sixteenth years and became a relatively popular boy. He felt comfortable in his body, which now matched his attitudes. When he was twenty-five, he married an older woman and adopted her children. He can now have coital orgasm. When asked as an adult why, as a child, he had not believed he was female, as his genitalia had then indicated, he said it was because "it did not feel right."

The John/Joan case clearly supports the notion that many male–female differences in sexual behavior are physiologically determined by pulses of hormones during prenatal development and are not simply under the control of hormonal states or social learning later in life. Those pulses may vary in their timing as a result of genetic or environmental factors, and that variation in timing may account for the rather wide range of gender attitudes found in human populations. What is now understood about the physiological basis of gender-related behavior makes it seem unlikely that people can simply "choose" to change their sexual orientation (the idea, for example, that homosexual individuals can be made heterosexual by persuasion seems to be simply an error).[247] Understanding of this biological result would seem to have policy relevance because people seem generally more tolerant of homosexuality when they believe that it is an accident of birth.[248]

The penis-accident anecdote unhappily doesn't tell us much about the ultimate reason why an individual becomes either heterosexual or homosexual or somewhere in between. There has been much discussion of a possible genetic basis for some homosexual behavior, but as might be expected for a characteristic so commonplace and difficult to define, results remain ambiguous.[249] In any case, recent claims of discovery of a "gene for homosexuality,"[250] weak as they are,[251] are almost beside the point.

If there were a simple genetic hetero–homo switch for sexual proclivity, then we would have to look even harder for the reason why the "homosexuality gene" is so favored despite its putative direct negative effect on fitness. I say *putative* because there are no reliable data on the actual reproductive performance of gay men. Many do marry and have children, and many may be very sexually active with both sexes. Selection against the allele (variant of the gene) predisposing to homosexual behavior might thus be very weak.[252]

On the other hand, if homosexual and heterosexual individuals do not differ genetically, then we still face the question of why closely related men or women living in the same society can have such different objects of sexual attraction—one sister attracted to men and another to women, for example. And if there is no significant hereditary component, why is homosexual behavior such a common result of cultural evolution? In addition, why is homosexuality sometimes integrated into the fabric of societies and considered normal, as it is said to have been in ancient China and in ancient and medieval Persia? The issue is clouded by the difficulties of analyzing available records and determining how widespread various practices actually were. After all, it's difficult enough to determine patterns of sexual behavior today. For instance, it is often claimed that homosexuality was considered normal in ancient Greece, yet several Greek writers declared it contrary to nature, and Aristophanes and Aristotle brutally condemned it. It appears actually to have been mostly an exercise of power frequently associated with warriors taking advantage of younger or lower-class males—an expression of what was considered a natural male sexual frenzy and thus condoned.[253]

In contrast, male homosexuality in various Melanesian societies is ritualized and compulsory. The Etoro of New Guinea have strong constraints on heterosexual intercourse because of the belief that a man dies when his store of semen is exhausted. That supply, according to local belief, is obtained in childhood by oral intercourse with adult males. The Etoro claim that "semen does not occur naturally in boys and must be 'planted' in them." In the process of nightly inseminations, all the properties of maleness are thought to be passed on.[254] Considering the great range of behavior lumped under the rubric of "homosexual," it is easy to doubt that there is one environmental "trigger" of homosexual behavior.[255]

Fertility Control and Incest

One possible reason why homosexuality apparently has evolved in some societies is fertility control. In a cross-cultural study of thirty-nine nonindustrial societies, attitudes toward homosexuality were related to perceived demographic needs. Seventy-five percent of societies classified as encouraging higher birthrates punished or disapproved of homosexual behavior. In societies classified as favoring the limiting of births, 60 percent encouraged or approved of it.[256] These data also show that attitudes toward homosexual behavior are modified by the social milieu; there is no sign of a universal horror of behavior thought by some to be "unnatural."

Fertility control, of course, is not just an invention of Margaret Sanger and the planned parenthood movement. Preindustrial societies, including hunter-gatherers, had many devices available to them for limiting reproduction, and they put them to use widely.[257] A study of the history of contraception brings home that often neglected point: human beings are unique in deliberately not maximizing their reproductive contribution to the next generation. Mostly, however, societies have lowered birthrates not by suppressing the urge to engage in sexual behavior, which clearly is under strong genetic control. Rather, they have done it by altering the consequences of that behavior, through abortion, infanticide, and contraception; by constraining it with penalties for various kinds of activities, such as extramarital and premarital intercourse or incest; through approval of homosexuality; or through pressure to raise the legal age for marriage. Constraints on marriage age can be an effective population-control mechanism in societies that have strong and enforced taboos against unwed motherhood.[258] Simply recommending abstinence has usually been a particularly unsuccessful method of preventing births. Curbing sexual appetites by fiat has always been difficult as well, as attested by the behavior of everyone from members of supposedly chaste religious orders to holders of high political office.

The single most discussed issue in the literature on evolution of human sexual behavior is incest. Sigmund Freud, for instance, saw incest taboos as a first step in civilization's campaign to restrict sexual life. It was an economic necessity, "since a large amount of the psychical energy which [civilization] uses for its own purposes has to be withdrawn from sexuality."[259] Although most of Freud's explanations of human behavior are of little utility today,[260] scientists still haven't agreed on an explanation of how the condemnation of incest evolved. We don't even know whether the taboo is a universal feature of human nature, and we don't know whether it is connected with the genetic problems of mating with close relatives, as is often claimed. Brother–sister incest is discouraged in most human societies, but surprisingly, it has been encouraged in some societies, notably among Egyptian pharaohs,[261] the royalty of ancient Persia, and the Incas.[262] In some areas in Roman Egypt, about one-sixth of all marriages were between brothers and sisters, and such matings may have been frequent beyond the Persian royalty of yore. The reasons for the evolution of incest approval seem to vary and include such factors as a shortage of acceptable mates, efforts to keep the "blood line" pure, and attempts to keep property in the family.[263]

The degree of inbreeding that is considered unacceptable varies from society to society and in different circumstances, but the Egyptian, Persian, and Incan attitude was (and is) unusual, and even in those cases it was probably restricted to a very small segment of society. Marriages between brother and sister, as well as between parent and child, have been rare in historical societies.

Nonetheless, it is not clear why incest, especially within nuclear families, most frequently is strongly disapproved of. The actual occurrence of incestuous behavior in modern societies is not known and may be relatively high.[264] One view is that it is only since the agricultural revolution that incest has been problematic in human societies. In hunter-gatherer groups, with wide spacing of births and with marriage and assumption of adult roles at puberty, unattached sexually mature brothers and sisters would rarely coexist, and mature offspring would not live in the parents' nuclear group. After the establishment of agriculture, as cultural evolution progressed, children began to become sexually mature before they were socially mature. By the time that began to happen, incest had been rare for so long that societies often reacted against it—it was a breach of custom.[265]

Another classic theory to explain the avoidance of incest was that of anthropologist Edward Westermarck, who argued that there is an inborn aversion to sexual contact between boys and girls who are raised together from an early age.[266] The counterargument was that incest is discouraged culturally because it disrupts family structure and is therefore bad for society: "the law only forbids men to do what their instincts incline them to do."[267] That argument claimed that the "primitive" desire to copulate with other members of the nuclear family had to be suppressed if more complex forms of society were to emerge.[268] It contended that if there were a genetic disinclination to mate with siblings, there would be no need for incest taboos.

On Westermarck's side is a demonstration by Stanford University anthropologist Arthur Wolf of what happens in "minor marriages" in Taiwan. In these marriages, the bride is taken into the future husband's household in infancy or as a young child and eventually marries the boy with whom she has been raised. Wolf notes that these couples' behavior differs from that of couples in "major marriages," in which the woman enters her future spouse's home as a young adult and often does not meet him until the day before the marriage: "Women who are forced to marry a childhood associate bear fewer children than those who marry a stranger. They are also more likely to leave their husband by divorce or avoid him in favor of other men."[269] The incestuous-seeming "minor" marriages are normally consummated on the eve of the lunar New Year. The reaction of young people when told they are no longer brother and sister but groom and bride is so aversive that one elderly man told Wolf "that he had to stand outside their room with a stick to keep the newlyweds from running away."[270] Wolf found that even the traditional jealousy of Chinese men did not carry over into these marriages. "In one case the husband and his mistress and the wife and her lover live next door to one another in the same compound without any apparent friction."[271] A similar pattern of disinterest has been reported among unrelated children raised in the extreme intimacy of the communal facilities of Israeli kibbutzes—including shared bathrooms and nudity into the teen years.[272] A notable lack of marriage has been reported among graduates from the same classes of such rearing programs.[273]

Anthropological reviews, however, have claimed that neither of these studies is definitive. The minor marriages are viewed by the Taiwanese as second-class marriages that could even be humiliating—so, as anthropologist Marvin Harris put it, it is "impossible to prove that sexual disinterest rather than chagrin and disappointment over being treated like second class citizens is the source of the couple's infertility."[274] In the kibbutz case, the anthropological counterclaim is that the number of marriages between alumni of the same kibbutz was actually higher than expected.[275] This is one of the reasons why anthropologists seldom resort to genetic explanations of incest taboos, even though there is widespread avoidance of incestuous mating among other animals.[276]

In all this confusion, one thing that seems reasonably certain is that incest avoidance did not evolve genetically in human beings because incest disrupted family structure and was thus unhealthy for society. Such an explanation is probably wrong because it invokes group selection, in this case the differential reproduction of societies rather than individuals. The special conditions that could permit group selection to be a powerful force seem unlikely to have been present often among populations of our ancestors. Genetically, then, the occurrence of incest should be influenced by a balance between two factors.[277] One factor is the physiological penalties that most animal species pay for inbreeding, which tends to make individuals genetically uniform (technically, to have identical alleles, forms of a gene, at each locus—that is, to be highly homozygous). When that occurs, deleterious alleles that are recessive—that do not change the phenotype when accompanied by a different, dominant allele (technically, when an individual is heterozygous at a locus)—are expressed and have a debilitating influence on the phenotype.[278] That is the case, for example, when a child receives two recessive alleles at a cystic fibrosis locus, producing a phenotype with that tragic disease.[279] The result is called inbreeding depression because highly homozygous individuals often have reduced viability, or reproductive capacity.

The other factor influencing the likelihood of incest is that incest may increase the presence of the genes of incestuous individuals in the next generation. If one mates with one's siblings (or other close relatives) instead of unrelated individuals and there is little inbreeding depression, the result will be an increase in inclusive fitness (the passing on of genes identical to one's own) as long as those relatives would not otherwise reproduce. Relatives share some identical genes, and unless there is a countervailing level of inbreeding depression, more copies of one's genes can be sent on to the next generation by mating with an otherwise nonreproducing relative than by mating with a stranger.[280] In the absence of the constraint of inbreeding depression, selection should generally favor incest concentrated within families. All of this, of course, begs the question of whether there actually is a significant genetic influence on whether or not one mates with relatives.

In human beings, inbreeding depression occurs in father–daughter matings and brother–sister matings[281] and, to a much smaller degree, in marriages between cousins (which are common in various Asian and African societies).[282]

Marriages among relatives increase the rate of genetic disorders, and if mortality rates from childhood infectious diseases continue to drop in the relevant developing countries, families will have more mature individuals of opposite sexes and the opportunities for marriage among relatives will increase. That, in turn, could cause a shift in the disease profile of those countries, with a decline in infectious diseases and an increase in those related to inbreeding, necessitating cultural evolution in the medical care delivery system.[283] Inbreeding depression may have been a fairly substantial problem in the relatively small groups of our early ancestors. But there is not enough information about it, about the costs of exogamy (mortality or reduced fecundity involved in migration to another group), and so on, to know exactly what balances were struck.[284] One question is whether incest taboos developed because our ancestors observed the deleterious effects of incest, because inbreeding depression (in the balance discussed earlier) selected for a genetic influence on mating behavior that caused people to avoid choosing close relatives as partners (as it has in other organisms), or both. Evolutionist John Maynard Smith suggested that "when symbolic communication, myth making and cultural transmission became the dominant modes of communication, [our ancestors'] earlier 'incest barriers' changed into 'incest taboos.'"[285]

Something like this may indeed have happened. A mathematical model developed by population geneticists Kenichi Aoki and Marcus Feldman examined what would happen if the mating pattern of parents affected that of their offspring—if, for example, the probability of a girl wanting to mate with her brother were a function of whether her parents were brother and sister. Aoki and Feldman's gene–culture coevolutionary model shows that if there is sufficient inbreeding depression, a purely *cultural* determination of mate preference may evolve that prevents brother–sister mating.[286] In any case, the existence of incest taboos (like the existence of kinship systems) shows that ideas of morality can be related to issues of genetic evolution.

We have seen repeatedly how complex some basic questions about behavioral evolution can be, even when the behavior has direct, obvious connections to fitness. A sociobiological approach, one that considers fitness costs and benefits, puts some constraints on the kinds of tales that can be woven. But it is still extremely difficult to develop wholly satisfactory answers because of the often overriding influences of cultural evolution, which is poorly understood. Reducing human problems to questions of gene frequency rarely provides a definitive solution for them. At the same time, over the past few decades we have clearly made progress toward understanding many aspects of the extremely complex natures of human beings, and the evolution of our sexual behavior is no exception. But hunter-gatherers did a lot more than eat and have sex, and as we will see, unraveling the evolution of phenomena such as violence and religious belief is no less challenging.

Chapter 9

THE DOMINANCE OF CULTURE

"Woman seems to differ from man in mental disposition, chiefly in her greater tenderness and less selfishness. . . . Owing to her maternal instincts [she] displays these qualities towards her infants in an eminent degree; therefore it is likely that she would often extend them towards her fellow-creatures."

—Charles Darwin[1]

"When you're wounded and left on Afghanistan's plains
And the women come out to cut up what remains,
Jest roll to your rifle and blow out your brains,
An' go to your Gawd like a soldier."

—Rudyard Kipling[2]

"One of the main functions of myths has always been to account for the bewildering and meaningless situation in which man finds himself in the universe."

—François Jacob[3]

We can be sure that our genes are involved in the human desires to eat and have sex: without those appetites, *Homo sapiens* would quickly disappear. Cultural evolution can certainly have substantial effects on the ways in which those appetites are expressed, but they are thoroughly grounded in biological evolution. Bacteria, plants, and butterflies take in nourishment and have sex; indeed, all living things do the first of these, and almost all do the second.

But when we start thinking about aspects of human natures such as violence, religion, and art, we're creeping over a subtle divide. Bacteria, plants, and butterflies don't make war, pray to gods, or compose operas. Only people do those things (although various other animals, from ants to chimpanzees, are sometimes accused of war-like behavior). What is certain, though, is that our discussion of human natures is now entering a realm in which the role of cultural evolution, because of the shortage of both genes and time for natural selection to act, clearly begins to overpower that of genetic evolution. There is no certain biological imperative for murdering, singing hymns, or painting pictures. Women can be tender or can dismember a living, wounded soldier; people can create and believe in stories that help them make sense of the world. And differences in people's genes have little or no effect on levels of empathy or the content of myths. When it comes to the intertwined behaviors associated with violence, religion, and aesthetics, cultural evolution thus moves toward center stage, and in these areas today, invoking our genetic heritage as the cause of various behaviors can be especially problematic.

Chimpanzee Politics

Evolutionist Richard Lewontin, one of the most determined critics of the popular literature on sociobiology, pointed out that to many who are exposed to this literature, "'man,' and the male of the species, in particular, is a violent and self-centred gene maximizer, the product of a deep biological core inherited from his ape-like ancestors."[4] Does this impression reflect a reasonable characterization of an important part of our natures? We share genes but not cultures with chimpanzees, so it makes sense to examine closely any of their complex behaviors that seem similar to those that crop up in human societies.

Chimps don't paint, sculpt, sing, or have rites of passage, but in chimp societies violence is common. Ethologist Frans de Waal vividly portrayed violent chimpanzee politics in a breeding colony of twenty-three individuals living in Arnhem Zoo in the Netherlands.[5] The zoo has a large, pleasant park-like enclosure where the chimps run loose in the daytime. A building with indoor cages features a second-story observation balcony with a good view of the outdoor interactions. In the chimp colony de Waal studied, individuals knew their place in society, which was structured as a dominance hierarchy established on the basis of winning or losing in aggressive interactions.[6] De Waal documented formalized dominance relationships in both sexes, though the patterns were less hierarchical and more stable in females than in males. Influence was not perfectly correlated with dominance, however. Even though males were rivals, they formed strong male–male bonds, and members of both sexes joined complex coalitions and mediated disputes. De Waal found that the Arnhem chimps were good at using other colony members as social instruments, manipulating their behaviors to benefit themselves. He also suspected that chimpanzees scheme beforehand to improve their dominance status. And he found that dominant males generally got more opportunities to copulate than did submissive ones, which points to one possible selective advantage of dominance, at least in the zoo environment.

Clearly, the zoo population was not living in typical chimpanzee surroundings. The Arnhem chimps may have had more time for politics than do many wild groups because they spent less time acquiring food and did not have to keep an eye out for predators or raiders from other groups. In the zoo, sneaking around sexually may be more difficult than it is in the wild. Nonetheless, many aspects of their behavior proved to be similar to those recorded in wild chimpanzees.[7]

One famous political sequence at Arnhem Zoo involved interactions among the dominant adult male, Nikkie, his older ally Yeroen, and Luit, pretender to the throne. Together, the wily Yeroen and strong Nikkie could defeat Luit, but individually Luit could beat either one. Whenever there was a falling-out between Nikkie and Yeroen, Luit would make his move, carrying out impressive stone- and branch-hurling displays of a sort that I was privileged to observe years ago at Gombe Stream Reserve.[8] Only a reunion of the Nikkie–Yeroen team to vanquish Luit would restore peace. Nikkie was a weak leader; not only could he be beaten in a fight by Luit alone, he also couldn't prevent Yeroen from being more successful with the female chimps. Eventually, Nikkie and Luit tired of being manipulated by Yeroen, made common cause against him, and subsequently ruled the roost.

As de Waal put it, "The balance in the male triangle, which had first been determined by Yeroen from below, was now determined by Nikkie from above."[9] Nikkie strove to prevent Luit and Yeroen from having contact, and, de Waal surmised, Yeroen became increasingly frustrated. He had supported Nikkie and helped him maintain dominance over Luit, but now Nikkie did not attempt to limit Luit's sexual successes, to Yeroen's cost. After some three-way squabbles over attempts by Yeroen and Luit to mate with a female, Krom, there was an unobserved nighttime fight in which both Nikkie and Yeroen were fairly seriously injured and Luit was barely scratched. Presumably Nikkie and Yeroen had had a violent falling-out and Luit had immediately become the dominant male.

The zookeepers isolated the three males from the rest of the group for a week, keeping them in separate cages at night. Then they reintroduced them to the group but kept them separated at night until that became difficult: Yeroen always tried to get into Nikkie's cage, and neither Yeroen nor Luit wanted the other to spend the night with Nikkie. Male chimps are powerful, and separating them became a real problem for the keepers. Eventually, the males were allowed to make their own nighttime arrangements. Nikkie was extremely submissive in Luit's presence, whereas Yeroen was much less so. Despite some signs of continuing tension among the males, harmony appeared to return to the colony as a whole. A couple of months after the Nikkie–Yeroen brawl, however, the two erstwhile enemies ganged up on Luit at night and injured him badly. The keepers found that toes had been bitten off both of Luit's feet, he had gashes over much of his body, and both testicles were missing. He died the next day.[10]

It seems clear that Yeroen and Nikkie simply put an end to the trouble Luit caused by putting an end to Luit as a threat. Whether they "intended" to kill him cannot, of course, be determined, but the obviously brutal and sustained nature of the attack makes one suspicious.[11]

Parallels to the violence that engulfed Nikkie, Yeroen, and Luit in the Arnhem colony can be seen in many, if not most, human societies. Indeed, disputes occur frequently in contemporary human hunter-gatherer societies, and, especially when men compete for women, disputes escalate into violence. Among the Netsilik Inuit in the middle of the twentieth century, a desire to steal another man's mate was the most frequent reason for murder or attempted murder.[12] Before the Canadian government suppressed the practice, murder of traveling individuals by local Inuit peoples often led to revenge raids by relatives of the slain, and even after suppression, murder rates remained much higher than in the United States. In one Copper Eskimo group of fifteen families, every adult male had been involved in a homicide.[13] From 1920 to 1955, murder rates among the !Kung Bushmen of southern Africa were twenty to eighty times higher than recent murder rates in industrial countries, and the Yahgan "canoe nomads" of Tierra del Fuego had a murder rate ten times that of the United States in the late twentieth century.[14] There is little reason to doubt that such levels of violence also occurred widely in prehistoric hunter-gatherer societies.

Chimp Violence, Human Violence

Violence seems to be more common among males than among females, as Harvard University anthropologist Richard Wrangham and many others have noted. The common roots of such behavior and the patriarchal structure of societies seem to be detectable in chimp behavior as well as our own.[15] After all, it wasn't two teenage girls who slaughtered the students at Columbine High School.[16] The story of Nikkie, Yeroen, and Luit gives, I think, some sense of the intricacy of relationships that can occur within the societies of our close living relatives.[17] Dominance interactions like those just described have been observed many times in wild chimpanzee communities, though no others are known to have resulted in the loser's demise—although adult males have been missing and presumed dead from undetermined causes.[18]

Although bonobos are not as well studied as chimpanzees, we do know that there are also many aggressive interactions among males of that species. Individuals hit, bite, slap, shove, grab, and otherwise abuse one another and use a variety of bluffing and charging displays.[19] Aggression by a dominant male is countered with various submissive reactions by lower-ranking males, including appeasing the aggressor by permitting him to mount. There is ample sexual competition among males, and dominant individuals appear to have more opportunities to mate. Even so, lower-ranking males can be quite successful,[20] perhaps because female bonobos have longer and more frequent periods of estrus than do chimpanzees.[21]

Chimps are also quite adept at making up after fights.[22] A short time after two males have had a screaming, rock- and stick-throwing display-battle, one may extend the hand of peace to the other, leading to a big hug and a peace sealed with a kiss.[23] Among bonobos, tensions are often relaxed by sexual activities, such as genital–genital rubbing between disputing female adults.[24] Such reconciliations appear to have evolved to preserve relationships that are valuable to the participants, protecting them from the corrosive effects of aggression. Former opponents, for example, preferentially make physical contact with each other rather than with third parties.[25] Reconciliation is now known to take place widely among our primate relatives, not just chimpanzees and bonobos. It's a mechanism for coping with conflict among social animals that cannot avoid the consequences of aggressive behavior simply by dispersing.[26] These interactions are part of what has been called Machiavellian intelligence, a component of human intelligence thought to be rooted in the selective value of using cunning, cooperation, and deceit to manipulate other members of a social group.[27]

Chimp violence doesn't take place only within groups; it can also occur between groups, where there may be no significant element of reconciliation. Relationships between different chimp groups occupying different territories may even involve the use of deadly force. In an incident at Gombe Stream Reserve, a mother from a chimp community other than the Kasakela community, the group of chimps that Jane Goodall had gradually accustomed to being observed by people, was brutally attacked, and the infant she was carrying was injured. The Kasakela males took the infant into a tree, killed it, and began to eat it. Then, according to the observers, the males "seemed to think they were doing something wrong," and one of them took the body some two miles to Jane's camp and left it on the laboratory porch.[28] I hesitate to interpret the last part of that behavioral sequence, but the deaths of infants in similar attacks on mothers and offspring by members of other groups occurred several more times at Gombe Stream. The attacks were primarily territorial, not infanticidal (i.e., not carried out by males to bring the female back into estrus), because the females were injured and no attempt was made to mate with them subsequently.[29]

About the time of that incident, in the early 1970s, the Kasakela community fragmented into northern (Kasakela) and southern (Kahama) communities.[30] The Kasakela community had eight males, six of them in their prime reproductive years and two of them old; the Kahama had seven males, one of them adolescent, four in their prime, one past his prime, and one old. Chimpanzees hold territories as groups, and they do so very aggressively.[31] As is typical of territorial chimps, members of the Kasakela and Kahama groups patrolled their territories—compact groups of males and, often, females in estrus moved silently and purposefully through peripheral areas. In 1971, during the early stages of fission, when Kahama males met Kasakela males there were sometimes charging displays by the southerners, but generally relations were peaceful. By 1974, though, encounters had taken a turn for the worse. Males from the Kasakela community began a series of violent forays toward the south featuring protracted, brutal attacks on members of the Kahama community. By 1977, the Kasakela males had wiped out the Kahama community, with all the healthy prime males killed (or missing and presumed dead). The restructuring of territories didn't end there; the destruction of the Kahama group removed a buffer between the Kasakela community and the Kalande community, which had nine males, farther south. The powerful Kalandes began to expand northward to the cost of the Kasakelas, who in turn moved further north, and the patrolling and tensions continued.

What about intent to kill in this case? On the basis of an analysis of the attackers' gruesome behavior, Jane Goodall thinks it is reasonable to conclude that the aggressors were deliberately homicidal (or, more correctly, panicidal?): "If they had had firearms and had been taught to use them," she commented, "I suspect they would have used them to kill."[32] However, it is not clear how widespread such nasty, aggressive behavior has really been in chimp groups in the wild. Chimp cultures can differ from place to place, and the Gombe Stream chimps were living in an area of recently and dramatically restricted habitat. In 1960, when Goodall first started her pioneering work at Gombe Stream, the forest stretched unbroken some sixty miles back from the shores of Lake Tanganyika. A decade later, it extended only about two miles to the rift escarpment, a ridgeline that more or less parallels the lakeshore, and most of the forest beyond the ridge had been cleared for cultivation. That drastic environmental change, conceivably leading to unusual crowding or resource shortages, may have been a factor in generating the intergroup strife at Gombe Stream.

The human-like complexities found in chimpanzee politics are truly impressive. Machiavelli would be proud of them.[33] Dominance depends not only on brute strength but also, to a degree, on coalitions and on acceptance by those dominated. Acceptance may be based on acquaintance. For instance, when the dominant male was removed from one experimental chimp group and two new adult males were introduced in his place, the nine females of that group (who were old friends) violently expelled the new males. Later, two other adult males were added; one was thrown out violently, but the other was allowed to stay. That male, Jimoh, was quickly but briefly groomed by two of the females, and one of them defended him vigorously against attacks by the dominant female. It was subsequently learned that those two females had been housed with Jimoh fourteen years earlier at another facility.[34] Despite his small size, Jimoh later achieved dominance over all the females, including some larger than he, through a combination of persistence and the occasional support of his female allies. However, on rare occasions when the females apparently perceived his behavior to be inappropriate, such as when he attacked juvenile males, they banded together and easily defeated him, a clear indication that he ruled by consent of the governed.

Hunter-Gatherer Politics

The ability of nondominant individuals to operate against those above them in the hierarchy and even to come to dominate them (a "reverse dominance hierarchy"),[35] thereby making a society more egalitarian, is easily found in human as well as chimp societies. It is memorialized in the many western movies in which the local citizens finally unite to help the hero overthrow the villains who have been terrorizing the community. The evolution of such behavior has been the subject of lively discussion in the anthropological community.[36] Sadly, we know nothing of the behavior of the common ancestor of chimpanzees and human beings. It seems probable that our ancestors, like most Old World primates, had social systems featuring moderate to strong dominance hierarchies.[37] Thus, in the transition to modern hunting and gathering *Homo sapiens*, there must have been a general trend toward softening of chimp-like dominance hierarchies, enabling evolution of the increasingly egalitarian, nonstratified societies that many scholars believe were characteristic of our hunter-gatherer ancestors.[38] In them, coalitions presumably would have limited the power of otherwise dominant individuals.[39] Furthermore, given that early hunting and gathering societies rarely had surpluses of food, the advantages of dominance would have been less than in agricultural societies. In the latter, as we will see, dominant individuals could appropriate the surpluses for themselves and begin the process of stratification and specialization that led to political states.

Although, like our chimpanzee cousins, historical hunter-gatherer groups raided one another,[40] it is clear that motivations for violence in chimps and in people can be very different. In our species, which has evolved intense consciousness, advance planning, language, and complex ethical systems, violence may be either socially sanctioned or socially disapproved—there is little reason to assume that genes are pushing us in one direction and culture trying to constrain us.[41] Historically, within what anthropologists Allen Johnson and Timothy Earle call family-level societies,[42] such as the Inuit groups, overly dominant, disruptive, or violent people could not be tolerated, and such behavior commonly resulted in socially approved homicide.[43] I well remember Tommy Bruce telling me, when I lived with the Aivilikmiut Eskimos in 1952, how a man who didn't fit in would be dealt with: "We would take him out seal hunting on a Peterhead [boat], and then when someone was shooting at a seal, the man's head would get in the way. *Ayornimut.*" (*Ayornimut* translated roughly as "Such things happen" or "Too bad; it couldn't be helped.")[44]

Such social sanctions, ranging from ridicule to assassination, against disruptive individuals are widely reported in hunter-gatherer societies.[45] This supports the hypothesized trend toward more egalitarian societies within these groups and a loss of overwhelming control by dominant males—people cooperate to dispose of those who possess what is judged to be too much power. But a very wide variety of behaviors, such as being domineering, chasing others' spouses, or behaving in unusual ways so as to be suspected of witchcraft, can be considered disruptive—egalitarian does not equate with peaceful. There can be heavy costs associated with sanctions that result in many people being killed. For example, among the Gebusi of New Guinea between 1980 and 1982, almost one-third of all mortality was the result of homicide, mostly the murder of people suspected of being witches,[46] and the small population was rapidly dying out as a result. Often, population declines and even extinctions appear to be at least partially caused by intragroup violence.[47]

The Roots of Warfare

The fact that such violence is virtually universal within human societies has given rise to a school of thought in which human beings are seen as having an innate drive or "military instinct"[48] that leads to aggression and is the root cause of warfare. For example, as zoologist Irenäus Eibl-Eibesfeldt wrote, "There is no conclusive proof of the existence of a primary aggressive drive, but there is strong circumstantial evidence that suggests it."[49] In part, the basis of this belief is how frequent *intra*community violence is among people, but *inter*community violence or warfare is nearly ubiquitous among human societies as well. War has been interpreted (parallel to Eibl-Eibesfeldt's comment) as a simple expression of humanity's genetically programmed aggressive nature. Belief that warfare is innate also traces in part to the observation that some other social animals, such as the Kasakela chimps, engage in intergroup violence. For example, among nonprimate mammals, female-led spotted hyena clans[50] in the Ngorongoro Crater of Tanzania defend well-marked territorial areas and kills of prey animals. This often leads to violent interclan battles in which individuals are badly mauled, sometimes killed, and on occasion also torn apart and eaten.[51] However, many social mammals don't engage in intergroup violence (peccaries, dolphins, elephants, and orangutans come to mind). Is this lack of violence evidence that human beings are innately peaceful and that warfare is a cultural adaptation that overrides a genetically based pacifism?

Warfare is less common among the great apes than are battles over food and mates within groups. But as demonstrated by the extermination of the Kahama chimp community by the Kasakelas, lethal intergroup violence among primates is not restricted to *Homo sapiens;* in both chimps and human beings, deadly conflict can occur even between bands made up of related males.[52] Intergroup violence appears to be much less common among bonobos than among chimpanzees;[53] indeed, on occasion, bonobo groups come together peacefully, with females initiating cross-community grooming and even copulation without objection from males.[54] The behavior of bonobos, unfortunately, is much less well known than that of chimps, and further study might reveal a darker side to their natures. Gorillas do not seem to be very territorial;[55] what intergroup violence does occur may be induced by battles among males over females or over the bounty of a fruiting tree.[56] Could the tendency to engage in warfare be a culturally evolved trait elicited by certain environmental conditions in chimps and in people as well? Remember, where the Kasakela and Kahama chimps were living, the area of available habitat had recently been greatly reduced.[57]

As with chimpanzees, warfare among human beings often is over territories (resources) or is caused, as in tribal conflict, by males fighting over females. Anthropologist Lawrence Keeley[58] points out that even though women may be common prizes of tribal warfare and sometimes its cause, their role is often as an economically valuable commodity rather than as sex objects. Such warfare over women is another aspect of humanity's more complex social interactions that, in my view, casts doubt on genetic determinist explanations of its origins.[59]

Intergroup violence among chimps and bonobos is carried out by the strongest individuals—primarily groups of young males. In addition, those males are ones that are closely bonded and, often, closely related.[60] Similar patterns of violence by bonded male youths can be detected even in modern human warfare. Men in combat generally fight not for principles but for their comrades. In World War II, for example, one major element in the impressive success of the badly outnumbered Wehrmacht was its policy regarding replacements. When an infantry soldier in the U.S. Army was killed or wounded, a new man was drawn at random from a replacement depot (a "repple depple," in the argot of the times). As military historian Stephen Ambrose noted, "Most of the replacements got to the most dangerous place in the world—unknown, unknowing, scared, bewildered."[61] Many were killed or wounded within a few days of joining units in the line. In contrast, the Nazi command carefully kept men from the same towns together in infantry units and replaced casualties with soldiers from home as well. The Wehrmacht appreciated the value of close bonding among men in combat (Kameradschaft) and tried to keep the young men who were doing the actual fighting in close-knit units reminiscent of those of the Kasakela chimps.

The idea that aggression, especially male aggression, is innate is too simple an explanation of the roots of war to do justice to the complexities of human behavior. The capacity to be aggressive appears to be virtually universal in human beings (as is the capacity to be friendly), but not all individuals or societies are aggressive. For instance, as author Barbara Ehrenreich points out, "throughout history, individual men have gone to near suicidal lengths to avoid participating in wars. . . . Men have fled their homelands, served lengthy prison terms, hacked off limbs, shot off feet or index fingers, feigned illness or insanity, or if they could afford to, paid surrogates to fight in their stead."[62]

The behavioral complexity associated with aggression was made clear by anthropologists Clayton Robarchek and Robert Dentan in their analysis of the widely cited myth that the Semai, a group of aboriginal swidden (slash-and-burn) farmers on the Malay Peninsula, were innately bloodthirsty.[63] The myth apparently traced in part to a report about the communist insurgency in Malaya in the 1950s. The Semai were famously peace-loving people.[64] But after some of their kinsmen were killed by the communists, Semai members of counterinsurgency forces took their vengeance. Dentan reported: "Taken out of their nonviolent society and ordered to kill, they seem to have been swept up in a sort of insanity they call 'blood drunkenness.' A typical veteran's story runs like this. 'We killed, killed, killed. The Malays would stop and go through people's pockets and take their watches and money. We did not think of watches or money. We thought only of killing. Wah, truly we were drunk with blood.' One man even told how he had drunk the blood of a man he had killed."[65]

The account is frequently cited to argue that even if, like the Semai, people are culturally trained to avoid violence, in an appropriate environmental circumstance an innate drive for violence will break through. As anthropologist Robert Paul put it: "Semai blood drunkenness is the eruption of a deeply repressed murderous passion which was certainly never taught or learned as part of Semai culture, but must have sprung from the depths of the psyche."[66]

In fact, the incident reported by Dentan was exceptional; in general, the Semai avoided lethal involvement in the war between the British and the Malayan communists. Furthermore, Dentan's translation, originally intended for a freshman-level college audience, was, as Dentan himself later pointed out, imprecise. Instead of "blood drunkenness," a better phrase would have been "blood light-headedness" or "blood nausea."[67] Rather than a breakthrough of a primitive genetic programming, the Semai incident is more reasonably interpreted as a result of disorientation on the part of frightened men who had been sickened by the sight of human blood. The experience of a firefight was very similar to what Semai mythology described as that of contact with obscene and violent supernaturals.[68] There is no need to postulate any deep, genetically programmed drive that had been suppressed until the conditions of combat released it, any more than we need to postulate a genetic urge to be peaceful that was environmentally suppressed in the course of the insurgency.

Whether or not the Semai were inherently bloodthirsty, it has become increasingly clear, as we saw earlier, that even relatively unstratified societies are not free of intra- and intergroup violence. As we will see in the next chapter, the Yanomamö, those Amazonian wife-abusers, suffer many deaths as a result of intergroup strife—but they are hardly exceptional in the level of casualties inflicted by tribal warfare. Indeed, in a careful review by Lawrence Keeley,[69] a final stake was driven through the heart of the Rousseauean myth that civilization corrupted a pacific humanity and led to the advent of warfare. Recent work indicating that bloody warfare over water and land plagued prehistoric North America, especially in the Southwest, and that resource shortages are involved in the generation of wars between pre-state groups[70] supports Keeley's demolition of the notion that war is basically a disease only of highly organized societies.[71]

A study of 132 cultures, however, gave some support to the idea that societies have gone to war more often as they have become more stratified and technologically sophisticated. In the study, the frequency of war was significantly related to certain aspects of social complexity (agricultural technology, size of settlements).[72] This makes some sense. Better agricultural technology would free more people for pursuits other than producing food, particularly soldiering, and bigger settlements would require more resources, which could be gained by violent appropriation of those belonging to other states. But the relationship in the study was between factors measuring the evolutionary stage of societies and the *frequency* of war, not its duration or the overall level of bloodshed (which may not increase with stratification). The correlation also does not demonstrate causation—more frequent warfare could have stimulated stratification and technological innovation. Furthermore, social complexity explained only a relatively small amount of the variation in frequency of war from society to society. War appears to have been a major and sanguinary human activity long before our ancestors got civilized. We'll return later to the topic of violent human behavior, especially highly organized warfare and genocide, features that have evolved in more complex societies than those of hunter-gatherers. But as you can see, the very complexity and variety of human behaviors related to violence give little support to the notion that we're genetically violent apes in sheep's cultural clothing or, for that matter, genetically saints that are good at occasionally donning wolves' clothing. And saints, too, can be involved in wars—religion is frequently intertwined with violence in many ways, often serving to reduce its occurrence and often serving as the motive for it.

Supernatural Relationships

Religion is sort of like love—it's difficult to define, yet everyone thinks he or she knows what it means. Unlike violence, chimps don't practice it, but we do. Religious or spiritual belief is clearly both a product and a critical part of most human natures. Such beliefs are a human universal, at least at the level of societies, if not of individuals.[73] For many people, religious belief and practice make up one of the most obvious facets of the complex jewel of their nature. Human self-consciousness has led people to feel a need to place themselves, as aware individuals, in time and space and to reflect on their position. There have been obvious limits on that placement. First, a knowledge of death, of the temporary nature of existence, limits it in time—one cannot be sure of what exists before or after an individual's life. Second, until recently, geographic constraints on knowledge meant that besides unknowable time, there were also potentially unknowable places—the end of the world off which to fall; dark forests, full of demons; an underworld. Self-consciousness and those constraints on it led to a sense of transcendence—the feeling that there are things that exist beyond ordinary experience.

Here, I will use the term *religion* simply to indicate a set of ideas about supernatural entities, agencies, and possibilities, some of which are held by the vast majority of human beings and by all significant groups. They are ideas that in any culture are used as aids in assembling a coherent worldview—what anthropologist Clifford Geertz called "conceptions of a general order of existence."[74] Examples of such beliefs are beliefs in the existence of gods, demons, and spirits, thought to influence natural events and perhaps to have been involved in the origins of society or the universe, and beliefs in the continuation of life after death. Religious ideas in this sense also include belief in the supernatural power of certain individuals or of carefully performed rituals[75] to communicate with or manipulate the behavior of unseen powers. Of course, the particulars of these beliefs vary widely from culture to culture.[76]

Marxism, capitalism, and science all have characteristics in common with religion, but I'll omit them from this discussion, even though they probably share some common evolutionary roots.[77] Science, above all, doesn't provide a father figure or other supernatural entities that can be prayed to for guidance and comfort. That makes science vastly different from most religious traditions and handicaps it as a means of orientation in a world in which comfort is often in very short supply.[78] Science tells us that we are creatures of accident clinging to a ball of mud hurtling aimlessly through space. This is not a notion to warm hearts or rouse multitudes.[79]

Religious ideas can be traced to the evolution of brains large enough to make possible the kind of abstract thought necessary to formulate religious and philosophical ideas. Not only do our brains permit us to outwit predators and colleagues and solve extraordinarily difficult problems; they also create problems by endowing us with intense consciousness and knowledge of our own mortality. This produces a load of anxiety that some feel is unbearable and that all of us attempt to find some way to reconcile. As the saying goes, life is tough, and then you die. In the face of life's tribulations, people seek a belief system that appears to justify those tribulations. Those who find such a system of orientation, it seems, tend to enjoy not only better mental health but also better physical health.[80]

Beliefs in the supernatural clearly have had—and continue to have—enormous influence on human behavior and the evolution of human societies. An examination of why religions evolved, the roles of religion, and the evolution of those roles is critical to understanding human natures and can be carried out without reference to the "truth" or "falsity" of the claims of any religion. As professor of religious philosophy Ninian Smart put it, we can "talk about worship and other activities in meaningful ways without having to comment on their validity, without having to comment on whether there is a Vishnu or a Christ. [We can] think of Vishnu as focus entering into the believer's life, dynamizing his feelings, commanding his loyalty, and so on."[81] The idea of "dynamizing" feelings is obviously a vague one. Encounters with certain people, thoughts, ceremonies, works of art or music, and the like can produce a surge of emotions that are often described by the word *thrilling*. For example, I am thrilled by experiences as diverse as catching the first sight of a granddaughter after a long separation, hearing a well-drilled band play "The Marines' Hymn" or "Scotland the Brave," pushing forward the throttles to launch an airplane into seemingly miraculous flight, seeing Vermeer's *The Milkmaid*, or even discovering a new connection in solving a scientific problem. Clearly, many people are thrilled by more specifically religious ideas and religious ceremonies. In the extreme, they may go into trances of religious ecstasy.[82] The pleasure of such thrills may be an important factor in religion's persistence,[83] some anthropologists have suggested, in addition to the solace many find in religion.

There is little direct evidence about the origins, original functions, and early evolution of religion, but one can construct reasonable hypotheses about them. The Neanderthals, for example, apparently buried their dead and may have been the first human beings to do so. The best evidence of this comes from three graves in southwestern France that appear to have been deliberately excavated and from the intact, articulated condition of skeletons from these graves, implying deliberate protection from natural decay processes and predators after death.[84] There is no dependable record of decoration of the corpses or inclusion of goods in the graves, as often characterized the burials of early *Homo sapiens.*[85] Nonetheless, it seems reasonable to conclude that the Neanderthal interments represented at least a strong awareness of death and a belief that there was something significant about the remains of group members that elicited a need to protect them from the elements or scavengers. The idea of a soul or spirit to be somehow honored or mollified could have been involved as well. Burials of anatomically modern *Homo sapiens* that include animal remains (offerings?) arguably go back almost 100,000 years.[86] Burials with certain kinds of art objects, almost all of them more recent than 50,000 years ago,[87] though, are the first solid clues we have to the presence of religious belief among early human beings.[88]

Religious ideas must have been especially diverse among early modern *Homo sapiens,* judging from historical hunter-gatherer cultures and the low population densities and wide dispersion of groups of our ancestors. Many different ideas about souls are held by hunter-gatherer cultures. For example, people need not have only one soul;[89] souls can wander, or not, during dreams;[90] souls are located in different organs (some Inuit groups thought that animal souls resided in the bladder[91] but human souls occupied the entire body of a living person, yet were immortal);[92] and so on. The diversity of notions about the soul and other beliefs about sacredness that characterized hunter-gatherer religions carries on to some extent from the state religions of antiquity to religions of today.[93]

Whenever religious ideas arose, and despite their likely early diversity, it seems reasonable to assume that early religion had the same two roles it serves today. One is explanatory and manipulative, designating which forces are driving mysterious-seeming events in the world and trying to influence them. The other is integrative and controlling—organizing groups to deal with those forces and justifying the power gained by some individuals over others within those groups. In the first instance, as our ancestors evolved ever greater intellectual capacities for understanding one another and their environments, questions not easily answerable were bound to arise. An early adaptive mode of thought, a recognition of simple cause-and-effect relationships (discussed in chapter 6), could have provided a way of addressing such tough questions.[94] Chimpanzees clearly understand that a blow from an appropriate rock against a palm nut placed on a rock anvil is a cause that will have the effect of opening the nut and making its contents accessible. Our early ancestors must also have spent much time trying to associate effects with causes even when those causes could not have been transparent. Why does the sun travel across the sky? What caused that sudden thunderstorm? Why did a member of the clan suddenly become so aggressive? Why did another stop breathing? When intense consciousness was evolving, that last question must eventually have led to the realization that the self was limited in duration.

Cause-and-effect interpretation of many observed sequences is very likely at least partially hard-wired into the human nervous system, as well as the nervous systems of other animals.[95] It seems reasonable to assume that as human powers of thought increased in the course of evolution, people would begin to invent causes of observed, immediately inexplicable effects, if for no other reason than to quell the anxiety that mysterious events often elicit.[96]

The idea of supernatural causes certainly came to our ancestors naturally. First of all, fear is an emotion that is "wired" into our human natures—fear of pain, fear of predators, fear of falling, fear of abandonment, fear of the unknown. Fear as a motivating factor appears to be a property of the nervous systems of not only human children and adults but also human infants and other animals,[97] although the objects or events that are feared and individuals' reactions to them, at least in *Homo sapiens*, are strongly influenced by environment. We also experience dreams,[98] trances, and hallucinations and see shadows and reflections of ourselves in still waters—all of which could suggest a separate existence, a subconscious, independent part of the individual that gives it motion but can depart from it, leaving the body lifeless. Death may well have inspired the early evolution of the concept of a soul. The recognition that life ends, combined with genetically evolved aversions to danger, eventually also must have led to the fear of death that philosophers since Thomas Hobbes have considered a powerful driving force in human behavior and thus critical in shaping human nature.[99]

And if people had souls animating them, why couldn't trees, rocks, and wind? If one accepts the existence of a variety of sentient forces of natural or supernatural beings, it would make sense for an individual to seek ways of manipulating them to the benefit of herself or her family or group. An organism that has developed a Machiavellian intelligence to deal with other members of its social group could quite easily move to negotiating with supernatural beings in various ways, through prayer, magic,[100] rituals, or other kinds of behavior, in attempting to achieve desired results. Effects with no clear causes; a desire to find causes; signs in dreams; death and other phenomena; and attempts, inspired by fear and pain, to influence causal agents—all these give us clues to the origin of religion.

Pioneering anthropologist Edward Tylor figured this much out well more than a century ago. As he put it, once "man" drew the conclusion that he had a soul, that idea "served as a type or model on which he framed not only his idea of other souls of lower grade, but also his ideas of spiritual beings in general, from the tiniest elf that sports in the long grass up to the heavenly Creator and Ruler of the world, the Great Spirit."[101] Many others have attempted to understand why human beings are involved with the supernatural, often taking an attitude similar to Tylor's. For instance, French sociologist Émile Durkheim surmised almost a century ago that religion allowed societies to deal with things that "surpass the limits of our knowledge."[102] Through the development of science, however, cultural evolution has been steadily expanding those limits of the known and changing human natures in the process.

The Roles of Religion

These explanations of the origins of religious thought still seem persuasive, but they do not address the integrative and controlling functions that cultural evolution has given religion in human cultures. For instance, the Machiguenga of the Peruvian rain forest are animistic, living in a spirit world that they believe permeates their lives.[103] The dominant idea of this religious view is that objects in nature, such as trees and rocks, are animated by spirits and that the world is inhabited by a variety of disembodied spirits. (The term *animism* comes from the Latin word for breath.) These spirits are especially active in impulse control—in limiting gluttony, aggression, and inappropriate sexual behavior. If an individual misbehaves, he or she will be punished or destroyed by the spirits. The spirits can be thought of as preventing sin, and control of sin (as in some organized religions) is exercised through fear.[104]

In the development of religion lies one of the first steps toward social stratification in human groups beyond the sexual stratification produced by biological evolution, that based in males' greater strength and females' greater reproductive responsibilities. Religion generated the emergence of shamans, or medicine women and men,[105] who took their place alongside other leaders such as headmen and chiefs. Each type of leader was a specialist who directed the group in special circumstances. Indeed, the emergence of individuals thought to be especially talented in contacting and dealing with supernatural entities is characteristic of the evolution of religions. A key aspect of family-level polities, in which the family was the basic economic unit and families were bound together in camps and regional networks, was that the authority of those with leadership roles was strongly circumscribed.[106] Headmen and shamans had separate domains. The shaman ordinarily did not make important decisions about the hunt or when to strike camp, although he or she surely was consulted. Small groups did not require the sorts of political control mechanisms that are needed to govern large, diverse populations. Although headmen may have been selected for their superior skills in planning, hunting, or negotiating, they had no intimate connections to powerful spirits. We can only guess at the special characteristics for which shamans were selected.

Stanford University neuroscientist Robert Sapolsky has beautifully elaborated historical speculation about the universal evolution of shamans and its possible connection with the universality of schizophrenia in human populations.[107] He speculates that people with so-called schizotypal behavior, which involves milder forms of the hallucinations and disordered thought that afflict schizophrenic individuals, were those who were chosen to be shamans.[108] They are "the forbidding, charismatic religious leaders in tribal life, the ones who sit and converse with the dead ancestors, who have solitary sojourns in the desert, whose huts sit separate from everyone else's, who spend the night transformed into wolves or bears or hyenas, the ones who lead the trance dances and talk in tongues and bring word of the wishes of the gods."[109]

Sapolsky also describes an interesting connection between obsessive-compulsive disorder (OCD) and the solace that many people find in religious rituals. OCD involves disabling compulsions to wash oneself, clean objects, count things, and so on. This disorder appears to involve rather minor deviations from genetically evolved "normal" connections among the neurons of the human brain; its link to religious rituals has been noted by people ranging from religious leaders themselves to Sigmund Freud.[110] In both schizotypal behavior and OCD, gene–culture coevolution may well be acting to shape the way human beings attempt to deal with the transcendent.

Schizotypal behavior may also be genetically related to schizophrenia in much the same way the sickle-cell trait is to sickle-cell anemia. Genes predisposing individuals to schizophrenic behavior (the evidence is quite good that such exist)[111] could be maintained in the population because schizotypal behavior is produced in heterozygotes (that is, individuals having two different alleles at the key locus), which have a selective advantage. The comparative reproductive contributions of shamans and ordinary members of societies are, of course, not known. There are anecdotal stories on both sides of the question—putatively celibate shamans in some societies, shamans with great sexual access in others.[112] Whether schizotypal behavior (and schizophrenia) was maintained among hunter-gatherers by selection favoring those who behaved in ways that increased their chances of being shamans cannot be determined—but it's an interesting idea. Whether OCD is a side effect of selection maintaining other critical brain functions is unknown. After all, "normal" people show compulsive behavior at times. Have you ever gone back to the mailbox and pulled the slot open a second time to be sure the letter went down, or found yourself unable to get some popular tune out of your head, or checked repeatedly to be certain your wallet was in your pocket or purse? There are numerous adaptive tasks that human beings do repetitively, and evolutionary design seemingly makes that easy, perhaps sometimes too easy.

Sapolsky has also noticed a most interesting correlation between major geographic factors in the environment and the evolution of religions.[113] Those religions that evolved in tropical rain forests tend to be polytheistic; those from deserts, monotheistic (or, in some cases, hierarchies in which subdeities exist as part of a supreme god).[114] The early peoples of rain forests lived in incredibly diverse, relatively benign environments, and perhaps as a consequence, some, such as the Mbuti Pygmies, imagined a diversity of gods. In contrast, deserts, in Sapolsky's view, "teach big singular things, like how tough a world it is, a world reduced to simple, desiccated, furnace-blasted basics in every realm. 'I am the Lord your God,' and 'There is but one god and his name is Allah' and 'There will be no gods before me's proliferate." Desert monotheisms also feature busybody gods, who continually meddle in the business of individual mortals. In Sapolsky's view, these religions provide explanations for the often catastrophic effects that deserts inflict on their dwellers. Lethal natural disasters on the order of unexpectedly dry water holes are rarer in rain forests; if one fruit crop fails, there's always another, or some succulent bugs to eat. Thus, religion's role of helping to orient people to their surroundings leads to the religions themselves being shaped by the environments in which societies are evolving. One might speculate that those early monotheistic desert religions lent themselves to the development of hierarchical societies ruled from above.

Some may consider today's Western religions to be an evolutionary advance from hunter-gatherer religions, but as with language, there is little basis for placing the religions of modern peoples on a scale from primitive to advanced. People tend to think of their own religion as the one true religion, of course, and adherents of the religions that have become "organized" as societies have become more complex tend to look down on those of hunter-gatherers. This was brought home to me during my Inuit summer. Father Rio was a Belgian Catholic Oblate missionary stationed at Coral Harbour in the Canadian Arctic, part of "the largest diocese on Earth." He was also battling with a native Anglican priest for the souls of the local Inuit. Father Rio made great raisin wine, and I spent many a pleasant evening helping him consume it. It was 1952, and in his view, the Inuit, with their "simple" religion, were "just like children." The "simple" Inuit religion was actually a form of animism based on a complex of spirits, ghosts, human and animal souls, and several major gods, employing shamans, numerous taboos, magic words, and the like. It was anything but simple.[115]

Religious ceremonies served important integrative functions in the lives of the Inuit, as they do in the lives of other hunter-gatherers and many people in industrial societies today. For hunter-gatherers, these functions include the assembly of groups of many families to engage in recreational pleasures, find mates, exchange information about the distribution of resources (game, fruit, raw materials, etc.), carry out cooperative hunting efforts, and participate in religious activities such as appeasing spirits and mourning the dead.[116] Many of our prehistoric ancestors presumably used religious ceremonies in similar ways.[117]

Interestingly, Father Rio's contempt for the religion of the people whom he sought to serve and convert was reciprocated. Once, when I was taking a language lesson with Tommy Bruce and several other Inuit, the talk turned to the feud between the Anglican priest and Father Rio, which had become intense. Why, Tommy wanted to know, if their religion was based on loving one's neighbors, did the priests shoot at each other's dogs with shotguns? Then he said,

"Do you know what Father Rio believes?" and regaled me with the story of the virgin birth. By the time he had finished, all the Inuit were laughing so hard that tears were running down their cheeks. Tommy was an Inuit of the old breed, a former shaman who had spent his life hating another Inuit because of a wife-trading deal gone bad in the distant past.[118]

The Inuit, when I knew them, were a people caught up in cultural evolutionary change induced by contact with Europeans. Their original religious beliefs, which were still strong, told them that game was supplied anew each year by spirits.[119] They, like members of most hunter-gatherer groups, had no concept of conservation;[120] it would not have made sense in the context of their worldview and traditional capabilities. Seal hunting, for example, was hard, dangerous work. It often involved standing motionless for hours in subzero winter weather over a seal's breathing hole in the sea ice, holding a harpoon at the ready. But when a thrust was made and a seal killed, its body was retrieved. So were those of seals killed at sea from boats. Inuit harpoons were ingeniously designed, with a bone point attached to a walrus-hide line.[121] After a strike, this harpoon head, still attached to the line, separated from the shaft and the latter floated to the surface, where it could be easily retrieved.[122]

The religious belief in spirit-supplied game had serious consequences when the environment changed. I was told that originally, about nineteen of every twenty seals killed ended up in the Inuit diet. But when the Inuit of Southampton Island were able to trap arctic foxes and sell the skins for a high price to the Hudson's Bay Company, they used the money to buy forty-foot engine-driven Peterhead boats from Scotland and high-powered rifles from the company. Seals were chased down and shot from the boats and then retrieved with harpoons. But in the spring, the density of seawater was reduced by abundant freshwater runoff from melting snow on land, and about nineteen of every twenty of the seals the Inuit shot and killed sank before they could be harpooned.[123] This cultural–evolutionary mismatch frequently led to some starvation as the Inuit depleted seal populations and caribou herds. The religious basis of the problem, the belief that game was annually renewed by spirits, was little understood by the Europeans. As one said, "It is a pity nobody has been able to teach them to conserve things properly: they are apt to 'blow' all they can get, kill all the caribou or seals they can, whether they need them or not. It is a strange kink for a generally intelligent race."[124]

When I returned briefly to Coral Harbour in 1988, I discovered that the previous Anglican–Catholic battle for Inuit souls had been largely won by evangelical Christian sects (which also have recently been making considerable inroads into the population of southern Costa Rica, where my research group now works and where hard feelings based on religious differences are on the upswing). The process of cultural evolutionary conversion had proceeded much further since the time when I had lived there, thirty-six years before. Hugh Brody, a research officer with Canada's Department of Indian Affairs and Northern Development, summarized the situation around the time I was in the Arctic: "If the Hudson's Bay Company may be said to have established an economic serfdom, then the missionaries sought to establish a moral serfdom."[125] The controlling effects of religion have come a long way since the emergence of shamans helped lay the foundations for the stratification of human societies. At Coral Harbour in 1988, however, I was struck by the growing movement among the younger Inuit to attempt to retrieve important elements of their original culture. Unhappily, they will be waging an uphill battle against the influence of Western consumerism, the most seductive and—at least temporarily—successful force for cultural evolution outside of religion the planet has ever seen. The human natures of the Inuit have been transformed over the past century or so, and it remains to be seen whether cultural evolution will allow some of the more traditional aspects of their natures to be restored.

Religion illustrates, perhaps better than any other aspect of human behavior, the way in which basic biological capabilities can be built into a vast cultural superstructure, influencing, indeed sometimes virtually determining, the natures of groups and individuals. It also shows how channeled some aspects of cultural evolution can be, constrained so that once a trajectory is established, environmental change has relatively little influence on it. Consider, for instance, the utter failure of the Soviet regime to extinguish religion in Russia despite three-quarters of a century of determined, often brutal effort, or, more generally, the persistence of major religions into an era of science in which many previous "mysteries" have been solved.

Art for Other Sakes

From its origins, art (in the broad sense, often combined with music and dance in rituals) has evolved in close association with religion and the use of symbols.[126] In preliterate societies, art often embodied communal values and knowledge, filling a role only partly replaced by the written word in literate societies.[127] Use of art, in one form or another, is a universal human characteristic and a major feature of all human societies. Its symbolic aspect is shared with other key features of human culture, including, of course, true language.[128] Art is thus an important product of human natures whose evolution we would like to understand. It is also a place where the plurality of our natures is most readily seen. Be they Makonde from Mozambique or Mexicans from Morelia, the variation in artistic natures among individuals and among groups is truly stunning. When I look at a Rembrandt painting or listen to a Beethoven piano concerto, I can't even imagine how those artists created their works, let alone do anything comparable. When one compares inro, decorated miniature lacquer containers for pills and other small objects crafted by Japanese artists, with the reliquary figures created by the Kota of Gabon, a difference in the natures of the craftsmen of Japan and central Africa seems apparent (see the figure).

The first physical traces of art in prehistory appeared quite suddenly in the archaeological record, starting around 40,000 years ago,[129] and included incised bones, small statues and figurines, clay bas-reliefs, and spectacular cave paintings.[130] The latter, exemplified by the extremely realistic figures portrayed on cave

Inro (left), three inches high, polished lacquer with stylized horse design inlaid with blue shell, about three hundred years old. Attached to the inro is a finely carved ivory netsuke. Kota skull guardian figure (right), twenty-two inches high, with a stylized human face, made of thin copper plates on a wooden core, more than a century old. All of these objects had functional roles in addition to their obvious aesthetic value. The inro is a medicine case divided into compartments for carrying pills; the netsuke served as a toggle or button to hold the case between the pocketless kimono and the kimono's sash. The guardian figure was placed on a basket of ancestors' remains to ward off evil that might desecrate the ancestral bones. Photographs by Cagan Sekercioglu.

walls at the Lascaux Grotto in southern France, the Altamira cave in Spain,[131] and similar sites, are compelling to anyone who sees them (see the figure).

Archaeologists remain uncertain of the motives of the people who created the art—they could leave no written record of their intentions. If the paintings move us, surely they moved their creators. One can easily imagine that much early art had a religious or ritual function,[132] designed to appease the spirits of game animals and the like in order to manipulate those undetectable causes of inexplicable effects. Even so, some or all of it may have been simply a type of aesthetic expression.[133] In fact, evolutionary psychologist Nicholas Humphrey recently proposed that creators of that magnificent art actually had premodern minds with little interest in communication: "Cave art, so far from being the sign of a new order of mentality, may perhaps be thought the swan song of the old."[134] He backs his conjecture by pointing out an eerie resemblance between the Lascaux animal paintings and the artwork of a young autistic girl in England in the 1970s. I have my doubts; this implies a lack of communication that seems quite unlikely 20,000 years ago, and autistic children, withdrawn and living in a fantasy world, seem a poor parallel for hunter-gatherers facing tough

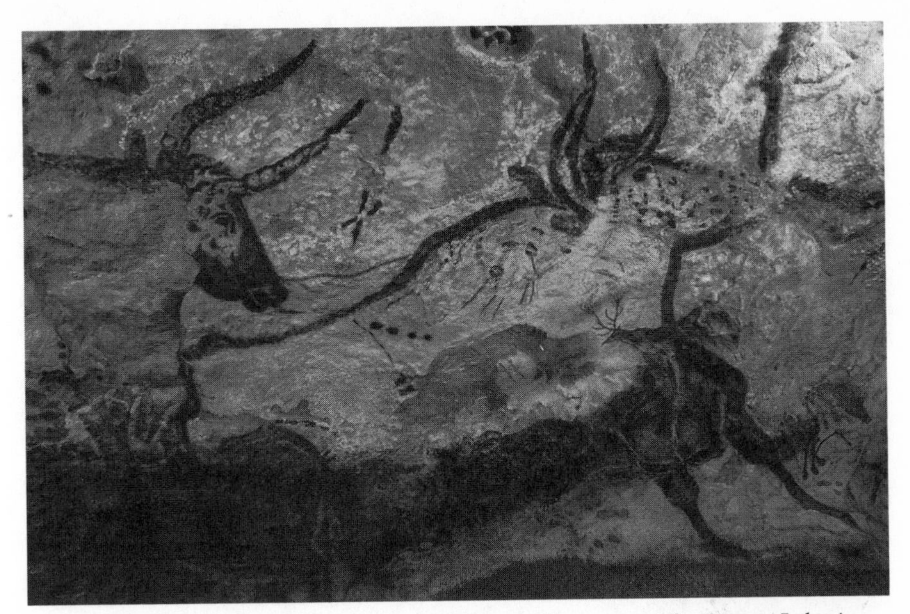

Cave paintings at Lascaux Grotto in southern France. Photo copyright Ferrero-Labat/ AUSCAPE International.

realities. It seems more likely that the explosion of art onto the human scene signaled a fundamental advance in communication.[135]

Because they are so often closely associated with one another and are so much a part of the fabric of historical societies,[136] it seems reasonable to speculate that music and dance may have developed at the same time as their close cousins, pictorial art and sculpture—that is, possibly around the time of the cultural Great Leap Forward.[137] Their beginnings may also be related to the origins of fashion. Much of early art, like some art today, was probably bodily adornment, including such things as body painting among Australian Aborigines and makeup and tattooing among those in Western societies. And much of it was quite likely used in connection with song and dance, as are masks among Africans, New Guineans, Native Americans, and others or jewelry and fancy clothes among pop singers. In preliterate societies, rhythmic movement virtually always accompanied song.[138] Bodily decoration, be it painting, feathers, or jewelry, doubtless often signaled status or skill (announcing, for example, that a male was skilled enough with darts to kill male birds of paradise and get their gorgeous plumes) and thus may have helped the wearer to gain sexual access.[139] Dance and art obviously also, like almost everything else in human society, are intertwined with sex. But the fact that aesthetic expression serves various functions in societies is not really an explanation for its arrival on the scene. Why should our ancestors have developed, apparently quite suddenly, a capacity for it?

That question is part of the puzzle of the Great Leap Forward. There is general agreement that the development of art did not coincide with any detectable physical change in our ancestors and that it probably occurred at the time of the Great Leap.[140] But, as one might expect from the lack of physical evidence for the origins of speech, there is no agreement about the relationship between human linguistic accomplishments, which were an exercise in symbolic communication, and art, much of which was similarly symbolic as well as iconic. (Again, symbols bear no resemblance to what they represent—the swastika representing the Nazis is an example. Icons bear a resemblance to their meaning, as when a silhouette of a walking child indicates a school pedestrian crossing.) But presumably language and art, because of their shared symbolism and iconicity, require similar mental gymnastics to use and appreciate. It's also possible that archaeologists are wrong in their estimation that art evolved at the time of the Great Leap. Was there a long period of artistic flowering expressed in, say, body painting and dance, which would not have left a "fossil" record? Are certain symmetries in stone tools that date back as much as half a million years signs of an aesthetic sense that might also have been expressed in temporary art forms?[141]

The Roles and Evolution of Art

The questions don't end there. Why did it take tens of thousands of years of symboling behavior before the glorious artistic abilities of Cro-Magnon *Homo sapiens* were transformed into pictorial symbols representing vocal ones? Writing came long after the Great Leap, developing only some five or six thousand years ago. It is generally believed to have begun in Mesopotamia and to have spread by both diffusion and independent invention—the latter in at least Mesopotamia, Mexico and Central America, and the Far East.[142] It appears to have evolved from art (iconic pictographs) to completely symbolic representation of the sounds of speech. Some believe that writing evolved from symbols representing first each word, then each syllable, and then each phoneme, but as one might suspect, it seems not to have been as simple as that.[143] The system of representing ideas and sounds used in the Chinese languages, for example, would not fit into such a scheme.[144] In fact, some scientists think that writing actually restructures the mental processes of literate peoples,[145] claiming, for instance, that the elements of a complex rational argument can't be carried in memory. First art and then writing added enormously to the human capacity to store extragenetic information; they allowed a whole new level of culture to be attained. In any case, writing is thought to have developed in close conjunction with religious practices, which required complex calendars for the timing rituals, and with economic affairs, which necessitated the keeping of financial records, some of the latter related to religious officialdom.[146]

Perhaps because art is entangled with almost all that is human, it is very difficult to define, and patterns in its evolution thus are difficult to detect. One approach to defining art is strictly aesthetic: something is art if it was created with the intention that it be aesthetically pleasing rather than merely functional.[147] Another approach is to view art as a kind of communication using especially effective visual images (or symbols).[148] But the images are highly culture bound and often are almost impossible for members of other contemporaneous cultures to interpret. For example, the Lega, a people of dense equatorial rain forests in central Africa, carve female figures with distended bellies. In museums and art galleries, these are often interpreted as fertility figures; in actuality, they evolved as warnings against committing adultery while pregnant. Among the Lega, cultural evolution produced art objects that help to ensure continuity of the social structure and to maintain social stratification. A central feature of Lega society is the Bwami ritual, in which a member bonds with kin and receives moral training in initiation ceremonies. Sorcery—the use of powers obtained from evil spirits—is controlled by the Bwami association. Various art objects are involved in the initiations, and there are strict rules about the use of the objects and who keeps them. They are also employed by older men to control the rise of younger men through the ranks.[149]

Art has evolved to play similar nonaesthetic roles in all human cultures. Among the Yolngu of Arnhem Land in northern Australia, each clan possesses a series of special designs that are traced back in legend to the Dreamtime, the ancestral past when the designs were received by the founders of the group. The clan designs are used in rituals to direct ancestral power to achieve certain goals as well as in a very complex system of clan relationships.[150] The aesthetic aspect of Yolngu art—the way it can thrill the Yolngu or make them just feel good or induce fantasies—is functional in religious rituals: as anthropologist Howard Morphy put it, "Yolngu artists claim to create a particular visual effect in their paintings . . . a shimmering quality of light which engenders an emotional response . . . interpreted as representing or being a manifestation of ancestral power"[151] (see the figure). Bushman rock art also has clear religious connotations, having to do, among other things, with the relationships of people with other living beings. These paintings clearly are not "gastronomic art," associated with hunted animals (as some European cave art may be) and gathered plants, because the animals most carefully depicted are not those commonly consumed, and plants are not depicted.[152]

Like art, music has in some cases evolved to serve social functions, as, for example, the funeral dirges on the Pacific island of Tikopia, where the singers have a specified kinship relationship to the deceased and are rewarded for their efforts.[153] Music also can be used to make routine work seem less of a burden or to synchronize it[154] as well as to stir up martial feelings among preliterate[155] or modern peoples (think of Australian troops marching off to the bloody disaster of Gallipoli to the tune of "Waltzing Matilda" or Nazi mobs singing the "Horst Wessel Song").[156]

Although those of us in Western societies tend to associate art first with its purely aesthetic role, it obviously plays practical roles in communication for us, just as it does for the Inuit, Lega, Yolngu, and Bushmen. It reinforces faith, supports government programs and dogmas (think of Soviet "tractor art" or war memorials and dramatic military recruiting posters in the United States), and, perhaps above all today, sells products. Even in the West, art today is still often also concerned with religion. Just consider the various elaborate religious ceremonies that make use of effigies, special costumes, candles, scrolls, hymns,

Yolngu painting on bark, depicting totemic animals (animals emblematic of the Dhalwangu clan, with which clan members claim affiliation due to common ancestry) at a sacred billabong (backwater; stagnant pool). Shown are long-necked turtles, fish, birds, and crayfish. The hatching represents weeds of the billabong; colors are primarily rich ochre, sienna, black, browns, and white, and the painting is thirty-seven inches high. The work was created in 1965 by one of the best-known Yolngu artists, Yangarriny Gumana. Photograph by Cagan Sekercioglu.

dances, and so on. Much art has been inspired by religion, and to a degree religion is a form of art.

Sex, violence, religion, and art are products of our natures whose evolution, beginning in our ancestors' long career as hunter-gatherers, has been tightly intertwined. A linkage of sex with violence probably has been around at least since our ancestors left the trees, whereas the association of religion with art has probably been limited to the last 50,000 years or so. Although human behaviors in these areas are built on a foundation constructed by biological evolution, the edifice itself—though constrained in some ways by the foundation—is a product largely of cultural evolution. There is no reason to believe that young men who commit violent crimes do so because they got an especially big dose of genes programming them for aggression. The biological–cultural evolutionary nexus has also been responsible for the evolution of *power* relationships in societies. Those relationships were critical in the cultural evolutionary development of complex social structures, and they receive all too little attention in many analyses of environmental and other social problems today. It's to those more complex social situations, which are now at the center of the human dilemma and in which the dominance of cultural evolution becomes virtually complete, that we turn next.

Chapter 10

FROM SEEDS TO CIVILIZATIONS

"The changes brought over the past 10,000 years as agricultural landscapes replaced wild plant and animal communities, while not so abrupt as those caused by the impact of an asteroid at the Cretaceous-Tertiary boundary some 65 million years ago or so massive as those caused by advancing glacial ice in the Pleistocene, are nonetheless comparable to these other forces of global change."

—Bruce D. Smith[1]

The millions of years our forebears spent roaming over the landscape as hunters and gatherers left important marks on our human natures. During that period, interacting genetic and cultural evolution affected everything from our sexual behavior and religions to our food preferences. But without the ensuing agricultural revolution and the sedentary life and divisions of labor it eventually made possible, cultural evolution could never have produced our complex modern civilization and the human natures that go along with it.[2] Without farming, which freed some people of the chore of wresting nourishment from the environment, there would be no cities, no states, no science, and no mayors, fashion models, professional soldiers, or airline pilots. In contrast to the periods of cultural stasis reflected in the Oldowan and Acheulean technologies—the latter, you will recall, a period of more than a million years with little sign of cultural change—the agricultural revolution led to a period of cultural evolution unprecedented in its rapidity and scale. That is the story of this chapter. It is a story that starts with the obtaining of food but returns us to two aspects of human behavior that, although present in hunter-gatherers, became even more important in sedentary groups—religion and violence. It's a story of cultural evolution, which at this stage of our development begins to swamp the more gradual processes of biological evolution.

Of Kings, History, and Mountain Ranges

Since the time of the agricultural revolution, and especially since the origins of writing, the diverse pathways of cultural evolution have been viewed through a variety of lenses, most of them some version of "history." Historians, political theorists, and other social scientists mostly examine changes within and among human societies in terms of human actors, motives, and actions—looking at what I like to call cultural *micro*evolution.[3] They could, somewhat unfairly, be characterized as taking primarily a "succession of governments, international conflicts, and changing economic systems" approach to cultural evolution. I say unfairly because there have been notable exceptions, such as Ferdinand Braudel, J. Donald Hughes, and Alfred Crosby.[4]

Most social scientists, however, tend to pay relatively little attention to the extrinsic environmental factors that help to shape the broad outlines of history. These include such things as whether mountain ranges or other barriers to the diffusion of innovations such as agriculture ran north–south or east–west (spread is easier at the same latitude); the availability of wild animals and plants suitable for domestication; and the distribution of resources or good harbors to shelter naval vessels. These are the sorts of factors that influence cultural *macro*evolution, changes in the trajectories of different human societies driven by environmental factors rather than by social, economic, and political machinations. The strong early influences of such macroevolutionary factors recently have been brought dramatically to public attention by Jared Diamond.[5] Although those who focus their interest on modern states also have paid little attention to environmental factors in their dynamics, that shortcoming may not always be as serious. Modern technologies, such as oceangoing vessels and fertilizers, have tended to reduce, but not to eliminate, the constraints of regional geography and thus of macroevolutionary forces.

The ideas and actions of individuals, their effects on society, and the institutions and cultural "spirits" of societies (such as attitudes toward work or corruption or the presence of some sort of cohesive social will) clearly play a critical and, at the level of microevolutionary detail, dominant role in determining the course of history. One need only look at the differences today between resource-rich Argentina and the United States, both of which were occupied by Europeans, but by Europeans with different cultural and economic histories, institutions, and relationships between church and state. Although there are environmental differences between the two countries, it's clear that these were not the key determinants of their different fates.

The extrinsic factors shaping cultural macroevolution were extremely important in establishing the conditions that prevailed at the dawn of history, the world at the first invention of writing. The Inuit, for example, could not invent farming because there were no plants suitable for domestication in their environment and the growing season was too short. People in the Middle East could, however, because they were surrounded by wild plants that could be so easily cultivated as to beg for domestication. Such factors remain as a background element shaping what is possible at any given time in any given society. For instance, cultural microevolution generated a need for liquid fuels in industrial societies. But macroevolutionary factors helped shape the course of World War II, when both Germany and Japan had to put enormous effort into obtaining and defending petroleum supplies because their own territories lacked them.

Such environmental factors varied greatly over the globe and have influenced the fates of groups that remained hunter-gatherers as well as those that underwent evolution toward more complex polities.[6] New Guinea's nearly impenetrable terrain of steep mountain ranges and dense tropical forests kept societies there quite isolated from one another and the outside world, making possible, among other things, the development of an amazing diversity of art forms and languages (about one thousand).

In Europe and North America, the ways in which history has unfolded seem more heavily influenced by cultural microevolution, although that in itself was partially determined by macroevolutionary factors—the absence of dramatic geographic isolating conditions or climatic extremes that so influenced the evolution of New Guinea and Inuit societies. Sociopolitical and economic considerations within each society, ranging from the emergence of charismatic leaders (Alexander the Great, George Washington, Winston Churchill) and the development of large-scale manufacturing (the early industrial revolution in England) to the creation of state bureaucracies (in the former Soviet Union), dominate our view of cultural evolution in the West. Cultural microevolutionary interactions among different groups have also greatly affected their fates—such things as the massive emigration from Europe to the United States or the spread of pop culture from the United States to Europe and the rest of the world. Sometimes both macro and micro factors have played strong roles. Some societies overstressed their environments and their cultures disappeared, often helped to oblivion by losses in wars with neighbors, for example.[7]

Factors guiding cultural macroevolution intervene periodically in the form of geographic and climatic conditions (e.g., the Russian winter, as Napoleon Bonaparte and Adolf Hitler learned to their distress), disease, climate change, resource depletion related to overpopulation, and so on. Although microevolutionary change has recently been at center stage in our historical accounts, it is important to note that such things as rapid climate change, depletion of underground water supplies, soil erosion, loss of biodiversity, and increasing chances of global epidemics may result in a reassertion of dominance by macroevolutionary factors that are waiting in the wings.[8]

Early historians did not take an evolutionary view of history; some modern ones have viewed history as basically cyclical;[9] and many people still view it simply as a sequence of kings and wars.[10] But beginning in the seventeenth century, scholars such as Giambattista Vico, Voltaire, Georg Hegel, Auguste Comte,[11] and others visualized history as an evolutionary process with a meaning that could be discovered, a process that was moving toward a goal. Karl Marx was one of the first great theorists of historical evolution. He built on the views of Hegel but substituted changes in the ways in which people extracted a living from their environments for Hegel's notion of historical evolution in spirit. Marx postulated that human history was a sequence of epochs based on modes of production—slavery, feudalism, and capitalism—each of which led to characteristic social relations. The force behind this historical evolution was humanity's drive to increase its powers of production; in the evolutionary sequence Marx anticipated, class struggle would lead to an eventual withering away of the coercive arm of the state and the start of a new history.[12] So far, it hasn't worked out that way, but that's beside the point. Marx's was a pioneering attempt to find an evolutionary pattern in history based on scientific analysis.[13]

That history should be viewed as an aspect of "cultural evolution" can be seen in a simple translation of that phrase: a gradual change, often into a more complex form, of the body of extragenetic information possessed by humanity. Historical trajectories show many features in common with those of biological evolution. They are constrained by the past—elephants are unlikely to evolve wings; successful technologies are rarely discarded.[14] They are influenced by isolation and exchange. For example, most speciation is promoted by geographic isolation, and populations may gain evolutionarily important variation from gene flow. Diversification of languages in New Guinea and the spread of exotic cuisines in what was once the culinary desert of Australia are similar phenomena. Trajectories in both kinds of evolution are subject to selective termination—most populations, species, dynasties, and civilizations go extinct. And the trajectories are not directed at apparent targets. The courses of both biological and historical evolution are notoriously difficult to predict. Viewing historical inquiry as a process of investigating aspects of cultural evolution encourages a search for generalities and also highlights the need to look simultaneously at other aspects of cultural evolution, such as changes in economic systems and marriage customs, when investigating the course of events. This is widely recognized among historians and is practiced by some of them, and a simplified evolutionary approach was a feature of the profession in the early 1920s. They just aren't often explicit now about their work being evolutionary—sometimes, I suspect, out of a desire to defend disciplinary boundaries that should long ago have vanished.

Social scientists other than historians also have embraced fundamentally evolutionary ideas—that is, they have become cultural microevolutionists—within their own disciplines. Some started long ago on the basis of a loose interpretation of Darwin's theories.[15] Anthropologists, of course, have a strong interest in the physical and cultural evolution of human beings.[16] Those who study patterns of population size and movement—demographers—have an evolutionary framework for the way birthrates and death rates change in response to industrial development—the theory of the demographic transition.[17] The theory is somewhat flawed, but it is indubitably evolutionary, given that first birthrates and then death rates are seen as gradually declining in response to changing economic conditions.[18] Perhaps the leading sociologist of the twentieth century, Talcott Parsons, took a deliberately evolutionary view of the development of society.[19] Other examples abound.[20] Even many thoughtful politicians see political evolution occurring. Some, for instance, view political systems worldwide as evolving away from nation-states and toward more global structures.[21] As the Czech Republic's president, Václav Havel, put it in 1999, "There is every indication that the glory of the nation state as the culmination of every national community's history, and its highest earthly value . . . has already passed its peak."[22]

Such contributions of social science (to say nothing of those from the humanities) are essential to a full understanding of our group and individual human natures and the ways in which they evolved. We are all creatures of our social backgrounds as much as of our biological backgrounds. The many common features of the natures of people in the United States were different after the Civil War than before; citizens of long-standing democracies have different human natures from those accustomed to living under dictatorships; Comanches have different human natures from !Kung Bushmen. Personal, group, and national histories are primarily products of cultural evolution. We as individuals and the societies in which we live are continuously evolving culturally, just as individuals and populations are continuously evolving biologically (although the cultural and biological evolution of individuals, from birth to senescence, is traditionally called development). It is often useful to take a reductionist approach and look only at sequences of historical events to understand part of cultural evolution, or to record changes in gene frequencies to understand part of biological evolution. But it is essential to view both processes and their interactions holistically as well.[23]

Evolutionary theory, primarily its biological aspects, converted biology into a coherent discipline. I have long hoped that in the same way, evolutionary theory, primarily its cultural aspects, could do the same for the social sciences. We badly need an evolutionary theory of culture that unites the social sciences (including history) into a coherent field of human behavioral science. It would be one in which scholars would still use diverse techniques but would always be aware that they are studying different aspects of the same problem—how do we account for both the unity and the diversity of human natures and their evolution through time? Although the basic problem of disciplinary fragmentation of the social sciences is well recognized,[24] I suspect it will take a revolution by some bright young scholars to develop such a unifying theory. Science tends to progress one funeral at a time.[25]

Some elements of such a theory may be foreshadowed by the rather coherent story of sociopolitical evolution from hunting and gathering through the establishment of states that can be put together today. It can be assembled without recourse to presumptions about genetics, such as the presumption that human beings are inherently hierarchical. That in itself is informative. The evidence for the dominance of cultural evolution in this sphere is clear: genetically similar *Homo sapiens* have created a vast diversity of social settings, both egalitarian and hierarchical. Even 250 generations of selection could hardly have changed our ancestors' genomes from "hunting and gathering" to "empire building," although that's all the generation time available between the onset of the agricultural revolution and the rise of empires. Furthermore, a child from a society that hunted and gathered just five generations ago easily learns to be a jet pilot, physician, or engineer. The change in human natures from those of hunter-gatherers in family groups to those of citizens in a modern state was clearly the result of cultural, not biological, evolution.

How did empires and modern states evolve in the first place? To begin to understand that, we must first look at the emergence of farming, a great sea change in the evolution of culture.

The Evolution of Agriculture

In the early 1970s, a team from the Institute of Archaeology at University College London explored the diets of the inhabitants of Abu Hureya on the Euphrates River in Syria. They were interested in what the residents had been eating sometime between 11,000 and 8,000 years before. But they weren't watching the locals shop—instead, they were laboriously floating charred plant remains from excavated debris and catching them in a cloth filter. They then sorted carefully through 712 samples, recovering, on average, about 500 seeds of some 70 kinds of plants. Using those results, they figured out what had been featured at those long-ago prehistoric meals. Of the total of 157 seed plants in the samples, one of the most frequent was wild einkorn wheat, the progenitor of a variety sometimes grown today on poor soils. The evidence gathered by the research team helped to establish that sedentary hunter-gatherers of that time relied in part on wild cereals that they harvested *before* those plants were brought under cultivation.[26] That evidence provided an important clue to *how* agriculture evolved. Much less is known, however, about *why*.

Considering the critical importance of agriculture in human history, it is unfortunate that the reasons for its first appearance remain hidden in the mists of time.[27] The origin of plant domestication, with both sowing and harvesting, apparently began over much of the world some 10,000 to 5,000 years ago.[28] Scholars believe that domestication started ten millennia ago in the Fertile Crescent, an arc of open pistachio woodlands and grasslands that stretches northward from what are now Israel and Lebanon through northern Syria (including Abu Hureya) and southern Turkey to northeastern Iraq.[29] The Fertile Crescent is limited on the west by the Mediterranean Sea and on the north and east by higher terrain and the Zagros Mountains; on the south, it is bounded by the arid Negev region and the Syrian Desert. The Crescent had an unusual abundance of plant species suitable for domestication.[30] Scholars believe it may have taken only a few centuries to produce changes in the form of the heads of wheat and barley that indicate the transition from wild to genetically modified domestic strains.[31] But, again, why agriculture evolved remains debatable.[32]

One theory is that a major (if sporadic) driving force throughout human social evolution was increasing population density. The needs of an increasing number of people in an area would severely deplete resources and trigger a demand to find ways to obtain more food per unit of effort.[33] In this reasonable and rather persuasive view, the rise of agriculture was simply a response to the need to intensify food procurement as population pressures grew. One can speculate that development of new "Great Leap" technologies, such as the spear-thrower or the bow and arrow, and the resultant intensification of hunting first increased food availability and then led to depletion of game herds, as suggested by the extinctions attributed to Pleistocene overkill.[34] As more efficient hunting techniques and changing climate reduced the numbers of large game animals in the late Pleistocene epoch (20,000–10,000 years ago), hunting became less rewarding and the gathering of plant foods became more important. Because women probably did most of the gathering, that could also have changed the social structure within groups in a more egalitarian direction and reduced male–male bonding, which had been important in the coordination of dangerous hunts.[35]

Population pressure, declining game supplies, and increased plant gathering make a good story about the origins of agriculture.[36] However, the evidence of food shortages in prehistory is largely circumstantial (although increasingly evidence is being uncovered of periodic nutritional stress in many hunter-gatherer societies),[37] and we must be careful about extrapolating from the behavior of historically known hunter-gatherers to that of physically modern *Homo sapiens* prior to the agricultural revolution. For much of the pre-agricultural existence of *Homo sapiens*, big game animals were still abundant.[38] There is also considerable difference of opinion about whether or not hunter-gatherer groups outside of polar regions[39] frequently outstripped the carrying capacity of their environments—that is, grew to the point at which their numbers were so large that the capability of the environment to sustain future populations was reduced.[40]

Indeed, anthropologist Marshall Sahlins has described tropical hunter-gatherers as "the original affluent society."[41] Sahlins's idea that hunters and gatherers were rarely in a desperate search for food seems to have been true for some groups.[42] His prime example of hunter-gatherer affluence, the !Kung Bushmen of southern Africa, maintained a low population density because of their long spacing between births, use of scattered resources, limited wants, and an occasional disastrous year. Given that low density, the Bushmen "preferred the foraging economy to its local alternative of pastoralism because of its cost advantage over domestication *on the average*."[43] Hunting and gathering simply required less effort than herding.

It may be, however, that the key to the spread of farming was not the *average* benign conditions hunter-gatherers enjoyed but the occasional periods of extreme scarcity—a variant of the general population pressure hypothesis.

Recent work suggests, for example, that some groups experienced seasonal changes that reduced nutritional intake by as much as 50 percent. As anthropologist Robert Foley put it, "The implications are all too clear: while some hunter-gatherers may have thrived in particular environments, many did not, but suffered seasonal and long-term starvation, and had to strive hard for survival."[44] It is possible that the motivation for herding and cultivation was to form a buffer against the uncertainties of hunting and gathering—as both insurance and a device to allow farmers to gain influence by provisioning others in time of shortage.[45] This could explain the rapid spread of agriculture from the Fertile Crescent, which might have been independent of population growth. It may have been that farmers and herders simply survived periods of hard times that hunter-gatherers did not, and therefore, with their nutritional and economic advantages, replaced them.[46]

There are other outstanding questions about the transition to farming. If food shortages promoted a switch to agriculture, why is it that many groups apparently did not experience such shortages and persisted in hunting and gathering, a few of them to this very day? Furthermore, hunter-gatherer societies presumably could avail themselves of a wide array of population-control methods, ranging from infanticide and prolonged lactation to contraception and abortion.[47] Couldn't they use them to keep their population sizes below the level at which resources became dangerously scarce?

In addition, although farming may have started independently in different places, why in so many areas did agriculture spread by diffusion[48]—people adopting it by imitating neighboring groups that had previously taken it up—if hunter-gatherers were leading affluent lives?[49] Perhaps it wasn't the technology itself that was so attractive. In Europe, there is evidence that agriculture did not spread solely as a matter of movement of ideas. Rather, the use of agriculture seems to have increased population growth rates of new farmers, who then produced surplus populations that migrated along with the new technology.[50] In this view, local overexploitation created foci of farming, which then spread because of population growth and migration—creating its own need as it went. On Pacific islands, there is abundant evidence that Polynesian settlers first polished off the local indigenous fauna before intensifying their use of domestic plants and animals.[51] Then at least some of them moved on to carry both extermination and farming practices to unoccupied areas.[52]

A second hypothesis, not necessarily contradictory to the first, is that agriculture developed in response to the rapid end of an ice age, when human beings had taken the Great Leap Forward and were intellectually ready to do the requisite planning and deal with the technical problems of raising crops, domesticating animals, and living together in permanent settlements. In the Near East about 13,000 years ago, climate change began to greatly expand the ranges of large-seeded grass species, potentially an enormous food resource.[53] Originally, people probably gathered these grasses by beating the ripened seeds into baskets, just as grass seeds have been gathered around the world even in recent times.[54] That worked well because the mature seed heads of wild wheat and barley had evolved to shatter—to break apart easily so that the seeds (the edible part) are dispersed. As conditions dried up about 12,500 years ago, the semi-sedentary exploiters of those grasses, already equipped to harvest and process grains, began to form villages and, about 10,000 years ago, to farm the grasses.[55] The grains were brought into cultivation and harvested by hand or sickle. Sowing (reseeding) was then added to harvesting. Inadvertent selection by the first farmers, as well as other genetic events involving chromosome duplication,[56] apparently soon produced strains with nonshattering heads—ones that held together when the plants were cut.[57] Individual plants with seed heads that tended to stay on as the plants were cut, rather than being dispersed and lost to the gatherers, were more likely to supply seeds for the farmers' subsequent sowing.

Such inadvertent selection by farmers also produced strains that germinated whenever farmers sowed them, that were more erect and thus easier to harvest, that had more and larger seed heads, and so on.[58] These and other patterns of unconscious selection were replaced by deliberate selective breeding perhaps as long as 10,000 years ago. That was an early example of "genetic engineering,"[59] a process now critical to maintaining agricultural productivity. The results of that selective breeding allow archaeologists and paleoethnobotanists (specialists in analyzing the relationships of prehistoric people to plants) to estimate when domestication started. They can separate the remains of wild-gathered feasts from those that originated in cultivated fields on the basis of physical characteristics of "fossil" seed heads.[60]

Whatever the cause of the emergence of agriculture, it is a mistake to assume that there was a series of "eureka events" as people discovered and developed its practice.[61] Early hunter-gatherer groups were, like modern ones, undoubtedly intimately familiar with the flora and fauna of their regions. They would not have been puzzled by the sprouting of seeds of their favorite foods in their kitchen middens or latrine areas and doubtless acted to encourage the production of favorite plant foods that they gathered—by weeding, pruning, and spreading seeds. Plant and animal domestication was generally a gradual process, moving from intimate knowledge through casual encouragement to full-scale control.[62] Nonetheless, within a region, the final stages of domestication could be quite rapid.

For example, in the southern Levant, the shift from casual use of grains to their systematic cultivation may have occurred in just a few hundred years,[63] sometime between 10,300 and 9,900 years ago. Selection appears to have quickly produced domesticated strains that could be harvested using the tools of the day—wooden sickles set with flint blades—with little loss of seeds. Those hard seeds were also relatively resistant to spoilage and thus could be stored for long periods—but they also needed to be husked and ground into flour with mortars and pestles to make them edible. Plant domestication in the Levant was followed by the domestication of cattle, goats, pigs, and sheep within a millennium or so (about 9,000–8,000 years ago).[64] In that area, the full shift to agriculture occurred in a few thousand years—an eye-blink in geological time, which is normally measured in millions of years. Globally, however, the shift was protracted in ecological time, taking some 400–500 generations to essentially finish the job.

The shift to agriculture probably happened in diverse ways in diverse places—independently in at least the Near East (Fertile Crescent), central Mexico, southern China along the Yangtze River, northern China (Yellow River), the south-central Andes, the eastern United States, sub-Saharan Africa, and perhaps New Guinea.[65] Some groups may indeed have been driven to take up agriculture by periods of hunger, when the need to intensify food procurement was clear and present—in some areas, it almost certainly was a successful response to population pressures.[66] Other peoples may have begun to farm to avoid shifting their homes as climate change altered the range of the game animals they traditionally hunted or because farming permitted a sedentary life that had other desirable attributes. Still others simply may have imitated neighboring groups or sought to increase their supplies of some favorite food previously gathered or hunted.[67] Agriculture was thus invented gradually, piecemeal, and quite probably sometimes reluctantly as groups changed time-honored lifestyles. In many areas, cultivation probably existed alongside hunting and gathering for millennia before true agricultural settlements were established.[68] But in a period of less than 10,000 years, agriculture became the most important source of food in all the major inhabited regions of Earth.

What Agriculture Changed

Although questions persist about why a substantial fraction of *Homo sapiens* settled down to farm, there is little doubt about the far-ranging consequences of that move. However it originated, agriculture started a positive feedback system that put humanity on the road to sociopolitical complexity. It first led to a regional (and eventually worldwide) expansion of the numbers of people that could be supported without decreasing an area's capacity to support people in the future.[69] It also reduced other natural constraints on population growth. For instance, agriculture made it easier for women to provide their infants with soft weaning foods. Although hunter-gatherer diets were generally of high quality—higher, in fact, than those of early agriculturalists—soft foods suitable for infants were usually in short supply.[70] Farmers, in contrast, could quickly move their babies to a diet of mush made from ground cereals. This reduced the length of lactation and its associated infertility. The sedentary lifestyle also permitted closer child spacing because women did not need to carry their offspring on long foraging trips or migrations.

Rising birthrates almost certainly accompanied the agricultural revolution. It seems unlikely, however, that the emergence of farming resulted in lower death rates. Indeed, they may have increased. As people became more numerous and colonized the world outside their African homeland, a variety of pathogens and parasites of other animals began to colonize them.[71] The establishment of settlements would have led to increases in the number and incidence of communicable diseases, many of which originated with domesticated animals.[72] Larger human populations allowed the persistence of diseases that otherwise would have died out,[73] and sedentary human lifestyles were made to order for worms and other parasites transmitted via feces. The interactions between human beings, urbanization, and the microbes causing disease have been both significant and complicated, with plagues often greatly increasing the human death rate.[74]

The different epidemiological (disease) environment of the agriculturalists, especially those with many domesticated animals, gradually enabled them (as opposed to the hunter-gatherers) to evolve some immunity to the new diseases. That immunity also made it possible for them to wipe out many of the hunter-gatherer populations they encountered—as exemplified by the decimation of indigenous New World populations of both hunter-gatherers and agriculturalists (who had relatively few domestic animals) after European contact.[75] The interactions that shaped the early epidemiological environment continue to this day as aggregations of human beings, sometimes poorly nourished, become increasingly dense; as intercontinental transport systems permit ever more rapid travel, allowing a quicker, more far-reaching spread of diseases; and as the evolution of antibiotic resistance in pathogens proceeds apace.[76]

There is some evidence that human health generally declined with the onset of agriculture. Infant and child mortality may have increased substantially because available weaning foods were nutritionally inferior to mothers' milk, and poor nutrition made the young especially susceptible to disease.[77] This conclusion about our past, like so many others, is not without controversy. It is based on examination of skeletal remains, which show an increased frequency of disease indicators after the agricultural revolution and also suggest a decline in average age at death. But traces of disease can also be interpreted as meaning that there was an increased tendency to survive certain illnesses that, if fatal, would have left no traces on the skeleton. Moreover, an apparent decline in life expectancy that is reflected in an increased proportion of youthful skeletons could also be the result of higher birthrates.[78]

Whatever the exact causes, birthrates became significantly higher than death rates, and population growth accelerated. That created a need for more food and thus pressure for technological change to make more intensive agriculture possible. One critical advantage of the settled life was that it permitted the development of technologies that were highly useful but not portable—a specialized "tool kit" containing items such as farm implements and irrigation systems, and gradually improved garden beds, that could not easily be carried from camp to camp. Once people became sedentary, they could invest their labor in such efforts as improving water flows around fields, preparing soil, weeding, planting and caring for tree crops, domesticating animals,[79] and constructing containers in which to store food against hard times. Intensification also gradually led to the ability to produce more food than a family needed, which permitted some individuals to turn to other occupations (and sometimes, as we'll see, to appropriate others' surpluses—i.e., taxation). Thus began the division of labor, not exclusively based on gender and age, that is so characteristic of more highly organized societies. Intensification and division of labor, in turn, required new social arrangements, and human societies underwent a sequence of changes through villages, clans, chiefdoms, and archaic states that led eventually to modern nation-states.[80]

The Rise of the State

During the agricultural revolution, family-level organization (bands or tribes with the family as the basic economic unit) probably predominated over much of the world. The exceptions were in areas of unusual resource concentration,[81] especially areas rich in seafood resources, where families could afford to aggregate in villages.[82] Why, then, weren't all early farmers, after they settled down to cultivate crops, satisfied with relatively relaxed, low-conflict, subsistence-farming lives? Most scientists believe that population growth, environmental deterioration, rising social pressures, or all three made those lives increasingly uncomfortable in many areas.[83] Population growth caused the environment to "fill up." That meant that forest agriculturalists had to return to fallowed lands too frequently to permit regeneration of the nutrients necessary for productive cultivation. It also meant conflicts over valued farm sites. Nutrient depletion, plant disease, pest problems, or combinations of such environmental changes reduced yields. Pressures from neighbors for food in exchange for tools or brides may have raised the demand for food, or threats from marauding pastoralists may have required farmers to aggregate for defense. In each case, a new level of intensification of food production was needed to solve the problem, and many groups of early agriculturalists responded to the challenge.[84]

Some common themes run through discussions of the evolutionary path from coalitions of farm families to states. States, as anthropologists Allen Johnson and Timothy Earle define them, are societies that cover entire regions and include hundreds of thousands or millions of people, often drawn from different ethnic groups and engaged in diverse economic activities.[85] They represent the culmination of humanity's rapid post-agricultural trend away from being a small-group animal whose social organization dealt exclusively with kin. States are also characterized by well-defined upper classes that profit from domination of the lower classes and by attempts to substitute pseudo-kin for the genuine article. This need for extended kinship was first reflected in legends of common descent of tribal groups. Its persistence is patent in the rhetoric of states about fatherland and motherland, leaders as "little fathers," "Uncle Joe" Stalin, "Uncle Sam," and so on.[86] The language of pseudo-kinship is widespread within states as well, used to bind together unrelated people in social or religious groups: for example, fraternity brothers, sorority sisters, the Holy Father, Mother Teresa, Jewish sisterhoods, and the term *family* used to describe everything from Japanese corporations to the Mafia.

Perhaps the most ubiquitous theme in discussions of the origins of states is what has become known as circumscription. It is an idea that can be traced back at least to nineteenth-century English social philosopher Herbert Spencer,[87] but its modern version owes its formulation to a classic paper published in 1970 by anthropologist Robert Carneiro of the American Museum of Natural History.[88] Carneiro's basic thesis of circumscription was simple: before a state system could evolve, something had to prevent future subjects from fleeing from their would-be rulers. When proposed in 1970, Carneiro's theory of circumscription stood in contrast to the view held by some that the creation of states was somehow automatic or voluntary and had nothing to do with people's ability to flee from prospective rulers. In that view, the state was seen as a stage in the evolution of society made possible by the production of surplus food by agriculture, a surplus that permitted specialization and greater division of labor within societies.[89] More specialization, in turn, permitted further agricultural intensification, for example, through the building of large irrigation systems,[90] which led to even larger populations and a demand for the more extensive managerial functions needed to maintain the state (whose power and control the rulers arrogated to themselves). In this view, whether or not the society was circumscribed was irrelevant.

Three kinds of circumscription were recognized by Carneiro and are supported by examples from the historical and archaeological record—geographic, resource, and social. Geographic (or environmental) circumscription obtains when barriers such as mountain ranges, oceans, or deserts limit the dispersion of farming peoples. Island societies and those confined to narrow mountain valleys are classic examples of geographic circumscription.[91] Resource circumscription describes the situation in which movement is limited by a sharp gradient in resource or environmental quality, as when people living on the rich soil of a river valley cannot spread to higher ground because soil quality there is poor. In social circumscription, other peoples already occupy the periphery of a society and limit its expansion. Social circumscription caused by the Philistines of the southern coastal lowlands of Palestine is one explanation for the rise of a monarchical Israelite state under David.[92] The Philistines kept Israelites from fleeing the imposition of a king who would control their lives.

According to Carneiro's theory, when the population of an uncircumscribed society grows, people can simply move into unoccupied lands. To a degree, North America was populated by people who had left Europe to find economic opportunity or to avoid governments they did not like; new technologies had made circumscription by oceans less effective. But when a circumscribed society grows to the point that there is no unused terrain, squabbles over land lead to warfare, and losers often are subordinated to winners because the former can no longer flee. This, Carneiro postulated, is the force that drives social stratification and political evolution, moving societies from the family-oriented band organization through tribes and chiefdoms to states.

This direction of political evolution toward the development of states is correlated with more frequent warfare,[93] and intensive warfare seems to be associated with the chiefdoms that were precursors of states. War captives, at first sacrificed[94] or even eaten,[95] eventually were enslaved and formed the basis of a lower class. Add in a group of successful warriors and other faithful followers of the chief who are awarded slaves, and the seeds of further stratification are sown—leading eventually from a status pyramid to clearly demarcated, more or less fixed classes or castes.[96] Political evolution also led to taxation, conscription, bureaucracies, and all the other features of states that subordinated people find difficult to tolerate. For all this to happen, the labor of the subordinated people had to be sufficient to produce goods beyond those needed for their own subsistence. Control of such surpluses by elites, which was a central focus of the great political economist Karl Marx,[97] is what made stratification (class divisions) possible.[98]

Circumscription in Action

Is there any way we can evaluate Carneiro's thesis of how cultural evolution led from the origins of agriculture to states? Maybe so, given that states evolved in some places and not in others. The islands of Polynesia form as good a test system for the theory as we are likely to find.[99] The populations occupying these islands originated rather recently (some 3,000 years ago) from a common ancestral population. Thus, similar people lived for various periods of time on an array of islands of different sizes, with different environmental characteristics and independent cultural evolutionary trajectories.

Patrick Kirch, an anthropologist at the University of California, Berkeley, has spent his career reconstructing the political and environmental histories of Pacific island microcosms. On those islands, the hard work of stratigraphic archaeology—detailed analysis of the sequence of past societies by the relative positions of their physical traces in geological strata—has gone on for only some fifty years. But the rate of investigation has been accelerating, and the results are impressive.[100] Careful excavation has produced a wide range of artifacts, including potsherds, intact pottery, stone tools and weapon points, collections of shells (some worked into tools or jewelry), the bones of extinct birds, pebbles used as net sinkers, shark teeth, accumulations of flakes from the working of flint-like rocks, the remains of crop trees, and the like. Paleoenvironmental investigation, especially that using cores of sediments, has also produced evidence of past floras from pollen, charcoal, and other plant materials and evidence of past faunas in the form of animal remains, including those of land snails and insects.[101] Analysis of this tediously gathered evidence by Kirch and others, with some contributions from linguistics, has in turn produced a most instructive picture of the colonization and transformation of the islands of the Pacific.[102] Back-breaking, dirty-hands work can pay off in science as elsewhere.

New Zealand is the largest of the Polynesian island groups, with a land area thirty times that of the Hawaiian Islands. Interestingly, New Zealand is the most recently colonized, and its temperate climate ironically restricted the possibilities for Polynesian agricultural development (which employed crops better suited for the Tropics). As a result, it still had a low population density at the time of contact with Europeans.[103] In light of Carneiro's theory, one would expect a low density of population over a wide geographic area to result in an absence of any highly complex sociopolitical organization. In fact, Europeans noted no centralized regional political structures among the Maori people of New Zealand like those found in Hawaii and Tonga. This might seem to be explainable simply by a lack of resource circumscription, but the story is more complex than that.

Conditions suitable for traditional Polynesian agriculture were found only in the northern part of the North Island of New Zealand, where the richest marine resources also were concentrated. That led people on the North Island, driven by competition for marine resources and rich land, to evolve a more complex political structure than that on the South Island, where there were far fewer people. New Zealand, then, illustrates both resource circumscription and, because expansion in the resource-rich areas was limited by the presence of hostile neighboring groups, social circumscription.[104] The short period of occupancy and relatively vast land area combined to limit the complexity of Maori

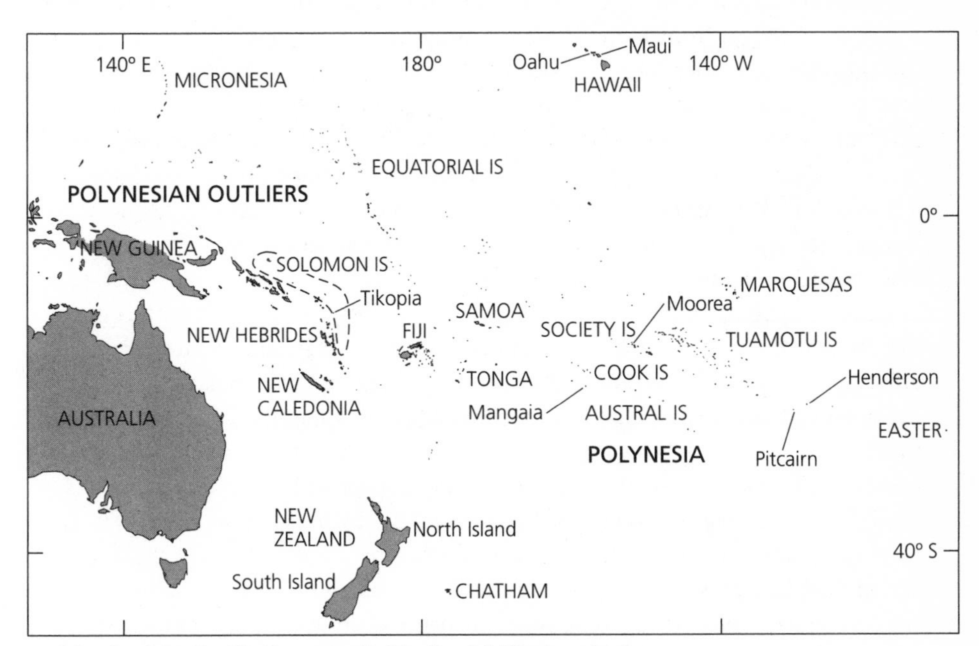

Islands of the Pacific Ocean studied by Patrick Kirch and others.

political organization, but at least on the North Island, society was beginning to evolve in the direction of statehood (see the map).

The "highest" level of Polynesian political evolution, essentially thoroughly stratified incipient states, was found on the Hawaiian Islands. The situation in the Hawaiian archipelago was very different from that in New Zealand. People arrived there about A.D. 400, which was 500–700 years earlier than the roughly A.D. 1000 arrival in New Zealand,[105] and occupied a much smaller island area, creating a population density in relation to arable land about ten times that of New Zealand. There was strong resource circumscription because there was considerable variation in the suitability of land for the wet (usually, irrigated pondfield) cultivation of taro.[106] Hawaii and eastern Maui, which lacked prime irrigable land areas, had terrain suitable for dry cultivation, and this terrain was used for a system based on farming of sweet potatoes.[107] Eastern Maui had "filled up" before European contact.[108] The eastern Hawaiian polities used their organization to attack those to the west in an attempt to acquire the valuable western pond-irrigation systems. Circumscription theory explains a lot about political evolution in Hawaii, but it doesn't explain all the details of such things as the possible role of conflicts over irrigation in generating intergroup competition and political evolution.[109]

The Tonga archipelago has only 4 percent of the land area of Hawaii, but it was colonized 1,500 years earlier (circa 1100–1000 B.C.) and thus had a longer period for its population to grow and its political systems to evolve than did either Hawaii or New Zealand. Like Hawaii, it had developed to the point of incipient statehood, with a secular and a sacred king residing in a "capital" on the main island of Tongatapu and with outlying districts and islands ruled by lower-ranked chiefs, who sent tributes to the king. At the time of European contact, population density was very high, although probably not quite as high as that of Hawaii. Apparently, Tongatapu had been brought under central political control about 1,000 years ago, following a period of population growth that had led to a "full land situation," in which no arable land was left unoccupied.[110] Tonga, then, appears to have been a classic case of resource circumscription, the resource again being arable land.

The Polynesian test system for Carneiro's theory also provides interesting insight into the degree to which circumscribed human societies can alter their environments and then follow disparate paths. Although in popular conception the islands of Polynesia are an unspoiled paradise, biologically nothing could be further from the truth. Despite their short tenure on New Zealand, the Maori managed to drive to extinction thirteen or more species of moa (flightless birds, some larger than ostriches) in less than 1,000 years by hunting them intensively for food[111] and destroying their preferred forest habitats.[112] In a northern valley on the island of Moorea in French Polynesia, the burning of vegetation, possibly in support of agriculture, converted primary forests into degraded fernlands and early successional forests. Erosion, probably connected with the loss of forest, deposited so much material on the valley floor that it was converted from a swamp into a relatively dry, flat area of alluvial soils.[113] Similar patterns of modification, combined with extinction of endemic species and introduction of exotics, have been documented as having commenced some 2,000 years ago on Mangaia Island, in the Cook Islands.[114]

The fate of extremely isolated, sixty-square-mile Easter Island is well known. The Polynesians colonized it around A.D. 400; they successfully introduced only chickens (and, inadvertently, rats). Pigs didn't make it, nor did breadfruit, coconut, or several fruit trees common to the Polynesian diet.[115] At first, the Easter Islanders had a rich source of protein in forest birds, nesting seabirds, fishes, and dolphins, which they hunted from large canoes. But as their population grew to a peak of some 7,000–10,000 people about A.D. 1400–1500, they completely deforested the island, where the principal forest tree was a slow-growing palm.[116] This ended the dolphin hunting and deep-sea fishing because there wasn't a single tree left to use in building canoes. The forest birds went extinct. The deforestation also ended the carving of the magnificent huge stone torsos *(moai)*, numbering nearly 1,000 and weighing as much as eighty tons, for which the island is famous (see the figure).[117] There was no longer a source of log rollers to use in dragging the *moai* from the quarry in the hills several miles to the shoreline, where they were to be erected, or levers to use in erecting them.

With the forests went much of the soil, swept away by strong, drying winds in the absence of trees to serve as windbreaks. Crop yields doubtless declined, and warfare over the dwindling resources became common. Many of the *moai* were toppled and destroyed, perhaps in connection with battles or insurrections, but the oral histories available to ethnographers are uninformative about this. In any case, warring tribes organized themselves into two loose, perpetu-

Giant heads and torsos *(moai)* on Easter Island. Left, re-erected statue, about twelve feet high and weighing about twelve tons; right, unfinished statue still in the quarry. Photographs by P. R. Ehrlich.

ally battling coalitions. Gradually, near slavery and predatory raiding took over, and a new cult arose, replacing the old-time religion.[118] Cannibalism apparently was common; a classic taunt was "The flesh of your mother sticks between my teeth."[119] By the time of European contact, in the eighteenth century, the population had declined to about 2,000 people. Thereafter, relations between outsiders and Easter Islanders, which included a Peruvian slave raid and introduction of exotic diseases to the island, depressed the population, which numbered only 111 people around 1900. As Patrick Kirch put it, "the culture of Easter Island went full circle—from a small band of Polynesian colonists on the shores of a new land, through remarkable heights of cultural brilliance, to a once again small and pitiful band striving to surmount yet another challenge to their island lifeways."[120] A few years ago, my wife, Anne, and I visited Easter Island, which is now a monument to overpopulation and disregard of conservation imperatives, with all its original rich forests gone and its endemic land birds extinct.[121] We had to focus our bird-watching on the three widespread, common birds that have been imported from the South American mainland: the Chimango caracara, the common diuca finch, and the rufous-collared sparrow.

Easter Island's story would seem to be just one more tale of ecological overexploitation and collapse—like those of the early civilization in the Tigris and Euphrates valleys, the classic Maya, the Anasazi, and the ancient Greeks.[122] Easter Island is different, though, in that it is one of that set of replicates that includes Hawaii, Tonga, and New Zealand—Pacific areas colonized by people of the same general culture from the same basic source at the same general time. Some other areas, such as the Marquesas Islands, northwest of Easter Island, followed similar patterns. They had rapid population growth, faced an array of environmental problems, and resorted to warfare, cannibalism, and construction of large monuments.[123] They never established the levels of stratification or rigid hierarchies of the incipient states found in Hawaii and Tonga—circumscription on Easter Island and similar islands did not lead to the development of states.

Kirch did an interesting analysis of the fates of those who colonized two other Polynesian islands: 20-square-mile Mangaia Island, southwest of the Marquesas Islands, and 1.8-square-mile Tikopia, the far western outlier of Polynesia, situated just east of the Solomon Islands.[124] The two islands differ in size by an order of magnitude, but Mangaia is much older geologically, has very poor soils (especially lacking in the critical nutrient phosphorus), and is located in the central Pacific Ocean, with a relatively unproductive coral reef system. Tikopia, in contrast, has rich soils and, being closer to the western Pacific centers of aquatic diversity, a relatively rich reef system. On both islands, geographically circumscribed societies developed.

Polynesians landed on Mangaia about 500 B.C., their arrival heralded in the sedimentary record by an abundance of charcoal particles, probably from the use of slash-and-burn cultivation. Pollen records show that at least a quarter of the island's landscape was rapidly deforested and the thin layer of nutrient-rich topsoil destroyed by a people accustomed to the young, rich soils of most of the volcanic islands they had previously colonized. By about A.D. 1200, the people had switched to irrigated taro agriculture in the soils deposited by erosion in the valley bottoms. The fields occupied less than 2 percent of Mangaia's surface and were the target of repeated tribal warfare. The Mangaians wiped out more than half of the island's native birds and decimated other elements of the fauna. They cultivated chickens, dogs, and pigs as food, but dogs and pigs (and perhaps chickens) were already gone when Captain James Cook showed up in 1777. Why they were eliminated is not certain, but one likely reason for the disappearance of the pigs was competition between them and people for vegetable foods. As pig and chicken bones faded from the late prehistoric fossil record and the native fauna declined, two new items showed up as major features of the Mangaian diet—rats and people. Broken and charred human bones not in graves tell a dramatic story of conflict, presumably over the dwindling resources. One recently excavated rock shelter appears to have been used exclusively for the ritual consumption of people.[125] By historical times, the Mangaian term for something tasty to eat was "as sweet as a rat."[126]

The politics of Mangaia changed as the resource situation deteriorated. Hereditary chieftainships were replaced with simple dominance by those with the military strength to overpower opponents and control the critical irrigation systems, which had become the focus of both politics and religion. Terror and human sacrifice became common, with offerings made to a dual god of war and irrigation. Population peaked at approximately 4,500 people and had declined to something more than 3,000 by the time of European contact. In many ways, Mangaia showed a pattern like that of Easter Island.

Tikopia, a solitary peak poking out of the Pacific Ocean, yielded quite a different story. In historical times, the tiny island maintained a population density almost five times that of Mangaia, even though the population was smaller in absolute size. The islanders managed this by developing an intensive system of arboriculture—orchard gardening—which covered the island with a multistory orchard of economically valuable plants, with fruit and nut trees shading yams and other crops. In the few places where there were no tree crops, there was intensive cultivation of yams and taro. Fisheries, carefully regulated by taboos, were the main source of protein. In addition, population-control mechanisms were in place—celibacy, contraception, abortion, infanticide, near-suicidal seafaring by young men (who were encouraged to leave on very dangerous expeditions), and, in some cases, expulsion of segments of the population. Zero population growth was incorporated into the Tikopian religion.

Why did the residents of tiny Tikopia find a way to sustainability whereas the Easter Islanders and Mangaians did not? The archaeological record suggests that for the first thousand years of occupation (900 B.C. to A.D. 100), the pattern of resource depletion found on the other two islands was repeated on Tikopia. In response, pig culture became more intense after about 200 B.C., but then between A.D. 100 and 1200 the sustainable arboricultural system was put in place. It was later refined by a deliberate decision to end pig production around 1600 because the pigs were too destructive to the farming system. The human population never crashed.

Differences in the productivity of the physical environment may have been partly responsible for the differences between Easter Island, Mangaia, and Tikopia. But people on the first two islands moved toward terror and cannibalism as they attempted to intensify production. The Mangaians, Kirch wrote, "chose a path that led in the end to terror: to the stalking of sacrificial victims in the night, to the incessant raiding of neighboring valleys, to a political system built on brute force."[127] The Tikopians took a different path. They paid a price for their sustainability, in brutal population control measures, but it appears to have been a mutually agreed-on price. Kirch doesn't know why the different paths were chosen, but he suspects that the key was the size of the islands and the (related) size of their populations. Easter and Mangaia Islands were large enough and had enough people for their populations to fragment into geographically defined groups: "us" in one valley, "them" in the next. On the other hand, the tiny size of Tikopia (one can circumnavigate the island on foot in less than a day) meant that everyone in the relatively small population must have been on rather close terms with everyone else, and this encouraged collective decision making. The size of the islands may well have been a major determinant of the different natures of their human populations.[128]

The Road to Inequality

The Polynesian test gives us some insight into why states evolved. It shows the importance of circumscription, but it also shows that cultural microevolution can produce different historical trajectories in the face of macroevolutionary constraints. Is there any other way to approach the problem of how political evolution may have proceeded in prehistoric times—before there was any written record? How was the egalitarian structure of hunter-gatherer groups transformed into the highly stratified societies of today? Are there recognizable stages in that evolution? There is a way to answer such questions, and interestingly, it is parallel to the way Ernst Mayr's model of speciation has been tested, as described in chapter 3. It is a technique that, for processes that go on for long periods, essentially substitutes changes in space for changes in time. In speciation, you will remember, when we look at a "snapshot" of the world in a single time-slice—the present day—we see populations with all degrees of differentiation, from geographically isolated but extremely similar populations to clearly distinct species living together without interbreeding. They seem to be in various stages of geographic speciation. And when we look at well-understood fossil records, such as those of *Homo sapiens*, we see the same thing. Similarly, when we look at present-day populations, all of them to some degree already circumscribed in a "full" world, or past populations that have very well understood archaeological records, we can detect societies in what can be interpreted as various stages of political evolution, from those with a simple family-based structure to one with complex social stratification. Let's take a look.

In economies based on subsistence agriculture, each family enterprise within a family-level society is organized to meet the basic needs of a self-sufficient household. Subsistence agriculture involves a division of labor based on sex and age,[129] and it does not ordinarily produce surpluses beyond necessary reserves against hard times. Furthermore, family organization probably is partially stabilized by the common genetic interests of family members, by the affinities people feel for their children, parents, and siblings,[130] and certainly by the many reciprocities that help bind families together. The family provides the necessary labor for domestication of plants and animals and manufactures the tools (such as sickles, digging sticks, cooking pots, and bows and arrows) needed to maintain the household.

Family-level agricultural societies characteristically existed first at low population densities (fewer than one person per square mile), given that because of resource limitations, the hunter-gatherer groups from which they evolved were necessarily thin on the ground. Early farming families were factors in other families' environments, but as in strictly foraging societies, interactions among families presumably were not usually intense. But families of some early agriculturalists formed settlements in the same areas, as do the modern Machiguenga who live in the Peruvian rain forest.[131] The Machiguenga get about one-third of their food by hunting and the remainder from highly productive gardens. The distribution of individuals varies greatly from time to time. People live mostly in semipermanent farming hamlets of a few related families that cooperate in limited ways (sharing food, clearing land) but retain quite strict family control over supplies, tools, and even plots within the communally cleared land. Periodically, the Machiguenga disperse to take advantage of periods of abundance of wild foods that are still considered essential components of the diet.

The Machiguenga practice slash-and-burn (swidden) agriculture, in which a plot of land is cleared, the remains of the vegetation are burned in the dry season, and crops are planted for a few seasons. The land is then allowed to revert to forest. Machiguenga population densities remain low because of the need for long fallow periods and the desire to produce yields by using the least labor in tilling the limited soils. The occasional occurrence of drought may be another factor: the Machiguenga population may be limited by the productivity of land in dry years but not in normal years, when a surplus is readily produced. Perhaps the most important constraint on population density is the relative scarcity of nonagricultural products that the Machiguenga need or desire: game, fish, firewood, and palm leaves for thatching huts.

Warfare (and thus group defense) is unknown among the Machiguenga. The scarcity of resources in their environment militates against their trying to hold territories—any occupied (and then defended) area would quickly be denuded of useful foods and materials. Thus, Machiguenga society would appear to be at the "resources too scarce and dispersed to be cost-effective to defend" end of the environmental–territoriality spectrum. Concentration of population would be counterproductive. The family is the fundamental unit of the Machiguenga's loose societal organization because there would be no reward for more complicated arrangements. Indeed, when a village of some 200 Machiguenga was formed under a Peruvian government program, it dissolved when local fish and game were depleted. The Machiguenga lacked social mechanisms to handle the tensions generated by shortages and the need for increased travel time to harvest wild foods.[132]

A modern group that may well represent a more complex organization found in early pre-statehood societies is the Yanomamö of Amazonia. Like the Machiguenga, the Yanomamö are horticulturalists (they raise fruits and vegetables in gardens), but they live in environments that are locally more densely populated.[133] They generally invest much effort in capital improvements to their gardens, which produce plaintains and peach palm fruits over many years. These more permanent gardens are resources worth fighting over, and fight the Yanomamö do. Indeed, it was the famous studies of the Yanomamö conducted by anthropologist Napoleon Chagnon[134] that helped shatter the view of some modern anthropologists that warfare in pre-state societies would be limited to times of stress, would be highly ritualized, and would generally result in light casualties. Chagnon found that some Yanomamö surprise attacks resulted in the deaths of about one-tenth of the enemy—a casualty rate far greater than those of World War II (an equivalent loss would have been the death of 13 million Americans at Pearl Harbor instead of some 2,200).[135] Over the course of a quarter-century study, Chagnon found that 44 percent of Yanomamö males had participated in killing someone, about 30 percent of adult males had met with a violent death, and nearly 70 percent of adults older than forty had lost a close relative to homicide.[136] The cultural evolution of complexity can carry a price.

The more complex sociopolitical organization among the Yanomamö seems to have developed largely in response to the need to defend productive gardens and hunting and fishing grounds in an increasingly densely populated landscape. The tribe was completely circumscribed: it was not possible for people to flee into the adjacent lowlands, so they were forced to be highly territorial, to form alliances, and to fight to retain their meager agricultural base.[137] The Yanomamö are dedicated traders because many of the goods they need or desire are unevenly distributed through the area the tribe occupies. Like that of hunter-gatherers and the Machiguenga, Yanomamö social structure is rooted in family and kinship, but it has started to branch out. Families live together in sizable groups, extended families of thirty to thirty-five people called *teri*. The *teri* tend to have individual leaders, although leadership is not formalized and several men may take leadership roles simultaneously. A *teri* has to be a formidable group or it risks having its valuable gardens or other assets appropriated by another *teri*.[138]

In response to conflict over (relatively) scarce resources, aggressive, fearless men become a great asset, to be integrated into the local group rather than, as among hunter-gatherers, considered dangerous outcasts or even executed. As population densities increased and social structures arose in which family autonomy was increasingly sacrificed, family or *teri*-level leaders and valued warriors, such as those found among the Yanomamö, probably gave way to clan leaders, "big men," warrior kings, and the like. That led to the evolution of further complexity—to chiefdoms, states, and empires that integrated larger and larger groups of peoples over greater areas.[139] One cause can be surmised from the global situation. There were only some 5–15 million human inhabitants of Earth at the time of the agricultural revolution,[140] but human numbers reached about 250 million 2,000 years ago. That growth, from about one person per square mile of arable land to about twenty-five, presumably created situations requiring ever more agricultural intensification and ever more complicated social arrangements.

When environmental pressure makes it necessary to integrate family- and village-level societies into regional economies, strong leadership becomes essential to organize the group's economy and to carry out negotiations with other groups. Relationships with other groups are needed to allow trade of valued items and to compensate for local unpredictability of food supplies. At that point, someone takes over the role of organizer and lives on the surplus created by the others. He (almost always he) ordinarily organizes his friends and relatives (who get their share) as proto–tax collectors to confiscate that surplus and proto-soldiers to deal with those who object to the confiscation. With such leadership, therefore, come the first stages of social stratification, with people consigned to different classes. At later stages of political evolution, this stratification becomes a hallmark of the state.

Public ceremonies may have evolved as devices to help define groups and their interrelationships and prepare the groups psychologically for territorial defense. Ceremonies and rituals originally could have functioned as replacements for the kinship bonds that maintain the unity of family-level societies. They bind together societies in which multifamily groups must work together and negotiate with other groups, despite the lack of close kinship to smooth the way. Eventually, ceremonies would also have served as material manifestations of ideologies—parts of the worldview of segments of societies that relate to their domination of or by other segments.[141] In more complex societies, those in which specialization into armies, priesthoods, and the like has occurred, ceremonies can serve to legitimize and make mysterious the social stratification, leadership structure, and extraction of surplus by the elites.

It is important to recognize that the hierarchical structure of more complex human societies such as chiefdoms and early states developed gradually from the relatively egalitarian structure of hunter-gatherer groups. In this respect, the state is not a lineal descendant of chimp-like dominance hierarchies; rather, it is a brand-new social invention. As anthropologists David Erdal and Andrew Whiten assert, the hierarchies occasioned by the evolution of states "are not merely reborn ape hierarchies but uniquely human in both their behavioural detail and cultural recognition."[142] So, no "hierarchical ape" label fits humanity. And unlike chimpanzee and hunter-gatherer interventions to keep societies more egalitarian, in state societies, despots typically are deposed from below only with the intention of finding a more satisfactory replacement despot. Nonetheless, it is clear that like chimps (and presumably our common ancestors), human beings are talented at such things as forming coalitions against dominant individuals (or against the weak, to keep them that way), making up after disputes, and generally manipulating one another. Whatever genetic predispositions there might be for people to dominate one another or to be egalitarian, these behaviors are utterly labile in response to environmental conditions, as clearly demonstrated by the wide differences among societies in space and time and the varieties of interpersonal relationships.

Origins of the State Revisited

When further intensification of agriculture was needed to support a growing population that could not be supported by a social organization based on extended families such as the *teri*, chiefdoms evolved. Remember that in order to permit such intensification, surpluses had to be created so that a farming family could feed more than itself, thus extending the previous gender- and age-based division of labor by freeing some adult males to take up nonagricultural specialties, including administrative posts. Control of surpluses made stratification (class divisions) possible. Chiefs, such as those who ruled in Hawaii, controlled regional polities of thousands of people, organizing the labor force for large-scale construction, especially for production of food and for its storage both against times of shortage and as a means of political control.[143] Large-scale technological improvements that increased the efficiency of production, but required investment beyond the means of local family-based groups, were made possible.[144] That means, as Timothy Earle put it, that the evolution of chiefdoms "[hinged] to a large measure on the ability to control or direct the flow of energy and other basic resources through a society as a means to finance new institutions."[145]

The financial resources of chiefdoms were garnered by taxation in kind; each household, for example, was taxed part of its crop.[146] Chiefdoms also provided a political structure that allowed violent acquisition of the resources of neighbors (and often added their populations to the chiefdom).[147] Furthermore, they supplied the organization and facilitated the development of transport technology to carry on extensive trade.[148] Positions of power in chiefdoms typically were a carryover from the family level of organization. Rank was (and in some places still is) heavily influenced by an individual's place in a kinship system—the relationship of a person to the chief.[149] Power was exercised over the allocation of labor by assignment of property rights,[150] and the entire system was legitimated in ideology—made manifest in ceremonies and sometimes in monuments (e.g., Stonehenge and the *moai* of Easter Island).[151] Specialists, once they emerged, often tended to concentrate geographically for efficiency in transport, and that concentration was one source of the nuclei of towns and cities. For instance, some 1,400 years ago in the first great imperial city of the Western Hemisphere, Teotihuacán, Mexico, about one-quarter of the households were involved in specialized craft production.[152]

Further intensification (and often higher population densities)[153] as well as increasing specialization doubtless were frequent root causes of the evolution of stratification. But there is evidence that the internal dynamics of societies—the sociopolitical environment—may in the long run play an equal or greater role in their cultural evolution than do changes in the ratio of population to resources. Overall, circumscription theory seems to provide some basic insight into the evolution of states, especially as indicated by interpretations of the archaeological record, the Polynesian test system, and the situation in modern societies. It seems a much more impressive explanation than the old theory that states developed automatically through voluntary association. But like scientific theories about many other phenomena, circumscription theory does not explain all the details.[154]

Dense human populations have been circumscribed or not circumscribed for different reasons, in different environments, with different resource endowments and different patterns of ritual exchange that in some cases bound people together more ideologically than economically, and so on.[155] Societies have found diverse ways to manage external aggression and risks of shortage and to solve economic problems through trade and development of more complex technologies. Each early state, for instance, evolved a unique economic blend of exchange through markets, trade, and taxation and tribute.[156] On Crete, the Minoan civilization that existed a couple of millennia before Christ was based on long-distance trading, whereas the Roman Empire concentrated more on extracting tribute from subject peoples. The Aztecs were more market than tribute oriented, whereas the Inca, like the Romans, relied more on tribute.[157] Not all societies concentrated specialized crafts in cities. In the ancient city of Monte Albán, also in Mexico, only 10–13 percent of people in the city were artisans, and craft production was dispersed among specialized villages. The natures of the peoples in all those societies would have evolved in somewhat different directions because of their different economic and social arrangements, just as those of Canadians and Mexicans have over the past few centuries.

But as we have seen, extrinsic factors such as the degree of circumscription have produced fundamentally very different trajectories for peoples in different locations because of geographic variation in resource distribution and climate. These ecological factors that are so influential in creating the broadest patterns of history, the determinants of cultural macroevolution, have not generally been given the attention they deserve. What accounts for the origins of states does not ordinarily account for their differential successes. Those may be based partly on environmental macroevolutionary factors, such as whether coal and iron ore were found close together within a state's boundaries, but they ordinarily also depend on such evanescent intrinsic factors as the presence or absence of competent leadership—factors in the potent force of cultural microevolution.[158] It's to a further examination of microevolutionary forces, such as capitalism and Napoleon, and their interactions with macroevolutionary ones, such as resource distribution and the Russian winter, that we now turn.

Chapter 11

GODS, DIVE-BOMBERS, AND BUREAUCRACY

> "Human social life cannot be understood apart from the deeply held beliefs and values that in the short run, at least, motivate and mobilize our transactions with each other and the world of nature."
>
> —Marvin Harris[1]

> "Even the most primitive of wars is a complicated, orchestrated, highly organized act of human imagination and intelligence, of which aggression is a necessary component but often not even the most important one. By the time we get to the large-scale wars of history, the aggressive component is even more reduced, and the logistical factors by far dominate the violent."
>
> —Robin Fox[2]

In A.D. 1519, Hernán Cortés, light skinned and bearded, appeared at what is now Veracruz, Mexico. That happened to be the exact year predicted for the birth of Quetzalcoatl (the Feathered Serpent), a deity backed by a powerful cult under the Mexican rulers of Yucatán. The Quetzalcoatl of legend was white and bearded, and Montezuma, Mexico's Aztec ruler, greeted Cortés with the honors due a god, laying himself and his powerful empire open to destruction. That, too, fit in with Aztec religious beliefs that successive universes were doomed to destruction, and it had doubtless made Montezuma pessimistic about his chances of avoiding annihilation by the Spaniard-gods.[3] Add in the chance joining with Cortés of a Cuban soldier infected with smallpox, and the Aztecs (and a large proportion of the indigenous population of the Americas), who had no immunity to the dread disease, were doomed.

History is replete with examples of seemingly minor events having tremendous consequences. If George Washington had died young in a battle with the French, the United States might never have achieved its independence. If John

Wilkes Booth had missed, the history of North–South relations in the United States after the Civil War might be entirely different. If Charles Darwin had not been patient and meticulous, his theory of evolution might not have been a major influence on the course of intellectual history in the late nineteenth and early twentieth centuries. If Martin Luther King Jr. had not been born (or if Henry David Thoreau had not invented civil disobedience and Mahatma Gandhi had not shown that it could work),[4] the course of the civil rights movement in the United States would have been radically different. History is often steered by individuals, particular incidents, and other microevolutionary forces. But the degree to which people make history, as opposed to history making people, is very difficult to sort out.

Cultural macroevolution—the shaping of cultural trajectories by environmental factors—has similarly momentous influences. For example, smallpox and many other diseases were acquired by *Homo sapiens* from domesticated animals—in the case of smallpox, probably cattle. As we have seen, the luck of the draw in domestication went to people in Eurasia, where there were more large animals with the appropriate characteristics than elsewhere. People in the Eastern Hemisphere also had more opportunity to acquire immunity to an array of diseases, even though they paid a very high price at numerous times and places.[5] Europeans thus unknowingly carried a potent weapon—epidemic diseases—when they invaded the Americas. The macroevolutionary ravages of disease, much of it taking place before large numbers of Europeans arrived,[6] interacted with religion and warfare to affect human natures over the entire hemisphere.

It is important to know what shapes the course of history, how that influences the evolution of our natures, and how that in turn feeds back on evolutionary history itself. We can see some of the patterns of interaction between cultural micro- and macroevolution in history by taking a close look at the development of states and their associated institutions of commerce, warfare, and religion. The need to understand interactions between cultural micro- and macroevolutionary forces becomes ever more urgent as the increasing scale of the human enterprise presses against global life-support systems, placing macroevolutionary constraints increasingly in the path of microevolution and setting up a collision that could threaten the viability of civilization.[7]

States not only are major actors in an increasingly problematic global drama but also in turn are powerful determinants of the human natures of their citizens. The natures of even such closely allied peoples as the British, Canadians, and Australians or the Costa Ricans, Nicaraguans, and Hondurans are quite different—from their accents and art to their political attitudes. States, through their political parties, religions, traditions, and institutions, tend on one hand to differentiate human natures and on the other to constrain cultural evolution and reduce its variation. Cultural evolution goes on both between and within groups, and the situation is made extremely complex and tough to analyze because of the human habit of belonging to many different groups simultaneously. The natures of members of any given group tend to be different from those of adherents of a different one, but members often conform to their group's norms and pass that conformity culturally to their offspring and successors. In our globalizing culture, wide-scale constraints can produce cultural microevolutionary hangovers analogous to the genetic evolutionary ones. An obvious example is the persistence of the war system long after technological evolution produced weapons quite capable of wiping out all of civilization.[8] Another is the urge to continue "business as usual" in the face of an escalating environmental crisis. Then again, the natures (e.g., preferences) of the individuals in a group, even the majority of the individuals, may not be reflected in the group's behavior. For example, even though most people are probably best described as peace loving, states spend a great deal of time at war.[9] And most people want to be "environmentalists" but continue to act in ways that lead to increasing deterioration of their life-support systems.

Some geographic variation in the environment that generates powerful macroevolutionary influences on the trajectories of cultural evolution can be explained by cultural microevolution. This was the case on Easter Island and many other Polynesian islands: human activities dramatically altered the environment, and this in turn changed the course of cultural evolution. As another example, some major historical shifts in population and power within China can be explained by shifting patterns of human-induced environmental degradation and recovery. During the Qin and Han Dynasties (221 B.C. to A.D. 220), for instance, much of China was deforested for the construction of great wooden cities and the Great Wall.[10] For the latter effort, wood was required for scaffolding and to fire brick kilns as well as for warmth, shelter, and cooking fires for the one-third of the nation's population needed to complete those projects. Forests in many areas were permanently destroyed. The regional depletion of the forests contributed to mass migrations and shifts in the center of political power. For example, in the late Western Han period (around the time of the birth of Christ), deforestation and increased flooding of the Yellow River, due to greater sediment loads traceable to the denuding of its watersheds, led to abandonment of the river's middle and lower watersheds as the center of Chinese civilization. Between 60 and 70 percent of the people who had lived along the river in those areas migrated south and east to the previously undeveloped lower Yangtze River. Later, greatly diminished population sizes due to migration and war allowed the ecosystems of northern China to recuperate, so the Yellow River watershed could once again support a larger population. But intercoupled political events and environmental deterioration produced a dramatic and complex dance of human population dynamics, ecosystem decline (and some recovery), and shifts of political centers over China for the next two millennia.

Religion and the Evolution of States

Two key constituents of the microevolution of state power are religion and violence, especially warfare between states.[11] Even though they are not primarily matters of political or economic organization, they are closely related to the evolution of stratification and to each other's evolution. Stratification is not confined to the presence of leadership hierarchies and social classes; it also manifests itself in professional priestly and warrior classes.[12] And intertwined, at least in part, with political, religious, and violent behavior is the fear of death that our intense consciousness keeps lurking in the background.[13] Human beings are the only animals that have developed religions, ethics, moral codes, and mutually agreed-on norms of conduct. These are human universals, but micro- and macroevolution have combined to produce a multitude of forms in different societies, as exemplified by the polytheistic rain-forest religions and monotheistic desert religions discussed earlier.[14]

A few thousand years ago, as some human groups evolved politically from family bands toward tribes and states, the roles of religion within them changed. Religion ceased to be solely a matter of personal and group belief, sometimes mediated by shamans, and of rules that help groups to function. Religion increasingly began to play another important role. It helped to sacralize—connect to the supernatural—codes of conduct that apparently made societies function more effectively, and it legitimated differences between classes of people (i.e., elites and commoners). Religious rituals not only presumably provided solace, pleasure, and a feeling of order to individuals; they also brought sanctity into previously nonreligious aspects of life, sanctity that elites believe societies need to allow the elites to remain in control.[15]

Sacralization may have first appeared in animistic societies, in which it must have been an easy step from assuming that disruptive individuals were inhabited by evil spirits to developing a religious rite for their removal. Equally, in such societies the unusual bad luck (or lack of skill) of some hunters could have been blamed on the operations of certain spirits, which could be appeased, and the good luck of others could be seen as the spirits' approval of their actions. Once rules of conduct and economic status had been sacralized, it was easy for religion to assume the role of sanctifying those rules and thus the status of the governing elites who enforced them. Investing arbitrary social conventions with sanctity made them seem natural—as if they were reflections of human nature—and this sanctification became a force justifying power relationships within societies. As anthropologist Roy Rappaport put it, "Our observation that sanctification presents to the individual as his own goals those of the society and thus replaces possible recalcitrance with compliance, remains true in contemporary societies."[16]

Organized religion thus seems to have evolved to help stabilize hierarchical social structure. That structure was necessary to carry out the complex tasks required to support ever larger populations on a limited, sometimes deteriorating, resource base. In Mesoamerica, the cultural region that extended from Mexico to northern Central America before the Spanish conquest, a complex and related set of polytheistic religions evolved hand in hand with the development of stratified states. One, for example, centered on the large city of Teotihuacán in the Valley of Mexico in the first 600 years or so A.D. The locations and plans of Mesoamerican cities themselves were based partly on religious considerations, something that a visit today to the ruins of the magnificent pyramids at Teotihuacán can confirm.[17] Even today, they seem clearly to have been designed to overawe. And although there is controversy as to which came first, the complex religions or the cities, there is little question that religious factors were entwined in the steps that led elites to build urban civilizations.[18]

Religion, reinforced by pseudoscience, continues to play a role in maintaining the status of elites today, for instance in justifying poverty and wealth as expressions of God's will. In the bad old days of heavy religion, we "knew" that the poor were as they were because they were iniquitous. The proof of their iniquity was their poverty. Now, with the deeper understanding of "science" (as represented by books such as *The Bell Curve* that attempt to associate class structure with innate intelligence), we "know" that the poor are as they are because they are stupid.[19] The proof of their stupidity is their poverty.

Religion has acquired many other roles in state societies, often, for instance, helping the poor and others who suffer the consequences of stratification.[20] Sometimes it has supported the economic and political establishment and sometimes it has sought to overturn it, the latter, for instance, when Christianity and other Jewish cults opposed the Roman government. And religious institutions can play a substantial role in the disintegration of states, through changes in their complex interactions with political and economic forces, as suggested by the preconquest decline of the Inca and Mexica[21] Aztec empires. Both empires

Teotihuacán, Mexico: Pyramid of the Sun in the foreground, Pyramid of the Moon in the background. The form and layout of the buildings, murals, carvings, and altars suggest religious significance. Photograph by P. R. Ehrlich.

started to run out of peoples to conquer, and some military defeats began to erode the peoples' self-confidence and trust in the empires. By the time of the conquest, ideological elements that had been critical to the initial successes of the empires were contributing to a general malaise, which was accelerating their decline. The basic difficulty was that Aztec and Inca state religions dictated that conquest and warfare continue forever. The Aztecs, for instance, needed a continuing supply of blood from human sacrifices to honor their militant patron god, Huitzilopochtli, or the universe would perish. But expansion of the empire to supply the needed victims was not possible politically, economically, or socially. The history of the Aztecs and the Inca suggests that religion can play just as fundamental a role in the fates of civilizations as do their economic and political institutions.[22]

Warfare and the Evolution of States

A religious apparatus and a military establishment are both prominent features of a state, so it is hardly surprising that religion is also closely connected to warfare. As societies evolved toward states, intercommunity violence continued to center mainly on control of territory and resources (and females of reproductive age). It evolved culturally into a much more organized, even ritualized, kind of conflict involving incorporation of the productive capacities of subjugated peoples into the resource flow of the conquerors, and it became entwined with and justified by religious ritual and belief.[23]

The question remains, however, of exactly why human beings persist in resorting to warfare in a wide variety of circumstances.[24] If it isn't the result of drives programmed into our genes, what is behind it? One way scholars have tried to understand war is by seeking to discover its functions.[25] War, among other things, serves as a mechanism for restructuring relationships within societies, such as rallying support around a faltering leader or redistributing labor or wealth. It also is used to grab another country's important resources, to occupy a strategic geographic position, or to spread religious or other ideas.[26] Some writers have suggested that it functions as a means of population control as well,[27] but if limiting numbers were the goal, it would be more effective to send teenage girls rather than boys into combat. Finally, historian Donald Kagan has persuasively argued that in a system of states, constant effort is required to maintain peace because "a contest among [the states] over the distribution of power is the normal condition and . . . such contests often lead to war."[28] In short, in Kagan's view, war, not peace, is the ordinary state of affairs among nations. He also believes that in both ancient and modern times, honor (defined as including such characteristics as fame, glory, prestige, and respect) has been an important reason why people go to war.[29]

If Kagan is correct about honor, that helps to explain the less than satisfactory results of attempts to understand the causes of war by searching for statistical correlations among possible "objective" factors such as population size, time since the last conflict, economic inequality, presence of disbanded soldiers looking for employment, availability of allies, sharing of geographic resources such as a major river system, cycles of relative power of nations, and so on.[30] These studies cast some light on human motivation to participate in wars, but they don't really supply what most would consider a satisfactory answer—as indicated by the lack of consensus on the issue.

In highly organized societies, religion can certainly be a prime cause of combat, as it was in the case of the Crusades. Those expeditions were in part a "solution" to the problem of a warrior class in Europe having overreproduced itself and the resultant scourge of local warfare. But there were other factors as well: population growth in general, deteriorating environmental conditions (loss of soil fertility), the need to occupy younger sons who had been pushed off the land by primogeniture (the law giving the eldest son the sole right to inherit family land), and a newly resurgent papacy that was starting to direct warfare in order to achieve dominance over the growing warrior aristocracy.[31] To see that population and environment were factors, one need only read a statement that, at the end of the eleventh century, was contemporaneously attributed to Pope Urban II as part of his preaching the First Crusade to the French nobility at the Council of Clermont:

> . . . since this land which you inhabit, shut in on all sides by the sea and surrounded by mountain peaks, is too narrow for your large population; nor does it abound in wealth; and it furnishes scarcely enough food for its cultivators. . . . Enter upon the road to the Holy Sepulchre; wrest that land from the wicked race and subject it to yourselves. That land which, as the Scripture says, "floweth with milk and honey," was given by God into the possession of the children of Israel. Jerusalem is the navel of the world; the land is fruitful above others, like another paradise of delights.[32]

Warfare has changed since the Crusades. Despite zoologist Irenäus Eibl-Eibesfeldt's notion that war is caused by an innate aggressive drive, individual aggressive propensities, violence, and war can become very much separated in modern states (as Robin Fox suggests in this chapter's second epigraph). A modern army, for example, is highly routinized; in its overall organization, it looks more like the Internal Revenue Service than a marauding band of Kasakela chimps. As historian D. L. Bilderback put it, "There have been two great leaps in increasing the power of the state, both in the seventeenth century. The first was getting people to shoot on command. The second was to get them to stop shooting on command: hostilities will cease at 11:00 A.M., November 11, 1918. Now state leaders could really direct coercion."[33] Another historian, Michael Howard, made a similar point: "The calculations of advantage and risk, sometimes careful, sometimes crude, that statesmen make before committing their countries to war are very remotely linked, if at all, to the displays of tribal *machismo* that we witness today in football crowds."[34]

Hormonally induced rage may lead to a barroom brawl; the ability to send surrogates to battle with calm deliberation can lead to war. Hormonal triggers are not required to incite mass violence. The pattern of old men in complete safety and with careful planning sending impressionable adolescent boys to kill and be killed persists into the twenty-first century. The root of the word *infantry*, one might remember, is the Latin *infans*, "infant." Connecting "genes for aggression" (if you'll excuse this lapse into the jargon of pop sociobiology) to the actions of warring governments is a bit of a stretch, just as would be connecting genes for conciliation to the deployment of United Nations peacemakers.[35] An inherited predisposition that in certain social environments biases people toward trying to manipulate others to their own benefit seems to me an adequate explanation for any "biological roots" of war. Indeed, genes that interact with environments to generate a phenotype with a problem-solving brain might be quite enough.

Violence in human societies has a much-neglected flip side as well. Fighting over the power to have first pick at the dining table and (among males) fighting over access to the sexual favors of females seems as natural to many social mammals as Mom's proverbial apple pie does to Americans. But remember, we apes are also specialists in reconciling and in forming coalitions for mutual aid and to maintain peace. One cannot help but be impressed by the way inner-city children form alliances in an environment in which the frequent lethal use of force is commonplace (although it is not clear whether the alliances reduce the exposure of members to violence). Equally compelling are the cases in which some individuals are able to see beyond their violent urban environments and dedicate their lives to peacemaking.[36]

A massive amount of data also indicates that it is surprisingly difficult to get most human beings to kill even during wars. In the Civil War in the United States and in both world wars, huge numbers of infantry soldiers—probably more than half of the men in combat—either did not fire their weapons or fired to miss. During the slaughter of British troops in the ill-fated 1916 offensive on the Somme, German machine gunners ceased firing rather than killing wantonly, allowing lightly wounded British survivors to retreat to their own lines.[37] Special training to kill was needed to greatly increase the rate at which U.S. infantrymen used their weapons in the Vietnam War.[38] Moreover, even some of those who were involved in some of the worst atrocities of the twentieth century took part only under extreme duress and deeply regretted it afterward.[39]

Are we to assume, then, that humanity has a "drive" toward kindness and reconciliation? Rather than being naturally "killer apes," could it be that we're genetically disposed to be "friendly apes" and that only extreme indoctrination, coercion, or both can create an environment in which murderous behavior can be elicited after thousands of generations of less violent people being more reproductively successful? We're short of data from virtually all societies to tell us whether the average violent man (or violent unemployed inner-city youth)[40] is more or less reproductively successful than the average nonviolent man. An exception is the Yanomamö, for whom Napoleon Chagnon reported preliminary data indicating that men who had killed were more reproductively successful than those who had not.[41] Being willing warriors may enable males to gain access to more females, either through sex appeal (on the basis of strength, ability to provide food, prestige) or through rape, but it may also lead to early death before their genes can be spread very far. And nothing at all is known about aggression and fitness in technologically complex societies.

Genocidal Natures

What about genocide—governmentally encouraged programs of mass murder directed at specific national, social, ethnic, racial, or religious groups[42]—a form of behavior that darkens our past, haunts our present, and clouds our future? Is it simply another unfortunate expression of an innate aggressive tendency? *Homo sapiens* has a long and undistinguished history of groups trying to exterminate other groups, often with considerable success. As an overview of genocidal behavior by Jared Diamond suggests, the practice has been so widespread that it might be considered a characteristic of our species.[43] Constraints on intragroup mayhem in social animals clearly have adaptive value, whether produced by individual selection, group selection, cultural evolution, or some combination of the three. In band societies, however, with periodic scarcity of resources, there could be substantial advantages to not allowing those constraints to extend to intergroup relations. There would be advantages to being able to define people in other groups as a different species—as "them," not "us." They could thus be placed outside the universe of obligation,[44] and it would then be psychologically easier to remove them indiscriminately as competitors. Such redefinition, at one level or another of consciousness, is clearly a part of the behavioral repertoire of chimpanzees (remember the total destruction of the Kahamas) and human beings but perhaps not of bonobos.[45]

Much of human genocide clearly has a resource–territoriality component, as seen in some of the treatment of Native Americans by Europeans or in the extermination of the Tasmanians by Australian settlers. The same can be said of the genocides Europeans committed in the process of acquiring the resources of Africa, and resource scarcity may have been an element in the recent intra-African genocides in Rwanda.[46] There was clearly a resource–territoriality element in the German and Japanese aggression during World War II, but that element hardly seems a major motivation in the attempted extermination of the Jews and Gypsies by the Germans or the Japanese army's rape of Nanking.[47] How does one explain "normal" grown men of the Nazi police walking little girls into the woods one after another and shooting them in the back of the head? According to one witness, "The executioners were gruesomely soiled with blood, brain matter, and bone splinters. It stuck to their clothes."[48] The Germans were notoriously cruel to Jews, humiliating them and often inflicting pain far beyond what would have been required to kill them, yet at the same time they took steps to prevent cruelty to dogs and cattle.[49]

The Japanese showed similar cruelty to the Chinese, but without the secrecy of the Nazi Holocaust. Apparently their goal was to lessen Chinese resolve through terror. Tens of thousands of women were not only raped but also, typically, mutilated and killed. Men were frequently forced to rape members of their own families. "Not only did live burials, castration, the carving of organs, and the roasting of people become routine, but more diabolical tortures were practiced, such as hanging people by their tongues on iron hooks or burying people to their waists and watching them torn apart by German shepherds."[50] The behavior of the Japanese was so atrocious that a Nazi living in Nanking at the time described the massacre as the operation of "bestial machinery."[51] Unlike the Germans, the Japanese have yet to come fully to grips with responsibility for the acts of their army in 1937, and those Japanese veterans who have faced up to their behavior and asked forgiveness have been targets of hate and violence from right-wing groups in Japan.

Genocide is clearly within the repertoire of human behavior, and I suspect it could be induced in virtually any group, given appropriate social conditions. Those conditions, unfortunately, can be quite diverse. For example, simple social pressure from people perceived as being in authority (scientists in white coats) caused ordinary Americans assigned the role of "teachers" in a famous experiment conducted by psychologist Stanley Milgram to deliver what they thought to be potentially lethal shocks to "learners" who were participating in the experiment with them. That occurred even though the learners (actually collaborators of the psychologist running the experiment) were pleading and screaming for the teachers to desist.[52] Journalist Iris Chang, who wrote the definitive work on the rape of Nanking, placed the blame for the genocidal actions of the Japanese army on indoctrination by "a dangerous government, in a vulnerable culture, in dangerous times, able to sell dangerous rationalizations to those whose human instincts told them otherwise."[53] She believed, as have many others, that the behavior of the Japanese troops traced in part to the brutality, often in the form of severe physical abuse, that the Japanese army of the time inflicted on its own soldiers and officers.[54] When turned loose on helpless civilians whom soldiers had been indoctrinated to believe were lower than pigs and ordered to massacre them, Chang wrote, it "is easy to see how years of suppressed anger, hatred, and fear of authority could have erupted in uncontrollable violence at Nanking."[55] Such behavior does not seem so hard to explain when people (the teachers) in a known experimental environment, subject to no abuse, indoctrination, or coercion and under no obligation to authorities, could be led to inflict punishment that might (in their minds) lead to the deaths of colleagues in an experiment.

Lest citizens of nations on the winning side of World War II conclude that atrocious behavior was confined to enemy countries, they should recall attitudes and actions in the United States, such as the vicious racial prejudice aimed at the Japanese in U.S. government propaganda. In an assembly held in the grade school I was attending just after the attack on Pearl Harbor, we ritually tore the

Japanese national anthem out of our songbooks. We were told, among other things, that the Japanese were inherently treacherous, had facial musculature that made them smile all the time, and had small tails. During the war, newsreel audiences in the United States cheered at pictures of burning Japanese soldiers running from bunkers that had been attacked with flamethrowers. U.S. soldiers were obviously profoundly influenced by racism. In one survey, 44 percent said they would "really like to kill a Japanese soldier," whereas only 6 percent expressed a similar desire to kill a German soldier.[56] Among civilian adults, racism "justified" brutalization of U.S. citizens of Japanese descent and the theft of their property. No similar campaign was mounted against the Germans or against U.S. citizens of German descent; in that case, it was not the German people themselves but the Nazis who were portrayed as our opponents. Unlike the situation in Europe, the United States and its allies pursued a near "take no prisoners" policy in the Pacific. Killing and maiming of Japanese military prisoners, survivors of ship sinkings, and sometimes civilians was routine and semi-official policy. Clean-cut American "boys" in large numbers collected Japanese ears and bones as souvenirs.[57]

That this was sometimes a response, in situations of great stress, to Japanese atrocities is partly true, but it certainly indicates that twentieth-century Americans were just as willing to define opponents as being outside the realm of humanity as had been their nineteenth-century counterparts in the U.S. Cavalry during the "Indian Wars." That the Japanese and Native Americans could also show genocidal behavior is neither a surprise nor an excuse—if there is a universal in genocide, it surely is the practice of dehumanizing the victims,[58] just as dehumanizing the enemy is a frequent feature of warfare. The ultimate horror inflicted in the Pacific war was the burning of Japanese cities and the use of atomic bombs against Hiroshima and Nagasaki, which killed hundreds of thousands of men, women, and children and made many millions homeless.[59] Here, the racist dehumanization of the Japanese, who were often portrayed as monkeys, was reinforced by the impersonal character of the weapons. Killing fellow human beings by pressing buttons to release bombs clearly requires a smaller psychological shift than does blowing their brains out in person. For example, it allowed the slaughter of hundreds of thousands of German civilians, who otherwise had not been dehumanized as the Japanese had been, in massive air raids on Hamburg, Dresden, and other German cities. And if the European war had continued longer, it seems likely that atomic bombs would have been used against the Germans as well.

Unhappily, of course, genocide did not pass from the human scene with World War II. The same old lessons can be drawn from the behavior of both Hutus and Tutsis in Rwanda and Burundi, who have been responsible for a series of genocidal events. The most horrifying of these occurred in 1994, when about 800,000 Tutsis were killed in Rwanda in a three-month period at the behest of the Hutu-led government.[60] There, the way was paved by a long and careful propaganda campaign, based largely on racial and class prejudices, that gave warrant—ideological legitimation—for the genocide.[61] The massive participation of the Hutu population in the slaughter, largely by machete, of their neighbors (and often former friends) once again demonstrated how readily unthinkable acts can be elicited. And, of course, so-called ethnic cleansing in the Balkans darkened the last decade of the twentieth century—again warranted by "legitimate" governments and demonstrating that the lessons of Adolf Hitler[62] had not been learned by many Europeans.

Violence, including warfare and genocide, certainly has not been excised from the human behavioral repertoire in the course of political evolution. Violent behavior was present in tribal societies, and it remains a feature of state societies. Perhaps the best summary of the evolution of human violence, including genocide, is that very smart primates have evolved (genetically and culturally) social and political systems that require manipulation of other individuals and groups. Those primates will at times resort to any tools they think will accomplish their goals. In some cases, those tools may be used to achieve goals that are immediate and personal: a blow with a club or a caress. In other cases, they may be long-term and impersonal: an apparently reluctant vote in a legislature to send troops to oppose a perceived aggressor or to initiate a welfare program. Today, statesmen's aggressive tools include intercontinental missiles tipped with hydrogen bombs and chemical and biological weapons with commensurately lethal effects. Without question, the development of weapons of mass destruction has created a potential for states to inflict casualties that proportionately would dwarf the brutalities of the Yanomamö and bring about the destruction of civilization.[63] Humanity has to work hard at its strained cultural devices for maintaining peace and averting genocide. What we can be certain of is that there is no possibility in the foreseeable future of changing our genes to make us more peaceful. It is senseless from any viewpoint for people to keep acting as if it were either possible or pertinent to determine whether human beings are "innately aggressive" or "innately pacific."

States of Inequality

Social stratification with religious and military components has been a feature of states from the formation of the first ones. The early development of states was associated with intensification of agriculture and trade, probably necessitated by increased population densities. State control was especially needed in connection with those aspects of economic intensification, such as the building and guarding of regional irrigation systems and road networks, that required large investments of capital and labor. In "hydraulic" civilizations[64]—those based on centrally directed irrigation—a substantial labor force is needed for maintenance, but periodically, in years when floods or other events create large-scale damage, it must be greatly augmented. How is the larger labor pool to be kept intact at other times? One suggestion is that it can be used to build ziggurats, pyramids, or other large and (from a practical point of view) useless projects whose actual completion is not important.[65]

In states, much or all of the land is ordinarily under government or elite ownership and control. Specialized institutions are required to organize production, collect taxes, manage construction projects, maintain internal order, defend against enemies, and conquer new subjects. Early in the development of states, elites formed hierarchies of bureaucrats and specialized guilds, a police apparatus, and priests in a religion that legitimated state rule and the privileges of the elites. Cultural anthropologist Leslie White considered religion and the original states to have been so tightly intertwined, for example, as to be "but aspects of something for which we have no better term than state-church."[66]

In the development of states, there was a trend away from taxation in kind—that is, farmers giving a portion of their produce to the government. Such "staple finance," the basis of states in the earlier stages of their evolution, such as that of the Inca, gave way to "wealth finance," in which objects considered valuable were produced and distributed under the control of the state—a move in the direction of the currencies now used.[67] Taxes in early states were levied at various times and places on individuals (head taxes) and households and on virtually all productive resources and economic activities (e.g., trade taxes, market taxes, herd taxes, land taxes).[68] According to anthropologist Frances Berdan, the power to tax derived from the idea "that the king owned everything in the land: every transaction to acquire wealth or produce goods had to be bought, by a tax, from the royal house."[69] That house, in return, was expected to meet a complex series of obligations to those ruled.

Stratification within and among societies often proceeded far beyond what was needed to support any intensification of production required by environmental scarcity. The surpluses generated suggest huge discrepancies in wealth between commoners and the ruling classes. Seven centuries before Christ, the royal cellars in the ancient kingdom of Urartu, roughly equivalent to today's Armenia, contained 55,000 gallons of wine,[70] quite enough to gladden the heart of any connoisseur. Temples from Babylon and Egypt to Cambodia amassed huge amounts of wealth, much of it in grain (which could be, and in times of shortage was, redistributed to the peasantry), but much also in the form of gold, silver, precious stones, statuary, ceremonial clothing, and the like, which indicated the appropriation of the efforts of thousands of people in support of the anointed.

Gigantic gaps between the rich and poor still exist between and within societies today. In 1997, for example, the per capita gross national product in Mozambique was $80, and in the United States it was $28,020.[71] Although at the extremes, such comparisons tend to exaggerate the gap because of the dominance of nonmonetary economies in very poor countries (most citizens of Mozambique don't shop at supermarkets), the more than 300-fold differential clearly represents an extraordinary international stratification. Similarly, in the United States in recent years, the top 1 percent of the population received more total income than did the bottom 40 percent.[72] The chief executive officer of one U.S. company made $127 million in a single year—almost 780,000 times the average per capita income of the poorest one-fifth of Earth's population.[73] Such preposterous levels of compensation are an aberration of recent cultural evolution in the United States; they are not found to the same degree in Europe or Japan (where differentials are nonetheless very substantial).

What factors in cultural environments repeatedly lead to disproportionate wealth, wealth far beyond that required to meet even exaggerated needs? At least in many societies, one factor obviously is a combination of fear and power: fear of losing wealth (and thus comfort or even one's life) leading to a quest for power to acquire and preserve ever more wealth. Wealth, after all, pays for bodyguards, priests, and ceremonies to impress the common folk and recruit the favors of gods, henchmen, and armies to loot wealth from others. It keeps poor people and nations relatively powerless to seek equality.

Another factor, if there is any significant hereditary component to wealth-acquiring behavior in states, would be the ancillary power to co-opt the services of members of the opposite sex (most frequently the services of women in harems) and ensure the survival of heirs. Certainly, genetic components of sexual behavior and care of offspring would be involved in these instances, but again the social environment may often override them. For example, cultural factors have led more than one king to kill an ambitious son, thwarting the ambitions of his own selfish genes; today, some of the world's richest and most powerful men may use their wealth to finance sexual escapades, but more to the point, others do not.[74]

Markets, Money, and Writing

One of the biggest driving forces of cultural evolution can trace its beginnings far back in prehistory, with the genesis of trade. Trade led to markets, markets to money and (scholars believe) writing. Markets, money, and literacy are, together with their offspring science and technology, primary shapers of our world and our natures.

The development of market economies over the past couple of centuries provided great impetus to the acquisition of wealth. It also stimulated technological innovation in agriculture and elsewhere, leading to the expansion of Earth's carrying capacity for people. At first, markets involved the exchange of a limited array of goods among hunter-gatherers—perhaps the exchange of meat from a recent mammoth kill for flints that made good spear tips or exchange of three dog skins for a pair of fancy fiber sandals. As groups became sedentary, markets became more formalized, often located at traditional places and conducted at traditional times. Some goods—cattle in Africa are a usual example—became standardized repositories of value and media of exchange. A small herd of cattle could be paid to a man as a bride price for one of his daughters. On some Pacific islands, modified shell artifacts (arm rings, necklaces) were used as a medium of exchange. In other places, beads, crops, feathers, gold, silver, stone, salt, and many other things have been brought into service as media of exchange.[75] Such items, known as commodity money, could also be used as standards of value (your daughter is worth 4 cows) and for storage and display of wealth (my herd has 200 head of cattle)[76] and thus have a great deal in common with money as we now understand it. When coins were invented and used in states, money had some additional attributes: it was generally portable, divisible into denominations, and under some form of government control. In modern society, the forms of money are continually and rapidly evolving as gigantic sums leap electronically between continents in microseconds.

Early portable forms of money made trade much easier, especially over long distances. It was more convenient to take a bag of gold coins to the next market town to seek a wife than to drive one's cattle herd there. Since its beginnings, money has always greased trade, and trade has always been a major promoter of specialization and thus an implement of stratification and intensification. Along with money, another human invention intimately connected with commerce has transformed the course of cultural evolution—writing. H. G. Wells summarized its role beautifully: "It put agreements, laws, commandments on record. It made the growth of states larger than the old city states possible. It made a continuous historical consciousness possible. The command of a priest or king and his seal could go far beyond his sight and voice, and could survive his death."[77]

Writing allowed complex information to be stored outside the human mind and transmitted over great distances in space and time.[78] It enabled both a complexification of economic arrangements (in the United States, the Federal Tax Code now occupies 40,000 printed pages)[79] and development of more effective ways to project power. Growing economies not only made the need for writing more critical but also led to new and better ways (printing, photocopying, e-mail) to distribute the written word. In short, writing and commerce coevolved. Writing also sowed some of the seeds for the development of modern states by reducing the need for rituals and personal contact with specific individuals who represented the power structure for organizing state society.[80]

Large states with complex economic arrangements seldom developed without writing (the Inca were a major exception),[81] and modern ones, based heavily on science and technology, could not exist without it. In fact, at least one theory of writing's origin is directly economic. Early Sumerian writing is thought to have evolved from a system of tokens that were often sealed inside clay containers incised with signs recording the tokens enclosed.[82] That system was used for accounting in Mesopotamia more than 5,000 years ago[83] and marked the start of the transition from prehistoric to historical time. Modern states also depend on a third crucial revolution in human communication, following on speech and writing: the invention of printing.

Money and writing remain promoters of specialization, stratification, and intensification today. In ancient Rome or fifteenth-century Europe, someone who was able to write could make a living as a scribe, writing for others. There are still scribes in Brazilian train stations today. Now, many people make their living as programmers, writing computer code that accelerates communication and interaction. Writing and its electronic descendants, and the flowering of the technology that writing made possible, have made global trade and the possibility of supporting 6 billion or more people, at least temporarily, a reality. With global free trade, it would be theoretically possible to operate the human economy at the global "production possibility frontier," defined as the various mixes of goods that theoretically could be produced by all human societies corporately if they operated efficiently.[84] This also could lead to a boring homogeneity of cultures—a trend that already is well under way with American-style fast food, films, and pop music in Saigon, Moscow, and Tokyo and Western-style consumerism rampant almost everywhere.[85]

Egalitarian hunter-gatherer societies never had a need for writing—it apparently is called forth by specialized functions, such as the keeping of accounts, or by the need for elites to promulgate decrees to keep distant subjects in line. For a long time, the haves wrote and the have-nots didn't. To a considerable degree, that condition persists. As Oxford University linguist Roy Harris put it, "It is not by chance that today the global maps of illiteracy and poverty so nearly coincide,"[86] although what is cause and what is effect may differ from place to place and may be difficult to untangle.[87] At the same time, improved communication brought about by printing has proven historically to be a great force for the removal of inequity through the diffusion of ideas such as liberty and equality. It remains so today.[88]

Cultural evolution has not produced the same kinds of states among all literate societies. For instance, it remains a matter of dispute why states in China and India took different courses from those in the West, not evolving toward a balance between a domain viewed as public and another domain viewed as protected and private.[89] Indeed, China seemed to be well on the way to controlling the world in the early 1400s. Admiral Zheng had taken a gigantic armada of some 300 ships and 28,000 sailors as far as East Africa decades before Christopher Columbus sailed in the *Niña*, *Pinta*, and *Santa María* with 90 men. Zheng's largest ships were 400 feet long and 160 feet wide, almost twice as wide as Columbus's largest ship was long (85 feet). His giant "treasure ships" had nine masts, many decks, and some cabins luxurious enough to include balconies. Zheng's armada had transports for troops and for horses, patrol boats, men-of-war, and tankers to hold freshwater;[90] it was the culmination of a great maritime era for China that stretched back in time for a millennium or more.

Why, with such a head start, was it not China, rather than Europe, that conquered the world? Part of the answer may lie with cultural macroevolution. The lack of geographic barriers within China allowed single rulers to control the entire area and prevented balkanization such as occurred in Europe: China, unlike Europe, did not subdivide into large numbers of separate states and cultural centers that competed with one another and became centers of innovation.[91] In addition, China as a whole was geographically isolated from other centers of civilization and thus could not benefit much from cross-fertilization of ideas. It was also relatively wealthy; the Chinese did not have as much of the wealth-seeking motive as did the Europeans.

Cultural microevolution—change driven by the internal dynamics of societies mixed with historical contingency and chance—helps to explain why China was at least partially eclipsed by European powers. Zheng was not a great navigator, as were Vasco da Gama and Columbus; wherever he went, he was guided by local pilots. A combination of Chinese politics and attitudes that valued inwardness[92] and denigrated "business" appears to have intervened to dampen the entrepreneurial spirit, which animated so much European exploration. That combination, in conjunction with the centralized bureaucratic control in China that committed a subcontinent-sized area to a single, inward-looking policy, seems to have handed the opportunity for global conquest to Europe.

What enabled Europe to seize the opportunity to extend its influence? A crucial factor was that when one country there lost the drive to explore and exploit, another could take up the slack.[93] Another important factor was the move among European nations away from the long-established pattern in agricultural states of bureaucratic control through monopolization of key aspects of the economy (e.g., water in hydraulic societies). European states may have been too small (controlling too few resources within their borders) and too weak to impose monopolies as technologies changed.[94] They were thus forced to allow economic competition. In other words, early weakness was an important factor in allowing them eventually to achieve a dominant position globally. The ideological transformation that led to the model of capitalism formalized by Scottish economist Adam Smith was also a critical factor. For a long time, the seeking of wealth had been condemned as avarice; gradually in the late eighteenth and nineteenth centuries, the view took hold that "men pursuing their private passions conspire unknowingly toward the public good."[95] Rather than a system in which the leaders controlled the people and made them behave themselves, Western societies moved to a system in which the "misbehavior" of people (greed, self-seeking) would produce an emergent stable and prosperous society (in Smith's terms, through the action of an "invisible hand").[96]

For Want of a Nail

In recent history, cultural microevolution—interactions between groups and the influences of nonenvironmental factors within them—has been at least as important as cultural macroevolution. Consider how different the world might be today had the leadership of nations involved in the second stage of the great twentieth-century world war[97] not included Winston Churchill, Adolf Hitler, Franklin D. Roosevelt, and Joseph Stalin. What if Roosevelt had not opposed Nazi Germany?[98] What might have happened if the Japanese leadership had listened to the brilliant commander of the Japanese navy, Admiral Isoroku Yamamoto, who warned against attacking the United States?[99] What if Hitler had not repeated Napoleon's error by attacking Russia? Such puzzles can sometimes be traced far back in history. What if Woodrow Wilson had not been greatly influenced by Immanuel Kant and had not taken the idea of the League of Nations from Kant[100]—an idea that still has a significant influence on events today? The role of gene–environment interactions in the development of individuals such as Kant, Wilson, Churchill, and Hitler and their subsequent shaping of human events is difficult to deny.

The what-if games played by all of us interested in history demonstrate the belief that individuals and the decisions they make are critical to the course of history. They resemble the what-if games played by evolutionists, such as "What if the comet had missed Earth 65 million years and the dinosaurs hadn't been wiped out?" Neither the genetic evolution of life nor human cultural evolution is deterministic, involving a straightforward unfolding of natural laws. Both depend to a large degree on all that has gone before (a point made frequently and eloquently by Stephen J. Gould).[101] All of our natures are a product of our histories, biological and cultural.

But that dependence on history doesn't mean that *perceptions* of cultural macroevolutionary factors, perceptions that themselves are a microevolutionary factor, were unimportant in the causation of World War II. Resource-poor Japan in the 1930s felt the need for access to Southeast Asian oil and other commodities,[102] but as that country demonstrated clearly in the decades following the war, a military solution was not required for it to gain access to the resources needed for prosperity (a cultural macroevolutionary factor). Cultural microevolution operated at many levels in the generation of the war. Would Japan have moved toward Southeast Asia had the United States continued to supply it with oil? If the Japanese military hadn't attacked Pearl Harbor, would the resources of the United States have been brought to bear so effectively against Nazi Germany? Politics and perceived environmental factors interacted in dramatic ways. Desire for the Soviet Union's potential agricultural productivity and oil reserves clearly was a factor attracting Nazi aggression, but those resources do not tempt aggressive behavior on the part of today's Germany. It is easier and cheaper to buy grain and oil. Technological advances and new trade and political arrangements—cultural microevolution—have made another German attack on Russia unthinkable.

World War II generated multitudinous examples of that microevolution in action. One cannot understand the genesis of that war, one of the most important international events in human history, by simply looking at environmental factors or broad patterns of cultural evolution. Instead, one must get into the nitty-gritty of the cultural evolution operating within nations at the time, the sweep of social and economic evolution leading up to it—especially the details of the course and resolution of its first act, World War I, which transformed the political landscape of Europe. That transformation, rooted in the unexpected slaughter of more than 9 million soldiers, can in part be blamed on mismatches in rates of cultural evolution. The destructive capabilities of military technologies had increased enormously, but technologies of communication had lagged slightly behind. Machine-guns, for example, made horse cavalry obsolete and gave an enormous advantage to the defense. But the way that weapon changed the balance of attack and defense was not appreciated by military strategists who had not adapted to the new technological realities of the war. Similarly, the potentials of radio were unappreciated, and lightweight portable radios that would allow the coordination of attacks had not yet been invented. The result, especially on the western front, was stalemate and slaughter.[103]

Chance events of other sorts shaped the outcome of World War II. If, for example, the aircraft carrier force of the United States had been present at Pearl Harbor at the time of the Japanese attack, the war and its aftermath might have been very different.[104] The history of World War II, like many other histories,[105] is replete with examples of pure luck changing the fates of hundreds of thousands or millions of people and the fates of powerful empires. One fine example of such events in cultural microevolution has always especially stuck with me. Interestingly, it seems to have stuck with my colleagues Jared Diamond and Peter Vitousek, who are war buffs, too, perhaps because the three of us spend our professional lives seeking broad principles to help explain biological phenomena replete with fascinating, chance-riddled detail. Or it could simply be that we're all interested in human behavior and realize that like other complex systems, human beings often show their characteristics most readily when under stress.

World War II started for the United States on December 7, 1941, with the destruction of the U.S. Pacific Fleet's battleship force by the Japanese air attack on Pearl Harbor. Months of uninterrupted and stunning Japanese victories followed. Then there was a heroic but militarily ineffective bombing of Japan by B-25 bombers, normally based on land but riskily launched from the aircraft carrier *Hornet* on April 18, 1942. That attack had an effect totally unanticipated by U.S. military planners. The implied threat to the emperor, then considered a deity, of the Doolittle Raid[106] settled a divisive issue within the Japanese high command. Japan's next major move would be to attack the most westerly of the Hawaiian Islands, Midway Island, instead of continuing its southward advance to cut Australia off from the United States. The idea was to bring the remnants of the U.S. fleet to decisive battle and to bar the United States from using the Midway "keyhole" as an opening through which to launch more air attacks on the home islands.[107] It was the proper strategy in any case, designed by Admiral Yamamoto, but it led to disaster because of an interlocked series of chance events—cultural evolutionary analogs of genetic drift.

The attack on Midway Island was launched in early June 1942; the critical events of the battle took place on the morning of June 4. The Japanese strike force commander, Admiral Chuichi Nagumo, when launching his aircraft against Midway, had kept ninety-three planes loaded with torpedoes and armor-piercing bombs in reserve against a possible encounter with the U.S. fleet. But when the need for a further strike at Midway became apparent, Nagumo, having had no signs of U.S. ships, ordered those ninety-three planes to reequip themselves with fragmentation bombs to drop on the island. When a scout plane reported the presence of U.S. Navy ships in the vicinity of Midway, Nagumo reversed his order and started rearming his planes with torpedoes and armor-piercing bombs. He could not launch those planes immediately, however, because his carriers first had to recover the planes returning from the initial attack on the island.

An hour later, just after the last of the Japanese aircraft from the first wave of attack had landed, there was a series of attacks on the powerful Japanese aircraft carrier force by antique U.S. torpedo bombers. Their attacks were utterly unsuccessful, thanks to skillful Japanese air defense and ship handling; not one torpedo exploded against a Japanese ship.[108] All fifteen planes of the U.S. carrier *Hornet*'s Torpedo Squadron 8, the first carrier-based planes to attack the Japanese, were shot down, and only one pilot[109] survived. The last fruitless American torpedo attack, by planes from the *Enterprise* and the *Yorktown*, ended at 10:24 A.M. For two minutes, it appeared to the Japanese that they had gained control of the central Pacific and essentially won the war. But even though their brave assaults were fruitless, the U.S. pilots' efforts were not, as it turned out, in vain.

At that key moment, the decks of Admiral Nagumo's four fleet carriers, the *Akagi*, *Kaga*, *Soryu*, and *Hiryu*, were covered with planes being refueled and rearmed. Luck then intervened on the U.S. side. Overhead at 14,000 feet appeared thirty-seven Douglas Dauntless dive-bombers from the carrier *Enterprise*, their leader having guessed the location of the Japanese from the course of a lone destroyer that had been spotted returning from an attack on a U.S. submarine. The nimble and feared Japanese "Zero" fighters assigned to give air cover to the carriers were at low altitude, having just finished off the U.S. torpedo bombers. The Zeros thus were unable to interfere with the U.S. dive-bombers. The sacrifice of the crews of the torpedo bombers was, by pure chance, crucial to what followed. At 10:26, there ensued a "beautiful silver waterfall"[110] of Dauntlesses, whose bombs scored the first U.S. hits of the battle. They fatally damaged the vulnerable *Akagi* and *Kaga*, the decks of which were laden with high explosives and laced with gasoline hoses.[111] That Japanese disaster was quickly followed by another Dauntless attack with planes from the *Yorktown*, which led to the sinking of the *Soryu*.[112] In six minutes, the doom of the Japanese empire was sealed by what probably was the most sudden and decisive defeat ever inflicted on a navy.[113] Japan ended up losing four of its best six large carriers and, more important, the cream of its naval aviators. The United States lost the *Yorktown* and relatively few airmen.[114]

Resource availability, environmental differences between the homelands of the combatants, or long-term differences in historical patterns in Japan and the United States caused by one being an island state and the other continental—no macroevolutionary explanation could have predicted the power differences in the assembled forces or the outcome of the battle. Planning, skill, bravery, and, above all, chance all seem to have more potent explanatory power in this case. A militarily ineffectual air raid designed to boost U.S. morale happened to come at a time to influence Japanese strategic decision making and send Japan's fleet toward Midway. And a complex series of human decisions, involving everything from the breaking of Japanese codes to inspired leadership at almost all levels of the U.S. command involved in the battle (in striking contrast to the incompetence that so often characterized U.S. leadership at the start of the war), combined with many more chance events to produce an unpredictable victory of weakness over strength. Factors of cultural macroevolution, especially the command of resources that permitted the United States to develop nuclear weapons, would almost certainly have led to an eventual U.S. victory. But if the cultural microevolutionary sequence had produced a Japanese triumph at Midway Island instead of one on the part of the United States, that victory might well have taken a very different form.[115]

It is, of course, impossible to demonstrate that the surprise U.S. victory at Midway resulted in a change in the natures of the Japanese or the Americans. But it was an important element in the pattern leading to Japan's loss of the war and a U.S. occupation under General Douglas MacArthur that imposed a new constitution on the Japanese people. There is no question, however, that the natures of many Japanese people changed greatly in response to the defeat and the revelation of Japanese war crimes.[116] The Japanese people today are different because of those events, and the same can be said of the people of the United States. The relatively benevolent U.S. occupation of Japan and abandonment of the racist propaganda campaign against the Japanese (stimulated, in part, by the country's desire to recruit Japan to its side in the cold war), along with the gradual realization of the atrocious wartime treatment by the United States of its own citizens of Japanese descent, has reduced anti-Japanese prejudice. Our inability to specify precisely how much the war changed the natures of the people involved is no surprise. Consideration of any significant aspect of history shows that numerous levels of analysis are usually necessary to get a reasonable grasp on causes and consequences, and even then, interpretations are tentative and often controversial. The interacting natures of millions of people have produced a complexity of events that, at the cultural microevolutionary level, frequently make it difficult to identify a few driving forces—especially because historians themselves often are participants and, at the very least, bring their own natures to the task.[117]

There certainly may be explicable regularities in cultural microevolution—in "history." For instance, the gradual trend away from despotisms and centrally planned economies and toward democratic governments and capitalistic economic systems may be traceable in part to superior information flow in the latter.[118] Historians, political scientists, and economists have created a huge literature that attempts, among other things, to document such regularities.[119] They (and we) have a long way to go.[120]

Cultural Evolution and History

Historians, like art critics and reviewers of novels (to say nothing of authors and composers), need have no fear that biological evolutionists will put them out of business. Smart evolutionists wouldn't even want to because they recognize that the social sciences and humanities should remain multivocal—that there is usually no one "proper" interpretation of motivation in a work of literature or in a novel historical sequence.[121] The vast body of culture produced by human minds provides an inexhaustible source of interest and challenge for scholars and others who wish to understand human natures.[122]

Scholars whose concerns are with culture itself inevitably will have a much greater interest in details, whether they consider themselves cultural evolutionists or not. They understand that there are constraints on the course of history—there can't be cities, for instance, if everyone must work full-time to produce food. But within those constraints, the courses of human history and evolutionary history can meander in ways that make prediction difficult even if they can often be explained post hoc. Although biology and anthropology have a lot to say about history and the possibilities of international cooperation, they don't have the last word, any more than neurophysiologists have the last word about pleasure or morals. We can't have a full appreciation of our history without knowing what kind of social animals we are, and we can't fully understand our sex lives or our ethics without understanding our brains. It is standard hubris on the part of those taking a more reductionist approach to a problem to think they have *the* explanation, and of those being more holistic to think, in contrast, that they have it. Understanding means different things—and historians have already shown us that reductionist explanations (e.g., that the fates of empires have been determined by biophysical environments) are incomplete at best.

With that said, can cultural evolutionary generalities be deduced about the progression from early states to modern ones? Max Weber, a founder of today's social science, distinguished earlier states from modern ones by the former's preoccupation with control of internal warring factions rather than defense of national territories from external threats. The problem of controlling internal factions was largely solved when leaders stopped ruling through lesser princes and replaced them with professional bureaucrats. Weber described the modern state, with its highly centralized power, as a system of "rule by human beings over human beings, and one that rests on the legitimate use of violence (that is, violence that is held to be legitimate). For the state to remain in existence, those who are ruled must *submit* to the authority claimed by whoever rules at any given time."[123]

The modern state has an executive entity that coordinates bureaucratic, military, and police organizations, supported by resources coerced from its subjects.[124] Perusal of the literature on the politics of the modern state[125] can give some notion of how far removed are the thoughts of most political theorists from consideration of the external environmental factors that may be shaping the state. Since the days of Thomas Hobbes, who, in the mid-seventeenth century, believed that states were formed to provide security,[126] these theorists have attempted to analyze the nature of the modern state in social, economic, and political terms and, in particular, to understand the tension between the presumed rights of individuals and rights of sovereign states.

The debates on how to balance the freedoms of individuals against the needs of communities become ever more poignant as a small-group animal, one that evolved genetically to deal with societies of a few hundred individuals at most, struggles to organize polities of hundreds of millions or even billions of people. For the past 200 years or so, the governance of nations has evolved steadily away from hereditary monarchies and toward either some form of democracy or authoritarianism. In recent years of this period, the most active intellectual debates have focused on the economic organization of society, on the freedoms related to production and consumption. They have especially centered on the role of centralized planning versus the role of market mechanisms—or, more crudely, on "capitalism" versus "socialism" or "communism." It is a debate that during the cold war threatened an end to civilization by nuclear holocaust.[127] The collapse and fragmentation of the Soviet Union, the explosive disintegration of Yugoslavia, and the general failure of centrally planned economies have illuminated what has long been obvious to many historians. Over the long run, it is hard to maintain large empires or even states containing many diverse ethnic groups, and it is virtually impossible for a government consisting of small-group animals to plan the economic activities of millions of individuals efficiently.

Conservatives and the business press in the United States and Europe have interpreted the breakdown of centrally planned economies as meaning that unrestrained capitalism is the only sensible way to organize the critical economic elements of society. They generally ignore other alternatives, downplay or actively oppose any attempt to consider the serious environmental consequences of such a system,[128] and view national environmental regulations and standards as a drag on economic growth and a barrier to free trade.

Such a view is clearly as senseless as was the idea that economies could (and should) be centrally controlled. That communism has failed does not speak to the obvious (and long-recognized) need to at least moderate the negative social and environmental effects of market economies. Interestingly, the glorification of unrestrained capitalism was not shared by even the most recent eloquent defender of individual choice,[129] the conservative Austrian-born English economist Friedrich August von Hayek. Today's conservatives and libertarians frequently cite Hayek's classic book *The Road to Serfdom*[130] in support of their views of the evils of government regulation, but one must suppose that many of these have never actually read his work. Although Hayek's central thesis—that government attempts to control the economy inevitably lead to totalitarian regimes and inefficient economies—could be claimed to have been borne out by history, those seeking comfort from his views in opposing environmental regulation or a social safety net clearly would be disappointed. Despite his usual label of "conservative," Hayek was no fan of the church-army-landholder conservatism of the nineteenth century or, say, of Generalissimo Francisco Franco in the twentieth.[131] As he wrote in the introduction to the 1956 edition of his magnum opus, "conservatism, though a necessary element in any stable society, is not a social program; in its paternalistic, nationalistic, and power-adoring tendencies it is often closer to socialism than true liberalism; and with its traditionalistic, anti-intellectual, and often mystical propensities it will never, except in short periods of disillusionment, appeal to the young and all those others who believe that some changes are desirable if this world is to become a better place."[132]

Hayek recognized that although central planning of how and for whom things are produced is an economically inefficient and politically disastrous course, the superiority of a market system in that sphere does not obviate the need for another type of government planning. That planning might be summarized today as the creation of a strong system of laws to provide a level (and relatively monopoly-free) playing field on which competition would be acted out and a system of controls to protect the public health, provide personal security, preserve ecosystem services,[133] and maintain an adequate social safety net.[134] The issue then becomes one of how to limit the wielding of government power to those "leverage points" in the legal-social system, where it is most clearly required and most effectively applied to those ends.[135]

Finding the best ways to moderate capitalism while avoiding too much central control is a key area for political debate—which is not surprising, given that all the issues are brand-new in evolutionary time (tens of thousands to millions of years) and many are novel even in ecological time (days to centuries). Human social organization has evolved, under the lash of necessity, to the state level only in the past few thousand years, and the scale of the human enterprise has transformed *Homo sapiens* into a global force only within the past century.[136] It is interesting, by the way, that Hayek clearly saw the negative social effects associated with the growing scale of the human population and of nation-states: "Least of all shall we preserve democracy or foster its growth if all of the power and most of the important decisions rest with an organization far too big for the common man to survey or comprehend."[137] He did not suffer as severely as did some of his admirers from the dissociation of political thought from the evolutionary history of human beings as small-group animals.[138]

The general failure of political theorists to consider biological and cultural evolution is easily understood because the connections of the evolution of modern states to agricultural intensification and resource scarcity are often obscured by social scientists' fascination with markets, industrialization, and national and international politics. The disconnect is reinforced by the relative isolation of political theorists, to say nothing of urban populations generally, from the natural systems that support the states, as well as the poor or nonexistent education most of us receive in hominid biological and cultural evolution and the ecosystem services on which our lives and the continuation of states depends.[139] The bottom line is that modern states should be viewed as one more stage in intensification of human activities, a stage in a long-term largely cultural macroevolutionary process—extracting ever more of the requisites of life from a changing and, in many respects, declining, resource base.

From my own perspective, much of the discussion of political theorists (i.e., all considerations of the formation of social contracts) has been based on a fallacious view of the origins of societies; questionable postulates about "human nature," such as egoism that prevents cooperation;[140] invented "rights" such as individuals inherently having the right to be free, equal, and autonomous;[141] and goals that are by definition impossible, such as seeking the greatest good for the greatest number.[142] A better foundation would be a realistic grasp of what is known about human natures, their evolutionary origins, and their future potentials.

Intensification and the Human Predicament

The production of agricultural surpluses, specialization, stratification, and the formation of cities and states has, after all, led to all that we familiarly know as history, and to the present issue of how to keep national and transnational societies that are organized around free markets from destroying humanity's life-support systems. Population growth has continued more or less unabated since our ancestors settled down to farm—with the number of human beings reaching about 200–300 million around the time of Christ, 500 million about 1650, 1 billion in the first half of the nineteenth century, 2 billion about 1930, 4 billion in 1975, and 6 billion at the turn of the twenty-first century.[143] In historical as in prehistoric times, intensification of food procurement and of the acquisition of other resources has evolved more or less apace—but like growth in the population itself, it has risen and fallen unevenly in space and time (moving through time in a pattern of spurts and relative stasis perhaps analogous to the punctuated equilibrium pattern in the geological record).

For instance, there were dreadful years in fourteenth-century Europe starting with the Black Death, which was followed by wars and turmoil. The Hundred Years' War (1337–1453) and bubonic plague had raised death rates and caused Europe's population to drop from some 85 million around 1300 to 60 million in 1500. It then climbed to perhaps 110 million at the start of the Thirty Years' War (1618–1648), only to fall some 10 million as a result of that conflict. The latter war was particularly bloody—in the storming of Magdeburg by Catholic forces in 1631, about 20,000 people died, and it is estimated that as many as one in three people in Germany and Bohemia perished as a direct result of the war.[144] After the Peace of Westphalia ended the Thirty Years' War, there followed a period of relative stability and tranquility in which population growth again surged.

At the same time, mercantilism was on the rise, state control was becoming more centralized, and agriculture was "modernized" by new land tenure arrangements that dispossessed serfs and peasants, creating larger, more efficient farms. Innovations were introduced to increase productivity, such as use of "green manure," clover to restore nitrogen to depleted fields. This intensification led to further population growth, European expansion (including the appropriation of lands in the Western Hemisphere), and the industrial revolution. That revolution laid the groundwork for the development of modern sanitary and medical technologies that greatly reduced death rates, especially when they spread to nonindustrial countries after World War II. The result was the unprecedented population explosion of the late twentieth century.[145]

Cultural evolutionary responses to the pressures of population growth have remained a mix of intensification of resource extraction, increased integration (e.g., in transport and banking systems), increased centralization and bureaucratic stratification (required to deal with ever larger numbers of people), the spread of birth-control practice and technologies, and migration. The "green revolution" is the best-known recent case of agricultural intensification,[146] and research is now under way to find ways to further intensify crop production.[147] Simultaneously, countries have vastly overinvested in technology-based intensification of fishing techniques, and this has led to unsustainable harvesting of the seas and widespread decline and collapse of fisheries.[148] That in turn has powered the creation of large, often unsustainable and environmentally destructive aquaculture operations to intensify seafood production.[149]

Although the role of external environmental factors in shaping the evolution of modern states may be controversial, there can be no controversy about the damage modern states are inflicting on the environment. The damage is so severe that the future of civilization is called into question. The total impact of the human enterprise on the ecological systems that support it multiplied almost twenty-five-fold between 1850 and 2000.[150] Now states are being united into a global economy by an instant global communications and finance system and the activities of multinational corporations (MNCs). MNCs seldom have any strong connection to a particular nation, and as political scientist David Held put it, "even when MNCs have a clear national base, their interest is above all in global profitability, and their country of origin may contribute relatively little to their overall financial position."[151] According to some observers, civilization is undergoing an information revolution that could rival the agricultural revolution in importance.[152] I have my doubts,[153] but there is no question that humanity increasingly can coordinate its intensification globally. Although many past civilizations failed to sustain themselves,[154] their collapses had only regional consequences. Now, with globalization, the power of nation-states to control their own destiny has seriously eroded,[155] and a plethora of nascent international regimes ranging from the United Nations to the Intergovernmental Panel on Climate Change still have not gained significant regulatory control over the way human beings treat one another or their rapidly deteriorating life-support systems.

As a species, we've left the trees far behind, and we've looked past our current dilemma toward the stars. But we still have natures that are primarily the products of interactions within groups of a few hundred acquaintances, leavened with inputs from many pseudo-acquaintances reaching us first through nationalist propaganda and more recently through films and television shows. We have barely begun to solve the problem with which cultural evolution has presented us: how to live in large groups, perpetually intensifying our activities, creating technologies few can understand and even fewer can control, without sowing the seeds of our own destruction.

Chapter 12

Lessons from Our Natures

"The natural history of our species is a fundamental beginning point for a new analysis of human disorders and degenerative disease, as well as the effects of culture on human environments."

—W. R. Trevathan, E. O. Smith, and J. J. McKenna[1]

We've delved into some of the intricacies of our past, tracing our biological and cultural evolution from early hominoid ancestors to the industrial societies of the present day. The evolution of human natures is a fascinating story as yet only partly told. We should all know the story, for the same sorts of reasons we should know our parents' stories. Just as we can gain a better understanding of ourselves from knowledge of our parents and their lives, so can we draw solace and a sense of orientation from knowledge of the genetic and cultural evolutionary processes that created our human natures and shaped humanity's long history and much longer prehistory.

But I believe that such knowledge can do even more. Our evolutionary history can provide important lessons to inform decisions today and suggest implications for our collective behavior in the future. So far in this book, I've attempted to present the consensus view of the scientific community, supplemented with my own interpretation of evidence where there is significant dispute. Now I'm going further out on a limb to explore a sample of lessons and the implications I draw from them. For convenience, I have rather arbitrarily divided these into lessons that are tactical (applicable to securing specific objectives), the subject of this chapter, and those that are strategic (useful in considering overall values and planning), the subject of chapter 13.

In many cases, my focus, explicitly or implicitly, will be on mismatches in human evolutionary rates. One mismatch involves the different paces of cultural and biological evolution. The incredible speed with which cultural evolution has altered the human environment, especially in the past century or two, has not allowed biological evolution enough time to make changes that could adapt us genetically to the new conditions. Another case involves mismatches in the rates of cultural evolution in the areas of technology and of values, ethics, and social organization. The increasing human ability to *do* things has outstripped the evolution of our ability to *understand* both what we should be doing and the full implications of what we are now doing.

Lessons from Coevolution

In some important areas, a knowledge of evolution is essential to the development of sane policies for our future well-being.[2] This is perhaps best seen in the coevolution of human beings with the pests that attack crops. Unhappily, our use of pesticides has proceeded largely as if Charles Darwin had never lived. The basic approach that humanity has used in its war against organisms that also wish to eat crops has been the broadcast spraying of pesticides. It's the real-world equivalent of the Animal House experiments with fruit flies in which Robert Sokal involved me, and the results have also been similar—resistance galore. Had those who deal with agricultural pests made use of evolutionary knowledge, we human beings would have been much cleverer in our coevolutionary race with our competitors. Pesticides, for example, would have been used only when other measures failed, and tactics would have been employed to retard the development of resistance, such as frequent moratoria on the use of each compound, with temporary substitutions of others.

Similarly, even though we may have (temporarily)[3] eradicated smallpox and have made great strides against polio and some other viral diseases, we cannot let down our guard in the struggle with microorganisms. The battle against agricultural pests is carried out by a relatively small group of people, but we're all in the fight against disease-causing viruses and bacteria every day. Perhaps the most important single tactical evolutionary lesson in this area is that there is a dynamic relationship between *Homo sapiens* and the microorganisms that attack it.[4]

The same sort of broadcast application featured in pesticide use is found in antibiotic use, with similar consequences for development of resistance. Penicillin, for example, was once given to virtually everyone who had anything worse than the common cold, with little attempt to distinguish viral diseases (against which antibiotics are ineffective) from bacterial diseases (against which they were originally extremely effective). The results were entirely predictable—the widespread and rapid evolution of resistance.[5] Penicillin is now useless in the treatment of many dangerous diseases, and new strains of bacterial and protozoan diseases have emerged that are resistant to all, or almost all, available drugs. Examples include the bacterium that causes tuberculosis and the organism that causes the most dangerous form of malaria.[6] In hospitals, there is now no reliable treatment for some strains of bacteria (enterococci) that cause serious infections of the heart valves and other parts of the body.[7] Yet indiscriminate use of antibiotics and other such evolutionarily ignorant practices continue to this day, and people who should know better still seem surprised as resistance threats mount.[8]

One of the most harmful practices is the widespread use of antibiotics as growth stimulants in animal feeds. Because this encourages the evolution of resistance by exposing ever more microorganisms to the antibiotics, it should be outlawed everywhere.[9] Discontinuing this use of antibiotics would have little effect on human diets, probably only influencing the cost of meat in societies in which diets are too high in fat to begin with. The same sorts of steps that should be employed to retard the development of pesticide resistance could be taken with antibiotics, although the ability of bacteria to swap resistance genes among themselves makes the job much tougher.[10]

There is a common but erroneous notion that a fixed number of infectious diseases exists to be conquered and that when they are vanquished, the threat of infection will essentially end. That idea would fade away if understanding of coevolution became widespread. And politicians educated about coevolution would pay more attention to the effects our practices have on the epidemiological environment—all the factors influencing the relationship of our species with disease-causing organisms.[11] The human immunodeficiency virus (HIV) was the first entirely novel pathogen to cause a global epidemic since the origin of modern medicine, but it may be only a warning of worse to come. If more people were aware of the continuing coevolution of human beings with bacteria, fungi, and viruses, they would realize that many aspects of the epidemiological environment are now deteriorating. As the human population grows, pushing larger numbers of people into contact with animal reservoirs of disease organisms,[12] and as transportation systems move more people around ever more rapidly, large-scale epidemics generated by novel strains of influenza or entirely new invaders of the human population may be in the cards. We will remain in peril even if precautions are taken, but the peril can be greatly lessened. Remember that bacteria's short generation times and their ability to pass genetic material among different bacterial species gives them an enormous advantage in their coevolutionary race with *Homo sapiens.* Scientists are trying to find ways to overcome the handicap humanity must face.[13] For instance, careful use of two antibiotics simultaneously rather than sequentially has been suggested to retard the evolution of resistance and encourage natural selection for susceptibility. And entirely new approaches to antimicrobial therapy involving drugs with very different modes of action from traditional antibiotics are likely to be seen in the coming years.

A greater understanding of evolution on the part of the general public would lead to the realization that evolving pathogens change not only their resistance to human defenses but also their virulence in response to changes in modes of their transmission. Disease-causing organisms that are commonly passed by personal contact usually evolve relatively low virulence so that host individuals are not confined to bed and immobilized but are able to move around and infect other people. Those that are carried by vectors (other organisms) or are transmitted in water supplies can be more virulent because the victims do not need to be mobile to allow the pathogens to infect a new host.[14] It is possible, for example, that HIV, the causal virus of acquired immune deficiency syndrome (AIDS), was present in a less virulent form in the human population for a long time. However, selection against virulence may have relaxed with changed patterns of sexual activity during the sexual revolution in the Western world, which was accompanied by extremely high levels of promiscuity in the male homosexual community. No longer did the human hosts need to survive for a long time to allow high probabilities of successful transmission, and viral strains that quickly became abundant in the host outcompeted others despite their tendency to kill their host relatively rapidly. If this scenario is correct, one can predict that carefully orchestrated changes in the virus's environment, such as reduced promiscuity among human hosts, use of condoms, and distribution of sterile needles to drug users, should lead not only to fewer infected people but also to the evolution of reduced virulence in HIV.[15]

The virulence of a pathogen also may evolve in response to selection pressures from other pathogens invading the same host, and knowledge of this may lead to new strategies in the coevolutionary war with our many microscopic enemies. Suppose, for example, that individuals in a population frequently infected by *Plasmodium falciparum*, the deadliest of the four malaria-causing protozoan species that infest humanity, were also likely to be infected by another protozoan, *Trypanosoma brucei*, which causes African sleeping sickness—virtually always a fatal disease. In that case, there would be little advantage to either parasite evolving restraint to keep its host alive. Better to evolve high virulence, use as large a share of the host's resources as soon as possible, and produce lots of offspring to be picked up by the next vector (*Anopheles* mosquito or tsetse fly, respectively) that comes along. Thus, programs to control one pathogen may have a second payoff in the reduced virulence of others. We must always remember that disease-causing organisms are coevolving with humanity and with one another; they are not fixed entities. If the human population in general and the medical community in particular were more aware of this, we could be much more successful in the struggles with our tiny but deadly adversaries, and we could alter our individual and corporate decisions accordingly.

A simple example of the value of tactical thinking about evolution can be drawn from a common self-medication practice—that of taking aspirin or some other medicine to return body temperature to normal when one has a minor fever. Doing that might not always be a good idea, knowledge of evolution suggests. Fever may be an evolved defensive reaction against disease-causing organisms,[16] a result of the coevolution of human beings with an array of pathogens. Allowing the fever to continue may shorten the course of a disease and even reduce the chances of death by making conditions difficult for pathogenic organisms. This, though, leads to another question: if a higher temperature of the body makes it an inferior habitat for, say, bacteria or viruses, why don't we "run a fever" all the time? One possible answer is that fever is energetically expensive to maintain. A fever of 103° F (40° C) can cause the body to use up nutrient reserves 20 percent faster than at normal temperatures, and it makes men temporarily sterile because the heat stops production of sperm. Another reason is that a higher "normal" temperature would increase the rate of aging as a result of increased production of oxygen free-radicals.[17] Furthermore, the body's temperature regulation mechanism is not precise, and if the normal temperature were higher, small upward excursions might push body temperature to the point where it could cause seizures and tissue damage. Finally, one must consider whether a higher body temperature would ultimately be an effective defense against organisms such as bacteria. Their generation times are so rapid that they might be able to evolve new temperature optima in a few days or less—much faster than human males could evolve heat-resistant sperm.

On the other hand, many fevers might be simply nonadaptive side effects of other defensive reactions or even responses triggered by pathogens to *improve* their environment. Medical researchers simply do not know, and unfortunately evolutionary theory is not part of the curriculum in most medical schools. An encouraging sign, though, is that physicians are beginning to give more recognition to the importance of evolution, gradually developing what has been called Darwinian medicine or evolutionary medicine.[18] There has been some research into the adaptive value of fevers, but so far it has been far from enough.[19] If the balance of evidence indicates that some fevers are defensive mechanisms, the wisest course may not always be to try to bring our temperatures back to normal. If we wish to keep an infection as brief as possible, we might decide to allow the fever to continue; if feeling better for some event is important, we might wish to suppress the fever even at the cost of some delay in total recovery. My own tendency is not to reach automatically for the aspirin or acetaminophen when I have a slight fever but to do so only if the fever starts making me feel really uncomfortable or I need to give a lecture that I can't cancel. But that's mostly a matter of guesswork—informed guesswork, but guesswork nonetheless. Obviously, in any case (and especially where a child is concerned), consultation with one's physician is a wise course of action. An understanding of evolution often allows us to ask the right questions, but it less often provides clear-cut answers, especially when some important research remains to be done.

Evolutionary Hangovers

Other health-related tactical lessons can be gleaned from knowledge of our evolutionary past. A classic example is the stress response. Mechanisms designed to deal with the periodic short-lived crises of our hunter-gatherer past are triggered too often by the near-constant stresses that may be encountered in modern life, contributing to a variety of "stress-related" diseases.[20] Here we have a true evolutionary hangover—an adaptation to past conditions that now causes great difficulties. In this case, as in many others, selection operates too slowly to adjust to changes caused by cultural evolution.

Other hangovers appear to be present in some of our basic food preferences. They and other nutritional adaptations evolved in the long period during which our ancestors were hunters and gatherers.[21] Their diets doubtless varied a great deal from group to group but were probably, on average, low in sugar and fat, neither of which is easily obtained from vegetable foods. Sources of both would have been prized.[22] It was not for fun that Cro-Magnons tackled mammoths and that *Homo ergaster* followed honeyguides, African birds that lead mammals to bees' nests in order to feast on beeswax after the honey-seeking mammal has torn open the comb.[23] The risks of being stamped or tusked to death or killed by the stings of thousands of bees were worth taking for the chance to obtain fat or sweets. Food preferences that evolved in hunter-gatherer societies, however, are not necessarily adaptive in environments polluted by hamburgers, hot dogs, soda pop, and candy bars.[24]

That modern diets might not be optimal for health was asserted as long as a century ago by English physician W. Roger Williams. He noted that the English registrar-general had repeatedly warned that each year the cancer death rate was "the highest on record." Between 1840 and 1896, the death rate from cancers had increased more than fourfold. Williams stated that in England,

> probably no single factor is more potent in determining the outbreak of cancer in the predisposed than high feeding. There can be no doubt that the greed for food manifested by modern communities is altogether out of proportion to their present requirements. Many indicators point to the gluttonous consumption of meat as likely to be especially harmful in this respect. Statistics show that the consumption of meat has for many years been increasing by leaps and bounds till it has now reached the amazing total of 131 lb. per head per year, which is more than double what it was half a century ago. . . . No doubt other factors coöperate and among these I should be especially inclined to name deficient exercise and probably also deficiency in fresh vegetable food.[25]

Williams's century-old statement has a curiously modern ring. Scientists do not yet fully understand the origins and development of cancer, but most today would agree that diet plays an important role in susceptibility to many of its forms. A recent study, one of many suggesting a connection between meat-eating and disease, reports that women who eat well-done meat—with a blackened, crisp crust—face more than four times the risk of cancer as do those who eat their meat cooked medium or rare.[26] The English are famous for eating their meat well done, which may explain some of the increase in cancer that Williams noted. In a related area, there is a suspicious correlation between fat consumption and the death rate from prostate cancer. Countries with high fat consumption, as in North America and most of western Europe, have high death rates from prostate cancer; on average, citizens of countries such as Thailand, El Salvador, and Japan consume about one-quarter of the fat and have about one-tenth the death rate from that cancer.[27] There are signs that eating soybean and tomato products can inhibit the growth of prostate tumors (though neither would have been part of our hunter-gatherer ancestors' diets), as can an ample intake of vitamin E and selenium. But dietary factors appear only to slow or speed the multiplication of cancer cells—postmortem microscopic studies of prostate glands, for example, indicate that "silent" microscopic cancers have about the same frequency regardless of diet.[28] Thus, dietary factors may not actually initiate cancerous growth but instead may play an important role in its pace. Few areas are as complex and controversial as the relationship between diet and health, but it is an area in which more thought about the connection between our evolutionary past and our present physical well-being again can encourage us to ask the right questions.[29]

Overall, there is ample evidence that modern diets have a negative influence on health, being implicated in heart disease, diabetes, and cancer, among other ailments.[30] About the only good news is that the damage done by bad diets can sometimes be reversed if an individual turns to healthier eating habits. The passing of several hundred generations since the agricultural revolution has not provided much time for selection to change food preferences.[31] The rich sources of energy that hunter-gatherers found rarely, and thus needed to exploit efficiently, are hardly needed by the overfed people of affluent countries today but are no less prized by them. Rich foods are almost certainly a source of disease today, but people do not usually die from cancer or coronary disease before they reproduce. So selection pressures that might make human beings more able to tolerate continuous high doses of fat and sugar, or that might predispose them to avoid these dangers, are likely to be relatively weak.

When selection pressures are stronger, however, human beings can clearly evolve in relation to dietary changes, as in the case of lactose intolerance (the problems some people have in digesting milk products, discussed in chapter 3). As another example, a diet high in fat and refined sugar has apparently created a relatively powerful selection pressure against certain genotypes in some populations—for example, those that make an individual susceptible to one form of diabetes.[32] On the Micronesian island of Nauru between the two world wars, sudden wealth brought about by phosphate mining transformed subsistence farmers into some of the world's most sedentary people, and they became hooked on junk food.[33] The incidence of diabetes at first skyrocketed (among some groups, it is over 60 percent) and has given the wealthy citizens of Nauru among the shortest of human life spans. Between 1975 and 1987, however, public health surveys showed the diabetes epidemic abating. The diet remained the same, but apparently there was genetic variation within the population in resistance to diabetes, and there were many fewer susceptible individuals left. Diabetes on Nauru strikes people in their prime reproductive years, and women with the disease have fewer than half as many babies as do those who are healthy. Thus, there would be substantial selection against genotypes that predispose individuals to diabetes.

The Nauru diabetes epidemic is just one of several epidemics that are now striking various parts of the developing world and that are associated with rapidly westernized diets.[34] A well-studied example is the set of health problems, such as susceptibility to diabetes and other chronic diseases, faced by Australian Aborigines who have recently switched to modern junk food diets.[35] Presumably, in the past, as diets in the Western world became richer, selection gradually reduced the frequency of susceptible genotypes in Western populations, given that these populations do not suffer the same high levels of diet-related ills as do new converts to rich diets.

But why would such susceptible genotypes exist in the population in the first place? One key possibility is that they provided an advantage in earlier populations at a time when opportunities to gorge on high-calorie foods were relatively rare and the ability to convert them rapidly into stored fat would be a substantial advantage. Those who managed to acquire those high-energy foods were more likely to be healthy and reproductively successful, so there may well have been selection favoring a desire to eat a great deal of such foods on the relatively few occasions when they were available—if the requisite genetic variation existed in hunter-gatherer populations. If that occurred, and if there has not been sufficient time for selection to change that desire, then junk food may rate as a "superoptimal stimulus" for many *Homo sapiens* today.[36]

Superoptimal stimuli are agents that produce greater responses than do related natural stimuli. A classic example of superoptimal stimuli involves the artificial lengthening by researchers of the tails of male Kenyan long-tailed widowbirds.[37] For female widowbirds, the superlong tails were superoptimal stimuli; even though these females had never before been exposed to them, they preferred males with superlong tails over those with normally long tails. There is little or no cost to a female widowbird of seeking to be inseminated by a very long-tailed male—such a male in nature would have to be an unusually strong or agile individual, able to escape predators despite his trailing encumbrance.[38] But in nature, the risks to a male with a tail too long, such as a higher probability of being devoured by a hawk, have presumably counterbalanced any sexual benefits an overexaggerated streamer-like appendage would bring. Selection, however, has simply programmed the female birds to choose as mates the males with the longest tails.

For me, a nearly raw, fatty steak smothered with buttery mushrooms or a hot-fudge sundae sprinkled with malted-milk powder may constitute superoptimal stimuli, just as males with extended tails do for those female widowbirds.[39] Selection may simply have resulted in people having a tendency to eat as much as they can when sweet and fatty foods are available.[40] A change in the environment (sudden great availability of high-calorie foods) converted this tendency into a disadvantage. So human problems with over-rich diets are traceable, at least in part, to that biological evolutionary hangover. Food preferences, of course, also undergo cultural evolution. Sushi restaurants are proliferating in the vicinity of Stanford University at the same time that McDonald's and Kentucky Fried Chicken franchises are invading Tokyo. I now rarely eat hot-fudge sundaes, no matter what my genes tell me, and like many people in the United States, I am, in general, consuming less beef and saturated fat. So evolutionary hangovers apparently are susceptible to cultural cures.[41]

An interesting related tactical question involves human acquisitiveness. Could there be a long-established biological evolutionary predilection to acquire things other than mates or certain foods? Could an abundance of consumer goods be a form of superoptimal stimulus? Hunters and gatherers normally had limited possessions, but meat from important kills, tools, weapons, and clothing have doubtless been highly valued for thousands of generations. That people have long paid careful attention to who procures what (and subsequently made it his or her property)[42] and how acquisitions are distributed among members of the community is more than amply documented in historical sedentary groups.[43] It seems certain that acquisitiveness was favored by cultural evolution, especially if an ability to provide goods (especially food) to others or simply to possess them had a payoff in prestige.[44] It probably also would have had a payoff in reproductive success, as seen in several preindustrial groups[45]—and could thus have a genetic evolutionary element. Control of resources by males, at least, certainly often appears to have had that effect historically. Remember Mulay Ismail the Bloodthirsty of Morocco and his reputed nearly 900 kids! Is it conceivable that in suitable environments, there is some contribution to acquisitiveness from our genes?

In considering the superconsumption characteristic of, say, society in the United States today, it does little good to focus on a genetic evolution "just so" story. There are no signs of "genes for consumption" and plenty of signs for cultural pressures for acquisitiveness. If consumption is to be controlled, it clearly can be done only by cultural evolution. There may be hope of that occurring because there is little indication that, beyond a certain point, consuming more and more produces more satisfaction. That point perhaps was reached sometime around the middle of the twentieth century for many people in the United States—at least, as per capita incomes were doubling between 1957 and 1992, the proportion of individuals reporting that they were "very happy" remained the same.[46] A similar trend has been recorded in Japan.[47]

People's feelings of satisfaction, in terms of income or consumption, derive substantially from their position in comparison with the positions of other people in a relatively limited reference group of acquaintances with whom a person identifies or has emotional affiliation. Presumably, in the studies just referred to, people in both the United States and Japan were keeping up with their neighbors. As neurophysiologist Donald Kennedy put it, "welfare detectors are disparity detectors."[48] That is, our sense of well-being depends in large part on the differences we perceive between ourselves and others. Sociologists call being less well-off than one's acquaintances relative deprivation[49]—deprivation with respect to others in the reference group. It's an idea that fits in well with the notion that a primary driver of human brain evolution was the need to deal with competing and cooperating members of social groups. "Competitive acquisition," as Harvard University economist Juliet Schor calls it,[50] has clearly been enhanced by the effects of television on our visually oriented psyches and by the advertising industry, which spends gigantic amounts of money to make those visual images as enticing as possible. But those very facts may give us clues to the cultural tools we might employ to control a level of consumption that is now totally out of control.

Every year, I spend a month or two doing fieldwork in Costa Rica. When I walk down the rural roads at night, I see television screens glowing from virtually every humble home. The ability of today's large populations in less developed regions to *see* fantasy versions of how the rich live has vastly inflated their aspirations. More and more people at home and abroad want to live like the super-rich on the screen, and their desire to consume is triggered by the portrayed affluence of others in a vastly expanded reference group. Wanting to keep up with Hollywood moguls, business tycoons, and rock, movie, and sports stars appears to be a prime mover of overconsumption.[51] Competitive acquisition is thus helping to drive one of the three great threats to the human environment, the middle factor of what John Holdren, Anne Ehrlich, and I have dubbed the $I = PAT$ equation,[52] which describes the impact of a human population on Earth's life-support systems.[53] In $I = PAT$, impact, I, is equal to population size, P, multiplied by affluence, A (measured as per capita consumption), in turn multiplied by the technologies, T (including social, political, and economic arrangements), that service the consumption. Thus, it may be that our past has built in evolutionary consumptive hangovers that do not serve us well in an overcrowded world in which natural capital—forests, soils, underground freshwater, and the like—is being rapidly depleted. An obvious cultural evolutionary strategy to reduce overconsumption would be to encourage behavioral change among members of the reference groups and alter the way that behavior is reported on television. For instance, humanity might be better off without programs such as *Lifestyles of the Rich and Famous.* How to accomplish the needed changes, however, is a much more difficult question.

Another area in which we may be dealing with evolutionary hangovers is in our struggles with bureaucracy. Anthropologists Lionel Tiger and Robin Fox argue that the reason for the frequent failure of bureaucracies is that they are inhuman: "In some sense they are human because they are human inventions. But it is one of the paradoxes of an animal endowed with intelligence, foresight, and language, that it can become its own animal trainer: it can invent conditions for itself that it cannot then handle because it was not evolved to handle them."[54] In the case of bureaucracy, the goal is putatively to eliminate favoritism, nepotism, love, hatred, and other "emotional" elements in human relationships—that is, to remove much of their humanity and put machine-like standardization in their place. This, in Tiger and Fox's view, causes stress for both the bureaucrats and those who must deal with them because our long evolutionary history has produced an emotional animal with a great propensity to favor kin. Producing an environment in which this biological evolutionary hangover could be overwhelmed would not be easy, even when and if it might be desirable. The frustration of the bureaucrat is often converted into an obsession with rules rather than the ends that the bureaucracy was designed to accomplish. And it frequently leads to "corruption" as bureaucrats tend to favor kin and pseudo-kin, as hominids have done for millions of years. Nepotism comes naturally, and those not related to the bureaucrat with whom they're dealing (or not part of her old-girl network) may suffer as a result. On the brighter side, though, our human flexibility shows through. Many bureaucrats are willing to bend or reinterpret the rules to make their organizations function smoothly and to maintain good relations with their clients.

In complex modern states, there seems to be no way to eliminate bureaucracies, but there may be ways to make them fit better with the natures of the human beings who must deal with them. Bureaucracies dealing with issues such as teenage pregnancy or mixed sex military units should recognize that trying to prevent sexual interaction between normal, sexually mature human beings is like trying to bail out the ocean with a thimble. Nothing is a more natural part of human natures, and there is abundant evidence that rules dictating abstinence will not be followed. Bureaucrats must instead try to promote conditions in which the *consequences* of the interactions are not inimical to the goals of society—although this can be incredibly difficult within bureaucracies themselves.

Of Skin Color and Science

The influence of the human evolutionary past and the results of ignorance of that past are nowhere more obvious than in the area of race, and few kinds of ignorance have had more pernicious effects in society. Racial prejudices stand, among other things, as a major barrier to the cooperation that will be required if the human predicament is to be resolved. Neither global nor regional and local environmental problems are likely to be solved if different groups are battling one another on the basis of imagined differences in human "quality."

Race and the related issue of group differences in intelligence deserve a close look, for in this area knowledge of the evolution of human natures provides some particularly important tactical lessons. One might claim that knowledge of how evolution works and how modern human beings evolved does not tell us much about racism (or sexism) that is not made obvious by direct observation of modern human beings or application of some simple and widespread moral concepts. After all, many people who are strongly opposed to racism and sexism are utterly ignorant of human evolutionary history. But I think that understanding more about our biological and cultural evolution will enable people to more readily see through and refute racist and sexist arguments based on evolutionary misapprehensions. In fact, the relevance of evolution to racial issues is acknowledged in popular culture in statements about, for instance, which "race" most recently "descended from the trees." The main basis for modern "racial" classification is skin pigmentation. That is, as we have seen, evolutionarily a very labile trait, responding to geographically varying selection pressures primarily related to the amount of ultraviolet radiation to which a group is exposed. Indeed, recent rapid human migrations have placed many people in environments to which their skin coloration is not adaptive. The coloration becomes an evolutionary hangover and creates public health problems.[55] Examples are dark-skinned peoples who have moved to northern areas, such as Indians and Pakistanis living in England, and light-skinned peoples who have migrated to sunnier areas, such as Europeans who have settled in Australia. The former are more susceptible to vitamin D deficiency than are their counterparts still in their homeland; the latter, to skin cancers.[56] Here again, some knowledge of evolution could help more people to behave in more sensible ways (e.g., taking vitamin supplements or covering up and using sunscreen) and could alert public health professionals to potential problems. Fortunately, this is beginning to happen.[57]

The idea that there are large groups of people who are differentiated from other groups by superficial characteristics and by putatively inherited mental qualities such as "intelligence" or "ambition" is one of the oldest myths cherished by people unfamiliar with evolutionary theory.[58] It persists even though virtually everyone is continually confronted with massive anecdotal evidence that people from all geographic regions and cultural groups, when adopted at birth into families from different areas or traditions, acquire the language and skills of their adopted cultures.

As is the case with other species, geographic variation in human beings does not allow *Homo sapiens* to be divided into natural evolutionary units. That basic point made by Edward O. Wilson and William L. Brown[59] has subsequently been demonstrated in a variety of organisms from butterflies[60] to *Homo sapiens*,[61] and use of the subspecies (or race) concept has essentially disappeared from the mainstream evolutionary literature. For human beings, whether we plot skin color, height, indices of nose or face shape, frequencies of genes controlling blood groups, or any other characteristic, the resulting maps are in most cases utterly different from one trait to the next (see the figure on pages 50–51).[62] Every geographic unit within our species has a unique combination of superficial attributes, but these do not fall into discrete subspecific groups of populations because the pattern of variation generally is discordant (having characteristics that change independently) rather than concordant (having different characteristics that change together from place to place).[63] This principle is illustrated in the figure on the next page. The only way people have managed to divide *Homo sapiens* into five (or as many as thirty)[64] putative races is to select arbitrarily one or a few characteristics (especially skin color) on which to base the classification—and then to throw in qualifiers such as "likely to be carrying

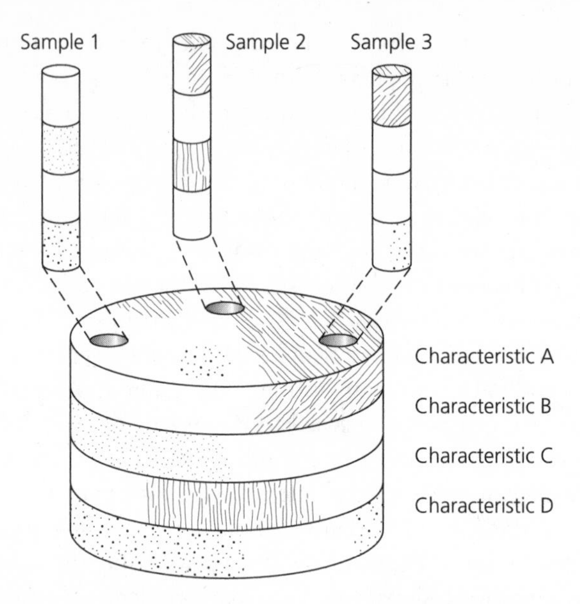

"Character layer cake," an explanatory model in which the horizontal axis represents a circular geographic region and the four "slices" show geographic variation in four characteristics (*A–D*) of a hypothetical organism distributed over the region. Each characteristic has substantial geographic variation over the region. Samples of organisms taken from different locations (three shown here as cylinders 1–3) will always be different from one another, but there is no way to divide the area logically into a series of subspecies. Geographic variation does not necessarily allow us to divide a species into subspecies that are logical (biological) units.

gene *X*" when saying how to determine the group to which an individual belongs. Pick a different set of characteristics and you get a different set of "races." Anthropologist Loring Brace and his colleagues put it succinctly after a detailed study of the characteristics of ancient Egyptians: "An assessment of 'race' is as useless as it is impossible."[65] As one recent report said: "[Government-mandated racial] categories could have enormous implications—from the distribution of government resources to political districting to demographic research. But as far as geneticists are concerned, they're meaningless."[66]

In the ignorance of evolution that has long prevailed, however, races are assumed to be coherent groups, and "race" has become a social and political issue—not just in the United States and South Africa but also in Europe, China, Japan, and elsewhere. A recent manifestation was the creation of a largely artificial "racial" divide based on putative physical differences between the Hutus and Tutsis, members of the same highly interbred ethnic group in central Africa.[67] That pseudo-difference, which reinforced a class distinction fostered by colonial powers, became the basis of one of the worst genocides of the twentieth century. There, as it has in many societies, "race" became a basis for social stratification, dominance of one group over another. Sadly, many scientists in such stratified societies have reinforced popular theories of purported racial superiority to support the views of those in the dominant stratum.[68]

At its most extreme, "race" has become an excuse for slavery.[69] That even great minds have accepted notions of racial inferiority as an excuse to exploit other people is instructive, if depressing. Consider the following statement: "The first difference [between the races] which strikes us is that of color. . . . Add to these, flowing hair, a more elegant symmetry of form, their [blacks'] own judgment in favor of the whites, declared by the preference of them, as uniformly as is the preference of the Oranootan [sic] for the black women over those of his own species. The circumstance of superior beauty, is thought worthy attention in the propagation of our horses, dogs, and other domestic animals; why not in that of man?"[70]

The author was not a dedicated racist gibbering on his or her Web page, or even the famous mid-twentieth-century bigot Senator Theodore Bilbo of Mississippi, but rather Thomas Jefferson. In the same book, Jefferson offered his analysis of racial differences, such as "In memory [blacks] are equal to whites; in reason much inferior."[71] He thought blacks were no good at art but better than whites at music. Indeed, his views—those of a highly educated man of his time—resemble those of many compassionate but ill-informed persons today. Jefferson, was, of course, one of the great liberal thinkers two centuries ago but, like all of us, a creature of his time and culture. He saw not only "political" barriers to emancipation but also "physical and moral" ones.[72] He commented that among Romans (whose slaves were white), "emancipation required but one effort. The slave, when made free, might mix with, without staining the blood of his master. But with us a second is necessary, unknown to history. When freed, he is to be removed beyond the reach of mixture."[73] There is a double irony here. Jefferson himself may have fathered children of a "black" woman,[74] Sally Hemings. And, after centuries of worry about "racial mixing," it now seems clear that the health dangers of too much inbreeding are real, whereas none have been documented for "racial mixture," even between members of the most genetically divergent human groups.[75]

Knowledge of genetic evolution provides a basic background for critical evaluation of the biological significance of race. The out-of-Africa 2 model discussed earlier in this book (the proposal that modern human beings trace to a relatively recent African exodus), if correct, demolishes once and for all one racist hypothesis, put forth in 1962 by anthropologist Carleton Coon. In a long and confused book titled *The Origin of Races*,[76] he wrote that "over half a million years ago, man was a single species, *Homo erectus*, perhaps already divided into five geographic races or subspecies. *Homo erectus* then evolved into *Homo sapiens* not once but five times, as each subspecies living in its own territory, passed a critical threshold from a more brutal to a more sapient state."[77] Guess who, in Coon's theory, got to be "sapient" before certain peoples with dark skins?[78] The scientists who favor the modern multiregional hypothesis—that geographic variants of *Homo erectus* evolved into *H. sapiens* in place—do not support this antique racist view.

Wilson and Brown had destroyed the concept of subspecies (synonymous with races) as evolutionary units almost a decade before Coon published his book. This sounds strange to many people, who claim to "know" races when they see them. The human emphasis on sight tends to override other evidence. Indeed, the urge to name populations that "look" different seems overwhelming to many taxonomists who work with nonhuman organisms, especially if the differences are in color.

The nonexistence of races as evolutionary units has never kept people from defining as inferior those who they perceive as belonging to different groups, often on the basis of color. Even Carolus Linnaeus, the father of modern taxonomy, in 1758 described the "white race" *(Homo sapiens europaeus)* as "nimble, of the keenest mind, [an] innovator" and the "black race" *(H. s. afer)* as "cunning, lazy, careless."[79] More than 100 years later, Thomas Henry Huxley—Darwin's liberal bulldog—wrote, "The highest places in the hierarchy of civilization will assuredly not be within the reach of our dusky cousins, though it is by no means necessary that they should be restricted to the lowest."[80] To Huxley's credit, however, he also wrote: "No human being can arbitrarily dominate over another without grievous damage to his own nature . . . no slavery can be abolished without a double emancipation, and the master will benefit by freedom more than the freed-man."[81] Huxley was clearly light-years ahead of Senator Bilbo, who as late as 1944 could write: "Historical and scientific research has established . . . beyond all controversy [that]: First: the white race has founded, developed, and maintained every civilization known to the human race."[82] Bilbo apparently didn't realize, among many other things, that China and Zimbabwe (in southern Africa) had thriving civilizations when northern Europeans were barbarians. But he certainly knew how to get elected in Mississippi in the middle of the twentieth century.

The attempt to find a scientific basis for racism and demonstrate that geographically based groups of human beings have different cognitive or emotional characteristics has been alive and well in recent decades. In 1973, William Shockley, a Stanford engineer who was awarded the Nobel Prize as a co-inventor of the transistor, launched himself on a crusade[83] to prove that dark-skinned people were inferior—going so far as to claim that people are "color-coded" for quality.[84] Shockley's grasp of the key issues was underlined a quarter of a century ago when, at the Stanford faculty club, he explained to me and geneticist Walter Bodmer that heritability could be studied in human beings by "mating together identical twins." He apparently was unaware that identical twins are always of the same sex. His amusing mistake in some ways foreshadowed the recent use of incorrectly analyzed data on twins in some studies as ammunition in overestimating the significance of genetic factors in human behavior.[85] The same racist theme—that some groups of human beings are innately inferior to others—is still periodically resurrected, most recently and more subtly by the late Richard Herrnstein and Charles Murray in *The Bell Curve: Intelligence and Class Structure in American Life.*[86]

The view has deep roots in a long history of discrepancies between the scores of black populations and white populations on standard intelligence tests—a discrepancy that is now declining.[87] The all-too-frequent interpretation of the scores has been that the poorer performance of blacks traces to genetic inferiority. But that is not the case. For one thing, the concept of intelligence is itself a difficult one and much disputed. Roughly, intelligence is considered to be the ability to understand the world, solve problems, and adjust to environmental changes. No single measure of intelligence is considered adequate to capture all its dimensions,[88] and what constitutes intelligence is a function of the attributes not only of individuals but also of the societies in which they are embedded. For example, an average Australian Aborigine might not do well on a translated Stanford-Binet intelligence test, and many people of European descent have historically flunked the Aboriginal intelligence test of surviving while traveling in the harsh Australian outback. Their failure is epitomized by a single phrase in a note left by the doomed William John Wills of the disastrous Burke and Wills expedition across the center of the continent in 1860–1861. After recording that the expedition's camels were dead and its provisions exhausted, he wrote: "We are trying to live the best way we can like the Blacks but find it hard work."[89]

As one might expect, it turns out to be exceedingly difficult to test any kind of intelligence cross-culturally. Different cultures generally speak different languages or dialects, so one culture is likely to be disadvantaged in any language-dependent test that attempts to compare people from two cultures or subcultures. For instance, the English spoken in inner-city black communities of the United States has evolved as a well-structured, normal language that conforms to the set of rules linguists call Black English Vernacular (BEV),[90] yet IQ tests are not given in inner-city English. If you did not grow up with the sort of language used in rap music, imagine how you might perform on a test given in that language. Cultural differences in perception—demonstrated, you will recall, in the differing susceptibilities of people from different environments to visual illusions—also may influence visual parts of some tests, in which the subject must identify one of a series of figures that represents a rotation of a sample figure.[91]

Scientists interested in measuring the degree to which genes are involved in an observed variation in a characteristic such as height, skin color, or score on an IQ test often attempt to calculate heritability. Properly calculated, heritability is a statistic that measures the degree to which some characteristic of the offspring in a population *in a given environment* will resemble that of their parents. Heritability is a tricky statistic to interpret because it describes the way a characteristic of a population responds in an environment, and it can vary enormously from environment to environment. It's also difficult to measure. Indeed, it is virtually impossible to assess the heritability of a trait in situations in which patterns of mating and the possibility of cultural transmission of traits from parents to offspring cannot be controlled. For that reason, the heritabilities of traits in human beings are frequently overestimated.[92] Two of the world's best quantitative geneticists, Marc Feldman and Richard Lewontin, summed up a technical discussion of the heritability situation concisely: "At present no statistical methodology exists that will enable us to predict the range of phenotypic possibilities that are inherent in any genotype. . . . Certainly the simple measurement of heritability . . . is nearly equivalent to no information at all for any serious problem of human genetics."[93] Use of the heritability statistic should probably be restricted to plant and animal breeding, in which it can be better measured and the measurements put to some practical use—such as in predicting the results of applying selection to increase the number of eggs laid by chickens.

But even if heritabilities were easy to calculate and useful, they'd tell us nothing about intergroup genetic differences—which, you will recall, are minor in human beings compared with genetic variation *within* populations.[94] Assume for a moment that there is general agreement about the meaning of the word *intelligence* and that some ideal test of "general intelligence" has been designed. Suppose further that a significant heritability of test score within groups has been demonstrated. That would mean that even if parents in a group did not raise their own children, the test scores of parents would be similar to those of their biological children who had been raised in another home. What would this tell us about the source of differences in the average test scores of different groups? Nothing at all. It would be a statistic that gave not a clue about whether one human group had evolved greater mental capacities than another.

To understand this failure of heritability to speak to group differences, consider a simple thought experiment. Suppose we were investigating a random sample of people from New York City. It would be a multicolored group and would have a high heritability of skin color, given that children's skin colors tend to resemble those of their parents. Then suppose that one winter, the group were randomly divided in two (so that both halves had the same average skin color), and one-half of the group went to Miami for a month to lie on the beach in the sun while the other half stayed in New York. When the groups came back together, there would be an average difference in skin color that would be entirely environmentally caused—the group that had been in Miami would be, on average, darker than the stay-at-home group. That would be the case even though the skin color difference was in a character previously determined to be very highly heritable.[95]

A key tactical lesson from all this is that a characteristic's high (or low) heritability gives no prescription for social action. *"Heritable" does not mean resistant to environmental change.*[96] Even if IQ *were* highly heritable in a population, it wouldn't mean that programs such as Head Start that supply intellectual stimulation to preschool children wouldn't work, any more than high heritability of height would protect babies from having their growth stunted by poor diets.

An even more important tactical conclusion is that it would be foolish to make social policy on the basis of group averages in characteristics, regardless of the reasons for the differences in those characteristics. Consider another thought experiment. Imagine that, counter to everything geneticists know, there is a unitary something that could be called "genetic IQ," and some way is discovered to assess it. Perhaps someone invents a sort of smart litmus paper on which, when the paper is placed on the forehead of a test subject, a number miraculously appears, faultlessly indexing his or her "genetic IQ." Imagine further that average "genetic IQ" litmus test scores tend to be somewhat higher in the black population, even though many whites score much higher than many blacks—some at the "genius" level. Would it then be good policy to give remedial aid to all whites and none to any blacks? Or would it be wiser to give additional help to those who have low scores, regardless of skin color? What, in fact, would be the reason for even bothering to calculate the group average IQ scores? Do we calculate them for populations differentiated on the basis of other characteristics, such as different blood groups? Would we want to know (or would we care) about litmus IQ average differences between those who have type AB blood and those with type O blood, between those who are tall and those who are short, or between those of normal weight and those who are overweight, or those with blue eyes and those with brown eyes?

It is only because people live in socially stratified societies and have a fascination with skin color (or nose shape or social class) that differences between certain groups are singled out for investigation and determination of their possible genetic causes. If average differences in IQ test scores were correlated with skin color in our society, would it make sense to try to decrease the incidence of low test scores by treating skin-color groups differently? Of course not, any more than it would make sense to attempt to lower the incidence of skin cancer (to which lighter-skinned people are more susceptible) by doling out sunscreen on the basis of IQ test score! We should be aiding individual students with low scores regardless of skin color.

On the cheery side, over the past few decades, belief in evolved genetic differences in "quality" between human groups has certainly declined. But it must decline further. Aside from the immorality of ill-treatment, too much potential intelligence is lost in the partial neglect of the enormous talents of those with the "wrong" skin colors—people whom, because of the evolved dominance of human visual systems combined with the vagaries of cultural evolution, societies have often defined as inferior and treated badly. There's an even more utilitarian reason as well: humanity faces critical problems that require unified action, and we can ill afford the divisive effects of racism.

Where Is a Woman's Place?

A woman's place is anywhere she wants to be. Sexism, much like racism, is a product of cultural evolution: a debilitating hangover of the past based on biological differences in strength and reproductive commitment. Sexism stands as another monument to the relatively slow rate of cultural evolution in social attitudes. As with racism, knowledge of genetic evolution permits no excuses for sexism, the all-too-common assumption that women are in some ways mentally, physiologically, or emotionally inferior to men and should therefore be barred from certain occupations and responsibilities. There is no evidence that differences in genomes make women less capable on average than men of carrying out any job or activity for which they have the basic physical strength.[97]

What we do know is that roles and attitudes associated with gender have biological roots (remember the case of John/Joan from chapter 8) but are very much influenced by culture. Evolutionist Bobbi Low showed, for instance, connections between the degree of stratification in polygynous human societies (those in which one man often bonds with more than one woman) and the education and training given to children in such societies.[98] Because in polygynous groups men would show greater variance in reproductive success (some men having many children, others going childless) than women, it seems reasonable that males and females should be trained differently, and indeed they are. Boys, for example, tend to be trained to show more fortitude, girls to be more sexually constrained.

But training differences go beyond that. Low found that sons in relatively unstratified societies are taught to be more aggressive. In those societies, successful men presumably have the best chances at high levels of reproduction. In more stratified societies, parents apparently do not so inculcate striving in their sons, presumably because a man's reproductive potential and social success are rather strongly set by his class or social position. Women, for example, are often able (indeed, encouraged) to "marry up" in class; men usually are not. These patterns of training *might* be partly explicable on the basis of predispositions established by selection, but the requisite data on reproductive success of individuals relative to their childhood training are not available even for current societies, let alone for past generations. After all, even for behaviors directly related to sexual activities (such as a preference for a variety of partners) it has been difficult to demonstrate that genetic predispositions exist. If they do (and I suspect they may because of the very different biological roles in reproduction assigned to men and women), those predispositions clearly are very labile in response to environmental influences.

Sexism has a long and dreary history in science (including the fields of evolutionary theory and ecology), which has long been dominated by white males.[99] Indeed, a classic 1963 book attempting to trace "the evolution of the scientific intellectual as a human type," which has an entire chapter on "the scientific revolution among the Jews," has no mention of the role of women in science and among its numerous cited scientists has no mention of Marie Curie or Lise Meitner.[100] What is less clear is exactly to what degree sexism has biased scientific research in various areas (my guess: a great deal in medicine and perhaps in anthropology, relatively little in nuclear physics).[101] This problem is at least on the road to solution. Both evolutionary biology and ecology are now populated with brilliant women who are doing world-class research even though they frequently have to contend with greater domestic commitments than do their male counterparts. As the younger female scientists mature, even the old men's club of the National Academy of Sciences (which is starting to recognize the cultural baggage carried by scientists of all stripes and persuasions) may become an old women's club as well.[102] Cultural evolution marches on, and in a tiny slice of evolutionary time it appears to be on its way to eliminating a division of labor rooted in biological evolutionary hangovers—behaviors no longer appropriate in modern environments.

Jumping the Tactical Gun

An understanding of the evolution of our natures clearly can help to abate the divisive forces of racism and sexism. But caution is necessary when considering more direct tactical interventions in areas in which our understanding of evolutionary factors is less clear-cut and the consequences of interventions are biologically and ethically more uncertain. For instance, evolutionary psychologists John Tooby and Leda Cosmides imply that knowledge of the ways in which genetic evolution has shaped the human brain opens the door to "understanding enough about human nature to eventually make possible intervention to bring about humane outcomes. Moreover, a program of social amelioration carried out in ignorance of human complex design is something like letting a blindfolded individual loose in an operating room with a scalpel—there is likely to be more blood than healing."[103] The key word in their statement is *eventually* because it's clear that current understanding of the gene–environment interactions involved in the development and functioning of the human brain is insufficient to support genetic "intervention" or even, at the moment, gene-replacement therapy. Moreover, the information does not now exist that would render plausible—even if not desirable—attempts to direct the course of evolution with artificial selection in order to "improve" people—that is, with eugenics. Indeed, the tactical proposals sometimes made for some form of human breeding program might well be analogized to plans to allow a maniac with an ax in that operating room. There are, however, some known genetic defects that cause mental and physical problems the occurrence of which could be prevented by genetic *counseling* about having children. Trisomy (Down syndrome—having three copies of chromosome 21) causing mental retardation is an example.[104] Another is neurofibromatosis, a disorder caused by a dominant gene that may cause disfiguring tumors and a variety of neurological problems, including blindness, dizziness, and deafness.[105] Even these situations produce serious ethical dilemmas, however, such as whether people should be advised not to have children if there is a one in four chance of a baby being seriously affected by a disorder.

Even the value, as a basis for social action, of achieving that understanding of human nature "eventually" must be examined carefully. Suppose we knew that there actually were a gene or genes that predisposed male carriers to homosexual behavior. What would that tell us? Would homosexual individuals then deserve less "blame" for their behavior? What policy recommendations would flow from that discovery? Would children with that gene be warned or given special training sessions?

Or consider the rare occurrence of infanticide in modern society. What if it were shown that a hereditary bias in human males made them slightly more likely to kill a stepchild or a girlfriend's offspring fathered by another male than their own biological child? What tactical steps should be taken? Should warnings be given to all remarried women who have children from a previous union and all unmarried mothers to keep a strict eye on mates who did not father the children, in order to keep their offspring maximally safe? Suppose that largely through genetic evolution, men actually have become more inclined than women to seek sexual variety. Estrogen injections for all of them? Would it be better or worse if they had become so inclined by cultural evolution?

The degree to which our past genetic evolution influences human behavior today is a fascinating topic for speculation, but even knowledge of exactly "how much" would not lead to immediate policy recommendations. Those who think that such knowledge would enable us to make sound tactical prescriptions tend to make some common errors. They tend to assume that genes are destiny—that selection often produces inflexible, DNA-based behavior patterns in human beings. They then conclude that any undesirable behaviors judged to be heavily influenced by genetic factors would best be altered by genetic intervention. But recall that a hallmark of our natures is incredible plasticity. That flexibility shows up even in the physiological mechanisms that underlie our behavior. Remember, for instance, that parts of the brain ordinarily devoted to helping us see can be dedicated to hearing in blind individuals, enhancing what is in essence a backup sensory system in animals genetically programmed to have sight as their dominant sensory mode. Knowledge of genetic influences is not only technically very difficult to obtain; even in theory it would rarely help us with policy decisions.[106]

Small-Group Animals

At the level of both tactics and strategies, I think the most important evolutionary lessons relate to our long history as a small-group animal. The evidence seems strong that the unique features of our brains evolved in large part to solve the problems of living and communicating in small communities of increasingly clever companions. Only recently in evolutionary time—a matter of a few hundred generations at most—has the opportunity even existed for interacting closely with more people than one would find living in a small village or employed in a small business. Most people, indeed, still have as friends and acquaintances a group of the same size (on the order of 150) as many hunter-gatherer bands. We still seem to bear many other traces of our small-group, gossiping-and-grooming past. Indeed, one study showed that about two-thirds of the conversation at a university refectory was basically gossip.[107]

Today, nearly one-half of all the people in the world, and three-quarters of those in the rich countries, live in urban areas—aggregations of more than 2,000 people.[108] The largest cities, Tokyo, São Paulo, New York, Mexico City, and Seoul, have populations of 15–25 million, and more than 100 metropolitan areas worldwide contain more than 2 million people.[109] In addition to having physical proximity to many more people, we have person-to-person contact with hundreds of others on a daily basis as a result of modern transportation systems, mail, and e-mail, and radio and television expand our circle of acquaintances—our pseudo-kin—much further.

As we search for tactics and strategies for reaching a sustainable society, it can be very useful to keep in mind that we are still primarily small-group animals, evolved genetically and culturally to operate in limited, homogeneous, largely closed societies.[110] Humanity is now challenged not only by the size of its component societies but also by their diversity. The problems of keeping many groups reasonably content and working toward common goals now exists globally as well as within states as humanity struggles to come to grips with issues such as ozone depletion, global warming, and the collapse of oceanic fisheries.

There are many consequences of our having evolved as a small-group animal and relatively suddenly having shifted to living in very large groups. People in small hunter-gatherer groups and farming villages have tight and usually stable cultural rules for dealing with one another and with the natural environment. Each person knows the environment intimately, interacts with crucial portions of it continually, and learns to synchronize with its patterns of change. Missing a game migration or planting a crop during the wrong season, for example, is an error with immediate, perhaps catastrophic, consequences. Everyone knows where her or his food comes from and what the major environmental dangers are.

Even though city dwellers tend to maintain about the same number of friendships as do villagers, they interact with a much larger number of people. But because most of the people they deal with in a given day do not belong to the same circle, they are less likely to share completely the same set of rules, a sharing that usually smooths social life in a village. And as we know, messages carried in the electronic media can overwhelm more parochial community values. There is great suspicion, simply to take the example most often discussed, that the high level of violence portrayed on television is a factor in generating crime.[111] In the United States, children in their elementary school years are exposed to some 8,000 murders and watch more than 100,000 acts of violence on television.[112] There is evidence that such routine childhood exposure to gory violence, including cartoons in which slaughtered people or animals apparently suffer no permanent consequences and simply come back to life, results in more violent behavior later.[113] Children become desensitized—habituated—so they tend to accept violence as a normal way of solving human problems, just as adults during the cold war habituated to the potential horrors that could flow from the nuclear arms race.[114] The ability of television in particular to induce changes in behavior is attested to by the willingness of giant corporations to pay millions of dollars for a half-minute commercial in a prized prime-time slot.[115]

The influence of television and other media, of course, isn't all negative.[116] The electronic media also tend to erode community values whose erosion many of us would consider beneficial. When those values include such widely shared attitudes as racial prejudice and the desirability of easy access to automatic weapons and armor-piercing bullets, television coverage often tends to bring disfavor on those who hold them. It seems likely that during the Vietnam War, the broadcast of mayhem on the evening news was a factor in convincing an influential segment of the U.S. population to abandon the community value of uniting behind the government in time of war.[117] But when American casualties are light, as in the Persian Gulf War, old chauvinistic community values remain in place.

One major problem, of course, is that communities differ in their values, and in today's large-group world, the diversity of values even *within* functional communities (which may be linguistic groups containing hundreds of millions of people) can be extreme—as is clearly demonstrated by the diversity of views in the United States on such issues as abortion, capital punishment, homosexuality, pornography, and gun control.[118] There are no easy ways to resolve such differences and establish broad values among such large groups. Problems arise even on the question of gratuitous violence in the United States, where there seems to be a growing consensus about the need to remove some of the depictions of violence from the public media. On one hand, reasonable arguments exist in favor of one controversial tactic—more stringent censorship for television programs, films, and video games aimed at children and teenagers in addition to the television receiver chips and film-rating system that give parents and society some control over what children see. On the other hand, some of these steps raise serious questions such as how to determine what is gratuitous and who to elect as censor. There would also be real problems in imposing small-group values on electronic media in the United States, especially in light of the difficulties of interpreting the important First Amendment guarantees of free speech and press.

Regardless of the evidence that small-group cultures persist in modern society, many small-group values may be going extinct, and thus human natures are evolving. In just the past half century or so, technology, communication, and especially automobiles have broken down many of the last vestiges of tight, limited neighborhood communities in the United States and in some other industrialized countries. Not only are people exposed to a much wider culture, but also their mobility has removed many cultural controls on behavior. We may have only a small group of acquaintances, as did hunter-gatherers, but the collection of people we know is a much less homogeneous and geographically delimited group and (with some exceptions in areas where there are very strong neighborhood traditions) a much less effective vehicle for developing and enforcing community values.

Being clever, human beings have, of course, evolved new values and other devices for dealing with changes in group size and mobility. Increasingly, stratification and organized coercion (supported by religion, art, and eventually written edicts and a near monopolization of within-group violence by the state itself) have replaced the less formal social coercion that regulated behavior in hunter-gatherer bands, early agricultural societies, and even the Old West of cinema fame. In the United States, at least, that trend can be seen to be continuing. Adults don't do as much of the disciplining of neighborhood children as they did half a century ago. The tendency now is to use the police powers of the state to deal with juvenile miscreants rather than to rely on constraints applied by parents and neighbors. That replacement for the more traditional personalized cultural controls is sometimes not very satisfactory. Those of us who remember the neighborhoods of the past or are fortunate enough to live in similar circumstances today may value a sense of neighborhood community. But adaptable as people are, I suspect that if current trends continue, in another few decades such values will have largely disappeared. Cultural evolution can be very rapid in some areas.[119]

We badly need to consider what, if anything, should be done about this decline of geographic community. Perhaps if people became more accustomed to taking a long evolutionary perspective on culture and entered into a systematic social dialog on cultural trajectories, society might steer a different course. We could seek strategies to restore the more stable and interactive communities into which we have evolved biologically and culturally to fit. One initial tactic would be to gradually redesign cities so that most people could live near their places of employment, to which they could travel primarily by foot or bicycle. That, combined with telecommuting, World Wide Web shopping, more carpooling,[120] and better public transportation, could enable the establishment of human-sized communities within cities and improve the health of individuals and the environment. It took about fifty years of deliberately designing the United States around the automobile to destroy communities; with another fifty years of continuous effort, doing such things as reversing the trend toward huge, distant shopping malls and planning neighborhood retail outlets for necessities such as food and medicines (if the Web should not suffice), these communities might be restored.[121] Cultural evolution, unlike biological evolution, is reversible.[122]

In many societies, groups exist that are trying to resist the changes brought about by rapid technological "progress." For example, powerful groups in many Islamic societies are determined to prevent Western values embodied in the mass media from destroying their sense of community and deeply held beliefs. One can be very sympathetic with that stance without endorsing such values as the suppression of women or the use of mutilation as punishment for crimes. Various fundamentalist (and "green") groups in Europe and the United States also are trying to resist what they perceive as threats to *their* community values. Our failure to understand the different reactions of change-embracing and change-resisting groups and ideologies underlines how little social scientists actually understand the process of cultural evolution.

What seems most desirable—and I think the diversity of more or less successful systems in our evolutionary past provides support for it—would be a sort of constrained, widely accepted ethical neopluralism.[123] By *ethical neopluralism* I mean a healthy mix of wide moral consensus and tolerance for a diversity of ethical positions within that consensus. Such a consensus must go far beyond issues such as mutual tolerance of religious beliefs to encompass equally or more contentious subjects such as how society could promote sustainable patterns of childbearing and consumption while retaining individual choice. For example, given the extent of present-day overpopulation and the resultant strain on our life-support systems, I think that all societies should move toward averaging fewer than two children per family. But I also believe that this would best be accomplished by people who are not particularly interested in reproducing remaining childless, which would enable some others with a strong interest in and talent for parenting to have more than two offspring. In order to obtain broad agreement on how to proceed on such issues, a wide-ranging discussion of what our purposes should be and how life should be lived—and what constraints nature puts on the latter—would be required. Evolving such a broad and tolerant consensus would in itself be a monumental task, but it is one that I suspect we have no choice but to address—despite the shadow of the Brave New World hanging over us. Perhaps we can avoid that world if we retain a sense of humor, don't take ourselves too seriously, and learn to live with substantial mystery and ambiguity, as Nietzsche tried to teach us to do more than a century ago.[124]

Chapter 13

EVOLUTION AND HUMAN VALUES

"The behavior of large and complex aggregates of elementary particles, it turns out, is not to be understood in terms of a simple extrapolation of the properties of a few particles. Instead, at each level of complexity entirely new properties appear, and the understanding of the new behaviors requires research which I think is as fundamental in its nature as any other."

—Philip W. Anderson[1]

"In the world of values, nature in itself is neutral. . . . It is we who create value and our desires which confer value. . . . It is for us to determine the good life, not for nature—not even for nature personified as God."

—Bertrand Russell[2]

I'm sitting at my computer having one of those introspective moments that I suspect afflict many who keep up with major developments in science. Researchers have learned so much about the evolution of human natures in the century and a half since *On the Origin of Species* was published that I wonder whether in another 150 years we'll understand it completely. Then I recall that phenomena such as perception, memory, learning, and consciousness involve, among other things, the changeable characteristics of neurons, their patterns of connection, the nature of the chemical systems that communicate between them, the gene-directed synthesis of proteins, and the overall chemical milieu of the nervous system. I think of the vast gap between knowing that and understanding how those miniature configurations and processes translate into such phenomena as my love of my wife, Anne, my fear of death, and the vivid picture of my environment resulting from an interaction between incoming electromagnetic radiation and the microstructure of my brain.

We've evolved biologically and culturally to the point at which we can ask big questions about issues such as how we "construct" our environments or maintain a sense of self, but it's not yet clear how far we're going to get, even in another 150 years, in finding big answers to them. In that quest, is science doomed to the sort of nonprogress that characterizes so much of thousands of years of philosophic discourse? Will I ever rationalize the seeming solidity of the steel desk on which my computer sits with the knowledge that the matter of which it is composed consists mostly of open space?[3] I'm not sure that some such big questions will ever be answered, but earlier ones such as "What is life, and how did its diversity come about?" have been yielding to determined effort, thanks in no small part to the pioneering work of Charles Darwin. And I am pretty sure that most of the great mysteries, if they are in some sense to be solved, will be solved within an evolutionary context.

That's one reason why I think that an understanding of evolution has strategic as well as tactical value. Evolutionary thinking goes far beyond figuring out how best to forestall all pests from evolving resistance to insecticides or even analyzing the concept of race or ways to restore small-group neighborhoods. It's to the more strategic areas that we now turn. They are ones in which consideration of the evolutionary past of our human natures can be helpful in thinking about sensible approaches to broad social, environmental, and philosophical problems.[4]

A modern knowledge of biological and cultural evolution helps us grasp what kinds of animals we are, where we came from, and how we fit into the natural world. It helps us to detect increasingly consequential mismatches between rates of biological and cultural evolution and mismatches between trends in cultural evolution in different areas of human endeavor. An understanding of evolution also can prompt us to consider the possible evolutionary significance of physical or behavioral traits when dealing with important problems.[5] It can help to explain important differences among societies—differences often misinterpreted as originating in innate differences in human qualities. It also helps us to interpret key uniformities within societies, ones misinterpreted as resulting from a uniform human nature. Those understandings should help us to think more wisely about ways to create a very different, more sustainable global society in the future, but they won't tell us what human life is for or how best to live with one another.

The Details Where the Devil Lives

Knowledge of the macroevolutionary big picture does not detract from the intricacy, interest, and importance of the details of cultural microevolution. Historical research is required to help us understand and appreciate the forces that create the patterns of change in which we are all continuously engulfed and with which we must live. Reduction, the attempt to explain larger phenomena in terms of smaller or more restricted ones,[6] has its place in history, as it does in biology. By looking back at individuals and events, historians can help untangle the complex internal dynamics and accidents that determine the course of momentous events, relatively free of influences from the physical and biological environments. These are the intertwined social, political, and economic changes of cultural microevolution.

Those microevolutionary interactions within groups have also greatly influenced issues at the very center of the human predicament—issues of how we behave toward one another and toward our environments, of our diverse values and ethics. Values and ethics first evolved during our past as a small-group animal living in hunter-gatherer bands, in situations in which every individual knew every other member of the group and the relationships of each. With that level of intimacy, elaborate kinds of social and political organization and codification of behavior were unnecessary. For instance, the difference between the herbal healer who almost always made people sicker and the herbal healer who often healed was well known to all, and thus the "market" solved the problem of quality control in the healing profession. There was no need for a "healer ethics" committee or indeed for any complex set of codified rules to protect the values of society. Rather than being codified, the "rules" were understood by all. Nonetheless, the concept of blame (and those of responsibility, resentment, and forgiveness) was necessary for even very simple human societies to function.[7]

The pace at which values evolved presumably increased as people started living in larger and larger groups and facing new dilemmas. Not all the changes in values have been adaptive. For example, appreciation of the natural environment has become less widespread within groups than in the days when people were hunter-gatherers or subsistence farmers. Although scientists now understand many details of the ways ecosystems work, most urbanized *Homo sapiens* do not value the services ecosystems deliver.[8] The average city dweller, for instance, has no idea what is involved in supplying his or her food and has a mental picture of environmental hazards that often ranks them in reverse order of their seriousness (as, for example, is suggested by great concern over roadside litter among some who are in the midst of producing large families and buying sport utility vehicles).[9] Because nature's services are simply unrecognized by most human beings today, remarkably few people are aware of the potential seriousness of the massive global changes now under way.[10]

It's not just the way people value the environment that has changed dramatically through cultural evolution over the past few hundred years. So have other aspects of human natures as the size of the human population and the scale of the human enterprise have exploded. And as both continue to grow, issues of lifestyles and ways in which people should organize their societies, central concerns of thoughtful people since before the time of Socrates and Plato, are sure to evolve further and to remain lively subjects of debate. The tension between intensification of resource extraction and liberty seems sure to heighten.

This cultural evolutionary trend confronts us with new dimensions to a long series of ancient dilemmas. How should subsocieties in a rapidly globalizing society divide up the portion of resources they "own"? How do they organize themselves to do it? How do they deal with public-access resources, those with no owners to exclude others from using them, such as oceanic fisheries? How about the resources shared with a neighboring society (such as rivers that flow through more than one state)? Or, more generally, how should people treat others and expect to be treated by them?

Evolutionary Ethics

That a knowledge of our evolutionary history can tell us about justice, and that ethics, morals, and norms can be explained on the basis of biology, is the most controversial claim made about evolution.[11] If it could, then humanity's age-old struggles with issues of ethics would largely have been an exercise in futility—evolution had the answers all along. That it did is suggested by statements of evolutionary psychologists such as ". . . the capacity for conscience may have been shaped by natural selection to promote and preserve reciprocity relationships. This explains why many moral principles require self-sacrifice for the sake of some relationship partner."[12] Thus, even if historians and music critics are safe, are evolutionists threatening to put philosophers and theologians out of business? Are ethics merely functions of a material brain that evolved gradually within groups of highly social primates, a view that can be traced back at least to Denis Diderot and his materialist contemporaries of the eighteenth-century French Enlightenment?[13] Or are ethics (as Kant thought)[14] derived from a supernatural or moral universe?[15]

My answer is a qualified yes to the question of whether ethics are functions of our material brains. We can explain the existence of ethics, morals, and norms without invoking the existence of that second moral universe. There is little to support a dualistic, mind-versus-matter view of existence. Human natures are clearly the result of biological and cultural evolution, and in some sense the ethical feelings and behaviors that are part of our natures must have arisen through these same processes. There is no sign that the other frequent postulate—that ethics are transcendental, that they exist without empirical explanation, beyond explanation by human history or human minds—is either valid or useful.

But I would respond with an unqualified no to an important related question: can the *choice* of the ethics, morals, and norms to which we ought to adhere be derived from an understanding of biological evolution?[16] In other words, the *capacity* to develop ethics is a product of biological evolution. It requires an ability to anticipate the consequences of one's actions; a critical capacity for empathy; a capability to internalize the moral standards of society and make value judgments (and thus feel guilt); and "free will," the capacity to make choices. There may even be a sensitive period of development during which a child has a special capacity to absorb the norms of his or her culture, including its prevailing values.[17] But the actual ethics, morals, and norms of a society—the products of that ethical capacity—are overwhelmingly a result of cultural evolution within that society.

There are, of course, certain uniformities in the ubiquitous development of ethical systems in human societies—in the words of anthropologist Robin Fox, the "universal moral discourse."[18] All societies have sets of rules associated with fairness, obligation, duty, self-control, respect for human life, and such. There is an analogy here with language. All societies have a language, and those languages share certain properties (vocabulary, grammar, syntax, etc.), but without special training, members of one society typically find the language of another unfathomable. Similarly, that all societies have moral discourse does not mean that members of different societies will understand or agree with the ethics of other societies.[19]

Although the capacity to develop ethics is a product of biological evolution, there is nothing in that evolution that tells us *what* we should do. The attempt to establish an ethical "ought" from a natural "is" deserves its old title, "the naturalistic fallacy."[20] What evolved is neither good or bad; it just is—including an evolved propensity to invent ethical systems. And the way we do behave or have behaved gives us no guidance about the way we ought to behave. That's what the moral discourse of our society is all about, and it's why that discourse is in flux. What we think of as the "oughts" at any given time are, though, largely products of cultural macro- and microevolution.[21]

There appears, then, to be no fountain of values buried in evolutionary theory. Knowledge of the evolutionary origins of attitudes or behaviors carries no normative message; if something evolved "naturally," it is not necessarily good,[22] and what is good is not necessarily natural.

That there is either a transcendental system of ethics or a set of evolutionary ethical values somehow "out there" seems especially unlikely because what is ethical or moral varies greatly from human nature to human nature within societies as well as between societies and over time. For instance, I think that euthanasia can be moral; people whose views I respect argue that it is always immoral. People in the West generally think that punishment of children is highly moral; the Inuit generally think it immoral.[23] Slavery was once widely considered ethical in many societies; in most, it no longer is. Adolf Hitler and Cambodian political leader Pol Pot thought genocide was ethically justified; most of the rest of human society had decided it was not. And the lack of a biologically evolved set of norms is what makes cross-cultural moral and ethical judgments such a contentious issue.

My view on the capacity for ethics—that it has its primary roots in biological evolution but that ethical standards themselves are based in cultural evolution—is far from universally held, especially among some of my evolutionist colleagues. They tend to see ethical norms as based in what might be thought of as genetic enlightened self-interest and tend to seek explanations in natural selection for most moral decisions.[24] Indeed, there is a field that can be thought of as evolutionary ethics,[25] but the gene-shortage problem, among other things, argues against its having much content.

My own conclusion is that organisms experience values, subjective rankings of "goodness," through such feelings as pleasure, pain, and hunger brought on in appropriate environments.[26] That is how selection encourages animals to behave in ways that lead to maximization of reproduction. To be ethical, animals need, among other things, to be able to experience rewards and punishments—that is, to have feelings (remember, "I feel, therefore I am").[27] Whereas the motivation to get our genes into the next generation may be the distant cause of much of our behavior, the immediate motivations are more familiar. We rarely mate to reproduce ourselves; we ordinarily mate because it feels good. We don't dodge an approaching car to preserve our ability to raise our children; we do it to avoid anticipated pain or death. We don't eat to gain energy; we eat to assuage hunger or for pleasure. These values, which are connected more or less directly to an animal's (including a human animal's) feelings, have been called perceived values.[28] Perceived values can be inferred in mammals as unlike human beings as rats. In simple experiments, rats not only learn to perform certain actions in order to obtain a desired result but also, as can be shown experimentally, learn that certain actions produce pleasant or painful consequences.[29] For instance, rats trained in an experimental arena[30] to press a lever to get food and to pull a chain to get sugar water are then removed from the arena. They are subsequently given sugar water followed by a sickening injection until they will no longer drink the sugar water. When returned to the arena, now rigged to give neither food nor sugar water, they press the lever but don't pull the chain—showing that they *know* the consequences of each action. They had not merely been conditioned to push the lever and pull the chain.[31]

Those values that we human beings develop indirectly by using our amazing brains to reason about behavior are called conceived values.[32] People have built on a biological background of perceived values, like the pleasure received from drinking sugar water, to create a cultural body of conceived values, such as feeling good about helping neighbors. Conceived values are first and foremost devices for dealing with the all-important social environment—ethics derive directly from our evolution as social mammals. In social mammals—everything from prairie dogs and vampire bats to wolves, chimps, and people—selection favors those that operate best within a group. Individual hominids managed to have more offspring when they could maneuver through the minefields that develop in any social organization. "Rules" often evolve within such organizations, even when they are organizations of nonhuman animals. Classic examples are dominance hierarchies, in which behavior of different individuals toward one another is determined at any given time by the individuals' relative positions in the hierarchy.

Human beings have developed the most ingenious ways of all to deal with social interactions. We have taken consciousness further than has any other animal—to the point of intense consciousness. We have combined it with an extraordinary awareness of our own being—of self—and use that understanding of self to interpret and respond to the behavior of others. As philosopher Mary Midgley noted, human beings developed ethical capacities that are "just what could be expected to evolve when a highly social creature becomes intelligent enough to become aware of profound conflicts among its motives."[33] Most of us want money, but most of us don't want to steal it (or are afraid to) and are concerned about the way others view our financial status and how we achieve it. That is, we use introspection as a route to dealing with the most important parts of our environments, the other members of our own group.[34]

The minds of chimpanzees probably provide a fair model for the sort of mind that our common ancestor possessed. There appears to be a sense of self in chimpanzees[35] and bonobos.[36] Along with orangutans[37] and gorillas,[38] they are the only other animals known to treat their mirror images as anything except another member of the same species.[39] They understand that they are looking at themselves, and as a consequence they will, for instance, use the mirror to examine portions of their bodies not normally visible, such as the tops of their ears. It seems clear that evolution of at least a strong consciousness, including a sense of self, is a necessary prerequisite for being able to impute mental states to other individuals. Intense consciousness, self-awareness, empathy, and social attribution (the capacity to project mental states onto others)[40] are hallmarks of humanity and prerequisites for developing ethical standards, although some experimental evidence suggests that chimpanzees also have these characteristics to some degree—remember the chimp mother I saw comforting her child when I first arrived at Gombe Stream Reserve.[41]

Empathy, in turn, provides the underpinnings of ethical systems. But the origin of ethics—culturally shared values that involve notions of right and wrong—cannot be traced to chimpanzees. Chimps have no way to share values; ethics had to await at least the evolution of language, of an efficient method of sharing the ideas that were presumably generated by notions of empathy. There thus appears to be an unbridgeable gap between the ethical capabilities of human beings and those of chimpanzees. If you don't believe it, try to imagine a chimp risking harm by intervening to help an unrelated individual from another group who is being attacked by one of the intervening chimp's relatives. No peaceniks were observed among the Kasakela chimps when that band was exterminating its rivals in the Kahama group.

How did natural selection allow ethical systems to develop in the hominid line in the first place? Darwin recognized a basic problem, as have many analysts since his day. If the unalloyed reproductive selfishness of natural selection is the basic engine of evolution, including human evolution, why did concerns for others arise? Why should a hungry person share a morsel of food with a starving brother or with the famished child of a stranger when a person's primary goal is supposed to be to maximize his or her own reproduction? Many biologists, as you will recall from chapter 2, thought that the explanation for altruism had been found in the discovery of inclusive fitness—in some cases, more copies of one's genes could be passed on to the next generation by helping relatives care for their children than in reproducing oneself. But that couldn't explain altruism toward nonrelatives. Altruism toward strangers could be explained, however, by the expectation of reward through reciprocity—later aid that improves reproductive prospects. That force would be especially potent if the altruistic acts exact small costs from the donor and confer great benefits on the recipient (as may often be the case).[42]

Unhappily, these answers are partial at best. Virtually nothing is known about any genes that altruistic human beings might promote by helping relatives, and even less is known about the possible reproductive differentials involved. Moreover, empathy and altruism often exist where the chances for any return to the altruist are nil. Indeed, careful psychological experiments suggest that much of human helping behavior is divorced from any real prospect of reproductive or other reward.[43] Work by psychologist Daniel Batson and his colleagues shows that helping can have as its ultimate goal *only* the improvement of the welfare of others without any physical, social, or psychological benefit whatsoever to the helper, including the benefit of pleasure in being altruistic. That is, there can be behavior that is truly altruistic,[44] and as far as we know, that behavior is unique to human beings.

Empathy would seem a necessary prerequisite for such altruism, and many of our empathetic feelings clearly are unrelated to personal advantage. Empathy comes from a combination of attribution and experience. As a child, I read stories of World War I aviators going to their doom in spinning airplanes. But I never really had a clue about their sensations and fear until, as a pilot in training, I first deliberately put an airplane into a spin. Even though I knew how to get the plane out of the spin, my first reaction was sheer terror—imagine what it would be like to drop down an elevator shaft in a spinning barrel. Some pilots are much less frightened the first time they spin an airplane; some are so paralyzed that they need to be saved by their instructors. I was in between, but now if I read of a crippled airliner spinning to destruction after a midair collision, I have a level of empathy for those inside that I never could have had before experiencing the spinning sensation myself. And my experiencing that empathy does not require that I have relatives on the flight or that there be any potential for reciprocity.

At a more dramatic level, empathy, where neither kinship nor the possibility of repayment existed, was the main motive for more than one-third of the people who helped Jews who had been scheduled for extermination by the Nazis.[45] Indeed, the rescuers' actions usually entailed great danger to themselves. A typical report from one such rescuer was: "Can you see it? Two young girls come, one sixteen or seventeen, and they tell you a story that their parents were killed and they were pulled in and raped. What are you supposed to tell them—'Sorry, we are full already'?"[46] When directly queried as to his motives, this man replied simply, "Human compassion."

It appears that development of empathy feeds into the emotional side of our existence—the side that makes the human brain the wonderful thinking device that it is. Goals and motives are important to the way we think and perform, and that means we must have values. Evolution of the capacity for empathy, which is so helpful in dealing with other members of our group, appears to have brought with it a value of caring for them. Apparently, we often see improvement of others' welfare as a reward in itself, not just a contribution to the goal of improving our own.

But if there is a strong genetic predilection for basic empathy, there is almost certainly some genetic tendency to place limits on the way in which we act on our empathy. As Mary Midgley put it: "Some degree of partiality is . . . built into our social nature. It shows itself, not just in favouring kin, but more widely in the way we form attachments or fail to form them, with all of the people who are of importance to us."[47] Mechanisms that permit discrimination of relatives from nonrelatives are widespread among nonhuman animals.[48] Although human beings probably do not have the ability found in some other animals to recognize relatives without prior clues (such as previous acquaintance with them),[49] the ubiquity of kinship systems suggests that there is a genetic predisposition for the establishment of such systems. During our hunter-gatherer past, children must have learned early who the members of their small group were (most or all of them were relatives) and learned to empathize with them in order to survive socially. There was little opportunity to carry that empathy over to individuals of other groups unless relations were constant and peaceful enough to allow it to develop.

Kinship systems would be pointless if people were not going to discriminate in their behavior toward others in a manner that affected the reproduction of copies of their own genes carried by relatives. Beyond any genetic predilections, it is clear that there is huge culturally evolved variation in the way empathy is deployed. Determination of exactly how large the universe of caring should be and what obligations this enlargement carries with it has shifted through time and still is a matter of substantial dispute in Western societies.[50] Should Eugenio Pacelli, Pope Pius XII, be condemned for putting what he viewed as the best interests of his church ahead of what others considered a moral duty to those being destroyed by the Holocaust, or should he be made a saint, as Pope John Paul II may do?[51] How much should be done by Englishmen to alleviate hunger in poor African countries? Should U.S. troops be risked to save the lives of Muslims threatened with genocide in the Balkans?

One area of environmental policy in which there has been debate about the proper size of the universe of caring has been that of foreign aid. The ethical issues involve consideration of the long-term results of people in industrial countries transferring to people in poor countries such technologies as antibiotics, pesticides for use against disease vectors, high-yielding crops, and synthetic fertilizers. Yet those technologies have lowered death rates and fueled the population explosion, creating a serious risk of severe problems for future generations. For example, between 1945 and 1955, following the introduction of malaria control with DDT, the death rate in Ceylon (now Sri Lanka) dropped by 50 percent and population growth spurted to approximately 3 percent per year.[52] Adoption of those "death-control" technologies was almost instantaneous—they went *with* the biological and cultural grain. Avoiding early death is good. But new birth-control technologies were not promoted vigorously or adopted enthusiastically in part because they go *against* the genetic and cultural grain—preventing children from being born is seen by many as bad. Would assistance in lowering death rates without parallel assistance in lowering birthrates be immoral, given that population growth will, sooner or later, lead to enormous suffering? Would sending aid to overpopulated areas damp out the signals that tell people in those areas that they have exceeded the carrying capacity of their homelands, or should it be sent anyway because people there are in such desperate need?[53]

This dilemma also involves the vexing issue of intergenerational equity—the degree to which the universe of concern should be expanded to include consequences for future generations. What duties did the donors have to people of the future? Should we forgo present pleasures to pay higher school taxes if we believe that will ensure better education for our children? How much should we care about maintaining environmental quality for them? Should we also care about the generation after that? How about the people ten generations from now? One thousand generations?

And what of other species? Should our universe of concern include them? For many or most people, it now does to some degree, although how far from *Homo sapiens* it should be extended is far from agreed on; consider the discussions of animal rights[54] and the rights of nature.[55] How much should we care about the well-being of pets, of livestock, of wild animals, of entire natural systems?[56] There is ongoing cultural evolution in this area, at present mostly in the direction of expanding those limits.

In the face of the dilemmas involved in defining a universe of concern, it seems that, as Midgley suggested, human beings have evolved some limits to the arousal of empathy—even though some individuals can conceive of ethics that are based on essentially limitless empathy. If the emotions of compassion, sympathy, pity, and the like were too strong or extended to too large a group, individuals would not care enough for themselves, and their reproductive performance would drop. Assuming that the necessary genetic variation were available, natural selection would gradually intervene over many generations to reduce the dimensions of compassion.

The tension between too much and too little empathy can readily be seen in personal experience. It is easy for most of us to become empathetically involved with fictional characters in literature or, especially, when they're visually presented in plays and films or on television. Similarly, empathy is aroused by stories about victims—few would not feel for the victims of mass murder waiting their turn to be subjected to the machete or for those awaiting their doom on a sinking *Titanic* or trapped in a high-rise fire. An urge to help can be generated even when one knows that the situation is in the past or entirely imaginary. The urge is particularly strong when the victims are depicted as individuals with whom one can identify rather than as anonymous individuals.

But evolution has apparently provided counterbalances to oversensitivity both in the human tendency to reserve the greatest empathy for those most like us—kin, members of the same ethnic or social group (however perceived), citizens of the same country—and in our capacity to habituate to repeated stimuli, including those that once induced empathy. Habituation can reduce empathy even for close relatives; think of such common statements as "I'm sick and tired of my brother's whining" or even the all-too-frequent occurrence of infanticide when "the baby just wouldn't stop crying." And even as modern visual communication technologies make possible the extension of empathy far beyond the sphere of influence in which it evolved (for example, extension of empathy on the part of people in the United States to starving African children or Balkan victims of rape), repeated exposure without opportunities to take action clearly also can lead to habituation. And, of course, violence within and between communities shows again that empathy is only one factor involved in human behavior toward others. At the extreme, history tells us only too clearly that when carried out with determination, cultural programs to dehumanize others—to provide that "warrant for genocide"[57]—can team up with habituation to permit holocausts.[58]

Of course, there is much more to ethics than altruism.[59] Although we can learn something from evolution about the origins of ethics and the capacity for it, the vast philosophical and theological literature suggests that a search for biological evolutionary answers to most ethical dilemmas is doomed to failure.[60] Human beings are able to generate conceived values because they acquired the brain capacity to do so while also becoming empathetic. But the values we conceive are for the most part, if not entirely, products of cultural evolution, and often they are products of diverse pathways of cultural evolution. Despite long philosophical searching, universal values have been difficult to discover. This is hardly surprising, considering that philosophers and sociobiologists alike are restricted to using their value-laden and emotional minds as tools for searching and analysis. We have great difficulty in keeping our own prejudices from coloring our attempts to study values.

The degree to which ethical judgments are culturally evolved and not inherited biological characteristics of the mind is exemplified by the evolution of ethics. The interaction of many minds is involved in the study of ethics, which makes it a form of what evolutionist Lawrence Slobodkin calls indeterminate scholarship—scholarship for which there is no way to test results empirically, scholarship that grows on itself.[61] This phenomenon is visible in almost any book about the philosophy of ethics. Open one and look at the index. You will probably find that most of the references will not be to facts, objects, or concepts; rather, they will be to names of people or book titles. For example, Mary Midgley's fine book *The Ethical Primate*[62] has an index containing 124 items, only 9 of which are not the names of people or books. A random index page of James Griffin's equally interesting *Value Judgement*[63] has 58 entries, of which all but 14 are people's names. Discussions of ethical issues are snapshots of cultural evolution in process.

Consider the sort of thought experiment that ethicists analyze. A group of six tourists is trapped by a landslide in a seashore cave, and the cave is rapidly being flooded by a rising tide. There is a glimmer of light at the top of the cave, and the tourists rush toward it. An overweight tourist reaches the hole first, struggles to get through it, and becomes hopelessly stuck. Those trapped below scream for help as the waters engulf them. Rescuers outside are faced with a dilemma. They have drills that can be used to free the overweight tourist, but that will take hours. The overweight tourist has the hours—he is in no danger of drowning, though the five poor souls below him will perish. But the rescuers also have a stick of dynamite that can be used to blow the overweight tourist out of the hole, killing him but saving the five below. He pleads for his life; his fellow travelers scream for theirs. What is the ethical course of action?[64] People may well differ in their answer to that question, as they do on the morality of abortion to save a mother's life or the morality of a mother in a group hiding from the Gestapo smothering her child to still the crying that would doom the entire group.[65]

It is impossible to find answers to these ethical dilemmas based on a standard traceable to either genetic or cultural evolution. There is no one human nature in these areas—feelings differ greatly among individuals and cultures. The waters of ethics are muddy indeed. Consider such uncomfortable questions as who, if anyone, was obliged to save members of persecuted minority groups in Bosnia or Kosovo from possible genocide in the 1990s; what ethical position should be taken by those in the United States or Africa on the issue of painful and dangerous female circumcision in central Africa (or even on that of the much less traumatic male circumcision in the United States);[66] and what environmentalists or animal rights activists should do about traditional whaling practices among Native Americans. What is the ethical position to assume if your child has a sore throat and a high fever and your physician suggests starting a course of antibiotics while laboratory tests are done to see whether the infection is bacterial or viral? Antibiotics are useless against viruses. Choosing to use the antibiotic in advance of the test results could provide a great benefit for your child but could help imperil society's ability to defend itself against pathogenic bacteria and also contribute to a possible deadly consequence for one of your grandchildren. If genetic intervention to improve the life of a child is ethical, what about something that might soon be possible—producing a child for a childless couple by culturing a body cell from the woman? In such cases, all too frequently some people's highly ethical position turns out to be other people's moral atrocity.[67]

Natural selection couldn't possibly operate at a sufficiently fine scale to promote neuronal connections that would program solutions to all such dilemmas.

As I've noted before, there aren't enough genes to code the various required behaviors, it's difficult for selection to do just one thing, and the required speed of evolution is often much too fast. Cultural evolution is the source of ethics, and it produces different answers to the ethical problems confronting different generations, different groups, and even different individuals within groups.[68]

Consciousness, Ethics, and Environments

Ethics, by definition, require consciousness so that moral choices can be made. Like many other people, although I am not a mind–matter dualist, my own version of human nature finds a strictly materialistic interpretation of the world unsatisfying. The original materialism, that the only reality was built of solid, irreducible atoms, has been destroyed by modern physics. We now believe that reality is made mostly of space, energy in its various forms, probabilistic distributions of subatomic particles, and similarly hard-to-picture stuff. I would take a naturalistic view that all that exists is phenomena that, in principle, can be approached by natural science.[69] But I am also pluralistic in thinking that many types and levels of explanation and "knowing" are possible. To me, some truths seem to be discovered (the mathematical concepts of pi and *e* are examples), whereas others are culturally evolved ("Thou shalt not kill"). There is no conflict between my "knowing" that the floor on which I am standing is mostly empty space and my "knowing" that it is solid. It is not difficult to see that science provides a more satisfactory way of explaining human problems and behavior than does a belief in astrology, without claiming that science is closer to some universal truth (or that there are in fact any universal truths).[70] That science is a better way of knowing many things seems to me correct even though certain aspects of reality are socially constructed; scientific views are influenced by our cultural evolution.[71] Our subjective impressions of the natural world are emergent properties of a complex interaction between that world and our physical organization (nervous systems, endocrine systems, etc.) and cultural attitudes that may forever resist satisfactory reductionist interpretation and about which communication will remain very difficult. This paragraph is just a small example of that difficulty!

To put it another way, we have the subjective views of our conscious selves—those imagined emotion-charged little persons sitting between our ears and looking out through our eyes, our principal windows on the world—as they enjoy pleasurable messages sent from, say, the food- and wine-tasting centers of our noses and tongues or the music-listening centers of our ears or are distressed by inputs that are boring, painful, or frightening. Of course, that little person, at least mine, wants to feel virtuous and has a sense of right and wrong.[72]

We also have the objective view that puts us in a real-world context—in a fine restaurant or a concert hall, in bed with a lover, giving money to a homeless pauper, or, on the other hand, in bed with a dull book, suffering from a bad sunburn, or hearing the sound of an intruder at the window. Which view is more "real"? I can't tell—can you? Can the joys of sex really be fully explained either by our being descendants of ancestors who were good at reproduction or by a fortuitous firing of neurons? Can an explanation of the neurochemistry of the human visual apparatus account for the beauty of a rainbow? Could knowledge about selfish genes and the electrochemical activities of our nervous systems allow someone who has never heard a Beatles song or a Beethoven piano concerto to predict what hearing them would be like? Can those who have heard them be sure that others experience them in the same way?[73] Scientists are trained to focus on the "reality" of the reductionist view, so much so that they often miss the origins of reductionism in holism. Understanding makes us feel good; our subjective reality leads us to think about an objective reality. Reductionist answers are called forth by holistic concerns and help—but only help—to resolve them.[74]

We seem to be always forced back to the larger view to find a degree of satisfaction not provided by dissection of a problem into its smallest parts. This quandary over levels of analysis—whether to seek answers in the arrangement of parts into a whole or in the parts themselves—is, of course, not restricted to issues related to our minds and feelings or even to biology. We understand a lot more about the structure of water than we do about the structure of our brains. And yet scientists cannot predict many properties of water by knowing only the properties of its constituent parts, hydrogen and oxygen. Until recently, chemists could not even predict water's freezing and boiling points by knowing its elementary constituents, let alone understand the beauty of a waterfall or a rainbow or the pleasure of sucking on an ice cube in a desert.[75]

How can we bring the richness of such subjective experience, which is really the basis of much of the disciplines known as humanities, into a scientific worldview that is so heavily reductionist—so determined always to explain things in terms of smaller and smaller elements? It may be impossible. We may be stuck with a practical dualism—that our minds cannot be satisfactorily explained in terms of neurons—even though we are convinced that the mind and the brain are as intimately related as a baseball pitch and a pitcher's arm. We may simply discover, as I said earlier, that we must accept that a satisfying reductionist explanation of many of our feelings, perceptions, and behaviors is beyond our grasp. I admit that I find the discussions of representations,[76] memory images and perceptual images,[77] parallel processing, neural networks, sensory feedback loops, and the like quite unsatisfactory as explanations of my conscious experience. There is, however, nothing to be lost by attempting to see how far reductionism can take us in our quest for satisfaction. That quest has come a long way in the century and a half since Phineas Gage set himself so far back and brain science so far forward by tamping too soon. Clearly, our level of reductionist understanding could increase even more over the next fifteen decades.

We should remember, however, that psychological and motivational explanations can be as real as physiological, cultural materialist,[78] and sociobiological[79] explanations. We are actors in our own destinies; we are intentional beings, and neither the external environment nor our genetic makeup directly controls many of our most important actions. We are conscious and able to make deliberate choices. I chose, for example, to have a vasectomy after fathering one child, despite my "selfish" genes and the social (environmental) pressures to avoid the putative problems of having an "only child." Yes, the operation of myriad environmental factors interacting with my genotype throughout my earlier life doubtless helped to shape that decision—may, indeed, have "determined" it. But that choice can still be seen as an act of my "free will." As philosopher Philip Kitcher put it: "Our actions are not simple reflexes. Nor are they the expressions of desires imposed on us by the surrounding culture. They are determined by the gene-environment interaction. If that interaction involves our appreciation of what is valuable and our modification of our desires in accordance with the appreciation, then the behavioral dispositions that result can lead us to free action."[80]

We may always have to deal with aspects of our human natures and experiences, as well as many of our feelings, ethical and otherwise, simply as epiphenomena of the characteristics of neurons in concert—the organization of our brains. We will undoubtedly be able to explain certain of their aspects and manipulate them, but there may well always be limits to our understanding of our experiences with them. Quite likely, there will remain limits to knowledge, and therefore we'll always need to be eclectic in choosing ways of "knowing." Perhaps those who, like me, are scientifically inclined will even learn to relax and enjoy a little mystery!

Human Natures and Nature's Services

Environmental ethics, the analysis of ethical problems created by the population-resource-environment predicament, provides instructive examples of the complexities of cultural evolution. It is a relatively new area, one that has been evolving—indeed, exploding—for only a few decades.[81] The ultimate issue in environmental ethics is what constitutes each individual's responsibility for maintaining the crucial natural services that those ecosystems supply to humanity. It is an area of ethics to which ecologists such as Charles Birch[82] and Garrett Hardin have been major contributors. This aspect of cultural microevolution, still largely confined to Western culture, has attracted thinkers from classically separate disciplines ranging from economics and ecology to political science and philosophy.[83] The rise of environmental ethics shows clearly how cultures respond to newly perceived ethical needs, and that rise could be considered a powerful argument against a basically transcendental basis for ethics. If the needed ethics were already "out there," we wouldn't need to invent them.[84]

The question of how much and what kind of human alteration of the environment is appropriate, especially in a context of preserving nature's services for future generations, was not widely considered an ethical issue until the 1970s. It deserves our attention because it will almost certainly be an increasingly central issue in the decades to come. Some already think they see an evolutionary ethical explanation of one aspect of environmental disruption. Recall that there is a school of thought that our ancestors were, because of genetic or cultural evolution or both, natural conservationists.[85] In this view, European and American industrial societies should look to the behavior of tribal peoples to learn how to be proper custodians of natural systems. The implication is that the current assault on biodiversity—the greatest extinction crisis in the past 65 million years—would not be occurring had modern people retained the ethics of their hunter-gatherer forebears instead of evolving new ones to suit urbanizing civilizations. As we have seen, however, there is little reason to believe that hunter-gatherer groups evolved automatic resource-conservation behavior,[86] and the degree to which the interests of indigenous peoples today are congruent with effective conservation policies is a matter of debate.[87]

Until recently, people have not paid much attention to the long-term environmental effects of their behavior but rather have focused on the satisfaction of their immediate needs. It appears that like the Inuit, our hunter-gatherer ancestors did not restrain themselves much in the exploitation of environmental resources. They could not afford the luxury of long-term planning. They changed their environments to the degree that their technological capabilities would permit—helping to exterminate many species of large animals at the end of the Pleistocene epoch and, in so doing, changing the biological communities of much of Earth.

The conservation record of peoples after the agricultural revolution is mixed. Control by forest dwellers, peasants, and nomadic herders ("ecosystem people")[88] of the local resources on which they depend often leads to superior husbandry of those resources,[89] in comparison with that of today's citizens of rich countries ("biosphere people"),[90] who are able to draw their resources from the entire biosphere. In contrast to ecosystem people, biosphere people receive little feedback about the status of the resource stocks they are tapping and thus have little incentive to conserve them. They "discount by distance,"[91] having less concern for possible depletion and degradation far away. Human history over the past ten thousand years has not been a story of sustainable management of resources so much as one of intensification of activities to support larger populations, which has in some cases led to ecological collapse.

If there is a lesson in this of value to those seeking overall strategies for maintaining our life-support systems, it is that global human society, which now dominates the biosphere, should be very cautious about further expanding its operations. Our husbandry of the ecosystems that supply society with essential services must be conscious and active lest we repeat the fate of the Easter Islanders on a global scale. We need to seek more contrasting cases, such as those of Mangaia and Tikopia, to try to understand the circumstances in which cultural evolution will lead to population stabilization and resource conservation and those in which it will lead to overpopulation and collapse. Few types of understanding would have more strategic value, given that our evolutionary future depends on it. People today, indigenous or industrial, must develop social constraints on resource use if their societies are to persist. And deciding what, where, and against whom constraints are necessary and how they are to be instituted presents monumental dilemmas. As the size of the global human enterprise shoots past the carrying capacity of Earth,[92] the ethical issues of both intergroup and intergenerational equity and the intimately connected ethics of the treatment of human life-support systems and their living components are now moving to the forefront.[93] These will almost certainly be among the great ethical issues of the twenty-first century.

Over the past five hundred years or so, the evolution of human technological capabilities hit an exponential phase as cultural developments built on previous advances in a positive feedback system. The invention of the printing press by Johannes Gutenberg in the fifteenth century allowed an explosion of scientific experimentation and invention, which led to steam engines, electric power grids, the recognition that bacteria and viruses cause many diseases, and the like. That explosion in turn led to the automobiles, airplanes, television sets, antibiotics, heart transplants, and nuclear and biological weapons of today. That vast array of technological advances, which would have stunned Gutenberg, shows how, in a positive feedback system, more produces even more. But during the past half millennium, the pace of technological change increased the mismatch between the evolutionary rates of technology and social organization and made the difference ever more problematic. For a substantial minority of humanity, the technological side of cultural evolution solved many of the age-old problems of physical existence. But it also led to the most lethal wars in history and opened the door to the possibility of civilization's destruction and the death of billions in a single cataclysm. Freedom from want has not been accompanied by freedom from fear.

Evolving human natures also permitted enlargement of the scale of the human enterprise to the point that it is destroying the life-support systems on which all our lives depend. They made it possible to condemn society to gradual extinction from loss of ecosystem services, to repeat on a global scale the fates of the civilizations of the Tigris and Euphrates valleys, Easter Island, the classic Maya, and the Anasazi. The technological advances combined with lagging social evolution of human natures have caused overpopulation and continuing population growth, overconsumption and continuing economic growth among the rich, and widespread use of environmentally malign technologies. Those problems have been exacerbated by inefficient, inequitable, and often iniquitous social, political, and economic arrangements. All of these factors are already denying another billion or more people a decent life and inexorably depleting the natural capital that sustains civilization.[94] Despite all the good things that have come out of human evolution, one thing is clear to me and to many of my colleagues who spend their time examining that predicament: our evolving human natures may be heading us toward the worst catastrophe in the history of *Homo sapiens.*

It is not too late for humanity to avert a vast ecological disaster and make the transition to a sustainable society, but the task will not be simple. The required strategic actions are evident. Population growth should be halted and a slow decline begun to a population size that, in a couple of centuries, might be environmentally sustainable. Such reduction toward an "optimum" population size might also help to ameliorate social problems.[95] Wasteful consumption in rich countries must be reduced to allow for needed growth in poor countries. Fortunately, a reduction in consumption accompanied by an *increase* in quality of life is technologically feasible. For instance, physicist John Holdren's scenarios, in which the rich become much more efficient and the poor consume more, offer a possible path toward more equitable and efficient patterns of energy use that could close the gap between rich and poor and reduce environmental damage compared with that which will result if current trends continue.[96] We might be able to reach those goals while temporarily supporting the substantially larger human population that is inevitable before growth can be halted. But technological feasibility is not enough. Our sociopolitical systems also must undergo dramatic revision in the direction of increasing equity at all levels if sustainability is to be achieved.[97] They will also have to deal with differences in the cultural attitudes and capabilities of *Homo sapiens*, which fuel the trend toward ethnic fragmentation worldwide at one level even as economic globalization is occurring at another—contrasting trends that are, at the least, disturbing.[98] The cooperation that will be needed to solve global environmental problems is unlikely to be achieved in a world divided into haves and have-nots and riven by ethnic antagonisms.

Evolution and Social Problems

What could a knowledge of our genetic and cultural evolutionary past tell us we might do about serious social problems such as violence, inequity, and environmental abuse? Animal behaviorist Stephen Emlen summarized his view of the evolutionary causality of human behavior, a view that I share: "The obvious importance of cultural influences on our behavior does not negate the probability that we humans also possess a set of biologically based predispositions for interacting with one another."[99] But geneticists have, at the moment, few clues to the genetic background of any behaviors aside from those known to be associated with severe hereditary defects.[100] Down syndrome, which is caused by having an extra copy of chromosome 21, affects approximately 1 in 750 children born alive and can be diagnosed by amniocentesis[101] a few months after fertilization. Fragile-X syndrome, resulting from a defect in the X chromosome, is more complex but is exceeded only by Down syndrome as a source of retardation. Phenylketonuria is another genetic defect that causes severe mental retardation, but in this case retardation can be avoided if the amino acid phenylalanine is rigorously excluded from the diet. Beyond having ideas about repairing a limited number of such defects, most geneticists do not look toward future eugenics programs, for technical as well as ethical reasons.

A basic strategic conclusion can be drawn from what geneticists know: trying to manipulate the human genome to "improve" human behavior by changing genetic predispositions would be a mistake. We have seen repeatedly how difficult it is to determine the relative contributions to human behavior of genes, environments, and gene–environment interactions. When it comes to genetic engineering as a tool for "improving" humanity, it is difficult to see how that could be done, technically, beyond finding ways to reduce the incidence of what almost everyone classifies as diseases or gross defects. Even there, the technical and ethical challenges are daunting.[102] For instance, one of the most identifiable (and thus investigated) negative mental conditions is schizophrenia.[103] The data suggest that multiple genetic factors and environmental triggers are involved, and use of genetic engineering to reduce the occurrence of the disease presents a formidable technical challenge indeed. Aside from the ethical constraints most of us would feel about going beyond the point of trying to avoid serious congenital problems, the potential for social disruption inherent in almost any proposed eugenics program makes the nontechnical issues at least as challenging as the technical ones.

Perhaps the most important practical reason for eschewing eugenics programs—even if humanity possessed the requisite knowledge—is that a never-ending argument over its goals would be virtually inevitable. Our diverse cultural evolution has guaranteed that. Chinese intellectuals might urge the selection of genes for community responsibility; Texan intellectuals might want to design genomes that produce a propensity for individualism. School administrators and master sergeants could thump for an increased willingness to accept discipline; scientists and artists might want to design people with more imagination. Football coaches would oppose efforts to produce a docile population. One could expand the list of such potential conflicts ad infinitum.

Furthermore, although there clearly are ethical concerns about large-scale genetic intervention, it's not just genetic engineering of human behavior that would be resisted. It seems certain that there would also be great opposition to any grand social-engineering plans to improve the human condition. Aldous Huxley's 1932 novel *Brave New World*,[104] with its directors of hatcheries and conditioning, neo-Pavlovian conditioning rooms, College of Emotional Engineering, soma, and the like, was an early cautionary tale about such engineering. But even people who have not read Huxley's great book would probably resist similar social engineering because of the distrust of science engendered by such technological triumphs as thermonuclear bombs and "gender reassignment" surgery. Resistance would come, too, from the large numbers of people who wisely distrust the competence of government planners in solving human problems. And, finally, there would be resistance to taking any kind of systematic approach to our dilemmas simply because some people have no taste for rationality.[105]

Distrust of government extends to what is taught in schools about values and beliefs—an increasingly difficult problem for large, diverse societies such as that of the United States. That distrust is prominently displayed in debates over whether to allow prayer in schools and—of direct pertinence to the theme of this book—debates over the teaching of evolution. In the United States, there is widespread and socially harmful acceptance of the claims of creationism. This persists even though it is widely recognized that there is no necessary conflict between religious belief and acceptance of modern science's view of evolution;[106] presumably, an omniscient and omnipotent deity could create a diversity of life through any mechanism it wishes.[107] Major media often treat creationism as a serious explanation of the origin of organic diversity,[108] and boards of education attempt to erase evolution from curricula. One recent such attempt, ironically, was in Kansas, where almost a half century ago I received superb training in evolution at the University of Kansas (which remains a fine educational institution).[109]

American neoconservatives promote creationism because, as their own statements reveal,[110] they apparently fear an educated population and see the theory of evolution as a threat to the religious beliefs they deem essential to keeping society from disintegrating. One strategic policy conclusion flows immediately from an understanding of evolution and its significance to people today. The mechanisms, results, and significance of evolution should be taught at all levels in schools in the United States, and the teaching of creationism as a "scientific" alternative to evolutionary theory should end.[111] The prices we pay in racism, antibiotic resistance, and pesticide resistance alone are too high to pay for a population ignorant of its evolutionary history and the processes that created humanity and all other organisms.

Is there anything in our evolutionary history that suggests ways in which tensions over such social issues could be reduced? Perhaps, although the most obvious lesson does not provide much insight into how to reduce the heat of the evolution–creation battle. We evolved as sight animals, and that does point to some ways to lessen the negative effects of differences in values among diverse communities. For example, aggressive or socially disapproved behaviors are sometimes referred to in the United States as "in your face" actions, with the implication that they are thrust aggressively into one's visual field. We don't describe aggression as "in your ear" or "in your stomach." Large multicultural societies might smooth some of their potential differences by avoiding "in your face" actions. For example, communities with theaters that specialize in showing very violent films or films with explicit sexual content might segregate these theaters in a single district and forbid them to display explicit advertisements. In that manner, those who wished to watch the films could do so, and those who were offended by them (and their children) would not have them "in their face." In a similar manner, the approval of RU 486 (the "morning-after" contraceptive pill) and similar drugs could, I hope, reduce tensions over the abortion issue. No longer would opponents of the procedure need to be confronted with the sight of young women going into abortion clinics. Of course, an understanding of evolution is not necessary to figure this out, but those who know human evolutionary history have the importance of our visual sense almost automatically in mind.

There seems to be one more fundamental policy implication that could be drawn from a consideration of these stressful social issues and the ethical dilemmas they generate. It is the need for a broad understanding of *both* cultural macroevolution (extrinsic shapers of culture) and cultural microevolution (internal dynamics of societies) if policies are to be adopted that will encourage our ability to *know* to catch up with our ability to *do*. As I've already indicated, macroevolutionary factors, such as the availability of animals suitable for domestication, appear to have been critically important early in the rise of complex societies. Then, for a long period, such macroevolutionary factors took a backseat, especially in the minds of historians and social scientists, to microevolutionary factors. Now, with the human predicament growing increasingly ominous, the force of macroevolution threatens a serious comeback, with environmental degradation increasingly constraining microevolutionary possibilities yet still demanding microevolutionary responses.

The environment will not, for instance, allow the physical human enterprise (populations, per capita material consumption) to grow perpetually (macroevolution), so the thinking of economists and business leaders will need to be directed toward ways to create healthy, physically nongrowing economies (microevolution).[112] Unhappily, however, the microevolution of the structure of our cultural apparatus for generating both understanding and responses is a lagging, not a leading, segment of cultural evolution. Nowhere in that apparatus is the lag more obvious than in higher education, especially in the way that it is divided into static, antique disciplines that actively work against badly needed interdisciplinary approaches to the most serious human problems.[113]

The universe has existed for some 13 billion years; over the past 800 years, academic disciplines have evolved to study it, influenced especially by the division in 1664 of the brand-new Royal Society of London into committees to deal with different academic areas.[114] When early inquiries and discoveries in science were being made and the universe appeared to be a neat, clockwork-style Newtonian place, the Western scientific disciplines—primarily physics, chemistry, geology, botany, and zoology—seemed reasonably coherent. But problems were not long in appearing. The boundaries between chemistry and physics and between botany and zoology soon began to break down.[115] The situation in the social sciences was similar but delayed by a century or two, the present disciplinary structure having crystallized in the nineteenth century.[116]

Considerable inertia can build up in institutions. Channels developed to direct flows of capital into university schools and departments; discipline-oriented infrastructure and reward systems were established. To those disciplines that had influence and money, more was given. The stronger disciplines became, the greater grew the career value of a disciplinary focus and the deeper became the channels controlling resource flows.

Thus, a conservative division of the world of scholarship has evolved. This conservatism influences virtually every aspect of scholarly inquiry, from the framing of research problems to the funding and carrying out of investigations to the publishing of findings in specialist journals and broader communication of the work's importance to policy makers and the public. Consider the framing of research problems. Since the Middle Ages, the process of cultural evolution has generated a body of nongenetic information sufficiently vast that no one person could hope to grasp more than a tiny fraction of it, even if given the necessary access. One more hominid joining the Java men and Neanderthals in extinction was the intellectual Renaissance man, the scholar with a command of many disciplines. If human beings are to learn more about how the world works and, better, to direct their collective understanding toward the long-term service of humanity, specialization is necessary. At the same time, because few significant human problems today lie strictly within the boundaries of current disciplines,[117] much more should be done to encourage interdisciplinary scholarship.[118] A disciplinary scholar does not need to be blind to other areas and should not consider disciplinary boundaries to be forever fixed. Rather, they should be viewed as eternally flexible and porous, and those who choose to tackle problems that cross the boundaries of the moment should not be punished, as they often are in academia today. The conservatism that was useful in the past is a luxury that society can no longer afford. Society also can no longer afford the split between the humanities and the sciences (the "two cultures" of physicist C. P. Snow)[119] or the marginalization of philosophy. In the cases of successful interdisciplinary collaboration with which I am familiar, four things have been required. The first is commitment on the part of senior people who have little to risk professionally and who are eager to involve bright junior people in the effort. The second is funders who are willing to back experimental efforts to foster interdisciplinary collaboration. The third is persistence—mutual understanding and cooperation build slowly. And the fourth is the creation of environments of social interaction and long-term association that allow friendships to develop.[120]

Considering the difficulty of fostering interdisciplinary research, it is very cheering that such research is gaining ground.[121] This could be a measure of the importance of the issues addressed. It is also encouraging that the public is interested in the "big picture" painted by science, and that picture is rarely painted by a single discipline. It is essential that scientists communicate with the public if humanity is ever to understand its roots and learn to husband and manage Earth's ecosystems sustainably and thus preserve nature's services.

Conscious Evolution

One step toward solution of the human predicament could be the creation of a more deliberate style of cultural evolution, one that would channel change in ways more beneficial to the majority of human beings. The need for a novel evolutionary approach can be seen in the most critical mismatch between biological and cultural evolution: the fact that the design of the human perceptual system makes it especially hard for people to recognize the most serious environmental problems. Unfortunately, few people are aware of this mismatch. Our perceptual system, remember, evolved to hold the environmental backdrop constant so that people could better detect, and react to, contrasting short-term events—the appearance of an edible animal or an individual of the opposite sex, or of a threat such as a lion or a rival. Recall that our visual system keeps the world still and that we habituate to constant sound. And "quick reflexes" are programmed into our nervous systems to allow us to react to short-term events. Biological evolution was unlikely to equip hominids with the ability to detect gradual alterations in their environmental backdrop—climate change, for example—because our ancestors generally were not capable of influencing or responding to such alterations. But today, threats ranging from global warming and loss of biodiversity to distribution of the materials and know-how to make nuclear and biological weapons are gradual changes in our environmental backdrop that humanity is causing and therefore can influence.

Our natures normally are poorly equipped to register such threats, but we can be trained to be aware of the biological evolutionary defaults that inhibit perception of certain key problems. People who have been so trained, scattered throughout society, have incorporated "slow reflexes" into their natures to accompany their quick ones. Those slow reflexes make it easier for them to detect threats that materialize not in seconds but in decades or centuries. Getting most people to develop slow reflexes and then to use them consistently would lay the groundwork for what psychologist Robert Ornstein and I call conscious evolution.[122] Part of the process of conscious evolution is already taking place on a small scale when groups get together to openly discuss and seek answers to gradually developing problems of society, such as the accumulation over a quarter century or so of massive arsenals of nuclear weapons mounted in independently targeted clusters on intercontinental ballistic missiles.[123]

The example of conscious evolution with which I am most familiar occurred in response to that civilization-threatening trend. It was the "nuclear winter" efforts of a broad collection of scholars, foundations, politicians, and representatives of the military from the United States and the Soviet Union in the early 1980s. Ecologists had long worried about the potential devastating environmental effects of a large-scale thermonuclear war,[124] and in 1983 a series of scientific meetings was held and collaborations begun to evaluate the potential consequences of such a conflict. Despite a wide diversity of views, a consensus was reached and scientific papers were prepared.[125] The issue was brought into public discussion by an intensive media campaign, including the first satellite-linked television broadcast of a transatlantic discussion among scientists in the United States and the Soviet Union. Many people believe that this effort to guide cultural evolution, by alerting people to the consequences of a trend of increasing destructive power too slow to be directly perceived, promoted subsequent political decisions that helped lessen the chances of an all-out war and contributed to the end of the cold war. The activities of the Intergovernmental Panel on Climate Change (IPCC), which involves hundreds of scientists from diverse disciplines who are engaged in a continuing evaluation of the global warming process in an effort to reach consensus on the technical issues related to that contentious topic and explore possible solutions, could be viewed in the same light.[126] So could relatively successful campaigns to change the natures of many people and reduce their intake of saturated fats and cigarette smoke.

Biological evolution obviously is much too slow to respond to the perils of nuclear arsenals or carbon dioxide emissions even when they are detected, and undirected cultural evolution often is too sluggish and moves in inappropriate directions when such threats arise. Thus, the mismatch between ability to *understand* and ability to *do* persists. In our approach to environmental problems, there is an enduring overemphasis on visible air and water pollution and an underemphasis on the decline of ecosystem services and the population and consumption factors driving that decline. Those faulty perceptions of the relative urgency of various environmental problems are caused in no small part by our evolution as sight animals and are exacerbated by the visual impact and immediacy of television. They result in much misallocation of effort in the battle to maintain environmental quality. Had we evolved the chemoreceptive capabilities of dogs (or male giant silk moths, which are sensitive to extremely low concentrations of certain chemicals), we would be much more concerned about issues such as the presence of hormone-mimicking synthetic chemicals in the environment.[127] If we could see certain frequencies of ultraviolet light, then the ozone crisis might never have occurred because corrective steps would have been taken much sooner.[128] If, in the late twentieth century, the sky had suddenly turned a sickly green when the concentration of carbon dioxide in the atmosphere climbed past 350 parts per million, then irresponsible elements of industry whose profits hinge on the heavy use of fossil fuels would not now be mounting a campaign of misinformation about global warming.[129] If, as long as the human population exceeds Earth's carrying capacity, a woman's face turned bright purple with the third pregnancy, people's reproductive behavior would change.

An answer to environmental misperceptions, if humanity could manage it, would be to create a conscious evolutionary process. Through such a process, the public would be encouraged to become more aware of the biological evolutionary defaults, and society would openly attempt to increase the rate of cultural evolution in the area of understanding our evolutionary background and the biases it produces. At the same time, a social dialog is needed, one designed to develop new criteria that would brake humanity's headlong "success" in the area of *doing*. In theory, the latter might be moderated simply by asking, "Doing to what end?" But it may be a much more difficult task in practice. For example, it is becoming increasingly clear that cities in countries as diverse as the United States, Mexico, England, Australia, Japan, and China have been or are being designed for the convenience of automobiles rather than people. The results are horrifyingly obvious and similar, but it has proven virtually impossible to generate a dialog about gradually rehumanizing those societies. Being able to get to work by foot, bicycle, or mass transit not only would greatly benefit our life-support systems but also would increase the quality and length of individual lives—not just by reducing pollution and accidental deaths and, in some cases, by increasing exercise but also by reducing the stresses of driving.[130] Nonetheless, when I was visiting car-clogged western Australia recently, politicians were still planning to build even more highways in order to "relieve traffic congestion," which, as experience around the world has shown, exacerbates rather than solves the problem.[131]

Bringing conscious evolution to the fore should involve broad-based efforts similar to the nuclear winter exercise but on a much larger scale. The idea would be to promote public discussion of crucial issues that are now largely ignored in order to redirect the malign trends now driving the human predicament into trajectories leading to a sustainable society. That the potential for conscious evolution exists is patent in the great social movements that societies have already experienced: in Western civilization, the abolition of slavery, the trend toward democratic government and freedom of individual expression, the rewarding of innovation rather than position, the labor movement, the woman suffrage and women's rights movements, the family planning movement, the civil rights movement in the United States, and the environmental movement all come immediately to mind. All of these were responses to gradual changes in thought and need that most people did not have the natures to detect but that eventually crossed thresholds that allowed relatively rapid change. Unhappily, we do not fully understand the cultural evolutionary processes that led to those rapid changes, but we must strive to do so if the rapid changes now required are to be agreed on and initiated. For instance, we should try to consider the cultural evolutionary effects in developing policy initiatives. When comparing a market-based mechanism (such as trade in emissions permits) with a command-and-control mechanism (such as limits on permitted emissions) to solve a pollution problem, the educational effect of the choice should be factored in. Thus, even though emissions trading might be more efficient, emissions limitations might have a greater influence on public perceptions of the seriousness of the problem.

The first thing that would be required in converting social movements into conscious evolution is a systematic, interdisciplinary consideration of the issues involved. This process should be transparent to all participants as well as to the general public and decision makers, goals toward which the IPCC works. It would take place in a framework in which possible default positions promoted by genetic evolution would be explicitly considered. For instance, time would not be wasted in discussions of programs to lower birthrates by planning to persuade people to have intercourse less often. The second requirement would be broad participation by informed nonscientists, ranging from ethicists to knowledgeable representatives of the public, more than that involved in the nuclear winter and IPCC examples. We certainly now have tools—global satellite television, the Internet, fax machines, conference calls—that make wide communication, debate, and consensus building feasible.[132]

In my view, it is highly unlikely that human beings will ever create a utopia, but I think it a counsel of despair to assume that we can't collectively do a lot better than we're doing today. Cultural evolution led many past civilizations to extinction. Our global civilization had better move rapidly to modify its cultural evolution and deal with its deteriorating environmental circumstances before it runs out of time. Whether the natures of most of us can be changed to establish better connections among diverse groups and to take more systematic control of our cultural evolution remains to be seen. One good starting point would be to drop the term *human nature* in the singular form from most of our discourse and learn to think automatically of the built-in genetic and cultural plurality of human beings. Our challenge is to learn to deal sensibly with both nature and our nature*s*—for all of us to learn to be both environmentalists and "people people." Utopian? Perhaps. I tend to be optimistic in thinking that we *can* do it but pessimistic about whether we *will* do it.

But maybe even the latter pessimism is unjustified. After all, grasping the basics of what is known about our natures is not all that daunting. Yes, genes do influence our behavior—we are not born with minds that are blank slates. But the gene-shortage problem tells us that the genes' influences on behavior must be limited and thus must apply mostly to very general capacities and propensities. Our genes can't be dictating most of our individual acts. And their influences will always depend, often heavily, on the environments in which they are expressed. Sorting out how gene–environment interactions influence human behavior is an extremely difficult chore that science has barely started to undertake—and an area in which much misleading nonsense has been published. If *Homo sapiens* is to improve its lot by manipulating human evolution, clearly it must do so by attempting to influence the course of *human cultural evolution*—and doing that with great care to avoid the abuses that could so easily occur and to preserve the diversity of natures that is such an important human resource.

So here we are, small-group animals trying to live, with increasingly rare exceptions, in gigantic groups—trying to maintain health, happiness, and a feeling of connectedness in an increasingly impersonal world in which individual natures are based on ever smaller fractions of society's culture. We've come a long way from those creatures snatching bugs from Paleocene bushes to the problems of the twenty-first century. We also have gone a long way away from one another. That distance and the degree of longing for reconnection that it has apparently created, at least in industrial societies, was symbolized for me recently by the public response to the tragic death of Diana, princess of Wales. About the same number of people turned out to watch her funeral procession as lived at the time of the agricultural revolution. Apparently, if people cannot find satisfactory social contacts in a small group, they attempt to compensate by forming pseudo-contacts with celebrities, who have been converted into super-optimal stimuli by the visual magic of television.

There is no easy formula for understanding the human past or today's human natures or for projecting the human future. We have clues about where we're going, but we can't tell for certain where we're going to end up. We know that we are apes, but we cannot be classified simplistically as "naked apes" or "killer apes" or "moral apes." We are products of a long and complex process of genetic and cultural evolution and gene–culture coevolution. Our past was complicated; so is our present, and so will be our future. Those who claim to have simple solutions for complex problems are most often wrong, but nonetheless the search for broad generalities is necessary. We'll never deal with the devils in the details unless we see the big picture.

NOTES

Preface

1. This notion was perhaps most strongly expressed by Niccolò Machiavelli in *The Prince* (Machiavelli 1981 [1513]). As he said of human beings generally, "They are ungrateful, fickle, dissembling, anxious to flee danger, and covetous of gain" (p. 60).
2. Quoted in Mayr 1997, p. 178. This is a big-picture statement—obviously, the aerodynamics of birds' wings, the functioning of the lens of an eye, and so on, "make sense" in a more limited context.
3. For a recent example, see Ghiglieri 1999.
4. Kahneman 1980, p. 190.
5. See, e.g., National Academy of Sciences USA 1993; Union of Concerned Scientists 1993.
6. Mencken 1922, p. 120.
7. Named after Pyrrho of Elis (ca. 360 B.C.–ca. 272 B.C.), the ultimate skeptic, who tried to behave as if there were nothing "out there" (because, he claimed, good arguments can be made for and against the existence of anything) and as a consequence had to be protected by his friends from walking into fires or falling into holes.
8. Hume 1977 (1777), p. 111 (I've removed a few commas for clarity).

Chapter 1: Evolution and Us

1. Jacob 1982, p. 23.
2. Darwin 1859, pp. 485–486.
3. For details about the altruistic actions of these non-Jews at the time of the Holocaust, see Oliner and Oliner 1988. It is estimated that the number of such selfless benefactors was actually between 50,000 and 500,000.
4. For a recent memoir by one such individual, see Opdyke and Armstrong 1999.
5. Lemonick 1999.
6. The article headlined was Gotthardt 1999.
7. Nesse and Williams 1995.
8. Wright 1994.
9. See, e.g., Barkow et al. 1992.
10. The human genetic endowment (genome) has about 3 billion nucleotides (building blocks of genes), of which about 120 million are in an estimated 60,000–70,000 functional genes, some 30,000 of which have now been mapped (DeLoukas et al. 1998). Another source gives a somewhat higher estimate, of as many as 142,000 genes (Travis 1999), yet if *that* number were multiplied by 10, or even 1000, the basic argument would not be changed.
11. Neurophysiologist Corey Goodman supplied me with recent neuron–synapse estimates, with the caution that the basis for them is still quite uncertain (personal communication, 23 July 1999). See also Kandel et al. 1995, p. 89.
12. Many social scientists who have become converts to strong genetic determinism appear to lack the nec-

essary background on gene–brain relationships to provide a balanced view of the crucial issues (see, e.g., Arnhart 1995).

13. Hamer and Copeland 1998, p. 25.
14. The narrowness of some reductionist scientists—brilliant as they often have been—must be observed to be appreciated. For a sense of such attitudes, see Edward O. Wilson's description of the molecular biology wars at Harvard University (Wilson 1994, chap. 12). Wilson's gracefully written autobiography will also give you insight into how scientists become interested in the critically important and difficult problems at higher levels of biological organization.
15. For a well-written and provocative example of the overly deterministic approach, supporting a clearly conservative agenda, see Ridley 1996b.
16. See, e.g., Herrnstein and Murray 1994.
17. See Lane 1995, in which the darker side of this issue is discussed, and also Tooby and Cosmides 1992, p. 40, which is a humane (if, in my view, very overoptimistic) consideration. For an example of popular discussion of the purported potential of genetic engineering to make people more intelligent, see Lemonick 1999.
18. Ehrlich and Holm 1963, pp. 322–323.
19. But not all; two-headed individuals (partial identical twins who did not separate) have been recorded.
20. In a 12-inch by 1-inch rectangle, does the length contribute more than the width? Doubling the latter to 2 inches doubles the area; doubling the former to 24 inches also doubles the area. One cannot say that one dimension contributes more than the other, nor can one define a chunk of area for which one dimension is primarily responsible.
21. Hamer and Copeland 1998, p. 11; see also Plomin 1989. For a less extreme, but still (in my view) overly deterministic, discussion based on probable overestimates of heritabilities, see Bouchard 1994.
22. Herrnstein and Murray 1994. For a recent critique, see Flynn 2000.
23. Barbujani et al. 1997.
24. Ehrlich et al. 1995, chap. 2. Predicting the effects of voluntary reproductive restraint over many more millennia is difficult; if there is genetic variation in the tendency to practice birth control, after very many generations genomes might arise that interact with most environments to produce individuals who desire very large families (or, rather, are especially incompetent at using birth control). But that might be difficult unless the cultural rewards of large families again became substantial.
25. Eaton and Mayer 1954. Old Colony Mennonite women in Mexico had an average of 9.5 children in the 1960s, almost as high a birthrate as the Hutterites sustained (Felt et al. 1990).
26. Peter 1987.
27. Details about Chang and Eng are from Wallace and Wallace 1978. The Wallaces' book contains much fascinating historical detail, most of it not pertinent to our story.
28. A wonderful essay on this topic relative to the causes of aggression can be found in Sapolsky 1997, pp. 149–159. It should be read by all "evolutionary psychologists."
29. That range of possibilities is the "norm of reaction" (Schlichting and Massimo 1998).
30. See, e.g., Falconer and Mackay 1996, pp. 132, 321–325; Maynard Smith 1998. In the latter, see the example on p. 95 (table 6.1), which shows that the growth rate of one genetic strain of mice is much more affected by nutrition than is that of another strain of mice. In contrast, two genetically different strains of flies show essentially the same reaction to changes in environmental temperature—there is little sign of gene–environment interaction. Technically, this effect is said to be additive. But the nature of gene–environment interactions limits such terms and conclusions. Explore a few new environments and the additivity may disappear.
31. There is abundant evidence that exposure to violence elicits violent behavior in children (Bandura et al. 1963; Comstock and Paik 1991; Paik and Comstock 1994) or makes them more tolerant of aggressive behavior in others (Molitor and Hirsch 1994).
32. Once again, I suggest Sapolsky 1997, pp. 149–159. Sapolsky's short book of essays is one of the great gems of biological literature aimed at nonbiologists. A longer but also accessible book specifically on gene–environment debates in human evolution and politics is Lewontin et al. 1984; see also Lewontin 1982a.
33. For an extended discussion of heritability designed for general readers, with references, see Ehrlich and Feldman 1977, pp. 105–118, 178–181.
34. For technical discussions of this much-abused measure, see Feldman and Lewontin 1975; Feldman and Otto 1997; and Feldman et al. 2000.
35. Clark 1987; Scharloo 1987.

36. For clear, short explanations of some non-Western traditions, such as the Confucian and Upanishadic Hindu views of human nature, see, e.g., Stevenson and Haberman 1998, chaps. 2, 3.
37. See, e.g., Betzig 1997; Degler 1991; Hume 1978 (1739); Lewontin et al. 1984; Stevenson and Haberman 1998; and Wilson 1978, as well as the title of the journal *Human Nature.*
38. For a fine overview of how some prominent thinkers of the past have viewed human nature, see Stevenson and Haberman 1998. For the views of a recent philosopher of science, see Hull 1986.
39. Breeding by human beings has made some domesticated species very variable. There is much more physical diversity among breeds of dogs, for example, than is shown in human beings, and considerable behavioral diversity among them as well.
40. One might argue that insects, bacteria, or plants are dominant—but we're a single species and influence the global environment, as far as we know, much more than does any other single species. *Homo sapiens* has created bigger changes over the past century than have all three of the other groups combined.
41. From the cartoon strip by Walt Kelly, 1971.

Chapter 2: Tales from the Animal House

1. Darwin 1859, pp. 80–81. Darwin here first mentions selection of variations useful to human beings and speculates about "variations useful in some way to each being in the great and complex battle of life." Then he begins the passage quoted here with, "If such do occur, can we doubt . . . ?" I dropped the first phrase in quoting the passage.
2. For geological details, see Press and Siever 1972, pp. 44–59.
3. From what's known about cosmic evolution, our origins actually can be traced back some 10–20 billion years, to the origin of the universe (Raven and Johnson 1999).
4. Darwin 1859. The first edition is generally considered the best. An excellent paperback facsimile edition published in 1964 by Harvard University Press, with an introduction by noted biologist Ernst Mayr, was still available in 1999. Darwin was stimulated to publish *Origin* by receipt of a manuscript from English naturalist Alfred Russel Wallace. Wallace's manuscript, titled "On the Tendency of Varieties to Depart Indefinitely from the Original Type," contained the idea of natural selection. Darwin and Wallace's work was presented simultaneously by two of Darwin's friends, botanist Sir Joseph Dalton Hooker and geologist Sir Charles Lyell, at a meeting of the Linnean Society of London on July 1, 1858, but Darwin's massive accumulated evidence, well known to his colleagues at the time, has resulted (properly) in his getting the major credit. See also Wallace 1870b. Wallace's views on human nature were interesting. For example, he believed that "man's naked skin could not have been produced by natural selection" because selection should have restored a furry coat as early human beings spread out from the Tropics (1870a, p. 348). His chapters on race (chap. 9) and the limits of selection in *Homo sapiens* (chap. 10) are well worth perusing for insight into the thinking of a very intelligent biologist 120 years ago.

   Our present scientific view of evolution traces to the activities of a small group of outstanding scientists in the middle of the twentieth century who in a series of books created what was called "the modern synthesis" (Huxley 1943). The foundational volume was Theodosius Dobzhansky's *Genetics and the Origin of Species* (Dobzhansky 1937), which went through three editions (biologists should check the mitotic figure on the spine of the first edition) and was replaced by *Genetics of the Evolutionary Process* (Dobzhansky 1970). Dobzhansky's classic work was quickly followed by Ernst Mayr's *Systematics and the Origin of Species* (1942), which then led to his *Animal Species and Evolution* (1963) and many other books. Next came George Gaylord Simpson's *Tempo and Mode in Evolution* (1944), which was replaced by *The Major Features of Evolution* (1953). Finally, there was Ledyard Stebbins's *Variation and Evolution in Plants* (1950), which extended the previously animal-oriented synthesis to plants. Other key books were Schmalhausen 1949; Darlington 1939; White 1945; and Rensch 1959 (1954). Providing an important background for the synthesis was the subdiscipline of modern population genetics, created by Ronald Fisher (1930), Sewall Wright (1931), and J. B. S. Haldane (1932). In two decades, the study of evolution itself underwent rapid cultural evolution.
5. I want to emphasize that despite the persistence of Darwin's central idea, evolutionary theory is not some sort of revealed truth to be believed without question. In the literature, in fact, I've been cited paradoxically as being both too enthusiastic about the theory of evolution (Peters 1991) and too critical of it (Stebbins 1977). In the past, colleagues and I have been critical of the substitution of untestable evolutionary speculation for careful investigation of contemporary explanations (Birch and Ehrlich 1967). And once, long ago, Richard Holm and I were so heretical as to suggest that the then-current evolutionary theory was not the only possible explanation of observed patterns in nature, "just the best explanation that has

been developed so far. It is conceivable, even likely," we wrote, "that what one might facetiously call a non-Euclidean theory of evolution, one that is beyond our current imaginations, lies over the horizon" (Ehrlich and Holm 1962). So far, no such new theory has been advanced. The basic structure of Darwinian evolutionary theory (now often called neo-Darwinian evolutionary theory) has held up pretty well for the thirty-five-plus years since Holm and I wrote that, despite some healthy fusses. There has been, for instance, a substantial debate over punctuated equilibrium; see Simpson 1953; Eldredge and Gould 1972; Charlesworth et al. 1982; Maynard Smith 1983; Ruse 1989; Futuyma 1998. Another debate is over so-called non-Darwinian evolution, or the neutral theory of molecular evolution, which focuses on the role of neutral variation in the genome; for a summary, see Futuyma 1998, pp. 628–629. Finally, there is the occasional complaint that natural selection is tautological. In the latter, it is pointed out that natural selection is often described as "the survival of the fittest" and those individuals that survive are defined as "most fit." This is a confusing oversimplification, as I try to show later in this discussion. See also Brady (1979, 1982) and Peters (1976, 1978).

6. Those organisms were not simple compared with contemporaneous nonliving material, but they are simple compared with the bacteria of today.
7. For recent work on *Drosophila*, see Powell 1997.
8. The developmental sequence in insects with complete metamorphosis (holometabolous insects) is, in technical jargon, ovum (egg), larva (maggot, caterpillar), pupa (chrysalis), and imago (adult). Most insects are holometabolous, and the clever way evolution has produced a stage adapted primarily for eating and growing (larva) and another primarily for reproducing and dispersing (imago) is probably a major reason why insects so vastly outnumber, in both species and individuals, all other land animals.
9. More technically, it was artificial selection. If selection is occurring in a natural population, it is called natural selection. If people are deciding which genotypes (genetic endowments) are to reproduce more, as when a pig farmer uses his heaviest hogs as breeders, it is called artificial selection. Artificial selection is indistinguishable in principle from natural selection and is the reason why we have hundreds of varieties of decorative roses, very heavy hogs, and cows that produce extraordinary quantities of milk.
10. One can alter this definition to say that selection simply is differential contributions of phenotypes (individuals with different characteristics), whereas the *response* to selection depends on genetic differences. I'm following the classic definition here (see, e.g., Dobzhansky 1951, p. 76), put succinctly by Ernst Mayr as "the differential perpetuation of genotypes" (1963, p. 183).
11. A technical caveat is needed here. If there is a relationship between the genotypes of different individuals in a population and the number of their offspring who become adults in the next generation, and that relationship is too strong to be attributed to chance, then there has been selection. Purely by chance, individuals with different genotypes will, on average, have somewhat different reproductive success. Statistical tests are required to determine whether differential reproduction can reasonably be credited to chance (genetic drift) or to a significant influence of genotypic differences on the survival and reproduction of the individuals carrying the genotypes.
12. Selection can produce results that at first seem impossible. One of the many interesting things I learned from Robert Sokal was how to evolve a strain of fruit flies that was *more* susceptible to DDT than the control strain. How could that possibly be done? After all, we couldn't use the flies that died as the parents of the next generation. But what we could do, it turned out, was divide eggs from a single female into two batches and raise one batch in a vial with DDT medium and the other in a vial with clean medium. We'd do that with a lot of females and then use as parents for the next generation the flies from clean vials that were brothers and sisters (siblings, or "sibs") of the flies that suffered the highest mortality in the DDT vials. Sibs of flies that carried genes for susceptibility and died easily were likely to carry the same genes. That they did so was made clear by the ease with which we were able to evolve a strain that died at the slightest whiff of DDT.
13. My discussion of evolutionary processes is simplified. They have been extensively analyzed by mathematical models, beginning especially with Fisher and Haldane in England and Wright in the United States (see, e.g., Fisher 1930; Haldane 1932; Wright 1931). For comprehensive modern treatments, see Wright 1968–1978; Roughgarden 1979; Falconer and Mackay 1996; Roff 1997; and Hartl and Clark 1997; and, at a more elementary level, Futuyma 1998; Ridley 1996a; and Freeman and Herron 1998. For a good sample of more advanced mathematical issues, see Feldman 1989.
14. Sokal and Hunter 1954.
15. Many environmental factors, such as the ambient temperature and water content of the medium, also influence pupation site (Sokal 1966; Sokal et al. 1960).
16. A single gene often influences more than one phenotypic characteristic. For example, as discussed later

in this chapter, the human gene that produces the form of hemoglobin that causes sickling of red blood cells also influences resistance to malaria. When heterozygous (see note 113), it influences reaction to low levels of oxygen in the environment; when homozygous (note 113), it causes a fatal anemia. This phenomenon of a single gene having more than one phenotypic effect is known technically as pleiotropy. Two other well-known phenomena complicating the connections among selection, genotypic change, and phenotypic change are linkage and epistasis. The latter obtains when nonallelic genes (those at different loci—different positions on the chromosome occupied by specific genes) interact to affect phenotypic expression (e.g., a gene at one locus influences the phenotypic expression of a gene at another locus).

17. The DNA in animal cells is combined with proteins in structures in the nucleus that can be stained with dyes (the root *chrom* comes from the Greek word for color) for microscopic examination. One can think of these structures as a physical manifestation of the genome. In human beings and other mammals, normally there are two sets of homologous chromosomes (chromosomes carrying the same loci) in the cell lines that, by meiosis (a process of cell division that halves the number of chromosomes), produce eggs and sperm (each with one chromosome from each homologous pair). See the discussion of chromosomes in Raven and Johnson 1999, pp. 207–246.
18. The members of the base pairs (nucleotides) are adenine, cytosine, guanine, and thymine. They are often referred to by their initial letters and pair only in the combinations A–T and G–C.
19. For an overview of molecular genetics, see Futuyma 1998. There is a finer level of detail in this process. Genetic information of organisms is used to direct the synthesis of polypeptide chains (chains of amino acid residues, one or more of which makes up a protein molecule). Triplets of the bases (e.g., CCC, GAT) code for the positions of particular amino acids or tell the cell where to terminate or initiate the assembly of a polypeptide. Proteins are both key structural elements of an organism and organic molecules (enzymes) that control its biochemical reactions. Polypeptide chains usually contain 100–1,000 amino acids, of which there are some twenty kinds. The genetic information of higher organisms is contained in the sequence of base (nucleotide) pairs of the DNA molecules. The nucleotide sequences are essentially coded blueprints for the synthesis of polypeptides; each triplet is a "word" (codon) of the code. The point-to-point relationship (colinearity) of the nucleotide sequence in the DNA and the amino acid residue sequence in the polypeptide was first shown in investigations of the tryptophan synthase gene in *Escherichia coli* by Charles Yanofsky (Yanofsky 1967; Yanofsky et al. 1964). For more detail, see Hartl and Jones 1998.
20. Confusingly, the term *phenotype* is used sometimes for the characteristics of the organism as a whole and sometimes for some particular trait. Thus, blue eyes might sometimes be described as a phenotype. Similarly, *genotype* is used sometimes to refer to the entire genome and at other times to refer only to the genetic background of a phenotypic characteristic under discussion. Behavioral as well as physical characteristics are part of the phenotype.
21. You may recall from a biology course that genes may be dominant or recessive. If gene *A* is dominant and gene *a* is recessive, individuals with *AA* genotypes and individuals with *Aa* genotypes will have the same phenotype (e.g., brown eyes), whereas those with *aa* genotypes will have a different phenotype (blue eyes). Because of the mixing of genes during reproduction, two *Aa* brown-eyed individuals can have an *aa* blue-eyed offspring; two *AA* brown-eyed individuals or an *AA* and an *Aa* cannot. For more detail, see Raven and Johnson 1999 or Hartl and Jones 1998.
22. In fact, every stage from fertilized egg to adult is a sequence of phenotypes, each resulting from an interaction between genotype, gene products, and various environments. Phenotypic characteristics also include such things as length of life, patterns of development and senescence (aging), number and timing of offspring produced, and the like (see, e.g., Charlesworth 1994; Roff 1992; Stearns 1992; G. C. Williams 1957). Most frequently, however, when people refer to "the phenotype," they mean the adult phenotype.
23. Most mutations involve a change in the nucleotide sequence caused by an error in replication that inserts the wrong nucleotide, but there are other kinds of mutations, such as those that delete or add a nucleotide. For details, see the discussions of DNA and mutation in a standard evolution or genetics textbook, such as Futuyma 1998; Hartl and Jones 1998; Price 1996; Ridley 1996a; or Freeman and Herron 1998. There are many sources of genetic variation that I do not cover here. Some of them, such as the evolution and behavior of transposable elements, DNA sequences that can move around in the genome (see McClintock 1948, for which she won the 1983 Nobel Prize), are still only partly understood (see, e.g., Futuyma 1998, pp. 269–271, 639–641).
24. Some mutational alterations of the DNA code do not change the amino acid sequences of the protein or

influence gene regulation and thus are "invisible" to selection. Many altered genes are recessive and have obvious phenotypic effects only when two copies of the mutant gene end up in the same genome—that is, in technical terms, when an individual is homozygous for the mutant change at that locus.

25. With a thousand or so DNA "rungs" in each gene, a gene often has different rungs at the same position. Substitution of one set of base pairs (rung) for another can lead to a different amino acid being inserted into a protein under construction by complex cellular machinery. Just a single amino acid difference out of the 100–1,000 amino acids in most proteins can make a gigantic difference to an individual if that amino acid plays a crucial role in the activity of that protein. For example, in a human being, it can make a life-threatening change in the hemoglobin of the blood.
26. See Futuyma 1998, pp. 18–19.
27. You can see why mutations are rarely helpful if you consider that the genetic message is read one triplet at a time, beginning at a fixed starting point. Thus, deleting a single nucleotide will cause all "downstream" triplets to change and result in a completely different set of amino acid residues in the resulting protein—a configuration that is hardly likely to allow that protein molecule to function.
28. There have been claims of directed mutation in bacteria under special circumstances (Cairns et al. 1988; Hall 1988), but these claims are controversial (e.g., Sniegowski and Lenski 1995). For a good overview, see Futuyma 1998, p. 285.
29. The following material can be a little daunting in its complexity. Further explanation can be found in the evolution and genetics texts referenced earlier.
30. In fewer than 300 generations, the entire universe would consist of chromosomes.
31. The mixing normally occurs only between homologous chromosomes; the phenomenon of exchange is known as "crossing over" (for a summary, see Futuyma 1998, chap. 3). Rates of recombination themselves evolve (see, e.g., Otto and Michalakis 1998).
32. The process of development in multicellular organisms is gradually becoming understood. The basic problem is how undifferentiated cells that originally have essentially identical genetic endowments become differentiated into a variety of cell types (say, cells that line the gut to help absorb nutrients or cells in the brain that transmit nerve impulses), tissues, and organs. The basic answer can be found in the differences in the microenvironments confronting each cell—such elements as the characteristics of neighboring cells and the chemical and energetic messages received from them (see, e.g., Müller 1997). More detailed explanations are now being found for such questions as how the fundamental three-dimensional organization of the cells of multicellular organisms is controlled—specifically through studies of so-called homeotic gene complexes (e.g., *HOM* loci in invertebrates and *Hox* loci in vertebrates) and homeoboxes (highly conserved nucleotide sequences coding for critical DNA-binding regions of proteins manufactured by homeotic genes); see Freeman and Herron 1998, pp. 455ff., for a summary. Classic papers include Scott and Weiner 1984 and Gehring 1986. More recent details can be found in McGinnis and Krumlauf 1992; Morata 1993; Gehring 1998; and Hirth and Reichert 1999. For a more elementary discussion of development, see Raven and Johnson 1999 and Purves et al. 1997.
33. The set of developmental pathways leading to an array of phenotypes that a given genotype will produce if exposed to all possible environments is known as the developmental reaction norm (Schlichting and Massimo 1998) or, more simply, the norm of reaction (Maynard Smith 1998, p. 95). The capacity of the environment to alter the expression of a given genotype is often referred to as phenotypic plasticity. For some of the complexities of genotypic variation being translated into phenotypic variation in an actual system, see Brakefield 1998. To see that phenotypic plasticity can cross generational gaps (maternally induced defense in the offspring of prey attacked by predators or herbivores), see Agrawal et al. 1999. Recognizing an organism–environment dichotomy is convenient for discussion, though it may overemphasize the distinction. Richard Lewontin has pioneered in pointing this out and has taken the position that "organisms construct every aspect of their environment themselves" (Lewontin 1985 [1983], p. 104); see also the discussion in Lewontin 1982b and Lewontin et al. 1984. The basic points that environments are defined by the characteristics of organisms and that organisms interact heavily with their environments, always modifying them to some degree, are important to emphasize. But Lewontin's position seems too extreme—for instance, although animals can move about and thus alter the microclimatic conditions they experience, they have little control over macroclimatic factors. A specific example is provided by the Bay checkerspot butterfly *(Euphydryas editha bayensis)*, for which the exact sequence of rainfall is an important element in the environment (Ehrlich 1992; Weiss et al. 1994) but which have precious little influence on that environmental element (although they can somewhat modify macroclimatic effects by changing location). For a pioneering analysis of environment by ecologists, see Andrewartha and Birch 1954. For a famous philosopher's view, closely related in many respects to Lewontin's, see the quotes

from John Dewey cited in Godfrey-Smith 1999 and Godfrey-Smith's discussion itself. For a mathematical treatment of the joint evolution of environment-altering traits and traits whose influence on fitness depends on those environments, see Laland et al. 1996.

34. Dawkins 1989 (1976). Dawkins is one of the great popularizers of science, and his analogy of the behavior of genes as "selfish" might be a useful heuristic device in some circumstances and at an elementary level. Dawkins certainly doesn't believe that this "selfishness" is the equivalent of the intentional selfishness (self-concern) of human beings, given that genes obviously are incapable of concern (Dawkins 1981). But he often writes as if he does. That can be particularly confusing to those who don't realize that neither DNA nor ribonucleic acid (RNA) can replicate itself, even though scientists often use the shorthand term *self-replicating* to describe these molecules. Unhappily, making such analogies, especially ascribing human motives to nonhuman entities, as in characterizing genes as playing out "tournaments of manipulative skill" and manipulating and shaping the world "to assist their replication" (Dawkins 1982, p. 5), can mislead. In response to a strong attack on his book (Midgley 1979), Dawkins correctly pointed out (1981, p. 557) that physicists, after all, describe a property of fundamental particles as "charm." But we do not associate our own behavior with particle physics, as we often do with evolutionary biology. Furthermore, Dawkins's idea that "looking at life in terms of genetic replicators preserving themselves by means of their extended phenotypes" (1982, p. 7) runs into deep trouble as soon as one considers linkage, pleiotropy, epistasis, and the inability of unaided and unsupported DNA to do anything at all (genes cannot replicate without organisms, and vice versa). Dawkins's writing also confuses some philosophers, who come away with the impression that "modern Darwinism . . . insists that the sole unit of selection is the gene" (Griffiths 1995, p. 125). For the views of a philosopher who tends to support Dawkins, read Dennett 1995; for those of one who tends to dissent, see Godfrey-Smith 1998.
35. Hartl and Jones 1998, p. 730. Genes and cultures interact in complex ways here. Although culture can reduce the deleterious effect of this particular genetic defect, it usually can't eliminate it because the dietary changes are expensive and require determination to adhere to. Thus, at the population level, one would still expect some selection against the deleterious allele (gene variant).
36. In population genetics, these are standardly called gene frequencies.
37. See, e.g., McKenzie 1996. A recent fine example of using insecticide resistance in mosquitoes both to understand microevolutionary processes and to suggest ways in which the development of resistance can be retarded is the work of Thomas Lenormand and his colleagues (Lenormand et al. 1998; Lenormand and Raymond 1998; Lenormand et al. 1999).
38. The word *adaptation* subsumes a host of complexities, as does the issue of what proportion of phenotypic features is the direct result of selection and the degree to which selection can produce characteristics that are in some sense optimal (Dennett 1983; Ehrlich and Holm 1962; Futuyma 1998; Gould and Lewontin 1979; Lewontin 1978; Rose and Lauder 1996; Taylor and Weibel 1981). For our purposes in this book, an adaptation can be viewed as any feature of an organism that increases the chances of its genes being represented in the next generation and that apparently has evolved because of this fitness benefit. Lewontin's views are more accurate than Dennett's in this area.
39. The vast majority of creative selection appears to occur among individuals within populations, but, as covered later, selection among groups of organisms (group selection) may account for a limited number of special traits, and selection among species (although probably not "creative" in the sense of producing new features that adapt organisms to their environments) has had substantial effects on the patterns of diversity of life.
40. Drift is actually always present at some level and can be seen as a sampling effect. For example, ignoring all other evolutionary forces, the sperm and eggs (gametes) produced by a sexually reproducing population can be viewed as containing a random sample of the genetic variation present in that population. The fertilized eggs (zygotes) that happen to be formed from those gametes contain a random sample of the genetic variation that was present in the gamete pool. The zygotes that mature into adults are a random sample of those originally formed. And so on. This process of repeated random sampling will cause greater changes in gene frequencies in a small population than in a large one. The reason is roughly the same as the reason why if you flip an honest coin a thousand times, the chance of a proportion of heads substantially different from one-half (say, 100 percent heads) is essentially zero, whereas if you flip the same coin twice, the odds of getting all heads are one in four. In populations of more than fifty reproducing pairs of individuals, the effects of drift are generally small enough to be ignored.
41. Selection and drift on average reduce the amount of variation, although if either selection or drift increases the frequency of a new mutant allele, the amount of variation increases.
42. For a less simplified view, see the treatment of the shifting balance theory in Futuyma 1998, pp. 408–409.

43. Treat 1957; Treat 1975.
44. See Busck 1912 for an account of what it's like to be infested with this interesting parasite, *Dermatobia hominis.*
45. Bates 1943. Sometimes *Dermatobia* oviposits on flies of the housefly family, on ticks, or even on leaves (Oldroyd and Smith 1973).
46. This includes both the common chimp *(Pan troglodytes)* and the fascinating pygmy chimpanzee, or bonobo *(Pan paniscus)*, which we'll meet later. After the original split, both the hominid and chimp lines split more times, with one representative of the hominid line and two of the chimp line surviving.
47. For an overview and discussion of techniques of studying selection in nature, see Endler 1986.
48. Reznik et al. 1997.
49. Tutt 1896; Kettlewell 1973; Owen 1997.
50. The latinized name of the peppered moth is *Biston betularia.* For a recent summary, see Brakefield 1987. I had this famous example stamped in my memory early in my career. In 1959, I had returned to the University of Kansas's entomology department as a postdoctoral fellow, and the university (like most others) was holding a Darwin lecture series to mark the centennial of the publication of *On the Origin of Species.* One evening, John Moore, a leading evolutionary developmental biologist of the era, called for the first slide of his lecture. On came a photograph of two moths, one dark and the other light but peppered with dark spots, sitting on a dark tree trunk. I and the rest of the large audience cracked up. Poor John was stunned; he had no idea why the audience was laughing. He studied the slide with puzzlement, peered at the audience, and looked to see if, perchance, he had left his fly open. No answer there either. Only later did he learn that each of the three lecturers who had preceded him, when they requested "first slide, please," had shown exactly the same photograph as if it were fresh to the audience. John Moore and I became friends, senior though he was, and occasionally we have reminisced about that long-ago evening when he was so puzzled. The coincidence of the moth slides, we agree, was no coincidence at all.
51. The exact frequencies cannot be determined from museum collections because in the early 1800s, the melanic individuals were considered rare aberrations. Captain N. D. Riley, keeper of entomology at the British Museum, told me proudly in 1960 that the museum had for more than a century been buying butterfly and moth collections, throwing out the "junk," and saving the aberrations.
52. Clarke et al. 1990. Many questions about the details remain unanswered (Brakefield 1987; Coyne 1998; Grant 1996; Majerus 1998), but the story still provides one of the best documented examples of the action of selection in nature.
53. See, e.g., Andres and Prout 1960; Roush and Daly 1990.
54. Roush and McKenzie 1987; Tabashnik 1990; Georghiou and Lagunes-Tejeda 1991; Pimentel and Lehman 1993. The complexity of evolutionary issues related to resistance can be seen in Tabashnik, Liu, Malvar, et al. 1997.
55. Resistance will also develop to the toxins now being genetically engineered into certain plants.
56. Georghiou 1986, p. 33; see also Tabashnik 1990. Of course, the pesticide industry might view this as a small price to pay for all the jobs and the portion of gross national product generated by the industry. But the social costs (in human and ecosystem health) are obviously very much underestimated in the billion-dollar estimate.
57. Daily and Ehrlich 1996b.
58. Baquero and Blázquez 1997.
59. Bacteria often contain plasmids, which are extrachromosomal molecules of DNA, and bacteriophages (*phages* for short), which are DNA or RNA viruses encapsulated in a protein coat. Both plasmids and phages can facilitate the transfer of genes between individual bacteria and are involved in the rapid evolution of antibiotic resistance (Levin and Lenski 1983).
60. My own adventures with the reality of the evolutionary process in nature took me from the Animal House to a converted bordello and involved the detection of selection in the absence of rapid environmental change. I had a postdoctoral research position at the Chicago Academy of Sciences in the late 1950s with the late Joseph Camin, and we did some research on evolution in water snakes *(Nerodia sipedon).* Joe and I collected the snakes on small islands in Lake Erie, where they were abundant. One of the islands, Middle Island, was owned by a man who was happy to have us remove snakes from his domain and use his large house there as our research base. The island was just inside Canada and had been a popular resort for swingers from the United States during Prohibition. Working from the house, Joe and I were able to demonstrate selection going on in snake populations without any rapid environmental change. We would sally forth and roll over rocks along the island's shoreline. Often, we would find four or five snakes, each about a yard long. We would grab them, and they would grab us back. Their mouths

were amply supplied with sharp teeth, and their saliva contained an anticoagulant. I well remember driving back to Chicago with Joe, bags of snakes in the trunk of the car and blood still oozing from bites on our arms.

That kind of water snake is common in swamps and bogs and along lake borders in the eastern United States, and most populations consist of individuals with dark transverse bands. But on the Lake Erie islands, most adults had little or no banding, and only a few were heavily banded (Camin et al. 1954).

We kept female snakes in our lab and allowed them to give birth (most snakes lay eggs, but water snakes' eggs hatch within the body and the young are born alive). The majority of the newborn individuals, unlike their parents, were heavily banded. We were thus able to measure selection by comparing frequencies of the genetically determined banding patterns in litters of very young snakes and in the adult population. Because the frequency of banded individuals was much higher in the litters than in the adult population, something appeared to be eliminating banded individuals from the population as they matured. There was obviously a rather strong selection pressure; snakes in the banded half of the banding spectrum had only about a quarter of the survival probability of those in the unbanded half (Camin and Ehrlich 1958; Ehrlich and Camin 1960).

A visual predator was likely to be the selective agent. The unbanded snakes were inconspicuous on the flat limestone rocks that lined the shores of the islands, the only suitable habitats there (the islands had no inland swamps), whereas banded snakes stood out. On the other hand, banded snakes were less conspicuous (to our eyes) against a background of complex swamp vegetation. Subsequent work supports the idea that the disruptive effect of banding (breaking up the outline of the snake, thus making it more difficult to see) gave snakes along a Kansas creek protection from visual predators (Beatson 1976). In our case, seagulls, which were abundant around the islands, were the probable predator, but we were unable to show that they were the actual source of the selection pressure.

It wasn't as if we didn't try. Joe and I attempted to test whether seagulls would differentially find and eat banded young water snakes on flat limestone rocks. Conveniently, the Chicago Academy of Sciences was right next door to the Lincoln Park Zoo. Marlin Perkins, star of the then-popular television show *Zoo Parade*, was director of the zoo and a friend of Joe's. He gave us the use of a big cage and lent us a herring gull, whom we promptly christened Herman. We made two identical mock-limestone rocks out of painted plaster of paris, used transparent tape to secure a banded young water snake to one and an unbanded one to the other, and introduced Herman to the cage simulation of a Lake Erie island shoreline. On the first trial, he stepped on the unbanded (camouflaged) snake on his way to grabbing and eating the banded (conspicuous) snake. Triumph! (Did you think that scientists really dispassionately test their hypotheses, trying to disprove them?) We should have quit while we were ahead. On the second trial, Herman figured out there was a snake on both rocks and tried to eat them both. Back to the drawing board.

We decided we'd better make sure that Herman could tell banded and unbanded snakelets apart. We presented him with the choice, with the snakes taped to tin dishes, and put a loud electric gong under the banded snake's dish to frighten Herman if he tried to choose banded. Sure enough, on the first trial he went for banded, and we triggered the gong. Herman paid no attention to the horrible racket and calmly started to devour the snake. We dashed into the cage, grabbed Herman, and snatched the snake from his beak. That taught him a lesson. From then on, if the gong went off, Herman grabbed the snake and ran to the farthest corner of the cage to eat it before we could catch him. Totally outwitted by a bird with a brain weighing a gram or so, we returned Herman to Marlin, and I swore off animal behavior experiments forever.

If the banded young had lower rates of survival, why hadn't the genes for the banded phenotype been eliminated over generations from the island populations? Joe and I suspected that the answer was to be found in observations of snakes swimming far out in the lake. We surmised—correctly, as later work demonstrated—that banded individuals from populations living in swamps around the mainland shore of the lake migrated occasionally to the islands. There, they mated with island snakes, and in the process added "banded" genes to the islands' gene pools. Thus, a persisting balance between gene flow and selection produced a situation that kept both types in the population and gave us a situation unusually amenable to study—a similar case in salamanders was recently described in Storfer et al. 1999. Recent work on the snakes supports the general picture Joe and I suggested, but with added complexities (King and Lawson 1995), showing how both scientific understanding and the environment evolve. For example, it appears that unbanded phenotypes are strongly favored among young-of-the-year, whereas banded phenotypes are favored among older (larger) snakes, which are better concealed against the island background than are the smaller banded individuals. But selection among the young snakes produces a net selection favoring the unbanded phenotypes (King 1992, 1993). This work also shows a general decline in population size over approximately fifty years. In 1937, it was possible to collect 20 snakes per hour. In the early 1950s, Camin and others got more than 130 adults on Middle Island in a single day. In 1957, he and I got about 11 per hour. In the 1980s, the rate dropped to about 1 per hour. In 1960, 84 adult snakes were collected on Middle Island in two days—three to five times the estimated population size on the island in the 1980s. The main reason for the population decline seems to be destruction of water snake habitat resulting from shoreline recreational use and construction of summer cottages (King 1986).

61. The snails are two species of the genus *Cepaea*, *C. nemoralis* and *C. hortensis.* Their shells may be banded, partially banded, or unbanded and colored yellow, brown, or any shade from pale fawn through pink and orange to red. The shell lip may be dark brown (normally) or black or, rarely, pink or white. The frequencies of various shell types vary enormously among local populations. Originally, it was claimed that differences were due largely to genetic drift (gene frequencies changing as a result of chance events) rather than to natural selection (Lamotte 1952, 1959).
62. Sheppard 1951, 1952; Cain and Sheppard 1954.
63. Technically, this mixture of types is called a polymorphism. A polymorphism requires two or more distinct genotypes (more rarely, phenotypes) persisting in a population in which the rarest phenotype is too numerous in the population for its presence to be explained by recurring mutations alone.
64. Jones et al. 1977.
65. Vicario et al. 1989; Mazon et al. 1990; Cameron 1992; Chang and Emlen 1993. Climate may also influence which proteins serve as biological catalysts in the living machinery of the snails' cells. The variations in the protein catalysts are usually referred to as enzyme polymorphisms—see, e.g., Guiller and Madec 1993. The situation in *Cepaea* is complicated. For example, we don't know whether or not there are coevolutionary interactions between *C. nemoralis* and *C. hortensis* (Arthur et al. 1993; Cowie and Jones 1987). In addition, there are signs of *frequency-dependent* selection, a situation in which the selection pressure changes as the frequency of different types in the population changes (Ayala and Campbell 1974; Haldane 1932; Heino et al. 1998; Hori 1993; Lewontin 1958; Wright 1969). Frequency-dependent selection will occur whenever the fitness of individuals with an inherited characteristic depends on the proportion of such individuals already in the population. For instance, the fitness of a male in a population that is 95 percent female is very different from that of one in a population that is 95 percent male. Thrushes attacking *Cepaea* can produce frequency-dependent selection pressures by forming search images, concentrating on finding and eating snails of one common shell type. Human beings often form search images as well; you may have discovered that after seeing the first of something you have been looking for (a camouflaged caterpillar, for example), finding subsequent ones is much easier. If thrushes disproportionately eat the snails with the commonest shell type, rare types are at an advantage and thus become more common (Allen et al. 1988; Tucker 1991). Eventually, the once-rare types become the common ones, and the birds switch to them. Such frequency-dependent selection can maintain two or more types of individuals in a population. A tool often used for analysis of frequency-dependent selection, especially in relation to optimal behavioral strategies, is that of evolutionarily stable strategy (ESS) (Maynard Smith 1978, 1982; Maynard Smith and Price 1973).

    Besides frequency-dependent selection, there is evidence of other complex patterns of selection acting in *Cepaea.* There are also signs of *disruptive selection,* in which phenotypes at opposite extremes of a spectrum are favored over phenotypes near the average. The direction of selection can change over very short geographic distances in the same population (not the purest form of disruptive selection, which would be different types being favored at the same time and place or at different times in the same location). There is also the possibility of *stabilizing selection* through removal of extreme forms or heterozygote advantage at loci (heterozygote more fit than either homozygote) with or without visible effects on shell characters (Cook and Gao 1996). Furthermore, there may be complex selection on interactions between loci, epistasis, so that the fitness of alleles is affected by the "genetic background" within which they occur (see, e.g., Otto 1997). Drift, Lamotte's original suggestion, against which Cain and Sheppard argued, is difficult to untangle from all the complex selective interactions (e.g., Cain 1989), but *founder effects* are probably important (Wilson 1995). For example, isolated populations of *Cepaea* at the limits of distribution in Scotland have low heterozygosity for shell polymorphism and are sometimes monomorphic (show no genetic variation).

    In this book, I sample only some aspects of selection. For instance, selection also operates to control life-history traits such as total life span (e.g., M. R. Rose 1991; Rose and Charlesworth 1981), which, among other things, influence the dynamics of populations. Selection pressures also often vary with population density and alter the growth rate and carrying capacity; in this respect there is density-dependent natural selection (Anderson 1971; Roughgarden 1971; Smouse 1976). An early experiment (Buzzati-Traverso 1955) showed a fruit fly population growing rapidly until it reached the carrying capacity *(K)* for the laboratory environment of the flies with the genetic composition of those used as founders for the experiment. After that, selection favored individuals better adapted to the particular laboratory environment, which led to a gradual increase in population size (increase in *K*) (see Ayala 1968). In contrast, in disturbed habitats and other nonequilibrium situations, selection may favor genotypes that maximize the rate of increase of the population *(r)*. Robert MacArthur (1962) called density-independent natural selection *r*-selection and density-dependent selection *K*-selection. Otto Solbrig first demonstrated how different phenotypes were favored by *r*- and *K*-selection in a study of dandelion plants at disturbed and undisturbed sites (1971). The two kinds of selection may be of some interest to human ecologists because there has been rapid growth in human population size—roughly a thousandfold increase over the past 10,000 years and a sixfold increase over the past two centuries—and the population is now apparently well above the planet's long-term carrying capacity (Daily and Ehrlich 1992; Daily et al. 1994). The degree to which *K*-selection may now be affecting human gene frequencies is unknown, but 500 generations of rapid growth should have been more than enough for us to expect some influence of *r*-selection on traits related to allocation of reproductive effort. The entire situation, however, is very difficult to analyze because of the interactions between genetic and cultural evolution.

66. For summaries of recent work on biological evolution on islands, see Grant 1998.
67. Carlquist 1974; Grant 1998.
68. Cody and Overton 1996.
69. There are many other examples of such rapid evolution in nature. In one such case, guppies living in a stream rich in large, predatory cichlid fishes that prefer to eat large guppies reproduced at a small size and had many young. When they were transferred to a stream that was free of the cichlids, in just four years (about seven generations) the guppies had shifted their life-history strategy to living longer, growing larger, and having fewer young (Reznik et al. 1997).

    Recent work by Peter Grant and Rosemary Grant has shown that the bills of Galápagos finches evolve in size and strength as rapid changes in climate require changes in diet. For instance, during one drought, a population of the medium ground finch *(Geospiza fortis)* was subjected to intense directional selection favoring large individuals with big bills. Only those birds could crack the large, tough fruits that remained when smaller fruits and seeds had been exhausted. Females also tended to mate with larger males, and as a result there was a detectable genetically based increase in size over a single generation (Grant 1986). For another island example, this one involving lizards, see Losos et al. 1997.
70. Two species of *Tribolium*.
71. *Escherichia coli*.
72. Lenski and Travisano 1994. For a semipopular overview of the work of Lenski's group (which is now up to more than 24,000 generations), see Appenzeller 1999.
73. Lenski et al. 1991.
74. There was no genetic variation within or among populations except for a single neutral marker allele used to identify the populations.
75. This was measured as increased competitive ability, which was determined by testing later generations against resurrected earlier generations.
76. See also Travisano et al. 1995. Fitness increases were similar but not identical after 10,000 generations. Population sizes were sufficiently large that many identical mutations must have appeared in the different populations. Repeated evolution has also been recorded in populations of viruses (Wichman et al. 1999) and stickleback fishes (Rundle et al. 2000).
77. Yoo 1980. As is frequently the case in experiments on artificial selection, progress in selection slowed after a while, and when selection was relaxed (Yoo stopped breeding from the flies with the highest bristle numbers), some of the gains were then lost, quite likely because without Yoo's intervention, flies with fewer bristles outreproduced those with more. In other words, natural selection returned bristle number toward the original value.
78. Interesting experiments on community evolution are also being carried out in bacterial systems (see, e.g., Bohannan and Lenski 1997; Rainey and Travisano 1998).
79. Gould 1997.
80. See, e.g., Grant 1986.
81. Tabashnik, Liu, Finson, et al. 1997; Thompson 1998. For other examples of rapid evolution, see Seeley 1986; Via and Shaw 1996.
82. Ehrlich et al. 1988, pp. 273–277.

83. All organisms, including viruses (which some might not define as organisms), have evolved mechanisms to protect their genetic material.
84. Porter 1997, pp. 296–297.
85. For an introduction to this topic, see Ehrlich and Ehrlich 1996, chap. 9.
86. Profet 1988.
87. Futuyma 1983, p. 212; see also Bernays and Chapman 1994, p. 48.
88. Colborn et al. 1996.
89. See, e.g., Harrell 1981, p. 818; Henderson et al. 1993; Eaton and Eaton III 1999, p. 255.
90. It is the immunoglobin-E (IgE) segment of the immune system. See Nesse and Williams 1995, chap. 12.
91. One evolutionary issue is sometimes raised in connection with the problem of controlling human population size. Today, people in better-educated, more prosperous portions of many human populations have smaller families than do those who are poorer and have less education. This has raised the fear that selection may be operating against intelligence and responsible reproductive behavior, that is, against limiting family sizes in a world that is already overpopulated (see, e.g., Hardin 1968a, p. 188; Loehlin 1998; Lynn 1998). There is substantial technical disagreement over whether this "dysgenic" trend actually exists (e.g., Preston 1999; Waldman 1999), aside from the vexing question of whether there is a single entity that can reasonably be called intelligence—a topic we'll visit later. If there were a persistent tendency for individuals to produce large families, because of either cultural values or genetic predisposition, this could be a factor retarding today's needed decline in human birthrates. But there is much variation in attitudes toward family size and little information on the heritability (if any) of those attitudes in diverse cultural environments. In addition, there is abundant evidence that cultural attitudes toward family size can change dramatically within a generation, as illustrated by substantial drops in fertility rates in many parts of the world in the past quarter century. Therefore, worry about the evolution of reproductive irresponsibility seems premature. Even if there were substantial genetic variation in preference for family size (amounting to selection in favor of wanting many offspring), it would be dozens of generations—several hundred years—before substantial behavioral changes could be expected. After all, even the very strong selection pressures involved in DDT resistance in *Drosophila* took several generations to produce significant results. The odds are that the population problem will be "solved" by a massive die-off, by the nearly universal achievement of limited family sizes, or by a combination of the two within a few generations (see, e.g., Ehrlich and Ehrlich 1990, chap. 9—when that was written, the chances of a large-scale nuclear war were declining; a decade later, they appeared to be increasing again); Bleek and von Hippel 1999; John Holdren, personal communication, 7 January 2000).
92. For an overview of the more general problem of the evolution of cooperation in animals, see Dugatkin 1997.
93. Hamilton 1964. Technically, inclusive fitness includes conventional fitness (as measured by the presence of offspring in the next generation), but inclusive fitness is often used (as in this book) to emphasize additional fitness gained by helping relatives other than one's own offspring.
94. That is, they are "identical by descent," meaning that they are copies of the same ancestral genes. Mutation, mitochondrial DNA (mtDNA), and Y-linked genes make this statement not quite precise, but it's close enough for our purposes. Inclusive fitness is of special importance in the evolution of social Hymenoptera (bees, ants, wasps). These insects are haplodiploid: males develop from unfertilized eggs and thus are haploid, whereas females come from fertilized eggs and thus are diploid. This results, among other things, in sisters sharing more genes than do mothers and daughters (Futuyma 1998, pp. 598–601) and in Hymenoptera showing extreme forms of altruism because often a worker can have higher fitness by rearing sisters that reproduce than by producing daughters herself (see, e.g., Michener 1974).
95. Calculation of relatedness and costs and benefits of altruism can be complex; see, e.g., Parker 1989.
96. See, e.g., Woolfenden and Fitzpatrick 1984; Koenig and Mumme 1987. Sometimes the helping is by nonrelatives, and the benefits to the helper may be in deferring the risks of breeding in order to develop the skills that will later give them greater chances of reproductive success (Brown 1987).
97. Trivers 1971; Axelrod and Hamilton 1981. Trivers concludes that "friendship, dislike, moralistic aggression, gratitude, sympathy, trust, suspicion, trustworthiness, aspects of guilt, and some forms of dishonesty and hypocrisy" may have evolved "to regulate the altruistic system" (p. 35). Axelrod and Hamilton point out that when individual recognition is not possible to ensure that potential reciprocators will not defect and become free riders, the problem can be circumvented by employing a fixed meeting place. Then it is not necessary to be able to discriminate individuals to ensure that the interaction is with the same individual. This is what happens at "cleaning stations" on coral reefs, where large fishes are picked over by smaller "cleaning fishes," which remove and eat the parasites of the larger species. "Defection"

can occur in this system. There are species of cleaner mimics that look just like cleaners but take bites out of the larger fishes. The latter cannot discriminate the mimics, but they do learn to avoid cleaning stations where the defectors hang out. For more recent work on this topic, modeling circumstances (especially levels of investment) under which cooperation should occur, see G. Roberts 1998 and Keller and Reeve 1998.

98. See, e.g., Wilkinson 1986. The bats are *Desmodus rotundus.*
99. Wilkinson 1984.
100. Wilson 1978, p. 153.
101. Richerson and Boyd 1978. For a classic anthropological treatment of kinship, see Fox 1983 (1967).
102. In some kinship systems, those considered the closest relatives of an individual are not those genetically closest—there is not a one-to-one mapping of kinship and genetic relatedness. However, that does not obviate this point. All that is required is that there be a significant correlation between kinship category closeness and actual genetic resemblance (see, e.g., Etter 1978).
103. Parents' caring for children is often considered just a result of ordinary natural selection rather than kin selection.
104. Group selection can also be carried out in laboratory experiments and has been shown to be capable of either reinforcing or countering the effects of individual selection (Wade 1977, 1979). Arguing over group selection models is a cottage industry in population genetics—see, e.g., Wynne-Edwards 1962; G. C. Williams 1966; Lewontin 1970; D. S. Wilson 1975, 1983; Wilson and Sober 1994; Uyenoyama 1979; Uyenoyama and Feldman 1980; Michod 1981, 1982; Wade 1985; Sober and Wilson 1998; Getty 1999; and Wade et al. 1999. This controversy is just a subset of a larger controversy over units of selection. For example, if the groups involved are species, then group selection can be called species selection (Stanley 1975) or taxon selection (G. C. Williams 1992). For a recent discussion of possible conflicts in evolution operating at different levels, see Keller 1999.
105. Passenger pigeons and woolly mammoths went extinct; city pigeons (rock doves) and African elephants persist—examples of species or genus selection.
106. For an excellent recent discussion, see Sober and Wilson 1998 and the review of it by Lewontin 1998.
107. Or, less interestingly, when deleterious mutations occur that lead to phenotypes that are unable to reproduce.
108. See, e.g., McNeill 1976, pp. 200ff.; Reff 1991.
109. McNeill 1976, p. 132.
110. Reff 1991, p. 118.
111. Carrington et al. 1999. Other work also indicates differential susceptibility of human genotypes to human immunodeficiency virus (HIV) infection (Lockett et al. 1999).
112. In red blood cells with both kinds of hemoglobin, infection by the malarial parasite causes sickling, which makes the cell a poor environment for growth of *Plasmodium* and often leads to the cell's destruction by the host's immune system.
113. Genes at the same position on each of a pair of chromosomes (that is, at the same locus) are called alleles. Members of a pair of alleles may be identical, in which case we say the individual is homozygous at that locus, or they may be different forms of the gene, in which case we say the individual is heterozygous at that locus. It is commonplace for geneticists to refer to individuals as homozygotes or heterozygotes, in which case it is understood that they are referring to the individuals' genotypes at a single locus (there are roughly 100,000 loci in the human genome).
114. Allison 1954a, 1954b, 1955, 1959, 1964; for more recent descriptions, see Bodmer and Cavalli-Sforza 1976; Diamond 1989a; and Diamond and Rotter 2000.
115. Diamond 1989a.
116. There are other interesting aspects to the sickle-cell example. When selection favors individuals carrying two different alleles, gene frequencies remain the same as long as the relative fitnesses of the different kinds of individuals do not change. In technical terms, the heterozygotes are favored over both homozygotes, and there is a stable gene frequency equilibrium determined by the selection coefficients against the two homozygotes. A simple mathematical model shows that this will result in both versions of the gene (both alleles) being preserved in the population—see Ehrlich et al. 1974, pp. 111–113, or Ridley 1996a, pp. 117–120. This situation is sometimes termed overdominance for fitness, and sickle-cell anemia is the classic case of the maintenance of genetic variability by selection favoring individuals with two versions of the same gene. This sort of mechanism is not unique to human evolution. A similar one has been elucidated in a series of elegant studies of the evolutionary ecology of sulfur butterflies by Ward

Watt (Watt 1977; Watt et al. 1983). He showed, among other things, that having two forms of a gene coding for a specific enzyme could produce a higher level of enzymatic activity than having just one form.

It is known that the difference between sickling hemoglobin and normal hemoglobin is a single amino acid substitution among the 287 amino acids from which the hemoglobin is constructed (Ingram 1956, 1959). In addition, many other changes in hemoglobins provide protection from malaria but also produce diseases in individuals with two copies of the responsible alleles. The distribution of these "genetic antimalarials" and the way they work and interact is a complex puzzle that is just beginning to be solved (Diamond 1989a; Flint et al. 1993; Weatherall 1987). The antimalarials are various forms of sickle-cell disease, a complex of other hemoglobin changes that cause diseases called thalassemias (Weatherall 1996), and a deficiency in the enzyme glucose-6-phosphate dehydrogenase (G6PD) (Ruwenda et al. 1995). There is substantial evidence that the responsible genes are descended from different mutations that occurred repeatedly and in different geographic areas (Kazazian Jr. et al. 1984; Nagel 1984; Kulozik et al. 1986; Chebloune et al. 1988).

117. See, e.g., Surtees 1970; for a fine popular summary, see Desowitz 1981, pp. 46ff.
118. Elliott 1972.
119. Wiesenfeld 1967.

Chapter 3: Our Natures and Theirs

1. Dobzhansky 1970, p. 3. The estimated number of species has risen considerably since he wrote.
2. I am concentrating on species-level biodiversity here, but other levels of biodiversity, especially the diversity of populations (Hughes et al. 1997), are also critical to human well-being.
3. Myers 1979; Ehrlich and Ehrlich 1981; Ehrlich and Ehrlich 1992; Wilson 1992; Ehrlich 1993; Heywood 1995; Hughes et al. 1997.
4. Darwin 1859, p. 399.
5. Darwin 1859, p. 396.
6. Mites of the family Tetranychidae.
7. McMurtry et al. 1970.
8. Blockstein 1998.
9. Mouse ticks, when older, attack deer, which thus are also factors in the epidemiology of Lyme disease.
10. Bacteria of the kind also responsible for syphilis.
11. Mayr 1942. Many of the basic ideas had been around for a long time (see, e.g., Dobzhansky 1937, chap. 8), but Mayr put them together, expanded them, and promoted them.
12. Genus is the taxonomic category within which all species are grouped. These two thrushes have been given the latinized specific names of *Catharus ustulatus* and *Catharus guttatus*, respectively. The first part of the name, *Catharus*, is the generic name and indicates the genus (plural *genera*) into which the bird has been classified. The second part, *ustulatus* or *guttatus*, is called the specific epithet, and the two together are the species name, following the system of binomial nomenclature developed by Carolus Linnaeus almost a quarter millennium ago (Linnaeus 1758) and used to this day. In the case of human beings, *Homo* is the genus and *Homo sapiens* the species. All species must belong not only to a genus but also to a family, an order, a class, a phylum, and a kingdom, in increasing order of inclusiveness. My favorite nonhuman organism, the checkerspot butterfly, *Euphydryas editha*, belongs to the family Nymphalidae ("four-footed" butterflies), the order Lepidoptera (scaly-winged insects—butterflies and moths), the class Insecta (insects), the phylum Arthropoda (joint-legged animals—including insects, spiders, lobsters, barnacles, etc.), and the kingdom Animalia (animals). *Homo sapiens* belongs to the family Hominidae, the order Primates (apes, monkeys, lemurs, etc.), the class Mammalia (mammals), the phylum Chordata (animals with backbones or notochords), and the kingdom Animalia (where they join *Euphydryas*).

    Finer details of relationships can be achieved by using nonobligatory categories such as subgenus, tribe, superfamily, and so on. Other taxonomists might give a somewhat different hierarchical classification, given that taxonomists persist in the hopeless task of trying to cram the multidimensional and continuous evolutionary relationships of organisms into an ever more refined hierarchy of discrete units, but overall they've done a pretty useful job of providing a handy classificatory structure for communicating about organisms.
13. See, e.g., M. B. Williams 1973.
14. Mayr 1940, p. 256; Mayr 1942, p. 120; Mayr 1982, pp. 273ff.; Mayr 1997, pp. 127ff.
15. Hybridization can itself be an important evolutionary force—for a recent overview, see Arnold 1997.

16. Especially among plants (Stebbins 1950, 1959), and perhaps in birds (Grant and Grant 1998) or other animals (Lewis 1980), hybridization may lead to the evolution of new species.
17. Ehrlich 1961; Mallet 1995; Brookes 1999.
18. See Ehrlich 1961 for more on "pretty well."
19. Ehrlich 1961; Sokal and Crovello 1970.
20. Reproductive isolation can evolve between two newly sympatric populations through selection against the tendency to hybridize (Dobzhansky 1951). In a classic set of experiments, Karl Koopman (1950) examined the tug-of-war between selection and swamping out of differences through interbreeding. Koopman experimented with artificial populations, mixing two similar species of fruit fly (*Drosophila pseudoobscura* and *D. persimilis*). He held them at low temperatures (16° C), at which sexual isolation is low and hybrids are more readily formed. He killed hybrids (identifiable by genetic markers) and in several generations found a marked decrease in the frequency of hybrids formed—selection rapidly produced a barrier that was shown to be at least partially sexual, individuals preferring mates of their "own" kind. For a similar result in nature, see Vamosi and Schluter 1999.
21. Van Gelder 1977. The problem of allopatry leads to endless debate and name changing as additional evidence is added to the taxonomic guesswork. For example, when sufficient intermediate (hybrid) individuals were discovered between the eastern Baltimore oriole and the western Bullock's oriole, they were lumped together into a single species, the northern oriole. The brilliant system of taxonomic nomenclature developed by the Swedish botanist Carolus Linnaeus, who wrote the foundational volume for zoological nomenclature (1758; for details, see Mayr et al. 1953), has served biologists well for almost 250 years but has not proven perfect (no taxonomic system could be). So we're stuck with trying to jam a continuum into boxes on the basis of morphological, behavioral, and genetic data to support our ideas about where best to cut the continuum of differentiation. Among other things, this has led to gigantic wastes of paper in trying to solve a "species problem" that had been recognized as insoluble long ago (Ehrlich 1961; Sokal and Crovello 1970). Part of the difficulty is the urge to view species as the basic evolutionary unit in sexually reproducing organisms. Usually, they are not. As expressed in Ehrlich and Raven 1969: "Modern evolutionary theory requires local interbreeding populations, far smaller groups than those normally called species, as evolutionary units in sexual organisms. It recognizes that such units will vary greatly in their genetic properties and may have a vast diversity of relationships with other such units. The evolution of larger . . . clusters—the species, genera, orders, and so forth, of taxonomists—is easily derived from the theory" (p. 1231). The whole situation has been made more complex by confusion inserted from the field of cladistics, the construction of classifications based on the length of time since the most recent common ancestor, which unfortunately tries to exclude from classification the crucial evolutionary dimension of morphological and genetic similarity. For details of the ongoing discussion, see, e.g., Ghiselin 1975; Mischler and Donoghue 1982; Cracraft 1983; Patterson 1985; De Queiroz and Donoghue 1988; Mallet 1995; Davis 1996; Harrison 1998; Sterelny and Griffiths 1999. For some insight into both the craze for creating largely useless hypothetical phylogenetic trees and a failure to understand what kinds of taxonomic information *need* to be transmitted to the general scientific community and the public, see Milius 1999; Cantino et al. 1999, and Pleijel 1999.

    The "species problem" in fossil organisms seems even more clearly self-inflicted (see, e.g., Foley 1991a; Kimbel 1991; Rightmire 1992). It's tough enough to decide whether organisms alive today are "potentially interbreeding"! Before the end of the nineteenth century, the Dutch physician-anatomist-turned-anthropologist Eugène Dubois's search for a missing link in human evolution was rewarded with the discovery of fossils of *Pithecanthropus* (now *Homo*) *erectus* along the Solo River near Trinil in central Java. "Java man" had a low, flat cranium with a brain capacity much smaller than that of *Homo sapiens* (the only human fossils known in Darwin's time were all Neanderthal people) yet stood fully erect. Subsequently, many fossils similar to Java man were discovered, one known as *Homo ergaster.* But how could scientists decide whether individuals of *Homo ergaster* (as represented by fossil skull KNM-ER 3733 from Kenya, age about 1.75 million years) could "potentially" have interbred with individuals of *Homo erectus* (as represented by specimen Trinil 2 from Java, the skullcap of Java man, age about 500,000 years)? The question is obviously absurd. With fossil material, species taxonomy (like the taxonomy of higher categories such as genera and families) becomes largely a matter of convenience: judging degrees of morphological unity and differentiation (Tattersall 1992) and attempting to communicate degrees of difference accurately. Often, there are serious problems in judging whether fossil parts found in the same general area and stratum belong to different populations or are just natural variants (e.g., different sexes) from the same population.

    When I discuss the *H. ergaster–H. erectus* case later in this book, I follow my colleague Richard Klein in recognizing *Homo ergaster* as a precursor of *H. erectus* because I think it convenient when discussing our history. Other distinguished paleoanthropologists prefer to refer to the same material as early (or African) *H. erectus*. All anthropologists should relax more about these taxonomic issues—there are so few examples of so few species and higher categories involved in their discipline that even if the viewpoint of extreme splitters is accepted, communication is not impeded and debate is largely a waste of time. To learn more about evolutionary, phenetic, and cladistic taxonomy, a good place to start would be the classic paper by my two Kansas mentors, Charles Michener and Robert Sokal (1957), which created the first revolution in taxonomy since Linnaeus and also laid the groundwork for modern cladistics. They could then avoid embarrassing statements such as "The logic behind cladistic analysis is simple: since evolution is the process of *change* over time, measures of overall similarity and distance between taxa (phenetics) can be deceptive because not all characters yield real information on evolutionary relationships" (Lieberman 1995, p. 160; but see Trinkaus 1995). Phenetic and cladistic information are both evolutionary, the first based on evolutionary rate × time and the second on time since divergence alone. Which relationships a scientist should investigate depends entirely on the question asked and the purpose to which the answer is to be put (for a brief discussion, see Ehrlich 1997, pp. 25–31). Linguist George Lakoff (1987) summarized the situation well: "Of all the current biological theories, only cladism might be interpreted as consistent with objectivist metaphysics on the issue of categorization—and then only by ignoring vital aspects of evolutionary biology" (p. 194).

Another vexing question of some interest to social scientists has been whether or not to recognize taxonomic entities below the level of species. The concept of subspecies as taxonomically recognizable populations below the species level was shown to be biologically virtually worthless almost fifty years ago by Edward O. Wilson and William L. Brown (1953), yet it soldiers on nonetheless. There is, of course, considerable interest in studying patterns of genetic variation within species (Avise 1994). But there is only one reason to name and describe new subspecies today. In some circumstances, populations in danger of extinction can be protected legally if they are designated subspecies. Because conservation of populations is critically important (Hughes et al. 1997), the subspecies still has utility on the policy front, but it should disappear from the evolutionary literature. See also the discussion in Howell 1996.

22. Brace 1964.
23. Ehrlich 1964.
24. Wilson and Brown 1953; Gillham 1965.
25. E. O. Wilson 1975. He is also the world's greatest expert on ants—his book written with Bert Hölldobler (Hölldobler and Wilson 1990) is one of the best books ever written on a segment of biodiversity. Wilson has been an eloquent spokesman for the preservation of biodiversity.
26. Wilson and Brown 1953. The late William L. Brown of Cornell University was, like his colleague, a distinguished student of ants—and a wonderful character.
27. George 1962; Coon 1962; Herrnstein and Murray 1994. Anthropologists Nina Jablonski and George Chaplin summarized a key point that makes the designation of such groups nonsensical—that the prime "racial" characteristic of skin pigmentation can respond quite rapidly to selection pressures as human populations migrate. "Because of its high degree of responsiveness to environmental conditions, skin pigmentation is of no value in assessing the phylogenetic relationships between human groups" (Jablonski and Chaplin 2000).
28. Owens and King 1999, p. 453.
29. Technical details about how such an estimate can be made can be found in Caccone et al. 1988. See also Miyamoto et al. 1987 and Goodman et al. 1990 and the references therein.
30. The variation is largely in the form of numbers of loci that are polymorphic (have different alleles) and gene frequencies at those loci or variation as measured by analogous patterns at the level of base pairs in the DNA code.
31. Lewontin 1982a, p. 123. That means that many loci are polymorphic with the same alleles within most populations; there are relatively few cases in which different populations are monomorphic for different alleles.
32. See, e.g., Dower 1986.
33. See, e.g., Cowles 1959; Roberts and Kahlon 1976. For a summary, see Ehrlich and Feldman 1977.
34. Robins 1991.
35. Folic acid is involved in the maturation of red blood cells and the manufacture of nucleotides, among other things. Recently, anthropologist Nina Jablonski made a case for dark skin being favored by selection because of melanin's protective effect on folate. Folate deficiency has been shown to be associated with serious birth defects in which the spine does not develop properly—including spina bifida, in which

the bony sheath around the spinal cord doesn't close (Jablonski 1992). If further research supports Jablonski's hypothesis, we'll understand another part of the selective basis of the relationship between skin color and solar radiation.

36. See, e.g., Branda and Eaton 1978; Harris 1997.
37. See, e.g., Diamond 1991, p. 100.
38. Robins 1991.
39. Loomis 1967; Neer 1975.
40. Beadle 1977. Although it is true that a light-skinned European farmer who spends a significant amount of time outdoors can synthesize enough vitamin D from a small area of exposed skin, this is not true of a large portion of light-skinned people living at the mid- to high latitudes of Europe (Thomas et al. 1998), and it is certainly not true of dark-skinned individuals at those latitudes (Henderson et al. 1987; Gullu et al. 1998; Douglas and Bakhshi 1998). The problems of analysis of vitamin D deficiency are exacerbated by relatively recent migrations and by practices such as purdah (the Hindu practice of secluding and veiling women), increasingly indoor lifestyles, and diets deficient in vitamin D.
41. Harris 1997, p. 82.
42. There is an extensive literature on the kinds of barriers to reproduction that are created by this independent evolution—it traces primarily to the greatest evolutionary geneticist of the twentieth century, Theodosius Dobzhansky (see, e.g., Dobzhansky 1937, pp. 231–232). Dobzhansky, Ernst Mayr, George Gaylord Simpson (paleontology), and G. Ledyard Stebbins (plant evolution) were the main architects of the "new synthesis" of evolution and dominated the field through the middle of the century.
43. Futuyma 1998; Ridley 1996a; Freeman and Herron 1998; Peterson et al. 1999; Wuethrich 1999.
44. It is also still unclear how important lack of genetic contact is in comparison with differences in selection pressures on isolated populations; see, e.g., Ehrlich and Raven 1969. For a recent comprehensive look at issues related to sympatric speciation, see Rice and Hostert 1993 and Howard and Berlocher 1998. Debate is also lively about the relative importance of major physical events (e.g., rapid climate change) as engines driving speciation and concentrating differentiation into relatively short periods of geological time (Eldredge and Gould 1972; Vrba 1995b).
45. Good examples of sympatric speciation are likely to be found where there is adaptation by insects to several different host plants (see, e.g., Bush 1969). It is certainly theoretically possible (Dieckmann and Doebeli 1999; Higashi et al. 1999); and there is experimental evidence that it can work (Rice 1988, 1990). The controversy tends to be generated by issues of how frequently conditions would permit differentiation in the presence of opportunity for substantial gene flow. But overall, there seems to be growing evidence that sympatric speciation may be important in some situations (Tregenza and Butlin 1999; Kondrashov and Kondrashov 1999); see also the material on lake fishes that follows.
46. Lenski and Travisano 1994.
47. The most important other speciation mechanisms are hybridization and polyploidy (having more than two sets of homologous chromosomes) (Futuyma 1998, pp. 504–510). Although much more common in plants, these mechanisms are also present in animals (Lewis 1980; Gallardo et al. 1999).
48. Galis and Metz 1998; for a popular overview with illustrations, see Stiassny and Meyer 1999.
49. These species flocks also are found, to a lesser extent, in other lakes in East and West Africa and even in Central America. I am particularly grateful to Les Kaufman for help with this section.
50. That genus was *Haplochromis*. They have now been divided into more genera and are generally described as the haplochromine cichlids, those now in the subfamily Haplochrominae of the family Cichlidae. The taxonomy of the group is very complex (Seehausen 1996).
51. Scale-snatching fishes of the genus *Perissodus* in Lake Tanganyika have asymmetrical mouths and apparently specialize in approaching their prey from one side or the other. Those, for instance, with mouths angled to the right attack the left side of the prey. In at least one species, *P. eccentricus*, there are individuals with mouths that face right and others with mouths that face left. The frequency of the two morphs remains near 50–50, apparently as a result of frequency-dependent selection. As one type becomes rarer, it also becomes fitter because the prey species are more alert to approaches from the side attacked by the more common morph (Hori 1993).
52. Zooplankton and phytoplankton.
53. Pharyngeal jaws.
54. Fryer and Iles 1972.
55. Meyer et al. 1990.
56. Greenwood 1974.
57. Kaufman et al. 1997.

58. Seehausen 1996.
59. Johnson et al. 1996.
60. See also the review in Orr and Smith 1998.
61. Owen et al. 1990.
62. See the discussion in Seehausen 1996 and Seehausen et al. 1999. The cichlid story is further complicated because several East African lakes are connected by rivers that ebb and flow with regional climate changes, first joining and then separating the lakes and their faunas. The number of possible histories is bewildering, and what is even less clear is how such high diversity within a single lake is maintained—that is, why routine extinction rates aren't much higher. Many of the Lake Victoria cichlids have evolved so recently and differentiated so little that hybrids between them are fully fertile (technically, there are no postmating barriers to successful hybridization; failure to hybridize is behavioral), though in nature they rarely hybridize. Older lakes, Lakes Malawi and Tanganyika, show faunas that are even more highly differentiated and that also have more complex histories (Dorit 1990). Fluctuations in the levels of those lakes alternatively submerged and exposed, separated and rejoined vast areas of shallow-water habitat, sometimes causing Lake Tanganyika to break up into three separate lakes.

    The brightly colored cichlid species apparently were kept separate by mate choice after sympatry was reestablished (if differentiation occurred allopatrically). Sexual selection appears to be strong enough to overcome the effects of natural selection against bright colors generated by predators, many of which hunt by sight in clear water, where the male's color signals are most conspicuous.
63. Bentzen and McPhail 1984; Bentzen et al. 1984; Taylor and Bentzen 1993; Rundle et al. 2000.
64. Greenwood 1974; Kaufman et al. 1997.
65. Water clarity, which influences the ability of females to select colorful males, and habitat diversity vary naturally as lake basins fill and empty during shifts between wet and dry climates. Africa—an especially high-and-dry continent—has been very volatile in this regard. Except for the deepest lakes, water is an ephemeral entity across most of the African landscape. This very volatility may have helped to fuel the evolutionary bursts of cichlids (Kaufman 1997), which must also have been interrupted by spectacular, rapid-fire extinctions as lakes oscillated between clear and murky, high and low, present and absent.
66. Disruptive selection is sometimes called diversifying selection.
67. Technically, it is a pharyngeal jaw apparatus. Liem 1974; Kaufman and Liem 1982; Liem and Kaufman 1984.
68. Dominey 1984; Galis and Metz 1998. The degree to which these pressures can produce sympatric speciation appears bound to remain controversial, but Galis and Metz seem convinced that sympatric speciation has been involved in production of the Lake Victoria species flocks.
69. The classic treatment of adaptive radiation is Simpson 1953, chap. 7; see also his earlier related discussion (Simpson 1944, chap. 7). For a fine discussion of a spectacular adaptive radiation in the distant past, see Paul Sereno's analysis of the radiation of the dinosaurs (1999).
70. Technically, the lake is becoming eutrophic. Eutrophic lakes are less clear than those that have fewer mineral nutrients (that are oligotrophic) because there are more algae in the water.
71. Seehausen, van Alphen, et al. 1997.
72. Kaufman 1991.
73. Seehausen 1999.
74. Seehausen, van Alphen, et al. 1997.
75. Kaufman 1991; Witte et al. 1992.
76. Cohen et al. 1996.
77. Kaufman and Ochumba 1993; Seehausen, Witte, et al. 1997.
78. Kaufman 1991.
79. See, e.g., Farrell 1998.
80. Lack 1947; Grant 1986.
81. Raikow 1977. Raikow concluded that the common ancestor of the extraordinary radiation in Hawaii was a finch with a standard finch-type bill.
82. See, e.g., Marshall and Ward 1996; Sharpton et al. 1996. Such collisions may have been involved in some or all of the other four great extinction events observable in the fossil record (Raup 1991). It has recently been claimed (Hedges et al. 1996) that continental breakup was largely responsible for the diversification of the mammalian orders, but even if this is correct, the demise of the dinosaurs doubtless would have opened many niches for groups at lower taxonomic levels.
83. Mayr 1982, pp. 433–434.
84. Darwin 1859, p. 317.

85. Darwin 1859, p. 320.
86. See, e.g., Ehrlich and Ehrlich 1981, p. 7; Futuyma 1998, p. 17.
87. For an interesting overview of the questions that evolutionists will be asking of the fossil record in the next wave of research, see Jablonski 1999.
88. The issue of how historical contingency constrains the action of natural selection was brought to the fore in a classic paper by Stephen J. Gould and Richard Lewontin (1979) that was heavily critical of the view that all features of organisms are adaptations optimized by natural selection (the Panglossian paradigm). A defense of the paradigm, but not of its naive use, can be found in Dennett 1983. For a more recent review, see Rose and Lauder 1996. A fine discussion of adaptationism from the viewpoint of a philosopher of biology can be found in Godfrey-Smith 1999 and, in greater detail, in Godfrey-Smith 2000, dissecting adaptationism from three perspectives.
89. Berkow 1992, p. 1362.
90. Eldredge and Gould 1972.
91. Eldredge and Gould 1972; Eldredge 1985; Eldredge 1999. The latter is an especially thoughtful discussion with which I often differ—I recommend it. For critiques, see Charlesworth et al. 1982 and Maynard Smith 1983.
92. The samples are biased for a number of reasons, not the least of which is that organisms that live in different habitats have different chances of becoming fossils. That's one cause of the relative completeness of the human fossil record *after* our ancestors moved into savannas, as opposed to the poor record of older members of our lineage who lived in trees. Forests are a poor place to live if your goal is to become a fossil.
93. Lenski and Travisano 1994; Elena et al. 1996.
94. An artifactual appearance of punctuation could be a product, technically, of a Markov process with absorbing barriers operating, especially when population sizes are relatively small; see Bergman and Feldman 1995.
95. See, e.g., Donald 1993; Bickerton 1995; but see Wills 1993.
96. If Eldredge and Gould are right that the punctuations in the punctuated equilibrium model are periods in which genetic change is concentrated in speciation events, then we must be in a period of punctuation for most groups because, as I indicated, there is abundant evidence of ongoing allopatric speciation today. Right or wrong, they have made an important contribution in raising an issue that has generated much research and discussion.
97. Dawkins 1997; Whitman et al. 1998.
98. Gould 1996.
99. Dawkins 1997.
100. Mayr 1997, p. 198. For a recent example of the lack of directional trends, see D. Jablonski 1997.
101. See, e.g., Ehrlich and Raven 1964; Wallace and Mansell 1976; Crawley 1983.
102. For a review of evolutionary theories of addictive behavior, see E. O. Smith 1999.
103. Judith Becerra (1997), for example, compared phylogenies (reconstructed from molecular data) of the chrysomelid (leaf) beetle genus *Blepharida* and plants of the genus *Bursera* (family Burseraceae) and profiles of the main defensive chemicals (terpenoids) of the Burseraceae. Terpenoids are often toxic to insect herbivores, and *Bursera* terpenoids reduce *Blepharida* growth rates and survival. Becerra used a chemogram (showing chemical similarity) of *Bursera* to compare with *Bursera* and *Blepharida* phylogenies. In *Bursera* and *Blepharida*, host shifts phylogenetically appear to correspond to patterns of chemical similarity.
104. Ehrlich 1970.
105. See, e.g., Ehrlich 1970; Ehrlich and Ehrlich 1973; Gilbert and Raven 1975; Roughgarden 1975, 1976; Anderson and May 1982; Nitecki 1983; Futuyma and Slatkin 1983; Roughgarden et al. 1983; Stone and Hawksworth 1986; Thompson 1989, 1993, 1999; Endler 1995; Ebert and Hamilton 1995.
106. See, e.g., Farrell et al. 1993; Farrell 1998.
107. Bohannan and Lenski 1997.
108. See, e.g., Deacon 1997; Durham 1991.
109. Schneider and Londer 1984. Climate evolves because of changes in Earth's orbit, in the concentration of greenhouse gases, and in the reflectivity of Earth's surface as plant communities evolve. It, of course, does not evolve by natural selection.
110. Schneider and Londer 1984.
111. Ehrlich and Holm 1963, pp. 285ff. Anthropologists and social scientists often formulate more complex definitions of culture (e.g., Holloway Jr. 1969; Waal 1999). They started long ago. The pioneering

anthropologist Edward B. Tylor defined it thus: "Culture or Civilization, taken in its wide ethnographic sense, is that complex whole which includes knowledge, belief, art, morals, law, custom, and any other capabilities and habits acquired by man as a member of society" (Tylor 1920 [1871], p. 1). Defining culture as humanity's store of nongenetic information is more general, and a lot shorter.

112. Ehrlich and Holm 1963; Cavalli-Sforza and Feldman 1981; Durham 1991.
113. Broad attempts to look at units of cultural evolution and to analogize cultural evolution with genetic evolution, including mine of long ago (Ehrlich and Holm 1963, pp. 285ff.), have been less than successful—see the discussion in Dawkins 1989 (1976), pp. 192ff., of "memes" as replicating cultural units of transmission. Dawkins and others (most recently and extremely Blackmore 1999) have produced a highly speculative literature on cultural evolution based on these postulated replicators—ideas, tunes, fashions, jokes, and so on that reproduce themselves by moving from mind to mind. The term *meme* is not very helpful analytically, but Dennett (1991, pp. 199ff.) does use it to illustrate some interesting points. Less ambitious, quantitative approaches to cultural evolution (e.g., Cavalli-Sforza and Feldman 1981) have been more profitable scientifically.
114. An interesting (if overemphatic) related discussion can be found in Harris 1998. See also Gardner 1998.
115. See, e.g., Richerson and Boyd 1989, p. 203.
116. Rogers 1995.
117. Platt 1973, p. 641; Costanza 1987.
118. See, e.g., Ehrlich and Ehrlich 1996, pp. 94ff. and chap. 8.
119. For an overview of mathematical modeling in this area, see Laland 1993.
120. See, e.g., Cavalli-Sforza and Feldman 1973; Feldman and Cavalli-Sforza 1976; Cavalli-Sforza and Feldman 1981; Lumsden and Wilson 1981; Durham 1991; Laland et al. 1995; Feldman and Laland 1996. For an interesting anthropological-psychological view that is somewhat at deviance with the foregoing, see Sperber 1985.
121. See, e.g., Shatin 1968; Feldman and Cavalli-Sforza 1989.
122. For an overview of the evolution of adult lactose absorption, which shows the complexity of issues in gene–culture coevolution, see Durham 1991, pp. 226–285.
123. The cells of genetic males have one X and one Y chromosome. The X and Y chromosomes carry the genes that determine the sex of the individual. Genetic females have two X chromosomes. Only one of each pair of chromosomes goes into each gamete (egg or sperm). Because the eggs of females can receive only an X chromosome, whereas sperm can have either an X or a Y, the father is the primary determiner of the sex of his children (I say primary because, for instance, a factor such as the acidity of the vagina may differentially select for or against Y-bearing sperm).
124. Dickeman 1975, 1979; Scrimshaw 1984.
125. Mathematical models indicate that the direction of that selection pressure depends on the cultural behavior of parents and can lead to either male or female bias in the ratio at birth—see, e.g., Nordborg 1992; Kumm et al. 1994; Laland et al. 1995; Kumm and Feldman 1997; Aoki and Feldman 1999. The situation is complex because the models indicate that changes in the frequency of alleles (different forms of the same gene) controlling the primary sex ratio will depend on the proportion of biased couples in the population, whether they compensate for the loss of female children, the loss of fitness due to the bias, and whether changes in the primary sex ratio alter the behavior of couples in the population (Aoki and Feldman 1999). In China, for example, where controls on reproduction make the fitness costs relatively small (final family size, a good measure of fitness, being much the same regardless of whether couples choose to reduce the number of daughters), one would expect the primary sex ratio to become more female-biased (to compensate for the reduction in the proportion of females in the reproductive population). On the other hand, evidence from burial sites and extrapolation from the behavior of modern hunter-gatherer groups indicate that the rate of prehistoric infanticide may have been high. Assuming that the victims were mostly daughters and that the practice involved substantial fitness loss, the primary sex ratio among those hunter-gatherers would have been male-biased (because parents who tended to produce males would have been more likely to pass on their genes) (Aoki and Feldman 1999).
126. This point is made in some detail by anthropologist Marvin Harris (1980, pp. 123–125).
127. For examples of maladaptive behaviors in tribal societies, see Edgerton 1992.
128. Cavalli-Sforza and Feldman 1981; Boyd and Richerson 1985.
129. See Cavalli-Sforza and Feldman 1981.
130. Ehrlich and Holm (1963, p. 287); Richerson and Boyd 1984; Aoki and Feldman 1999.
131. See, e.g., Cavalli-Sforza 1986.
132. Binford 1963; Cavalli-Sforza and Feldman 1981.

133. Ehrlich and Holm 1963, p. 291.
134. Diamond 1978.
135. I include the weasel word *largely* because I would approve of intervention to correct genetic defects (such as phenylketonuria, caused by a rare recessive gene) or to abort fetuses with serious genetic defects that cannot be cured—and such actions would, in a sense, modify behavior.
136. One should note that knowledge of cosmology and some other areas of physical science is at least as incomplete.
137. The continuing discussion of the importance of random changes in the proportions of different kinds of genes in populations and of how much genetic variation is normally under the control of selection—a discussion that was once framed as the neutrality controversy (Ayala 1974; Futuyma 1998; King and Jukes 1969; Martin and Palumbi 1993; Price 1996)—illustrates this. So do debates over such topics as the importance of sympatric speciation, the meaning of punctuated equilibrium, and so on. Our knowledge is still incomplete.
138. A good example is the expanding work on the major histocompatibility complex (MHC), an extremely variable cluster of genes that codes for cell surface proteins that was first discovered in connection with the rejection of skin grafts (in human beings, the MHC is known as the HLA—human leukocyte antigen—complex). The complex not only is involved in providing signals within an animal's body that determine what is self and what is nonself (especially pathogenic nonself such as viruses and bacteria) but also may be involved in such things as mate selection in mice and human beings (Ober et al. 1997). Because of its medical importance, the evolution of the MHC is under intense scrutiny (see, e.g., Edwards and Hedrick 1998).

Chapter 4: Standing Up for Ourselves

1. Harris 1989, p. 2.
2. I've told this story to classes for years, and I thought it came from William Irvine's classic book *Apes, Angels, and Victorians* (1955), but in a quick reperusal of the book I was unable to find it. Even so, that book is a great read. It is possible that Irvine recounted the story to me when I first arrived at Stanford University in 1959 and started teaching a course in evolution. A version of it, attributing the statement to the wife of the canon of Worcester Cathedral, can be found in Ashley Montagu's *The Concept of Race in the Human Species in the Light of Genetics* (1964, p. 2). Darwin's basic idea of evolution by natural selection was quite well accepted by the educated public of Victorian England when *Origin* was published. *The Descent of Man and Selection in Relation to Sex* (Darwin 1871) and *The Expression of Emotions in Man and Animals* (Darwin 1998 [1872]) were much more controversial.
3. Huxley 1863.
4. The skulls were purchased from a company called Skullduggery Inc., at 624 South B Street, Tustin, CA 92780; (800) 336-7745; Internet: http://skullduggery.com. In 1999, that set cost a little more than $1,000, about the price of a single computer. One set should be available in every American high school. (I have no connection whatsoever with the company.)
5. Family Hominidae (limited in this book to the genera *Ardipithecus*, *Australopithecus*, *Paranthropus*, and *Homo*).
6. Most people pronounce the name *bah*-no-bo, but some have claimed that it should be bah-*no*-bo. Ethologist Jan van Hooff told me (27 September 1998) that the latter traced from his student Frans de Waal jokingly claiming that the second pronunciation was correct when he was presenting a scientific paper at the San Diego Zoo. One story of the name's origin, van Hooff told me, is that it derives from a misspelling of the name *Bolobo*, which was on a crate containing a bonobo when it arrived at a German zoo. Bolobo is a town in the Democratic Republic of Congo, presumably the source of wood for the crate (see also Waal 1997, p. 7b).
7. Tsutsumi et al. 1989.
8. *Pan troglodytes* and *Pan paniscus*, respectively.
9. *Gorilla gorilla.*
10. Horai et al. 1995.
11. Many more generations (7 million years' worth) separate chimps and gorillas from their most recent common ancestor than separate us from chimps (only 5 million years' worth).
12. Linnaeus 1758.
13. That is, the bonobos are probably closer to us phenetically (in overall similarity), including behaviorally

(see, e.g., Takeshita and Walraven 1996). Chimps and bonobos are their own closest relatives phenetically and cladistically (in time of most recent common ancestry) and are equally related to us cladistically.

14. Kano 1992; Wrangham et al. 1996.
15. Zihlman 1996.
16. Matsuzawa 1985a, 1985b, 1990.
17. Gannon et al. 1998.
18. In addition to chimps and people, tool use is found in (at least) insects, birds, and mongooses.
19. See, e.g., Sugiyama 1994.
20. Sugiyama 1997.
21. Wrangham and Peterson 1996.
22. Limongelli et al. 1995; Povinelli et al. 1990; Povinelli, Nelson, et al. 1992. Evidence suggests that monkeys are not capable of such reasoning; see, e.g., Povinelli et al. 1991; Povinelli, Parks, et al. 1992; Visalberghi 1993; Visalberghi and Limongelli 1994.
23. On reconciliation, see Waal and van Roosmalen 1979 and Kappeler and van Schaik 1992. Gorillas are also adept at forming alliances to protect relatives from harm, gain access to otherwise unattainable resources, and the like (Harcourt and Stewart 1989).
24. Boesch 1996. For overviews, see Heyes 1993 and Gibson and Ingold 1993; see also the discussion of apes' tool use in McGrew 1989.
25. For an overview, see Whiten et al. 1999.
26. Matsuzawa 1996.
27. Boesch 1991b.
28. McGrew 1992. For a summary, see Whiten et al. 1999.
29. Sugiyama and Koman 1979.
30. Matsuzawa 1991, 1996.
31. See, e.g., Toth 1985.
32. McGrew and Marchant 1996. See also MacNeilage et al. 1987.
33. Lateralization, the preference for using one paw or the other, and other anatomical and behavioral asymmetries are widespread in animals (Bradshaw 1991), and some species seem to show a tendency to prefer the right side (Ornstein 1997, pp. 18–24).
34. Chimps do not, however, show strong right- or left-hand preferences in a population as a whole (McGrew and Marchant 1996).
35. Daly and Wilson 1984; Tudge 1997.
36. Work on the significance of chimp–human genetic differences traces to geneticist Mary-Claire King and biochemist Allan C. Wilson (1975), who concluded that "the average human polypeptide is more than 99 percent identical to its chimpanzee counterpart" and "a relatively small number of genetic changes in systems controlling the expression of genes may account for the major organismal differences between humans and chimpanzees." King and Wilson may have been correct that most of the relevant differences are located in regulatory genes. No one knows, in part because the significance of the claimed 2 percent difference depends on the distribution of the 2 percent of base pairs that are different. That is, in theory, enough to alter every single gene in the genome. Even single changes in the amino acid residue sequence in a protein, caused by a single base pair substitution, can dramatically change that protein's functioning. A classic example is the single substitution that changes one amino acid residue in hemoglobin and, when the affected allele is homozygous, causes sickle-cell anemia. That relatively few genetic changes form the basis of chimp–human differences is even more remarkable because only about 4 percent of the nucleotide sequence actually codes for proteins (Hartl and Jones 1998). Many of the differences between the chimp and human genomes thus are probably meaningless differences in the "junk DNA," most of which has no known function, although functions for some parts of it, such as DNA "microsatellites," are being uncovered (Moxon and Wills 1999). Meaningful genetic differences between human beings and the other great apes, and how so few changes in the DNA can be responsible for the differences in kind between us and our nearest relatives, are now being investigated intensively (Gibbons 1998e; Muchmore et al. 1998). How genetic differences are translated into phenotypic differences remains perhaps the most critical piece of the evolutionary puzzle (see, e.g., Schlichting and Massimo 1998).
37. Short 1994; Waal 1995; Waal 1997, p. 185.
38. Tomonaga et al. 1993.
39. Other organisms may limit their reproduction physiologically when environmental conditions deteriorate, but this is a physiological rather than a conscious behavioral response.
40. Ehrlich et al. 1995.

41. An unfortunate side effect of our large brains has been human decimation of chimpanzee populations, threatening them with extinction in the wild. In addition, we often treat our nearest relatives brutally as food, entertainment, and experimental animals. For a sad photographic record of their treatment, see Nichols and Goodall 1999; if you want to help, contact the International Primate Protection League, P.O. Box 766, Summerville, SC 29484; (803) 871-2280.
42. Chimps, 350–400 cubic centimeters (cc); modern people, a spread of about 1,200–1,700 cc.
43. See Brown 1991 for an excellent summary. One of the most famous studies of cross-cultural universals was carried out by psychologist Paul Ekman and his colleagues on the expression of emotions by facial expressions, based originally on similarity of predictions about emotions by people from New Guinea, Borneo, Brazil, Japan, and the United States after examining photographs of faces (Ekman and Friesen 1978; Ekman and Friesen 1986; Ekman et al. 1969). For an interesting discussion of the problems of doing such work and its possible interpretations, see Fridlund 1991. This work builds on a classic book by Charles Darwin, *The Expression of Emotions in Man and Animals*, which has just been reissued in a fine edition with excellent and interesting contributions—introduction, commentaries, and afterword—by Ekman (Darwin 1998 [1872]). For an early view on universals that make it possible for people to live together in culturally organized social groups, see Wallace 1961.
44. Those objects and actions can be either existing in the real world or imagined. There are also many universals and near universals in the details of language—for example, there are great regularities in the way folk taxonomies of plants and animals develop from one culture to the next (Berlin 1992; Berlin et al. 1973; C. H. Brown 1977, 1979).
45. Mitani 1996. Nonetheless, there are similarities between the acoustic patterns shown in vocal communication among nonhuman primates and those found in human speech (Maurus et al. 1988).
46. Waal 1996b.
47. Tooby and Devore 1987.
48. Boyd and Richerson 1996.
49. My wife, Anne Ehrlich, worked with me on the butterfly research; we first reported on our Gombe experiences in Ehrlich and Ehrlich 1981, pp. 3–5.
50. In looking at our evolutionary history, I assume at least one universe, life, and mammals. For a tour of the history of all life in a single excellent volume, I recommend Richard Fortey's *Life: A Natural History of the First Four Billion Years of Life on Earth* (1997). A more technical treatment, full of interesting ideas about the ecology of ancient organisms, is Vermeij 1987. A recent semipopular account of the problems of early life can be found in Kerr 1999. For more detail about broad evolutionary patterns, especially the problems of determining the base structure of the "tree of life," see Doolittle 1999, 2000 and Knoll and Carroll 1999.
51. Darwin 1859.
52. If the extreme version of solipsism (the notion that only the self and the contents of one's own consciousness exist—or can be known) were taken as correct, one could not do science.
53. A scientific theory can be defined as a coherent framework for an entire field, one ordinarily so well supported by empirical data that scientists act as if it were "proven." Once established, theories are treated as "true" (I use quotation marks because truth itself is not a scientific concept; all scientific ideas are subject to revision) as long as they are useful—generally until and unless some scientist demonstrates them to be wrong. Any position that is based on authority and is not subject to revision cannot be accurately described as scientific.
54. Darwin 1859, especially chaps. 6, 9. Darwin did not mention the Neanderthal fossils, which had just been discovered—and, as we shall see, they were not really missing links.
55. Seven specimens are now known. In more detail, *Archaeopteryx* had the snout, teeth, and long tail of a reptile and lacked the keeled sternum (breastbone) that anchors the flight muscles of modern birds. Yet it did have another avian characteristic associated with strong flight muscles, a large furcula (wishbone), and delicate extended "hand" bones that seem designed to support a wing but were not fused like the hand bones in modern birds. Above all, the "arms" of *Archaeopteryx* and its long reptilian tail were covered with feathers, their details exquisitely reproduced in the limestone. Feathers are the defining feature of birds, and each flight feather of *Archaeopteryx* was asymmetrical and indistinguishable from those of modern flying birds (flightless birds tend to have feathers that are symmetrical). For more details, see Feduccia and Tordoff 1979 and Feduccia 1996.
56. See, e.g., Sereno and Rao 1992; Feduccia 1996; Hou et al. 1996; Dingus and Rowe 1997; Chatterjee 1997; Burke and Feduccia 1997; Padian and Chiappe 1998; Shipman 1998; Forster et al. 1998; Burgers and Chiappe 1999.

57. For an excellent elementary textbook–level overview of human physical evolution, see Lewin 1998.
58. Shrews are small insect-eating mammals resembling mice with pointed noses. Today's tree shrews (Tupaioidea) are generally shrew- and squirrel-like primates that are active insectivores in tropical forest undergrowth and that may be living lives somewhat like those of our distant ancestors (Klein 1999, p. 87).
59. Ehrlich and Raven 1964; Farrell 1998.
60. Nowak 1991, vol. 2, pp. 490–492; Fleagle 1999, pp. 11ff.; Klein 1999, pp. 66ff.
61. Klein 1999.
62. Fleagle 1994; Ciochon and Etler 1994.
63. Ehrlich and Holm 1963, p. 282. I thought, as did many other evolutionists of the time, that sharp, frontally directed eyes were evolved mainly to provide binocular vision to allow leaping about in the trees (where the typically keen mammalian sense of smell would be a poor guide to the location of the next branch). Considering the success of squirrels, however, which have eyes directed more to the sides but still have binocular vision (John Allman, personal communication, 1 November 1999), which is not necessary for distance judgment (Dobbins et al. 1998), the scenario I present in what follows is now the consensus; see, e.g., Jones et al. 1992. For an overview of primate evolution, see Martin 1990 and also Cartmill 1972.
64. John Allman, personal communication, 1 November 1999.
65. The details of the phylogenetic origins of the anthropoids are still a contentious subject, as, e.g., in Bloch et al. 1997 and Kay et al. 1997. For recent information, see Gebo et al. 2000.
66. Dating of the geological epochs of the Cenozoic era is as follows: Paleocene, 65–56.5 mya; Eocene, 56.5–35.4 mya; Oligocene, 35.4–23.3 mya; Miocene, 23.3–5.2 mya; Pliocene, 5.2–1.64 mya; Pleistocene, 1.64 mya–10,000 years ago; Holocene, or Recent, 10,000 years ago to the present (Jones et al. 1992).
67. Begun et al. 1997; Fleagle 1999.
68. See, e.g., Andrews 1992; Klein 1999, pp. 118ff.
69. Technically, both convergent and parallel evolution are homoplasy—the possession of similar character states not derived from a common ancestor. Convergence occurs when two unrelated evolutionary lines acquire similar characteristics, as have the shapes of whales and fishes. Parallelism is the acquisition of similar characteristics in related lines, presumably by alteration of the same developmental pathway, as in the evolution of "eye spots" in different families of butterflies. For a detailed discussion of both convergence and parallelism in subterranean mammals and other organisms, see Nevo 1999.
70. Köhler and Moyà-Solà 1997b.
71. This paragraph is based largely on personal communications from paleoanthropologist Richard Klein in late 1999.
72. Partridge, Bond, et al. 1995; Partridge, Wood, et al. 1995; Klein 1999.
73. The Ruwenzori range, called the "Mountains of the Moon," whose highest peak is Mt. Stanley (in the Democratic Republic of Congo, called Mt. Ngaliema).
74. Pickford 1993.
75. Heroic efforts are being made to reconstruct Pliocene–Pleistocene climatic events and correlate them with faunal and evolutionary changes in Africa. See, for example, Pickford 1990; Vrba 1995a; White 1995; and other articles in Vrba et al. 1995. The question of whether there was a distinct turnover in the mammalian faunas in East Africa between 2.8 and 2.5 mya or whether the change was primarily a more gradual one between 2.5 and 1.8 mya (Behrensmeyer et al. 1997) remains unsettled.
76. See, e.g., Coppens 1994; Vrba 1995b; Boaz 1997. The exact importance of the move into the savanna and subsequent climatic variability is still under extensive discussion (see, e.g., Potts 1998).
77. Andrews 1992.
78. Coppens 1994.
79. Skybreak 1984, p. 84. On the other hand, there is evidence that among chimps, some of the behaviors that are most hominid-like are most developed in those chimp populations that live in forests rather than in more open habitats (see, e.g., Boesch-Achermann and Boesch 1994). For a general discussion, see Strum 1987; Susman 1987; Wrangham 1987; and Rodseth et al. 1991, as well as the discussion that follows.
80. The state of knowledge in the 1960s indicated that the earliest members of the human line belonged to the genus *Australopithecus.* They were upright, bipedal hominids that were estimated to have lived in Africa beginning some 1.75 million years ago. Although they had small brains (a brain volume of 450–600 cc), they were thought (probably incorrectly) to have used stone tools. The australopithecines were replaced by the larger-brained (775–1,200 cc) *Homo erectus* some 600,000–500,000 years ago, and they in

turn gave rise to *Homo sapiens* (including the variant then known as *H. sapiens neanderthalensis*) 200,000–100,000 years ago, with a brain volume averaging about 1,350 cc. All this is summarized in Ehrlich and Holm 1963. The story has subsequently become more complex, but overall that picture assembled by paleoanthropologists by the 1960s was pretty close to the mark. They had been fortunate enough to make a series of spectacular finds.

81. *Neandertal* in today's German, with *valley (Tal)* spelled *Thal* in the 1850s (Klein 1999).
82. Tattersall 1995; Klein 1999.
83. Bonobos were not recognized at the time.
84. Tattersall 1995, pp. 32ff. At the time, it was assumed that only modern human beings had reached Europe, and therefore Dubois had to look elsewhere.
85. The name had originally been proposed by German biologist Ernst Haeckel to describe a hypothetical human ancestor.
86. Jean-Jacques Hublin, personal communication, 12 May 1999.
87. An australopithecine skull found by Broom in 1936 at a lime-mining site at Sterkfontein (near Pretoria) was originally named *Australopithecus transvaalensis*. Broom later placed it in a new genus, *Plesianthropus*, but it is now generally considered synonymous with *A. africanus*. The famous skull was nicknamed Mrs. Ples, although it is now considered to have been the skull of a male (Conroy et al. 1998). Sadly, Mrs. Ples, the first australopithecine skull discovered since the Taung child, apparently was nearly intact before it was discovered, "but the quarryman's blast blew out the endocranial cast and most of the skull, leaving the imprint of the skull top and part of the cranial roof itself in the rock wall" (Gregory and Hellman 1939, p. 560).
88. Gregory and Hellman (1939) stated that "these South African Pleistocene man-apes were in both a structural and a genetic sense the conservative cousins of the contemporary human branch" (p. 564).
89. Tattersall 1995, p. 60.
90. Klein 1999.
91. Mayr 1944; Klein 1999.
92. In the early 1930s, a mining engineer uncovered twelve skulls in Java's Solo Valley with endocranial capacities of 1,035–1,255 cc, as compared with about 400 cc for a chimpanzee and an average of about 1,350 cc for modern human beings. The Solo people were christened *Homo (Javanthropus) soloensis* (Tattersall 1995, pp. 63–64). These appear to be a late form of *H.* (then *P.*) *erectus* (Klein 1999). A few years later, work in Java by Ralph von Koenigswald turned up more remains of earlier *H. erectus.* (Von Koenigswald paid by the piece for fossil finds, and he eventually discovered that local people were finding almost intact skulls and using rocks to smash them into many more pieces to greatly increase their cash value.)
93. The transfer was proposed in 1950 by Ernst Mayr (Mayr 1950). That move was implied as early as 1939 in a paper by von Koenigswald and Weidenreich (1939) commenting on the close relationship of "Pithecanthropus" and "Sinanthropus," in which they carefully printed those names in roman (normal) type while italicizing *Homo modjokertensis* as one did a formal latinized species name—the implication being that they were unsure about the generic assignment of the material that has come to be known as *H. erectus* (Le Gros Clark 1964; *Sinanthropus* was officially sunk into *Pithecanthropus* in the 1955 edition of this work); see also Tattersall 1995.
94. These discoveries began with one by Broom.
95. Some paleoanthropologists believe that *H. erectus* survived in the Far East until it was replaced by *H. sapiens,* advancing from Africa about 50,000 years ago (Klein 1999).
96. Current views given on the human fossil record lean heavily on the work of Richard Klein; for details, see Klein 1999.
97. There are about 6.5 cc to a cubic inch, so those brains would have been about 180 cubic inches in volume—the equivalent of a cube a little more than five and a half inches on a side.
98. The skull measured was the famous Mrs. Ples (Conroy et al. 1998).
99. Ruff et al. 1997.
100. Brain volume is usually considered synonymous with endocranial capacity, which is the quantity normally measured. Recent work suggests that the brain volumes of all early hominids should be reexamined with a new, more accurate technique than was available for the original measurements (Falk 1998).
101. See, e.g., Geertz 1966; Boyer 1994.
102. For an overview, see Richard Klein's superb and balanced treatment (1999).
103. The latinized name derives from the Afar Triangle, the part of Ethiopia containing Hadar and marking the juncture of three geological rifting systems, where the fossil was discovered. The triangle is home to

the Afar people, a nomadic Islamic people who take great pride in Lucy and now believe that their tribe was ancestral to all of humanity (Johanson and Edgar 1996, p. 124). A recent cladistic analysis has led to the conclusion that *A. afarensis* should be confusingly renamed *Praeanthropus africanus* (Strait et al. 1997). One can only hope that paleoanthropologists will be slow to make this change, which seems to trace to the usual misunderstanding of evolution by cladistic taxonomists.

104. Tuttle et al. 1991.
105. Recent *A. afarensis* fossils from the Afar Triangle support these conclusions and reinforce the idea that Lucy represented a single, rather widespread bipedal species with substantial sexual dimorphism. The question of whether or not *A. afarensis* was partially adapted to arboreal locomotion is not settled (Kimbel et al. 1994; White et al. 1993).
106. The two recently discovered species are *Australopithecus anamensis* (Leakey et al. 1995; Leakey et al. 1998) and *Ardipithecus ramidus* (White et al. 1994).
107. *A. anamensis* was upright and probably lived mostly in relatively open habitats with nearby forests (Leakey et al. 1995); this is possibly in contrast with *A. ramidus*, which may have lived in the forests themselves (WoldeGabriel et al. 1994). *A. anamensis* had chimp-like jaws, teeth, hands (possibly), and (possibly) a gorilla-like substantial sexual dimorphism (male–female difference in size) (Leakey et al. 1998). We and the apes are sexually dimorphic in other characters, but in this book, the unadorned term refers only to size.

Just before *A. anamensis* was described, fragmentary remains from Ethiopia of an even older (approximately 4.4 mya) australopithecine that apparently lived in woodlands were found and named *Ardipithecus ramidus* (White et al. 1994; WoldeGabriel et al. 1994). It, too, apparently was bipedal, but lower limb bones of a fragile partial skeleton, which await description, will determine that with certainty (Richard Klein, personal communication, mid-1999). The remains are mostly dental, and the fossil apparently is so close to the point of divergence between the lineage leading to us and that leading to the chimps and bonobos that paleoanthropologist Bernard Wood stated that *Ardipithecus ramidus*' "attribution to the human line is metaphorically—and literally—by the skin of its teeth" (Wood 1994). The original describers named it *Australopithecus ramidus*, but then they almost immediately reconsidered its generic status and decided that *A. ramidus* was sufficiently distinct to place it in a separate genus, *Ardipithecus* (White et al. 1995).

Interestingly, recent restoration of previously unpublished fossil material of an Italian fossil hominoid (a member of the superfamily Hominoidea, apes and people, as opposed to the family Hominidae, just people) 7–9 million years old shows that bipedalism had independently evolved on at least one occasion among earlier apes. The hominoid was *Oreopithecus bambolii* (Harrison and Rook 1997). Careful study of its anatomy suggests that *Oreopithecus* walked upright (Köhler and Moyà-Solà 1997a). The morphology of its foot was unlike that of any other great ape, including *Homo sapiens*, with the big toe sticking out at an angle of about ninety degrees from four other short toes. The Spanish paleontologists who did the research suggest that this ape expanded its foraging to be more terrestrial because its island habitat was free of predators and it could avoid energetically expensive and risky climbing activities. The new evidence shows that among the apes, bipedalism was not restricted to the australopithecines and the genus *Homo;* but it is, of course, possible that *Oreopithecus* still spent a portion of its time in the trees. In any case, it was clearly not a direct ancestor of human beings, and its bipedalism is unrelated to ours. Most recently, it has been suggested that the *Oreopithecus* had a hominid-like precision grip capability in its hands (Moyà-Solà et al. 1999), but this conclusion is not universally accepted (Bower 1999c).

There is some variation in modern treatment of our relatives, but the two great superfamilies of Old World higher primates (that is, excluding tarsiers, lemurs, and their relatives) are the Cercopithecoidea (monkeys, baboons, etc.) and the Hominoidea (great apes and people). Together, they make up the so-called catarrhines, a name that refers to their narrow noses with downward-facing nostrils. (New World monkeys are platyrrhines and have flat noses with outward-facing nostrils). Within the Hominoidea (which, unlike the cercopithecoids, lack tails), the family Hominidae contains the australopithecines and the genus *Homo* and, in some classifications (e.g., Harrison and Rook 1997), the other living great apes except for gibbons (Hylobatidae). In other classifications (e.g., Fleagle 1999), the chimps, gorillas, and orangutans are placed in a separate family, the Pongidae. The differences are based largely in the different philosophies of cladistic, phenetic, and evolutionary classification (Ehrlich 1997).

108. The species represented by the skull was first named *Zinjanthropus boisei* by the Leakeys.
109. I use the term *cranium* to indicate the portion of the skull, including the face, that encloses the brain and supports the jaws. I use *skull* simply to indicate the skeleton of the head, including the jaws. The original *Zinjanthropus* specimen lacked an associated mandible (lower jaw).

110. Some scientists still consider it to have belonged to the genus *Australopithecus*, in which case *Paranthropus* would be a subgenus—it is largely a matter of taste, but overall I think that using *Paranthropus* aids in communicating about our ancestors.

111. It is possible that Broom's and the Leakeys' discoveries are just geographic variants of the same species. *Paranthropus boisei* is now known to have occurred from Tanzania to Ethiopia, and the morphology of the recent Ethiopian finds reopens the question of whether there was just one widespread robust australopithecine species; see, e.g., Delson 1997 and Suwa et al. 1997. The final word on that may never be in because, as mentioned earlier, the taxonomic problems of distinguishing species are even more difficult in fossils than they are in living organisms. See Mayr 1942; Mayr et al. 1953; Ehrlich 1961; Sokal and Crovello 1970; Cracraft 1983; Patterson 1985; Tattersall 1992. For recent discussions of taxonomic problems in the genus *Homo*, see Wood 1996 and Stringer 1996.

112. Robinson 1963; Grine 1981, 1987.

113. Grine 1985; Klein 1999.

114. The original discovery of *A. aethiopicus* was a toothless lower jaw found by two French anthropologists, Camille Arambourg and Yves Coppens (1967), but most information about this species is based on the famous "black skull" (Kenya National Museum WT-17000) found by Alan Walker and Richard Leakey; the discovery is well described in Leakey and Lewin 1992, pp. 121–134. Manganese salts had penetrated the skull and turned it a beautiful blackish bronze. *A. aethiopicus* was originally described by the French as being in a new genus, *Paraustralopithecus*, but Walker, Leakey, and their colleagues considered it an early form of *Paranthropus boisei* (Walker et al. 1986). Here, I follow McHenry 1996 and Klein 1999 in considering it a case of parallel jaw evolution with the robust australopithecines in response to a common diet rather than viewing *A. aethiopicus* as an ancestor of *P. boisei* and *P. robustus*, as suggested by some cladistic analyses (Strait et al. 1997). The question of whether or not the jaw evolution was parallel is basically unresolved (McCollum 1999).

115. It was an East African fossil skull with an ape-like lower face, dating from about 2.5 million years ago. *Australopithecus garhi* was a gracile species that combined characteristics of australopithecines (a typically small brain capacity of some 450 cc) with those of early *Homo* (some of the same dental characteristics). It had, however, remarkably large molars, even bigger than those of *Paranthropus* but structurally different from theirs and showing a wear pattern characteristic of a more varied, less coarse diet than that of the latter (Asfaw et al. 1999). Richard Klein commented to me that *A. garhi* "closely resembles *A. afarensis* in the shape and small size of the brain case and in the degree of subnasal prognathism. If we didn't have the dentition, I think we would call it a very late surviving *A. afarensis*, or we might worry that the 2.5 million year date was mistaken and that it was actually more than 3 million years old" (personal communication, 15 May 1999). If nothing else, the new fossils will keep cladists busy in trying to determine the exact branching of the hominid family tree with inadequate data (Strait and Grine 1999). Those fascinated with such details would probably get a better return for their effort by helping in the search for more and better pertinent fossil material. The *garhi* find was from the same formation from which the arm and leg bones of another individual were recovered a year earlier. The upper leg bone was very much like those of some modern *Homo sapiens*, but the forearm was long, like those of apes and australopithecines (Heinzelin et al. 1999). It is not certain, however, that the skull and postcranial skeleton are from the same species. To further tantalize scientists, the same formation yielded, only a meter away from the arm and leg bones, the remains of some large mammals. These included bones that had been broken open (presumably to extract marrow) and showed numerous cut marks made by stone tools (Heinzelin et al. 1999). If the finds go together, we have an australopithecine with rather modern-appearing legs, rather ape-like arms, and a strange dentition a bit reminiscent of that of the first representatives of *Homo*. This creature may have had a substantial component of meat in its diet (because the vegetarian portion leaves no traces, the relative contribution of meat to its overall nutrition is unknown).

116. The oldest *Homo* with a reasonably secure date, represented by a maxilla from Hadar, lived about 2.3 mya (Richard Klein, personal communication, 15 May 1999).

117. Clarke 1996. The time of their disappearance is still unsettled (Klein 1988, 1999).

118. Jean-Jacques Hublin, personal communication, 12 April 1999. Because of their dietary differences, it is quite likely that the conservative robusts did not compete much with the *Homo* species that evolved from the graciles, at least at first (Grine 1985). *Paranthropus* succumbed without leaving descendants (Suwa et al. 1997), possibly as a result of global climate changes or increased competition from *Homo*. By the time *Paranthropus* disappeared, *Homo* had been making tools for about a million years (and perhaps controlling fire, although there is no sound evidence of this latter accomplishment for about another half

million years), an accomplishment that would have allowed them to utilize the tough vegetation that comprised important *Paranthropus* resources (Klein 1988; Vrba 1988; but see White 1995).

119. Strassmann and Dunbar 1999.
120. See, e.g., Bartholomew and Birdsell 1953.
121. Aiello 1981; Foley 1995, p. 117.
122. E.g., see articles in Coppens and Senut 1991.
123. See, e.g., Washburn 1960; Hewes 1961; Lovejoy 1981.
124. Kortlandt 1980.
125. Fifer 1987.
126. Another hypothesis is that the knuckle-walking of the chimp, though an elegant compromise between the need for terrestrial locomotion and the need for arboreal locomotion, is much less efficient than bipedalism for moving across the land surface, even if there is no need to use forelimbs to carry things (Rodman and McHenry 1980; Steudel 1994). Substantial evidence suggests, however, that the energetic differences between the two postures in movement are at most small, and considerations of energetic efficiency are unlikely to have been the major factor in the evolution of an upright, bipedal lifestyle (R. M. Alexander 1991; M. D. Rose 1991). For an excellent summary, see Jablonski and Chaplin 1993.
127. See, e.g., Jolly 1970; Rose 1976; Wrangham 1980; Hunt 1994.
128. The savanna- and desert-dwelling baboons are in the genus *Papio*, subgenus *Papio*.
129. The forest-dwelling baboons are in the genus *Papio*, subgenus *Mandrillus*.
130. The mandrill, which weighs as much as 54 kilograms (120 pounds), is the largest monkey.
131. Geladas belong to the genus *Theropithecus*.
132. DeVore and Hall 1965; Altmann and Altmann 1970.
133. Dunbar 1984.
134. Crook 1966; Dunbar 1984; Jablonski 1993.
135. Iwamoto 1993.
136. Jolly 1970.
137. The bipedal shuffle is used for about 15–30 percent of the distance geladas cover.
138. Our ancestors might well have been preadapted for such a transition if they were similar to ancient proto-gibbons, which are assumed to have used some bipedal branch-walking for moving around in trees (Tuttle 1994).
139. Leuteneggar 1987.
140. Wheeler 1984, 1991.
141. Chaplin 1994; but see also Wheeler 1992.
142. Bonnefille 1994; Klein 1999.
143. Jablonski and Chaplin 1993.
144. A related idea was first put forward in Livingston 1962.
145. Upright displays were common in the pre–machine gun days of human warfare—with foot soldiers wearing everything from feathered headdresses to giant fur hats to make themselves look taller and intimidate the enemy. The importance of height carries over into our language. A U.S. Navy official, discussing the navy's support of Robert Ballard's exploration of the wreck of the *Titanic*, said that the point was to show the Soviets how skilled the U.S. military was at underwater operations so the Soviets would believe "we were not merely 10 feet tall but 20 feet tall" (Sontag and Drew 1998, pp. 322–323). Whether there is any connection between such sayings and the possibility of bipedalism being connected to display behavior is, of course, completely unknown. Great apes as well as many other animals erect hair or feathers and change posture to make themselves look larger when attempting to threaten other individuals.
146. McHenry 1992a. Difficulties include those of assigning parts to individuals, avoiding circularity by assuming that smaller individuals are females, and being sure that individuals of different sizes are members of the same species.
147. Kano 1992; Waal 1997.
148. Lovejoy 1981.
149. This is not to denigrate the social importance of meat-sharing in patriarchal primate societies (see, e.g., Stanford 1999).

Chapter 5: Bare Bones and a Few Stones

1. Allman 1999, pp. 2–3.
2. The average brain size for modern *Homo sapiens* is about 1,350 cc (Ruff et al. 1997).

3. Ehrlich and Holm 1963, p. 285, on the basis of tools found in association with *Paranthropus (Zinjanthropus)* that are now believed to have been made by *H. habilis*. The modern view is given in Klein 1999.
4. See, e.g., Steele 1999; Roche et al. 1999.
5. Schick and Toth 1993.
6. The hand structure of *Paranthropus* was appropriate for tool use (Susman 1994). See also McGrew 1993.
7. The robust australopithecines persisted for a long time alongside evolving *Homo* while that genus began to produce Acheulean artifacts, an advance on the Oldowan industry, and the Acheulean tradition continued unabated after those australopithecines became extinct. There is no evidence for a separate contemporaneous tool tradition that disappeared when *Paranthropus* did. There is also some question of whether the hand bones analyzed by Susman actually came from *Paranthropus* (Klein 1999). It is widely assumed that early *Homo* produced the oldest stone tools, but animal bones with stone-tool percussion and cut marks are associated with *Australopithecus garhi* (Heinzelin et al. 1999), and thus the issue of who made the first stone tools is still open.
8. Leakey et al. 1964. The placement of *H. habilis* (and *H. rudolfensis*) in the genus *Homo* has recently been questioned on the basis that those creatures are more like australopithecines than the later *H. ergaster* (Wood and Collard 1999). It is and will remain largely a matter of taste; for ease of communication, I follow the conventional taxonomic treatment.
9. One must be careful not to overinterpret brain size differences when samples are small, in view of the very large spread of sizes (about 1,200–1,700 cc) in modern *Homo sapiens*.
10. Wood 1991, 1992; but see Rightmire 1995. There is a thermal connection between enlarging brains and bipedality. The brain needs an abundant blood supply to both nourish and cool it, and an upright posture makes that something of an engineering challenge for natural selection—a challenge that it met in the hominid line (see Falk 1990 and the comments that follow her paper).
11. Trinkaus 1992; these are long opposable thumbs with a special joint at the base that allows not only a "power grip" for holding a hammer but also a "precision grip" for delicately manipulating a needle or pencil (Napier 1964).
12. This refers especially to the famous *H. rudolfensis* skull, ER-1470; see Tattersall 1998a, p. 133. The asymmetry is presumably related to handedness and is possibly related to the origins of language (MacNeilage et al. 1984). However, more or less similar anatomical and behavioral asymmetries are widespread in other animals (Bradshaw 1991; Bauer 1993), including chimpanzees, in which the asymmetry involves the same areas that are specialized for language reception in modern human beings (Gannon et al. 1998).
13. Toth 1985; Schick and Toth 1993, p. 142.
14. Ornstein 1997.
15. Shipman 1984; Klein 1999, p. 229.
16. The period characterized by this technology and the Acheulean tradition, which succeeded it, is often referred to as the Lower Paleolithic or Early Stone Age. It stretched from about 2.5 mya to 250,000 years ago (Klein 1999).
17. Schick and Toth 1993.
18. Schick and Toth 1993, pp. 133–134.
19. Wright 1972.
20. Schick and Toth 1993, pp. 136ff.; Richard Klein, personal communication, late 1999.
21. Toth et al. 1993.
22. Tattersall 1998a, pp. 128ff.
23. See, e.g., Binford 1985; Shipman 1986; Blumenschine 1987.
24. Schick and Toth 1993, p. 208. Because the ratio of animal hunting to gathering of plant materials as a source of nourishment often favors the latter, some have suggested using the term *gatherer-hunter*. For uniformity, I'm sticking with *hunter-gatherer*, with no implication for division of activities.
25. Roche et al. 1999.
26. Tattersall 1995, pp. 187ff. Not all paleoanthropologists consider *H. ergaster* a species distinct from *H. erectus*. For a detailed biometric analysis, see Bräuer 1994. For a fine discussion by a scientist who separates *H. ergaster* but considers the evidence too incomplete to make the position of *H. ergaster* certain, see Klein 1999, pp. 287–295.
27. Within modern populations, differences between males and females include facial characteristics, such as the browridges being generally more strongly developed in males. But in a mix of skulls from different populations, such comparisons won't work—for example, many Australian Aboriginal women have stronger browridges than many European men. Overall size can also be important. Pelvises are of little use in determining the sex of australopithecines because the need for a wide birth canal to allow the deliv-

ery of big-brained infants had not yet arisen. Lucy is thought to have been a female because she was only about three feet tall. In what was presumably a strongly sexually dimorphic primate, if Lucy were a male, the females would have been truly tiny. Similarly, if the Turkana boy had actually been a girl, then *Homo ergaster* men would have been gigantic.

28. Johanson and Edgar 1996, p. 182.
29. Ruff et al. 1997; John Allman, personal communication, 1 November 1999.
30. Ruff and Walker 1993.
31. Wheeler 1993.
32. Klein 1999, p. 250.
33. Jablonski and Chaplin 2000.
34. Chaplin 1994.
35. Steudel (1994).
36. Folk Jr. and Semken Jr. 1991; Wheeler 1992; Klein 1999, p. 292.
37. C. D. Michener, personal communication, 3 August 1999.
38. Richard Klein (personal communication, late 1999) comments that cleavers were most commonly made from intractable materials such as basalt and quartzite, and it may be that they were mainly failed hand axes.
39. Named after a site at St. Acheul, near the northern French city of Amiens. On the Oldowan–Acheulean transition, see also Gowlett 1986 and Wynn 1993.
40. The dispersal into Europe may have coincided with a general dispersal into Europe of large mammalian species from the east. In Europe, early human beings may have faced severe competition for carcasses from two large hyena species, which may have limited their scavenging possibilities (Turner 1992). In any case, it suggests that meat was a substantial component of the dietary resource base of European *H. ergaster/erectus* (Turner 1994).
41. Gibbons 2000; Yamei 2000.
42. Dean and Delson 1995; a mandible found in Dmanisi, in the Caucasus Mountains, by a joint Georgian-German expedition has been assigned to African-type *Homo erectus (H. ergaster)* and dated to 1.8–1.6 mya (Gabunia and Vekua 1995). If these interpretations are correct, the Dmanisi mandible would be one of the earliest signs of our ancestors living outside Africa. In 1998, two nearly complete skulls were found at Dmanisi that look very much like those of *H. ergaster* (Gabunia et al. 2000). Recent datings of *H. erectus* in Java suggest that hominids had reached that area at about the same time (Swisher III et al. 1994), but the dating is in dispute (de Vos and Sondaar 1994; Swisher III 1994).
43. Bar-Yosef 1995.
44. Stringer and Gamble 1993; Klein 2000b.
45. Klein 1999, p. 256.
46. Klein 1999, pp. 535–536.
47. Busvine 1948. The taxonomic treatment of the two forms as races or species is largely a matter of taste.
48. What may be a more important discovery was made very recently and could significantly alter ideas of the relationship of *Homo erectus* to *Homo sapiens.* That was a skull uncovered in the Northern Danakil (Afar) Depression of Eritrea, in northeastern Africa. The well-preserved adult skull is estimated to be about a million years old and thus falls in the temporal gap between the last known African *H. erectus,* dated at about 1.4 mya, and the earliest known fossils assigned to *H. sapiens,* about 0.6–0.5 mya (Abbate et al. 1998). The skull shows a mix of *H. erectus* and *H. sapiens* characteristics. Preliminary estimates of brain volume were 750–800 cc, which, combined with a long braincase and heavy browridges, suggests *H. ergaster/erectus.* But the widest part of the skull is high, with the sidewalls (parietal walls) converging slightly lower down. That is characteristic of *H. sapiens* and may indicate that the transition to our own species began earlier than previously surmised. This could be crucial because a long time during which natural selection was differentiating modern people from *H. erectus*-like creatures could help to explain our special characteristics. The find is also viewed as reinforcing the importance of the Horn of Africa as a critical area in human evolution, although it may be just a place where luck has been on our side as far as fossil ancestors being preserved and our finding them (Gibbons 1998c).
49. Here, the three-part name indicates that Neanderthals were then considered a subspecies, a European geographic variant of *H. sapiens.*
50. Ehrlich and Holm 1963.
51. Mitochondria, like the chloroplasts of plant cells, are thought to have originated as separate organisms that had a mutualistic relationship with the cells they inhabited. They are the only organelles of cells that

have their own genetic systems, including both DNA and the RNA that functions to "translate" the genetic information encoded in DNA. For details of this fascinating story, see Gillham 1994.

52. Recent work casts some doubt on the assumption of inheritance only from the mother (Awadalla et al. 1999). If confirmed, dates based on mtDNA analyses will need reexamination.
53. Sometimes referred to as early archaic *Homo sapiens*, e.g., in Bräuer et al. 1997, or *Homo heidelbergensis*, e.g., in Tattersall 1998a.
54. Bräuer et al. 1997; Klein 1999, p. 277.
55. Klein 1999; Tattersall 1992, 1998b; P. Lieberman 1998. On the basis of skull shape, Daniel Lieberman (1998) would also exclude what are usually referred to as archaic *Homo sapiens* from our species. It's a matter of taste, and keeping the archaics in our species seems fine to me at the moment. My taxonomic conservatism is based in part on the "rule of obligatory categories" (Ehrlich and Murphy 1981), the idea that categories to which all organisms must be assigned (species, genus, family, etc., as opposed to superspecies, subgenera, subfamilies, etc.) should be kept inclusive for ease of communication to the general public. The long separation and probable partial sympatry of *H. sapiens* and *H. neanderthalensis* makes splitting them seem sensible; similar logic would argue against a new name for archaic *H. sapiens*. Recent molecular evidence has led to the suggestion by cladists that both of the genera *Homo* and *Pan* (chimps and bonobos) be lumped into *Homo*, with *Homo* and *Pan* as subgenera (Goodman 1999). This would be a double mistake—first, in using purely cladistic rather than evolutionary (rate × time) standards in classification where there is especially strong reason to include the phenetic dimension, and second, in upsetting a long-standing taxonomic arrangement, widely used in scientific and popular communication, solely to emphasize the temporal sequence of phylogenetic splitting.

    Recently, the suggestion has been made that Neanderthal morphology is the result of iodine deficiency, based on a morphological comparison between the skeletons of cretins (sufferers from extreme iodine deficiency) and Neanderthal skeletons (Wilford 1998). The idea is that the problem was solved either by improved trade bringing iodine-rich foods to interior Europe or by a new mutation making Neanderthals' thyroid glands better able to absorb iodine. I find the idea that the Neanderthals were cretins far-fetched, given that the Neanderthal morphology stretched all the way to the Middle East. Cretinism persisted in the Alps into recent historical times—indeed, the term *cretin* has its roots in *Christian* in French dialect and was coined by medieval liberals trying to point out the basic humanity of individuals who were mentally retarded as a result of iodine deficiency. Interestingly, the famous Venus figures from Cro-Magnon sites, obese women with pendulous breasts and conspicuous vulvae, resemble female cretins. This could indicate that after the Neanderthals disappeared, cretins among the modern people were considered either paragons of beauty or shaman-like individuals to be revered or appeased.
56. Krings et al. 1997; see also Lindahl 1997.
57. See, e.g., Hublin 1996.
58. Klein 1999.
59. Mellars 1998; F. H. Smith, Trinkaus, et al. 1999. The issue of the timing of overlap of Neanderthals and *H. sapiens* in Europe remains a crucial element in the debate over why the Neanderthals went extinct (F. H. Smith, Trinkaus, et al. 1999).
60. See, e.g., Trinkaus 1986, 1989.
61. Jones et al. 1992, p. 247; Tattersall 1998a, p. 151.
62. Arsuaga, Carretaro, et al. 1997.
63. Aiello 1993.
64. The main modern proponent of this hypothesis is Milford Wolpoff (1984, 1989, 1999). His views are summarized in Wolpoff and Caspari 1997.
65. See, e.g., Ehrlich and Holm 1963, p. 284.
66. It also allowed a racist interpretation of human history to be developed (e.g., Coon 1962), which, of course, does not speak to the scientific validity of the basic hypothesis.
67. Nei 1995.
68. See, e.g., Nei and Roychoudhury 1993; Nei 1995. Most genetic differentiation in modern *Homo sapiens* is within, rather than between, populations (Lewontin 1982a).
69. Pritchard et al. 1999; Thomson et al. 2000. For a balanced summary covering both pros and cons of this model, see Klein 2000b.
70. Stringer and Andrews 1988. For a more recent summary by Christopher Stringer, one of the leading advocates of the hypothesis, see Stringer and McKie 1996. Richard Klein (1998) makes a persuasive case that it wasn't until major behavioral changes occurred, some 50,000–40,000 years ago, within the same

morphological context that archaeologically (behaviorally) *and* anatomically modern people were able to displace the Neanderthals.

71. Dean et al. 1994; Hublin 1996; Arsuaga, Martínez, et al. 1997.
72. See, e.g., Bräuer 1984; Aiello 1993. There are also those who, accepting that Africa was the source of important features of modern *Homo sapiens*, deny the overall importance of migration and accept many features of the multiregional model, e.g., F. H. Smith, Falsetti, et al. 1989.
73. Klein 2000b.
74. Stoneking et al. 1986; Cann et al. 1987; Stoneking 1993; Relethford 1995; Manderscheid and Rogers 1996; Harpending and Relethford 1997.
75. Templeton 1992, 1993; Long 1993; Ballard and Kreitman 1995; Loewe and Scherer 1997.
76. Horai et al. 1995; Penny et al. 1995; Zischler et al. 1995; but see Wills 1995. Wills estimates that Eve lived between 806,000 and 436,000 years ago.
77. See, e.g., Cavalli-Sforza et al. 1988; Nei and Roychoudhury 1993; Batzer et al. 1994; Castiglione et al. 1995; Dorit et al. 1995; Goldstein et al. 1995; Hammer 1995; Martinson et al. 1995; Nei 1995; Pääbo 1995.
78. The term *haplotype* is often used in the literature of human genetics to refer to a DNA sequence that differs from sequences at homologous sites on one or more other chromosomes—equivalent to a series of alleles at two or more loci determined molecularly.
79. Technically, if alleles are found together more often than would be expected under an assumption of random association, there is linkage disequilibrium. Linkage disequilibrium can be caused by genetic drift and tends to decay with time. Thus, high levels of disequilibrium in non-African populations could indicate small size of founder migrant groups, recency of origin, or both—if the patterns are not being caused by selection.
80. Tishkoff et al. 1996; see also Wood 1997.
81. Gibbons 1997c. Other recent DNA studies of people from China and the Andaman Islands also support the out-of-Africa 2 hypothesis (Mukerjee 1999). Remember, some recent work on the Y chromosome suggests that the out-of-Africa date could be as recent as 60,000 years ago (e.g., Pritchard et al. 1999). The causes of the spread are not clear but could well have been a combination of population growth and curiosity to explore new foraging grounds—a population diffusion rather than a directed march. A rate of a few miles per year would have spread modern people all over Eurasia in a few thousand years.
82. Valladas et al. 1988.
83. Valladas et al. 1987. It appears that *H. neanderthalensis* and *H. sapiens* alternately occupied some Middle Eastern sites as the climate warmed (favoring an African biota and *H. sapiens*) and cooled (favoring a boreal biota and *H. neanderthalensis*).
84. Hublin et al. 1996.
85. Stringer 1992.
86. But it would be valuable to have a more ample record of early *Homo sapiens* fossils from Africa to establish fully the connection of the Israeli archaics with African populations (F. H. Smith 1992).
87. My summary in this paragraph is based largely on Klein 1999. There remains abundant controversy on the adaptiveness of Neanderthal characteristics; see, e.g., Graves 1991 and accompanying comments. One open question about the differences between *H. sapiens* and *H. neanderthalensis* is whether the former had domesticated dogs and the latter didn't. A suggested explanation is that our species had a competitive edge because of the presence of guard dogs, which had excellent senses of smell and hearing (Allman 1999, pp. 204ff.). The actual time of domestication of dogs is a matter of debate, with some recent genetic evidence suggesting that it occurred several times independently and more than 100,000 years ago (Vilà et al. 1997), but archaeological evidence suggests it may have occurred only about 14,000 years ago (e.g., Morell 1997; Federoff and Nowak 1997 and response). It is important to note that *Homo sapiens* could not have domesticated dogs until after it left Africa because the wolf progenitors of the dog do not occur there. See also Brisbin Jr. and Risch 1997.
88. Holliday 1997; Churchill 1998.
89. It had been claimed that features of the Neanderthal pelvis indicate it may have enclosed a larger birth canal than that characteristic of *H. sapiens* (see Brace 1988). Unfortunately for this hypothesis, the only complete Neanderthal pelvis we have suggests that despite pubic elongation and other peculiarities, the Neanderthals did not have especially large birth canals (Tague 1992).
90. Churchill 1998.
91. Mellars 1996.
92. From the type site at Le Moustier, in the Dordogne region of France. The term *Mousterian* is often

applied to similar collections of tools of the same antiquity from Africa and Asia. The technology is also called Middle Stone Age, which is the sub-Saharan African equivalent of the European, western Asian, and northern African Mousterian or Middle Paleolithic. Middle Paleolithic, or Middle Stone Age, assemblages first appeared about 250,000 years ago and lasted until 50,000–35,000 years ago, depending on the locality.

93. Bar-Yosef 1994; Mellars 1996; Klein 1998.
94. Schick and Toth 1993.
95. Trinkaus and Shipman 1993; Berger and Trinkaus 1995. The first really ancient *Homo sapiens* remains found in Europe, about 30,000 years old, were recovered from a site in the Dordogne region of western France. A series of skeletons was unearthed in a shallow rock shelter in a cliff known locally as Cro-Magnon. That cliff gave its name to all ancient European *Homo sapiens* (Tattersall 1995).
96. See Hayden 1993; Brainard 1998; Marean and Kim 1998.
97. See, e.g., Crosby 1986; Diamond 1997a; but see also d'Errico, Zilhão, et al. 1998.
98. Zubrow 1989.
99. This is a point amply documented by Jared Diamond 1997a.
100. Mellars 1996, 1998; Klein 1999.
101. There is a recent report of a *Homo sapiens–H. neanderthalensis* hybrid specimen, a child who was buried north of Lisbon, Portugal, an estimated 24,500 years ago (Bower 1999d). Its status as a hybrid is open to question, and its significance, if it was a hybrid, would remain questionable—especially because there would be no way to determine whether the individual would have been fully fertile as an adult. See also the discussion in Norris 1999.
102. Stiner et al. 1999. It is unfortunate that we have so little information about prehistoric hunter-gatherer group sizes and population densities; that information would be helpful in sorting out various genetic and cultural scenarios.
103. See, e.g., Graves 1991 and accompanying comments.
104. See, e.g., Mellars 1989; Hublin et al. 1996; but see d'Errico, Zilhão, et al. 1998.
105. Called the Châtelperronian.
106. Named for the Aurignac rock shelter in the Pyrenees of France.
107. Goebel et al. 1993; Otte 1995. It is possible, however, that the Great Leap Forward actually started in Africa, where carefully shaped barbed points have been found that apparently date to 155,000–90,000 years ago (Brooks et al. 1995; Yellen et al. 1995).
108. Diamond 1989b. See the discussion in Sahlins 1968, p. 85.
109. See, e.g., Mellars 1991.
110. See, e.g., Stringer 1995.
111. The scenario of a series of waves seems most reasonable to me, and it was suggested by the first mtDNA study (Cann et al. 1987), but more recent work seems to support something close to a single, explosive dispersal event (Tishkoff et al. 1996).
112. Pope 1995; Swisher III et al. 1996.
113. Gibbons 1997c.
114. See, e.g., Rightmire 1992.
115. Morwood et al. 1998.
116. Quoted in Thwaites 1998 and in Gibbons 1998a. As far as I know, however, there is no good evidence of the use of "well-crafted spears" by *H. erectus*, although it is certainly possible. The suggestion is made on the basis of spears that may be 400,000–350,000 years old excavated from the depths of a coal mine in Germany (Bower 1997). Because there are no associated human remains, the makers of the spears, which may have been throwing (as opposed to thrusting) weapons, could have been *H. neanderthalensis* or archaic *H. sapiens*. There is also the possibility that stone tools attributed to *Homo erectus* on Flores were flaked by natural processes (Richard Klein, personal communication, 29 December 1999).
117. Klein 1999, p. 329; see also C. Stringer, quoted in Thwaites 1998.
118. Straus et al. 1993.
119. Dayton 1999; Roberts et al. 1994. The dates are uncertain because this time is close to the limit of radiocarbon dating, and the results of other dating techniques are subject to various uncertainties (O'Connell and Allen 1998). The date of 60,000 years ago seems unlikely in view of other evidence regarding the spread of human beings after the Great Leap, but there is some evidence that it could have been some 50,000 years ago (Miller et al. 1999). Dating of rock art in a shelter in the Northern Territory of Australia to more than 100,000 years ago (Fullagar et al. 1996) has since proven to be a result of problems with the dating method (Gibbons 1998f).

120. Allen et al. 1989; Stringer and Gamble 1993; Klein 1999. For a history of human occupation of Australia, New Guinea, and New Zealand, see Flannery 1994.
121. Evidence from rock shelters shows that people reached the Solomon Islands and vicinity more than 30,000 years ago (Wickler and Spriggs 1988; Allen et al. 1989). Interestingly, there is no sign of substantial sea travel elsewhere until about 13,000 years ago, from the Mediterranean Sea (Jared Diamond, personal communication, March 1999).
122. See, e.g., Gibbons 1996.
123. Morell 1998.
124. Gibbons 1998b.
125. See, e.g., Simons 1989; Stringer 1992, 1996; Aiello 1993; Lieberman 1995; Klein 1992, 1995, 1999. There is still, however, strong paleoanthropological dissent, e.g., Frayer et al. 1993; Otte 1995; Brace 1997; Wolpoff and Caspari 1997.
126. Gibbons 1997b; Reich and Goldstein 1998; Marshall 1998; Stefanini and Feldman 1999; Pritchard et al. 1999; although some of the genetic evidence is consistent with both hypotheses or even favors the multiregional one (e.g., Ayala et al. 1994; Bower 1999b; Harding et al. 1997). See also discussions in Harpending 1994 and Sherry et al. 1994.
127. Klein 1999.
128. Kimbel 1995. Some sense of the excellence of the fossil record can be gained by perusing David Brill's magnificent photographs of fossil hominids in Johanson and Edgar 1996.
129. Their work provides an excellent example of how historical events can be reconstructed by combining meticulous field research, sophisticated dating techniques, data from molecular genetics, field experiments, careful use of several kinds of analysis of fossil morphology, intelligent speculation, and the kind of healthy debate that characterizes most active areas of science. The field experiments have been in taphonomy, the science of determining how distorted are the remains of dead organisms, including the fossil record. With fossils, this is done in part by examining the possibility that they have been transported away from the communities in which the organisms actually perished. One approach to this has been to experiment with the fates of the remains of modern organisms under different conditions (e.g., Schick and Toth 1993).
130. Diamond 1989b.

Chapter 6: Evolving Brains, Evolving Minds

1. Damasio 1994, p. xvi.
2. See, e.g., Lakoff 1987. Lakoff writes that in cognitive models that are embodied, "meaning is understood via real experiences in a very real world with very real bodies," as opposed to objectivist models, in which "such experiences are simply absent. It is as though human beings did not exist, and their language and its (not *their*) meanings existed without any beings at all" (p. 206). See also Damasio 1999.
3. For an interesting discussion of the evolution of the human mind in relation to life-history strategies, brain structures, complexity of technology, and other features, see Foley 1996.
4. Barton et al. 1995; Barton 1997; Hoffman 1998.
5. Other parts of the cerebral cortex include the olfactory cortex, the entorhinal cortex, the hippocampus, and the cingulate cortex.
6. Allman 1999, p. 44.
7. For more detail about the possible mechanisms behind this, homeotic mutants (ones that transform one homologous body structure into another—such as transforming the third body segment of a fly's thorax into the second), and gene duplication and how these may operate in brain evolution, see Allman 1999, chap. 3. More basic information can be found in Hartl and Jones 1998. Some background on homeobox genes can be found in Ruddle and Kappen 1995.
8. For a summary, see Holloway 1995. The volume in which that appears (Changeux and Chavaillon 1995) contains a number of very interesting chapters on brain evolution. Another interesting book about the evolution of the brain, which ties it to language evolution, is Deacon 1997.
9. Thomas Hobbes thought that fear of violent death is a driving motivation for much of human behavior—and is in itself central to human nature (Hobbes 1997 [1651]). Thoughtful fear of one's own death certainly demands intense consciousness. A confusing but much-discussed philosophical analysis of levels of consciousness is that of existentialist Jean-Paul Sartre (1905–1980), who had a Cartesian (dualist, mind-versus-matter) view and focused on the relationship between consciousness and self-consciousness (1960

[1956]). His ideas converged to a degree with those of Sigmund Freud, with non-self-conscious consciousness partially playing the role that in Freud's theory is attributed to the subconscious mind.

10. Dewey 1988, p. 80.
11. See, e.g., Griffin 1976, 1992; Calvin 1996b, p. 146; Pinker 1997, pp. 131ff.; Zimbardo and Gerrig 1999, chap. 5.
12. See, e.g., Humphrey 1992. See also Gallup Jr. 1985 and a recent fine review by Antonio Damasio 1999.
13. Ornstein 1988.
14. Jackendoff 1987. These activities obviously include monitoring of physiological states. Here, I do not use the term *unconscious*, which traces back to the great neurologist Sigmund Freud (1856–1939), the father of psychoanalysis. In Freud's view, the unconscious portion of the mind, which could not be observed directly, was more extensive than the conscious portion. He greatly modified the Cartesian dilemma of "How does one know other minds?" by shifting the focus to "How does one know one's own unconscious mind?" He developed an elaborate theory of a three-part structure of the unconscious mind, which, he postulated, consisted of the *id*, a complex of primitive desires and demands; the perceiving and reasoning *ego;* and the *superego*, the center of moral and ethical behavior and restraint. The ego, he postulated, mediates between the id and the superego. The scheme has partial parallels to Platonic notions of appetite, reason, and spirit and Christian ones of devil, individual, and God. Freud also explained the meaning of dreams and posited that slips of the tongue ("Freudian slips") were one avenue to knowledge of the unconscious mind. Unfortunately, despite Freud's brilliance, creativity, and influence, there are few data to support most of his original ideas, which have little influence in the scientific community today and must be categorized as "interesting even if not true." That does not mean that psychoanalysis cannot be helpful for someone with psychological problems, just as a conversation with an intelligent friend can be. For an introduction by Freud to his own writing, see Freud 1977 (1909) and 1961 (1930); see also Storr 1989; Ornstein 1986b, 1988, especially the material on dreams, p. 264; and Stevenson and Haberman 1998.
15. Dennett 1991, p. 308. Another example is the phenomenon of "blindsight," which occurs in some individuals who have suffered certain types of damage to the brain's visual cortex (Weiskrantz 1986; Cowey and Stoerig 1991). These people are conscious of being blind, and they have no sensation of sight. Yet experiments show that they nevertheless receive and evaluate visual information. For instance, these subjectively blind patients can reach in the direction of visual stimuli that they cannot see or can make reasonable decisions about objects in their visual field of which they have no conscious awareness. See, e.g., Weiskrantz 1990, 1995; see also a good discussion in McGinn 1999, pp. 147ff.
16. Humphrey 1992.
17. Humphrey 1992, p. 179.
18. The distinction traces to Scottish philosopher Thomas Reid (1710–1796) (1785, essay 2), who was a famous critic of the premier Scottish philosopher David Hume. He developed a philosophy based on "common sense."
19. "I think, therefore I am" (Descartes 1970 [1637], p. 101).
20. Humphrey 1992, p. 180. I am bypassing here the difficult but important topic of the function of emotions, their relationship to the nervous system (Ekman et al. 1983; Ekman 1992), and how they interact with consciousness and cognition. For some of the diverse views on this vexing issue, see Hohmann 1966; Lazarus 1982; Zajonc 1984; Johnson-Laird and Oately 1992; Ekman and Davidson 1993, 1994. Work by Paul Ekman shows that composing the face into an expression of emotion can cause one to feel the corresponding emotion (e.g., smiling produces a "feeling" of happiness). There is still substantial debate about the degree to which emotional responses are strongly and similarly genetically influenced in all human beings and the degree to which they respond to differences in the cultural environment (e.g., Bower 1998).
21. Kundera 1991, p. 225.
22. I haven't really done justice to Humphrey's arguments here—for those with an interest in the relationship of mind, brain, and body, his book is well worth reading. For a further sampling, see Dennett 1991; Crick and Koch 1992; Searle 1997; Damasio 1999; and Crick 1994—don't miss Crick's entertainingly annotated "Further Reading," pp. 281–291.
23. See, e.g., Nagel 1974; Dennett 1991, pp. 441ff. Nagel claims that we can never know what it's like to be a bat; Dennett thinks we can come pretty close. See also Dennett 1996; Griffin 1992; Searle 1997; and articles in Beckoff and Jamieson 1996. Dewey said, "With the animals, an experience perishes as it happens, and each new doing or suffering stands alone" (1988, p. 80), but I think this clearly underestimates

the level of consciousness of some of our closer relatives, as the material on chimpanzee politics in chapter 9 of this book and the effects of experience on chimpanzees' later behavior may convince you.

24. See the discussions of "Consciousness-1, -2, and -3" in Bickerton 1995, pp. 126–135, and the fine brief review by Pennisi 1999a.
25. McGinn 1999. This is the best work I have read on the topic, even though I don't agree with all of McGinn's views (which he admits are highly speculative).
26. Dennett 1991; see also Dennett 1996.
27. Minsky 1985; this is related to Bickerton's notion of primary and secondary representational systems (1990). It does not imply that human beings were the "goal" of the evolution of consciousness, but it does imply that the evolution of consciousness was progressive in the sense of Richard Dawkins's views (see chapter 3) (Dawkins 1997). For an interesting early discussion of consciousness by an evolutionist, see Rensch 1959 (1954), chap. 10.
28. Cheney and Seyfarth 1990b, p. 193. The quote continues, "What remains to be determined is whether they are also adept at understanding each other's *minds*" (emphasis in the original).
29. Humphrey 1976, quote on p. 316. See also Jolly 1966. Jolly wrote, "The social use of intelligence is of crucial importance to all social primates . . . social integration and intelligence probably evolved together, reinforcing each other in an ever-increasing spiral" (p. 504). A root of this idea can be seen in Trivers 1971. Trivers wrote about altruism (just one aspect of social problem solving), "Given the psychological and cognitive complexity the [human altruistic] system rapidly acquires, one may wonder to what extent the importance of altruism in human evolution set up a selection pressure for psychological and cognitive powers which partly contributed to the large increase in hominid brain size during the Pleistocene" (p. 54).
30. Sawaguchi and Kudo 1990. The human neocortex is a couple of millimeters thick; if unfolded, it would make a fair-sized napkin of some 200,000 square millimeters (about 17 inches square) (Allman 1999).
31. See, e.g., Dunbar 1993, 1998a. The assumption of growth in size of groups in the hominid line is based on regression analysis to determine the relationship between the sizes of groups in modern nonhuman primates and hunter-gatherers and the ratio of volume of neocortex and volume of the rest of the brain. This issue is controversial—for a sense of the technical questions involved, see the open peer commentary and response at the end of the first reference cited. John Allman emphasizes the human attribute of being able to "participate in a large variety of different social networks, each with its own rights and obligations"—families, kinship groups, regional trading webs, and the like—and the fact that "behaviors that are expected in one social context must be withheld in others" (personal communication, 24 September 1999). He points out that the parts of the brain that seem to be especially involved in navigating through these different social situations, dealing, for instance, with emotional arousal and response selection, are the anterior cingulate cortex and the adjacent frontal lobe (e.g., Lane et al. 1998; Nimchinsky et al. 1999; Posner and Rothbart 1998).
32. The statement is vague because, unfortunately, we have essentially no evidence about when the scattered groups of our ancestors began to reach critical size and how much contact they had with other groups. I'll return to this issue later.
33. And within tribes, moieties (two subdivisions of a tribe that exchange women in a certain pattern). Fox 1983 (1967), pp. 180ff.
34. See, e.g., Chomsky 1980.
35. In a PET scan, a patient is injected with water labeled with a rapidly decaying isotope of oxygen, ^{15}O, which emits positrons. The labeled water shows up in the brain in about a minute, accumulating wherever the blood flow (and thus brain activity) is highest. The positrons are then captured by a "camera" that produces an image indicating the degrees of activity of different parts of the brain. For details about this process and illustrations of images, see Posner and Raichle 1994.
36. Magnetic resonance imaging (MRI) is based on the tendency of many atoms to line up in response to a magnetic field and then emit, in reaction to radio-wave pulses, radio signals that reveal the number of particular atoms at a given position and features of their chemical environments. With this technique, it is now possible to follow functional changes in the brain by monitoring changes in blood oxygen concentration (Posner and Raichle 1994, pp. 18–19, 228–231; Wagner et al. 1998; Brewer et al. 1998).
37. Synapses are connections between neurons across which the neurons communicate electrochemical nerve impulses, either by the passage across a narrow physical gap of one of perhaps a few hundred chemical neurotransmitters or by direct electrochemical connections. They are crucial parts of the wiring of the brain and the rest of the nervous system, and understanding how they function in thought processes is a major challenge in brain research. (There is also growing interest in chemical signals in the brain travel-

ing relatively slowly through the cerebrospinal fluid [Zoli et al. 1998; Anonymous 1999b].) Recent work on mice shows how genetic changes can influence a characteristic of synapses that is involved in memory and learning (Tang et al. 1999). For a rather detailed summary of the physiology of the nervous system, see Kandel et al. 1995. For the increasingly understood role of malfunctioning neurotransmitters in mental illness, see Barondes 1999; in addictions, Restak 1995. Both books cover fascinating material—as a former smoker who still craves the weed more than forty years later, I found Restak's discussion of nicotine (pp. 116–121) especially interesting.

38. Harlow 1868, quoted in Damasio et al. 1994. Interestingly, free-roaming rhesus monkeys living in an experimental colony on Cayo Santiago (an island off Puerto Rico) were surgically damaged in a manner similar to that Gage suffered. They, too, retained their motor functions, but their ability to engage in normal social behavior was destroyed (Myers 1975). The morality of such experiments is, to me, questionable.
39. Ornstein 1988; Cytowic 1993; Damasio et al. 1994.
40. Parts of Gage's prefrontal cortex were destroyed in the accident, as was some of the underlying "white matter," especially in the left hemisphere (Damasio et al. 1994). The cortex is made up mostly of the layered bodies of nerve cells (neurons) and appears dark, and therefore it is often referred to as "gray matter." The underlying white matter is made up mostly of axons (nerve fibers) of the cells whose bodies are in the gray matter (Damasio 1994).
41. Researchers have compared paintings done by artists before and after they suffered a stroke in one brain hemisphere. A well-known painter, Lovis Corinth, suffered severe damage in the left hemisphere of his brain from a stroke. Afterward, his work showed, among other things, misplacement of details and deficiency in the presentation of texture, especially on the left sides of drawings. This example is discussed in a classic semipopular work about the effects of brain damage, Gardner 1975. In another example, intricacies of the way the brain handles language have been illustrated by studies of the speech deficits (aphasias) that result from different kinds of injuries.
42. See, e.g., Gardner 1975; Ornstein 1997; Gazzaniga 1998; Doricchi and Incoccia 1998.
43. Strauss 1998.
44. Ornstein 1997, pp. 66–67.
45. Ornstein 1997, pp. 151ff., 175.
46. Fodor 1983, 1985; but see the diversity of views in the discussion following the latter and in Gazzaniga 1989.
47. See, e.g., Piaget 1952; Gruber 1977; Beilin 1989.
48. For other neo-Piagetian nonmodular views, see, e.g., Fischer 1980 and Pascual-Leone 1987.
49. See comments in Ornstein 1986a, p. 17.
50. Hirschfeld and Gelman 1994.
51. Allman and Kaas 1971; Allman 1999.
52. Evidence that there is a separate module for recognizing faces can be found in Behrmann et al. 1992.
53. See, e.g., Kling 1986.
54. See many articles in Barkow et al. 1992.
55. Fodor 1983, pp. 52–53.
56. Fodor 1983, p. 53.
57. See, e.g., Donald 1991; Hirschfeld and Gelman 1994; Karmiloff-Smith 1992.
58. Pinker 1997, p. 524; see also Pinker 1991. As an example, see Hirschfeld 1989 on domain specificity and the acquisition of kinship terms.
59. Neurophysiologists would probably describe these specialized modules as complex circuits of neurons. Neurons are the basic signal-transmitting units of the nervous system—most of them can send signals to and receive signals from thousands of other neurons.
60. The most persuasive use of the parallel-processing computer analogy I have seen is in Dennett 1991.
61. Searle 1997. For some sense of the variety of views on brain analogies, see also Jackendoff 1987; Dennett 1991; Churchland and Sejnowski 1992; Crick 1994.
62. P. Lieberman 1998, pp. 99–100.
63. The classic paper associating computers and brains is Turing 1950. This is the paper in which the famous Turing test was proposed, wherein a computer could be said to think if it could answer questions in such a way that the interrogator could not distinguish it from a human respondent. In the domain of problem solving, which is often seen as the most basic aspect of intelligence, it is still not clear whether brains solve problems in the same way computers do, although in both human beings and nonhuman animals, problem solving often appears to include analogs to computation; see, e.g., Real 1991. The classic counterar-

guments to Turing's "computation equals thinking" equation are the views of John Searle (1980, 1992, 1993, 1997). The most famous and controversial of these is Searle's "Chinese room" argument. He points out that without understanding Chinese, he could sit in a room and follow instructions that assigned Chinese ideograms in response to incoming ideograms that asked questions, giving answers to the questions also in the form of ideograms. The questions and answers would be displayed outside the room, where someone literate in Chinese would see the questions being answered. Inside the room, however, there would be no way that Searle could, simply by following the programmed instructions, gain an understanding of Chinese. Computers work by computing without understanding, but understanding is the essence of intelligence. For contrary views, see, e.g., Bridgeman 1980 and Dennett 1991. I tend to think that Searle's criticism is pertinent, but that does not necessarily mean that the creation of a machine with understanding is impossible or that nothing can be learned from computational models of the mind.

64. Damasio 1989. Even the issue of how time is experienced in relation to "objective" time is far from settled—see, e.g., Dennett 1991, chap. 5.
65. Edelman 1992.
66. A concise description of the cortex can be found in Calvin 1996b, pp. 116ff. For a well-illustrated popular description of the brain, I recommend Ornstein and Thompson 1984. For more detail about brain structure and function, see Kandel et al. 1995. Each hemisphere is conventionally divided into four lobes: occipital, at the rear; temporal, near the temples; parietal, on the sides above the temporal and in front of the occipital; and frontal, at the front. The hemispheres, as mentioned earlier, are connected by a cable of nerve fibers called the corpus callosum.
67. Churchland and Sejnowski 1992.
68. To get some idea of the diversity of views on this question, sample Ornstein 1986b; Dennett 1991; Searle 1993, 1997. There is, of course, a great deal of neural activity in our brains of which we are not consciously aware—from activities such as regulation of body temperature to extremely rapid decision making that involves intelligent processing of inputs (Dennett 1991, p. 308).
69. Classic work showing that protein synthesis is required for the functioning of long-term memory is Agranoff and Klinger 1964. For more recent views on memory, see Goldman-Rakic 1992; Malenka and Nicoll 1997, 1999; Rugg 1998; Brewer et al. 1998; Wagner et al. 1998; and especially Squire and Kandel 1999. For a discussion of how timing tricks may be essential to creating an "illusion" of spatial integration and continuity in a retrieved event, see Damasio and Damasio 1993. Another issue related to memory is how the formation of habits differs physically from mechanisms involved in recalling episodes; see, e.g., Hebb 1949; Sherry and Schacter 1987; Calvin and Ojemann 1994; Calvin 1995, 1996a, 1996b.
70. See, e.g., Calvin 1987; Pinker 1997.
71. See, e.g., Stein and Levine 1987; Levenson 1994; Damasio 1999, pp. 40ff.
72. This is because a single gene often affects more than one characteristic (pleiotropy) and because of linkage and interactions of genes at different loci (epistasis).
73. For a wonderful, brief, nontechnical description of the morphology and organization of the brain, as well as interesting comments about the mind, see Edward O. Wilson's *Consilience* (1998), pp. 102ff.
74. His reasoning, which seems a little antique now, is that we can be certain that we have a mind but not that we have a body. Thus, the two could not be identical; ergo, dualism. But his conjecture led to the mind–body "problem"—how the physical body could affect the mind (e.g., dropping a rock on your foot causing the mind to register pain) and vice versa (the thought of raising your hand leading to its raising). See the discussion in Dennett 1991, pp. 33ff.
75. Ryle 1949.
76. Damasio 1994. See also his later book (1999).
77. Calabrese et al. 1987; Hosoi et al. 1993; Damasio 1994.
78. The point of there being obvious mind–body connections, of course, traces at least as far back as French physician and philosopher Julien La Mettrie (1709–1751) (1996 [1747]).
79. See, e.g., Laitinen and Livingston 1973.
80. See, e.g., McEwen 1991.
81. Ehrlich 1964.
82. Ornstein 1997.
83. Kohn and Dennis (1974).
84. Robert Sapolsky, personal communication, late 1999; Jay and Becker 1995.
85. Neville 1977.
86. Lessard et al. 1998. In cats, early blindness expands the part of the cortex devoted to hearing

(Rauschecker and Korte 1993) and may provide the neural basis for improved auditory localization in visually deprived cats (Rauschecker and Kniepert 1987).

87. See, e.g., Jenkins et al. 1990; Wang et al. 1995; Buonomano and Merzenich 1995, 1998. For an interesting application of what has been learned from studies of cortical plasticity in primates, see Merzenich et al. 1996.
88. However, there certainly are genetic influences on various aspects of brain morphology (Oppenheim et al. 1989; Tramo et al. 1995).
89. See, e.g., Pinker 1997; see also Spelke et al. 1992.
90. Ornstein 1988, pp. 191ff.
91. Zimbardo and Gerrig 1999, chap. 4.
92. See, e.g., Levenson 1994.
93. See the discussions in Pinker 1997, chap. 6, and Damasio 1999, chap. 2.
94. The definition of emotion is classically difficult; mine is based on, among others, those of Panksepp 1994; Clore 1994; and Ledoux 1994.
95. See, e.g., Panksepp 1994.
96. It is interesting to note that the growing interest in the role of emotions in cognition mirrors a move away from sterile Platonic-Christian-Kantian philosophies, which emphasized the importance of duty and pure reason, a move represented by the very diverse existentialist philosophies of Søren Kierkegaard (1813–1855; a critic of Christianity from within); Friedrich Nietzsche (1844–1900; a critic from without—one of his most famous aphorisms was "In truth, there was only *one* Christian, and he died on the cross"); Martin Heidegger (1889–1976); Albert Camus (1913–1960); Jean-Paul Sartre; and many others. Their philosophies emphasized an individual's subjective, personal experience, choice making, commitment, and the need to live a passionate life—not necessarily externally, but internally, with a "passion of the spirit" (Nietzsche 1974 [1887], p. 74). Belief that duty should override emotions played a role in the behavior during the Holocaust of people such as Adolf Eichmann and Heinrich Himmler, the former of whom actually invoked the beliefs of German philosopher Immanuel Kant (1724–1804)—incorrectly, given that Kant saw duty as relating to transcendental moral law and believed that, morally, other individuals must be treated as ends, not means—to exculpate himself for his role in the slaughter of millions (Singer 1993, pp. 185–186).
97. Damasio 1994, chap. 3. Current views of the role of emotions bear considerable resemblance to those pioneered by Thomas Hobbes in *Leviathan* (1997 [1651]). In the mid-seventeenth century, Hobbes contested the Platonic view of people as rational beings, instead emphasizing people's emotional (passionate and sensual) aspects. Scientists wishing to understand human behavior, even those aspects of it that are considered rational, ignore emotions at their intellectual peril.
98. See, e.g., Michotte 1965.
99. See, e.g., Avis and Harris 1991, which gives support to the notion of a universal "mentalistic" framework adopted by young children—the notion that they make assumptions about the mental states of others in interpreting their behavior. For a recent overview, see Frith and Frith 1999.
100. Daniel Povinelli reported experiments further supporting this at the 1999 meeting of the Association for the Study of Animal Behaviour. His results indicated that unlike young human children, chimps could not associate seeing something with knowing about it, and it seems doubtful that chimps can "read" each other's thoughts or experience empathy (Anonymous 1999a).
101. See, e.g., Damasio 1999, pp. 228–229.
102. Birnbaum 1955.
103. Samelson 1981. Behaviorism itself is the doctrine that empirical data about what people say and do, how stimuli produce behaviors, should be the essence of psychology. Watson was an intellectual descendant of English philosopher John Locke (1632–1704), who thought that the mind started out as a blank slate ("white paper") (Locke 1975 [1690], p. 59). Watson thought that introspection (the predominant way in which psychologists studied the mind in the early twentieth century) was ineffective because researchers could not agree about how the mind worked. Instead, he proposed the study of behavior, taking the reasonable position that knowledge comes largely from experience (Ornstein 1988, pp. 5–6). In the process, though, he underestimated the amount of preprogramming that evolution built into the brain and, almost certainly, the amount of genetic variation in intellectual capacity among individuals. The idea of the almost infinitely malleable human individual who is shaped by culture traces back at least to two Frenchmen, physician Jean-Marc-Gaspard Itard (1775–1838), who attempted to educate the "wild boy of Aveyron," who had been found living in a forest, and Émile Durkheim (1858–1917), one of the fathers of modern sociology. Durkheim stated that the natures of individuals "are merely the indeterminate

material that the social factor molds and transforms. Their contribution consists exclusively in very general attitudes, in vague and consequently plastic predispositions. . . ." (Durkheim 1962 [1895], pp. 105–106). The roots of a more modern approach to brain and mind, which tries to tie behavior to brain functioning, owe much to the work of psychologist Donald Hebb (Hebb 1949; Milner 1993).

104. Watson 1970 (1924), p. 104.
105. Geertz 1973, p. 49.
106. See, e.g., Wilson 1984, 1988, 1992.
107. E. O. Wilson 1975, p. 4. For a full discussion of the sociobiology controversy, see Segerstråle 2000.
108. For a discussion of this "standard social science model," see Barkow et al. 1992, pp. 24ff.
109. For a clear exposition of the complex history of this movement, see Degler 1991.
110. See, e.g., Ehrenreich and English 1979.
111. It wasn't only the notion that people's behavior could simply be made to conform to good socialist ideals through proper training, regardless of any genetic propensities, but also the notion that wheat plants could be turned into good, high-yielding, socialist wheat plants regardless of their genomes—as so clearly demonstrated by the deadly Lysenko affair, in which the first-rate geneticists in the Soviet Union were destroyed (see, e.g., Hull 1988, p. 61; Joravsky 1970).
112. When I tell this story to biology students, I always add that I'm sure that "peoples' photosynthesis" must go on in the "rubroplasts" of the leaves. Today, those radical students are doubtless tenured deconstructionist professors in departments of English language and literature.
113. For a recently analyzed example, see Brunner et al. 1993.
114. See chapter 1, notes 10 and 11.
115. See, e.g., Pinker 1997, pp. 386–389.
116. As Jared Diamond (1993, pp. 264ff.) points out, "If there is any single place in the world where we might expect an innate fear of snakes among native peoples, it would be in New Guinea, where one third or more of the snake species are poisonous and certain non-poisonous constrictor snakes are sufficiently big to be dangerous." Yet there is no sign of such innate fear among the indigenous people, and children regularly "capture large spiders, singe off the legs and hairs, and eat the bodies." The people there laugh at the idea of an inborn phobia for snakes, and they account for the fear in Europeans as a result of Europeans' stupidity in being unable to distinguish which snakes might be dangerous. There is reason to believe that fear of snakes in other primates is largely learned as well (Mineka et al. 1981 and references therein).
117. See, e.g., Buss 1994; Hamer and Copeland 1998; Geary 2000. Ridley 1999 is one of the best of this genre, informative even if (in my opinion) overenthusiastic.
118. Whether or not people have free will is basically a badly posed philosophical question. Because people think they are making free choices, and other people think that those choices are being made freely (a view that is inherent, for example, in justice systems that assign responsibility), then in my opinion there is free will. For an erudite consideration of the issue, which illuminates the complexity of the philosophical issues and draws conclusions with which I sometimes agree, see Kane 1998.
119. At a practical level, causes of behavior can often be discerned, and often behavior is coerced. But the issue of free will centers on uncoerced choices where causes of decisions made are not obvious.
120. Obviously, many animals also make choices, but "free will" is normally thought of in an ethical context and thus is probably pretty much limited to *Homo sapiens* today.
121. Allman 1999. Allman ties human brain expansion to environmental variability, especially variability in climate (pp. 192ff.); see also Potts 1996a, 1996b and the comments in Klein 1999, p. 251.
122. For a pioneering work on "natural categories," suggesting that our perceptual systems differentially recognize certain color and form categories, see Rosch 1973. For a comprehensive overview that builds on the way we categorize things (e.g., lions as dangerous) and calls into doubt many standard ideas of the mind and abstraction, see Lakoff 1987. Lakoff summarizes research in this area by saying it "shows clearly . . . that human categories are very much tied to human experiences and that any attempt to account for them free of such experience is doomed to failure" (p. 206). See also Berlin et al. 1973, 1974 for interesting discussions of categories in folk taxonomies of plants.
123. See the discussion of the "activity-independent period" in Tessier-Lavigne and Goodman 1996.
124. Meister et al. 1991; Katz and Shatz 1996. Interestingly, it appears that the sense of smell is under greater genetic control than the visual sense and the olfactory system does not require the input fine-tuning seen in the visual system (O'Leary et al. 1999), perhaps because olfaction provides less precise information.

125. This is the activity-dependent period of Goodman and Shatz 1993. For textbook overviews of what is known about the way the nervous system is assembled, see Müller 1997 and Kandel et al. 1995.
126. Kandel et al. 1995, pp. 688ff.
127. Instincts are usually viewed as complex behavioral responses, largely genetically programmed, to environmental cues. The term was important in early attempts to understand animal behavior, using a dichotomous "instinctive or learned" or "innate or intelligent" approach (e.g., Lorenz 1937, 1950, 1966; Tinbergen 1951). For early criticism of the dichotomous approach, see, e.g., Schneirla 1952; Lehrman 1953; Hebb 1953; and the discussion in Hinde 1966, pp. 315–321. The terms *instinct* and *innate* have faded from the technical literature (e.g., Krebs and Davies 1997) but hang on in popular parlance.
128. Ornstein 1988, p. 237; Hoffman 1998, pp. 17–19.
129. Kandel et al. 1995, pp. 470ff.; Hata and Stryker 1994.
130. Allman 1999, pp. 138ff.
131. Shatz 1992.
132. Allman 1999, p. 177.
133. Hebb 1949; Edelman 1987.
134. Edelman 1987; Edelman 1992, chap. 9.
135. Waddington 1960, p. 126. See also Kohlberg 1984, pp. 7ff.
136. For an overview of what is known about the cellular mechanisms of learning and memory—a complex and intriguing story of neurotransmitters, receptors, changes in neuron sensitivity, growth and loss of synapses, and so on—see Kandel et al. 1995, chap. 36. There is a vast literature on learning. Foundational works include Pavlov 1927 and Skinner 1938, and a wide array of approaches have been taken subsequently (e.g., Bandura 1986; Dewhirst and Berman 1978; Garcia and Ervin 1968; Gelman 1988; Johnston 1981; Seligman 1970; Tierney 1986). Teaching plays a very special role in learning by human beings (see, e.g., Premack 1984). For broad overviews of learning in *Homo sapiens*, see Leahy and Harris 1985; Ornstein 1988; Zimbardo and Gerrig 1999.
137. This is because a single gene often affects more than one characteristic (pleiotropy) and because of linkage and epistasis.
138. However, tastes and smells may be important to us in some circumstances, especially in connection with food choices and sexual behavior (see, e.g., Stoddart 1990).
139. The great philosopher Immanuel Kant, building on Plato's ideas, postulated that the universe has a dual character, divided between a "phenomenal" world, one made up of space and time, which are available to sense perception, and a "noumenal" world, one made up of objects that are not accessible to the senses but about which we are able to reason. It is interesting to speculate whether Kant would have considered the extension of human sensory abilities by instruments as transforming noumena into phenomena. Of course, no one has yet devised an instrument that would allow us to detect and decode transcendental moral codes, which were one of Kant's central concerns.
140. These are called proprioceptors.
141. That is, anyone raised in a "carpentered" world.
142. Gregory 1973, 1980.
143. Gestalt laws were formulated by Gestalt psychologists, who rejected a reductionist approach to the investigation of behavior and perception, instead studying responses to the entire set of elements into which people automatically organize their perceptions. *Gestalt* is a German word that does not translate easily into English but means roughly "overall integrated form." Thus, one might say, "He identified the bird from its gestalt rather than from the red spot on the nape of its neck."
144. For a good overview of the Gestalt laws, see Bruce et al. 1996, pp. 106–110.
145. It is difficult to know how we perceive things in real time but even more difficult to know how evolution has shaped our brains to store our perceptions—to form part of the crucial memories that so shape our individual natures. Images that are not being actively viewed seem to be stored in idealized form as fragments of real images (try to remember the details of the arrangement of your car's instrument panel—exactly where each gauge is and how it presents information). And there is good evidence that most thoughts are not stored as images, although a lot of us, me included, certainly do much of our thinking visually. For example, try to visualize the image that in your mind would stand for "hunger." Did you picture a starving individual? Which sex? Who? Did you see your own stomach empty? Or did you find an image for the abstract concept of hunger? (I can't.) And if that was easy, try "ravenous" or the more difficult "no longer ravenous but thirsty."

Steven Pinker says: "Pictures are ambiguous, but thoughts, virtually by definition, cannot be ambiguous. . . . If a mental picture is used to represent a thought, it needs to be accompanied by a caption, a set of instructions for how to interpret the picture—what to pay attention to and what to ignore" (Pinker 1997, p. 297). On the other hand, John Allman makes the point that all of the brain's representations of reality are inherently ambiguous, whether they take the form of pictures, words, or equations: "Reliance on verbal descriptions may be a particularly ineffective way of thinking about spatial structure or dynamic physical processes. Verbal labels may actually impede thought in these and many other areas. (The false dichotomy between the verbal labels 'nature' and 'nurture' is an example. The term 'module' as applied in cognitive function may be another.)" (Personal communication, 24 September 1999.) He notes how easily we receive complex messages from facial expressions for which there are not good verbal descriptors: "Words can enhance the meaning of images, but by the same token images can greatly enhance the meaning of words. In each case, more information is added to facilitate understanding." Once again, it is clear that our mental processes, so central to our natures, are barely beginning to be elucidated.

146. Hume 1977 (1777). David Hume (1711–1776) based his notions of causality on the recognition of the "constant conjunction" of objects or events. He did not deal with the now-famous problem that "correlation is not causation," a point emphasized by his contemporary Thomas Reid 1970 (1764), who theorized that our notions of causation trace to the early realization that we ourselves can cause things—the understanding that we have, as Reid called them, "active powers" (pp. 42ff., 293). Causality remains an active topic of philosophical discussion.
147. Michotte 1965, translated from the 1946 French edition. More specifically, Hume and Michotte were interested in "event causation" (Searle 1997, p. 7), not the kind of causation considered when we ask "What causes this table to seem solid?"
148. Beasley 1968; Boyle 1960.
149. Experience, though, can alter judgments of perceived causal events; see, e.g., Schlottmann and Shanks 1992; Schlottmann and Anderson 1993.
150. Leslie 1982, 1994; Leslie and Keeble 1987.
151. The research focuses especially on the way we recognize three-dimensional objects, often regardless of the angle from which we view them (and thus almost regardless of the two-dimensional stimulation of our retinas); see, e.g., Marr and Nishihara 1978; Biederman 1987; Biederman and Gerhardstein 1993; Bülthoff and Edelman 1992; Tarr 1995; Tarr and Bülthoff 1995; Lawson and Humphreys 1996. One fascinating and controversial idea is that complex shapes are in part recognized as combinations of some thirty-six idealized components, which psychologist Irving Biederman called "geons" (by analogy with protons, neutrons, and electrons, which are components of atoms). See Biederman 1987; the illustrations in the article are very informative. See also Biederman and Gerhardstein 1993; the authors argue that geons, not viewpoint, are critical to object recognition—a view that is now quite controversial (see the following). Thus, both a cup and a pail are seen as a combination of a cylindrical geon and a curved-tube geon, the difference being whether the tube is attached to the side of the cylinder (cup handle) or to its top (pail handle). This hypothesis has recognition being relatively viewpoint-independent, and it may well account for much of the way in which, at least at first, the apparently two-dimensional "snapshot" representations (e.g., Bülthoff and Edelman 1992; Tarr 1995) of objects in our brains are translated into perceptions (and recognition) of three-dimensional objects. Despite the evidence for this hypothesis, considerable evidence now also indicates that recognition can be heavily dependent on viewpoint as well. In this area, a possible consensus seems to be emerging that there are multiple routes to object recognition, as discussed in Tarr 1995; Tarr and Bültoff 1995; and Lawson and Humphreys 1996. See also the popular summary in Pinker 1997. I recommend Pinker's book even though I disagree with much that is in it—which is hardly surprising because there is so much in it. My major reservation is that Pinker seems to think that a relative handful of genes more or less directly specifies a gigantic amount of interconnected brain wiring and behavior. But his book is an excellent exposition of an entire genre in which, I think, the same unfortunate leap of faith occurs. And it is written with a wonderful sense of humor and lots of interesting anecdotes. For a more technical summary of geons and related issues, see Bruce et al. 1996, pp. 222–227.
152. Cohen 1999; Parr and Waal 1999.
153. Bruce and Humphreys 1994; Bruce et al. 1996, pp. 227–231.
154. Bruce and Humphreys 1994.
155. Etcoff et al. 1991.
156. Behrmann et al. 1992. Evidence from observations of another brain-damaged patient indicates that there

are separate visual and language-based representations of the characteristics of physical objects and that there is a submodule that specializes in recognition of animals (Hart Jr. and Gordon 1992).

157. Cohen 1999; Parr and Waal 1999.
158. In theory, that could trace to human social groups a million or so years ago having been large compared with those of, say, the ancestors of chimps or bonobos. Again, the assumptions of group size among our ancestors are based on indirect evidence (Dunbar 1993) and must be taken with a grain of salt.
159. Hudson 1960.
160. Turnbull 1961, pp. 304–305.
161. Segall et al. 1963, 1966.
162. Interestingly, Segall and co-workers (1966) showed appropriate concern about whether willingness to comply or fear might have contaminated the responses of colonial peoples being given tests by those perceived as their superiors or oppressors. Segall and colleagues attempted to control for that problem (for a discussion of related issues, see Segall 1963).
163. This theme is developed in more detail in Ornstein and Ehrlich 1989.
164. See note 94 in chapter 13 for some references to this literature.
165. Evolutionists often call speculations about adaptation "just so" stories, after a set of stories of the same name (including "How the Whale Got His Throat" and "How the Rhinoceros Got His Skin") by the great poet of the British Empire, Rudyard Kipling (1902).
166. See, e.g., Luria 1980, p. 179; Laughlin Jr. and d'Aquili 1974, p. 116. Another example is the dyadic structure of myths; see, e.g., d'Aquili and Laughlin Jr. 1975.
167. For an excellent summary of what may or may not be universal in human cultures and behavior, see Brown 1991.
168. On the other hand, the Gombe chimps are much more closely related to the Wamba bonobos of the Democratic Republic of Congo than they are to us—and the bonobos' forest habitat is certainly more similar to that of Gombe Stream than it is to Palo Alto. Nonetheless, some aspects of bonobo behavior (such as frequent copulation in the "missionary" position) is more like that of human beings than that of chimps (Tratz and Heck 1954; Waal 1997).
169. One popular notion that is quite pervasive is that we cannot find out much about human behavior from that of our living relatives because we're so much smarter than they are. Most people think that an unbridgeable "intelligence gap" between human beings and other animals has been created by the great expansion of our brains. In some instances, there is such a gap: no other animal can do higher mathematics or can design, build, and fly a 747. If, however, intelligence is defined as the ability to solve certain kinds of problems, then there is little sign of a gap even though in other areas the gap is immense. (A related issue is the degree to which teaching is found in nonhuman animals. For an overview, see Caro 1992.)

    Fishes, pigeons, dogs, apes, and people alike are able to solve a problem in which they are presented with a tray holding two food wells, one with a circular cover and one with a triangular one. At each trial, the food reward is always under the triangular cover, and subjects learn to respond to it, not to the position of the well. In more complex tests involving "oddity," trays with three food wells have two wells covered by identical covers and a third covered by an odd cover. The reward of food is always under the odd cover (which may be either a circle or a triangle). This problem cannot be solved by pigeons, rats, or dogs or by some young children. Another level of ambiguity is added when the color of the tray signals whether the food reward is under the odd object or under one of the matched objects. Nonprimates can't handle this one, but some monkeys and apes do it easily and can solve even more complex oddity problems. These tests are of the sort originally designed to measure human mental abilities, and many human beings cannot solve them. Psychologist Harry Harlow concluded on the basis of these results that defining human beings (as famous evolutionist Theodosius Dobzhansky once did) as possessing mental abilities that occur in other animals at most in rudimentary forms "must of necessity disenfranchise many millions of United States citizens from the society of *Homo sapiens*" (Harlow 1958). Nonetheless, when it comes to other areas of intelligence, some other animals can be just plain dumb. Daniel Dennett gives a good example: "People are less fond of telling tales of the jaw-dropping stupidity of their pets, and often resist the implications of the gaps they discover in their pets' competences. Such a smart doggie, but can he figure out how to unwind his leash when he runs around a tree or lamppost?" (Dennett 1996, p. 115).

    Indeed, I think Dobzhansky's statement is correct—but the gap in mental abilities between people and other animals is best seen not in human problem-solving abilities but in our possession of two characteristics unique among living animals that, among other things, allow us to generalize from our small intellectual triumphs to a much greater extent than any other living animals: intense consciousness and a language with syntax.

Chapter 7: From Grooming to Gossip?

1. Wilson 1998, p. 132.
2. This tribe of Inuit is sometimes also called Aivilik Eskimo or Aivilingmiut (Balikci 1970). My transliteration follows Sutton 1932.
3. The transliteration of Tommy Bruce's name is after Copeland 1960; after half a century, I give no guarantees about my transliterations of various other Inuit terms. I had learned about Bruce from the writings of George Miksch Sutton, a distinguished biologist who had done fieldwork on the island in 1929–1930. Sutton was primarily an ornithologist, but he had collected butterflies of the genus *Erebia*, a group that had fascinated me as a teenager (my collection of them is now in the American Museum of Natural History). Sutton described Tommy as both intelligent and helpful (Sutton 1932, p. 5), as indeed he was.
4. For a recent overview of some of the contentious issues in language evolution, see Hurford et al. 1998. I'm especially grateful to Derek Bickerton for comments on many of these issues. He and I disagree on some of them, but he has helped me greatly in sharpening my thinking. His recently published book, *Lingua ex Machina: Reconciling Darwin and Chomsky with the Human Brain*, written with a neuroscientist, William H. Calvin (Calvin and Bickerton 2000), is a fine exposition of their well-informed views.
5. Järvi and Bakken 1984. For further avian examples, see Ehrlich et al. 1988, pp. 591–593, 611–615.
6. For a detailed but popular overview of neo-Chomskyan linguistics (so called because they are rooted in the pioneering linguistic work of Noam Chomsky), with definitions of terms and thought-provoking material about the evolution of language, I recommend Pinker 1994. For a taste of the technical difficulty of linguistics from the old master himself, try Chomsky 1980, or, if you read Spanish, Chomsky 1988, which is in English but uses many Spanish examples. For a different view, one not focused on innate rules of grammar, see P. Lieberman 1998.
7. Liska 1994.
8. This phenomenon is sometimes called generativity (Corballis 1992).
9. Chomsky 1957, p. 15.
10. There is ample evidence that young children are genetically predisposed to recognize "most or all of the more than two dozen consonant sounds characteristic of human speech, including consonants not present in the language they normally hear" (Gould and Marler 1987).
11. See, e.g., Chomsky 1971, 1975.
12. Fernald 1992.
13. See Crain and Nakayama 1986 but also Crain 1991 and the discussions following it that contest various of his views. For other evidence that has been interpreted to mean that children are genetically programmed to use structure-dependent grammar rules, see Read and Schreiber 1982; Crain and McKee 1985; Crain and Thornton 1991. Unfortunately, in this work it is not possible to differentiate with assurance genetic programming from early learning.
14. Crain and Nakayama 1986. The grammatical errors were concentrated in the most difficult sentences—with many more correct phrasings when asking Jabba whether "the boy who is unhappy is watching Mickey Mouse" than whether "the boy who was holding the plate is crying."
15. See, e.g., Gordon 1985. Gordon concludes that children follow built-in word-formation rules "independent of the input received" (p. 73)—that is, that the rules are unlearned. For a nontechnical view of what is known about the way the brain is structured in relation to language, see Damasio and Damasio 1992.
16. Pinker 1994, e.g., p. 51. For insight into the ongoing debate over exactly how children learn grammar, see Rumelhart and McClelland 1986; Pinker and Prince 1988; Pinker 1991, 1999; and McClelland and Seidenberg 2000.
17. Van Petten and Bloom 1999.
18. Saffran et al. 1996; Bower 1999e; Marcus et al. 1999; Steinhauer et al. 1999. For general discussions of the capacities of young children, with frequent reference to language skills, see Mehler and Fox 1985.
19. This critical period is similar to the period in early development in which visual experiences "teach" the brain to see properly. Children blinded by cataracts at birth who have their sight restored some years later never are able to see properly (Newport 1990; Hurford 1991; for a popular account, see Hubel 1988, chap. 9). Although it is easy to postulate a selective advantage to early and rapid language acquisition, it is not clear exactly why the ability to learn languages decays with time. One appealing idea (Hurford's) is

that selection pressures favoring retention of the capacity decline rapidly after puberty. Another (Newport's) is that the development of other, nonlinguistic mental abilities interferes with the capacity to easily acquire language. Mixed in with both is the issue of the possible adaptive value of the long period of infancy in *Homo sapiens*, in which sensory systems develop sequentially, perhaps to avoid interference (Turkewitz and Kenny 1982). Those interested in these issues may wish to consult the literature on the evolution of senescence, which deals with the parallel problem of why selection should favor aging when longer life expectancies can be produced in selection experiments on fruit flies and mechanisms for flawless repair of cell and tissue damage could theoretically exist; see, e.g., Medawar 1952; G. C. Williams 1957; Hamilton 1966; Rose 1984; Rose and Charlesworth 1981; Partridge and Fowler 1992; Partridge and Barton 1993; Roper et al. 1993; Charlesworth 1994; Nesse and Williams 1995. For a summary, see Freeman and Herron 1998.

20. Bickerton 1985, pp. 137, 141.
21. See, e.g., Bickerton 1985; Bickerton 1990, p. 169; Jackendoff 1994, pp. 134–135; S. J. Roberts 1998.
22. Bickerton 1995, p. 39. Bickerton (1985) believes, contra Chomsky, that there is a single, genetically determined model available to a child, not a range of them. "It is only in pidgin-speaking communities, where there was no grammatical model that could compete with the child's innate grammar, that the innate grammatical model was not eventually suppressed. The innate grammar was then clothed in whatever vocabulary was locally available and gave rise to the creole languages heard today" (p. 147). Thus, creoles are easier to learn than other languages, and children do learn a "native creole" at first. Then, under pressure from the speakers around them, they change the grammar of the creole "until it conforms to that of the local language."
23. See, e.g., Metter et al. 1987. Technically, the prefrontal cortex is the cortex of the frontal lobe, in front of the premotor cortex. For an overview of the physical spreading of language functions, see P. Lieberman 1998, pp. 100ff.
24. Calvin 1996b, p. 141.
25. Personal communication, 19 February 1999.
26. See, e.g., Berko 1958; Katz et al. 1974; Hirsh-Pasek and Golinkoff 1991; Macnamara 1982; Crain 1991.
27. Jackendoff 1994. There is some confusion about whether deaf children "babble" in sign language (MacNeilage 1998; Meier and Newport 1990; Petitto and Marentette 1991.
28. Goldin-Meadow and Feldman 1977; Goldin-Meadow 1978, 1982; Goldin-Meadow and Mylander 1984. Susan Goldin-Meadow thinks that language probably coevolved with toolmaking and points out: "The deaf children in our studies, while lacking conventional language, nevertheless had access to the artifacts which evolved along with language and which could have served as supports for the child's invention of a language-like system for communicating both within and beyond the here-and-now" (Goldin-Meadow 1993, p. 82).
29. See, e.g., Goldin-Meadow and Mylander 1990; Goldin-Meadow et al. 1994.
30. Jackendoff 1994, pp. 146–151.
31. Gopnik and Crago 1991.
32. Kuhl et al. 1992. There are similarities in the discrimination of speech sounds by people and other animals (Kuhl 1981).
33. I could, of course, be wrong here. Differences of opinion among well-informed people such as Chomsky, Pinker, Lieberman, and Bickerton are standard in science and are one of the things that makes science so much fun.
34. Pinker 1994, p. 57.
35. The examples are from Pinker 1994, pp. 79, 452.
36. The example is based on Pinker 1994, p. 79.
37. This example is also based on Pinker 1994, pp. 79–80. For more examples of the differences between representations in the brain and what is communicated by language, see pp. 80–81.
38. Pinker 1994, chap. 3.
39. Roe 1953, p. 146.
40. An example can be found in the perspectival writings of Friedrich Nietzsche (e.g., 1996 [1878], pp. 18–19).
41. Wittgenstein 1921, 1953.
42. See, e.g., Hoijer (1954).
43. Whorf 1956, p. 221.
44. Alford 1978; Lucy 1985.
45. Or, more specifically, not only English but also French, German, and other European languages, which

he lumped into a category called Standard Average European (SAE) because "with respect to the traits compared, there is little difference [between them]" (Whorf 1941).

46. Malotki 1983. One putative example of language shaping worldview is the plethora of words the Inuit have for different kinds of snow—an example that is not central to the Whorfian hypothesis (after all, biologists have many names for different kinds of organisms) and has generally confused the issue (Martin 1986). For an amusing essay on Inuit vocabulary, see Pullum 1991, pp. 159–171.
47. Lenneberg 1953; Pinker 1994. Berlin and Kay (1969) mistakenly thought that their work refuted Whorf's—see Alford 1978—but they disproved a version of the hypothesis not traceable to Whorf himself. Whorf's own views were complex and not easily categorized (see, e.g., Lakoff 1987, pp. 324–325).
48. See, e.g., Rumsey 1990; Slobin 1990, 1991; Lucy 1997.
49. Lucy 1992a. Some of Lucy's early work related to the Sapir-Whorf hypothesis was on the issue of how people divide up the color spectrum (Lucy and Shweder 1979)—an area dominated by the work of Berlin and Kay (1969), which demonstrated the cross-cultural universality of a limited number of basic color terms, similarity in ways in which terms are added to the lexicon, and a close matching of actual colors chosen to match the basic terms (Sahlins 1976). For a new formulation of the Sapir-Whorf hypothesis, see Lucy 1992b.
50. For a brief summary, see Foley 1997.
51. See, e.g., Brown 1957; Brown and Lenneberg 1954; Kay et al. 1991; Gumperz and Levinson 1991. See also the debate on differences in use of counterfactual statements in English and Chinese (Au 1983, 1984; Bloom 1984). The late Richard Holm and I were fascinated by the Sapir-Whorf hypothesis forty years ago (Ehrlich and Holm 1962, 1963).
52. See, e.g., Witkowski and Brown 1982. After an analysis of color nomenclature, they concluded that their findings were "consistent with both a universalist interpretation of color terminology and a Whorfian hypothesis which asserts that language exerts an active influence on thought and behavior" (p. 419).
53. Davidoff et al. 1999.
54. Despite some cultural differences that might be traceable to the diversity of human languages, there are many human universals and near universals. Universals are traits or complexes, as anthropologist Donald Brown says, "present in all individuals (or all individuals of a particular sex or age range), all societies, all cultures, or all languages—provided that [they are] not too obviously anatomical or physiological or too remote from the higher mental functions" (Brown 1991, p. 42). The existence of religious beliefs and (to date) male dominance of politics are two examples of such universals. So many universals are tightly tied to language that despite the influence of different languages on worldviews, those views must retain a fundamental similarity, even if the Sapir-Whorf hypothesis holds in some instances (Brown 1991). For a more technical discussion of "the common basis for all human language," see Greenberg 1975. Greenberg points out, among other things, that languages universally have a dominant word order. The three most common sequences of word order are, in order of frequency among languages, subject-object-verb (SOV), subject-verb-object (SVO), and verb-subject-object (VSO). (The dominant word order in the English language is SVO.) Of the three other possible sequences, VOS occurs very rarely, but OSV and OVS are unknown. Knowledge of the dominant sequence of a given language can enable one to predict other of its characteristics. See also Greenberg 1987 for an interesting discussion of marking as a linguistic universal (*he* is the unmarked default term of a pair, and *she* is the marked version, which cannot be substituted if sex is unknown or ambiguous; *steward* is unmarked, *stewardess* marked) and Friedrich 1975 for a discussion of the "natural iconicity" of the coding of shapes. Interestingly, the complexity of the terms for some concepts of property rights and social sanctions to protect them makes it difficult to compare them across cultures (see, e.g., Hallowell 1943) and even to develop a reasonable discourse about such rights within our own society (see, e.g., Bromley 1991). Just how similar they are in various dimensions can be determined only with more careful research to enable us to fully understand "how language shapes, and is shaped by, the nature of our knowledge" (Hill and Mannheim 1992, p. 401).
55. Pinker 1994, p. 232; Ruhlen 1994, pp. 4, 62. There is a contentious literature on language evolution, focusing in part on how to determine whether sources of similarity are the result of what might be thought of as parallel evolution within languages or the result of borrowing after cultural contact (see, e.g., Thomason and Kaufman 1988).
56. Klicka and Zink 1997.
57. Nichols 1998, p. 128.
58. Cultural evolution of languages is so rapid that most linguists believe that linguistic descent cannot be traced back more than 10,000–12,000 years and, in most cases, not more than 6,000 years (Nichols 1998); see also Renfrew 1998. However, this is not a universal view (Bengtson and Ruhlen 1994).

59. Bengtson and Ruhlen 1994.
60. Bengtson and Ruhlen 1994; Ruhlen 1994; Merritt Ruhlen, personal communication, 28 March 1999.
61. See, e.g., Livingstone 1963; Cavalli-Sforza et al. 1988; Barbujani and Sokal 1990; Barrantes et al. 1990; Barbujani and Sokal 1991.
62. There is a single reported child fossil interpreted as a hybrid between *Homo neanderthalensis* and *H. sapiens* (Duarte et al. 1999).
63. For an overview of the issue and an introduction to the variety of modern opinions, see Pinker and Bloom 1990, the commentaries following it, and the papers in Wind et al. 1992.
64. See, e.g., Davidson and Noble 1989; Mellars 1991; Noble and Davidson 1991.
65. See, e.g., Bradshaw 1991; Greenfield 1991; Foley 1991b; Aiello and Dunbar 1993; Pinker 1994; P. Lieberman 1998. Vocal communication in monkeys has acoustic patterns in common with human vocal communication (Maurus et al. 1988), and there is much other evidence to suggest that human language was assembled evolutionarily from building blocks widespread in living animals (Beynon and Rasa 1989; Bradshaw 1991; Cheney and Seyfarth 1990a; Seyfarth et al. 1980). Philosopher Thomas Reid made the observation that before what he called "artificial languages" could evolve, there had to be "natural languages" (e.g., cries, tears, facial expressions, gestures) that allowed communication of agreement on the social conventions that make up the natural language (Reid 1970 [1764], pp. 55ff.).
66. Nottebohm 1993; Bauer 1993. This, of course, doesn't mean that the vocalizations of those organisms are homologous to those of human beings.
67. Seyfarth et al. 1980; Cheney and Seyfarth 1990a, pp. 102–110. The different alarm calls produced different responses—leopard alarms caused the monkeys to climb trees; eagle alarms caused them to dash into bushes; snake alarms caused them to search the ground around them. Similar communication has been recorded in mongooses (Beynon and Rasa 1989).
68. Bickerton 1990. Bickerton, who is responsible for the term in its modern sense, himself contends that a protolanguage would have no syntax at all. Some workers have questioned the utility of intermediate or primitive forms of grammar and thus whether grammar could evolve. Such questioning seems to be grounded on a misunderstanding of evolution, communication in general, and language in particular (see, e.g., Bates et al. 1991). For a recent discussion of the vexing issue of the origins of syntax, see Hurford et al. 1998.
69. Savage-Rumbaugh 1991; Savage-Rumbaugh and Rumbaugh 1993. Kanzi also learned to make stone tools by imitating human beings—once he was shown how useful they could be (especially sharp-edged flakes he could use to cut a cord to get into a box to obtain a reward). He never, however, achieved the ability to make recognizable artifacts at the level of Oldowan technology (Schick and Toth 1993, pp. 136–140).
70. But he didn't seem to grasp the meaning of prepositions—his knowledge of the world was sufficient to divine that *go, get, microwave,* and *tomato* could not mean "Get the microwave out of the tomato" (P. Lieberman 1998, p. 44).
71. Coren 1994, pp. 114–115.
72. Brannon and Terrace 1998; Carey 1998.
73. Beynon and Rasa 1989.
74. Classic work here was done by a husband-and-wife team, Beatrix T. Gardner and R. Allen Gardner, starting when they adopted a baby chimpanzee, Washoe (Gardner and Gardner 1971; Gardner and Gardner 1984; Gardner et al. 1989). Unfortunately, the Gardners' work lost support after it was erroneously attacked on the basis of another attempt to communicate with a chimp, Project Nim (Terrace et al. 1979). See the critique of the latter work in P. Lieberman 1998, pp. 38–39. Lieberman gives details of the experimental protocol of the Terrace group, which are not presented in the necessarily abbreviated description of Project Nim in the *Science* article just cited.
75. See, e.g., Chomsky 1972; Piattelli-Palmarini 1989.
76. For a discussion of the many changes during hominid evolution that are related to the emergence of speech, see Wind 1983.
77. Darwin 1859, p. 191.
78. P. Lieberman 1998, p. 139.
79. For a good discussion of laryngeal anatomy, see P. Lieberman 1998, pp. 137ff.
80. It is true that Rhodesian man (archaic *H. sapiens*) had essentially the modern degree of basicranial flexion, but Neanderthals had very flat cranial bases, as far as can be told from the three specimens in which the degree of flexion can be established. What can be questioned is whether only those with the modern

degree of flexion could produce full human vocal language (Richard Klein, personal communication, late 1999).

81. Arensburg et al. 1990; Duchin 1990; P. Lieberman 1998, p. 137. Earlier reports cast doubt on the vocal ability of Neanderthals (Lieberman and Crelin 1971; Lieberman et al. 1972).
82. Kay et al. 1998.
83. Pinker 1994, p. 354. My own view is that much of the focus on the anatomy of the vocal tract (e.g., Duchin 1990) as an indicator of the origin of language is an evolutionary red herring and may not even be all that important at some levels in vocal communication (as the quote from Pinker implies). Refinement of speech is a different issue (P. Lieberman 1998).
84. Crawford 1937; Miles 1991; Mitchell and Miles 1993.
85. See, e.g., Savage-Rumbaugh 1991.
86. Savage-Rumbaugh et al. 1996. Bees, and perhaps some other animals, can signal directions to other individuals of their groups.
87. An icon is not entirely arbitrary, as is a pure symbol—for example, a line in the sand to indicate direction would be an icon; four hoots to indicate the same direction would be a symbol. Icons transmit information in part by sharing characteristics with what they represent.
88. Boesch 1991a.
89. Discussion of the significance of gestures is scattered throughout the vast literature on the origins of language—see, e.g., Hewes 1973b; McNeill 1985; Davidson and Noble 1989; Corballis 1992, 1999; Iverson and Goldin-Meadow 1998.
90. Hewlett and Cavalli-Sforza 1986.
91. Aoki and Feldman 1989.
92. This made-up example borrows from the language spontaneously developed by deaf children (Goldin-Meadow and Feldman 1977). A key issue, though, is whether the gestures of people who already have a spoken language tell us anything about those of deaf youngsters who have not been exposed to signing adults.
93. For an interesting discussion of gestures in relation to language, see Kendon 1997.
94. See, e.g., Savage-Rumbaugh and Rumbaugh 1993, p. 88.
95. Goldin-Meadow 1993.
96. Hewes 1973a; see also Wynn 1993 and Armstrong et al. 1994 and the discussion following it. For an overview of the connection between tools and language, see Gibson 1991. A discussion of human nonverbal communication, which downgrades the possibility that it was involved in the evolution of language, can be found in Burling 1993b; see also the comments that follow it directly as well as Davidson and Noble 1993; P. Lieberman 1993; Ragir 1993; and Burling 1993a. Unhappily, it is very difficult to confirm a connection of gestures and protolanguage through the fossil record (see, e.g., Dibble 1989). Indeed, this problem plagues attempts to reconstruct not only the origins of language (e.g., Davidson 1991) but also the origins of many aspects of human behavior.
97. Calvin and Bickerton 2000.
98. Calvin 1993, 1996b.
99. Reynolds 1993.
100. It has been suggested that a switch from gestural to vocal communication accompanied the appearance of *Homo sapiens* in Africa (Corballis 1994).
101. Bickerton 1995.
102. The implication is that the vertebrate eye could not have evolved by small, incremental steps. But in fact, a single light-sensitive spot is, for most animals, better than no light sensitivity whatsoever. Computer modeling of the evolution of the eye by a process of natural selection is easy and shows that there has been superabundant geological time for eyes such as ours to have evolved by natural selection. Indeed, theoretically there would have been enough time for vision to have evolved, to have been lost by natural selection, and then to have evolved again tens of thousands of times—see Nilsson and Pelger 1994. (Selection favors eyelessness where there is no light, as in cave fishes that are descended from sighted fishes. For such fishes, eyes would be useless organs, vulnerable to damage and infection.) If eyes did evolve more than once in an evolutionary line, they would be expected to have different forms each time, as do the eyes of dragonflies, squid, and mammals today. For some mysteries (and answers) about the evolution of eyes, see Goldsmith 1990.

    The notion of rapid reorganization also recalls the ideas of geneticist Richard Goldschmidt, who postulated a sudden mutation-like reorganization of the entire genome to produce a new species (Goldschmidt 1940). Goldschmidt thought (correctly) that such reorganizations would result in "hopeful monsters," most of which would be inferior, but he believed (incorrectly) that a significant proportion would be the principal source of entirely new species. Goldschmidt was a distinguished geneticist, well known for his work on the gypsy moth genus *Lymantria.* For a refutation of Goldschmidt's ideas about speciation, see Mayr 1942, pp. 137–138; Stebbins 1950, p. 194; and Dobzhansky 1951, pp. 202–204. Goldschmidt's ideas were originally very popular with paleontologists (Mayr 1997, p. 194).

103. See, e.g., Davidson and Noble 1989 and the discussion that follows it and Byers 1994 and discussion. After struggling through one anthropologist's especially prolix and jargon-laden argument, which was nearly devoid of concrete examples (Byers 1994), I was relieved to read the comment of another anthropologist (a self-described semiotician—one whose work is concerned with signs and symbols in languages): "I must confess I had difficulty following [the] argument, even after multiple close readings and careful comparison with an earlier paper . . . that makes many of the same points" (Parmentier 1994, p. 388). For a similar comment about another mind-paralyzing exercise, see Black 1989.
104. See, e.g., Newmeyer 1991; Foley 1997, pp. 66ff.
105. Bickerton 1995, p. 51. Some linguists believe that protolanguages consisted of large lexicons with no syntax. Bickerton has an interesting discussion of protolanguages based in part on his interest in pidgin and creole languages. Remember that children of pidgin users sometimes develop creoles as their first language. This progressive development of syntax may mimic aspects of the evolution of human language abilities as a whole.
106. Calvin and Bickerton 2000.
107. For further debate on these issues, see Bickerton 1986 and Premack 1986. Premack's argument (1985, p. 282) that "human language is an embarrassment for evolutionary theory because it is vastly more powerful than one can account for in terms of selective fitness" simply reflects a misunderstanding of selection. Obviously, better ability to communicate can provide reproductive advantages in social animals. It is more difficult to explain the relative paucity of communication ability in chimps and bonobos than to explain its refinement in human beings—a "choice point" problem (see the discussion later in this chapter, under "Why Us and Not Them?").
108. Dunbar 1992. Dunbar's data pretty much dispose of explanations of increased brain size in higher primates based on a series of ecological hypotheses, such as the hypothesis that fruit-eating primates need more neocortical volume than do leaf-eaters in order to monitor a resource that is comparatively sparse and scattered in space and time (Clutton-Brock and Harvey 1980) or the hypothesis that it is primarily a matter of needing more neocortex to store mental maps of larger home ranges or to exploit food items (such as nuts in hard shells) that are embedded in a matrix (see, e.g., Gibson 1990; Mace and Harvey 1983). Even though the reasons for early neocortex growth may have been ecological, later growth was probably driven by selection pressures related to social behavior. In any case, the human neocortex is about three times the size that would be expected for a nonhuman primate of equivalent body size, and it also differs in patterns of anatomical specialization (Passingham 1973).
109. For example, Allman and Hasenstaub (1999) carefully analyzed the Stephan database (Stephan et al. 1984, 1987) and found weak, statistically insignificant correlations between size of social group and size of neocortex and other brain structures. Allman (personal communication, 1 November 1999) is dubious about Dunbar's (e.g., 1998a) analysis of the relationship between neocortex and group size, in part because of the way Dunbar measured the visual cortex in an attempt to refine his study. Group size tends to be larger in highly visual primates and smaller in those that rely more heavily on their olfactory abilities. Allman and Hasenstaub also note that Dunbar did not include the orangutans, which have a large neocortex and small group size (Allman, personal communication, 1 November 1999).
110. Dunbar (1991) showed that frequency of social grooming correlated with group size in primates but not with body size, suggesting that grooming could not have a purely hygienic function (but the correlation with body size was higher in the New World, or platyrrhine, monkeys). See also Reynolds 1981; Reynolds argues that grooming is the primate behavior from which cooperative tool construction in the human line evolved.
111. Muroyama and Sugiyama 1994, p. 169.
112. Aiello and Dunbar 1993; Dunbar 1993.
113. Range 90–220, mean 153. The significance of human group size has long fascinated social scientists (see, e.g., Simmel 1902), but there has been little investigation of even such elementary questions as how crowding affects human behavior; see Ehrlich and Freedman 1971 and Freedman et al. 1971.
114. Killworth et al. 1984. In contrast, intimate circles tend to be more in the range of 10–15 persons (Dunbar and Spoors 1995).
115. Dunbar 1991.

116. There also appears to be physical limits on verbal grooming in the absence of technical aids such as telephones and e-mail. Dunbar and colleagues have assembled evidence that the maximum size of freely forming conversational groups is about four persons (Dunbar et al. 1995).
117. For a good popular summary of this hypothesis, see Dunbar 1996.
118. It should not be assumed that this hypothesis necessarily involves group selection. Individuals who spent less time physically grooming and more time verbally gossiping could, at this point, outreproduce persistent groomers.
119. Iwamoto and Dunbar 1983.
120. Aiello and Dunbar 1993, p. 187.
121. Aiello and Dunbar 1993, p. 190.
122. These points were emphasized to me by brain scientist John Allman in personal communications on 24 September and 1 November 1999.
123. The explanations that follow are based on Aiello and Dunbar 1993.
124. Hill and Lee 1998.
125. Aiello and Wheeler 1995. The energy demand of the brain is a function of the energetic needs of individual neurons, which require a great deal of energy to maintain their resting membrane potential (the ability to transmit a nerve impulse rapidly) (Kandel et al. 1995) as well as to carry out the operations involved in altering the arrangements of neurons in learning and memory (e.g., changing the dendritic processes, strengthening and weakening synapses). As a result, even though the human brain accounts for only some 2 percent of the body by weight, it uses about 20 percent of the body's energy (Aiello and Wheeler 1995; Passingham 1973; Potts 1996b).
126. Energetically expensive nonbrain organs in human beings, such as the liver and the gastrointestinal tract, have been compensatorily reduced as the brain has grown, allowing brains to increase in size without a rise in basal metabolic rate (Aiello and Wheeler 1995). For an overview of the current debate about energetic constraints on brain size, see Gibbons 1998d.
127. This does not mean group selection, although it is conceivable that the rather restrictive conditions required for group selection to work (see chapter 2) pertained.
128. Mellars 1996, pp. 398ff.
129. There is a large literature on the idea that human intelligence evolved primarily to handle the problems of living in more or less permanent groups, in which a balance between cooperation and competition must be maintained. That kind of intelligence has been coined Machiavellian intelligence (Byrne 1996; Byrne and Whiten 1988; Byrne and Whiten 1992; Whiten and Byrne 1997). Other workers, however, emphasize natural history intelligence, which is designed to help an otherwise relatively helpless animal deal with factors outside the group—see the discussion in Fox 1997, pp. 79ff. M. R. A. Chance (1967, 1975, 1962), looking at the "attention structure" of primate societies (the amount of attention subordinates pay to dominant individuals), concluded that those Old World primates, such as baboons and Japanese macaques, that concentrated much of their attention on dominant males were less able than chimpanzees (and presumably our ancestors) to develop natural history intelligence. Chance extends his conclusions to *Homo sapiens*, stating that "peer groups create rank-ordered social relations wherever the culture permits it" (1975, p. 104).
130. Bickerton 1990, 1995.
131. Bickerton (1998) believes that developing protolanguages have no syntax whatsoever.
132. Sahlins 1972.
133. The argument can be made that great apes in general had entered that niche (Byrne 1994).
134. Schick and Toth 1993, p. 128.
135. See, e.g., Clark 1994; Hublin et al. 1996; Mellars 1996; Howells 1997; Conroy 1997; Tattersall 1998a; Klein 1999. There are signs, though, that less energy-efficient hunting techniques might have put the Neanderthals at a disadvantage in some areas (D. Lieberman 1993).
136. Roberts 1992.
137. It has also been argued that the Great Leap "was not the 'origin' or 'beginning' of art" (Marshack 1997, p. 53). See also Marshack 1972.
138. An argument that there was no brain reorganization can be found in Wynn 1991.
139. See, e.g., Lindly and Clark 1990 (see also the commentaries following it); Milo and Quiatt 1994; Klein 1998.
140. Average brain size, however, seems to have reached its maximum well before the Great Leap (Allman 1999, p. 194; Ruff et al. 1997).
141. Klein 2000a.

142. The issue of just how innovative our ancestors were and we are today is the subject of an interesting if inconclusive literature; see, e.g., van der Leeuw and Torrence 1989, especially the chapter by Stephen Shennan (1989). For an especially stimulating essay, see Fox 1997, chap. 3. Some of the best examples of innovation and, especially, failure of innovation can be extracted from the literature on warfare (e.g., Keegan 1993, 1998; Rosen 1991).
143. The issue of when *Homo sapiens* reached Australia bears on this reconstruction. Current suggestions that it occurred as much as 60,000 years ago seem unlikely, but if the extinction of the large flightless bird *Genyornis newtoni* was caused by human beings, a date of 50,000 years ago may be possible (Miller et al. 1999), implying a very rapid spread of post–Great Leap people. See also Klein 2000a.
144. The story is further complicated because part of the cultural transition could be accounted for by a statistical artifact of the sort that may be involved in generating biological patterns of punctuation in the fossil record (Bergman and Feldman 1999).

Chapter 8: Blood's a Rover

1. Lee 1968b, p. 3.
2. Jacob 1982, p. 5.
3. Housman 1951, p. 12.
4. Tooby and Cosmides 1992, p. 69; Symons 1979, p. 36.
5. Foley 1987; Potts 1996b; Potts 1998, pp. 242–243; Foley 1999.
6. Hunting and gathering can be viewed as a mode of production, which can be compared with other modes (pastoralism, agriculture) in terms of efficiency: the yield of food relative to the effort invested. See the articles in Ingold et al. 1991 (1988), especially that by Ingold in vol. 1 (pp. 269–276).
7. See, e.g., Pälsson 1988.
8. This assumes an average of roughly twenty years per generation over 5 million years; generation times of modern *Homo sapiens* are generally assumed to be about twenty-five years—see, e.g., Rogers and Jorde 1995. Fossil evidence suggests that australopithecines rarely survived beyond that age (Dumond 1975), and today, female chimpanzees begin reproducing at the age of about fourteen or fifteen (Goodall 1986, p. 81).
9. There is some question as to whether or not it was only agricultural peoples that managed to colonize the rain forests (Bailey et al. 1989).
10. One example of relatively rapid evolution in human beings might be the effect of female infanticide on sex ratio (Laland et al. 1995).
11. Eventually, microscopic examination of patterns of wear from use might change this to a degree. See, e.g., Hayden 1979.
12. A classic example is the hand ax, which probably was a multipurpose cutting and smashing tool.
13. Foley 1988, pp. 219–220.
14. Ingold 1991 (1988); Ellen 1994.
15. Dart 1957.
16. Washburn 1960.
17. Kuroda et al. 1996; Basabose and Yamagiwa 1997. Chimpanzees have a different suite of prey available to them in their mostly forest habitats, as opposed to the largely savanna areas that presumably were home to the australopithecines.
18. Ihobe 1997.
19. Stanford 1999.
20. Even gorillas, the most herbivorous of the great apes, both purposely and inadvertently consume insects and snails (Harcourt and Harcourt 1984).
21. Harding 1975.
22. Bartlett and Bartlett 1961; Altmann and Altmann 1970; Harding 1975; Hamilton 1987.
23. In hominoids (apes and people) in general, there seems to be a positive relationship between meat-eating and tool use (McGrew 1989).
24. Klein 1978, 1982, 1987; Bunn and Kroll 1986 and the commentaries following it.
25. Bunn and Kroll 1986; Shipman 1986; Klein 1999. Bunn and Kroll have good pictures of fossil bones that apparently show signs of having been worked with stone tools. For the complexities of interpreting the material and determining how much meat was consumed and how it was obtained, see the excellent discussion in Klein 1999, pp. 239–248. The long debate over the degree to which our various ancestors scavenged as opposed to hunted (e.g., Binford 1985; Chase 1988) seems far from settled; for a recent discus-

sion, see Klein 2000a. For some 5 million years, our ancestors roamed, first over Africa's savannas (although some scientists think that our most distant ancestors started taking on human characteristics first in forests—see, e.g., Boesch-Achermann and Boesch 1994) and eventually, beginning just a few tens of thousands of years ago, over almost all the planet. The australopithecines and earliest members of the genus *Homo* may themselves have been confined to the warmer parts of Africa in part because they could not control fire to warm themselves. It has been claimed, however, that *H. habilis* may have occurred in southern China as well (Huang et al. 1995), but the evidence presented in this paper is very fragmentary, based on a few teeth and jaw fragments of an "indeterminate" species that has affinities "with *H. habilis* and *H. ergaster*."

26. Heinzelin et al. 1999.
27. Richard Klein, personal communication, 1 May 1999.
28. See, e.g., Klein et al. 1999.
29. Kaplan and Hill 1985a and discussion following it.
30. Sharing can be a complex business, however, and it occurs in forms that do not involve generosity in the same sense as in modern Western societies (Peterson 1993).
31. Isaac 1978; Potts 1984; Rapoport 1994; Klein 2000a.
32. Tattersall 1995, p. 136.
33. Brooks 1996; Fruth and Hohmann 1994.
34. The relationship of technological changes captured in the archaeological record to increasing brain volume in evolving hominids is, you will recall, not clear. At one extreme is the view that new cultural activities and improved techniques may have developed gradually along with enlarging brains and that apparent revolutions in implements and practices are just a result of accidents of sampling and other coincidences. Such coincidences are sometimes technically referred to as statistical artifacts. These explanations are similar to some (using the statistical theory of Markov chains) of periods of stasis and punctuation in the fossil record of life (Bergman and Feldman 1995). At the other extreme, the technological changes could have been associated with a series of population explosions for which the causes are not known but that could have been the result of rapid increases in mental abilities that increased survivorship; see, e.g., Polgar 1972. Whether the technological changes facilitated the population explosion by permitting the intensification of production or whether they were driven by a need for more efficient production created by the explosion is a contentious issue, especially when considering the agricultural revolution. See also the discussion in Cowgill 1975. One revolution could have occurred around the time *Homo habilis* appeared, another with *H. ergaster/erectus* (leading to the first out-of-Africa dispersal), another with the evolution of archaic into modern *H. sapiens* (the second out-of-Africa event and the preagricultural Great Leap Forward). Brain volumes increased significantly at the first two of those stages, with essentially modern volumes reached only within the past 200,000 years or so.
35. See Harris 1989. Harris believes that *Homo erectus* would have been unable to resist the lure of abundant meat in the form of the game herds on the savanna (p. 46). This book, like his others, makes a good read. Modern *H. sapiens* hunters do use thrusting spears, but they are generally restricted to special circumstances and smaller game, such as peccaries (Churchill 1993).
36. See, e.g., Klein 1999.
37. Defleur et al. 1999; Culotta 1999. There are putative other cases of prehistoric cannibalism stretching back almost 1 million years (Klein 1999, pp. 360–361; Bower 1999a).
38. Dart (1948) described a new species associated with what he took to be evidence of fire use as *Australopithecus prometheus*, in honor of the heroic Greek who recaptured fire from Zeus and returned it to Earth.
39. Clark and Harris 1985; Bellomo 1994; Johanson and Edgar 1996, pp. 96–97. But see Klein 1999, pp. 237–238.
40. It is very difficult to obtain evidence about human use of fire in the distant past (Bellomo 1993). Charcoal deposited in the open disappears from the fossil record; ancient caves tend to collapse; and it is often difficult to distinguish anthropogenic from naturally occurring fire. There is some evidence that fire was used as far back in time as 1.5–1.0 million years ago (Brain and Sillen 1988), so its use by *H. ergaster/erectus* is certainly possible.
41. Straus 1989; James 1989 and comments that follow it; Schick and Toth 1993; Klein 1999, pp. 237–238, 350–354; Richard Klein, personal communication, late 1999. Doubt has recently been thrown on the textbook case of the first proven use of fire (by *H. erectus*) in China (Weiner et al. 1998).
42. Ehrlich and Raven 1964; Ehrlich 1970. Avoiding toxicity of plants was a constant challenge for hunter-gatherers, and some of them doubtless ate various clays that bind toxins (Johns 1989). In the Amazon Basin, parrots eat clay at certain cliffs, presumably also to help them deal with toxic plant foods. For a

complex hypothesis about the role of cooking in human evolution, suggesting that cooking goes back almost 2 million years and greatly influenced male–female relationships and reduced sexual size dimorphism, see Wrangham et al. 1997 and the discussion that follows it. Especially interesting are the comments in Brace 1999. Brace sees fire having been brought under control much more recently, less than 300,000 years ago, in connection with permanent occupation of the Temperate Zone and the problems of dealing with the frozen remains of big game animals. He thinks the decrease in dimorphism with the evolution of *Homo* from *Australopithecus* was caused by the increased size of females required to enable them to carry fetuses for nine months and bear large-brained babies.

43. Lieberman 1987.
44. O'Connell et al. 1999.
45. For an extended discussion, see Kaplan et al. 1999.
46. Aiello and Wheeler 1995; Pennisi 1999b.
47. Wrangham et al. 1997; O'Connell et al. 1999.
48. See, e.g., Stanford 1999. The original work emphasizing the importance of hunting in human evolution was Lee and DeVore's (1968b) edited book *Man the Hunter*, especially a chapter titled "The Evolution of Hunting" (Washburn and Lancaster 1968).
49. Heinzelin et al. 1999.
50. For a synthetic view on the evolution of human diets, in which meat plays a key role in addition to vegetable matter, see Milton 1999.
51. See, e.g., Brooks 1996.
52. Beyries 1988; Shea 1989; Klein 2000a.
53. Diamond 1989b, 1991.
54. The first evidence is indirect, but needles and beads and eventually direct evidence in burials showed up in the Gravettian culture, 28,000–21,000 years ago (Klein 1999, pp. 535–536); see also Barber 1994.
55. Conkey et al. 1997. Establishment of what were the earliest art objects has proven quite difficult (see, e.g., d'Errico and Villa 1997), as has been their interpretation (d'Errico and Cacho 1994).
56. See, e.g., Chase and Dibble 1987.
57. Klein 1999, pp. 490–491, 537ff., 548ff.
58. Pringle 1998a.
59. Kuttruff et al. 1998.
60. Peterkin et al. 1993.
61. See Martin and Klein 1984. This fine edited book gives a balanced treatment of the climate-change versus overkill hypotheses. See especially the interesting chapter by Jared Diamond (1984), which uses information about historical extinctions to cast light on prehistoric ones. Despite further data, the relative importance of climate and hunting remains in doubt as far as many continental extinctions are concerned (Barnosky 1989). Still, it seems likely that human hunting has often been a major factor—a case argued ably by Tim Flannery (1994, 1999) for the disappearance of the Australian megafauna soon after the arrival of *Homo sapiens* (see also Levy 1999; Miller et al. 1999). The dominant role of human hunting seems clear in island episodes such as the destruction of the moas of New Zealand (A. Anderson 1984; Trotter and McCulloch 1984), whereas the extinction of the Irish elk appears to be a result of climate change (Barnosky 1986). My guess, as indicated, is that both hunting and climate change were factors in almost all continental megafaunal extinctions, with the relative roles varying geographically and among species (see, e.g., Mithen 1993). Populations that were reduced in size (or contained weakened individuals) because of climate change might easily have been pushed past the point of recovery by hunters. There were more megafaunal extinctions in North America (where people had recently invaded) than there were in Europe, where *Homo* had been present for tens of thousands of years. That might be explained by the prey having had much greater evolutionary experience with human hunting in Europe, coevolving defenses against gradually improving human skills.
62. Martin 1967, 1984. See also Diamond 1991, pp. 339–348.
63. Kay 1994, 1995.
64. There have been many attempts to reconstruct the behavior of prehistoric hunters based primarily on detailed examination of animal remains, but evidence is sparse and difficult to analyze, and the behavior of the hunters doubtless was variable in both space and time. This leads to brave attempts at deriving behavioral conclusions from very limited data. For instance, Daniel Lieberman (1993) concluded that in the Levant, "Neanderthals may have used habitats in a different, *less* seasonally mobile manner than modern *Homo sapiens*" (p. 216). This may have increased their energetic cost of foraging and might have been

one factor that put them at a competitive disadvantage with our ancestors. Then again, the foregoing may boil down to a cascading set of assumptions. See also Lieberman and Shea 1994.

65. See, e.g., Frison 1986.
66. Balikci 1970, pp. 44–45.
67. Hames 1991; Alvard 1993a, 1993b; Kay 1994, 1995; Low 1996; Winterhalder and Lu 1997; Alvard 1998. The same can be said of pastoralists, who frequently turn huge areas into wastelands by overgrazing.
68. J. M. Diamond 1993, p. 267.
69. See, e.g., Terborgh 1988.
70. Aiello and Dunbar 1993; Dunbar 1993. See also Birdsell 1968 and the references therein.
71. Just a few generations ago, most of humanity still lived in those small groups, and there is no reason to believe that in the short time that has elapsed since then, selection of behavioral traits in response to increasing group size has ever been as intense as, say, selection for DDT resistance in the fruit flies of the Animal House—or that it has had much, if any, genetic evolutionary effect.
72. See, e.g., Kirch 1991b; Weisler and Kirch 1996; Mellars 1996, chap. 5 and pp. 398–401.
73. Cashdan 1989, pp. 42ff.
74. Berdan 1989, p. 104. Trade may also have been an important facilitator of intertribal marriage and thus a mechanism for restraining intertribal warfare (Podolefsky 1984).
75. Hamilton et al. 1990.
76. See, e.g., van den Berghe and Frost 1986.
77. Selection pressures may also change in response to the frequency of types in populations—with, for example, less frequent hair colors being favored (do blondes have more fun?). Sexual selection pressures thus can vary with the frequencies of certain genes in a population—that is, they can be frequency-dependent. Frequency-dependent selection occurs when the fitness of different genotypes depends on their frequency in a population. Some predators (including human beings) form "search images" (Tinbergen 1960) that allow them to find prey with a certain pattern (say, a snail with a striped shell as opposed to an unstriped one) more readily than those with other patterns. Then predation tends to be more severe on the commonest genotypes in the population, which are thus reduced in number until another genotype becomes more frequent. The predators then shift to them. Such a system of frequency-dependent selection (sometimes called apostatic selection) can maintain genetic variability in a population (Ayala and Campbell 1974). In some cases, it may be the commonest genotype that is favored (see, e.g., Mallet and Barton 1989).

    Frequency dependence explains why there is about the same number of men as women in a population—selection that favors less common types until they increase in frequency usually produces a roughly even sex ratio in animals such as *Homo sapiens.* If for some reason the genes causing the production of females become more frequent in a population, males will become rarer. On average, then, males will reproduce more (because on average there will be more females for each one to inseminate). The females carrying genes that favor the production of males (presumably by a mechanism such as one that adjusts vaginal conditions to favor sperm carrying the male-determining chromosome) would then have more grandchildren, and those genes would then become more frequent and shift the sex ratio back toward equality. (This explanation was originally set forth by Charles Darwin [1871] but is usually attributed to Sir Ronald Fisher [1930].) The situation is actually somewhat more complicated than this, but the details need not concern us here. The point is that selection favors genes leading to production of the rarer sex.

    The existence of two sexes rather than just one is one of the major mysteries in biology, on which there is an extensive literature. It is a puzzle because asexual reproduction is the easiest way to maximize the presence of one's own genes in the next generation (rather than getting only a 50 percent representation by sharing reproduction with a sexual partner). For a recent overview, see Wuethrich 1998; Barton and Charlesworth 1998; and the other articles in the same issue of *Science*, especially the one on genomic imprinting (Pennisi 1998), for a strange twist on the genetics of sexual reproduction.

78. There is plenty of evidence to support this notion (e.g., Cavalli-Sforza and Bodmer 1971; Harcourt 1995; Harcourt and Gardiner 1994; Vogel and Motulsky 1996). Recently, there have been claims that people actually avoid mating with individuals who have the same genotypes at major histocompatibility complex (MHC) loci (human leukocyte antigen, or HLA, loci) (Ober et al. 1997); but see Hedrick and Black 1997.
79. Bamberger 1974, p. 266. However, power relationships between the sexes can be very complex and can vary with environmental circumstances (Hayden et al. 1986; Langness 1974; Turnbull 1982), and women are often more powerful than men in certain circumstances (Gewertz 1981).
80. Bamberger 1974, p. 280.
81. Ortner 1974, pp. 67–68.

82. For an interesting discussion, see Ralls 1977.
83. E. O. Wilson 1975.
84. Kano 1992; Waal 1997, p. 24.
85. Martin et al. 1994.
86. Reynolds 1994. However, it has not been possible so far to positively link sexual dimorphism in body weight to degree of competition for females.
87. McHenry 1992b.
88. McHenry 1992b; Short 1994; Martin et al. 1994; Klein 1999, p. 192.
89. See, e.g., Symons 1982; Donald Symons, personal communication, 25 January 1999.
90. For an interesting discussion of male supremacy, see M. Harris 1977, chap. 6. Anthropologists have long realized that the pattern of patriarchy goes back to prehistory; see, e.g., Rivers 1924, among many others.
91. Chagnon 1992, p. 147.
92. See, e.g., Herdt 1982, 1984; Harris 1989, pp. 292–293.
93. The groups are then said to be patrilocal. In human beings, kinship systems can be complex (for explanations, see Harris 1997), but roughly 70 percent of known societies are patrilocal in the sense used here (Murdock 1967).
94. Matrilocality. See, e.g., Reynolds 1994.
95. When lineages become too large to exploit local resources effectively or otherwise become difficult to manage, they split, but often the daughter lineages continue to recognize their consanguinity and collectively become units that anthropologists describe as clans.
96. Bar-Yosef and Belfer-Cohen 1991, p. 182.
97. Ehrlich and Roughgarden 1987.
98. See, e.g., Gill and Wolf 1975.
99. See the discussion in Dyson-Hudson and Smith 1978.
100. Keeley 1996, pp. 108ff., 86ff.
101. Divale and Harris 1976; Hamilton 1982; Harris and Ross 1987a, p. 56.
102. Gregor 1990.
103. Palmer 1989. There has been substantial debate on the causes and meaning of rape in Western societies (Backman and Backman 1997; Brownmiller 1975; Groth and Burgess 1977a, 1977b; Kanin 1957; Malamuth 1986; Scully and Marolla 1985; Symons 1979; Zillmann 1984). Many analysts emphasize the violent aspect of rape over the sexual, often grounding it in the patriarchal structure of human societies. Others, seeing a connection to the relatively small parental investment of males and the connection of sex with violence in nonhuman primates and other mammals, propose that there is some genetic predisposition to rape in male primates (Ellis 1989; Ghiglieri 1999; Smuts and Smuts 1993); for an overview, see Pavelka 1995. I suspect that a strong sexual urge, male physical and cultural dominance, and lack of appropriate social learning and cultural restraints for some individuals all are contributing factors. Here, violence and sex seem completely intertwined. "More directly violent" excludes such things as relations between male slave owners and female slaves in the antebellum South, which were largely condoned.
104. Altman 1975; Malmberg 1980; Sack 1987; Taylor 1988; Rapoport 1994.
105. I'm indebted to John Allman for suggesting this interpretation (personal communication, 1 November 1999).
106. The same can be said for "personal space," the little "bubble" of space that each of us maintains between ourselves and other members of society. Nor does the culture-versus-genes question seem terribly important, given that both territoriality and personal space are clearly very environmentally labile—for example, concepts of personal space differ greatly among modern cultures. Arabs very closely approach individuals to whom they are speaking, bathing them with their breath. As anthropologist Edward T. Hall, who pioneered the study of personal space, put it (from an Arab's perspective): "To smell one's friend is not only nice but desirable, for to deny him your breath is to act ashamed. Americans on the other hand, trained as they are not to breathe in people's faces, automatically communicate shame in trying to be polite" (Hall 1966, p. 160).
107. Idani 1991b; Waal 1995, 1997.
108. I hope you don't need a reference for this, but if you do, see Diamond 1997b, or go to the movies and listen to popular music.
109. Lack of dependable results makes it difficult, for instance, to discover such things as the relationship between frequency of copulation and fertility; see, e.g., Nag 1972. See also Berk et al. 1995.
110. Schröder 1992.

111. Betzig 1989.
112. Petrie and Kempenaers 1998.
113. Even though gibbon pairs mate for years, there is plenty of sexual activity outside of the pair bond (Anonymous 1998). Much complexity and variation are buried in the term *monogamous* (Palombit 1994, 1996).
114. E.g., Diamond 1997b.
115. Baker and Bellis 1995; Dunbar 1998b.
116. Ehrlich et al. 1988.
117. Watts 1996.
118. Wilson 1978, p. 125.
119. Short 1994, p. 13. Extreme forms of polygyny do occur, with husbands having six or more wives, in which case it can encourage homosexual behavior on the part of young unmarried males and on the part of wives who have limited sexual access to their husbands (Evans-Pritchard 1970).
120. Information about harems and dimorphism is from Diamond 1991, pp. 60–61.
121. Martin et al. 1994. Two other common mating systems, polyandry and promiscuity, are not found as the dominant system in any human society. Polyandry is much rarer than monogamy or polygyny, especially in primates, among which it is found only in tamarins of the genus *Saguinus* (Garber 1997).
122. Harcourt et al. 1981; Goodall 1986; Waal 1997.
123. Short 1979.
124. Human males are possibly outdone in that department only by bonobos (Waal 1995). Some believe that large penises, like fatty breasts underlying mammary glands, are basically secondary sexual characteristics, ones not necessary to "get the job done" (Sheets-Johnstone 1989). Both are clearly characteristics that are largely genetically controlled, and both have been presumed to function to enhance pair-bonding (Short 1994, p. 13). There is actually little evidence for that role, however, aside from attempts in many cultures to enhance these characteristics, as in males' use of penis sheaths (in New Guinea) and codpieces (Europe in the early modern period) and females' use of breast enhancement. At least in Western culture, women seem unimpressed with the size or other characteristics of penises. In one survey, women were asked to name the parts of the male anatomy they most admired. Only 2 percent expressed interest in penises, whereas 39 percent were interested in buttocks (R. L. Smith 1984, p. 630).
125. R. L. Smith 1984, p. 630.
126. Baker and Bellis 1995.
127. Martan and Shepard 1976; Dewsbury 1984.
128. Baker and Bellis 1995. It has been suggested that in some primates, perhaps even in *Homo sapiens*, sperm are a limiting resource for females—see Small 1988 and comments following it. Soon after I left the Animal House and began teaching at Stanford University, Pat Labine, one of my best graduate students, discovered sperm competition in a checkerspot butterfly (*Euphydryas editha*, a species that our group now has been studying for more than forty years). Pat showed that when mating, which takes an hour or more, males of the butterfly transfer to the females a nutrient-rich structure called a spermatophore, which contains their sperm. The spermatophore is often finished with a plug that, once hardened, blocks the female's vagina, preventing a second male from mating with her (Labine 1964). In a difficult and ingenious series of experiments, Pat, among other things, sterilized males with radiation and then mated them with fertile females that had previously been mated to normal males and had their plugs removed. The results showed "sperm precedence" because most of the eggs laid after the second mating were not fertile. Her work showed that if two fertile males mated with the same female, the sperm from the last male won the competition and fertilized the eggs (Labine 1966). Thus, male checkerspot butterflies appear to have evolved the plugging mechanism to protect their genetic investment.
129. Eliasson and Lindholmer 1972.
130. Ford and Beach 1951; Donald Symons, personal communication, 25 January 1999.
131. Baker and Bellis 1995, pp. 166–167.
132. Stoddart 1990, pp. 68–70. Stoddart points out that the situation is complex and reactions to the odor may be largely subconscious (and, in some societies, highly negative when the odor is rancid), but he nonetheless supports the attraction hypothesis.
133. Ovulation in chimpanzees occurs around the end of maximal tumescence, and if there is no fertilization, menstruation begins about nine days after detumescence starts. It lasts about three days (Goodall 1986), although there is some variation among chimp populations. The cycle is similar in bonobos (Takahata et al. 1996).
134. There have been claims that the degree of concealment is overstated (e.g., Small 1993, 1996), but at the

very least, evolution appears to have reduced the level of advertisement of ovulation to both sexes (Alexander and Noonan 1979).

135. Doty et al. 1975.
136. McClintock 1971; Graham and McGrew 1980; Russell et al. 1980; Weller et al. 1995; see also Whitten 1999.
137. Stoddart 1990, p. 106.
138. I say "more or less" because there is some evidence that women are not as continuously receptive as is sometimes claimed. One study found a statistically significant monthly peak in sexual activity "for all female-initiated behavior, including both autosexual and female-initiated heterosexual behavior" (Adams et al. 1978). See also Manson 1986; Manson found that sexual arousal peaked twice, once around ovulation and once before menstruation.
139. Chimpanzees and bonobos also have lactational (postpartum) amenorrhea. Chimps, on average, do not cycle for more than four years after giving birth; bonobos resume cycling in less than a year (Takahata et al. 1996).
140. Strassmann 1997. For more of a popular account, see Small 1999. In fact, the high frequency of menstruation in modern women may be related to the high frequency of such problems as reproductive cancers and endometriosis (see, e.g., Maynard Smith et al. 1999).
141. Harrell 1981, p. 817; see also Derr 1982.
142. Profet 1993.
143. Strassmann 1996b.
144. See, e.g., Strassmann 1996a. Some scientists (e.g., Finn 1998) believe that menstruation is a nonadaptive consequence of selection operating on other aspects of the uterine environment.
145. Broude and Greene 1976; M. Harris 1977; Herdt 1982; Gregor 1985; Harris 1997, pp. 334–335. See also Buckley 1982 and Child and Child 1985. Child and Child suggest that menstrual odors, if transferred to men by sexual contact, might make it more difficult for the men to approach game animals while hunting.
146. Strassmann 1991; Small 1999.
147. Strassmann 1996c.
148. Johnson 1983. The material here is from a personal communication on 16 August 1999.
149. See also M. Harris 1977, pp. 85–86.
150. Evolutionists, remember, are rightly nervous about invoking selection operating among groups, as opposed to individuals, because of the rigorous conditions required for differential survival of groups to override the force of individual selection within groups, e.g., possibly favoring men who spend their time trying to inseminate more females instead of helping to provision offspring.
151. See the quotes in Symons 1979, p. 107.
152. For example, a male might pass more of his genes to the next generation by consorting with many females, whereas each female might be more reproductively successful if she monopolized the male's help. Parents and offspring—indeed, even mothers and fetuses (see, e.g., Haig 1993)—do not have entirely congruent evolutionary interests.
153. Alexander and Noonan 1979.
154. Anthropologist Donald Symons, in his classic but controversial treatise on human sexual evolution (1979), describes two other scenarios for how natural selection may have caused the loss of estrus. Symons's scenarios and the one described in the text lean heavily on the need for males to be confident that if they put effort into caring for a female (and, directly or indirectly, her offspring), they will be promoting their own genes. Both of Symons's scenarios are also tied to the development of a division of labor in which males hunted and females gathered. One is the meat-for-sex scenario, which neatly brings together eating and sexual activity. This scenario was suggested by the observation that female chimps in estrus were more successful in getting meat from males than were females not in estrus (for an overview, see Stanford 1999, pp. 199ff.). Among hominids, male hunters who obtained a surplus of meat might trade it to females in estrus in return for copulations, thus increasing the male's fitness, or they might give it to relatives, in which case it could increase their inclusive fitness. Loss of estrus benefited females by ending male confidence that copulations would be likely to produce offspring, shifting the balance of power so that males would have to provision females continuously in return for a long-term sexual contract with one or more females—the origins of marriage. The females gain by getting meat more reliably over time, and the males respond with devices that allow them to monopolize the reproductive capacity of females and that contribute to the viability of those females and their offspring.

    The other Symons scheme is a sex-on-the-side scenario. In this scenario, pair bonds evolved while females still experienced estrus, promoted by the advantages of a sexual division of labor once males became proficient hunters. The most desirable males could be married to only a few women at most—so the females' choices for a male mate with whom to form a pair bond were limited. By the evolutionary loss of estrus, females increased their opportunities to be fertilized on the sly by males more desirable than their bonded mates. The loss made the problem of mate-guarding for males a difficult and perpetual chore. The male previously could limit his vigilance to the period of estrus, leaving time for hunting expeditions and the like. In the absence of estrus, however, avoidance of provisioning other males' offspring became much more difficult.

155. Diamond 1997b.
156. Hrdy 1977, 1981, 1983. As this book goes to press, Hrdy has just published a provocative book examining motherhood. It includes, among other things, an examination of female infanticide and of the importance of the notion of partible paternity, wherein some human groups believe that babies can have multiple fathers, receiving contributions from various men who mate with a pregnant woman (Hrdy 1999).
157. Sugiyama 1965; Steenbeek et al. 1999.
158. See, e.g., Fossey 1983.
159. Interestingly, it is possible that female baboons, at least, fake estrus after a male takeover, possibly to keep the males from killing their offspring to bring on genuine ovulation (Motluk 1997).
160. Andelman 1987.
161. Hausfater and Hrdy 1984.
162. Goodall 1986.
163. Dixson 1998, p. 70 (emphasis in the original).
164. Daly 1984; Hill and Kaplan 1988; Tudge 1997.
165. Dixson 1998, p. 71. An exception to this rule of nonparent–consort infanticide is the tendency in some societies, such as those of China and India, for female children to be killed or to die of neglect because parents favor sons. Sen (1992) includes general neglect as probably more important than overt infanticide in producing the 80 million women estimated to be "missing" in India and China. See also Vines 1993 and Ehrlich et al. 1995, p. 48. Dickeman (1979) suggests that production of more male children in these societies results in enough more descendants in later generations to counterbalance the selective costs of female infanticide or neglect. The issue remains unresolved (see Alexander 1988; Kitcher 1985).
166. Burley 1979.
167. Some women today claim to be able to detect their own ovulation (Small 1993, p. 195).
168. There is another possible reason why female human beings would deceive themselves about their time of ovulation. Animal behaviorists Richard Alexander and Katherine Noonan (1979) suggested that it is simply because social deception may be more likely to succeed if the deceiver is also self-deceived. But there is a contrary view. Symons suggested that it may actually be easier to lie when one knows he or she is lying. As he put it: "The truth fits seamlessly into the world, and doesn't require managing. Lies don't, and constantly need superintending so that other supporting lies can be told (the tangled web we weave when first we practice to deceive)" (Symons, personal communication, 25 January 1999). I usually add a further layer of complexity to the tangled-web argument with "But if we practice day and night, we learn to weave each strand just right"! Human behavior rarely yields to simple explanations.
169. The most convincing demonstration has been of female–female communication through a pheromone or pheromones produced in the armpits of donors that change the length of the ovulatory cycle of recipient women (Stern and McClintock 1998); see also the discussion in Weller 1998.
170. Robert Sapolsky, personal communication, 24 February 1999.
171. Sillén-Tullberg and Møller 1993.
172. This assumes that concealed ovulation indeed serves a direct function, which I believe is likely.
173. For a recent discussion, see Pawlowski 1999 and the commentaries following it.
174. I cannot help but wonder whether any role was played by those related hallmarks of human evolution—the interacting expansion of the brain (especially the prefrontal cortex) and development of enhanced intellectual skills, a complex culture, and, especially, language. A greater capacity to analyze the world could have added an element of the "not for me" scenario as some females, finally understanding the connection between copulation and pregnancy, tried to avoid copulation around the time of ovulation. This, of course, would have amounted to a countervailing selective force against the evolution of great intelligence (and, perhaps, understanding of early language) because the females who hadn't figured out the consequences of copulation would have had more children. But in populations in which language could be used to spread the news of the "sex causes babies" connection, most females might have the desire to limit their reproduction below the maximum level, and any selective difference between those who had

figured out the connection on their own and those who hadn't would be reduced. Therefore, it seems unlikely that the "not for me" scenario provides the chief explanation for concealed ovulation. Once concealed ovulation was fixed in a population in which the possible results of copulation around the time of ovulation was well understood, selection against females who were better at detecting their own ovulation and avoiding intercourse at those times could have helped to maintain the concealment (just as selection would have favored men who somehow were better at detecting ovulation in females). But again, the selection pressure would be weak, as suggested by the high failure rate of the "rhythm" method of birth control. This assumes that women often would want to avoid getting pregnant, which, considering the manifest social benefits of motherhood in most societies, would not often be the case, especially early in their reproductive careers.

If, on the other hand, ovulation had become largely cryptic under pressure from males committing infanticide, an increase in female intelligence could have been a force keeping that trait at the concealment end of the spectrum. Smart females who recognized the male threat and could detect their fertile periods presumably might attempt to copulate often and with numerous males when they were not ovulating, reducing their consorts' confidence in paternity while maintaining some control over how many offspring to have and with whom. That, again, would create a situation of selection for concealment (its presumed effect on intelligence would depend on what assumptions are made), but it involves postulation of a lot of reasoning about an area in which people even today tend not to be altogether rational.

175. Donald Symons, personal communication, 25 January 1999. Symons's book *The Evolution of Human Sexuality* (1979) is a landmark volume in this area.
176. A puzzle related to loss of estrus, for which various explanations have been advanced, is why ovulation ends rather early in a woman's life. That is, why did menopause evolve? The most likely explanation is that as women age, not only does childbearing grow increasingly dangerous but also they are playing a very important role in rearing the children they have already borne. Remember, fitness is measured not by the number of children a woman bears but by the number who survive to reproduce in the next generation. Therefore, a woman who ceases reproduction before old age and concentrates on raising her children may prove much more fit than one who continues to reproduce. This behavior may be rooted in mother–child food-sharing, in which grandmothers can contribute more of their genes to posterity by helping to feed the daughters who are producing their grandchildren. This grandmothering hypothesis is discussed in detail in Hawkes et al. 1998. For an opposing view, see Packer et al. 1998.

    Like many other animals, human beings must balance the fitness costs and benefits of trying to produce and rear large broods against those of producing and rearing small broods. Furthermore, in preliterate societies, older women are important repositories of cultural knowledge. Because these societies are structured around family groups, by surviving into old age and grandmotherhood a woman may increase her inclusive fitness (which, you may recall, is promotion of one's genes not just by reproducing but also by aiding the reproduction of relatives with identical genes). For excellent overviews of this topic, see Diamond 1997b, chap. 6, and, especially, Sherman 1998. See also Hill and Hurtado 1991; Austad 1994; Perls et al. 1997; Hawkes et al. 1997 and the discussion following it.
177. Ford and Beach 1951. Ford and Beach also report a few rather minor exceptions.
178. Large body size may have lost out evolutionarily in allocation of energy to the development of energy-hungry brains. Sexual size dimorphism seems evolutionarily related to harem-keeping (as in gorillas), which would be an increasingly unsuccessful reproductive strategy for males in a society in which ovulation signals were disappearing and intelligence was increasing.
179. Grooming remains an important human activity where lice are prevalent—for an interesting account of grooming and gossip in the Pyrenees, see Ladurie 1979, p. 10.
180. Schröder 1992.
181. Gagneux et al. 1997.
182. This study reminded me of the way DNA studies surprised ornithologists when they revealed that offspring in the nests of "monogamous" birds, those presumed paragons of sexual fidelity, often have multiple fathers and sometimes multiple mothers (females may lay eggs in the nests of other females) (see, e.g., Petrie and Kempenaers 1998).
183. See, e.g., Manderson 1995.
184. More technically, ventro-ventral.
185. Takahata et al. 1996.
186. Waal 1995, p. 48.
187. Small 1993, pp. 139–140.
188. See, e.g., Burton 1971; Goldfoot et al. 1980.

189. Small 1993, p. 149.
190. Williamson and Nowak 1998. Among other things, that tissue serves to clamp the urethra shut during intercourse, possibly to block the movement of bacteria toward the bladder.
191. Although the presence of a single gene on the Y chromosome (the SRY gene of the short arm) determines whether testes will develop, other genes influencing sexual characteristics are scattered throughout the genome and are "turned on or off" depending on the hormones secreted by testes or ovaries (Gilbert 1997, chap. 20).
192. See, e.g., Trivers 1972; Symons 1979.
193. Small 1993, p. 28.
194. Symons 1979.
195. Bermant 1976.
196. See, e.g., Bermant et al. 1969.
197. Hrdy 1983.
198. Or one publisher of a "men's magazine" who allegedly copulated with more than 2,000 young women in a twenty-year period (Nobile 1974).
199. It is important to remember, however, that the *average* number of heterosexual contacts must be the same for men and women—only the *distributions* can differ. In a population of 100 men and 100 women, if the average number of different contacts for a man is 5, the average number for a woman must be 5 as well. But, of course, every man may have had sex with 5 women, whereas 90 women may have been monogamous and 10 prostitutes may each have had sex with 41 men.
200. Symons 1979, pp. 292ff. See also Bell and Weinberg 1978.
201. Clark and Hatfield 1989.
202. Small 1992a, 1992b, 1993.
203. Whyte 1978.
204. See, e.g., Hayes 1975; Edgerton 1992. Female genital mutilation differs in degree and significance, the latter of which is sometimes far removed from any goal of sexual repression (Ellen Gruenbaum, personal communication, 22 March 1999).
205. See, e.g., Ghiglieri 1987; Freedman 1979; Schröder 1992.
206. See, e.g., Harcourt 1981; Wrangham 1979.
207. Ghiglieri 1987, p. 331.
208. Symons 1979, p. 203; Hrdy 1983; Buss 1989; Small 1992a, 1992b, 1993; Schaik and Paul 1996. Chimpanzee females also exercise a degree of mate choice (see, e.g., Nishida 1997).
209. The women accomplished this by squatting over 250-milliliter glass beakers and, when necessary, sneezing or urinating to induce the flow.
210. Gould 1987; Konner 1990, pp. 181ff.
211. Morris 1967; Baker and Bellis 1993b. The term traces to the idea of dropping "as if pole-axed."
212. Fox et al. 1970; Singer 1973; Baker and Bellis 1993b.
213. Much of the material in this paragraph is based on Baker and Bellis 1993b.
214. Englert et al. 1986.
215. Baker and Bellis 1993a, 1995.
216. One interesting evolutionary result, if Baker and Bellis are correct, is that there should be a powerful selective pressure for men who can easily and quickly bring women to orgasm and then ejaculate immediately—especially where there is a chance of sperm competition (this was suggested by Donald Symons in a personal communication on 14 January 1999). Anecdotal evidence seems to suggest that if there is a genetic bias toward such performance, it has been culturally overridden in many men.
217. For more about sperm competition in primates, see Harcourt et al. 1995 and Harcourt 1997.
218. Baker and Bellis 1995, p. 161. Furthermore, a recent report suggests that women may actually change their mate preferences during different stages of their menstrual cycle to enhance their chances of being fertilized by an especially healthy, very masculine male (while remaining bonded to a less masculine male who is more likely to cooperate in child rearing) (Penton-Voak et al. 1999).
219. Bellis and Baker 1990.
220. Berk et al. 1995.
221. Baker and Bellis 1989, 1993a.
222. That might mean a greater chance of extra-pair copulations.
223. Baker and Bellis 1993a.
224. Brown 1991, p. 109.
225. See, e.g., Brecher 1969, p. 257. The nature and extent of the sexual revolution is not well documented,

and some claim that it had minimal effect (e.g., Laumann et al. 1994); but they depend on the results of surveys, yet the value of data based on interviews is difficult to evaluate—see also Buss 1994.

226. As usual, there are too many factors for them to really be sorted out. For instance, the situation regarding sexually transmitted diseases has changed, especially with the appearance of AIDS, and even though abortions are legal, they are increasingly difficult and unpleasant to obtain in the United States because of the terrorist activities of some anti-abortionists.
227. It is not possible to give a precise figure because definitions of homosexuality vary and survey results on attitudes and behavior are often unreliable (M. Diamond 1993; Fay et al. 1989; Laumann et al. 1994). Because of biases against homosexuality, the level of underreporting may be substantial. There are many varieties of homosexual behavior, including differences in preferred sex role (e.g., "inserter" or "insertee"), which seem to be strongly culturally influenced (Carrier 1977). Homosexuality is condemned by a substantial minority of people, who believe that it is a matter of choice, as opposed to a matter of genetic determination (Aguero et al. 1984; Whitley Jr. 1990).
228. See, e.g., Evans-Pritchard 1970; Broude and Greene 1976; Herdt 1988, 1984; Whitam and Mathy 1986; Fulton and Anderson 1992.
229. Vines 1999; Bagemihl 1999. The genetic control of homosexual behavior in fruit flies is gradually being uncovered by detailed experimental analysis, with very interesting results for the hierarchical genetic control of brain and behavior (see, e.g., Ryner et al. 1996).
230. Furuichi and Ihobe 1994; Waal 1997, pp. 66–67.
231. Waal and Roosmalen 1979; Goodall 1986; Waal 1989; Hashimoto and Furuichi 1994.
232. Eibl-Eibesfeldt 1977.
233. Waal 1995.
234. Herdt (1984).
235. The idea that people are conceived sexually neutral and are made heterosexual or homosexual traces back to the first "gay activist," the German judge Karl Heinrich Ulrichs, who crusaded for gay rights in the middle of the nineteenth century. Similar views were held by Sigmund Freud. For details, see the excellent summary in LeVay 1996.
236. The cells of genetic males have one X and one Y chromosome; females have two X chromosomes. Because the eggs of females can have only an X chromosome, whereas sperm can have either an X or a Y, the father is the primary determiner of the sex of his children (I say primary because, for instance, a factor such as the acidity of the vagina may differentially select for or against Y-bearing sperm).
237. LeVay 1991 and the discussion in Gilbert 1997, pp. 787–788. See also Zhou et al. 1995.
238. There is a large literature showing the influence of hormonal changes early in development on later sexual behavior in mammals (e.g., Pomerantz et al. 1986).
239. Roughgarden 1999. The frequency of intersexuality is probably underestimated, and the problems of intersexuals have been too long ignored (see, e.g., Fausto-Sterling 1993).
240. See, e.g., Phoenix et al. 1959; Beatty 1992.
241. Goy et al. 1988.
242. Zhou et al. 1995.
243. The penis accident story is based on Diamond and Sigmundson 1997 and Colapinto 1997.
244. See, e.g., Money and Ehrhardt 1972.
245. Duckett and Baskin 1993.
246. Perlmutter and Reitelman 1992.
247. Signs of gender preference have been reported in children as young as three months of age, according to Anne Campbell and colleagues at the University of Durham in England (reported in Anonymous 1997).
248. Ernulf and Innala 1989.
249. See, e.g., Pattatucci and Hamer 1995. For a recent careful discussion, see Dixson 1998, pp. 164ff.
250. Hamer et al. 1993; Hu et al. 1995.
251. Wickelgren 1999; Rice et al. 1999.
252. Another possibility is some form of heterozygote advantage (in which individuals carrying both alleles of a gene are favored over those carrying just one), as in the case of the sickle-cell trait discussed in chapter 2 (see, e.g., MacIntyre and Estep 1993), but the possible selective advantage of heterozygotes for a homosexuality allele is not known. What evidence there is about the reproductive output of homosexual women does not support the view that any genes for that trait are selected against; homosexual women seem to be about as reproductively successful as heterosexual women. In addition, homosexual men begin their sexual activities early and may gain competence by practicing together (Davenport 1965) for later bisexual roles (exclusive homosexuality is relatively rare) or heterosexual roles—see also Baker and Bellis

1995. Baker and Bellis go beyond this to weave an ingenious story about the selective value of bisexuality that I find too complex to be credible. Finally, one might ask whether homosexual individuals' helping of relatives provides them with a high level of inclusive fitness even if they don't personally reproduce much. Again, there are no data to support such a conjecture.

253. Harris 1989, pp. 240–245; Thornton 1997; personal communication, Bruce S. Thornton and Victor Hanson, 22 March 1999. The received wisdom of the Greeks' having considered all homosexual behavior natural is simply not accurate—men who allowed themselves to be penetrated were despised.
254. Kelly 1976. This can also be viewed as men imitating womens' nursing (McElvaine 2000).
255. When one considers the great difficulties of determining whether there is a genetic component that biases individuals toward homosexual behavior (a characteristic with an apparently direct effect on fitness), it is easy to see why we must remain agnostic about issues such as the selective effect of various aspects of sexual "attractiveness," a subject on which there is a growing but inconclusive literature. See, e.g., Langlois et al. 1987; Langlois and Roggman 1990; Buss et al. 1990; Buss et al. 1992; Buss and Schmitt 1993; Buss 1994; Jankowiak et al. 1992; Singh 1993a, 1993b, 1994; Perrett et al. 1994; Perrett et al. 1998; Thornhill and Gangestad 1996; Yu and Shepard 1998 and comments on their work (*Nature* 399:214–216, 1999); Enquist and Ghirlanda 1998. Much of this work implies a degree of genetic control of behavior that is unsupported by what is now known about the size of the human genome, the structure of the brain, and the potential power of natural selection. Nonetheless, the accumulation of data has intrinsic interest quite outside the issue of the degree of genetic influence (if any) on many of the behaviors studied. An example is the discovery that romantic love is a near universal in human cultures (Jankowiak and Fischer 1992).
256. Werner 1979.
257. Harris and Ross 1987a; Ehrlich et al. 1995, chap. 2.
258. Harris and Ross 1987a; Ehrlich et al. 1995, chap. 2.
259. Freud 1961 (1930), pp. 59–60. Freud's view could be charitably interpreted as simply saying that incest was forbidden for the good of society.
260. See note 15 in chapter 6.
261. Hopkins 1980.
262. May 1979; Harris 1989, pp. 198–199.
263. Barnard 1994.
264. May 1979. The frequency of incestuous births (resulting from matings between parent and offspring or between brother and sister) in Western populations is thought to be 1 in 100,000 (Cavalli-Sforza and Bodmer 1971) to 1 in 10,000 (Adams and Neel 1967) or even higher. Consanguineous marriages (those between individuals who are related as second cousins or closer) are very frequent, with on the order of one-sixth of all human beings living in societies in which 20 percent or more of marriages are consanguineous (Bittles et al. 2000).
265. Slater 1959. Interestingly, Slater cites (p. 1058) Hoebel (1949, p. 195) to the effect that "the Plains Indians think incest neither a crime nor a sin, but impossible . . . they have no taboo, but simply an absence of incest." See also Fox 1983, pp. 13–14.
266. Westermarck 1891.
267. Frazer 1910.
268. Sahlins 1960.
269. Wolf 1970, p. 503. See also Wolf 1995 and, for a review of the work, Lieberman and Symons 1998.
270. Wolf 1970, p. 508.
271. Wolf 1970, p. 508.
272. Spiro 1958.
273. Talmon 1964; Shepher 1971, 1983.
274. Harris 1989, p. 201.
275. Hartung 1985; see also Harris 1989, p. 201. There are serious questions about actual type and duration of contact in the kibbutz and, possibly, about how many of those who married had entered the kibbutz after the critical period.
276. See, e.g., Sade 1968; Emlen 1995. Japanese quail raised with their siblings prefer to mate with a first cousin of the opposite sex (Bateson 1982), showing the sort of low-level incest avoidance that may also have been characteristic of hominids when they lived in small, isolated bands. As another example, inbreeding appears to be avoided in chimpanzees in part by lessened sexual attraction between individuals who were close when they were young (Pusey 1980). Anthropologists think the taboo is overemphasized because there are so many different levels and forms of incest avoidance in various societies. They

have explained the taboos in terms of the culturally perceived need for hunter-gatherer bands to form alliances with other bands (Tylor 1888). This is accomplished by enforced exogamy (marriage outside the group) (Harris 1989, p. 204). Because individuals were so valuable to small hominid groups, there would have been a temptation "to keep sons and daughters and brothers and sisters at home to enjoy their economic, sentimental, and sexual services" (Harris 1989, p. 205). Therefore, cultural taboos presumably arose to ensure that alliances could be formed that benefited everyone. As population densities increased and people settled down, the need for maintaining trade relations presumably put an even higher premium on exogamy, and the taboo was extended to more distant relatives. More than exogamy may have been involved in the need to form alliances so that bands could move out of areas where there were environmental stresses. Ecstatic religion could have originated as part of the intergroup bonding process (Hayden 1987). Now that money—commercial arrangements—is playing much of the role formerly filled by exogamy, the taboos against incest are predicted to narrow in terms of the degree of relatedness covered by the taboo. This hypothesis was first put forward by M. N. Cohen (1978) and is generally supported in terms of taboos outside the nuclear family (Leavitt 1989). But the taboos would narrow only so far, given that brother–sister and parent–offspring incest are thought to be potentially much more disruptive to society (especially the latter, interfering as it does with the marriage bond). That makes those taboos unlikely to disappear.

277. Wilson 1976. See also van den Berghe 1983 and the discussion following it.
278. Consider two forms of a gene—that is, two alleles—that control the number of bristles on a fruit fly's behind; we'll arbitrarily call these alleles *A* and *a*. Because there are two copies of each gene in the fly's cells (this is a common and necessary simplification), one on each of a pair of chromosomes, a fly can have one of the following three genotypes: *AA*, *Aa*, or *aa*. If *AA* flies have 30 bristles, *Aa* flies 20 bristles, and *aa* flies 10 bristles, there is no dominance. If the numbers are, respectively, 30, 25, and 10, there is partial dominance. And if both *AA* and *Aa* flies have 30 bristles and *aa* flies 10, there is full dominance—the *A* allele is completely dominant over the *a* allele. Inbreeding will result in flies that are homozygous for more genes, and more recessive genes will be expressed, which, in animals that do not normally inbreed, will usually result in lowered fitness. See also Bengtsson 1978; Maynard Smith 1978; Ralls et al. 1979; Ralls et al. 1987.
279. Hartl and Jones 1998, pp. 637–638.
280. See, e.g., Aoki and Feldman 1999.
281. Adams and Neel 1967; Bittles and Neel 1994. It is difficult to assess the overall effect of inbreeding in human populations. In many developing regions, consanguineous marriages are common and often result in increased fertility as well as increased mortality and morbidity—two counterbalancing forces (Bittles et al. 1991).
282. Bittles et al. 1993; Bittles 1995; Grant and Bittles 1997; Hussain and Bittles 1998.
283. Bittles et al. 1993; Nelson et al. 1997.
284. R. M. May 1979.
285. Maynard Smith 1978, p. 143.
286. Aoki and Feldman 1997.

Chapter 9: The Dominance of Culture

1. Darwin 1871, p. 583.
2. This was cited in Ehrenreich 1997, p. 129, but not quite as I remembered it. Librarian Colleen Bilderback located the original for me in Kipling 1912. The entire poem, "The Young British Soldier," can more conveniently be found today in Whitehead 1995, pp. 58–60.
3. Jacob 1982, p. 12.
4. Lewontin 1999, p. 728. For the view from sociobiologists, which makes this characterization rather close, see Wilson 1978, chap. 5, or Hamer and Copeland 1998, chap. 3.
5. Waal 1982.
6. The first dominance hierarchy to have been studied was the "pecking order" of barnyard hens (Schjelderup-Ebbe 1935).
7. See, e.g., Goodall 1986; Boehm 1994a.
8. I recall a spectacular display by a chimp named Humphrey in the presence of Jane Goodall; Anne Pusey; my wife, Anne Ehrlich; two female students; and me. Humphrey hurled rocks and then grabbed a palm frond and thrashed Jane, Anne Pusey, and the students. Anne Ehrlich stepped behind me as Humphrey approached us, and I raised my shoulders to make myself look as large as possible and looked him in the

eyes. Humphrey raised the frond, threatened me, and then wandered off. My utterly unscientific conclusion was that he recognized me as a male and thought better of it—thanks to Jane's wise hands-off policy (there was never any physical contact between researchers and chimps), which prevented the chimps from learning how easily their enormous strength could overpower an adult male *Homo sapiens.*

9. Waal 1989, p. 59.
10. The account of the chimp male triangle is based entirely on Waal 1989.
11. Recently, some evidence has indicated that wild chimpanzees may actually coerce other, initially unwilling individuals to participate in aggression (Arcadi and Wrangham 1999).
12. Balikci 1970, p. 179.
13. See, e.g., Steenhoven 1959, pp. 55–57; Keeley 1996, p. 29.
14. Keeley 1996, pp. 29ff.; see also Knauft 1987, table 2.
15. Wrangham and Peterson 1996.
16. In the United States, more than six times as many males as females are arrested for violent crimes (Macionis 1997). However, the relationship of testosterone to male aggression is complex—there is no simple "more testosterone produces more aggression" relationship (Sapolsky 1997). S. Bhasin and co-workers (1996) gave men weekly doses of 600 milligrams of testosterone, six times more than the dose ordinarily given as therapy to men suffering from lack of the hormone. Despite that, no differences in mood or behavior or in scores on a standardized test, the Multidimensional Anger Inventory (Siegel 1986), were seen between that group and a control group given a placebo. Indeed, even the relationship between levels of testosterone and male sexual activity is not straightforward, despite the "popular monkey-gland legend of geriatric lechery, and the widely-held clinical belief in testosterone as the remedy for impotence" (Money 1981, p. 398).
17. Chimpanzee communities (like those of their closest relatives, the bonobos) are described as fusion-fission societies (Goodall 1986; Stanford 1999, p. 57; White 1996). The membership of chimp groups is in constant flux, with individuals joining (fusion) and leaving (fission) in response to the composition of the group (number of males, number of sexually mature females, changes in dominance relations or relations with other groups, etc.). Male chimpanzees are more gregarious than females, preferring other males' company except when in the presence of females in estrus. Females tend to associate primarily with their offspring, except when in estrus.
18. Nishida 1983; Goodall 1986, p. 417.
19. Kano 1992. Nonetheless, examination of skeletal remains suggests a lower level of aggressive interactions among bonobos than among chimpanzees (Jurmain 1997).
20. Kano 1996.
21. Common chimpanzees average 9.6 days (25 percent of a cycle of 37.2 days) (Tutin 1979). In contrast, bonobos usually are in estrus for more than 20 days, or more than 40 percent, of a 46-day cycle (Kano 1992, p. 160).
22. Waal 1989, 1996a.
23. Waal 1996a, p. 161.
24. Waal 1997, pp. 108ff.
25. Waal and van Roosmalen 1979.
26. Kappeler and van Schaik 1992.
27. Byrne and Whiten 1988; Byrne 1996; Whiten and Byrne 1997. There seems to be little question that part of our intellectual capacity can be traced to selection for such social skills and related attributes as rapid face recognition. On the other hand, remember that our ancestors are also thought to have entered the cognitive niche to help themselves survive in a hostile environment with few physical specializations. Lacking fangs, claws, spines, or thick hides, our ancestors needed what has been called natural history intelligence in order to survive and prosper. See Cachel 1994; Fox 1997, pp. 79ff.; and also Chance 1967, 1975.
28. This occurred while Anne and I were at Gombe Stream. I have not seen accounts of this specific event published, but reports of similar incidents (without accounts of transport of dead infants to the lab) can be found in Bygott 1972 and Goodall 1986. Anne and I took the baby chimp's remains to Dar es Salaam to be autopsied by Professor Msangi at the university there. The long trip resulted in a decaying body, and Msangi could not meet us because of a death in his family. We ended up in a rather fancy hotel with a very odoriferous package—and a comedic sequence ensued that eventually ended with the corpse overnighting in a freezer.
29. Goodall 1986, p. 523.
30. Bygott 1979.

31. Goodall et al. 1979; the group territoriality of chimps is no doubt a factor in their endangerment because when a group's habitat is made untenable by human activities, there is little chance of chimp refugees being integrated into a neighboring group (Goodall 1994).
32. Goodall 1986, p. 530.
33. Niccolò Machiavelli (1469–1527) was an Italian political theorist noted for his advocacy of the clever use of power and manipulation to maintain the state (Machiavelli 1979).
34. Waal 1996a.
35. Boehm 1993.
36. See, e.g., the comments following Boehm 1993; Erdal and Whiten 1994; Boehm 1994b; Knauft 1994.
37. E. O. Wilson 1975, p. 283; Jolly 1972.
38. See Leacock 1978 and the discussion following it; Knauft 1991 and the discussion following it.
39. Just as the power of dominant individuals may have lessened in *Homo ergaster/erectus* (where sexual dimorphism first appears to have been reduced). See, e.g., Chagnon 1979; Harcourt and Waal 1992; Boehm 1993 and the discussion following it; Erdal and Whiten 1994 and the discussion following it.
40. Balikci 1970, p. 183.
41. For a contrary view by an anthropologist whom I often find thought-provoking, see Fox 1991; Fox 1994, chap. 5. Fox sees us as having a series of "drives" (including sexual and aggressive ones) that one function of the burgeoning neocortex was to inhibit (see, e.g., Fox 1994, p. 79). His view is that we're as unlikely to abolish aggression as to abolish sex and that in the case of aggression, the job of cultural evolution is to inhibit or ritualize it. My view is that human beings clearly have a capacity for aggression in many environments, just as they have a capacity for appeasement, and that creating environments in which aggression is minimized would be a lot simpler than creating ones in which sexual desires are not expressed. Most people feel frustrated when they are deprived of sex, but most don't seem to build up a desire to be aggressive if they haven't punched or shot anyone for a while. Behavioral phenotypes, I think, are best viewed in a "norm of reaction" context (see the discussion in chapter 2) rather than in terms of expression of drives, a concept that harks back to early hydraulic analogies developed by famed ethologist Konrad Lorenz for the origins of behavior in the brain (Lorenz 1937, 1950). These models have long been considered unrealistic because they don't correspond to what is known about the neurophysiology of the brain (see, e.g., Lehrman 1953). The difference is subtle but real.
42. Johnson and Earle 1987.
43. Hoebel 1964; Balikci 1970, p. 189.
44. In the American slang of my youth, the best translation of *ayornimut* would have been "That's the way the cookie crumbles." (Again, my transliterations of Inuit terms are from memory, with no guarantees as to accuracy.)
45. See, e.g., Woodburn 1982; Knauft 1987; Boehm 1993.
46. Knauft 1987; Edgerton 1992, p. 172.
47. Edgerton 1992, pp. 180ff.
48. See, e.g., James 1880. In a letter to Albert Einstein, Freud explained war as being based on an instinct (in Bramson and Goethals 1964, pp. 73–74); Holloway Jr. 1967.
49. Eibl-Eibesfeldt 1979 (1975), p. 114. Other early examples are Ardrey 1961, 1966 and Lorenz 1966. A more recent example from the sociobiological literature is Wilson and Daly 1985. There was a related idea that animals subjected to pain or frustration would be more likely to act aggressively—and that human beings, when thwarted, would tend to show aggressive behavior, including going to war (Berkowitz 1962; Dollard et al. 1939).
50. The hyena clans are groups that have social bonds but are not necessarily related to one another (Kruuk 1972).
51. Van Lawick and van Lawick-Goodall 1971, pp. 201–207; Kruuk 1972, pp. 251–265.
52. Cause and effect can be complex here (see, e.g., Harris and Hillman 1989, pp. 318–321). See also Divale and Harris 1976.
53. White 1996.
54. Idani 1991a.
55. Schaller 1963; Tutin 1996, pp. 66–67.
56. Tutin 1996.
57. Perhaps the similarities between intergroup conflict among people and that among other great apes is largely coincidental. In human beings, warfare often involves rational, long-term calculation that is beyond the capabilities of chimps, bonobos, and gorillas. Thus, one possibility is that genetic predispositions related to intergroup violence evolved differently in the evolutionary lines leading to the other

African great apes and the one leading to *Homo sapiens* because of those different capabilities. Another possibility is that the genetic tendency that makes warfare possible may well have been present in the chimp–bonobo and hominid lines since our common ancestor. After all, fighting between groups is known in other primates, such as rhesus monkeys (see, e.g., Southwick et al. 1965), where it was recorded in a disturbed environment. Cultural evolution could have led first to the suppression of intergroup violence in the hominid line, perhaps when population densities were quite low, and then to its reappearance and flourishing as population growth necessitated intensification of food procurement efforts and caused competitive tensions among some early human societies (Knauft 1991). Some modern human groups, such as the Mrabri of northeastern Thailand, do not appear to engage in intergroup conflict at all (Hewes 1994), which may indicate (as is also suggested by the Kasakela chimp episode) that environmental triggers are critical to generating intergroup conflict in our primate cousins and in hunter-gatherer (if not industrialized) human groups.

58. Keeley 1996, pp. 86–87.
59. See, e.g., Ferguson 1995.
60. Goodall 1986; White 1996. See also Wrangham and Peterson 1996, e.g., p. 24.
61. Ambrose 1997, p. 276.
62. Ehrenreich 1997, p. 10.
63. Robarchek and Dentan 1987.
64. Their murder rate, however, was actually some three times that of the United States today (Knauft 1987, p. 458).
65. Dentan 1968, pp. 58–59.
66. Paul 1978, p. 77.
67. Robarchek and Dentan 1987, p. 360.
68. Robarchek and Dentan 1987, p. 361; Robarchek 1989.
69. Keeley 1996.
70. This applies especially to fear of future shortages (Ember and Ember 1992), but the relationship can be tenuous (Ferguson 1989).
71. Pringle 1998b.
72. Leavitt 1977.
73. Brown 1991, p. 139.
74. Geertz 1973, p. 90.
75. For one evolutionary and neurobiological explanation of rituals, see d'Aquili and Laughlin Jr. 1975.
76. See, e.g., Boyer 1994.
77. For a discussion of evolutionary biology in this context, see Midgley 1985.
78. Philosophers have long been fountainheads of not comfort but pessimism, as exemplified by the writings of Arthur Schopenhauer (1788–1860), who believed that life consists of suffering (Higgins 1996).
79. Marxism was a religion of sorts to the Bolsheviks of the 1920s and 1930s; unrestrained capitalism is a religion of many current conservatives, although both systems largely lack the power to impart a sense of transcendence. A noncritical belief in the omnipotence of science is shared by many today, and some people have been tempted to view science as a replacement for religion. It may, in fact, be useful to see it as another form of religion. Science is, so far, the most successful system for assigning causes to observed "mysterious effects." And it is, so far, the most successful system for finding ways to manipulate those forces. Yes, followers of religion must take on faith such things as the existence of spirits or gods. But followers of science must take on faith a wide variety of things, such as the existence of a reality outside of the mind—a reality in which certain rules operate uniformly in both space and time, effects have causes, and samples can be used to induce valid generalities. Scientists are also inclined to take on faith the materialist position that science, religions, ethics, and morals are creations of human minds interacting in human societies—although certain ideas of science (especially in areas such as quantum mechanics and cosmology) can provide, for some, a sense of transcendence. Many reject, as I do, the transcendentalist position of religious thinkers and some lay philosophers that ethical and moral precepts would exist even if there were no human beings. Edward O. Wilson has a thoughtful discussion of the materialist and transcendentalist positions (Wilson 1998, chap. 11). I would be less inclined than he is to emphasize genetic elements in the origins of the materialist view, but the issue is really an open one.

    Science, which at least has roots in religion even if it is not to be classified as one, has the advantages of built-in skepticism and a tradition of using observations of nature or experiments as courts of last resort. But science has failed, and I suspect will always fail, to provide answers to many ultimate questions: What was there before the origin of the universe? What would there be if the universe didn't exist?

Can we be sure it exists? Does it exist only in the human mind? Philosophers (if they exist) have puzzled over such questions for millennia. For that reason, the explanatory role of religion remains important for many people; few seem able to live with persistent uncertainty. On the other hand, inserting a creator as an ultimate explanation for the universe simply raises the same questions about the creator as previously existed for the universe itself.

80. See, e.g., Maton 1989; Benson 1996; Roush 1997.
81. Smart 1994, p. 9.
82. For an interesting discussion of the neurobiology of trances in an Indian society, see Gell 1980.
83. Nadel 1954.
84. Mellars 1996, pp. 375–381.
85. See Gargett 1989 and the commentaries following it, including Kooijmans et al. 1989; also Mellars 1996, pp. 375–381.
86. Vandermeersch 1970; Mellars 1996, p. 380.
87. One example is a triple Czech burial dated about 27,000 years ago (Bahn 1988) that embodies enigmatic but clearly complex notions.
88. Chase and Dibble 1987.
89. Harris 1989, pp. 400–401.
90. Torrance 1994, p. 138.
91. Ray 1967, p. 37.
92. Balikci 1970, pp. 198ff.
93. Smart 1994, pp. 61–66. A classic work on religion, the sacred, rituals, and myths throughout history is Eliade 1957.
94. Those familiar with the history of philosophy will know that there is nothing at all simple about notions of cause and effect, the modern discussion of which traces in no small part to the writings of David Hume (see, e.g., Russell 1945, pp. 664ff.).
95. Remember the work of A. E. Michotte (1965), discussed in chapter 6 of this book, and those who have succeeded him, e.g., A. Leslie (1982).
96. There is an interesting psychological literature on causal inference (e.g., Arkes and Harkness 1983; Crocker 1981; Schustack and Sternberg 1981), but as far as I can see, it has not been employed much in the discussion of the origins of religion.
97. A good example is the apparent hard-wiring of fear of heights (see, e.g., Gibson and Walk 1960), although, of course, the emotion is inferred.
98. One theory of dreams, based on the neurobiology of dreaming sleep, is that they are basically the result of the brain reorganizing information acquired during wakefulness (see, e.g., Hobson 1994; Hobson and McCarley 1977).
99. Hobbes 1997 (1651).
100. Prayer is normally thought of as an attempt to influence the gods to change the forces of nature; magic is an attempt at direct influence. It is a fine line to draw.
101. Tylor 1920 (1871), vol. 2, p. 196. Tylor believed there was an evolutionary sequence from animism through "polydemonism" and polytheism to monotheism. A more modern view is that there has long been a complexity in human beliefs and no sign in what we know of a directional evolution—for an early analysis, see Blanc 1962.
102. Durkheim 1965 (1915).
103. Material on the Machiguenga is from Allen Johnson (personal communication, 16 August 1999).
104. Ideas traceable to animism, such as that of the existence of a soul separable from the body, persist in many religions today. An interesting discussion of this and other aspects of religion, which has been influential in my thinking, can be found in La Barre 1954, chap. 14.
105. Shamans are as often women as men (Harris 1997, p. 373).
106. Johnson and Earle 1987.
107. Sapolsky 1997. The brain biochemistry behind schizophrenia and other disorders is slowly being unraveled (see, e.g., Barondes 1999).
108. Early discussions of this can be found in Ackernecht 1943 and Silverman 1967. The suggested connection between shamanism and mental disorders has been hotly debated (e.g., Boyer 1969; Noll 1983; Silverman 1967). I am persuaded by Sapolsky's analysis.
109. Sapolsky 1997, pp. 243ff. This is one of the best books I've ever read on a biological topic—highly recommended. See also Jaynes 1976. Jaynes ties religion to schizophrenia (pp. 412ff.) in the course of inno-

vative (or far-out, depending on one's viewpoint) speculation about consciousness in its present form having evolved only some 3,000 years ago.

110. Sapolsky 1997, pp. 256ff.; see also Rapoport 1989. For an attempt to tie brain structure to the universality of rituals and myths, see d'Aquili and Laughlin Jr. 1979.
111. Some evidence suggesting that there is a hereditary component to the tendency to develop schizophrenia can be found in Heston 1966; Kringlen 1967; Gottesman and Shields 1972; and Gottesman 1991.
112. Sapolsky 1997, p. 250.
113. Sapolsky 2000. The quotations that follow are from his article.
114. Sapolsky took his data from the monumental work by R. B. Textor (1967).
115. Cf. Balikci 1970.
116. See, e.g., Johnson and Earle 1987, pp. 35ff.
117. Religious ceremonies are acts or series of acts designed to persuade supernatural forces to produce some result or results in the natural world—or simply to appease them. Religious ritual is the required order in which a ceremony is to be performed, or the entire body of ceremonies characteristic of a religion. The term *rite* is often used synonymously with *ceremony* ("fertility rite"). Early hunter-gatherers had complex art forms connected with the religious aspects of the ceremonies. A well-studied case is the complex relationship of rock art to the religious beliefs of Australian Aboriginal cultures (Layton 1992, chap. 2; Crumlin and Knight 1991). Another example is the Inuit of Alaska's nineteenth-century use of diverse masks, which expressed the fundamental unity of animals and people in the mythical past, to guarantee successful hunts, among other things (Oosten 1992; Ray 1967, p. 37).
118. The story of Tommy is told in Copeland 1960. I learned many years later that his overall view of the *kabloonak* ("big eyebrows," the Inuit name for Europeans) was not complimentary. "He is not an Eenook (Inuit). . . . He cannot build an igloo. He cannot guide a man across the tundra without becoming lost. He cannot endure the dangers of hunting the wild animal on land and sea. He cannot live by the skill of his hands in a land where all but the Eenook would perish." Tommy and his friends thought white men were fools (Copeland 1960, pp. 123–124). This doubtless included the young entomologist whom he and his companions christened *Tah-kah-lee-koo-tah-chee*, "Butterfly catcher."
119. This is doubtless an oversimplification of their views. For an interesting discussion of hunter-gatherer relationships to the natural agencies that support them, see Bird-David 1992.
120. Alvard 1998; Winterhalder and Lu 1997.
121. See, e.g., Balikci 1970, p. 68; see also Dumond 1977.
122. Driftwood was the only source of wood for Inuit who lived on the treeless tundra, so the shafts were precious. The harpoon head remained in the seal, which could be pulled from the water with the line.
123. This information is from a personal communication from Bert Swaffield, Hudson's Bay Company factor at Coral Harbour, Northwest Territories, summer 1952.
124. Polunin 1949, p. 78.
125. Brody 1975, p. 23.
126. Lindly and Clark 1990. For the role of art in nonverbal communication in a New Guinea society, see Eilers 1977.
127. Dissanayake 1988.
128. Chase and Dibble 1987.
129. The earliest claim I've come across is about 50,000 years ago, for a flint plate engraved with four nested semicircles surrounded by parallel slanting lines (Marshack 1996).
130. See, e.g., White 1989. For recent overviews, see Conkey et al. 1997 and Appenzeller 1998.
131. These were recently dated to 13,200–12,800 years ago (Valladas et al. 1992).
132. Marshack 1996.
133. See, e.g., Halverson 1987.
134. Humphrey 1998.
135. Delporte 1995.
136. Seeger 1994.
137. The earliest artifacts that are known with certainty to have been musical instruments are flutes found in a French cave that date from the Aurignacian period (about 40,000–30,000 years ago) (d'Errico, Villa, et al. 1998).
138. Dissanayake 1992, p. 117. The origins of chanting and singing, not unexpectedly, have been subject to extensive speculation.
139. Diamond 1991, pp. 159–161.
140. Chase and Dibble 1987; Lindly and Clark 1990 and comments following it.

141. See, e.g., Schick and Toth 1993, pp. 282–283.
142. See, e.g., Street and Besnier 1994.
143. Harris 1986.
144. Street and Besnier 1994.
145. Ong 1982; see also Harris 1986 and Donald 1993. Donald discusses writing in terms of the "externalization" of memory. For a strong claim about the extent to which literacy influences thought, see Olson 1986.
146. Street and Besnier 1994.
147. Layton 1991.
148. See, e.g., the discussion in Mithen 1996, pp. 154–167.
149. Biebuyck 1973; Layton 1991, p. 27.
150. Morphy 1988.
151. Morphy 1992, p. 196.
152. Guenther 1988.
153. Layton 1991, p. 42.
154. Bücher 1909.
155. Glaze 1981.
156. For a sense of the impact of the latter, see Klemperer 1998.

Chapter 10: From Seeds to Civilizations

1. B. D. Smith 1995, p. 16.
2. The settling down of humanity was also related to exploitation of marine resources (Pälsson 1988), but it is highly unlikely that the sequence described in this chapter that followed the agricultural revolution could have been supported solely by exploitation of marine resources. Although the formation of sizable communities was roughly simultaneous with the development of agricultural systems to support them, excavation of the large community of Çatalhüyük in Turkey, home to perhaps 10,000 people some 9,000–7,500 years ago, raised serious questions about the motivation for settlement (Balter 1998, 1999). Among other interesting features, the settlement has not yet revealed the signs of division of labor or construction of communal buildings that one would normally expect in so large an agglomeration. For a model of the transition from mobile band to sedentary community, see Harris 1977b.
3. I was hesitant to describe levels of cultural evolution, which are somewhat different from the similarly labeled levels of genetic evolution, but it seemed a convenient shorthand and preferable to complete neologisms. Other divisions of micro and macro investigation are discussed in the social realm (e.g., Iggers 1997, p. 14).
4. See, e.g., Braudel 1993; Hughes 1975; Crosby 1986. Braudel represents the Annales group of French historians, dating from the early 1930s, which in essence forms an entire school of exceptions (Iggers 1997, chap. 5). That school paid more attention to "structures," things such as geography and crop yields, than to leaders and battles. There is also a developing area of investigation into what might be called environmental cultural microevolution—examination of the history and sociology of concern for the environment. See, e.g., White Jr. 1967; McHenry and Van Doren 1972; Worster 1994, 1995; Petulla 1980; Fox 1981; Thomas 1983; Nicholson 1987; Shabecoff 1993; Soulé and Lease 1995.
5. Jared Diamond is the leading analyst of cultural macroevolution today, and his Pulitzer Prize–winning book *Guns, Germs, and Steel* (1997a) is a brilliant, innovative, and entertaining introduction to the subject. Consideration of macroevolutionary factors goes back at least as far as the work of French political philosopher Baron de La Brède et de Montesquieu (1689–1755), who might be viewed as the father of modern political science. Montesquieu (1989 [1748]) discussed the roles of such things as climate, soil fertility, and natural barriers in shaping societies and politics. See also Lowenthal 1987.
6. Polities are politically organized social units.
7. See, e.g., Hughes 1975; Tainter 1988; Yoffee and Cowgill 1988.
8. Postel et al. 1996; Schneider 1989, 1995, 1997; Daily and Ehrlich 1996a.
9. See, e.g., Spengler 1996 (1923).
10. For a brief overview, see Gilderhus 1996.
11. Giambattista Vico (1668–1744), considered by some the first modern historian, who saw history as a process of evolutionary development but retained elements of the Greco-Roman view that history was basically cyclical; Voltaire (François-Marie Arouet, 1694–1778), who examined the evolution of Christianity in Europe; Georg Hegel (1770–1831), whose complex and obscure ideas contained the notion of

discerning laws governing history (see, e.g., Hegel 1991 [1821]) and concluded that it consisted of the evolutionary development of the human spirit or universal soul (Geist); and Auguste Comte (1798–1857), a brilliant but unbalanced historian, philosopher, and sociologist who had a theory of social evolution. More recently, the French intellectual Michel Foucault (1926–1984) followed partially in Hegel's historical (and Friedrich Nietzsche's radical-historical) footsteps, examining in a historical context such subjects as the evolution of the designation and treatment of the "insane" (Foucault 1988) and "criminal" (Foucault 1995) elements of society. The basic issue of objectivity in history remains as alive and well in historiography as it is in science (see, e.g., Gilderhus 1996; Iggers 1997; Novick 1988). The issue was recently brought into sharp focus by the appearance of "historical revisionists" who claim that the Holocaust was a hoax (for details, see Lipstadt 1993).

12. For instance, his early statement: "In broad outline, the Asiatic, ancient, feudal and modern bourgeois modes of production may be designated as epochs marking progress in the economic development of society. The bourgeois mode of production is a last antagonistic form of the social process of production . . . but the productive forces developing within bourgeois society create also the material conditions for a solution of this antagonism. The prehistory of human society accordingly closes with this social formation" (Marx 1970 [1859], pp. 21–22). This was written eight years before his magnum opus, *Das Kapital* (Marx 1967 [1867]). Marx referred to prehistory because he believed that when classes were gone, true human history would begin.
13. Gilderhus 1996, p. 59. There is still a discipline of "Marxist historiography," a method of studying history in the tradition of Marx and Engels, especially from an economic determinist perspective (Novick 1988).
14. An instructive exception is the renunciation of the use of firearms by Japan from 1543 to 1879. It is described in detail in an important book by Noel Perrin (1979).
15. See, e.g., Wallerstein et al. 1996.
16. For an informative discussion of the development of anthropological ideas about evolution, see Harris 1968.
17. See, e.g., Notestein 1945; Wilson 1999.
18. Teitelbaum 1975; Ehrlich and Ehrlich 1990, pp. 214–216; Ehrlich et al. 1995, pp. 65ff. An important flaw in the theory is that it assumes an automatic equilibration of birthrates and death rates even though no mechanism capable of accomplishing that exists (Boserup 1981; Lee 1987; Lesthaeghe 1980; Malthus 1970 [1798]; Wilson and Airey 1999; Wrigley 1978), a fact that has substantial policy implications.
19. See, e.g., Parsons 1964. Unfortunately, Parsons also pioneered the writing of gibberish in the social sciences: "An element of a shared symbolic system which serves as a criterion or standard for selection among the alternatives of orientation which are intrinsically open in a situation may be called a value. . . . But from this motivational orientation aspect of the totality of action it is, in view of the role of symbolic systems, necessary to distinguish a 'value orientation' aspect. This aspect concerns, not the meaning of the expected state of affairs to the actor in terms of his gratification-deprivation balance but the content of the selective standards themselves" (Parsons 1951; see also Mills 1959, chap. 2). This tradition is all too often carried on today by deconstructionists in disciplines such as anthropology.
20. Legal scholars, for instance, have a deep interest in the evolution of laws and legal structures. For a readable and classic example, see Rakove 1996. Today, an evolutionary view is also embraced by archaeologists, who have taken to computer modeling of the evolution of ancient civilizations (Wright 1997), a process about whose utility I must admit to having doubts. To see how it's done and judge for yourself, see Epstein and Axtell 1996. Sociologists (e.g., Nolan and Lenski 1999, chap. 3) and political scientists are also on the bandwagon. A basically evolutionary view is seen in detailed analyses of such things as the development of states (Gouldner 1977). It is also implicit in discussions of the development of political ideas in standard textbooks and in the role of views of human nature in their evolution (e.g., Scott 1997, pp. 46–47). For interesting readings, see Strauss and Cropsey 1987.
21. See, e.g., Rieff 1999.
22. Havel 1999, p. 3. Others point out that the demise of the nation-state system has been prematurely announced many times. Whatever their views on that issue, historians have remained fascinated by the evolution of nations and nationalism (see, e.g., Gellner 1983). They look at such issues as the origins of what Benedict Anderson called the "imagined communities" of nationality (Anderson 1991) and their ethnic roots (e.g., A. D. Smith 1986), the roles of ethnic and civic nationalism in political evolution and revolution and in decolonization (e.g., Chatterjee 1986; Hobsbawm 1992; McPherson 1999), and so on. Anderson addresses such topics as why people become patriots (or even chauvinists), why multicultural

empires could not persist, and why nationalism is such a potent force in developing nations—all pertinent questions for understanding differences and similarities among human natures today.

23. Much of what we can understand of cultural evolution before the invention of writing depends on extrapolation. In trying to reconstruct prehistoric events such as the beginning of cultivation, social scientists lean on what is known of historical and contemporary societies. In reading brief summaries such as those in the previous chapter and this one, one should remember that the stories are based on a mixture of fragmentary archaeological evidence, often biased written accounts (left by later societies), and speculation. Archaeological evidence is almost always fragmentary because, most importantly, evidence is often differentially preserved (i.e., bones and pottery do better than bark paintings and feather headdresses). Written evidence may mix myth and fact and normally is not analytical by modern standards. The abundance of material remains of older cultures, in contrast to the paucity of clues about the *ideas* of those cultures, tends to bias scientists toward a materialist (economic) interpretation of cultural evolution rather than, say, one that focuses more on ideological or spiritual elements. As a result, conclusions about prehistoric cultural evolution must be considered tentative. Of course, all scientific conclusions are tentative, but some are more tentative than others.
24. Wallerstein et al. 1996.
25. The problem is made more complex by the inroads of deconstruction, especially in anthropology (Kuznar 1997). Deconstruction (the term traces to attempts to see how structures and ideas have been constructed) focuses entirely on texts, not on the "real" world or even on the intentions of the authors of texts (Derrida 1976). For an attempt at a balanced view of the battle between those who believe there is a reality totally detached from human experience and those who believe there are only social constructions, see Hacking 1999.
26. This story is based on material from Bruce D. Smith's fine book (1995) about the origins of agriculture.
27. Reed 1977b; Ellen 1994.
28. Diamond 1997a; Ellen 1994, p. 207; Cowan and Watson 1992.
29. B. D. Smith 1995, p. 50.
30. For a discussion of the Fertile Crescent's botanical riches and suitability for agriculture, see Diamond 1997a, chap. 8.
31. B. D. Smith 1995, pp. 72–73.
32. See, e.g., Balter 1998; Pringle 1998c.
33. Data on actual population sizes are not available, although early settlements left archaeological remains and burials from which some demographic data could be derived. One can guess that early hominids required large home ranges to support themselves, especially in the presence of environmental stresses such as droughts. Such events could have greatly reduced the carrying capacity of a group's environment.
34. Modern events support such a view. We know that once firearms were available to them, some historical hunter-gatherers intensified their hunting and dramatically reduced the availability of game—not just on arctic shores, as did the Aivilikmiut, but also in moist tropical forests (Kay 1995; Robinson 1996). The role of firearms in the hands of Native Americans in reducing buffalo herds prior to the final slaughter by Europeans is disputed because their bows and arrows were powerful and accurate weapons that persisted for many decades alongside firearms. But it is likely that hunting with guns contributed to at least local shortages on the fringes of the range of *Bison bison* (see, e.g., Roe 1970).
35. The emphasis on gathering might even have reduced the sexual size dimorphism. The latter could happen genetically or, more likely, as a result of changes in diet or as a result of both; when dietary quality drops, the difference in size between the sexes is reduced (Foley 1988, p. 220). Five hundred generations would not have been much time for genetic change unless selection pressures were substantial. Late Pleistocene modern human beings were more sexually dimorphic than early post-Pleistocene ones (p. 221). The transition to agriculture might also have contributed to the creation of incest taboos to encourage marital exchanges that promoted cooperative land clearing, digging of irrigation ditches, and the like, where a temporary concentration of labor was required (Harris 1989).
36. See, e.g., Cohen 1977a; Reed 1977a. For a sense of the problems involved in working out details of the population pressure theory, see Rosenberg 1990, 1991; Graber 1989, 1991, 1992; McCorriston and Hole 1991; and Hole and McCorriston 1992. Unfortunately, almost nothing is known about densities of pre-agricultural hunter-gatherer populations, but data on variations in small game hunting (Stiner et al. 1999) and genetic evidence (Harpending 1994; Long 1993; Rogers and Harpending 1992; Sherry et al. 1994; Reich and Goldstein 1998) both suggest that human populations expanded greatly and rapidly following the cultural Great Leap Forward.
37. Foley 1999.

38. Foley 1988, pp. 219–220.
39. Central Arctic peoples did not have a broad array of plant foods to fall back on if game became scarce (see, e.g., Lee 1968, pp. 40–41). They did, however, have relatively dependable supplies of fish and marine mammals. In any case, what are thought of as typical Inuit may in fact be quite atypical because most Inuit in historical times lived south of the Arctic Circle (Laughlin 1968).
40. Harris 1977a; Roosevelt 1984.
41. Sahlins 1968, p. 84; but see Lorna Marshall's comment on Sahlins following it (p. 94); Sahlins 1972. Sahlins's idea seems to have roots in the writings of John Locke, who believed that humanity's original state of nature was benign. Thomas Hobbes, in contrast, is famous for having commented that the original state of human life was "solitary, poore, nasty, brutish, and short" (Hobbes 1997 [1651], p. 70).
42. Subsequent work has shown that hunter-gatherers, especially those in "immediate return" economies (living basically hand to mouth, with little food storage or processing), can obtain their subsistence with relatively few hours of labor per day (Barnard and Woodburn 1988). For uncertainties of yield, see the discussion in Bird-David 1992.
43. Johnson and Earle 1987, p. 46.
44. Foley 1999. Foley was summarizing work by M. Jenicke, R. Pennington, A. Fromont, and others reported at a conference in Durham, England, in May 1999.
45. B. D. Smith 1995, p. 83.
46. B. D. Smith 1995; John Allman, personal communication, 1 November 1999.
47. Harris and Ross 1987a; Ehrlich et al. 1995.
48. Zohary and Hopf 1988.
49. Many of these issues are discussed in Rindos 1984.
50. See, e.g., Ammerman and Cavalli-Sforza 1973; Sokal et al. 1991.
51. See, e.g., Diamond 1995; Diamond 1997a, p. 110.
52. Steadman 1977, pp. 68–69.
53. Van Zeist 1986.
54. Jardin 1967; Harlan 1989.
55. Bar-Yosef and Kislev 1989; B. D. Smith 1995.
56. The original grasses were diploid, whereas the cultivated strains were tetraploid and hexaploid.
57. Zohary and Hopf 1988.
58. Harlan et al. 1973; Zohary and Hopf 1988. In the wild in the Mediterranean area, germination is often inhibited so that it is delayed until the onset of favorable conditions even if the seed is wetted.
59. Alteration of plants and animals by selective breeding is just as much genetic engineering as are the gene-transfer techniques now featured under that rubric.
60. Zohary and Hopf 1988. In recent decades, there have been substantial advances in the techniques for retrieving and studying plant remains from archaeological sites, especially the use of flotation to separate plant remains from the rest of archaeological sediments, which sink (Pearsall 1989; Streuver 1968).
61. Cowan and Watson 1992.
62. For a brief, insightful summary of animal domestication, see Clutton-Brock 1992.
63. Hillman and Davies 1990. See also Bar-Yosef and Belfer-Cohen 1991. Genetic studies suggest that maize domestication in southwestern Mexico took at least several hundred years and, interestingly, that the artificial selection pressures placed on that grass by ancient cultivators was quite strong (Wang et al. 1999).
64. Bar-Yosef 1991; Bar-Yosef 1994, pp. 52–55.
65. B. D. Smith 1995; Diamond 1997a; MacNeish 1992. MacNeish recognizes two schools of explanation: one by cultural ecologists, who saw agriculture as a response to environmental (climate) change, and the other by materialists, who saw it coming from population pressures, population growth, or both (p. 10).
66. Cohen 1977b. For those interested in more details of the adoption of agriculture, what is known is summarized in Diamond 1997a. More recent details of the issues can be found in Pringle 1998c and B. D. Smith 1995.
67. For an interesting collection of papers on this issue, see Harris and Hillman 1989.
68. Pringle 1998c.
69. Daily and Ehrlich 1992.
70. Cohen 1994, p. 289.
71. Cohen 1989; Porter 1997.
72. McNeill 1976; Cohen and Armelagos 1984; Armelagos et al. 1991.
73. For a comprehensive overview, see Anderson and May 1992.
74. See, e.g., McNeill 1976.

75. See, e.g., McNeill 1976; Reff 1991; Diamond 1997a.
76. A critical but often neglected point is that human beings are subject to a constantly evolving epidemiological environment (Armelagos and McArdle 1975; Ehrlich and Ehrlich 1970, p. 148). For a discussion of the current significance of this point, see Daily and Ehrlich 1996b.
77. Scrimshaw et al. 1968; Cohen and Armelagos 1984; Armelagos et al. 1991; Diamond 1991.
78. Milner et al. 1989; Wood et al. 1992. Birthrates higher than death rates cause the population to grow and also shift its age structure toward the younger age classes, and this can lead to a proportionate increase in the remains of young people.
79. Nomadic pastoralists, of course, did this as well.
80. That broad outline may be accurate, but uncovering the exact cultural evolutionary forces that led from the agricultural revolution to jet aircraft and home computers is fraught with difficulties. For much of the critical early period, there is either no written record or a very inadequate record from which data on such key issues as resource scarcity, political organization, and intergroup relations can be extracted. Even where substantial information is available, it is clear that all societies are subject to a variety of internal and external forces whose relative influences are difficult to sort out, both because of lack of relevant information about certain factors and because of the complexity of interactions among them. One need only consider the controversial nature of broad-scale attempts to deal with *current* sociopolitical arrangements, such as Samuel Huntington's suggestion that there are cultural "fault lines" between civilizations today (Huntington 1993, 1996; Matlock Jr. 1999), to see what students of prehistoric politics are up against. Nonetheless, a systematic approach can be taken in examining the issue of how and why states evolved; the story naturally starts with the agricultural revolution because hunting and gathering cannot support agglomerations of hundreds of thousands or millions of people in limited geographic regions. The structure of the following discussion of the transition from family economy to nation-state owes much to Johnson and Earle 1987.
81. Ellen 1994.
82. Yesner 1980.
83. See, e.g., Boserup 1965; Ellen 1994, pp. 218ff.; Cohen 1994.
84. Defense of resources such as prime agricultural land, long-yielding tree crops, rich foraging territories, and flocks of domestic animals was necessitated when the benefits of violent seizure began to exceed the costs. Communal effort was required to deploy technologies beyond the capabilities of individual families. One family cannot readily construct and operate a canoe big enough for use in hunting whales or build and maintain a complex irrigation system. Trade in food and specialized tools became ever more important as the need to increase the efficiency of food production grew. Economics as we know it had arrived on the scene. And despite all the intensification, the risk of starvation escalated as foods that once were gathered only in times of shortage became part of the staple diet and could no longer serve as a buffer against failure of other sources. Communal food storage became an important strategy. These developments led to suprafamilial organization because lifestyles and circumstances had changed to such a degree that the challenges and risks faced by individual families made the benefits of forming interfamily groups greatly exceed the costs.
85. Johnson and Earle 1987, p. 246.
86. Fox 1997, pp. 67–71.
87. He was the most famous social Darwinist and one of the fathers of sociology. The connection is pointed out in Graber and Roscoe 1988 and in Spencer 1891 (1860), pp. 281–282.
88. Carneiro 1961, 1970. For an expanded overview of the evolution of political complexity, see Flannery 1972.
89. See, e.g., Harris 1968; see especially the work of V. Gordon Childe (e.g., 1936). Of course, Hobbes (1997 [1651]) and many other early modern theorists in the seventeenth and eighteenth centuries (e.g., Rousseau 1762) had similar ideas of a voluntary social contract in which individuals in a "state of nature" bound themselves together into a society or state by accepting obligations to one another in return for security or other benefits—and then formed a government (or acquired a king) to enforce the obligations. Even though we know, contrary to that theory, that hominids have been social for millions of years, social contract theory can sometimes be a useful fiction. A recent social contractarian version is found in the important work on social and distributive justice of John Rawls (1971). Rawls's work is best known for the idea that ethical standards should be determined as if from behind a "veil of ignorance," wherein the parties "do not know how the various alternatives will affect their own particular case and they are obliged to evaluate principles solely on the basis of general considerations" (pp. 136ff.)—that is, as if they knew nothing of their status or situation in the society in which they would live. Rawls's ethical system

permits differences in wealth as long as the differences benefit the least advantaged in society. His views are not utilitarian but transcendentalist (justice originates outside of experience in Rawls's concept of fairness) and are in the long tradition of those who assert the existence of rights (Ehrlich 1968, pp. 187–188). For a libertarian critique of Rawls, see Nozick 1974; for criticism from the opposite (communitarian) side, see Sandel 1988. Major controversy surrounds what the theoretical people behind the veil would know about such things as their own abilities, whether they could be risk takers (e.g., prefer to take their chances in a less equitable system), and the inevitability of differences in wealth leading to differences in power that would reduce equity (basically a Marxian critique). Sandel's criticism and the fundamentally theological one of MacIntyre (1984) fit better with the theme of this book because they more fully recognize the crucial importance to all moral judgments of the cultural milieu in which individuals are embedded. Rawls (1996) responded to his critics and subsequently modified his views to less transcendental ones. See also Walzer 1984. It is a tradition that includes the founding fathers of the United States, and the assertion of rights can serve what, in my view, are worthy goals (e.g., Dasgupta 1993). The growth in awareness and pressure for respecting human rights over the past few decades is simply part of the ongoing process of cultural evolution.

90. See, e.g., Wittfogel 1957.
91. Carneiro 1981, p. 64.
92. Hauer 1988.
93. Leavitt 1977.
94. Apparently the sacrifices were, for the most part, carried out in religious ceremonies to propitiate gods. For an overview, see Harris 1989, pp. 422–425; for a recently discovered example, see Pringle 1999.
95. There has been an interesting debate over whether or not cannibalism following ceremonial sacrifice was undertaken in Aztec society in part out of a need for high-quality protein (see, e.g., Harner 1977; Ortiz de Montellano 1978). Marvin Harris comes down on the side of the dietary importance of the process (1989, pp. 432–436). Increasingly, the evidence indicates that cannibalism has been an ancient and widespread practice (see, e.g., Gibbons 1997a).
96. Carneiro 1981. For another take on stratification, which looks at how elite minorities originally get control of wealth, see Webster 1990 and the comments following it.
97. See, e.g., Marx 1967 (1867), vol. 3, pp. 818ff. See also Plattner 1989b.
98. For an interesting recent discussion of this, see Earle 1997, chap. 3.
99. Kirch 1988, 1997.
100. Kirch and Weisler 1994; Chang and Goodenough 1996; Kirch 1996a.
101. See, e.g., Athens 1997.
102. Kirch and Hunt 1997.
103. Kirch 1984, p. 98.
104. Vayda 1961.
105. Kirch 1984; see the map on p. 78 for dispersal patterns of Polynesians.
106. Kirch 1994. The two central root crops were *Colocasia* (taro; wet) and *Dioscorea* and *Ipomoea* (yam and sweet potato, respectively; dry). The Hawaiian Islands were formed as a geological plate passed southeastward over a plume of upwelling molten rock, which built volcanoes from the ocean floor. The weathered soils, permanent streams, and valley alluvium of the geologically younger northwestern islands of Kauai, Oahu, Molokai, and western Maui allowed for major areas of pondfield irrigation. Eastern Maui, around the base of the Haleakala volcano, and Hawaii had major areas of intensive dryland agriculture (with some pondfield agriculture on Hawaii's windward coast) (pp. 251–253). A seamount southeast of the island of Hawaii is building and will eventually penetrate the sea's surface and form a new Hawaiian island.
107. *Ipomoea batatas.* Polynesians also cultivated several kinds of yam (*Dioscorea* spp.), and although sweet potatoes dominated, yams were a secondary crop on Hawaii and eastern Maui.
108. Kirch 1990, pp. 333–334.
109. Kirch 1988.
110. Kirch 1984.
111. Trotter and McCulloch 1984.
112. A. Anderson 1984.
113. Lepofsky et al. 1996.
114. Kirch 1996b.
115. Kirch 1984.
116. Dransfield et al. 1984; Brand and Taylor 1998.

117. Van Tilburg 1994.
118. Englert 1970; McCall 1980; Kirch 1984.
119. Diamond 1995.
120. Kirch 1984, p. 278.
121. Diamond 1995.
122. Jacobsen and Adams 1958; Culbert 1973; Abrams et al. 1996; Sabloff 1971; Hughes 1975. Factors other than overexploitation may well have been involved in some of these collapses. For instance, both warfare (Appenzeller 1994) and cultural macroevolution (climate change) (Hodell et al. 1995; Sabloff 1995) may have been involved in the collapse of the classic Maya. For an interesting analysis of collapses focused on a declining marginal return to complexity, see Tainter 1988.
123. Kirch 1991a.
124. Kirch 1997. The account that follows is based primarily on his paper.
125. Patrick Kirch, personal communication, 16 July 1999.
126. J. Williams 1837, pp. 244–245.
127. Kirch 1997, p. 38.
128. Finally, there are at least twelve Polynesian islands, mostly small and limited in resources, where circumscribed societies just didn't make it. Henderson Island (fifteen square miles), Raoul Island (twelve square miles), and Pitcairn Island (two square miles) are examples—islands that were occupied for at least several hundred years before their human populations disappeared for unknown reasons (Kirch 1984, pp. 89–92). Why they lost their entire populations, whereas Tikopia, Mangaia Island, and Easter Island did not, can only be guessed.
129. Sahlins 1972.
130. This point is based largely on first principles but is supported by such things as observations of the ubiquity of kinship systems.
131. Johnson 1983.
132. M. Baksh 1984, reported in Johnson and Earle 1987.
133. Regional densities of Yanomamö populations are roughly the same as those of the Machiguenga, but fertile soils are limited to relatively small pockets, leading to more intensification and defense of those pockets (Allen Johnson, personal communication, 16 August 1999).
134. Chagnon 1983, 1988; Keeley 1996.
135. Keeley 1996, p. 68; Morison 1947–1962.
136. Chagnon 1988.
137. Johnson and Earle 1987, p. 129.
138. There has been considerable debate over the degree to which Yanomamö combat has its roots either in hunger and the need for resources or in the simple desire to carry on feuds (e.g., Lizot 1977).
139. For examples of the earlier, less familiar stages of this political evolution, see Johnson and Earle 1987.
140. These population sizes are guesses based on population densities of historical hunter-gatherers, which rarely exceeded one person per 3 square kilometers and were often much lower—as low as one person per 300 square kilometers (Lee and DeVore 1968a, p. 11). There are some 133 million square kilometers of ice-free land, so if the average hunter-gatherer density over that land (deserts and mountain ranges included) were one person per 10–25 square kilometers, that would coincide with the estimates given. See also Ehrlich et al. 1977; Armelagos et al. 1991; Cohen 1994.
141. Thompson 1990; Earle 1997. A more general definition of an ideology is simply "a value system or belief system accepted as fact or truth by some group" (Sargent 1993). This definition, which overlaps completely with religion but is more inclusive, seems excessively general to be useful. But it is common in political science (e.g., Scott 1997) and is close to the dictionary definition.
142. Erdal and Whiten 1994, p. 178.
143. D'Altroy and Earle 1985; Earle 1991b.
144. Gilman 1981.
145. Earle 1991a, p. 71.
146. This is technically called staple finance (Johnson and Earle 1987).
147. Carneiro 1970; Webster 1975.
148. Burton 1975.
149. Earle 1997.
150. The rights were either individual or collective—there is a substantial debate over the role of property rights in, for example, the origins of agriculture and the tension between individual and collective land

management regimes (see, e.g., Demsetz 1967; North and Thomas 1977). See also the discussion in Bromley 1991, pp. 136ff.

151. Earle 1991a.
152. Berdan 1989, p. 93.
153. Johnson and Earle 1987, p. 244.
154. For a brief critical review, see Schacht 1988.
155. See, e.g., Webb 1988; Feinman 1991.
156. For a discussion of Mesopotamian states, see Yoffee 1995.
157. Berdan 1989, p. 80.
158. Service 1978.

Chapter 11: Gods, Dive-Bombers, and Bureaucracy

1. Harris 1989, p. 398.
2. Fox 1994, p. 119.
3. Coe 1981, pp. 161–162, 170–171.
4. See Thoreau 1986 (1849, 1854) and J. Brown 1977. Thoreau's essay on civil disobedience, or "peaceable revolution" (p. 399), much impressed Gandhi, who in turn profoundly influenced King.
5. See, e.g., McNeill 1976.
6. Crosby 1986, p. 200; Reff 1991; Denevan 1992. There is substantial controversy about the course of events in North America north of Mexico. See, e.g., Snow 1995 and Diamond 1997a, pp. 210ff. and the reference he cites.
7. Union of Concerned Scientists 1993; National Academy of Sciences USA 1993.
8. Ehrlich et al. 1983.
9. This sort of phenomenon is discussed at length in Schelling 1978. In this fascinating book, Schelling considers the aggregate results of individual choices and how they can vary from what the participating individuals would, on average, prefer.
10. Information about China in this paragraph is from Chang-Qun et al. 1998.
11. I use the term *religion* even though it carries, probably falsely, the implication that non-Western cultures recognize an entity equivalent to "religion" in the Western sense (Wax 1984).
12. For an interesting evolutionary discussion of the role of religion in relation to social conventions and hierarchies, see Rappaport 1971.
13. Or, if you agree with Thomas Hobbes (1997 [1651], e.g., p. 71), more in the foreground.
14. Tylor 1920 (1871), vol. 1, p. 418; Kidder 1940, pp. 534–535; Sperber 1985, p. 85; Brown 1991, p. 139. There are near universals within organized religion. An example, pointed out by English anthropologist Sir Edmund Leach, can be seen in the myths of virgin births that are a common feature of religions (Leach 1967). Leach emphasized the broad unity of humanity's religious notions with the following statement: "It is high time we finally abandoned the traditional distinction between the stupidity of savages and the theology of civilized men. Stories about the ignorance of paternity among primitive peoples are of the same kind as stories about the virgin birth of deities in the so-called higher religions. And in neither case are the story-tellers stupid. If we are to understand such stories we need to consider them all together as variations on a single structural theme" (Leach 1967, p. 46).

    The origins of religions appear to be rooted in the human penchant to create explanations for natural phenomena and the human desire to find ways to control fate. But religions have evolved into a massive and slowly changing superstructure with many subdivisions espousing diverse views. Consider the variation in ethics and morals among cultures and over time. Slavery may have originated in religion, that is, in temple organizations (White 1959, p. 308). Another view is that slavery evolved from the taking of prisoners in warfare. Instead of being eaten, the prisoners were made to do productive work. In most societies, slavery was considered morally acceptable until fairly recently (and in a few places it still is). In the past, abortion was not considered particularly immoral even by the Catholic Church, yet today its morality is a matter of impassioned dispute in the United States and even within the church itself (Schwarz 1998). Note that there is a debate over whether a fetus should be considered equivalent to an infant by the Catholic Church, which does not consider a miscarriage a death and mark it with ceremony.

    Genital mutilation of women is considered moral in some African societies; genital mutilation of male babies is widely practiced in North American and European communities; murder of women who have been "unfaithful" (even by being raped) is more or less condoned in some Muslim societies (Beyer 1999). Indeed, keeping women subservient is a moral imperative in various societies and groups today, even some in the United States. Homosexuality has been ritualized in a few societies and strongly disapproved of in many others. Some societies practice polygyny, others serial monogamy. In Australia, exposure of women's breasts was indecent in the early 1960s, yet topless beaches were common in the 1970s.

As these random examples make clear, human religious, ethical, and moral views—human values—are evolving rapidly today. Values are in flux because the body of nongenetic information possessed by humanity, our culture, is growing. They are in flux because one area of that culture, science, which has not only its own values but also enormous influence on other value-laden areas, is undergoing a period of disproportionately rapid growth. They are in flux because individual religions, like other institutions in our culture, are evolving, in part because people of different cultures are in contact with one another as never before in history. And they are in flux because many cultures, along with their values, are going to extinction.

The development of organized religion in human societies has long fascinated anthropologists, including (or especially) anthropologists who were themselves religious. Very often, the latter have viewed religions as evolving toward higher forms—theirs, of course, being already at the peak. Nineteenth-century English anthropologist Sir John Lubbock, for example, was convinced that "primitive savages" lacked anything that could be described as religion—they were "atheists" who lacked any ideas on the subject of a deity (Lubbock 1870, pp. 118 ff.). This sort of culture-bound attitude was and is widespread. The brilliant French mathematician and scientist Blaise Pascal (1623–1662) wrote, in *Pensées* (Pascal 1995 [1670], p. 74), a spirited defense of Christianity: "Heathen religion has no foundations today." Pascal was interested in probability and argued that lacking evidence in either direction, one should believe in God because there was much to be gained if one was right and little to be lost if God did not exist. Known as "Pascal's wager" (Pascal 1995 [1670], pp. 121–127), this has triggered long and fruitless debate among philosophers. Pascal's views were, of course, not reflected by later secularist thinkers of the Enlightenment, such as Julien de La Mettrie (1709–1751), David Hume (1711–1776), Denis Diderot (1713–1784), and Baron d'Holbach (1723–1789). La Mettrie wrote, characteristically (La Mettrie 1996 [1747], pp. 22–23): "This does not mean that I question the existence of a supreme Being; on the contrary, it seems to me that the greatest degree of probability is in its favor. But as this existence does not prove the need for one religion any more than another, it is a theoretical truth which serves very little practical purpose. So that, as we can say after so much experience that religion does not imply perfect honesty, the same reasons allow us to think that atheism does not preclude it." It is a view refreshingly enlightened in contrast to the bigotry of Lubbock more than a century later or, indeed, that of many members of the United States House of Representatives two and one-half centuries later. As Marvin Harris put it, Lubbock had "an infuriating certainty about his own brand of superstition" (Harris 1968, p. 200).

Social scientists have barely started on the road to understanding the religious side of human natures. One thought-provoking (if not compelling) theory to explain the near ubiquity of religious notions and their persistence and replication has been put forth by psychologist Pascal Boyer. It is simply that they are counterintuitive (Boyer 1994; the material in the next two paragraphs is based on this work). The existence of angels; the possibility of virgin births; the persistence of the personality after death; the existence of an all-powerful, all-loving deity that permits millions of innocent children to die horrible deaths in holocausts, visits lethal plagues on homosexuals, severely punishes for eternity those who do not follow its rules to the letter, and so on, are counterintuitive to many, if not most, people in Western culture. Why did the Deity provide us with free will if just about all one can do with it is make bad choices and end up in hell? Self-contradictory notions abound in Christianity. The Trinity, combined with the dual nature of Christ, is such a notion, bringing together the opposites of unity and division. The ways of God are indeed mysterious, and they have puzzled philosophers in the Western tradition as long as there have been philosophers.

Similar counterintuitive notions exist in cultures that have not developed into states. For example, most of the Fang people of Cameroon believe that witches are persons with an extra organ called the *evur.* This powerful appendage is believed to be able to leave the body at night, fly on banana leaves, murder fetuses, and cause a person's blood to thicken and turn black. The Fang do not find these features "natural" or "normal"; indeed, they get attention because they are counterintuitive.

This counterintuitive aspect of religion could also explain why science is so appealing to the human mind, and so close to religion. What could be more counterintuitive than flying machines, television sets, or the notions that Earth circles the sun and the book you are reading is made largely of empty space? Boyer believes that the human mind is predisposed to build certain abstract representations. Those representations result from the interactions of the human nervous system with external objects about which we have intuitive expectations (e.g., the different expectations of behavior of a chicken and a rock, because the former is known to be alive). *Intuitive* here does not necessarily imply any genetic bias—just that reasoning isn't required to know something about a thing or event immediately. It is possible that there may be some genetically based tendency to recognize a difference between living and nonliving, given that expectations about different behavior of living and nonliving things appear to be present in infants (e.g., Richards and Siegler 1986). In religious ideas, the counterintuitive and intuitive are linked, as they are in Boyer's view, and that aids in their transmission. And this gets us back to A. E. Michotte and the invention of counterintuitive causes for mysterious effects, as discussed in chapter 6.

15. Rappaport 1971, 1979.
16. Rappaport 1971, p. 39.
17. Bloomgarden 1998.
18. Coe 1981, p. 170.
19. This point was made to me by historian D. L. Bilderback (personal communication, 23 August 1999). The book is Herrnstein and Murray 1994.
20. For a thoughtful discussion of the role of religion in democratic states, see Greenawalt 1995, or, for a summary, Greenawalt 1998.
21. The Mexica (Nahuatl) are a group of peoples from southern Mexico and Central America, including the Aztecs.
22. Conrad and Demarest 1984, pp. 185–186.
23. After all, virtually every state that marches off to battle claims to have God on its side. (An exception would be the Soviet army in World War II, which thought it had History on its side.) The U.S. Navy, for example, at the end of World War II employed 2,787 chaplains to serve 4 million personnel, and all of the chaplains were from religions that support the commandment "Thou shalt not kill" (Drury 1984).
24. For a varied overview, see Nettleship et al. 1975.
25. I am ignoring an old argument that intergroup violence evolved for population control—that its fundamental function was avoiding overexploitation of natural resources (see, e.g., Russell and Russell 1973)—because even using group selection or wild cultural speculation, one would have a hard time putting together a scenario for that. In chimpanzee and human societies, it seems reasonable to look for ecological factors that give advantages to aggressive individuals or groups (see, e.g., Harris 1984; Vayda 1961).
26. See, e.g., Leeds 1963.
27. See, e.g., Divale and Harris 1976.
28. Kagan 1995, p. 569.
29. Kagan 1995, p. 8. Richard Wrangham and Dale Peterson, in their interesting book about male violence in people and chimpanzees (1996, p. 191), make a connection between male chimp violence and human emotions that might be described as pride or arrogance (or, perhaps, honor).
30. See, e.g., Richardson 1960; Doran and Parsons 1980; Ember 1982.
31. A population–resource component in the generation of wars even in the twentieth century has been detected by careful analyses, as well as being explicitly claimed, for example, when Adolf Hitler announced Germany's need for living space (Lebensraum) in the East prior to his invasion of the Soviet Union. For analyses, see Choucri and North 1975; for a summary, see Ehrlich et al. 1977, pp. 908–910. Religion is often involved in the generation of war, as it was in the ethnic strife that followed the breakup of Yugoslavia and, much earlier, in the Crusades. In the former Yugoslavia, Eastern Orthodoxy, Catholicism, and Islam clashed, leading to "ethnic cleansing." It is astonishing, but the total victory in 1389 of the invading Muslims over Orthodox Serbians at the Field of Blackbirds in Kosovo seems as recent to the Serbs today as the Holocaust does to the Jews (Catherwood 1997, p. 59). But in this case, the involvement of religion is perhaps more of a secondary cause or an excuse and is inextricably intertwined with ethnicity. Ethnic fragmentation is a powerful force in the world today, and one could draw the conclusion that some states (such as Kosovo) will never be stable until they consist of just one group—although how that could be humanely achieved is a very difficult question.
32. Quoted in Burns 1963, p. 358.
33. Personal communication, 4 April 1999.
34. Howard 1984, p. 7.
35. Indeed, human cultures, including those of hunter-gatherers (see, e.g., Turnbull 1982), have evolved many institutional mechanisms for conflict reduction. Organizations such as the United Nations clearly have deep roots.
36. See, e.g., Canada 1995.
37. Keegan 1998, p. 296.

38. This was first brought to light by the famous military historian S. L. A. Marshall (1978); see also the discussion in Keegan 1976, pp. 73ff. For detailed coverage, see Grossman 1995. Some of Grossman's explanations of reticence to kill, especially the Freudian ones, I find less than convincing. But his book, though somewhat repetitive, is very interesting and does document the reticence quite thoroughly. Firing rates (percentages of men who actually fired their weapons at the enemy) appear to have been very low throughout World War I and were documented among U.S. troops to have been about 15–20 percent in World War II and 55 percent in the Korean War. They rose to a surprising 90–95 percent in the Vietnam War, apparently as a result of new training methods designed specifically to condition men to shoot and kill. Grossman makes the most interesting point (pp. 54ff.) that it seems to be not fear of death that leads to the high rate of psychiatric problems among infantrymen but fear of killing what are undeniably fellow human beings. Civilians, prisoners of war, and naval personnel subject to horrendous and persistent dangers in wartime show negligible levels of combat fatigue and the like.
39. See, e.g., Chang 1997, pp. 56–59.
40. Wilson and Daly 1985.
41. Chagnon 1988.
42. I've modified here the definition in the Convention on the Prevention and Punishment of the Crime of Genocide approved by the United Nations General Assembly on 9 December 1948. See also Horowitz 1997.
43. Diamond 1991, pp. 250–278; see also Chalk and Jonassohn 1990.
44. Fein 1979.
45. Idani 1991a.
46. See, e.g., Sven Lindqvist's fascinating brief account of European treatment of Africans (1992). The Hutu genocide against the Tutsis in ecologically devastated Rwanda (Holdren and Ehrlich 1984; Ehrlich et al. 1995, pp. 10–11) is described in Gourevitch 1998.
47. Gilbert 1985; Chang 1997. The massive horror of atrocities committed on Jews in the Holocaust overshadowed the suffering inflicted by the Japanese (especially on the Chinese) and the Nazis' genocidal killing of some 250,000 Gypsies and millions of Russian prisoners of war. The Japanese slaughter of more than a quarter of a million Chinese people, mostly civilians, and rape of 20,000–80,000 women took place over just a few weeks (Chang 1997).
48. Goldhagen 1997. See especially chap. 7; the quote is from p. 218.
49. Goldhagen 1997, p. 269. Although Goldhagen's work has been supremely controversial, especially with respect to the uniqueness of German anti-Semitism and the full extent of culpability among the Germans (see, e.g., Finkelstein and Birn 1998), the point made here is essentially uncontested. A much less controversial but equally horrifying account is Browning 1998. Biologists might wish to examine the dismal record of many of their colleagues under the Third Reich (Deichmann 1996). A fascinating recent work that examines what the United States and Great Britain knew of the genocide, and the role of anti-Semitism in their inaction, is Richard Breitman's book *Official Secrets* (1998). The issue of how prejudice influences the reaction of nonperpetrators to genocides was alive in the 1990s during events in Rwanda and the Balkans. What the United States and European nations should have done in these cases is a very difficult ethical question.

    An attempt has been made to explain Nazi behavior in revised Freudian psychological terms as a distortion of a drive by Germans to kill their own children, connecting it to the widespread practice of infanticide and Jewish opposition to it (Kestenberg and Kestenberg 1987). The claim is that "infanticide and cruelty to children were revived in pure culture not only in the Nazi aggression against alien children but also against their own children" (p. 139). The same authors also state (p. 152): "The impulse to kill one's children and commit genocide on one's own people is not easy to conquer. . . . Ridding themselves of the Jews, as saviors of children, freed them to commit mass murder on their own." Considering the frequency with which genocide is practiced and the diversity of practitioners, I find this Freudian line of reasoning less than compelling, especially because the analysis does not speak to the origins of the "impulse" or in fact to its existence, beyond pointing out correctly that infanticide and child abuse (according to current moral standards) have historically been widespread. Freudian "explanations" of behavior are largely of interest today as another illustration of the persistence of counterintuitive ideas (the existence of a "drive" to kill one's children) among a basically religious following. For a more informative and brilliantly crafted literary look at what it meant to be "different" in Nazi Germany, see Hegi 1997. For a stunning set of diaries by a Jew who lived through the Nazi period in Dresden, see Klemperer 1998, 2000.
50. Chang 1997, p. 6.
51. Christian Kröger, "Days of Fate in Nanking" (unpublished diary), in the collection of Peter Kröger.

52. Milgram 1974.
53. Chang 1997, p. 220.
54. In contrast, Daniel Goldhagen, who wrote the controversial book *Hitler's Willing Executioners* (1997), believed that the long tradition of virulent anti-Semitism in Germany, combined with relentless Nazi propaganda and dehumanizing treatment of the Jews, was responsible for the widespread (but far from universal) willingness of Germans to participate in genocide. Goldhagen blamed a long-developed negative image of Jews that was an exceptional part of German culture for allowing the Holocaust and for its cruelty. Other scholars, I think correctly, disagree—disputing both his selection of evidence and his conclusions (Oberman 1997; Browning 1998; Finkelstein and Birn 1998). Cruel holocausts are not simply expressions of German culture, as clearly demonstrated by those in Rwanda (Gourevitch 1998), Cambodia, and Croatia. Goldhagen's critics point out that systematic dehumanization of the victims was important in creating circumstances in which "ordinary Germans" would behave with extraordinary cruelty as well as in making possible the atrocious behavior of many non-German co-perpetrators from eastern Europe and elsewhere (Browning 1998). Breitman 1998, in my opinion, is a balanced overview of this sad controversy. See also Kaplan 1998, e.g., "The history of Jewish daily life shows how the Nazi government—through indoctrination, bribery, and coercion—turned antisemitic prejudices into a mass movement" (p. 4), a view congruent with Breitman's. What is clear from the writings of both Goldhagen and his critics is that very many Germans, faced with dehumanized victims and working in wartime in an officially sanctioned program, were willing participants in the extermination of fellow human beings, even though they had not been subjected to the sort of abuse that was inflicted on Japanese soldiers.
55. Chang 1997, p. 218.
56. Cited in Grossman 1995, p. 162.
57. Dower 1986.
58. Interestingly, during World War II, even the most liberal and antiracist anthropologists in the United States participated in efforts to denigrate Japanese culture and (at least by implication) the Japanese as a people; see, e.g., Bateson and Mead 1942, p. 263. For a more understanding contemporary view of the issues from the old Freudian perspective that was once so popular, see Silberpfennig 1945.
59. Keegan 1990.
60. Gourevitch 1998. This is a brilliant and thought-provoking book.
61. For an analysis of warrant for genocide, see Kuper 1981, p. 84.
62. But *not* of the Ottoman Empire's actions against Armenians during World War I, which did not fit a reasonable definition of genocide, although often cited as such. The Armenians revolted against the Ottoman Empire when that empire was embroiled in World War I and the British had invaded the Dardanelles. The Ottoman Empire's resulting program to relocate the Armenian population of Asia Minor was accompanied by much brutality, and as many as a million people may have died, some by murder and some of starvation and exposure. But incompetent as the empire was, it attempted to control excesses in what it considered an act of military necessity. The government moved to punish crimes against the Armenians, and there were almost 1,400 courts martial during which Ottoman military and civilian personnel were sentenced, sometimes to death, for crimes against the Armenians (Lewis 1995, pp. 339–340). That contrasted sharply with the behavior of the Nazis (or of the Hutus in the most recent Rwandan genocide). There is considerable dispute about the actual numbers of deaths, with one careful estimate being a horrifying 600,000 (McCarthy 1983). It was a vast human tragedy, regardless of exact numbers or labels.
63. Ehrlich et al. 1983.
64. Steward 1949; Wittfogel 1957; Mitchell 1973.
65. D. L. Bilderback, personal communication, 4 April 1999.
66. White 1959, p. 303.
67. Brumfiel 1980.
68. Berdan 1989, p. 96.
69. Berdan 1989, p. 96.
70. Ozguc 1973, p. 51.
71. Population Reference Bureau 1998.
72. Korten 1995, p. 108.
73. Anonymous 1993.
74. The connection between sex and power is well illustrated in pre-state social arrangements, so it is not absurd to think that the number one message of the genes—reproduce as much as possible—would play a role in the acquisition of wealth or, in the case of women legally denied direct control of wealth, con-

sortship with the rich. Accumulation of wealth was not physically or socially feasible in hunter-gatherer societies, but maneuvering by both men and women within certain culturally set rules to gain access to the opposite sex was often (perhaps always) as high on the social agenda, just as it is in modern society. The most parsimonious explanation of our complex behavior related to power and sex is that whatever genetic evolution there might be of a disposition to plan, plot, and scheme to maintain or improve one's position in society (and thus one's reproductive output) is shaped by cultural evolution in different environments into many different human natures.

75. Plattner 1989a.
76. Berdan 1989.
77. Wells 1946, p. 49.
78. See Harris 1986 and the discussion in d'Errico 1995.
79. Senator John McCain on *Face the Nation* (CBS), 4 July 1999.
80. Gellner 1994, p. 191.
81. Diamond 1997a, p. 360.
82. For an interesting discussion of the evolution of writing, see Lecours 1995.
83. Schmandt-Besserat 1978.
84. If humanity produced only two commodities, wheat and rice, the production possibility frontier (PPF) would be a line on a graph with wheat on one axis and rice on the other. The line would be a convex curve representing the maximum amount of wheat that could be produced if a given amount of rice were produced, and vice versa, as land that could be used to produce either was switched from one to another. If the economy were producing a combined amount of the two commodities that was within (to the lower left of) the PPF, it would be economically inefficient. All this assumes, of course, that all social costs would be properly captured in market prices, which in practice they never are.
85. Ehrlich 1997, chap. 1.
86. Harris 1986, p. 24.
87. I could add that there is nothing more fundamental to maintaining the inequality of the sexes and the promotion of runaway population growth than high rates of illiteracy among women (Koblinsky et al. 1993; United Nations 1991b). For the situation in one developing country, see World Bank 1991. In 1990, about 20 percent of men, but about 33 percent of women, were illiterate worldwide (Ballara 1991; United Nations 1991a). Besides their tragic direct effect on women's lives, high birthrates are an unfortunate demographic correlate of female illiteracy (Ehrlich et al. 1995, chap. 3). One positive response to sexism has been to tie the need for equity between the sexes to the environmental destruction produced by societies (all of which are patriarchal). That has occurred in a social movement called ecofeminism. The basic ecofeminist issue is the degree to which environmental degradation is connected to the inferior status of women—the literature on which consists of an interesting mix of good ideas, viewpoints from "radical ecology" (with which I sometimes agree), and occasionally atrocious scholarship and distortion. For an introduction to that literature, see Shiva 1989; Merchant 1992, chap. 8; Mies and Shiva 1993; Mellor 1997; and Sturgeon 1997. Ecofeminism has an interesting view of the cultural evolution of science, one that I think has some merit: "The scientific revolution in Europe transformed nature from *terra mater* into a machine and source of raw material; with this transformation it removed all ethical and cognitive constraints against its violation and exploitation" (Shiva 1989, p. xvii). As is the case with most social movements, however, ecofeminism sometimes draws unwarranted conclusions in order to support its worldview. For example, assuming that concern about overpopulation is basically an attack on women and claiming that environmental scientists disregard other sources of environmental degradation is purest nonsense—as the wide use of the $I = PAT$ equation clearly illustrates (e.g., Bradotti et al. 1994, p. 143). To judge the silliness of these assertions, see Ehrlich and Ehrlich 1990, p. 51, a cited source for Bradotti et al. 1994. Systematic discussion of the relationship of population and other factors in causing environmental degradation goes back at least as far as Ehrlich and Holdren 1971. Sexism and racism exist in the environmental movement just as they exist in many other areas of society. That, however, is no justification for sloppy scholarship on the subject.
88. The fax machine has been a serious enemy of despotism (although it can also be a tool of repression).
89. Heller 1997.
90. Details about the armada are from Kristof 1999.
91. Diamond 1997a, pp. 411–417. Of course, events in Europe had other important cultural *macro*evolutionary components, such as the abundance of fine natural harbors along Great Britain's coasts, as opposed to their paucity on the western shores of the European continent. That was an important cause

of Britain's success as a naval power, which shaped the destiny of the entire world (Keegan 1988, pp. 14–15).

92. For the inward, world-renouncing orientation of China, see Needham 1956 and Feuer 1992.
93. Kristof 1999.
94. I'm leaning here on some ideas of Thomas Heller (personal communication, 18 August 1999).
95. Hirschman 1977, p. 10. Note that Adam Smith considered that this pursuit of private passion would go on in a basically moral setting, as expressed in his earlier (and then very famous) book *The Theory of Moral Sentiments* (1974 [1759]).
96. A. Smith 1976 (1776). Smith stated (p. 477) that every individual "intends only his own gain, and he is in this, as in many other cases, led by an invisible hand to promote an end which was no part of his intention."
97. World War II was essentially a continuation of World War I; see, e.g., Keegan 1998, chap. 1.
98. For a superb new analysis of Roosevelt's policies, see Kennedy 1999a.
99. Harvard-trained Yamamoto knew the Americans well, admired them, and had advised his government against attacking the United States. When his advice was not followed, however, he prosecuted the war with amazing success and fulfilled his own prediction that he would be able to run wild in the Pacific for six months. He was well aware of the macro imbalances that made it highly unlikely that Japan would prevail if the United States were provoked into enthusiastic prosecution of the war.
100. Kant 1996 (1797), pp. 114–115: "The elements of the right of nations are these . . . (3) A league of nations in accordance with the idea of an original social contract is necessary, not in order to meddle in one another's internal dissensions but to protect against attacks from without."
101. See, e.g., Gould 1989, pp. 280ff.
102. In 1941, Japan needed to import 40 percent of its steel, 60 percent of its aluminum, 80 percent of its oil, 85 percent of its iron ore, and 100 percent of its nickel (Keegan 1990, p. 103).
103. Keegan 1998; see especially pp. 22–23.
104. It is interesting to speculate about what the U.S. response might have been had the Japanese simply invaded French, Dutch, and English colonies in Southeast Asia without bombing Hawaii. Japan thought it had to neutralize the U.S. presence in the Philippines to protect its flank if it was to take over Malaya and Indonesia, given that the United States had made it clear with oil and steel embargoes that there were limits to its tolerance of Japanese expansionism. But whether military intervention by U.S. forces would have been politically possible without a provocation such as the invasion of Pearl Harbor is an open question.
105. The wide recognition of the role of chance is found in the sequence "For want of a nail, the shoe was lost; for want of a shoe, the horse was lost; for want of a horse, the general was lost; for want of a general, the battle was lost; for want of a battle, the war was lost," versions of which date back to at least 1651 (Evans 1968, p. 475).
106. Named after Colonel James Doolittle, who led the attack—seventy-one members of the aircrew that launched the precarious enterprise survived, including Doolittle (1896–1993), who received the congressional Medal of Honor for his exploit, was promoted to the rank of lieutenant general, and went on to have a long and successful career in aviation.
107. For a summary, see Keegan 1990, pp. 270–272. A more extended account by this brilliant military historian can be found in *The Price of Admiralty* (1988), marred slightly by some confusion about the names of U.S. torpedo bombers and dive-bombers of the era. See also Prange 1982.
108. Historian David Kennedy (1999b) has a recent short account of the battle in his fine article "Victory at Sea."
109. Ensign George H. Gay, who described the events in his book *Sole Survivor* (1980). I have remembered his name ever since reading a magazine article shortly after the battle, titled "Torpedo 8," after the designation of his doomed squadron.
110. Kennedy 1999b, p. 64.
111. The *Kaga* sank at 7:25 that evening; the *Akagi* was scuttled at 5:00 the next morning (Morison 1947–1962, p. 158).
112. The *Soryu* sank five minutes before the *Kaga.*
113. That afternoon, planes from the *Hiryu* severely damaged the *Yorktown*, which was abandoned. But revenge was obtained by Dauntlesses from the *Enterprise*, including planes that were refugees from the *Yorktown.* They found the *Hiryu* and inflicted fatal wounds on the ship, which was scuttled the next morning, completing the U.S. sweep of Japan's carrier force.
114. Ever since I first read a detailed account of those momentous events, one bit of trivia that strangely (and

nonadaptively) seems stored in the active memory of my brain has been the name of Lieutenant Commander Wade McCluskey, who led those world-changing thirty-seven dive-bombers from the *Enterprise*.

115. For instance, the United States might not have been able to help the Allies against Germany to the degree that it did, Russia might have been defeated, the atomic bomb might first have been used in the European theater, and on and on.
116. See, e.g., Dower 1999, pp. 496–508. Especially interesting is Dower's description of the transformation of one of Japan's most distinguished philosophers, Hajime Tanabe. He was steeped in Hegelian and Kantian philosophy and had studied under the famous existentialist Martin Heidegger (1889–1976), who, interestingly enough, gave outspoken praise to the Nazi regime. Heidegger did not intervene when his professor, the founder of phenomenology, Edmund Husserl (1859–1938), was silenced by the Nazis because Husserl was Jewish. Tanabe went through a period of self-flagellation and declared the Western philosophical tradition (which he had represented in Japan) to be inferior. Furthermore, as a sign of his change of nature, he turned against Emperor Hirohito, declared the emperor's great responsibility for the war, and urged that the wealth of the imperial family be transferred to the poor (pp. 497–501).
117. The question of how objective history can be is under constant debate. See, e.g., Novick 1988; the material about the role of the pre–World War II interventionist–isolationist debates among historians is especially interesting (pp. 247–249). In any case, major microevolutionary forces often can be detected—as in the role of the Treaty of Versailles, which ended World War I, in generating World War II (Keegan 1998, pp. 423–424).
118. This was suggested to me by John Allman (personal communication, 1 November 1999).
119. For a sample that I have dipped into and enjoyed even when I've disagreed with the authors, see Braudel 1993; Doran and Parsons 1980; Huntington 1996; Kennedy 1987; Landes 1998; Melko 1994; McNeill 1992; Modelski 1978; Quigley 1961; Sanderson 1995, 1998; Toynbee 1934–1961 (no, I haven't read it all!).
120. One broad pattern that emerges from a study of cultural evolution is that nation-states are temporary arrangements, even though they may persist for many human lifetimes (Bronson 1988). This could be a salutary lesson for those now happily, if often unconsciously, fashioning the first global state as well as for those concerned that the globalizing world is simultaneously in a state of ethnic fragmentation. There is little reason to assume that either of the centrifugal and centripetal tendencies now visible in global politics will lead to a kind of stasis, whether Plato's ideal city or some dreadful alternative. Politics will continue to evolve, and those concerned about the shape of future political arrangements would do best to focus on ways to guide that evolution in ethically desirable directions rather than seeking single-minded utopian "solutions" to the problem of how people should organize their societies. It may be true that the term *United Nations* is virtually an oxymoron and that nations "exist to be disunited with one another, only in coming together in temporary alliances out of self-interest" (Fox 1997, p. 63). But the same trend toward intensification that created such "unnatural units" as states when there were no alternatives should be able to develop relatively permanent international regimes to deal with pressing problems of the global commons, such as regulating the quality of the atmosphere, managing oceanic fisheries, and avoiding large-scale nuclear war.
121. See, e.g., Iggers 1997.
122. Consider how one's view of human natures is stretched and enriched by reading, for instance, Johann von Goethe's *Faust* or the radical analyses of Michel Foucault (1972, 1988, 1994, 1995).
123. Based on a lecture given in Munich in 1919; first published as a brochure, *Politik als Beruf*, in that year, and reprinted in a collection of Weber's political writings (Lassman and Speirs 1994, p. 311). According to Weber, acceptance of state authority is generally based on custom (tradition), the existence of appropriate legal statutes, or the presence of a charismatic leader.
124. Skocpol 1979; Held 1989.
125. A useful and concise summary can be found in Held 1989, especially chap. 1.
126. That is, men form commonwealths (states) with the goal "of getting themselves out from that miserable condition of Warre" (Hobbes 1997 [1651], p. 93). Hobbes here was referring, of course, to his concept of the war "of every man, against every man" (p. 70), which he conceived as humanity's natural condition. Needless to say, issues of state politics go back far beyond those of the modern state, to the times of Plato and Confucius.
127. Ehrlich et al. 1983. As of this writing (February 2000), the threat of disastrous use of nuclear weapons has hardly ended. Large inventories of arms still exist, the problems of weapons proliferation are becoming increasingly serious, and there is a danger of the United States reigniting a nuclear arms race with Russia and starting one in Asia by deploying a national missile defense system.

128. Ehrlich and Ehrlich 1996.
129. Hayek was a liberal in the original sense of John Locke, Adam Smith, and others who generally opposed government intrusion into individual choices and (tracing from roots in classical civilization) valued highly the emergence of relatively unconstrained individuals within a society. Recently, liberalism has acquired exactly the opposite connotation.
130. Hayek 1944, p. 257. Of course, the case for having small political units in which a sense of community can persist, as well as arguments over democracy as a political system, go back to Plato (ca. 427–347 B.C.) and before (Kraut 1992; Plato 1991 [ca. 387]). For an introduction to current discussions of the problems of the "worst political system—except for all the rest," see Cronin 1989; Dahl 1989, 1998; Sandel 1996. On the related and increasingly critical issue of the role of the U.S. presidency, see Cronin 1998; its bibliography gives good access to the extensive literature.
131. The Rush Limbaughs and Tom DeLays of the world are not conservative in the traditional sense, although, as made clear in the following quote, they do share certain attitudes with conservatives in the older sense.
132. Hayek 1944, p. xxxvi. He goes on to point out that conservatives always try to defend privilege and push governments to help them do so. In short, Hayek would have seen through today's so-called conservatives who oppose government regulation except when it comes to the things *they* want the government to regulate—such as abortion, the teaching of religion and creationism in schools, and any other behavior they want promoted or discouraged.
133. Daily 1997.
134. See, e.g., the 1994 Fiftieth Anniversary edition of *The Road to Serfdom*, pp. 41–43, 133–135.
135. See Pirages and Ehrlich 1974, pp. 57ff., on macroconstraints and microfreedoms.
136. See, e.g., Crosby 1986. The Polynesian expansion took rats and pigs around the Pacific, the European expansion started the spread of weeds and diseases, and so forth. Over the past 150 years, of course, the global influence of *Homo sapiens* has enormously increased (Ehrlich 1968, 1995; Ehrlich et al. 1977; Ehrlich and Holdren 1971; Holdren and Ehrlich 1974; Holdren 1991; Marsh 1874; Myers 1979, 1996a, 1996b; Turner II 1990; Vitousek, Aber, Howard, et al. 1997; Vitousek, Aber, Howarth, et al. 1997; Vitousek et al. 1986; Vitousek, Mooney, Lubchenco et al. 1997).
137. Hayek 1944, p. 257.
138. Hayek, in fact, had a considerable interest in cultural evolution late in his career (Hayek 1979, pp. 153ff.).
139. Daily 1997.
140. Hobbes 1997 (1651).
141. Except, in earlier days, for slaves, women, and the poor.
142. A double maximization. In any physical sense, the greatest good is a quantity of something, and beyond the need for sufficient people to produce or mobilize that something, the greater the number of people who have access to it, the less of the good there is for each. One cannot maximize the amount of gold or grain possessed by each individual simultaneously with maximizing the number of people possessing that quantity.
143. Ehrlich et al. 1977, chap. 5; Ehrlich and Ehrlich 1996.
144. Ehrlich et al. 1977, p. 187.
145. Ehrlich et al. 1977; Ehrlich and Ehrlich 1990.
146. Dalrymple 1976; Ehrlich et al. 1977, chap. 7 and references thereto.
147. Ehrlich et al. 1993, 1995; Waterlow et al. 1998.
148. See, e.g., Safina 1997.
149. Naylor et al. 1998.
150. Holdren 1991, updated.
151. Held 1989, p. 230. There is an ongoing discussion of the significance of the roles of MNCs and the globalization of the economy, and considerable disagreement, but there is general agreement that the day of the nation-state as it has existed for the past few hundred years is coming to an end. See, e.g., Korten 1995; Mander and Goldsmith 1996; Caldwell 1997.
152. Chichilnisky 1997.
153. For a detailed discussion of the "knowledge revolution" and its environmental effects, see Ehrlich et al. 1999.
154. See, e.g., Jacobsen and Adams 1958; Culbert 1973; Hughes 1975; Yoffee and Cowgill 1988.
155. Held 1989, p. 228.

Chapter 12: Lessons from Our Natures

1. Trevathan et al. 1999, p. 3.
2. Some of the tactical evolutionary lessons about human behavior would be the same even if we thought people were the products of special creation. There is reason to believe, for instance, that whatever predispositions people might have for language acquisition, languages are most easily acquired early in life. Exposing children to a great deal of appropriate aural stimulation early on, including foreign languages, would probably benefit the educational process. Language, mathematics, art, and music appear to have been tightly interwoven by evolution; all involve symbolism, and all may have similar developmental windows of opportunity. It would seem wise to encourage their expression early in life. Some claim that the influence of the environment in the first two years of life on subsequent adolescent and adult behavior and capability is greatly overestimated (see also Bruer 1999; for a summary, see Kagan 1998). However, one should not underestimate the potential for damage in those years either, for at least in rats and monkeys, early emotional events can have long-lasting effects on the neurochemistry of the brain and on later mood and behavior (Mlot 1998). Interestingly, evidence is building that much of children's socialization comes not through the efforts or behavior of parents but via peer groups (Harris 1995, 1998).
3. Even if the last cultures of smallpox are destroyed, there is no guarantee that a mutated animal virus with the same effects won't again invade the human population (as far as we know, that was the original source of smallpox).
4. There are also interesting evolutionary implications for health that do not involve biological and cultural coevolution with microorganisms. Instead, they concern genetic and cultural responses to substances in the environment (including the epidemiological environment) that cause severe allergic reactions, such as asthma (see, e.g., Barnes et al. 1999; Finn 1992; Gamlin 1990; Hurtado et al. 1999; Profet 1991).
5. Levy 1992.
6. Levin and Anderson 1999.
7. Beers and Berkow 1999.
8. A recent example was the disbelief expressed by some experts at the spread of drug-resistant *Staphylococcus aureus* bacteria (Stolberg 1999, p. A17) from hospitals and nursing homes into the general population—a totally predictable occurrence.
9. Levy 1992, pp. 138ff. The issue of use in pets is more delicate.
10. Levy 1992, chap. 4.
11. That has long been a central concern of mine and of other biologists; see, e.g., Ehrlich 1968; Ehrlich and Ehrlich 1970, pp. 148ff.; Daily and Ehrlich 1996a. For a reconstruction of humanity's early epidemiological environment, see Cockburn 1971 and the commentaries following it.
12. It seems likely that chimpanzees are the reservoir for HIV (Gao et al. 1999). Hunting of chimpanzees by human populations has reduced their numbers, making them endangered. Ironically, it may be that the chimps have had a grim revenge.
13. Levin and Anderson 1999; Ebert 1999. Unhappily, the most advantageous strategies may involve forgoing short-term gain for long-term benefit.
14. Ewald 1988, 1994. Some of Ewald's ideas are controversial, but the notion that virulence is an evolved characteristic of pathogens is not.
15. Nesse and Williams 1995, p. 61. The example in the next paragraph is also suggested by Nesse and Williams (p. 58). There is considerable debate over whether the evolution of virulence is actually well enough understood to support the making of public health recommendations based on it (Ebert 1999; Levin and Anderson 1999).
16. Kluger et al. 1997. A strongly sociobiological viewpoint on evolutionary aspects of medicine, with references, can be found in Nesse and Williams 1995. It is rather biased to the view that various symptoms and syndromes are adaptive (the result of natural selection), but it is an interesting read nonetheless. A more recent, balanced, and technical treatment of many of the issues can be found in Stearns 1999.
17. Highly reactive forms of oxygen that can damage large, biologically important molecules. See also Sohnle and Gambert 1982.
18. Nesse and Williams 1995; Trevathan et al. 1999.
19. Doran et al. 1989; Graham et al. 1990.
20. See Sapolsky 1998, which is a great read.
21. Eaton et al. 1996. For an overview of human dietary choices, see Pyke 1986.
22. Honey may be the food with the most calories per unit mass found in nature, and dispersing reproductive forms of termites and ants can have a higher caloric yield than meat—and both were very likely important parts of our ancestors' diets (McGrew 1999).

23. The best-known honeyguide, *Indicator indicator*, has a chattering call that attracts attention to itself. It makes short flights toward the bees' nest, drawing a person, honey badger, baboon, or other mammal to the nest. After the mammal breaks into the nest, the thick-skinned honeyguide feasts on the wax, impervious to the enraged bees.
24. Milton 1993. For an overview of human dietary habits, see Harris and Ross 1987b.
25. W. R. Williams 1896. Williams was aware of the possibility that higher cancer rates could have reflected greater life expectancy and was able to rule that out.
26. Raloff 1998.
27. Cited in Garnick and Fair 1998. However, the data are from 1975, so some changes may well have occurred.
28. Garnick and Fair 1998.
29. For a recent review, see Eaton et al. 1999.
30. See, e.g., O'Dea 1991.
31. See, e.g., Rendel 1970.
32. Non-insulin-dependent diabetes mellitus (NIDDM) (Zimmet 1991; Diamond 1992).
33. The Nauru story is based on Diamond 1992 and McDaniel and Gowdy 2000.
34. See, e.g., Weiss et al. 1984.
35. O'Dea 1991.
36. See, e.g., Hamilton 1987.
37. Andersson 1982. For an illustration and a discussion of female choice, see Small 1992b.
38. The widowbirds are *Euplectes progne*. The "little or no" refers to possible costs to the female of superlong male tails. Because the males build nests for the females or help the females build them, postcopulatory predation on the males would influence female fitness.
39. See also the discussion in Nesse and Williams 1995, pp. 147–151.
40. On the other hand, my father also loved nearly raw beef and "dusty road" sundaes, so there could be a large element of cultural transmission. In sorting this out, there is obviously a substantial problem of distinguishing proximate from ultimate causes. Interestingly, there may be a connection between a sugar hangover and human addiction to alcohol. Honey may have been the first substance to be fermented and consumed (O'Dea 1991; E. O. Smith 1999).
41. Cultural prejudices about food can be very difficult to change. I can draw two examples from my personal experience. During the Great Depression, just after I was born, my father lost his job; although he soon found another, throughout my childhood a roast beef on Sunday, cooked rare, was a sign that all was still well. As I said, I still crave rare beef, and my reduced consumption requires substantial force of will. On the negative side, even though cockroaches are eaten in China and Thailand and by Kalahari Bushmen and Australian Aborigines (Abrams 1987), the very thought makes me want to vomit. When I was a child, my mother was horrified by cockroaches, and I clearly picked up that cultural trait from her.
42. This idea traces back at least to John Locke (1978 [1689])—see the discussion in Ingold 1986, pp. 226–227.
43. See, e.g., Dentan 1968, p. 491; Gould 1982; Cashdan 1989; Jonaitis 1991. There is a substantial and argumentative literature on sharing in nonstratified and stratified societies—see Ingold 1986; E. A. Smith 1991 (1988); Lee 1991 (1988).
44. Economist Thorstein Veblen posited the latter a century ago when he wrote about "conspicuous consumption" (1967 [1899]).
45. See, e.g., Kaplan and Hill 1985b; Flinn 1986; Betzig 1988. In the rural village of Grande Anse on the isolated northern coast of Trinidad, for example, individuals with bigger landholdings had more children, men with more land had greater mating success, and young people whose parents were residents of the village (as opposed to living elsewhere) had higher reproductive success (Flinn 1986).
46. Myers and Diener 1995.
47. Kenneth Arrow, personal communication, Askö, Sweden, September 1998.
48. Chocolate Group Seminar, Center for Conservation Biology, Stanford University, 5 March 1999.
49. Merton 1968; Townsend 1987.
50. Schor 1998, p. 3.
51. A demonstration of this in formal economic terms is being developed—Lawrence Goulder and Geoffrey Heal, "Do We Overconsume?" Chocolate Group Seminar, Center for Conservation Biology, Stanford University, 5 March 1999. For a thoughtful book about overconsumption that expands on this theme, see Frank 1999; for a diverse group of interesting general essays on our consumer culture, see Rosenblatt 1999.

52. Ehrlich and Holdren 1971; Holdren and Ehrlich 1974; Ehrlich and Ehrlich 1990.
53. See, e.g., Daily 1997.
54. Fox 1997, pp. 27–54, is a wonderful recent discussion of this (the quotation is from p. 31). For the original, see Tiger and Fox 1971.
55. Henderson et al. 1987; Kaidbey et al. 1979.
56. Jablonski and Chaplin 2000.
57. Race could become a social issue only when some people decided that members of other groups or tribes, different from themselves in various ways and living in other areas, were actually members of their own species rather than nonhuman. Human beings seem to "discount by distance" (Daily and Ehrlich 1992) as well as by time. Even today, to many, the remoteness and perceived differences of some people make them of less concern than people who are more similar culturally or live closer (e.g., compare the different reactions in the United States to atrocities in Rwanda and Kosovo). There is a substantial and ancient debate over the role of emotion in general and empathy in particular in prosocial (as opposed to antisocial) behavior. For a review, see Eisenberg and Miller 1987.
58. This is a myth that is often taken very seriously; see, e.g., Browne 1994; Snyderman 1994.
59. Wilson and Brown 1953; see also Brown 1959.
60. See, e.g., Gillham 1965.
61. Brace 1964.
62. Cavalli-Sforza et al. 1994; but see Sokal et al. 1999.
63. The most concordant of commonly used characteristics in human racial classification are skin color and hair type (Ehrlich and Feldman 1977).
64. Coon et al. 1950.
65. Brace et al. 1993, p. 1.
66. Marshall 1998, p. 654.
67. Gourevitch 1998, pp. 47–48.
68. See, e.g., Ehrlich and Feldman 1977, pp. 8–12. There are abundant other examples not detailed in that source.
69. Slavery, contrary to some popular opinion, has not always been a race-based institution.
70. Jefferson 1801, pp. 204–205.
71. Jefferson 1801, p. 206.
72. Jefferson 1801, p. 204.
73. Jefferson 1801, p. 214.
74. Culturally defined; she had one black grandparent.
75. Hamilton 1999.
76. Coon 1962.
77. Coon 1962, p. 657.
78. Coon's book is replete with egregious material. On the different times and "levels" of the putative races, see also especially pp. vii, x, 427, 655 (first printing; some later printings have different pagination). Coon's material on speciation and "racial intermixture" (e.g., p. 661) was primitive even for his day.
79. Linnaeus 1758, pp. 20–22 (translation from the original Latin courtesy D. L. Bilderback).
80. Reprinted in T. H. Huxley 1910, p. 115.
81. T. H. Huxley 1910, p. 116.
82. Quoted in Degler 1971, pp. 120–121.
83. Shockley 1972.
84. Quoted in Daniels 1973.
85. For an excellent critique, see Goldberger and Kamin 1998.
86. Herrnstein and Murray 1994; see reviews in Goldberger and Manski 1995 and Brace 1996.
87. Neisser 1998; in that volume, see especially Hauser 1998.
88. In statistical analyses of test results, however, there are signs of some sort of general intelligence in individuals (for two careful analyses coming to somewhat different conclusions, see Carrol 1997 and Kagan 1998, pp. 50ff.), although there is an increasingly widespread view that intelligence has many different facets. For a popular summary, see H. Gardner 1999.
89. Moorehead 1963. A photograph of the actual note can be found following p. 68.
90. Labov 1969; Pinker 1994.
91. To get a sense of IQ tests and their limitations, see Ehrlich and Feldman 1977, chap. 4.
92. See, e.g., Jensen 1969; Bouchard et al. 1990; Herrnstein and Murray 1994; Plomin et al. 1994. For critiques, see Feldman and Lewontin 1975; Ehrlich and Feldman 1977; Lewontin et al. 1984.

93. Feldman and Lewontin 1975, p. 1168.
94. Barbujani et al. 1997.
95. Lying in the sun would in fact have *reduced* the heritability of skin color in the Miami group! There is a thought experiment, more difficult to construct, that shows how interpopulation differences can be genetic with a zero heritability of a character. Think of two groups—in this case, each group being a pair of young identical twins. One pair has very dark brown skins, having had parents with dark brown skins. The other has pale skins, having had pale-skinned parents. Both groups live in the same environment, having been adopted by a loving, tan-skinned couple. Within each group, the heritability of skin color is zero; by definition, there is no genetic variation within pairs of identical twins. But the difference between the two groups in this characteristic that has zero heritability is caused entirely by a difference in their genetic endowment. (Mutation can make this statement slightly oversimplified, and, of course, the entire example is artificial because groups consisting of single pairs of identical twins cannot perpetuate themselves. A more realistic example would be highly inbred dark-skinned and light-skinned populations that lived in the same environment and did not interbreed. There, heritability of skin color would be very low but nonetheless the interpopulation color differences would be almost entirely due to genetic differences.)
96. Lewontin 1982a.
97. Although on average women are somewhat smaller and less physically strong than men, there is abundant overlap in these categories. The basic physical differences related to their reproductive role means that women may need to avoid certain activities around the time of childbirth, but increasingly, among the better-off in the industrial world, society is recognizing that child rearing is a joint responsibility for which men as well as women should be allowed leave from their jobs. A skeptic like my friend Loy Bilderback would say that the shift is anchored in the need for the woman's income to facilitate the continuance of our orgy of consumerism (Loy, in fact, did say that—personal communication, 28 August 1999).
98. Low 1989.
99. See, e.g., Harding 1991. Her book is well worth reading, though I differ with some of the ideas in it—especially that peasants or factory workers are likely to carry less cultural baggage than aristocrats or managers. Certainly white male biases influence public support of scientific disciplines. If the goal of medical research is to maximize human health, for example, then the research agenda of the U.S. biomedical community is badly distorted toward the sorts of diseases that frighten middle-aged male congressional representatives. But that does not mean that the research results themselves, subject to the scientific adversary system, are similarly distorted. For a thoughtful recent treatment of the effects on women of the science of genetics, see Mahowald 2000.
100. Feuer 1992. The quote is from p. vii; the chapter is 10.
101. An interesting discussion of Harding's views can be found in Kuznar 1997. See also Fedigan 1986 for sexual and other biases in anthropology.
102. National Academy of Sciences USA 1989.
103. Tooby and Cosmides 1992, p. 40.
104. Hartl and Jones 1998, p. 274.
105. Beers and Berkow 1999, p. 1496; Kitcher 1996.
106. Both aspects are illustrated by the dilemmas associated with schizophrenia (see, e.g., Gottesman 1991).
107. Dunbar 1993, table 4.
108. Population Reference Bureau 1998.
109. See, e.g., Johnson 1997. Estimates vary.
110. The issue of how to define *society* is a nontrivial one and has a rich history in sociology going back at least to Émile Durkheim. His notion was that societies are defined by shared values and that deviance (e.g., crime) holds societies together by promoting debate about values that enhance solidarity (Durkheim 1962 [1895]). See also Merton 1938 on anomie and, for an overview, Macionis 1997.
111. Ornstein 1988, pp. 609–610.
112. Hamburg 1992, p. 192.
113. See, e.g., Staub 1986, p. 153; Huesmann and Eron 1986; Singer and Singer 1990.
114. The conditioning to accept violence as normal or even desirable may be even more effective because, for example, viewing of violent films in theaters and on television is often associated with pleasant things: friends, dates, popcorn, candy, and so on (Grossman 1995), pp. 308ff.
115. According to Reuters/Variety (2 December 1999), ABC-TV was selling thirty-second-long commercial spots during the January 2000 Super Bowl broadcast for $3 million. Television advertisements are also

clearly a major driver of the rampant consumerism that is helping to destroy society's life-support systems.

116. Children, for example, have also been shown to learn helping behavior from television (Friedrich and Stein 1975; Singer and Singer 1990).
117. These speculations are most applicable to television and radio; the influence of the Internet, which features, among other things, many "hate sites," on community values is less well understood.
118. Of these topics, less seems to be known systematically about the content and meaning of pornography than about the others (see, e.g., D. D. Smith 1976).
119. These ideas have a long history in sociology, tracing back to the Chicago School (Macionis 1997, pp. 576–577) and Ferdinand Toennies (1855–1936), who focused on the social problems of urbanization. Toennies (1971) stressed the contrast between Gemeinschaft (values of a rural community based on religion, custom, and personal relationships) and Gesellschaft (urban society based on instrumental values, public opinion, and laws).
120. G. Gardner 1999.
121. One must note, however, that designing decent cities requires much more than dealing with the problems created by automobiles. As Jane Jacobs pointed out long ago in her classic book *The Death and Life of Great American Cities* (1993 [1961], p. 10), "the destructive effects of automobiles are much less a cause than a symptom of our incompetence at city building."
122. For instance, starting around 1650, Japan renounced the use of firearms for more than two centuries, after using them very successfully for about 100 years (Perrin 1979).
123. To contrast it with its partial ancestor, the pluralism of William James (1975 [1907, 1909]), see, e.g.: "Philosophy has often been defined as the quest or the vision of the world's unity. We never hear this definition challenged, and it is true as far as it goes, for philosophy has indeed manifested above all things its interest in unity. But how about the *variety* in things. Is that such an irrelevant matter? . . . What our intellect really aims at is neither variety nor unity taken singly, but totality" (pp. 64–65). In the interests of disclosure, I admit to having streaks of ethical relativism and existentialism in me and believe that the meaning of life is the meaning an individual gives to it (part of the message of Sartre 1960 [1956]). For me, life is best lived passionately in every sense. One motto I use, which I'm sure I got from Weston La Barre (1954, damned if I can find the page), is "Only those who haven't lived long for immortality." For an interesting discussion of the problem of private and public morality, a continuing tension in pluralistic societies, see Robert H. Kane's (1999) series of tapes available from The Teaching Company ([800] 832-2412). Although I do not agree with all that Kane says, his lectures raise a series of important questions to which he gives thoughtful answers.
124. With plenty of ambiguity, as was his wont. See, for instance, his wonderful book *The Gay Science* (Nietzsche 1974 [1887]).

Chapter 13: Evolution and Human Values

1. Anderson 1972, p. 393.
2. Russell 1957, pp. 55–56.
3. The Newtonian view of matter as inert lumps is dead, having been replaced by counterintuitive quantum views of matter as "weird excitations and vibrations of invisible field energy" (Davies and Gribbin 1992, p. 14). For the views of a series of distinguished physicists on quantum weirdness, as well as a fine brief summary, see Davies and Brown 1986. But at a more philosophical level, why don't I find the "philosophy of organism" of Alfred North Whitehead (1978 [1929])—the notion that the entities of the universe, from protons to people, should be considered not substances but processes—helpful in explaining my experience?
4. A historical example may make this clear. The ideas of a social contract as proposed by Thomas Hobbes, Jean-Jacques Rousseau, and other early modern theorists—of people having voluntarily come together to form societies—would never have been developed had those theorists possessed today's evolutionary understanding that human beings and their primate ancestors have lived in societies for tens of millions of years. Today's understanding of this stretches back at least to Charles Darwin: "Judging from the habits of savages and of the greater number of the Quadrumana [primates excluding human beings], primeval men, and even their ape-like progenitors, probably lived in society" (1871, p. 64). Had Great Britain's former prime minister Margaret Thatcher understood that, she would not have made the ultimate reductionist claim that society doesn't exist (Birch 1999, p. 36). But knowing that social contract theory (use-

ful as it may have been in some contexts) was based on a complete fiction and that Thatcher was just plain silly does not give us policy prescriptions for ways to organize societies today.

5. One strategic question that bears on topics such as city planning but that I will not discuss here is the degree to which our evolutionary past, starting in savannas as hunter-gatherers, has led to genetic influences on human habitat preferences—environments in which people feel more comfortable. This notion is sometimes called the biophilia hypothesis. Those interested in the issue should see Kellert and Wilson 1993. See also Iltis et al. 1970; Wilson 1984; Kaplan 1987; Orians and Heerwagen 1992.
6. That is, methodological reductionism.
7. Strawson 1962; Alexander 1987; Fox 1994, pp. 255ff.
8. For an excellent overview of ecosystem services, see Daily 1997.
9. At the strategic level, the environment that shapes our values is also being rapidly shaped by them in a complicated feedback process that we only partially understand, a point made long ago by neuroscientist Roger Sperry (1972, p. 117). Failure to understand that process makes finding robust answers to crucial questions about sustainability all the more difficult.
10. Vitousek, Aber, Howard, et al. 1997; Vitousek, Mooney, Lubchenco, et al. 1997.
11. For an account of early post-Darwin views on this issue, centered on the thinking of "Darwin's bulldog," Thomas Henry Huxley, see Paradis and Williams 1989.
12. Nesse and Lloyd 1992, p. 615.
13. See Copleston 1964, pp. 1–58. Among the most prominent materialists were Denis Diderot (1713–1784), who might be called a proto-evolutionist, realizing as he did that the biblical story of creation is mythical; Julien de La Mettrie (1709–1751); and Baron d'Holbach (1723–1789). All were atheists (Diderot not until late in his development) who believed there was no supernatural source of moral guidance.
14. Kant's philosophy assumed that God had to exist as a basis for the realization of justice and morality. Remember that Kant viewed the cosmos as dualistic, split between a Platonic noumenal world, one of objects of awareness (such as morals) that are not produced by sensory awareness, and a phenomenal world of space and time, available to sense perception. For insight into the intricacy of Kant's thought by a great philosopher who was not terribly sympathetic with Kant's ideas, see Bertrand Russell's *A History of Western Philosophy* (1945), pp. 704ff. One of the works I most enjoyed reading as a teenager was Russell's 1927 essay "Why I Am Not a Christian" (Russell 1957, pp. 3–23). In it, he wrote regarding Kant, "He was like many other people: in intellectual matters he was skeptical, but in moral matters he believed implicitly in maxims that he had imbibed at his mother's knee." Russell summarized Kant's argument as saying that unless God existed, there would be no right or wrong, and then he proceeded to demolish it (p. 12). Russell, a determined positivist, shared his culturally relativistic (or subjectivist) view of ethics with Jean-Paul Sartre, the most famous existentialist. Neither believed that there are absolute criteria by which to judge moral systems. People like to believe that there are cross-cultural absolute ethical standards—that, for example, the behavior of the Nazis would be universally seen as immoral (Hatch 1983). However, I think that the dismaying ease with which genocidal behavior can be elicited argues against that view—remember, Adolf Eichmann thought he was doing his duty, which he evidently thought had a higher ethical value.
15. The monist view is that the mind and ethics can be explained eventually in terms of neurophysiology—that the mind and the brain are the same thing. That's where my sympathies tend to lie. The dualist view posits a separate world of the mind; it is counterintuitive for many, if not most, scientists. For a stimulating recent discussion, which comes down on the somewhat out-of-fashion dualist side, see Stent 1998. Maybe Stent is right.
16. My view here is quite close to that of Ernst Mayr, who wrote a fine overview of evolutionists' thoughts about the genesis of ethics (Mayr 1988, chap. 5, pp. 75–89). See also Simpson 1969 and Ayala 1987.
17. Waddington 1960, pp. 151, 156ff. See also Kohlberg 1984.
18. Fox 1997, pp. 128ff.
19. I owe this analogy to Robin Fox (1997, pp. 129–130).
20. The basic point traces to David Hume (1978 [1739]); see also Moore 1903; Taylor 1992; Taylor 1992 (1989). For an early opposing view from Edward O. Wilson (a view about which I obviously have my doubts), see Wilson 1978, e.g., pp. 4–5. For a later one, see Wilson 1998, pp. 248–251. Wilson himself seems to have modified this view somewhat over time (Wilson 1984, in which on p. 119 he combines "emotion with the rational analysis of emotion" to develop a better conservation ethic), just as a fine scientist should with the acquisition of new data or reconsideration of old. Nonetheless, he remains more optimistic than I am about the degree to which neurobiology can help us understand ethical issues (Wilson 1998, chap. 11). Even though we disagree, I for one hope he's right in his optimism—better to under-

stand our deep motivations than to cling to untested assumptions about them. For examples of some of the intense discussion of such issues that has developed since Wilson wrote *Sociobiology* (1975), see Pugh 1977; Campbell 1979; Singer 1981, 1986; Lumsden and Wilson 1981, 1983; Ruse 1984, 1986, 1987, 1993; Ruse and Wilson 1985, 1986; Kitcher 1985; Richards 1986a, 1986b; Thomas 1986; Trigg 1986; Alexander 1988; Slobodkin 1993; G. C. Williams 1993; P. A. Williams 1993.

21. Clearly, a great deal of chance was factored into those evolutionary processes. I added *largely* to my statement because some ethical "oughts," such as incest taboos, may be related to genetic predispositions. Darwin (following Aristotle and others) supposed that human social sense and morals germinated in the long and close bond between parents and offspring: "The feeling of pleasure from society is probably an extension of the parental or filial affections, since the social instinct seems to be developed by the young remaining for a long time with their parents; and this extension may be attributed in part to habit, but chiefly to natural selection" (Darwin 1871, p. 108). Morals then flowered as a result of the need for cooperation as groups became ever larger and more complex (pp. 126ff.). Chimps, for example, show close and enduring parent–offspring bonds and can cooperate but show precious little moral sense by our standards, whereas upper-class Englishmen often fail to bond with their sons, who nonetheless generally grow up with the morals of upper-class Englishmen. I can't help but think that morals and ethics are primarily a product of complex cultural evolution not easily subject to a universal reductive explanation. For a contrary view, see Ruse 1984 and Ruse and Wilson 1986.
22. There is an extensive, if frustrating, philosophical literature on the meanings of *natural* and *good.* For an introduction to this and related topics, see Honderich 1995.
23. Inuit child rearing is very permissive (see, e.g., Balikci 1970, pp. 108–109, for the Netsilik; P. R. E., personal observation of the Aivilikmiut, 1952). Child-rearing practices vary greatly from society to society. For instance, in Tibet a mother may fellate a five-year-old boy on request, whereas in other cultures babies may be prevented even from touching their own genitals (Harris 1997, pp. 356–357).
24. For example, Edward O. Wilson has stated that all human behavior is in its roots adaptive, consisting of ways of promoting the continuity of our genetic material: "Morality has no other demonstrable function" (1978, p. 167). See also Alexander 1987.
25. Nitecki and Nitecki 1993.
26. Cabanac 1979.
27. Humphrey 1992.
28. Elzanowski 1993.
29. Mackintosh 1983, 1987.
30. A standard "Skinner box."
31. Mackintosh 1987, pp. 114–115.
32. Elzanowski 1993.
33. Midgley 1994, p. 3.
34. Evolutionary traces of that may be imputed from the way children's awareness of other individuals' capacities gradually develops (Miller and Aloise 1989; Strayer 1987). Youngsters at the age of three often do not understand how others come to know things. For example, they do not connect the act of their older sister's looking into a box with her knowledge of what's in the box (Povinelli and Godfrey 1993, p. 281). The ability to make that connection develops a little later. Around the age of four or five, young children can modify their behavior in light of the way they size up others' mental capabilities. For instance, they realize that babies are best addressed in simplified language and adjust their speech accordingly (Sachs and Devin 1976). One can suspect that in this particular case, development more or less recapitulates evolutionary history (although in many instances it does not), with abilities that developed gradually in the human evolutionary past also developing gradually in human children, but we cannot show this to be the case.
35. Gallup Jr. 1970.
36. Hyatt and Hopkins 1994.
37. Lethmate and Dücker 1973; Miles 1994.
38. Swartz and Evans 1994; Patterson and Cohn 1994. Gorillas show many other signs of self-awareness, as these articles suggest.
39. Gallup Jr. 1979, 1985. Interestingly, it was not at first possible to demonstrate that gorillas recognize themselves in mirrors (Gallup Jr. 1983; Suarez and Gallup Jr. 1981), but see Patterson and Cohn 1994 and Parker and McKinney 1999. Mirrors test self-recognition rather than self-awareness, and it seems reasonable to suppose that some primates that cannot recognize themselves in a mirror nevertheless have a degree of self-awareness—see Povinelli 1987; Povinelli and Godfrey 1993; Swartz et al. 1999.

40. Povinelli and Godfrey 1993. For an excellent recent summary regarding the human capacity to understand and manipulate the mental states of others, see Frith and Frith 1999.
41. A classic experiment that can be thus interpreted was that of Premack and Woodruff 1978. It is, however, open to other interpretations (Savage-Rumbaugh et al. 1978). For a useful discussion that makes clear that these issues remain unsettled, see Tomasello and Call 1997, chap. 10. See also Povinelli et al. 1990; Povinelli, Nelson, et al. 1992; Povinelli, Parks, et al. 1991, 1992; Heyes 1993; and Swartz et al. 1999.
42. Trivers 1971. See also the model by Simon 1990, which depends on there being genetic variance for "docility," adeptness at learning from others (social learning).
43. Batson et al. 1983; Batson et al. 1988; Batson et al. 1989.
44. Batson 1990. Buried in this problem is the one of free will because one can hardly be ethical unless one can choose without constraint to do so. Determinism is the notion that all acts have predetermined causes—are part of a fixed chain of causes and effects that extends both into the distant past, to the origin of the universe, and into the distant future, to its end. Some assume that within our own lives, determinism erases the possibility of free will—we do what we have been fated to do by the original configuration of the universe. That view of a clockwork-like Newtonian universe led French astronomer and mathematician Pierre-Simon de Laplace (1749–1827) to state that a sufficiently intelligent being, if made aware of the position and momentum of every particle in the universe, would know the entire past and future. The more statistical scientific view of the world that accompanied the development of quantum mechanics has not solved this age-old philosophical problem but rather, from a physicist's point of view, has added complexity to it (see, e.g., Davies and Brown 1986). For a brief overview of this problem, see Roy C. Weatherford's entry "Determinism" and connected entries in Honderich 1995. Ruse 1986 has an interesting related discussion. For a sociobiological view of the problem, see Alexander 1987, p. 253; for an excellent response to his position by a philosopher, see Kitcher 1985, pp. 406ff.
45. Other motives were a sense of obligation to the norms of behavior of a social reference group in response to religious or (as in the case of many Danes) national standards or a belief that the treatment of the Jews violated central principles around which the rescuer organized his or her life—the principle of justice, for example (Oliner and Oliner 1988).
46. Oliner and Oliner 1988, p. 197.
47. Midgley 1994, p. 148.
48. See, e.g., Holmes and Sherman 1983; Fletcher and Michener 1987. For mechanisms of kin recognition, see Sherman 1991; R. D. Alexander 1991; Parr and Waal 1999.
49. Wells 1987.
50. It is also a matter of dispute in the social science community. See, e.g., the discussion of Richard Rorty's concept of ethnocentrism in Festenstein 1997.
51. See the recent book *Hitler's Pope* by Catholic author John Cornwell (1999). In the view of another Catholic analyst, "*Hitler's Pope* makes it clear that if Pacelli . . . is to be canonized now, the Church will have sealed its second millennium with a lie and readied its third for new disasters" (Carroll 1999, p. 112). Cornwell basically concludes that Pacelli's universe of caring did not include Jews, apparently because he linked Jews and communists in a plot to destroy Christendom (pp. 295ff.). To others, Cornwell's conclusions seem biased (see, e.g., Jones 1999; Rubinstein 2000), and the debate is certain to continue. Overall, the issue underlines the complex ethical dilemmas faced even by those to whom society looks for ethical leadership.
52. Davis 1956; Ehrlich et al. 1977, pp. 196–197.
53. This is basically a recent round of the old debate over the ethical obligations of the rich toward the poor and what constitutes social justice. Ecologist Garrett Hardin has thought long and hard about that ethical issue and others related to population growth, environmental deterioration, international migration, foreign aid, and intragenerational and intergenerational equity (see, e.g., Hardin 1968a, 1968b, 1977, 1993, 1996). Hardin tries to consider the long-term consequences, especially environmental consequences, of human activities, and in the process he has generally come down on the side of what might be called "ecological tough love." A characteristic passage from his work on "lifeboat ethics" expresses this (Hardin 1996, p. 292):

> We are all descendents of thieves, and the world's resources are inequitably distributed, but we must begin the journey to tomorrow from the point where we are today. We cannot remake the past. We cannot, without violent disorder and suffering, give land and resources back to the "original" owners—who are dead anyway.
>
> We cannot safely divide the wealth equitably among all present peoples, so long as

> people reproduce at different rates, because to do so would guarantee that our grandchildren—everyone's grandchildren—would have only a ruined world to inhabit.

At the opposite ethical extreme is philosopher Peter Singer, who believes that Englishmen should care as much about starving Bengalis as about destitute Britons. He asserts that "if it is in our power to prevent something bad from happening, without thereby sacrificing anything of comparable moral importance, we ought morally to do it. By 'without sacrificing anything of comparable moral importance' I mean without causing anything else comparably bad to happen, or doing something that is wrong in itself, or failing to promote some moral good, comparable in significance to the bad thing that we can prevent" (Singer 1972).

54. See, e.g., Singer 1975; Regan 1983.
55. See, e.g., Stone 1974; Ehrenfeld 1978; Devall and Sessions 1985.
56. For an overview of many relevant ethical issues, see Rolston 1988.
57. Cohen 1967; Kuper 1981, chap. 5. For an interesting discussion of denial of identity, dehumanization, and the sanctioning of massacres, see Kelman 1973.
58. Even in such cases, however, ordinarily there are untoward effects on some of the perpetrators in whom, presumably, empathetic feelings are stronger. As an example, there were relatively high suicide rates among Nazi units of "order police" involved in mass executions of Jews in eastern countries, and Heinrich Himmler was concerned about the mental health of the SS men who he had commanded to carry out mass murder personally. The Nazi switch from execution by gunfire to the use of gas chambers was in part motivated by this concern (see, e.g., Breitman 1998).

    Pol Pot is another example. He murdered more than 1 million of his countrymen in Cambodia between 1975 and 1979 (McCarthy 1998), but he was reported to have been a doting father (Chandler 1999). In contrast, President Bill Clinton was, in the view of many, if not most, U.S. citizens, "immoral" in his personal life. He was hardly a prototypical family man, although many viewed him as a good parent. But unlike Pol Pot, Clinton worked hard for the underdogs of society.
59. See the discussion of altruism in Sober 1988, 1993. Sober contends that group selection is necessary for the development of evolutionary altruism. For an extended discussion, see Sober and Wilson 1998.
60. For a brief sampling of the depth and complexity of ethical issues from a philosopher's viewpoint, I recommend Griffin 1996. The writings of Mary Midgley are also well worth perusing (e.g., Midgley 1985, 1994).
61. Slobodkin 1993, p. 344.
62. Midgley 1994.
63. Griffin 1996, p. 178.
64. The example is one of several interesting ones in Griffin 1996, chap. 8.
65. Kant might have had trouble with that one, given that under his system of the "categorical imperative" (basically, assuming that one would want any moral rule under which one operates to be a universal law of nature—Kant 1996 [1797]), he might have decided that it was unethical to take the life of an innocent human being. Hume could have handled this situation more readily because he wanted to base a universal ethical system on shared feelings, or "sentiments" (Hume 1978 [1739], pp. 470ff.), and might have concluded that everyone would believe the mother's actions proper because the baby also would have died had the group been discovered. As suggested earlier, I tend to take a rather relativistic view of ethics, one that is also utilitarian (see, e.g., Mill 1998 [1863]). Utilitarians believe that ethical decisions must be judged largely on the basis of their consequences, a position that traces to the father of utilitarianism, Jeremy Bentham (1748–1832) (Bentham 1988 [1789]). He defined utility as "that property in any object, whereby it tends to produce benefit, advantage, pleasure, good or happiness . . . or . . . to prevent the happening of mischief, pain, evil, or unhappiness to the party whose interest is considered" (p. 2). Utility is to be maximized globally. As has frequently been pointed out, utilitarianism can make the creation of moral codes very difficult because differences of degree are almost always involved. ("Would you make love with me for a billion dollars?" "Well, I guess so." "For one dollar?" "What do you think I am?" "We've determined that—what we're haggling about now is price.") Strict utilitarianism also requires that no consideration be given to self, friends, or relatives in the ethical calculus—if I could improve the human condition more by spending $1,000 to feed hungry people halfway around the world than by helping a granddaughter through college, then the moral thing would be to do so. I cannot fault the logic, nor can I live up to the standard. For a much stricter modern utilitarian view, see Singer 1972.
66. In Europe, circumcision was done largely for religious purposes; in the United States, it was promoted in the late nineteenth century to help prevent masturbation and reduce the sex drive. In 1999, the Amer-

ican Academy of Pediatrics stated that the evidence "is not sufficient to recommend routine neonatal circumcision"; see Ayton 1999. Under what circumstances it might be justifiable to inflict useless pain on male infants, who do not have the power to give or withhold their consent, is an interesting ethical dilemma.

67. All of these cases also highlight Mary Midgley's point about a moral sense evolving in response to awareness of the profound conflicts among motives.
68. See, e.g., Slobodkin 1993, p. 346.
69. I agree in many respects with the pragmatism that traces to Charles Sanders Peirce (1839–1914) (Peirce 1955); William James (1842–1910) (James 1975 [1907, 1909]); and John Dewey (1859–1952) (Dewey 1988) and that developed by many subsequent philosophers, especially Willard Van Orman Quine (b. 1908) (e.g., Quine 1977) and Richard Rorty (b. 1931) (e.g., Rorty 1982). I use the term *pragmatism* to mean several things. One is that rather than being concerned with whether or not beliefs are some sort of representation of an independent "reality," we all, scientists and others, bring our worldviews with us when we approach a practical or theoretical question—we are part of the reality. In addition, I think the validity of our answers depends largely on their usefulness (in a very broad sense of usefulness for the community of concern, including the putting together of a worldview I find coherent)—similar to the view that Edward O. Wilson has made famous as "consilience" (Wilson 1998). Pragmatists are supposed to think that what is true or right is "what works"; a counterpragmatic argument has been that Nazi policies such as persecution of Jews "worked" (Habermas 1979; Habermas claims that ethics cannot be based on "nothing more than what is anyhow contained in the everyday norm consciousness of different populations" [p. 202]). The technical complexities of philosophical pragmatism in theory and practice can be appreciated by a perusal of Festenstein 1997.
70. Aside from those that are basically statements of definition or synonymy in sentences: "All triangles have three sides." Modern consideration of the issue traces especially to Kant. For a more recent technical discussion, see Ayer 1952 (1946) on analytic and synthetic truths (pp. 78ff.) and the contrary views of W. V. Quine (1977, 1992) or Quine as explained by C. Hookway (1988).
71. For a recent attempt by a philosopher to deal with the contentious issues of social construction—of whether there is a reality "out there" that is completely independent of human attempts to study it (e.g., that an electron "really" can have both position and momentum at the same time)—see Hacking 1999.
72. Kagan 1998, p. 190.
73. These are basically the same issues that stimulated the romantic response to Enlightenment science that flowered in the late eighteenth and, especially, early nineteenth centuries. This romantic response represented a preference for feeling, intuition, imagination, and self-expression over rational analysis and intellect. It is associated with such writers and thinkers as German philosopher Friedrich von Schelling (1775–1854) and Johann von Goethe, Johann von Schiller, Samuel Taylor Coleridge, and Percy Bysshe Shelley. Philosophically, it traces to Kant.
74. See, e.g., N. Williams 1997.
75. For an excellent discussion of the general issue, see Anderson 1972.
76. See, e.g., Medin and Ortony 1989.
77. See, e.g., Fodor 1998.
78. Cultural materialists attempt to explain differences in attitudes and behavior among groups of people in terms of constraints in their biophysical environments.
79. Sociobiology involves the search for explanations of those differences in the genetic evolutionary history of populations.
80. Kitcher 1985, p. 411.
81. See, e.g., Cooper and Palmer 1992; Regan and Singer 1989.
82. See, e.g., Birch 1976, 1990, 1993a, 1993b; Birch and Cobb 1981.
83. For a sampling of the developing literature of environmental ethics and related topics (including "deep ecology"), see Daly and Cobb Jr. 1989; Dasgupta 1993; Dasgupta and Mäler 1995; Devall 1980; Devall and Sessions 1985; Eckersley 1992; Fox 1984, 1989, 1990; Leopold 1966; Murdoch 1980; Naess 1973, 1989; Nash 1989; Pirages 1977, 1996; Pirages and Ehrlich 1974; Sessions 1981; Soulé and Lease 1995; Stone 1987; Tobias 1985; Zimmerman 1986.
84. One might, of course, deny that ethics themselves evolve, suggesting instead that we simply improve our access to the eternal contents of a Kantian moral universe. That, too, would be an example of cultural evolution.

    Ethics is just one complex cultural area that has arisen through and been shaped by evolution—religion, art, music, storytelling, and literature are other examples, as we have seen. Evolution clearly provided the necessary mental attributes to allow these areas to emerge and both promoted and constrained their development. The practices of magic, music, and art may originally have provided substantial biological evolutionary advantages (bringing approval and mates to their best practitioners), and they obviously developed under strict evolved constraints on the human perceptual system. There has been no flowering of music in frequency ranges beyond the evolved perceptive abilities of the human ear; no art has been created that can be "illuminated" only by X-rays. Other areas of the arts had to await further cultural evolution. Storytelling may have promoted the reproduction of raconteurs, but there was no literature until cultural evolution led to writing (although much early literature, such as the Gilgamesh epic and the *Iliad*, circulated orally for long periods before being immortalized in writing). The art of the film (and of television) obviously had to await the harnessing of electricity.

We seek in vain for explanations based on natural selection for many cultural "choices," however. Was some sort of "genetic preference" for classical music expressed in one culture in one general time period and a "genetic preference" for rock and roll expressed in the same culture at another time? Can we expect the legendary sexual exploits of rock stars to increase future interest in rock music? Will the ability to write poetry dwindle because it is virtually impossible to make a living by writing poetry and thus poets might have fewer children? Does having a celibate clergy tend to remove genes for piety from a population? These are extreme examples, perhaps, but a needed antidote to the vision the uninformed have of those selfish little DNA sequences steering our lives.

85. See, e.g., Bodley 1976; Martin 1978; Clad 1984; van Lennep 1990; International Working Group on Indigenous Affairs 1992. Much of this attitude among environmentalists probably traces to a fictitious environmental version of Chief Seattle's 1854 Treaty Oration that originally appeared in the *Seattle Sunday Star* on 29 October 1887, in a column by Henry A. Smith (Internet: http://www.halcyon.com./arborhts/chiefsea.html). For information about the environmental version by Ted Perry, see http://www.geocities.com/Athens/2344/chiefs3.htm#speech and Jones and Sawhill 1992.
86. See, e.g., the material on the Inuit in chapter 9 of this book; Hames 1991; Alvard 1993a, 1993b, 1998; Kay 1994, 1995; Low 1996; Winterhalder and Lu 1997.
87. Tisdell 1989; Redford and Stearman 1993a, 1993b; Alcorn 1993; Soulé 1995a; Spinage 1998; Attwell and Cotterill 2000.
88. See, e.g., Gadgil 1991; Bawa and Gadgil 1997.
89. See, e.g., Dasgupta 1993; Gadgil et al. 1993; Ehrlich et al. 1995.
90. Bawa and Gadgil 1997, p. 295.
91. Daily and Ehrlich 1992.
92. At least in terms of current and foreseeable behaviors and technologies (Daily and Ehrlich 1992; Ehrlich and Ehrlich 1990, pp. 67–70, 92).
93. See, e.g., Birch 1993a, pp. 107–110; Birch 1993b.
94. I have not dealt with what is known about the human predicament and what the scientific community thinks should be done to solve it because those issues are amply covered in a diverse literature. For a sample emphasizing environmental aspects, see Arrow 1996; Cairns and Bidwell 1996; Colborn et al. 1996; Daily et al. 1998; Daily and Ehrlich 1990, 1996b; Daly 1973; Dasgupta 1993; Eldredge 1995; Heywood 1995; Holdren 1991; Holdren and Ehrlich 1974; Holling and Meffe 1996; Homer-Dixon and Percival 1996; Hughes et al. 1997; Leopold 1966; Levin 1999; Lubchenco 1998; Matson et al. 1987; May 1993; Myers 1979, 1993, 1996b; Pimm 1991; Pirages and Ehrlich 1974; Postel et al. 1996; Raven 1980, 1990, 1993; Recher et al. 1986; Ricketts et al. 1999; Root and Schneider 1995; Safina 1997; Saunders et al. 1993; Schneider 1989, 1997; Schulze and Mooney 1993; Simberloff and Cox 1987; Sisk et al. 1994; K. Smith 1987; Socolow et al. 1994; Soulé 1995b; Soulé and Wilcox 1980; Soulé et al. 1979; Terborgh 1988; Tilman et al. 1996; Vitousek et al. 1986; Vitousek et al. 1996; Vitousek, Aber, Howard, et al. 1997; Vitousek, Mooney, Lubchenco, et al. 1997; Wackernagel and Rees 1996; Wilcove 1994; Wilson 1992.
95. See, e.g., Daily et al. 1994. There are many possible definitions of *optimum;* the one used in this study considered the preservation of broad environmental and social options for individuals.
96. Holdren 1991.
97. Daily and Ehrlich 1996c; Ehrlich et al. 1995.
98. Barber 1995. Ethicists must accelerate their consideration of how to deal with differences in cultural norms. Our collective ethics will continue to evolve, and in my view people should try to take as pluralistic a view as they can of what is morally acceptable. Some of the cross-cultural questions will be especially difficult. But the problems of analyzing them may be exaggerated by anthropological emphasis on cultural differences that may not be greater than the variety of behaviors seen among individuals (Goldschmidt 1960). To take a rather extreme example, were the youngsters engaged in ritualized fellatio in

New Guinea societies being abused? My gut says yes, but my brain isn't so sure. Could that behavior have been considered nonabusive in those rain-forest societies when they were isolated from other cultures but then transformed into abusive when the societies were exposed to a global Western culture that had its religious roots in the desert?

99. Emlen 1995.
100. For more details of those defects, see Hartl and Jones 1998.
101. Amniocentesis is the analysis of cells extracted with a fine needle from the sac of fluid around the fetus.
102. See, e.g., Kitcher 1996.
103. See, e.g., Gottesman 1991; Gottesman and Shields 1982.
104. Huxley 1932.
105. Scientists like me always tend to make the subconscious assumption that people would like to find rational solutions to human problems, despite massive evidence of a widespread preference for irrational (or nonmaterial) approaches to the world. One need only look at the ubiquity of gambling, the ease with which swindlers bilk the unwary with "too good to be true" schemes, the large audiences for "suspend your disbelief" motion pictures or television shows such as *The X-Files*, the belief in angels by more than two-thirds of people in the United States, and the popularity of horoscopes to see that human natures often show an affinity for the irrational.

Thus, many people tend to seek their answers to the problems of life in ways that don't lean heavily on rational analysis. Organized religion flourishes in the United States. As Gustav Niebuhr, senior religious correspondent for the *New York Times*, put it: "This is a fundamentally religious nation. The United States over the last 30 years, has not followed the trajectory of many western European countries toward an intrinsically secular society. . . . The theological proposition that God was absent or dead never took root here" (Niebuhr 1998). Nonetheless, the character of religion in the United States has been changing. The numbers of participants in mainstream—Presbyterian, Methodist, Catholic, Jewish—religions are decreasing while those of a variety of newer congregations, many classed as fundamentalist, are increasing (Lindner 1998; see especially table 3). This is a continuation of a decades-long trend; as one commentator put it, "Later in the [twentieth] century and into the contemporary period, the cultural hegemony of classical Protestant churches has been challenged by an aging and declining membership, new prominence of 'younger' churches, especially Evangelical and Pentecostal, and growing religious pluralism" (Lindner 1999, p. 9). See also the discussions in Kelley 1972 and Rosten 1975 and, especially, the thoughtful analysis of Robert Wuthnow (1988).

In 1947, 6 percent of people in the United States said that they had no religion; by 1987, that number had risen to 9 percent (statistics in this paragraph are from Gallup Jr. and Castelli 1987). At the same time, membership in churches and synagogues was declining, from 76 to 65 percent. But even though confidence in mainstream religion may be declining, there is little sign of a decline in a general faith in God—94 percent of people in the United States said they believe in God, 90 percent pray to God, and 88 percent believe that God loves them. Furthermore, some 71 percent believe in life after death, a proportion that has held steady for decades. People in the United States combine high levels of education with deep religious convictions.

People in the United States also show a high level of belief in other supernatural phenomena. Twenty-five percent believe in ghosts, 10 percent believe they have had contact with ghosts, and 10 percent think they have talked with the devil. One in four believe in horoscopes, one in six say they've been in contact with the dead, and one in seven say that they've seen an unidentified flying object (UFO) (Gallup Jr. and Newport 1991). Stock market experts frequently consult astrologers (CNN, 26 August 1998). Large numbers of people believe that extraterrestrial beings not only visit Earth regularly but also frequently molest people sexually (Sagan 1995, chap. 4).

That people hold beliefs such as these—that they prefer the "spiritual" to the "rational," if you will—is not necessarily inimical to the functioning of society. We badly need to harness the spiritual side of existence in the battle to solve the human predicament, as in, for example, a widespread quasi-religious conversion to the valuing of other organisms (Ehrlich and Ehrlich 1981). In recent years, various religious groups have entered the battle to save Earth's nonhuman life-forms, recognizing them as part of the creation. A recent example is the successful effort of His All Holiness Ecumenical Patriarch Bartholomew I (Eastern Orthodox Church) to bring various groups together to help preserve biodiversity (Hobson and Lubchenco 1997). The World Council of Churches has long been active in examining issues related to the environment, justice, and the human future (see, e.g., World Council of Churches 1980). For a thoughtful discussion of religion and ecology, see Haught 1995, chap. 9. Efforts such as that of the patriarch should be encouraged by people of other traditions. It also seems important to note that dialogs about ethics are probably more concentrated in religious communities than anywhere else in modern society, and such discussions should be welcomed by all.

106. Thoughtful scholars, such as John Greene (1999a), fully support the modern scientific view of evolution but object to the attempt to draw ethical conclusions on an evolutionary basis. Although I do not share Greene's religious views, I am largely in agreement with his reservations. To get an idea of Greene's differences with the great evolutionary biologist Ernst Mayr, see Greene 1999b. I am largely in accord with Mayr's conclusion that "all the achievements of the human intellect were reached with brains not specifically selected for these tasks by the Darwinian process" (Mayr 1997)—I would just change *all* to *many* because it is not clear exactly which achievements are being referred to and (if one of them, for example, is the acquisition of syntactical language) when and how they evolved. My basic point, again, is that ethics and the like are largely epiphenomenal in a brain whose basic structure evolved for other reasons.
107. See, e.g., Ruse 1998.
108. E.g., PBS, *NewsHour with Jim Lehrer*, 21 April 1998.
109. Belluck 1999; for analysis, see Larson and Witham 1999. The ignorance of science inherent in the creationist movement is indicated by one newspaper subheading: "Christians Who Believe God Created Human Beings Fully Formed Say Darwinian Theories Are Unproven" (Rosin 1999). Of course, science "proves" nothing—including that Earth circles the sun—scientific results are perpetually open to revision. The classic battle over the teaching of evolution in schools led to the Scopes trial in 1925. For a fine recent view of the trial and its milieu, as well as its continuing deep significance, see Larson 1997. Creationist opposition to teaching it had largely disappeared from the intellectual community by the 1880s, but as universal high school education spread in the twentieth century, the exposure of the general public to the idea of evolution led to a recrudescence. Larson's book explains why creationism is not likely to disappear soon.
110. Bailey 1997.
111. The intellectual reason is that there is no science in either the procedures or the conclusions of creationism. In general outline, evolutionary theory is utterly uncontroversial among scientists (National Academy of Sciences USA 1998). But creationism thrives because many people find it difficult or impossible to accept that human beings and all other organisms are modified descendants of organisms of the past, a past stretching back billions of years. Curiously, the phenomenon of creationism is largely confined to the United States and Australia; most educated people in Europe, for example, consider the fact of evolution as being no less acceptable than the fact that Earth is round.

    Why, then, is creationism not restricted to those who are uneducated or who belong to fundamentalist religions? Neoconservatives with college degrees and no fundamentalist affiliations have launched frequent attacks on the theory of evolution. Neoconservative ABC-TV commentator William Kristol has asserted that Darwinism is on the way out, "no longer accepted so easily by [many] biologists and scientists," because of a new focus on "the old-fashioned argument from design." Yet *all* competent biologists accept that today's organisms are modified descendants of organisms that lived billions of years ago; there is no scientific dispute whatsoever on this point. As the National Academy of Sciences USA recently put it (1998, p. 4): ". . . there is no debate within the scientific community over whether evolution occurred, and there is no evidence that evolution has not occurred. Some of the details of how evolution occurs are still being investigated. But scientists continue to debate only the particular mechanisms that result in evolution, not the overall validity of evolution as the explanation of life's history." The notion that the occurrence of complex living entities implies a conscious designer was argued famously by William Paley (1803) at the start of the nineteenth century. That argument doesn't hold water.

    People are often confused by the argument from design, which is indeed, as Kristol noted, old-fashioned—as old-fashioned as it is ignorant. It most often appears in a form such as "What good is half an eye?" As suggested in the notes to chapter 7, there is no need to assume a designer to explain the universe in general or eyes in particular—evolutionary theory does that just fine. See, e.g., Jacob 1977; Dawkins 1986.

    Mathematician David Berlinski wrote an article, much praised by neoconservatives, in which he resurrects an ancient claim that evolutionary theory is constructed in such a way that it cannot be falsified by any observations (Berlinski 1996; see also Orr et al. 1996). This, of course, follows the now-outdated Popperian view that creating falsifiable hypotheses is the only way to do science (Popper 1968 [1935]; Popper 1968). It is also incorrect (M. B. Williams 1973). Evolutionary theory predicts that many things will *not* be observed. For example, behavior would not evolve by natural selection in one species the sole effect of which was to enhance the success of another species. Another is that human footprints would not be found in undisturbed geological formations from the age of dinosaurs. Still another is that the more children blue-eyed people had, the rarer blue eyes (a largely genetically determined attribute) would become in the population. A final one is that two isolated populations, one living for thousands of generations in an environment in which a substantial portion of the individuals in each generation were killed by a poison and the other living in a poison-free environment, would at the end of that time show no difference in their resistance to the poison. Observation of any of these phenomena would cast serious doubt on the theory of evolution by natural selection. It is sometimes claimed that the theory of natural selection is not a truly scientific theory because it can explain *any* observation—it cannot be falsified. The foregoing examples falsify that claim.

Evolution also predicts things that *will* be found, in particular links that are missing from the fossil record. Creationism has always seized on a putative absence of "missing links" to boost its case. The fact that, as predicted by evolutionary theory, link after link has been found has had little effect on creationist thought. In the minds of creationists, each missing link that is found creates the need for two more (e.g., one between reptiles and *Archaeopteryx* and one between *Archaeopteryx* and birds). But the creationists' task of obfuscation and denial has become increasingly hopeless as the fossil record has been filled in.

In the face of all the evidence, why would people as well educated as Kristol, Berlinski, and United States Supreme Court nominee Robert Bork be so opposed to one of the best-established and most important scientific frameworks—one with enormous explanatory power? Bork stated (1997, pp. 294–295), for example, that "the fossil record is proving a major embarrassment to evolutionary theory." He naively accepts the long-discredited argument that the universe must have been created, given that it has just the right characteristics to support life. That notion was first put forward in detail by L. J. Henderson (1913). Henderson's view of the "fitness of the environment" is an example of a scientific contribution that was important not because it was correct but because of the discussion it stimulated. For an early refutation, see Lillie 1913; a recent, sophisticated demolition is provided in G. C. Williams 1993. Aspects of Henderson's idea persist in the discredited strong form of the Gaia hypothesis (see, e.g., Craik 1989; Ehrlich 1991). In fact, of course, what we know is that life has evolved to fit into a tiny corner of the universe.

One possible explanation for creationist silliness on the part of the educated, well supported by the statements of neoconservatives themselves (Bailey 1997), is that they believe it is necessary to promote religion in order to maintain social order and evolutionary theory undermines religion. Bork claims, perhaps correctly, that "a world without religion . . . is likely to be very unpleasant," and he believes that "a major obstacle to a religious renewal" is Darwinism. Other neoconservatives are concerned that a more widespread realization that Genesis is just a fable might undermine religion and destabilize society. For example, Leon Kass, a member of the University of Chicago's prestigious Committee on Social Thought, believes that evolutionary theory represents a threat to social order: "The creationists and their fundamentalist patrons . . . sense that orthodox evolutionary theory cannot support any notions we might have regarding human dignity or man's special place in the whole. And they see that western moral teaching, so closely tied to scripture, is also in peril if any major part of scripture can be shown to be false" (quoted in Bailey 1997, p. 24). Perhaps Kass, Kristol, and the others support what Plato regarded as the noble lie—in this case, preserving the faith of the common people in Genesis and thus in social order. Stratification rears its head as usual, and neoconservatives apparently take their cue from Karl Marx in treating religion as the opiate of the people.

Strangely, it seems to be only in the United States that neoconservatives are concerned about evolution. Nonetheless, creationism is a factor in worsening the human predicament. The decline in the teaching of evolution in the schools of the United States cuts most people off from knowledge about their origins and the origins of the ecosystems in which they operate, and it generally encourages a feeling of human power and exceptionalism that makes it difficult to deal with environmental problems. In a recent survey of 534 newspaper editors, only about half disagreed strongly with the statement "Dinosaurs and humans lived contemporaneously"; in 1999, the majority whip of the U.S. House of Representatives was Tom DeLay of Texas, a creationist; and in the presidential primary campaign in 1996, Pat Buchanan, a candidate for the presidency, proudly declared that he was a creationist.

Although we tend to think of the Victorians as hostile to evolutionary ideas, their thinking on the topic was actually much more advanced than DeLay's or Buchanan's (Robinson 1977). For example, a review of *On the Origin of Species* in the *Dublin Review* (Morris 1860), which one might expect to be vitriolic, ended, ". . . we owe Mr. Darwin many thanks for the steadiness of investigation, the thorough knowledge of natural history, and the comprehensive grasp of his subject which characterizes the very remarkable book before us" (p. 81).

112. See, e.g., Ehrlich 1997, chap. 5.
113. They have often worked against progress in much more narrow areas of scholarship as well, as they have (Ehrlich and Holm 1962) and still do (Ehrlich 1997, pp. 17ff.) in the artificial disciplinary divisions between ecology, evolution, systematics, and behavioral biology.
114. This material on interdisciplinary research is slightly modified from Daily and Ehrlich 1999, with Gretchen Daily's permission. For a brief discussion of eighteenth-century science and the burgeoning of scientific academies at that time, see Frängsmyr 1999.
115. Scientists realized that explanations for much of chemistry lay in the realm of physics, and the similarities of structure and process in plants and animals became increasingly clear. By the middle of the twentieth century, the line between chemistry and biology (a descendant of botany and zoology) had lost its sharpness. Among other things, the functioning of enzymes was being uncovered and DNA was rearing its double spiral head.
116. Wallerstein et al. 1996.
117. A question such as "How did consciousness evolve, and how does it relate to emotions?" might be considered to belong primarily in the arenas of neurobiology and philosophy, but important dimensions clearly also lie in fields such as genetics, endocrinology, and behavior. Similarly, a problem such as "How can the harmful environmental effects of human activities be greatly reduced?" might seem squarely situated in demography, ecology, and economics, but further consideration quickly takes one into fields of engineering, sociology, psychology, anthropology, political science, law, and ethics, to name just a few. We've already seen how the nearly ridiculous structure of social science disciplines hinders attempts to develop an integrated science that focuses on cultural evolution as a unitary phenomenon. Here again is that critical mismatch in rates of cultural evolution. The disciplines that study evolution—tools for knowing—developed over centuries, are isolated from one another, and change slowly. Technologies—tools for doing—have proliferated in recent decades, are evolving with extraordinary rapidity, and have created a need for rapid changes in the way the world of knowledge is structured.
118. How could a more interdisciplinary approach to addressing human problems be fostered? It is, of course, always theoretically possible to assemble transdisciplinary teams (groups of cooperating disciplinarians), but that often presents a difficult challenge in finding people willing and able to work together. Given the way rewards flow down disciplinary lines, very senior people are the only ones in a position to participate. Nonetheless, with effort, such collaborations are possible and may be very successful. At Stanford University, for instance, ecologists, economists, businesspeople, and legal scholars have been able to work together on transdisciplinary problems such as the following: How can ecosystem services be valued? What sorts of legal structures are needed to protect the areas that supply the services? How efficient would population limitation be in comparison with carbon taxes in limiting the flow of carbon dioxide into the atmosphere? Similar collaboration has emerged among a group of economists, ecologists, and other scholars centered at the Beijer International Institute of Ecological Economics of the Royal Swedish Academy of Sciences, and slowly but surely, answers to such important questions are being developed.
119. Snow 1998 (1959, 1964). Interestingly, in view of the role of "literary criticism" in the postmodern movement, Snow in 1959 called one of the two cultures "literary intellectuals" and the other "scientists" (p. 4). Although postmodernists would have us believe that science is just one more set of texts or ways of knowing among many, actually the scientific culture is thriving while the postmodern culture is thrashing about, and the gap between the two is, sadly, growing. Science has found out how to be more and more effective, whereas literary intellectuals have discovered how to become more and more ineffective. From the viewpoint of one who thinks that science doesn't have all the answers, the situation seems far from healthy.
120. The last seems particularly central to the success of these enterprises, much more than I would have expected at the outset. Professors are small-group animals, too, and for these extradisciplinary activities, a group of colleagues have bonded naturally into an interdisciplinary group of "intellectual hunter-gatherers."

    The biggest job at the beginning of any transdisciplinary enterprise is establishing communication. Originally, ecologists and economists were separated by differences in worldview clearly related to differences in their training. Considerable mutual instruction, translation of jargon, and patience were required before the problems themselves were clear to both groups and a joint attack on them could be launched. For example, before economists could get interested in the valuation and safeguarding of ecosystem services, they had to become informed of the nature and importance of those services. Before ecologists could participate productively in the effort, they had to learn about economic principles and methods of valuation as well as the role of institutions in mediating human interactions with the environment. The explanation of the limitations of that knowledge has been as important as its communication. But gradually, in the cases I've observed, a common ground on which to build has become established.

121. In my own experience, students are keenly interested in interdisciplinary problems and will eagerly undertake research on them with the slightest encouragement (as long as there is a prospect of a job at the end of the tunnel). I think there is a brilliant future for such research, both because we have so far to go and because more and more people are beginning to realize that we must make the trip.
122. Ornstein and Ehrlich 1989.
123. A sort of conscious evolution was seen by sociologist Anthony Wallace (1956) in a study of revitalization movements that he defined as "a deliberate, organized, conscious effort by members of a society to construct a more satisfying culture . . . the persons involved in the process of revitalization must perceive their culture . . . as a system" (p. 265). Revitalization movements included the rise of Christianity and the Ghost Dance of Native Americans.
124. Ehrlich 1968, p. 77; Ehrlich and Ehrlich 1970, pp. 191–193.
125. Turco et al. 1983; Ehrlich et al. 1983.
126. See, e.g., Intergovernmental Panel on Climate Change (IPCC) 1996. The IPCC provides only a partial model because it does not make a major attempt at dialog with the public at large and no consideration is given to why the public is evolutionarily disinclined to worry about a gradually increasing level of carbon dioxide in the atmosphere, detectable only as a squiggly line on a graph at an observatory on Mauna Loa in Hawaii.
127. Colborn et al. 1996.
128. Roan 1989; Ehrlich and Ehrlich 1991, chap. 4.
129. Ehrlich and Ehrlich 1996, chap. 8; for excellent overviews of the global warming issue, see Schneider 1989, 1997.
130. U.S. General Accounting Office 1989; Freund and Martin 1993, pp. 33–41.
131. Freund and Martin 1993, pp. 20–21.
132. Although the tools are available, the main one, television, at least in the United States, is being used less and less for educating the public and generating informed discussion. One need only note the decline in international news coverage following the end of the cold war (with correspondents withdrawn from many countries) and the inordinate attention paid to the O. J. Simpson trial and the Monica Lewinsky affair to see the unhappy trend from serious news coverage to "infotainment" in process.

REFERENCES

Abbate, E., A. Albianelli, A. Azzaroli, M. Benvenuti, et al. 1998. A one-million-year-old *Homo* cranium from the Danakil (Afar) Depression of Eritrea. *Nature* 393:458–460.

Abrams, E. M., A. Freter, D. J. Rue, and J. D. Wingard. 1996. The role of deforestation in the collapse of the late classic Copán Maya state. In L. E. Sponsel, T. N. Headland, and R. C. Bailey, eds., *Tropical Deforestation: The Human Dimension.* Columbia Univ. Press, New York.

Abrams, H. L., Jr. 1987. The preference for animal protein and fat: A cross-cultural survey. Pp. 207–223 in M. Harris and E. B. Ross, eds., *Food and Evolution: Toward a Theory of Human Food Habits.* Temple Univ. Press, Philadelphia.

Ackernecht, E. H. 1943. Psychopathology, primitive medicine, and primitive culture. *Bulletin of the History of Medicine* 14:30–67.

Adams, D. B., A. R. Gould, and A. D. Burt. 1978. Rise in female-initiated sexual activity at ovulation and its suppression by oral contraceptives. *New England Journal of Medicine* 299:1145–1150.

Adams, M. S., and J. V. Neel. 1967. Children of incest. *Pediatrics* 40:55–62.

Agranoff, B. W., and P. D. Klinger. 1964. Puromycin effect on memory fixation in the goldfish. *Science* 146:952–953.

Agrawal, A. A., C. Laforsch, and R. Tollrian. 1999. Transgenerational induction of defences in animals and plants. *Nature* 401:60–62.

Aguero, J. E., L. Bloch, and D. Byrne. 1984. The relationship among sexual beliefs, attitudes, experience, and homophobia. *Journal of Homosexuality* 10:95–107.

Aiello, L. C. 1981. Locomotion in the Miocene Hominoidea. Pp. 63–98 in C. B. Stringer, ed., *Aspects of Human Evolution.* Taylor and Francis, London.

Aiello, L. C. 1993. The fossil evidence for modern human origins in Africa: A revised view. *American Anthropologist* 95:73–96.

Aiello, L. C., and C. Dean. 1990. *An Introduction to Human Evolutionary Anatomy.* Academic Press, London.

Aiello, L. C., and R. I. M. Dunbar. 1993. Neocortex size, group size, and the evolution of language. *Current Anthropology* 34:184–193.

Aiello, L. C., and P. Wheeler. 1995. The expensive-tissue hypothesis: The brain and the digestive system in human and primate evolution. *Current Anthropology* 36:199–221.

Alcorn, J. B. 1993. Indigenous peoples and conservation. *Conservation Biology* 7:424–426.

Alexander, R. D. 1987. *The Biology of Moral Systems.* Aldine, Hawthorne, NY.

Alexander, R. D. 1988. Evolutionary approaches to human behavior: What does the future hold? Pp. 317–341 in L. Betzig, M. B. Mulder, and P. Turke, eds., *Human Reproductive Behavior: A Darwinian Perspective.* Cambridge Univ. Press, Cambridge.

Alexander, R. D. 1991. Social learning and kin recognition. *Ethology and Sociobiology* 12:387–399.

Alexander, R. D., and K. M. Noonan. 1979. Concealment of ovulation, parental care, and human social evolution. Pp. 436–453 in N. A. Chagnon and W. Irons, eds., *Evolutionary Biology and Human Social Behavior.* Duxbury Press, North Scituate, MA.

Alexander, R. M. 1991. Characteristics and advantages of human bipedalism. Pp. 255–266 in J. M. V. Rayner and R. J. Wootton, eds., *Biomechanics in Evolution.* Cambridge Univ. Press, Cambridge.

Alford, D. K. H. 1978. The demise of the Whorf hypothesis. *Proceedings of the Fourth Annual Meeting of the Berkeley Linguistics Society* 4:485–499.

Allen, F. J., C. Gosden, and J. P. White. 1989. Human Pleistocene adaptations in the tropical island Pacific: Recent evidence from New Ireland, a Greater Australian outlier. *Antiquity* 63:548–561.

Allen, J. A., D. L. Raymond, and M. A. Geburtig. 1988. Wild birds prefer the familiar morph when feeding on pastry-filled shells of the landsnail *Cepaea hortensis* (Müll.). *Biological Journal of the Linnean Society* 33:395–401.

Allison, A. C. 1954a. The distribution of the sickle-cell trait in East Africa and elsewhere, and its apparent relationship to the incidence of subtertian malaria. *Transactions of the Royal Society of Tropical Medicine and Hygiene* 48:312–318.

Allison, A. C. 1954b. Protection afforded by sickle cell trait against sub-tertian malarial infection. *British Medical Journal* 1:290–292.

Allison, A. C. 1955. Aspects of polymorphism in man. *Cold Spring Harbor Symposia on Quantitative Biology* 20:239–255.

Allison, A. C. 1959. Metabolic polymorphisms in mammals and their bearing on problems of biochemical genetics. *American Naturalist* 93:5–16.

Allison, A. C. 1964. Polymorphism and natural selection in human populations. *Cold Spring Harbor Symposium for Quantitative Biology* 29:137–149.

Allman, J. M. 1999. *Evolving Brains.* Scientific American Library, New York.

Allman, J., and A. Hasenstaub. 1999. Brains, maturation times, and parenting. *Neurobiology of Aging* 20:447–454.

Allman, J., and J. Kaas. 1971. A representation of the visual field in the caudal third of the middle temporal gyrus of the owl monkey *(Aotus trivirgatus). Brain Research* 31:84–105.

Altman, I. 1975. *The Environment and Social Behavior.* Brooks/Cole, Monterey.

Altmann, S. A., and J. Altmann. 1970. *Baboon Ecology: African Field Research.* Univ. of Chicago Press, Chicago.

Alvard, M. S. 1993a. Conservation by native peoples. *Human Nature* 5:127–154.

Alvard, M. S. 1993b. Testing the "ecologically noble savage" hypothesis: Interspecific prey choice by Piro hunters of Amazonian Peru. *Human Ecology* 21:355–387.

Alvard, M. S. 1998. Evolutionary ecology and resource conservation. *Evolutionary Anthropology* 7:62–74.

Ambrose, S. E. 1997. *Citizen Soldiers: The U.S. Army from the Normandy Beaches to the Bulge to the Surrender of Germany.* Simon and Schuster, New York.

Ammerman, A. J., and L. L. Cavalli-Sforza. 1973. A population model for the diffusion of early farming in Europe. Pp. 343–357 in C. Renfrew, ed., *The Explanation of Culture Change.* Duckworth, London.

Andelman, S. 1987. Evolution of concealed ovulation in vervet monkeys *(Cercopithecus aethiops). American Naturalist* 129:785–799.

Anderson, A. 1984. The extinction of moa in southern New Zealand. Pp. 728–740 in P. S. Martin and R. G. Klein, eds., *Quaternary Extinctions: A Prehistoric Revolution.* Univ. of Arizona Press, Tucson.

Anderson, B. R. 1991. *Imagined Communities: Reflections on the Origin and Spread of Nationalism.* Rev. ed. Verso, London.

Anderson, E. 1984. Who's who in the Pleistocene: A mammalian bestiary. Pp. 40–89 in P. S. Martin and R. G. Klein, eds., *Quaternary Extinctions: A Prehistoric Revolution.* Univ. of Arizona Press, Tucson.

Anderson, P. W. 1972. More is different. *Science* 177:393–396.

Anderson, R. M., and R. M. May. 1982. Coevolution of hosts and parasites. *Parasitology* 85:411–426.

Anderson, R. M., and R. M. May. 1992. *Infectious Diseases of Humans: Dynamics and Control.* Oxford Univ. Press, Oxford.

Anderson, W. W. 1971. Genetic equilibrium and population growth under density-regulated selection. *American Naturalist* 105:489–498.

Andersson, M. 1982. Female choice selects for extreme tail length in a widowbird. *Nature* 299:818–820.

Andres, L. A., and T. Prout. 1960. Selection response and genetics of parathion resistance in the Pacific spider mite, *Tetranychus pacificus. Journal of Economic Entomology* 53:626–630.

Andrewartha, H. G., and L. C. Birch. 1954. *The Distribution and Abundance of Animals.* Univ. of Chicago Press, Chicago.

Andrews, P. 1992. Evolution and environment in the Hominoidea. *Nature* 360:641–646.

Anonymous. 1993. Executive pay: The party ain't over yet. *Business Week* (16 April): 56–64.

Anonymous. 1997. Boys will be boys. *New Scientist* (22 November): 29.

Anonymous. 1998. "Monogamous" gibbons really swing. *Science* 280:677.

Anonymous. 1999a. Animal behavior: Chimps or chumps? *Economist* (11 December): 75.

Anonymous. 1999b. Signals that go with the flow. *Trends in Neuroscience* 22:143–145.

Aoki, K., and M. W. Feldman. 1989. Pleiotropy and preadaptation in the evolution of human language capacity. *Theoretical Population Biology* 35:181–194.

Aoki, K., and M. W. Feldman. 1997. A gene-culture coevolutionary model for brother-sister mating. *Proceedings of the National Academy of Sciences USA* 94:13046–13050.

Aoki, K., and M. W. Feldman. 1999. Theoretical aspects of the evolution of human social behaviour. Pp. 1–30 in A. Kazancigil and D. Makinson, eds., *World Social Science Report.* UNESCO, Rome.

Appenzeller, T. 1994. Clashing Maya superpowers emerge from a new analysis. *Science* 266:733–734.

Appenzeller, T. 1998. Art: Evolution or revolution? *Science* 282:1451–1454.

Appenzeller, T. 1999. Test tube evolution catches time in a bottle. *Science* 284:2108–2110.

Arambourg, C., and Y. Coppens. 1967. Sur la découverte, dans le Pléistocène inférieur de la vallée de l'Omo (Éthiopie) d'une mandibule d'Australopithécien. *Comptes Rendus de Séances de l'Académie des Sciences (Paris)*, ser. 2, 300:227–230.

Arcadi, A. C., and R. W. Wrangham. 1999. Infanticide in chimpanzees: Review of cases and a new within-group observation from the Kanyawara study group in Kibale National Park. *Primates* 40:337–351.

Ardrey, R. 1961. *African Genesis.* Columbia Univ. Press, New York.

Ardrey, R. 1966. *The Territorial Imperative.* Atheneum, New York.

Arensburg, B., et al. 1990. A reappraisal of the anatomical basis for speech in middle Palaeolithic hominids. *American Journal of Physical Anthropology* 83:137–146.

Arkes, H. R., and A. R. Harkness. 1983. Estimates of contingency between two dichotomous variables. *Journal of Experimental Psychology: General* 112:117–135.

Armelagos, G. J., and A. McArdle. 1975. Population, disease, and evolution. *American Antiquity* 40:1–10.

Armelagos, G. J., A. H. Goodman, and K. H. Jacobs. 1991. The origins of agriculture: Population growth during a period of declining health. *Population and Environment* 13:9–22.

Armstrong, D. F., W. C. Stockoe, and S. E. Wilcox. 1994. Signs of the origin of syntax. *Current Anthropology* 35:349–368.

Arnhart, L. 1995. The new Darwinian naturalism in political theory. *American Political Science Review* 89:389–400.

Arnold, M. L. 1997. *Natural Hybridization and Evolution.* Oxford Univ. Press, New York.

Arrow, K. J. 1996. Economic aspects of environmental challenges. Pp. 29–31 in H. W. Kendall et al., eds., *Meeting the Challenges of Population, Environment, and Resources: The Costs of Inaction.* World Bank, Washington, DC.

Arsuaga, J. L., J. M. Carretero, C. Lorenzo, A. Gracia, et al. 1997. Size variation in middle Pleistocene humans. *Science* 277:1086–1087.

Arsuaga, J. L., I. Martínez, A. Gracia, and C. Lorenzo. 1997. The Sima de los Huesos crania (Sierra de Atapuerca, Spain): A comparative study. *Journal of Human Evolution* 33:219–281.

Arthur, W., D. Phillips, and P. Mitchell. 1993. Long-term stability of morph frequency and species distribution in a sand-dune colony of *Cepaea. Proceedings of the Royal Society of London*, ser. B, 251:159–163.

Asfaw, B., T. White, O. Lovejoy, B. Latimer, et al. 1999. *Australopithecus garhi:* A new species of early hominid from Ethiopia. *Science* 284:629–635.

Athens, J. S. 1997. Hawaiian native lowland vegetation in prehistory. Pp. 248–270 in P. V. Kirch and T. L. Hunt, eds., *Historical Ecology in the Pacific Islands.* Yale Univ. Press, New Haven.

Attwell, C. A. M., and F. P. D. Cotterill. 2000. Postmodernism and African conservation science. *Biodiversity and Conservation*, in press.

Au, T. K. 1983. Chinese and English counterfactuals: The Sapir-Whorf hypothesis revisited. *Cognition* 5:155–187.

Au, T. K. 1984. Counterfactuals: In reply to Bloom. *Cognition* 17:289–302.

Augusta, J., and Z. Burian. N.d. *Prehistoric Animals.* Spring Books, London.

Austad, S. N. 1994. Menopause: An evolutionary perspective. *Experimental Gerontology* 29:255–263.

Avis, J., and P. L. Harris. 1991. Evidence for a universal conception of mind. *Child Development* 62:460–467.

Avise, J. 1994. *Molecular Markers, Natural History, and Evolution.* Chapman and Hall, New York.

Awadalla, P., A. Eyre-Walker, and J. Maynard Smith. 1999. Linkage disequilibrium and recombination in hominid mitochondrial DNA. *Science* 286:2524–2525.

Axelrod, R., and W. D. Hamilton. 1981. The evolution of cooperation. *Science* 211:1390–1396.

Ayala, F. 1968. Evolution of fitness. II. Correlated effects of natural selection on the productivity and size of experimental populations of *Drosophila serrata. Evolution* 22:55–65.

Ayala, F. J. 1974. Biological evolution: Natural selection or random walk? *American Scientist* 62:692–701.

Ayala, F. J. 1987. The biological roots of morality. *Biology and Philosophy* 2:235–252.

Ayala, F. J., and C. A. Campbell. 1974. Frequency-dependent selection. *Annual Review of Ecology and Systematics* 5:115–138.

Ayala, F. J., A. Escalante, C. O'hUigin, and J. Klein. 1994. Molecular genetics of speciation and human origins. *Proceedings of the National Academy of Sciences USA* 91:6787–6794.

Ayer, A. J. 1952 (1946). *Language, Truth, and Logic.* Dover, New York.

Ayton, P. 1999. Clear cut. *New Scientist* (11 December): 47.

Backman, E. L., and L. R. Backman. 1997. Sexual harassment and rape: A view from higher education. Pp. 133–155 in R. F. Levant and G. R. Brooks, eds., *Men and Sex: New Psychological Perspectives.* Wiley, New York.

Bagemihl, B. 1999. *Biological Exuberance: Animal Homosexuality and Natural Diversity.* St. Martin's Press, New York.

Bahn, P. G. 1988. Triple Czech burial. *Nature* 332:302–303.

Bailey, R. 1997. Origin of the specious: Why do neoconservatives doubt Darwin? *Reason* (July): 22–28.

Bailey, R. C., G. Head, M. Jenike, B. Owen, et al. 1989. Hunting and gathering in tropical rain forest: Is it possible? *American Anthropologist* 91:59–82.

Baker, R. R., and M. A. Bellis. 1989. Number of sperm in human ejaculates varies in accordance with sperm competition theory. *Animal Behaviour* 37:867–869.

Baker, R. R., and M. A. Bellis. 1993a. Human sperm competition: Ejaculate adjustment by males and the function of masturbation. *Animal Behaviour* 46:861–885.

Baker, R. R., and M. A. Bellis. 1993b. Human sperm competition: Ejaculate manipulation by females and a function for the female orgasm. *Animal Behaviour* 46:887–909.

Baker, R. R., and M. A. Bellis. 1995. *Human Sperm Competition: Copulation, Masturbation, and Infidelity.* Chapman and Hall, London.

Baksh, M. 1984. Cultural ecology and change of the Machiguenga Indians of the Peruvian Amazon. Ph.D. diss., Univ. of California, Los Angeles.

Balikci, A. 1970. *The Netsilik Eskimo.* Natural History Press, Garden City, NY.

Ballara, M. 1991. *Women and Literacy.* Zed Books, London.

Ballard, J. W. O., and M. Kreitman. 1995. Is mitochondrial DNA a strictly neutral marker? *Trends in Ecology and Evolution* 10:485–488.

Balter, M. 1998. Why settle down? The mystery of communities. *Science* 282:1442–1445.

Balter, M. 1999. A long season puts Catalhöyük in context. *Science* 286:890–891.

Bamberger, J. 1974. The myth of matriarchy: Why men rule in primitive society. Pp. 263–280 in M. Rosaldo and L. Lamphere, eds., *Women, Culture, and Society.* Stanford Univ. Press, Stanford.

Bandura, A. 1986. *Social Foundations of Thought and Action.* Prentice Hall, Englewood Cliffs, NJ.

Bandura, A., D. Ross, and S. A. Ross. 1963. Imitation of film-mediated aggressive models. *Journal of Abnormal and Social Psychology* 66:3–11.

Baquero, F., and J. Blázquez. 1997. Evolution of antibiotic resistance. *Trends in Ecology and Evolution* 12:482–487.

Barber, B. R. 1995. *Jihad vs. McWorld.* Ballantine Books, New York.

Barber, E. W. 1994. *Women's Work: The First Twenty Thousand Years.* Norton, New York.

Barbujani, G., A. Magagni, E. Minch, and L. L. Cavalli-Sforza. 1997. An apportionment of human DNA diversity. *Proceedings of the National Academy of Sciences USA* 94:4516–4519.

Barbujani, G., and R. R. Sokal. 1990. Zones of sharp genetic change in Europe are also linguistic boundaries. *Proceedings of the National Academy of Sciences USA* 87:1816–1819.

Barbujani, G., and R. R. Sokal. 1991. Genetic population structure of Italy. II. Physical and cultural barriers to gene flow. *American Journal of Human Genetics* 48:398–411.

Barkow, J. H., L. Cosmides, and J. Tooby. 1992. *The Adapted Mind: Evolutionary Psychology and the Generation of Culture.* Oxford Univ. Press, New York.

Barnard, A. 1994. Rules and prohibitions: The form and content of human kinship. Pp. 783–812 in T. Ingold, ed., *Companion Encyclopedia of Anthropology: Humanity, Culture, and Social Life.* Routledge, London.

Barnard, A., and J. Woodburn. 1988. Property, power, and ideology in hunter-gathering societies: An introduction. Pp. 4–31 in T. Ingold, D. Riches, and J. Woodburn, eds., *Hunters and Gatherers.* Vol. 2, *Property, Power, and Ideology.* Berg, Oxford.

Barnes, K. C., G. J. Armelagos, and S. C. Morreale. 1999. Darwinian medicine and the emergence of allergy. Pp. 209–243 in W. R. Trevathan, E. O. Smith, and J. J. McKenna, eds., *Evolutionary Medicine.* Oxford Univ. Press, New York.

Barnosky, A. D. 1986. "Big game" extinction caused by Late Pleistocene climatic change: Irish elk *(Megaloceros giganteus)* in Ireland. *Quaternary Research* 25:128–135.

Barnosky, A. D. 1989. The Late Pleistocene event as a paradigm for widespread mammal extinction. Pp. 235–255 in S. K. Donovan, ed., *Mass Extinctions: Processes and Evidence*. Columbia Univ. Press, New York.

Barondes, S. H. 1999. *Molecules and Mental Illness*. Scientific American Library, New York.

Barrantes, R., P. E. Smouse, H. W. Mohrenweiser, H. Gershowitz, et al. 1990. Microevolution in lower Central America: Genetic characterization of the Chibcha-speaking groups of Costa Rica and Panama, and a consensus taxonomy based on genetic and linguistic affinity. *American Journal of Human Genetics* 46:63–84.

Bartholomew, G. A., Jr., and J. B. Birdsell. 1953. Ecology and the protohominids. *American Anthropologist* 55:481–498.

Bartlett, D., and J. Bartlett. 1961. Observations while filming African game. *South African Journal of Science* 57:313–321.

Barton, N. H., and B. Charlesworth. 1998. Why sex and recombination? *Science* 281:1986–1990.

Barton, R. A. 1997. *Visual Specialisation, Brain Evolution, and Behavioural Ecology in Primates*. Blackwell, Berlin.

Barton, R. A., A. Purvis, and P. H. Harvey. 1995. Evolutionary radiation of visual and olfactory brain systems in primates, bats, and insectivores. *Philosophical Transactions of the Royal Society of London*, ser. B, 348:381–392.

Bar-Yosef, O. 1991. The Early Neolithic of the Levant: Recent advances. *Review of Archaeology* 12:1–18.

Bar-Yosef, O. 1994. The contributions of Southwest Asia to the study of the origin of modern humans. Pp. 23–66 in M. H. Nitecki and D. V. Nitecki, eds., *Origins of Anatomically Modern Humans*. Plenum Press, New York.

Bar-Yosef, O. 1995. The role of climate in the interpretation of human movements and cultural transformations in western Asia. Pp. 507–523 in E. S. Vrba, G. H. Denton, T. C. Partridge, and L. H. Burckle, eds., *Paleoclimate and Evolution, with Emphasis on Human Origins*. Yale Univ. Press, New Haven.

Bar-Yosef, O., and A. Belfer-Cohen. 1991. From sedentary hunter-gatherers to territorial farmers in the Levant. Pp. 181–202 in S. A. Gregg, ed., *Between Bands and States*. Southern Illinois Univ., Edwardsville.

Bar-Yosef, O., and M. E. Kislev. 1989. Early farming communities in the Jordan Valley. Pp. 632–642 in D. R. Harris and G. C. Hillman, eds., *Foraging and Farming: The Evolution of Plant Exploitation*. Unwin Hyman, London.

Basabose, K., and J. Yamagiwa. 1997. Predation on mammals by chimpanzees in the montane forest of Kahuzi, Zaire. *Primates* 38:45–55.

Bates, E., D. Thal, and V. Marchman. 1991. Symbols and syntax: A Darwinian approach to language development. Pp. 29–65 in N. A. Krasnegor, D. M. Rumbaugh, R. L. Schiefelbusch, and M. Studdert-Kennedy, eds., *Biological and Behavioral Determinants of Language Development*. Erlbaum, Hillsdale, NJ.

Bates, M. 1943. Mosquitoes as vectors of *Dermatobia* in eastern Colombia. *Annals of the Entomological Society of America* 36:21–24.

Bateson, G., and M. Mead. 1942. *Balinese Character: A Photographic Analysis*. New York Academy of Sciences, New York.

Bateson, P. 1982. Preferences for cousins in Japanese quail. *Nature (London)* 295:236–237.

Batson, C. D. 1990. How social an animal? *American Psychologist* 45:336–346.

Batson, C. D., J. G. Batson, C. A. Griffitt, S. Barrientos, et al. 1989. Negative-state relief and the empathy-altruism hypothesis. *Journal of Personality and Social Psychology* 56:922–933.

Batson, C. D., J. L. Dyck, J. R. Brandt, J. G. Batson, et al. 1988. Five studies testing two new egoistic alternatives to the empathy-altruism hypothesis. *Journal of Personality and Social Psychology* 55:52–77.

Batson, C. D., K. O'Quin, J. Fultz, and M. Vanderplas. 1983. Influence of self-reported distress and empathy on egoistic versus altruistic motivation to help. *Journal of Personality and Social Psychology* 45:706–718.

Batzer, M. A., M. Stoneking, M. Alegria-Hartman, H. Bazan, et al. 1994. African origin of human-specific polymorphic *Alu* insertions. *Proceedings of the National Academy of Sciences USA* 91:12288–12292.

Bauer, R. H. 1993. Lateralization of neural control for vocalization by the frog *(Rana pipiens)*. *Psychobiology* 21:243–248.

Bawa, K. S., and M. Gadgil. 1997. Ecosystem services in subsistence economies and conservation of biodiversity. Pp. 295–310 in G. C. Daily, ed., *Nature's Services: Societal Dependence on Natural Ecosystems*. Island Press, Washington, DC.

Beadle, P. C. 1977. The epidermal biosynthesis of cholecalciferol (vitamin D_3). *Photochemistry and Photobiology* 25:519–527.

Beasley, N. A. 1968. The extent of individual differences in the perception of causality. *Canadian Journal of Psychology* 22:399–407.

Beatson, R. R. 1976. Environmental and genetical correlates of disruptive coloration in the water snake *Natrix s. sipedon*. *Evolution* 30:241–252.

Beatty, W. W. 1992. Gonadal hormones and sex differences in non-reproductive behaviors. Pp. 85–128 in A. A. Gerall, H. Moltz, and I. L. Ward, eds., *Handbook of Behavioral Neurobiology*. Vol. 11, *Sexual Differentiation*. Plenum Press, New York.

Becerra, J. X. 1997. Insects on plants: Chemical trends in host use. *Science* 276:253–256.

Beckoff, M., and D. Jamieson, eds. 1996. *Readings in Animal Cognition*. MIT Press, Cambridge.

Beers, M. H., and R. Berkow, eds. 1999. *The Merck Manual of Diagnosis and Therapy*. 17th ed. Merck Research Laboratories, Whitehouse Station, NJ.

Begun, D. R., C. V. Ward, and M. D. Rose, eds. 1997. *Function, Phylogeny, and Fossils: Miocene Hominoid Evolution and Adaptations*. Plenum Press, New York.

Behrensmeyer, A. K., N. E. Todd, R. Potts, and G. E. McBrinn. 1997. Late Pliocene faunal turnover in the Turkana Basin, Kenya and Ethiopia. *Science* 278:1589–1593.

Behrmann, M., G. Winocur, and M. Moscovitch. 1992. Dissociation between mental imagery and object recognition in a brain-damaged patient. *Nature* 359:636–637.

Beilin, H. 1989. Piagetian theory. *Annals of Child Development* 6:85–131.

Bell, A. P., and M. S. Weinberg. 1978. *Homosexualities: A Study of Diversity among Men and Women*. Simon and Schuster, New York.

Bellis, M. A., and R. R. Baker. 1990. Do females promote sperm competition? Data for humans. *Animal Behaviour* 40:997–999.

Bellomo, R. V. 1993. A methodological approach for identifying archaeological evidence of fire resulting from human activities. *Journal of Archaeological Science* 20:525–553.

Bellomo, R. V. 1994. Early Pleistocene fire technology in northern Kenya. Pp. 16–28 in S. T. Childs, ed., *Society, Culture, and Technology in Africa*. Museum of Applied Science, Center for Archaeology, Philadelphia.

Belluck, P. 1999. Board for Kansas deletes evolution from curriculum. *New York Times* (12 August): A1, A13.

Bengtson, J. D., and M. Ruhlen. 1994. Global etymologies. Pp. 278–337 in M. Ruhlen, ed., *On the Origin of Languages: Studies in Linguistic Taxonomy*. Stanford Univ. Press, Stanford.

Bengtsson, B. O. 1978. Avoid inbreeding: At what cost? *Journal of Theoretical Biology* 73:439–444.

Benson, H. 1996. *Timeless Healing: The Power and Biology of Belief*. Scribner, New York.

Bentham, J. 1988 (1789). *Introduction to the Principles of Morals and Legislation*. Prometheus Books, Amherst, NY.

Bentzen, P., and J. D. McPhail. 1984. Ecology and evolution of sympatric sticklebacks *(Gasterosteus)*: Specialization for alternative trophic niches in the Enos Lake species pair. *Canadian Journal of Zoology* 62:2280–2286.

Bentzen, P., M. S. Ridgway, and J. D. McPhail. 1984. Ecology and evolution of sympatric sticklebacks *(Gasterosteus)*: Spatial segregation and seasonal habitat shifts in the Enos Lake species pair. *Canadian Journal of Zoology* 62:2436–2439.

Berdan, F. F. 1989. Trade and markets in precapitalist states. Pp. 78–107 in S. Plattner, ed., *Economic Anthropology*. Stanford Univ. Press, Stanford.

Berger, T. D., and E. Trinkaus. 1995. Patterns of trauma among the Neandertals. *Journal of Archaeological Science* 22:841–852.

Bergman, A., and M. W. Feldman. 1995. Question marks about the period of punctuation. Working Paper No. 96-02-006. Santa Fe Institute, Santa Fe.

Bergman, A., and M. W. Feldman. 1999. On the population genetics of punctuation. Pp. 1–25 in J. Crutchfield and P. Schuster, eds., *Towards a Comprehensive Dynamics of Evolution: Exploring the Interplay of Selection, Neutrality, Accident, and Function*. Addison-Wesley, New York.

Berk, R., P. R. Abramson, and P. Okami. 1995. Sexual activities as told in surveys. Pp. 371–386 in P. R. Abramson and S. D. Pinkerton, eds., *Sexual Nature/Sexual Culture*. Univ. of Chicago Press, Chicago.

Berko, J. 1958. The child's learning of English morphology. *Word* 14:150–177.

Berkow, R., ed. 1992. *The Merck Manual of Diagnosis and Therapy*. 16th ed. Merck Research Laboratories, Rahway, NJ.

Berkowitz, L. 1962. *Aggression: A Social Psychological Analysis*. McGraw-Hill, New York.

Berlin, B. 1992. *Ethnobiological Classification*. Princeton Univ. Press, Princeton.

Berlin, B., D. E. Breedlove, and P. Raven. 1973. General principles of classification and nomenclature in folk biology. *American Anthropologist* 75:214–242.

Berlin, B., D. E. Breedlove, and P. H. Raven. 1974. *Principles of Tzeltal Plant Classification*. Academic Press, New York.

Berlin, B., and P. Kay. 1969. *Basic Color Terms.* Univ. of California Press, Berkeley.

Berlinski, D. 1996. The deniable Darwin. *Commentary* 101:19–29.

Bermant, G. 1976. Sexual behavior: Hard times with the Coolidge effect. Pp. 76–103 in M. H. Siegel and H. P. Zeigler, eds., *Psychological Research: The Inside Story.* Harper and Row, New York.

Bermant, G., M. T. Clegg, and W. Beamer. 1969. Copulatory behaviour of the ram, *Ovis aries. Animal Behaviour* 17:700–705.

Bernays, E. A., and R. F. Chapman. 1994. *Host-Plant Selection by Phytophagous Insects.* Chapman and Hall, New York.

Betzig, L. 1988. Mating and parenting in Darwinian perspective. Pp. 3–20 in L. Betzig, M. B. Mulder, and P. Turke, eds., *Human Reproductive Behavior: A Darwinian Perspective.* Cambridge Univ. Press, Cambridge.

Betzig, L. 1989. Causes of conjugal dissolution: A cross-cultural study. *Current Anthropology* 30:654–676.

Betzig, L., ed. 1997. *Human Nature: A Critical Reader.* Oxford Univ. Press, New York.

Beyer, L. 1999. The price of honor. *Time* (18 January): 55.

Beynon, P., and O. A. E. Rasa. 1989. Do dwarf mongooses have a language? Warning vocalisations transmit complex information. *South African Journal of Science* 85:447–450.

Beyries, S. 1988. Functional variability of lithic sets in the Middle Paleolithic. Pp. 213–224 in H. L. Dibble and A. Montet-White, eds., *Upper Pleistocene Prehistory of Western Eurasia.* Monograph 54. Univ. of Pennsylvania Museum, Philadelphia.

Bhasin, S., T. Storer, N. Berman, C. Callegari, et al. 1996. The effects of supraphysiologic doses of testosterone on muscle size and strength in normal men. *New England Journal of Medicine* 335:1.

Biasutti, R. 1967. *Le Razze e Popoli Della Terra.* 4th ed. Unione Tipografico, Turin.

Bickerton, D. 1985. Creole languages. Pp. 134–150 in V. P. Clark, P. A. Eschholz, and A. F. Rosa, eds., *Language: Introductory Reading.* St. Martin's Press, New York.

Bickerton, D. 1986. More than nature needs? A reply to Premack. *Cognition* 23:73–79.

Bickerton, D. 1990. *Language and Species.* Univ. of Chicago Press, Chicago.

Bickerton, D. 1995. *Language and Human Behavior.* Univ. of Washington Press, Seattle.

Bickerton, D. 1998. Catastrophic evolution: The case for a single step from protolanguage to full human language. Pp. 341–358 in J. R. Hurford, M. Studdert-Kennedy, and C. Knight, eds., *Approaches to the Evolution of Language.* Cambridge Univ. Press, Cambridge.

Biebuyck, D. 1973. *The Lega: Art, Initiation, and Moral Philosophy.* Univ. of California Press, Berkeley.

Biederman, I. 1987. Recognition by components: A theory of human image understanding. *Psychological Review* 94:115–147.

Biederman, I., and P. C. Gerhardstein. 1993. Recognizing depth-rotated objects: Evidence and conditions for three-dimensional viewpoint invariance. *Journal of Experimental Psychology: Human Perception and Performance* 19:1162–1182.

Binford, L. 1963. "Red ochre" caches from the Michigan area: A possible case of cultural drift. *Southwestern Journal of Anthropology* 19:89–108.

Binford, L. R. 1985. Human ancestors: Changing views of their behavior. *Journal of Anthropological Archaeology* 4:292–327.

Birch, L. C. 1976. *Confronting the Future. Australia and the World: The Next Hundred Years.* Penguin Books, Ringwood, Victoria.

Birch, L. C. 1990. *On Purpose.* Univ. of New South Wales Press, Sydney.

Birch, L. C. 1993a. *Confronting the Future.* Rev. ed. Penguin Books, Ringwood, Victoria.

Birch, L. C. 1993b. *Regaining Compassion for Humanity and Nature.* Univ. of New South Wales Press, Sydney.

Birch, L. C. 1999. *Biology and the Riddle of Life.* Univ. of New South Wales Press, Sydney.

Birch, L. C., and J. B. Cobb. 1981. *The Liberation of Life: From the Cell to the Community.* Cambridge Univ. Press, Cambridge.

Birch, L. C., and P. R. Ehrlich. 1967. Evolutionary history and population biology. *Nature* 214:349–352.

Bird-David, N. 1992. Beyond "the original affluent society": A culturalist reformulation. *Current Anthropology* 33:25–47.

Birdsell, J. B. 1968. Some predictions for the Pleistocene based on equilibrium systems among recent hunter-gatherers. Pp. 229–240 in R. B. Lee and I. Devore, eds., *Man the Hunter.* Aldine, Chicago.

Birnbaum, L. C. 1955. Behaviorism in the 1920s. *American Quarterly* 7:15–30.

Bittles, A. H. 1995. When cousins marry: A review of consanguinity in the Middle East. *Perspectives in Human Biology* 1:71–83.

Bittles, A. H., J. C. Grant, and S. A. Shami. 1993. Consanguinity as a determinant of reproductive behaviour and mortality in Pakistan. *International Journal of Epidemiology* 22:463–467.

Bittles, A. H., W. M. Mason, J. Greene, and N. A. Rao. 1991. Reproductive behavior and health in consanguineous marriages. *Science (Washington, DC)* 252:789–794.

Bittles, A. H., and J. V. Neel. 1994. The costs of human inbreeding and their implications for variations at the DNA level. *Nature Genetics* 8:117–121.

Bittles, A. H., H. S. Savithri, H. S. Venkatesha Murthy, G. Baskaran, et al. 2000. Human inbreeding: A familiar story full of surprises. In H. Macbeth and P. Shetty, eds., *Ethnicity and Health.* Taylor and Francis, London. In press.

Black, L. T. 1989. Comments. *Current Anthropology* 30:138.

Blackmore, S. 1999. *The Meme Machine.* Oxford Univ. Press, Oxford.

Blanc, A. C. 1962. Some evidence for the ideologies of early man. Pp. 119–136 in S. L. Washburn, ed., *Social Life of Early Man.* Methuen, London.

Bleek, P. C., and F. N. von Hippel. 1999. Missile defense: A dangerous move. *Washington Post* (12 December): B09.

Bloch, J. I., D. C. Fisher, P. D. Gingerich, G. F. Gunnell, et al. 1997. Cladistic analysis and anthropoid origins. *Science* 278:2134–2135.

Blockstein, D. E. 1998. Lyme disease and the passenger pigeon. *Science* 279:1831.

Bloom, A. H. 1984. Caution—the words you use may affect what you say: A response to Au. *Cognition* 17:275–287.

Bloomgarden, R. 1998. *Guía Fácil de Las Piramides de Teotihuacan.* Editur, S.A., DF, Mexico.

Blumenschine, R. J. 1987. Characteristics of an early hominid scavenging niche. *Current Anthropology* 28:383–407.

Boaz, N. T. 1997. *Eco Homo.* Basic Books, New York.

Bodley, J. 1976. *Anthropology and Contemporary Human Problems.* Benjamin-Cummings, Menlo Park, CA.

Bodmer, W. F., and L. L. Cavalli-Sforza. 1976. *Genetics, Evolution, and Man.* Freeman, San Francisco.

Boehm, C. 1993. Egalitarian behavior and reverse dominance hierarchy. *Current Anthropology* 34:227–240.

Boehm, C. 1994a. Pacifying interventions at Arnhem Zoo and Gombe. Pp. 211–226 in R. W. Wrangham, W. C. McGrew, F. de Waal, and P. G. Heltne, eds., *Chimpanzee Cultures.* Harvard Univ. Press, Cambridge.

Boehm, C. 1994b. Reply to Erdal and Whiten. *Current Anthropology* 35:178–180.

Boesch, C. 1991a. Symbolic communication in wild chimpanzees? *Human Evolution* 6:81–90.

Boesch, C. 1991b. Teaching among wild chimpanzees. *Animal Behaviour* 41:530–532.

Boesch, C. 1996. The emergence of cultures among wild chimpanzees. Pp. 251–268 in W. G. Runciman, J. Maynard Smith, and R. I. M. Dunbar, eds., *Evolution of Social Behaviour Patterns in Primates and Man.* Oxford Univ. Press, Oxford.

Boesch-Achermann, H., and C. Boesch. 1994. Hominization in the rainforest: The chimpanzee's piece of the puzzle. *Evolutionary Anthropology* 3:9–16.

Bohannan, B. J. M., and R. E. Lenski. 1997. Effect of resource enrichment on a chemostat community of bacteria and phage. *Ecology* 78:2303–2315.

Bonnefille, R. 1994. Palynology and paleoenvironment of East African hominid sites. Pp. 415–427 in R. S. Corruccini and R. L. Ciochon, eds., *Integrative Paths to the Past: Paleoanthropological Advances in Honor of F. Clark Howell.* Prentice Hall, Englewood Cliffs, NJ.

Bork, R. H. 1997. *Slouching toward Gomorrah: Modern Liberalism and American Decline.* HarperCollins, New York.

Boserup, E. 1965. *The Conditions of Agricultural Growth.* Aldine, Chicago.

Boserup, E. 1981. *Population and Technological Change.* Univ. of Chicago Press, Chicago.

Bouchard, T. J. 1994. Genes, environment, and personality. *Science* 264:1700–1701.

Bouchard, T. J., D. T. Lykken, M. McGue, N. L. Segal, et al. 1990. Sources of human psychological differences: The Minnesota study of twins reared apart. *Science* 250:223–226.

Bower, B. 1997. German mine yields ancient hunting spears. *Science News* 151:134.

Bower, B. 1998. Getting a feel for emotions. *Science News* 154:190–191.

Bower, B. 1999a. Cave finds revive Neandertal cannibalism. *Science News* 156:213.

Bower, B. 1999b. DNA's evolutionary dilemma: Genetic studies collide with the mystery of human evolution. *Science News* 155:88–90.

Bower, B. 1999c. Fossil ape's grasp gets two thumbs way up. *Science News* 155:23.

Bower, B. 1999d. Fossil may expose humanity's hybrid roots. *Science News* 155:295.

Bower, B. 1999e. Speech insights sound off in the brain. *Science News* 155:68.

Boyd, R., and P. J. Richerson. 1985. *Culture and the Evolutionary Process.* Univ. of Chicago Press, Chicago.

Boyd, R., and P. J. Richerson. 1996. Why culture is common, but cultural evolution is rare. Pp. 77–93 in

W. G. Runciman, J. Maynard Smith, and R. I. M. Dunbar, eds., *Evolution of Social Behavior Patterns in Primates and Man.* Oxford Univ. Press, Oxford.

Boyer, L. B. 1969. Shamans: To set the record straight. *American Anthropologist* 71:307–309.

Boyer, P. 1994. *The Naturalness of Religious Ideas: A Cognitive Theory of Religion.* Univ. of California Press, Berkeley.

Boyle, D. G. 1960. A contribution to the study of phenomenal causation. *Quarterly Journal of Experimental Psychology* 12:171–179.

Brace, C. L. 1964. A non-racial approach towards the understanding of human diversity. Pp. 103–152 in A. Montagu, ed., *The Concept of Race.* Free Press of Glencoe, New York.

Brace, C. L. 1988. Comments: The functional significance of Neanderthal pubic length, by Karen Rosenberg. *Current Anthropology* 29:607–608.

Brace, C. L. 1996. Review of *The Bell Curve. Current Anthropology* 37:S156–S161.

Brace, C. L. 1997. One human line, two million years. *Natural History* (September): 5–6.

Brace, C. L. 1999. Comments: The raw and the stolen. *Current Anthropology* 40:577–579.

Brace, C. L., D. P. Tracer, L. A. Yaroch, J. Robb, et al. 1993. Clines and clusters versus "race": A test in ancient Egypt and the case of a death on the Nile. *Yearbook of Physical Anthropology* 36:1–31.

Bradotti, R., E. Charkiewicz, S. Haüsler, and S. Wieringa. 1994. *Women, the Environment, and Sustainable Development: Towards a Theoretical Synthesis.* Zed Books, London.

Bradshaw, J. L. 1991. Animal asymmetry and human heredity: Dextrality, tool use, and language in evolution—ten years after Walker (1980). *British Journal of Psychology* 82:39–59.

Brady, R. H. 1979. Natural selection and the criteria by which a theory is judged. *Systematic Zoology* 28:600–621.

Brady, R. H. 1982. Dogma and doubt. *Biological Journal of the Linnean Society* 17:79–96.

Brain, C. K., and A. Sillen. 1988. Evidence from the Swartkrans cave for the earliest use of fire. *Nature* 336:464–466.

Brainard, J. 1998. Giving Neandertals their due: Similarities with modern humans shift the image of the caveman brute. *Science News* 154:72–74.

Brakefield, P. M. 1987. Industrial melanism: Do we have the answers? *Trends in Ecology and Evolution* 2:117–122.

Brakefield, P. M. 1998. The evolution-development interface and advances with the eyespot patterns of *Bicyclus* butterflies. *Heredity* 80:265–272.

Bramson, L., and G. Goethals, eds. 1964. *War.* Basic Books, New York.

Brand, J. A., and M. S. Taylor. 1998. The simple economics of Easter Island: A Ricardo-Malthus model of renewable resource use. *American Economic Review* 88:119–138.

Branda, R. F., and J. W. Eaton. 1978. Skin color and nutrient photolysis: An evolutionary hypothesis. *Science* 201:625–626.

Brannon, E. M., and H. S. Terrace. 1998. Ordering of the numerosities 1 to 9 by monkeys. *Science* 282:746–749.

Braudel, F. 1993. *A History of Civilizations.* Penguin Books, New York.

Bräuer, G. 1984. The "Afro-European *sapiens* hypothesis" and hominid evolution in East Asia during the Late Middle and Upper Pleistocene. *Courier Forschungsinstitut Senckenberg* 69:145–165.

Bräuer, G. 1994. How different are Asian and African *Homo erectus*? *Courier Forschungsinstitut Senckenberg* 171:301–318.

Bräuer, G., Y. Yokoyama, C. Falguères, and E. Mbua. 1997. Modern human origins backdated. *Nature* 386:337–338.

Brecher, E. M. 1969. *The Sex Researchers.* Little, Brown, Boston.

Brecher, J. 1993. Global village or global pillage? *Nation* (6 December): 685–688.

Breitman, R. 1998. *Official Secrets.* Hill and Wang, New York.

Brewer, J. B., Z. Zhao, J. E. Desmond, G. H. Glover, et al. 1998. Making memories: Brain activity that predicts how well visual experience will be remembered. *Science* 281:1185–1187.

Bridgeman, B. 1980. Brains + programs = minds. *Behavioral and Brain Sciences* 1980:427–428.

Brisbin, I. L., Jr., and T. S. Risch. 1997. Primitive dogs, their ecology and behavior: Unique opportunities to study the early development of the human-canine bond. *Journal of the American Veterinary Medical Association* 210:1122–1126.

Brody, H. 1975. *The People's Land: Eskimos and Whites in the Eastern Arctic.* Penguin Books, Harmondsworth.

Bromley, D. W. 1991. *Environment and Economy: Property Rights and Public Policy.* Basil Blackwell, Cambridge, MA.

Bronson, B. 1988. The role of barbarians in the fall of states. Pp. 196–218 in N. Yoffee and G. L. Cowgill, eds., *The Collapse of Ancient States and Civilizations.* Univ. of Arizona Press, Tucson.

Brookes, M. 1999. Live and let live. *New Scientist* (July): 32–36.

Brooks, A. S. 1996. Behavior and human evolution. Pp. 135–166 in W. E. Meikle, F. C. Howell, and N. G. Jablonski, eds., *Contemporary Issues in Human Evolution.* California Academy of Sciences, San Francisco.

Brooks, A. S., D. M. Helgren, J. S. Cramer, A. Franklin, et al. 1995. Dating and context of three Middle Stone Age sites with bone points in the upper Semliki Valley, Zaire. *Science* 268:548–553.

Broude, G. J., and S. J. Greene. 1976. Cross-cultural codes on twenty sexual attitudes and practices. *Ethnology* 15:409–429.

Brown, C. H. 1977. Folk botanical life-forms: Their universality and growth. *American Anthropologist* 79:317–342.

Brown, C. H. 1979. Folk zoological life-forms: Their universality and growth. *American Anthropologist* 81:791–817.

Brown, D. E. 1991. *Human Universals.* McGraw-Hill, New York.

Brown, J. 1977. *Gandhi and Civil Disobedience: The Mahatma in Indian Politics, 1928–1934.* Cambridge Univ. Press, Cambridge.

Brown, J. L. 1987. *Helping and Communal Breeding in Birds: Ecology and Evolution.* Princeton Univ. Press, Princeton.

Brown, R. W. 1957. Linguistic determinism and the part of speech. *Journal of Abnormal and Social Psychology* 55:1–5.

Brown, R. W., and E. H. Lenneberg. 1954. A study in language and cognition. *Journal of Abnormal and Social Psychology* 49:454–462.

Brown, W. L. 1959. Some zoological concepts applied to problems in the evolution of the hominid lineage. *American Scientist* 46:151–158.

Browne, M. W. 1994. What is intelligence, and who has it? *New York Times Book Review* (16 October): 3, 41.

Browning, C. R. 1998. *Ordinary Men: Reserve Police Battalion 101 and the Final Solution in Poland (with a New Afterword).* HarperCollins, New York.

Brownmiller, S. 1975. *Against Our Will: Men, Women, and Rape.* Simon and Schuster, New York.

Bruce, V., P. Green, and M. A. Georgeson. 1996. *Visual Perception: Physiology, Psychology, and Ecology.* Psychology Press, Hove, East Sussex.

Bruce, V., and G. W. Humphreys. 1994. Recognizing objects and faces. *Visual Cognition* 1:141–180.

Bruer, J. T. 1999. *The Myth of the First Three Years: A New Understanding of Early Brain Development and Lifelong Learning.* Free Press, New York.

Brumfiel, E. 1980. Specialization, market exchange, and the Aztec state. *Current Anthropology* 21:459–478.

Brunner, H. G., M. R. Nelen, P. van Zandvoort, N. G. G. M. Abeling, et al. 1993. X-linked borderline mental retardation with prominent behavioral disturbance: Phenotype, genetic localization, and evidence for disturbed monoamine metabolism. *American Journal of Human Genetics* 52:1032–1039.

Bücher, K. 1909. *Arbeit und Rhythmus.* Teubner, Leipzig.

Buckley, T. 1982. Menstruation and the power of Yurok women. *American Ethnologist* 9:47–90.

Bülthoff, H. H., and S. Edelman. 1992. Psychophysical support for a two-dimensional view interpolation theory of object recognition. *Proceedings of the National Academy of Sciences USA* 89:60–64.

Bunn, H., and E. Kroll. 1986. Systematic butchery by Plio-Pleistocene hominids at Olduvai Gorge, Tanzania. *Current Anthropology* 27:431–452.

Buonomano, D. V., and M. M. Merzenich. 1995. Temporal information transformed into a spatial code by a neural network with realistic properties. *Science* 267:1028–1030.

Buonomano, D. V., and M. M. Merzenich. 1998. Cortical plasticity: From synapses to maps. *Annual Review of Neurosciences* 21:149–186.

Burgers, P., and L. M. Chiappe. 1999. The wing of *Archaeopteryx* as a primary thrust generator. *Nature* 399:60–62.

Burke, A. C., and A. Feduccia. 1997. Developmental patterns and the identification of homologies in the avian hand. *Science* 278:666–668.

Burley, N. 1979. The evolution of concealed ovulation. *American Naturalist* 114:835–858.

Burling, R. 1993a. On the evolution of language. *Current Anthropology* 34:168–170.

Burling, R. 1993b. Primate calls, human language, and nonverbal communication. *Current Anthropology* 34:25–53.

Burns, E. M. 1963. *Western Civilizations: Their History and Their Culture.* 6th ed. Norton, New York.

Burton, F. B. 1971. Sexual climax in female *Macaca mulatta*. *Proceedings of the Third International Congress of Primatology* 3:180–191.

Burton, R. 1975. Why do the Trobriands have chiefs? *Man* 10:544–558.

Busck, A. 1912. On the rearing of a *Dermatobia hominis* Linnaeus. *Proceedings of the Entomological Society of Washington* 14:9–12.

Bush, G. L. 1969. Sympatric host race formation and speciation in frugivorous flies of the genus *Rhagoletis* (Diptera, Tephritidae). *Evolution* 23:237–251.

Buss, D. M. 1989. Sex differences in human mate preferences: Evolutionary hypotheses tested in thirty-seven cultures. *Behavioral and Brain Sciences* 12:1–49.

Buss, D. M. 1994. *The Evolution of Desire*. Basic Books, New York.

Buss, D. M., M. Abbott, A. Angleitner, A. Asherian, et al. 1990. International preferences in selecting mates: A study of thirty-seven cultures. *Journal of Cross-Cultural Psychology* 21:5–47.

Buss, D. M., R. J. Larsen, D. Westen, and J. Semmelroth. 1992. Sex differences in jealousy. *Psychological Science* 3:251–255.

Buss, D. M., and D. P. Schmitt. 1993. Sexual strategies theory: An evolutionary perspective on human mating. *Psychological Review* 100:204–232.

Busvine, J. R. 1948. The "head" and "body" races of *Pediculus humanus* L. *Parasitology* 39:1–16.

Buzzati-Traverso, A. 1955. Evolutionary changes in components of fitness and other polygenic traits in *Drosophila melanogaster* populations. *Heredity* 9:153–186.

Byers, A. M. 1994. Symboling and the Middle-Upper Palaeolithic transition: A theoretical and methodological critique. *Current Anthropology* 35:369–399.

Bygott, J. D. 1972. Cannibalism among wild chimpanzees. *Nature* 238:410–411.

Bygott, J. D. 1979. Agonistic behavior, dominance, and social structure in wild chimpanzees of the Gombe National Park. Pp. 405–428 in D. A. Hamburg and E. R. McCown, eds., *The Great Apes*. Benjamin-Cummings, Menlo Park, CA.

Byrne, R. W. 1994. The evolution of intelligence. Pp. 223–265 in P. J. B. Slater and T. R. Halliday, eds., *Behaviour and Evolution*. Cambridge Univ. Press, Cambridge.

Byrne, R. W. 1996. Machiavellian intelligence. *Evolutionary Anthropology* 5:172–180.

Byrne, R. W., and A. Whiten. 1988. *Machiavellian Intelligence: Social Expertise and the Evolution of Intellect in Monkeys, Apes, and Humans*. Clarendon Press, Oxford.

Byrne, R. W., and A. Whiten. 1992. Cognitive evolution in primates: Evidence from tactical deception. *Man* 27:609–627.

Cabanac, M. 1979. Sensory pleasure. *Quarterly Review of Biology* 54:1–29.

Caccone, A., R. DeSalle, and J. R. Powel. 1988. Calibration of the change in thermal stability of DNA duplexes and degree of base pair mismatch. *Journal of Molecular Evolution* 27:212–216.

Cachel, S. 1994. The natural history origin of human intelligence. *Social Neuroscience Bulletin* 7:25–30.

Cain, A. J. 1989. Persistence and extinction in some *Cepaea* populations. *Biological Journal of the Linnean Society* 38:183–190.

Cain, A. J., and P. M. Sheppard. 1954. Natural selection in *Cepaea*. *Genetics* 39:89–116.

Cairns, J., Jr., and J. R. Bidwell. 1996. Discontinuities in technological and natural systems caused by exotic species. *Biodiversity and Conservation* 5:1085–1094.

Cairns, J., J. Overbaugh, and S. Miller. 1988. The origin of mutants. *Nature* 335:142–145.

Calabrese, J. R., M. A. Kling, and P. Gold. 1987. Alterations in immunocompetence during stress, bereavement, and depression: Focus on neuroendocrine regulation. *American Journal of Psychiatry* 144:1123–1134.

Caldwell, L. K. 1997. Implications of a world economy for environmental policy and law. Pp. 220–237 in P. Dasgupta, K. Mäler, and A. Vercelli, eds., *The Economics of Transnational Commons*. Clarendon Press, Oxford.

Calvin, W. H. 1987. The brain as a Darwin machine. *Nature* 330:33–34.

Calvin, W. H. 1993. The unitary hypothesis: A common neural circuitry for novel manipulations, language, plan-ahead, and throwing. Pp. 230–250 in K. R. Gibson and T. Ingold, eds., *Tools, Language, and Cognition in Human Evolution*. Cambridge Univ. Press, Cambridge.

Calvin, W. H. 1995. Cortical columns, modules, and Hebbian cell assemblies. Pp. 269–272 in M. Arbib, ed., *Handbook of Brain Theory and Neural Networks*. MIT Press, Cambridge.

Calvin, W. H. 1996a. *The Cerebral Code*. MIT Press, Cambridge.

Calvin, W. H. 1996b. *How Brains Think: Evolving Intelligence, Then and Now*. Basic Books, New York.

Calvin, W. H., and D. Bickerton. 2000. *Lingua ex Machina: Reconciling Darwin and Chomsky with the Human Brain*. MIT Press, Cambridge.

Calvin, W. H., and G. A. Ojemann. 1994. *Conversations with Neil's Brain.* Addison-Wesley, Reading, MA.
Cameron, R. A. D. 1992. Change and stability in *Cepaea* populations over twenty-five years: A case of climatic selection. *Proceedings of the Royal Society of London,* ser. B, 248:181–187.
Camin, J. H., and P. R. Ehrlich. 1958. Natural selection in water snakes (*Natrix sipedon* L.) on islands in Lake Erie. *Evolution* 12:504–511.
Camin, J. H., C. Triplehorn, and H. Walter. 1954. Some indications of survival value in the type "A" pattern of the island water snakes in Lake Erie. *Chicago Academy of Sciences, Natural History Miscellanea* 131:1–3.
Campbell, D. T. 1979. Comments on the sociobiology of ethics and moralizing. *Behavioral Science* 24:37–45.
Canada, G. 1995. *Fist Stick Knife Gun.* Beacon Press, Boston.
Cann, R. L., M. Stoneking, and A. C. Wilson. 1987. Mitochondrial DNA and human evolution. *Nature* 325:31–36.
Cantino, P. D., H. N. Bryant, K. De Queiroz, M. J. Donoghue, et al. 1999. Species names in phylogenetic nomenclature. *Systematic Biology* 48:790–807.
Carey, S. 1998. Knowledge of number: Its evolution and ontogeny. *Science* 282:641–642.
Carlquist, S. 1974. *Island Biology.* Columbia Univ. Press, New York.
Carneiro, R. L. 1961. Slash and burn cultivation among the Kuikuru and its implications for cultural development in the Amazon Basin. Pp. 47–68 in J. Wilbert, ed., *The Evolution of Horticultural Systems in Native South America: Causes and Consequences.* Editorial Sucre, Caracas.
Carneiro, R. L. 1970. A theory of the origin of the state. *Science* 169:733–738.
Carneiro, R. L. 1981. The chiefdom: Precursor of the state. Pp. 37–79 in G. D. Jones and R. R. Kautz, eds., *The Transition to Statehood in the New World.* Cambridge Univ. Press, Cambridge.
Caro, T. M. 1992. Is there teaching in nonhuman animals? *Quarterly Review of Biology* 67:151–174.
Carrier, J. M. 1977. "Sex-role preference" as an explanatory variable in homosexual behavior. *Archives of Sexual Behavior* 6:53–65.
Carrington, M., G. W. Nelson, M. P. Martin, T. Kissner, et al. 1999. HLA and HIV-1: Heterozygote advantage and B*35-Cw*04 disadvantage. *Science* 283:1748–1752.
Carrol, J. B. 1997. Theoretical and technical issues in identifying a factor of general intelligence. Pp. 125–156 in B. Devlin, S. E. Fienberg, D. P. Resnick, and K. Roeder, eds., *Intelligence, Genes, and Success: Scientists Respond to "The Bell Curve."* Springer-Verlag, New York.
Carroll, J. 1999. The Holocaust and the Catholic Church. *Atlantic Monthly* (October): 107–112.
Cartmill, M. 1972. Arboreal adaptations and the origin in the order Primates. Pp. 97–212 in R. Tuttle, ed., *The Functional and Evolutionary Biology of Primates.* Aldine-Atherton, Chicago.
Cashdan, E. 1989. Hunters and gatherers: Economic behavior in bands. Pp. 21–48 in S. Plattner, ed., *Economic Anthropology.* Stanford Univ. Press, Stanford.
Castiglione, C. M., A. S. Deinard, W. C. Speed, G. Sirugo, et al. 1995. Evolution of haplotypes at the DRD2 locus. *American Journal of Human Genetics* 57:1445–1456.
Catherwood, C. 1997. *Why Nations Rage: Killing in the Name of God.* Hodder and Stoughton, London.
Cavalli-Sforza, L. L. 1986. Cultural evolution. *American Zoologist* 26:845–855.
Cavalli-Sforza, L. L., and W. Bodmer. 1971. *The Genetics of Human Populations.* Freeman, San Francisco.
Cavalli-Sforza, L. L., and M. W. Feldman. 1973. Cultural versus biological inheritance: Phenotypic transmission from parent to children (a theory of the effect of parental phenotypes on children's phenotype). *American Journal of Human Genetics* 25:618–637.
Cavalli-Sforza, L. L., and M. W. Feldman. 1981. *Cultural Transmission and Evolution: A Quantitative Approach.* Princeton Univ. Press, Princeton.
Cavalli-Sforza, L. L., P. Menozzi, and A. Piazza. 1994. *The History and Geography of Human Genes.* Princeton Univ. Press, Princeton.
Cavalli-Sforza, L. L., A. Piazza, P. Menozzi, and J. Mountain. 1988. Reconstruction of human evolution: Bringing together genetic, archaeological, and linguistic data. *Proceedings of the National Academy of Sciences USA* 85:6002–6006.
Chagnon, N. A. 1979. Kin selection and conflict: An analysis of a Yanomamö ax fight. Pp. 213–238 in N. A. Chagnon and W. Irons, eds., *Evolutionary Biology and Human Social Behavior: An Anthropological Perspective.* Duxbury Press, North Scituate, MA.
Chagnon, N. A. 1983. *Yanomamö: The Fierce People.* 3rd ed. Holt, Rinehart and Winston, New York.
Chagnon, N. A. 1988. Life histories, blood revenge, and warfare in a tribal population. *Science* 239:985–992.
Chagnon, N. A. 1992. *Yanomamö: The Last Days of Eden.* Harcourt, Brace, San Diego.
Chalk, F., and K. Jonassohn. 1990. *The History and Sociology of Genocide: Analyses and Case Studies.* Yale Univ. Press, New Haven.

Chance, M. 1967. Attention structure as the basis of primate rank orders. *Man* 2:503–518.

Chance, M. 1975. Social cohesion and the structure of attention. Pp. 93–113 in R. Fox, ed., *Biosocial Anthropology*. Halsted Press, New York.

Chance, M. R. A. 1962. Natural and special features of the instinctive social bond of primates. Pp. 17–33 in S. L. Wasburn, ed., *Social Life of Early Man*. Methuen, London.

Chandler, D. P. 1999. *Brother Number One: A Political Biography of Pol Pot*. Westview Press, Boulder.

Chang, H.-W., and J. M. Emlen. 1993. Seasonal variation of microhabitat distribution of the polymorphic land snail *Cepaea nemoralis*. *Oecologia* 93:501–507.

Chang, I. 1997. *The Rape of Nanking: The Forgotten Holocaust of World War II*. Basic Books, New York.

Chang, K.-C., and W. H. Goodenough. 1996. Archaeology of southeastern coastal China and its bearing on the Austronesian homeland. Pp. 36–56 in W. H. Goodenough, ed., *Prehistoric Settlement of the Pacific*. American Philosophical Society, Philadelphia.

Changeux, J. P., and J. Chavaillon, eds. 1995. *Origins of the Human Brain*. Oxford Univ. Press, Oxford.

Chang-Qun, D., G. Xue-Chun, J. Wang, and P. K. Chien. 1998. Relocation of civilization centers in ancient China: Environmental factors. *Ambio* 27:572–575.

Chaplin, G. 1994. Physiology, thermoregulation, and bipedalism. *Journal of Human Evolution* 27:497–510.

Charlesworth, B. 1994. *Evolution in Age-Structured Populations*. 2nd ed. Cambridge Univ. Press, Cambridge.

Charlesworth, B., R. Lande, and M. Slatkin. 1982. A neo-Darwinian commentary on macroevolution. *Evolution* 36:474–498.

Chase, P. G. 1988. Scavenging and hunting in the Middle Paleolithic: The evidence from Europe. Pp. 225–232 in H. L. Dibble and A. Montet-White, eds., *Upper Pleistocene Prehistory of Western Asia*. Univ. of Pennsylvania Museum, Philadelphia.

Chase, P. G., and H. L. Dibble. 1987. Middle Paleolithic symbolism: A review of current evidence and interpretations. *Journal of Anthropological Archaeology* 6:263–296.

Chatterjee, P. 1986. *Nationalist Thought and the Colonial World: A Derivative Discourse?* Univ. of Minnesota Press, Minneapolis.

Chatterjee, S. 1997. *The Rise of the Birds*. Johns Hopkins Univ. Press, Baltimore.

Chebloune, Y., J. Pagnier, G. Trabuchet, C. Faure, et al. 1988. Structural analysis of the 5′ flanking region of the BX-globin gene in African sickle cell anemia patients: Further evidence for three origins of the sickle cell mutation in Africa. *Proceedings of the National Academy of Sciences USA* 85:4431–4435.

Cheney, D. L., and R. M. Seyfarth. 1990a. *How Monkeys See the World: Inside the Mind of Another Species*. Univ. of Chicago Press, Chicago.

Cheney, D. L., and R. M. Seyfarth. 1990b. The representation of social relations by monkeys. *Cognition* 37:167–196.

Chichilnisky, G. 1997. The knowledge revolution. *Journal of International Trade and Economic Development* 7:39–54.

Child, A., and J. Child. 1985. Biology, ethnocentrism, and sex differences. *American Anthropologist* 87:125–128.

Childe, V. G. 1936. *Man Makes Himself*. Watts, London.

Chomsky, N. 1957. *Syntactic Structures*. Mouton, The Hague.

Chomsky, N. 1971. *Problems of Knowledge and Freedom*. Pantheon Books, New York.

Chomsky, N. 1972. *Language and Mind*. Harcourt, Brace, New York.

Chomsky, N. 1975. *Reflections on Language*. Pantheon Books, New York.

Chomsky, N. 1980. *Rules and Representations*. Columbia Univ. Press, New York.

Chomsky, N. 1988. *Language and the Problems of Knowledge*. MIT Press, Cambridge.

Choucri, N., and R. C. North. 1975. *Nations in Conflict: National Growth and International Violence*. Freeman, San Francisco.

Churchill, S. E. 1993. Weapon technology, prey size selection, and hunting methods in modern hunter-gatherers: Implications for hunting in the Palaeolithic and Mesolithic. Pp. 12–24 in G. L. Peterkin, H. M. Bricker, and P. Mellars, eds., *Hunting and Animal Exploitation in the Later Palaeolithic and Mesolithic of Eurasia*. American Anthropological Association, Washington, DC.

Churchill, S. E. 1998. Cold adaptation, heterochrony, and Neandertals. *Evolutionary Anthropology* 7:46–61.

Churchland, P. S., and T. J. Sejnowski. 1992. *The Computational Brain*. MIT Press, Cambridge.

Ciochon, R. L., and D. A. Etler. 1994. Reinterpreting past primate diversity. Pp. 37–67 in R. S. Corruccini and R. L. Ciochon, eds., *Integrative Paths to the Past: Paleoanthropological Advances in Honor of F. Clark Howell*. Prentice Hall, Englewood Cliffs, NJ.

Clad, J. 1984. Conservation and indigenous peoples: A study of convergent interests. *Cultural Survival Quarterly* 8:68–73.

Clark, A. G. 1987. Genetic correlations: The quantitative genetics of evolutionary constraints. Pp. 25–45 in V. Loeschcke, ed., *Genetic Constraints on Adaptive Evolution.* Springer-Verlag, Berlin.

Clark, G. A. 1994. Comments. *Current Anthropology* 35:382.

Clark, J. D., and J. W. K. Harris. 1985. Fire and its roles in early hominid lifeways. *African Archaeological Review* 3:3–27.

Clark, R. D., and E. Hatfield. 1989. Gender differences in receptivity to sexual offers. *Journal of Psychology and Human Sexuality* 2:39–55.

Clarke, C. A., F. M. M. Clarke, and H. C. Dawkins. 1990. *Biston betularia* (the peppered moth) in West Kirby, Wirral, 1959–1989: Updating the decline in f. *carbonaria. Biological Journal of the Linnean Society* 39:323–326.

Clarke, R. J. 1996. The genus *Paranthropus:* What's in a name? Pp. 93–104 in W. E. Meikle, F. C. Howell, and N. G. Jablonski, eds., *Contemporary Issues in Human Evolution.* California Academy of Sciences, San Francisco.

Clore, G. R. 1994. Why emotions require cognition. Pp. 181–191 in P. Ekman and R. J. Davidson, eds., *The Nature of Emotions: Fundamental Questions.* Oxford Univ. Press, New York.

Clutton-Brock, J. 1992. How the wild beasts were tamed. *New Scientist* (15 February): 41–43.

Clutton-Brock, T. H., and P. H. Harvey. 1980. Primates, brains, and ecology. *Journal of Zoology (London)* 190:309–323.

Cockburn, T. A. 1971. Infectious diseases in ancient populations. *Current Anthropology* 12:45–62.

Cody, M. L., and J. M. Overton. 1996. Short-term evolution of reduced dispersal in island plant populations. *Journal of Ecology* 84:53–61.

Coe, M. D. 1981. Religion and the rise of Mesoamerican states. Pp. 157–171 in G. D. Jones and R. R. Kautz, eds., *The Transition to Statehood in the New World.* Cambridge Univ. Press, Cambridge.

Cohen, A., L. S. Kaufman, and R. Ogutu-Ohwayo. 1996. Anthropogenic impacts and conservation efforts on the African Great Lakes: A review. Pp. 575–624 in T. C. Johnson and E. O. Odada, eds., *The Limnology, Climatology, and Palaeoclimatology of the East African Lakes.* Gordon and Breach, Amsterdam.

Cohen, M. N. 1977a. *The Food Crisis in Prehistory: Overpopulation and the Origin of Agriculture.* Yale Univ. Press, New Haven.

Cohen, M. N. 1977b. Population pressure and the origins of agriculture: An archaeological example from the coast of Peru. Pp. 135–177 in C. A. Reed, ed., *Origins of Agriculture.* Mouton, The Hague.

Cohen, M. N. 1978. The disappearance of the incest taboo. *Human Nature* 1:72–78.

Cohen, M. N. 1989. *Health and the Rise of Civilization.* Yale Univ. Press, New Haven.

Cohen, M. N. 1994. Demographic expansion: Causes and consequences. Pp. 265–296 in T. Ingold, ed., *Companion Encyclopedia of Anthropology.* Routledge, London.

Cohen, M. N., and G. G. Armelagos, eds. 1984. *Paleopathology at the Origins of Agriculture.* Academic Press, Orlando.

Cohen, N. 1967. *Warrant for Genocide.* Harper and Row, New York.

Cohen, P. 1999. Face values. *New Scientist* (19 June): 19.

Colapinto, J. 1997. The true story of John/Joan. *Rolling Stone* (11 December): 55–97.

Colborn, T., D. Dumanoski, and J. P. Myers. 1996. *Our Stolen Future.* Dutton, New York.

Comstock, G., and H. Paik. 1991. *Television and the American Child.* Academic Press, San Diego.

Conkey, M. W., O. Soffer, D. Stratmann, and N. G. Jablonski, eds. 1997. *Beyond Art: Pleistocene Image and Symbol.* California Academy of Sciences, San Francisco.

Conrad, G. W., and A. A. Demarest. 1984. *Religion and Empire: The Dynamics of Aztec and Inca Expansionism.* Cambridge Univ. Press, Cambridge.

Conroy, G. C. 1997. *Reconstructing Human Origins: A Modern Synthesis.* Norton, New York.

Conroy, G. C., G. W. Weber, H. Seidler, P. W. Tobias, et al. 1998. Endocranial capacity in an early hominid cranium from Sterkfontein, South Africa. *Science* 280:1730–1731.

Cook, L. M., and G. Gao. 1996. Test of association of morphological variation with heterozygosity in the snail *Cepaea nemoralis. Heredity* 76:118–123.

Coon, C. S. 1962. *The Origin of Races.* Knopf, New York.

Coon, C. S., S. M. Garn, and J. D. Birdsell. 1950. *Races: A Study of the Problems of Race Formation in Man.* Thomas, Springfield, IL.

Cooper, D. E., and J. A. Palmer, eds. 1992. *The Environment in Question: Ethics and Global Issues.* Routledge, London.

Copeland, D. M. 1960. *Remember, Nurse.* Ryerson Press, Toronto.

Copleston, F. 1964. *A History of Philosophy*. Vol. 6, *Modern Philosophy from the French Enlightenment to Kant.* Doubleday, New York.

Coppens, Y. 1994. East Side story: The origin of humankind. *Scientific American* (May): 88–95.

Coppens, Y., and B. Senut, eds. 1991. *Origine(s) de la Bipédie chez les Hominidés.* Éditions du CNRS, Paris.

Corballis, M. C. 1992. On the evolution of language and generativity. *Cognition* 44:197–226.

Corballis, M. C. 1994. Comments. *Current Anthropology* 35:360.

Corballis, M. C. 1999. The gestural origins of language. *American Scientist* 87:138–145.

Coren, S. 1994. *The Intelligence of Dogs.* Free Press, New York.

Cornwell, J. 1999. *Hitler's Pope: The Secret History of Pius XII.* Viking, New York.

Costanza, R. 1987. Social traps and environmental policy. *BioScience* 37:407–412.

Cowan, C. W., and P. J. Watson, eds. 1992. *The Origins of Agriculture: An International Perspective.* Smithsonian Institution Press, Washington, DC.

Cowey, A., and P. Stoerig. 1991. The neurobiology of blindsight. *Trends in Neuroscience* 14:140–145.

Cowgill, G. L. 1975. On causes and consequences of ancient and modern population changes. *American Anthropologist* 77:505–525.

Cowie, R. H., and J. S. Jones. 1987. Ecological interactions between *Cepaea nemoralis* and *Cepaea hortensis:* Competition, invasion, but no niche displacement. *Functional Ecology* 1:91–97.

Cowles, R. B. 1959. Some ecological factors bearing on the origin and evolution of pigment in the human skin. *American Naturalist* 93:283–293.

Coyne, J. A. 1998. Not black and white. *Nature* 396:35–36.

Cracraft, J. 1983. Species concepts and speciation analysis. *Current Ornithology* 1:159–187.

Craik, J. C. A. 1989. The Gaia hypothesis: Fact or fancy? *Journal of the Marine Biological Association of the United Kingdom* 69:759–768.

Crain, S. 1991. Language acquisition in the absence of experience. *Behavioral and Brain Sciences* 14:597–650.

Crain, S., and C. McKee. 1985. Acquisition of structural restrictions on anaphora. *Proceedings of the North Eastern Linguistic Society* 16:94–110.

Crain, S., and M. Nakayama. 1986. Structure dependence in grammar formation. *Language* 63:522–543.

Crain, S., and R. Thornton. 1991. Recharting the course of language acquisition. Pp. 321–337 in N. A. Krasnegor, D. M. Rumbaugh, R. L. Schiefelbusch, and M. Studdert-Kennedy, eds., *Biological and Behavioral Determinants of Language Development.* Erlbaum, Hillsdale, NJ.

Crawford, M. P. 1937. The cooperative solving of problems by young chimpanzees. *Comparative Psychology Monographs* 14:1–88.

Crawley, M. J. 1983. *Herbivory: The Dynamics of Animal-Planet Interactions.* Univ. of California Press, Berkeley.

Crick, F. 1994. *The Astonishing Hypothesis: The Scientific Search for the Soul.* Simon and Schuster, New York.

Crick, F., and C. Koch. 1992. The problem of consciousness. *Scientific American* 267:111–117.

Crocker, J. 1981. Judgment of covariation by social perceivers. *Psychological Bulletin* 90:272–292.

Cronin, T. 1989. *Direct Democracy.* Harvard Univ. Press, Cambridge.

Cronin, T. E. 1998. *The Paradoxes of the American Presidency.* Oxford Univ. Press, New York.

Crook, J. H. 1966. Gelada baboon herd structure and movement: A comparative report. *Symposia of the Zoological Society of London* 18:237–258.

Crosby, A. 1986. *Ecological Imperialism: The Biological Expansion of Europe, 900–1900.* Cambridge Univ. Press, Cambridge.

Crumlin, R., and A. Knight. 1991. *Aboriginal Art and Spirituality.* HarperCollins, North Blackburn, Victoria.

Culbert, T. P., ed. 1973. *The Classic Maya Collapse.* Univ. of New Mexico Press, Albuquerque.

Culotta, E. 1999. Neanderthals were cannibals. *Science* 286:18–19.

Cytowic, R. E. 1993. *The Man Who Tasted Shapes.* MIT Press, Bradford Books, Cambridge.

Dahl, R. A. 1989. *Democracy and Its Critics.* Yale Univ. Press, New Haven.

Dahl, R. A. 1998. *On Democracy.* Yale Univ. Press, New Haven.

Daily, G. C., ed. 1997. *Nature's Services: Societal Dependence on Natural Ecosystems.* Island Press, Washington, DC.

Daily, G. C., P. Dasgupta, B. Bolin, P. Crosson, et al. 1998. Food production, population growth, and the environment. *Science* 281:1291–1292.

Daily, G. C., A. H. Ehrlich, and P. R. Ehrlich. 1994. Optimum human population size. *Population and Environment* 15:469–475.

Daily, G. C., and P. R. Ehrlich. 1990. An exploratory model of the impact of rapid climate change on the world food situation. *Proceedings of the Royal Society of London,* ser. B, 241:232–244.

Daily, G. C., and P. R. Ehrlich. 1992. Population, sustainability, and Earth's carrying capacity. *BioScience* 42:761–771.

Daily, G. C., and P. R. Ehrlich. 1996a. Global change and human susceptibility to disease. *Annual Review of Energy and the Environment* 21:125–144.

Daily, G. C., and P. R. Ehrlich. 1996b. Impacts of development and global change on the epidemiological environment. *Environment and Development Economics* 1:309–344.

Daily, G. C., and P. R. Ehrlich. 1996c. Socioeconomic equity, sustainability, and Earth's carrying capacity. *Ecological Applications* 6:991–1001.

Daily, G. C., and P. R. Ehrlich. 1999. Managing Earth's ecosystems: An interdisciplinary challenge. *Ecosystems* 2:277–280.

Dalrymple, D. 1976. *Development and Spread of High-Yielding Varieties of Wheat and Rice in the Less Developed Nations.* U.S. Department of Agriculture, Foreign Agricultural Service, Washington, DC.

D'Altroy, T., and T. Earle. 1985. Staple finance, wealth finance, and storage in the Inca political economy. *Current Anthropology* 26:187–206.

Daly, H. E., ed. 1973. *Toward a Steady-State Economy.* Freeman, San Francisco.

Daly, H. E., and J. B. Cobb Jr. 1989. *For the Common Good: Redirecting the Economy toward Community, the Environment, and a Sustainable Future.* Beacon Press, Boston.

Daly, M., and M. Wilson. 1984. A sociobiological analysis of human infanticide. Pp. 487–502 in G. Hausfater and S. B. Hrdy, eds., *Infanticide: Comparative and Evolutionary Perspectives.* Aldine, New York.

Damasio, A. R. 1989. The brain binds entities and events by multiregional activation from convergence zones. *Neural Computation* 1:123–132.

Damasio, A. R. 1994. *Descartes' Error: Emotion, Reason, and the Human Brain.* Avon Books, New York.

Damasio, A. R. 1999. *The Feeling of What Happens: Body and Emotion in the Making of Consciousness.* Harcourt, Brace, New York.

Damasio, A. R., and H. Damasio. 1992. Brain and language. *Scientific American* 267:89–95.

Damasio, A. R., and H. Damasio. 1993. Cortical systems underlying knowledge retrieval: Evidence from human lesion studies. Pp. 233–248 in T. A. Poggio and D. A. Glaser, eds., *Exploring Brain Functions: Models in Neuroscience.* Wiley, New York.

Damasio, H., T. Grabowski, R. Frank, A. M. Galaburda, et al. 1994. The return of Phineas Gage: Clues about the brain from the skull of a famous patient. *Science* 264:1102–1105.

Daniels, N. 1973. The smart white man's burden. *Harper's* 244:25.

D'Aquili, E. G., and C. D. Laughlin Jr. 1975. The biopsychological determinants of religious ritual behavior. *Zygon* 10:32–58.

D'Aquili, E. G., and C. D. Laughlin Jr. 1979. The neurobiology of myth and ritual. Pp. 152–182 in E. G. d'Aquili et al., eds., *The Spectrum of Ritual: Structural Analysis.* Columbia Univ. Press, New York.

Darlington, C. D. 1939. *The Evolution of Genetic Systems.* Cambridge Univ. Press, Cambridge.

Dart, R. A. 1948. The Makapansgat proto-human *Australopithecus prometheus. American Journal of Physical Anthropology* 6:259–284.

Dart, R. A. 1957. *The Osteodontokeratic Culture of Australopithecus prometheus.* Transvaal Museum Memoirs, Pretoria.

Darwin, C. 1859. *On the Origin of Species.* Murray, London.

Darwin, C. 1871. *The Descent of Man and Selection in Relation to Sex.* Murray, London.

Darwin, C. 1998 (1872). *The Expression of Emotions in Man and Animals.* Oxford Univ. Press, Oxford.

Dasgupta, P. 1993. *An Inquiry into Well-Being and Destitution.* Oxford Univ. Press, Oxford.

Dasgupta, P., and K.-G. Mäler. 1995. Poverty, institutions, and the environmental resource base. Pp. 2371–2463 in J. Behrman and T. Srinivasan, eds., *Handbook of Development Economics.* Elsevier, Amsterdam.

Davenport, W. 1965. Sexual patterns and their regulation in a society of the southwest Pacific. Pp. 164–207 in F. A. Beach, ed., *Sex and Behaviour.* Wiley, New York.

Davidoff, J., I. Davies, and D. Roberson. 1999. Colour categories in a Stone-Age tribe. *Nature* 398:203–204.

Davidson, I. 1991. The archaeology of language origins: A review. *Antiquity* 65:39–48.

Davidson, I., and W. Noble. 1989. The archaeology of perception. *Current Anthropology* 30:125–155.

Davidson, I., and W. Noble. 1993. On the evolution of language. *Current Anthropology* 14:165–166.

Davies, P. C. W., and J. R. Brown, eds. 1986. *The Ghost in the Atom.* Cambridge Univ. Press, Cambridge.

Davies, P. C. W., and J. Gribbin. 1992. *The Matter Myth: Dramatic Discoveries That Challenge Our Understanding of Physical Reality.* Simon and Schuster, New York.

Davis, J. I. 1996. Phylogenetics, molecular variation, and species concepts. *BioScience* 46:502–511.

Davis, K. 1956. The amazing decline of mortality in underdeveloped areas. *American Economic Review* 46:305–318.

Dawkins, R. 1981. In defence of selfish genes. *Philosophy* 56:556–573.

Dawkins, R. 1982. *The Extended Phenotype: The Long Reach of the Gene.* Oxford Univ. Press, Oxford.

Dawkins, R. 1986. *The Blind Watchmaker: Why the Evidence of Evolution Reveals a Universe without Design.* Norton, New York.

Dawkins, R. 1989 (1976). *The Selfish Gene.* Oxford Univ. Press, Oxford.

Dawkins, R. 1997. Human chauvinism. *Evolution* 51:1015–1020.

Dayton, L. 1999. Aging fast: An early Australian just got 30,000 years older. *New Scientist* (29 May): 13.

Deacon, T. W. 1997. *The Symbolic Species: The Coevolution of Language and the Brain.* Norton, New York.

Dean, D., and E. Delson. 1995. *Homo* at the gates of Europe. *Nature* 373:472–473.

Dean, D., J.-J. Hublin, R. Ziegler, and R. Holloway. 1994. The Middle-Pleistocene pre-Neanderthal partial skull from Reilingen (Germany). *American Journal of Physical Anthropology* 18 (supp.): 77.

Defleur, A., T. White, and P. Valensi. 1999. Neanderthal cannibalism at Moula-Guercy, Ardàche, France. *Science* 286:128–131.

Degler, C. N. 1971. *Neither Black nor White.* Macmillan, New York.

Degler, C. N. 1991. *In Search of Human Nature: The Decline and Revival of Darwinism in American Social Thought.* Oxford Univ. Press, New York.

Deichmann, U. 1996. *Biologists under Hitler.* Harvard Univ. Press, Cambridge.

DeLoukas, P., G. D. Schuler, G. Gyapay, E. M. Beasley, et al. 1998. A physical map of thirty thousand human genes. *Science* 282:744–746.

Delporte, H. 1995. Man's intelligence as seen through Palaeolithic art. Pp. 200–210 in J. P. Changeux and J. Chavaillon, eds., *Origins of the Human Brain.* Oxford Univ. Press, Oxford.

Delson, E. 1997. One skull does not a species make. *Nature* 389:445–446.

Demsetz, H. 1967. Toward a theory of property rights. *American Economic Review* 57:347–359.

Denevan, W. L., ed. 1992. *The Native Population of the Americas.* 2nd ed. Univ. of Wisconsin Press, Madison.

Dennett, D. C. 1983. Intentional systems in cognitive ethology: The "Panglossian paradigm" defended. *Behavioral and Brain Sciences* 6:343–390.

Dennett, D. C. 1991. *Consciousness Explained.* Little, Brown, Boston.

Dennett, D. C. 1995. *Darwin's Dangerous Idea: Evolution and the Meanings of Life.* Simon and Schuster, New York.

Dennett, D. C. 1996. *Kinds of Minds: Toward an Understanding of Consciousness.* Basic Books, New York.

Dentan, R. K. 1968. *The Semai: A Nonviolent People of Malaya.* Holt, Rinehart and Winston, New York.

De Queiroz, K., and M. J. Donoghue. 1988. Phylogenetic systematics and the species problem. *Cladistics* 4:317–338.

Derr, B. W. 1982. Implications of menstruation as a liminal state. *American Anthropologist* 84:644–645.

D'Errico, F. 1995. A new model and its implications for the origin of writing: The La Marche antler revisited. *Cambridge Archaeological Journal* 5:163–206.

D'Errico, F., and C. Cacho. 1994. Notation versus decoration in the Upper Paleolithic: A case-study from Tossal de la Roca, Alicante, Spain. *Journal of Archaeological Science* 21:185–200.

D'Errico, F., and P. Villa. 1997. Holes and grooves: The contribution of microscopy and taphonomy to the problem of art origins. *Journal of Human Evolution* 33:1–31.

D'Errico, F., P. Villa, A. C. Pinto Llona, and R. Ruiz Idarraga. 1998. A Middle Palaeolithic origin of music? Using cave-bear bone accumulations to assess the Divje Babe I bone "flute." *Antiquity* 72:65–79.

D'Errico, F., J. Zilhão, M. Julien, D. Baffier, et al. 1998. Neanderthal acculturation in western Europe? A critical review of the evidence and its interpretation. *Current Anthropology* 39:S1–S44.

Derrida, J. 1976. *Of Grammatology.* Johns Hopkins Univ. Press, Baltimore.

Descartes, R. 1970 (1637). *The Philosophical Works of Descartes.* Vol. 1. Cambridge Univ. Press, New York.

Desowitz, R. S. 1981. *New Guinea Tapeworms and Jewish Grandmothers.* Norton, New York.

Devall, B. 1980. The deep ecology movement. *Natural Resource Journal* 20:299–322.

Devall, B., and G. Sessions. 1985. *Deep Ecology: Living as if Nature Mattered.* Peregrine Smith Books, Salt Lake City.

DeVore, I., and K. R. L. Hall. 1965. Baboon ecology. Pp. 20–52 in I. DeVore, ed., *Primate Behavior: Field Studies of Monkeys and Apes.* Holt, Rinehart and Winston, New York.

De Vos, J. D., and P. Sondaar. 1994. Dating hominid sites in Indonesia. *Science* 266:1726–1727.

Dewey, J. 1988. *Reconstruction in Philosophy.* Vol. 12, *1920, Essays.* Southern Illinois Univ. Press, Carbondale.

Dewhirst, J. R., and J. S. Berman. 1978. Social judgments of spurious and causal relations between attributes and outcomes. *Journal of Experimental Social Psychology* 14:313–325.

Dewsbury, D. A. 1984. Sperm competition in muroid rodents. Pp. 547–571 in R. L. Smith, ed., *Sperm Competition and the Evolution of Animal Mating Systems.* Academic Press, Orlando.

Diamond, J. M. 1978. The Tasmanians: The longest isolation, the simplest technology. *Nature* 273:185–186.

Diamond, J. M. 1984. Historic extinctions: A Rosetta stone for understanding prehistoric extinctions. Pp. 824–862 in P. S. Martin and R. D. Klein, eds., *Quaternary Extinctions: A Prehistoric Revolution.* Univ. of Arizona Press, Tucson.

Diamond, J. M. 1989a. Blood, genes, and malaria. *Natural History* (February): 8–18.

Diamond, J. M. 1989b. The Great Leap Forward. *Discover* 10:50–60.

Diamond, J. M. 1991. *The Rise and Fall of the Third Chimpanzee.* Radius, London.

Diamond, J. M. 1992. Diabetes running wild. *Nature* 357:362–363.

Diamond, J. M. 1993. New Guineans and their natural world. Pp. 251–271 in S. R. Kellert and E. O. Wilson, eds., *The Biophilia Hypothesis.* Island Press, Washington, DC.

Diamond, J. M. 1995. Easter's end. *Discover* (August): 63–69.

Diamond, J. M. 1997a. *Guns, Germs, and Steel: The Fates of Human Societies.* Norton, New York.

Diamond, J. M. 1997b. *Why Is Sex Fun? The Evolution of Human Sexuality.* Basic Books, New York.

Diamond, J. M., and J. Rotter. 2000. The evolution of human genetic diseases. In R. A. King, J. I. Rotter, and A. G. Motulsky, eds., *The Genetic Basis of Common Diseases.* 2nd ed. Oxford Univ. Press, Oxford. In press.

Diamond, M. 1993. Homosexuality and bisexuality in different populations. *Archives of Sexual Behavior* 22:291–310.

Diamond, M., and H. K. Sigmundson. 1997. Sex reassignment at birth: Long-term review and clinical implications. *Archives of Pediatrics and Adolescent Medicine* 151:298–304.

Dibble, H. L. 1989. The implications of stone tool types for the presence of language during the Lower and Middle Palaeolithic. Pp. 415–431 in P. Mellars and C. Stringer, eds., *The Human Revolution: Behavioural and Biological Perspectives on the Origins of Modern Humans.* Edinburgh Univ. Press, Edinburgh.

Dickeman, M. 1975. Demographic consequences of infanticide in man. *Annual Review of Ecology and Systematics* 6:107–137.

Dickeman, M. 1979. Female infanticide, reproductive strategies, and social stratification: A preliminary model. Pp. 321–373 in N. Chagnon and W. Irons, eds., *Evolutionary Biology and Human Social Behavior: An Anthropological Perspective.* Duxbury Press, North Scituate, MA.

Dieckmann, U., and M. Doebeli. 1999. On the origin of species by sympatric speciation. *Nature* 400:354–357.

Dingus, L., and T. Rowe. 1997. *The Mistaken Extinction: Dinosaur Evolution and the Origin of Birds.* Freeman, New York.

Dissanayake, E. 1988. *What Is Art For?* Univ. of Washington Press, Seattle.

Dissanayake, E. 1992. *Homo Aestheticus: Where Art Comes from and Why.* Free Press, New York.

Divale, W. T., and M. Harris. 1976. Population, warfare, and the male supremacist complex. *American Anthropologist* 78:521–538.

Dixson, A. F. 1998. *Primate Sexuality: Comparative Studies of the Prosimians, Monkeys, Apes, and Human Beings.* Oxford Univ. Press, Oxford.

Dobbins, A. C., R. M. Jeo, J. Fiser, and J. M. Allman. 1998. Distal modulation of neural activity in the visual cortex. *Science* 281:552–555.

Dobzhansky, T. 1937. *Genetics and the Origin of Species.* Columbia Univ. Press, New York.

Dobzhansky, T. 1951. *Genetics and the Origin of Species.* 3rd ed. Columbia Univ. Press, New York.

Dobzhansky, T. 1970. *Genetics of the Evolutionary Process.* Columbia Univ. Press, New York.

Dollard, J., L. W. Doob, N. E. Miller, O. H. Mowrer, et al. 1939. *Frustration and Aggression.* Yale Univ. Press, New Haven.

Dominey, W. J. 1984. Effects of sexual selection and life history on speciation: Species flocks in African cichlids and Hawaiian *Drosophila.* Pp. 231–249 in A. A. Echelle and I. Kornfeld, eds., *Evolution of Fish Species Flocks.* Univ. of Maine Press, Orono.

Donald, M. 1991. *Origins of the Modern Mind: Three Stages in the Evolution of Culture and Cognition.* Harvard Univ. Press, Cambridge.

Donald, M. 1993. Précis of *Origins of the Modern Mind: Three Stages in the Evolution of Culture and Cognition. Behavioral and Brain Sciences* 16:737–791.

Doolittle, W. F. 1999. Phylogenetic classification and the universal tree. *Science* 284:2124–2128.

Doolittle, W. F. 2000. Uprooting the tree of life. *Scientific American* 282:90–95.

Doran, C. F., and W. Parsons. 1980. War and the cycle of relative power. *American Political Science Review* 74:947–965.

Doran, T. F., C. De Angelis, R. A. Baumgardner, and E. D. Mellits. 1989. Acetaminophen: More harm than good for chickenpox? *Journal of Pediatrics* 114:1045–1048.

Doricchi, F., and C. Incoccia. 1998. Seeing only the right half of the forest but cutting down all the trees. *Nature* 394:75–78.

Dorit, R. L. 1990. The correlates of high diversity in Lake Victoria haplochromine cichlids: A neontological perspective. Pp. 322–353 in R. M. Ross and W. D. Allmon, eds., *Causes of Evolution: A Paleontological Perspective.* Univ. of Chicago Press, Chicago.

Dorit, R. L., H. Akashi, and W. Gilbert. 1995. The absence of polymorphism at the ZFY locus on the human Y chromosome. *Science* 268:1183–1185.

Doty, R. L., M. Ford, G. Preti, and G. R. Huggins. 1975. Changes in the intensity and pleasantness of human vaginal odors during the menstrual cycle. *Science* 190:1316–1318.

Douglas, A. S., and A. S. Bakhshi. 1998. Does vitamin D deficiency account for ethnic differences in tuberculosis seasonality in the U.K.? *Ethnic Health* 3:247–253.

Dower, J. W. 1986. *War without Mercy.* Pantheon Books, New York.

Dower, J. W. 1999. *Embracing Defeat: Japan in the Wake of World War II.* Norton, New York.

Dransfield, J., J. R. Flenley, S. M. King, D. D. Harkness, et al. 1984. A recently extinct palm from Easter Island. *Nature* 312:750–752.

Drury, C. M. 1984. *The History of the Chaplain Corps, United States Navy.* Vol. 2, *1939–1949.* U.S. Government Printing Office, Washington, DC.

Duarte, C., J. Maurício, P. B. Pettitt, P. Souto, et al. 1999. The early Upper Paleolithic human skeleton from the Abrigo do Lagar Velho (Portugal) and modern human emergence in Iberia. *Proceedings of the National Academy of Sciences USA* 96:7604–7609.

Duchin, L. E. 1990. The evolution of articulate speech: Comparative anatomy of the oral cavity in *Pan* and *Homo. Journal of Human Evolution* 19:687–697.

Duckett, J. W., and L. S. Baskin. 1993. Genitoplasty for intersex anomalies. *European Journal of Pediatrics* 152 (supp. 2): 580–584.

Dugatkin, L. A. 1997. The evolution of cooperation. *BioScience* 47:355–362.

Dumond, D. E. 1975. The limitation of human population: A natural history. *Science* 187:713–721.

Dumond, D. E. 1977. *The Eskimos and Aleuts.* Thames and Hudson, London.

Dunbar, R. I. M. 1984. *Reproductive Decisions: An Economic Analysis of Gelada Baboon Social Strategies.* Princeton Univ. Press, Princeton.

Dunbar, R. I. M. 1991. Functional significance of social grooming in primates. *Folia Primatologica* 57:121–131.

Dunbar, R. I. M. 1992. Neocortex size as a constraint on group size in primates. *Journal of Human Evolution* 20:469–493.

Dunbar, R. I. M. 1993. Coevolution of neocortical size, group size, and language in humans. *Behavioral and Brain Sciences* 16:681–735.

Dunbar, R. I. M. 1996. *Grooming, Gossip, and the Evolution of Language.* Harvard Univ. Press, Cambridge.

Dunbar, R. I. M. 1998a. The social brain hypothesis. *Evolutionary Anthropology* 6:178–190.

Dunbar, R. I. M. 1998b. Your cheatin' heart. *New Scientist* 160:29–32.

Dunbar, R. I. M., N. D. C. Duncan, and D. Nettle. 1995. Size structure of freely forming conversational groups. *Human Nature* 6:67–78.

Dunbar, R. I. M., and M. Spoors. 1995. Social networks, support cliques, and kinship. *Human Nature* 6:273–290.

Durham, W. H. 1991. *Coevolution: Genes, Culture, and Human Diversity.* Stanford Univ. Press, Stanford.

Durkheim, E. 1962 (1895). *The Rules of Sociological Method.* Free Press, Glencoe, IL.

Durkheim, E. 1965 (1915). *The Elementary Forms of Religious Life.* Free Press, New York.

Dyson-Hudson, R., and E. A. Smith. 1978. Human territoriality: An ecological reassessment. *American Anthropologist* 80:21–41.

Earle, T. 1991a. Property rights and the evolution of chiefdoms. Pp. 71–99 in T. Earle, ed., *Chiefdoms: Power, Economy, and Ideology.* Cambridge Univ. Press, Cambridge.

Earle, T., ed. 1991b. *Chiefdoms: Power, Economy, and Ideology.* Cambridge Univ. Press, Cambridge.

Earle, T. 1997. *How Chiefs Come to Power: The Political Economy of Prehistory.* Stanford Univ. Press, Stanford.

Eaton, J., and A. J. Mayer. 1954. *Man's Capacity to Reproduce: The Demography of a Unique Population.* Free Press, Glencoe.

Eaton, S. B., and S. B. Eaton III. 1999. The evolutionary context of chronic degenerative diseases. Pp. 251–259 in S. C. Stearns, ed., *Evolution in Health and Disease.* Oxford Univ. Press, Oxford.

Eaton, S. B., S. B. Eaton III, and M. J. Konner. 1999. Paleolithic nutrition revisited. Pp. 313–332 in W. R. Trevarthan, E. O. Smith, and J. J. McKenna, eds., *Evolutionary Medicine.* Oxford Univ. Press, New York.

Eaton, S. B., S. B. Eaton III, M. J. Konner, and M. Shostak. 1996. An evolutionary perspective enhances understanding of human nutritional requirements. *Journal of Nutrition* 126:1732–1740.

Ebert, D. 1999. The evolution and expression of parasite virulence. Pp. 161–172 in S. C. Stearns, ed., *Evolution in Health and Disease.* Oxford Univ. Press, Oxford.

Ebert, D., and W. D. Hamilton. 1995. Sex against virulence: The coevolution of parasitic diseases. *Trends in Ecology and Evolution* 11:79–82.

Eckersley, R. 1992. *Environmental and Political Theory: Toward an Ecocentric Approach.* Univ. College London Press, London.

Edelman, G. M. 1987. *Neural Darwinism: The Theory of Neuronal Group Selection.* Basic Books, New York.

Edelman, G. M. 1992. *Bright Air, Brilliant Fire: On the Matter of Mind.* Basic Books, New York.

Edgerton, R. B. 1992. *Sick Societies: Challenging the Myth of Primitive Harmony.* Free Press, New York.

Edwards, S. V., and P. W. Hedrick. 1998. Evolution and ecology of MHC molecules: From genomics to sexual selection. *Trends in Ecology and Evolution* 13:305–311.

Ehrenfeld, D. 1978. *The Arrogance of Humanism.* Oxford Univ. Press, New York.

Ehrenreich, B. 1997. *Blood Rites: Origins and History of the Passions of War.* Holt, New York.

Ehrenreich, B., and D. English. 1979. *For Her Own Good: One Hundred Fifty Years of the Experts' Advice to Women.* Pluto Press, London.

Ehrlich, P. R. 1961. Has the biological species concept outlived its usefulness? *Systematic Zoology* 10:167–176.

Ehrlich, P. R. 1968. *The Population Bomb.* Ballantine Books, New York.

Ehrlich, P. R. 1970. Coevolution and the biology of communities. Pp. 1–11 in K. L. Chambers, ed., *Biochemical Coevolution.* Oregon State Univ. Press, Corvallis.

Ehrlich, P. R. 1991. Coevolution and its applicability to the Gaia hypothesis. Pp. 19–22 in S. H. Schneider and P. J. Boston, eds., *Scientists on Gaia.* MIT Press, Cambridge.

Ehrlich, P. R. 1992. Population biology of checkerspot butterflies and the preservation of global biodiversity. *Oikos* 63:6–12.

Ehrlich, P. R. 1993. Biodiversity and ecosystem function: Need we know more? Pp. vii–xi in E.-D. Schulze and H. A. Mooney, eds., *Biodiversity and Ecosystem Function.* Springer-Verlag, Berlin.

Ehrlich, P. R. 1995. The scale of the human enterprise and biodiversity loss. Pp. 214–226 in J. H. Lawton and R. M. May, eds., *Extinction Rates.* Oxford Univ. Press, Oxford.

Ehrlich, P. R. 1997. *A World of Wounds: Ecologists and the Human Dilemma.* Ecology Institute, Oldendorf/Luhe.

Ehrlich, P. R., and J. H. Camin. 1960. Natural selection in Middle Island water snakes (*Natrix sipedon* L.). *Evolution* 14:136.

Ehrlich, P. R., D. S. Dobkin, and D. Wheye. 1988. *The Birder's Handbook: A Field Guide to the Natural History of North American Birds.* Simon and Schuster, New York.

Ehrlich, P. R., and A. H. Ehrlich. 1970. *Population, Resources, Environment: Issues in Human Ecology.* Freeman, San Francisco.

Ehrlich, P. R., and A. H. Ehrlich. 1973. Coevolution: Heterotypic schooling in Caribbean reef fishes. *American Naturalist* 107:157–160.

Ehrlich, P. R., and A. H. Ehrlich. 1981. *Extinction: The Causes and Consequences of the Disappearance of Species.* Random House, New York.

Ehrlich, P. R., and A. H. Ehrlich. 1990. *The Population Explosion.* Simon and Schuster, New York.

Ehrlich, P. R., and A. H. Ehrlich. 1991. *Healing the Planet.* Addison-Wesley, Reading, MA.

Ehrlich, P. R., and A. H. Ehrlich. 1992. The value of biodiversity. *Ambio* 21:219–226.

Ehrlich, P. R., and A. H. Ehrlich. 1996. *Betrayal of Science and Reason: How Anti-Environmental Rhetoric Threatens Our Future.* Island Press, Washington, DC.

Ehrlich, P. R., A. H. Ehrlich, and G. C. Daily. 1993. Food security, population, and environment. *Population and Development Review* 19:1–32.

Ehrlich, P. R., A. H. Ehrlich, and G. C. Daily. 1995. *The Stork and the Plow: The Equity Answer to the Human Dilemma.* Putnam, New York.

Ehrlich, P. R., A. H. Ehrlich, and J. P. Holdren. 1977. *Ecoscience: Population, Resources, Environment.* Freeman, San Francisco.

Ehrlich, P. R., and S. S. Feldman. 1977. *The Race Bomb: Skin Color, Prejudice, and Intelligence.* New York Times Book Co., New York.

Ehrlich, P. R., and J. Freedman. 1971. Population, crowding, and human behavior. *New Scientist and Science Journal* (1 April): 10–14.

Ehrlich, P. R., J. Harte, M. A. Harwell, P. H. Raven, et al. 1983. Long-term biological consequences of nuclear war. *Science* 222:1293–1300.

Ehrlich, P. R., and J. Holdren. 1971. Impact of population growth. *Science* 171:1212–1217.

Ehrlich, P. R., and R. W. Holm. 1962. Patterns and populations. *Science* 137:652–657.

Ehrlich, P. R., and R. W. Holm. 1963. *The Process of Evolution.* McGraw-Hill, New York.

Ehrlich, P. R., and R. W. Holm. 1964. A biological view of race. Pp. 153–179 in A. Montagu, ed., *The Concept of Race.* Free Press, New York.

Ehrlich, P. R., R. W. Holm, and D. R. Parnell. 1974. *The Process of Evolution.* 2nd ed. McGraw-Hill, New York.

Ehrlich, P. R., and D. D. Murphy. 1981. Butterfly nomenclature: A critique. *Journal of Research on the Lepidoptera* 20:1–11.

Ehrlich, P. R., and P. H. Raven. 1964. Butterflies and plants: A study in coevolution. *Evolution* 18:586–608.

Ehrlich, P. R., and P. H. Raven. 1969. Differentiation of populations. *Science* 65:1228–1232.

Ehrlich, P. R., and J. Roughgarden. 1987. *The Science of Ecology.* Macmillan, New York.

Ehrlich, P. R., G. Wolff, G. C. Daily, J. B. Hughes, et al. 1999. Knowledge and the environment. *Ecological Economics* 30:267–284.

Eibl-Eibesfeldt, I. 1977. Patterns of greeting in New Guinea. Pp. 209–247 in S. Wurm, ed., *Language, Culture, Society, and the Modern World.* Australian National Univ. Press, Canberra.

Eibl-Eibesfeldt, I. 1979 (1975). *The Biology of Peace and War: Men, Animals, and Aggression.* Viking, New York.

Eilers, F.-J. 1977. Non-verbal communication in Northeast New Guinea. Pp. 249–259 in S. A. Wurm, ed., *Language, Culture, Society, and the Modern World.* Pacific Linguistics, Canberra.

Eisenberg, N., and P. A. Miller. 1987. The relation of empathy to prosocial and related behaviors. *Psychological Bulletin* 101:91–119.

Ekman, P. 1992. Facial expressions of emotion: New findings, new questions. *Psychological Science* 3:34–38.

Ekman, P., and R. J. Davidson. 1993. Voluntary smiling changes regional brain activity. *Psychological Science* 4:342–345.

Ekman, P., and R. J. Davidson, eds. 1994. *The Nature of Emotion: Fundamental Questions.* Oxford Univ. Press, New York.

Ekman, P., and W. V. Friesen. 1978. *Unmasking the Face.* Prentice Hall, Englewood Cliffs, NJ.

Ekman, P., and W. V. Friesen. 1986. A new pan-cultural facial expression of emotion. *Motivation and Emotion* 10:159–168.

Ekman, P., R. W. Levenson, and W. V. Friesen. 1983. Autonomic nervous system activity distinguishes among emotions. *Science* 221:1208–1210.

Ekman, P., E. R. Sorenson, and W. V. Friesen. 1969. Pan-cultural elements in facial displays of emotion. *Science* 164:86–88.

Eldredge, N. 1985. *Time Frames: The Evolution of Punctuated Equilibria.* Princeton Univ. Press, Princeton.

Eldredge, N. 1995. *Dominion: Can Nature and Culture Co-exist?* Holt, New York.

Eldredge, N. 1999. *The Pattern of Evolution.* Freeman, New York.

Eldredge, N., and S. J. Gould. 1972. Punctuated equilibria: An alternative to phyletic gradualism. Pp. 82–115 in T. J. M. Schopf, ed., *Models in Paleobiology.* Freeman, Cooper, San Francisco.

Elena, S. F., V. S. Cooper, and R. E. Lenski. 1996. Punctuated evolution caused by selection of rare beneficial mutations. *Science* 272:1802–1804.

Eliade, M. 1957. *The Sacred and the Profane: The Nature of Religion.* Harcourt, Brace, San Diego.

Eliasson, R., and C. Lindholmer. 1972. Distribution and properties of spermatozoa in different fractions of split ejaculates. *Fertility and Sterility* 23:252–256.

Ellen, R. 1994. Modes of subsistence: Hunting and gathering to agriculture and pastoralism. Pp. 197–225 in T. Ingold, ed., *Companion Encyclopedia of Anthropology.* Routledge, London.

Elliott, R. 1972. The influence of vector behavior on malaria transmission. *American Journal of Tropical Medicine and Hygiene* 21:755–763.

Ellis, L. 1989. *Theories of Rape: Inquiries into the Causes of Sexual Aggression.* Hemisphere, New York.

Elzanowski, A. 1993. The moral career of vertebrate values. Pp. 259–276 in M. H. Nitecki and D. V. Nitecki, eds., *Evolutionary Ethics.* State Univ. of New York Press, Albany.

Ember, C. R., and M. Ember. 1992. Resource unpredictability, mistrust, and war. *Journal of Conflict Resolution* 36:242–262.

Ember, M. 1982. Statistical evidence for an ecological explanation of warfare. *American Anthropologist* 84:645–649.

Emlen, S. T. 1995. An evolutionary theory of the family. *Proceedings of the National Academy of Sciences USA* 92:8092–8099.

Endler, J. A. 1986. *Natural Selection in the Wild.* Princeton Univ. Press, Princeton.

Endler, J. A. 1995. Multiple-trait coevolution and environmental gradients in guppies. *Trends in Ecology and Evolution* 10:22–29.

Englert, P. S. 1970. *Island at the Center of the World.* Scribner, New York.

Englert, Y., F. Puissant, M. Camus, M. Degueldre, et al. 1986. Factors leading to triplonucleate eggs during human in-vitro fertilization. *Human Reproduction* 1:117–119.

Enquist, M., and S. Ghirlanda. 1998. The secrets of faces. *Nature* 394:826–827.

Epstein, J., and R. Axtell. 1996. *Growing Artificial Societies.* MIT Press, Cambridge.

Erdal, D., and A. Whiten. 1994. On human egalitarianism: An evolutionary product of Machiavellian status escalation. *Current Anthropology* 35:175–178.

Ernulf, K. E., and S. M. Innala. 1989. Biological explanation, psychological explanation, and tolerance of homosexuals: A cross-national analysis of beliefs and attitudes. *Psychological Reports* 65:1003–1010.

Etcoff, N. L., R. Freeman, and K. R. Cave. 1991. Can we lose memories of faces? Content specificity and awareness in a prospagnosic. *Journal of Cognitive Neuroscience* 3:25–41.

Etter, M. A. 1978. Sahlins and sociobiology. *American Ethnologist* 68:160–169.

Evans, B. 1968. *Dictionary of Quotations.* Delacorte Press, New York.

Evans-Pritchard, E. E. 1970. Sexual inversion among the Azande. *American Anthropologist* 72:1428–1434.

Ewald, P. W. 1988. Cultural vectors, virulence, and the emergence of evolutionary epidemiology. *Oxford Surveys of Evolutionary Biology* 5:215–245.

Ewald, P. W. 1994. *Evolution of Infectious Disease.* Oxford Univ. Press, Oxford.

Falconer, D. S., and T. F. C. Mackay. 1996. *Introduction to Quantitative Genetics.* 4th ed. Longmans, Essex.

Falk, D. 1990. Brain evolution in *Homo:* The "radiator" theory. *Behavioral and Brain Sciences* 13:333–381.

Falk, D. 1998. Hominid brain evolution: Looks can be deceiving. *Science* 280:1714.

Farrell, B. D. 1998. "Inordinate fondness" explained: Why are there so many beetles? *Science* 281:555–559.

Farrell, B. D., and C. Mitter. 1993. Phylogenetic determinants of insect/plant community diversity. Pp. 253–266 in R. E. Rickelefs and D. Schluter, eds., *Ecological Communities: Historical and Geographical Perspectives.* Univ. of Chicago Press, Chicago.

Fausto-Sterling, A. 1993. The five sexes. *Sciences* (March–April): 20–24.

Fay, R. E., C. F. Turner, A. D. Klassen, and J. H. Gagnon. 1989. Prevalence and patterns of same-gender sexual contact among men. *Science* 243:338–348.

Federoff, N. E., and R. M. Nowak. 1997. Man and his dog. *Science* 279:207.

Fedigan, L. M. 1986. The changing role of women in models of human evolution. *Annual Review of Anthropology* 15:25–66.

Feduccia, A. 1996. *The Origin and Evolution of Birds.* Yale Univ. Press, New Haven.

Feduccia, A., and H. B. Tordoff. 1979. Feathers of *Archaeopteryx:* Asymmetrical vanes indicate aerodynamic function. *Science* 203:1021–1022.

Fein, H. 1979. *Accounting for Genocide.* Free Press, New York.

Feinman, G. M. 1991. Demography, surplus, and inequality: Early political formations in highland Mesoamerica. Pp. 229–262 in T. Earle, ed., *Chiefdoms: Power, Economy, and Ideology.* Cambridge Univ. Press, Cambridge.

Feldman, M. W., ed. 1989. *Mathematical Evolutionary Theory.* Princeton Univ. Press, Princeton.

Feldman, M. W., and L. L. Cavalli-Sforza. 1976. Cultural and biological evolutionary processes, selection for a trait under complex transmission. *Theoretical Population Biology* 9:238–259.

Feldman, M. W., and L. L. Cavalli-Sforza. 1989. On the theory of evolution under genetic and cultural transmission with application to the lactose absorption problem. Pp. 145–173 in M. W. Feldman, ed., *Mathematical Evolutionary Theory.* Princeton Univ. Press, Princeton.

Feldman, M. W., and K. N. Laland. 1996. Gene-culture coevolutionary theory. *Trends in Ecology and Evolution* 11:453–457.

Feldman, M. W., and R. C. Lewontin. 1975. The heritability hang-up. *Science* 190:1163–1168.

Feldman, M. W., and S. P. Otto. 1997. Twin studies, heritability, and intelligence. *Science* 278.

Feldman, M. W., S. P. Otto, and F. B. Christiansen. 2000. Genes, culture, and inequality. Pp. 61–85 in K. Arrow, S. Bowles, and S. Durlauf, eds., *Meritocracy and Economic Inequality.* Princeton Univ. Press, Princeton.

Felt, J. C., J. C. Ridley, G. Allen, and C. Redekop. 1990. High fertility in Old Colony Mennonites in Mexico. *Human Biology* 62:689–700.

Ferguson, R. B. 1989. Game wars? Ecology and conflict in Amazonia. *Journal of Anthropological Research* 45:179–206.

Ferguson, R. B. 1995. *Yanomami Warfare: A Political History*. School of American Research Press, Santa Fe.

Fernald, A. 1992. Human maternal vocalizations to infants as biologically relevant signals: An evolutionary perspective. Pp. 391–428 in J. H. Barkow, L. Cosmides, and J. Tooby, eds., *The Adapted Mind: Evolutionary Psychology and the Generation of Culture*. Oxford Univ. Press, New York.

Festenstein, M. 1997. *Pragmatism and Political Theory: From Dewey to Rorty*. Univ. of Chicago Press, Chicago.

Feuer, L. S. 1992. *The Scientific Intellectual: The Psychological and Sociological Origins of Modern Science*. Transaction, New Brunswick, NJ.

Fifer, F. C. 1987. The adoption of bipedalism by the hominids: A new hypothesis. *Human Evolution* 2:135–147.

Finkelstein, N. G., and R. B. Birn. 1998. *A Nation on Trial: The Goldhagen Thesis and Historical Truth*. Holt, New York.

Finn, C. A. 1998. Menstruation: A nonadaptive consequence of uterine evolution. *Quarterly Review of Biology* 73:163–173.

Finn, R. 1992. John Bostock, hayfever, and the mechanism of allergy. *Lancet* 340:1453–1455.

Fischer, K. W. 1980. A theory of cognitive development: The control and construction of hierarchies of skills. *Psychological Review* 87:477–531.

Fisher, R. A. 1930. *The Genetical Theory of Natural Selection*. Oxford Univ. Press, London.

Flannery, K. V. 1972. The cultural evolution of civilizations. *Annual Review of Ecology and Systematics* 3:399–426.

Flannery, T. F. 1994. *The Future Eaters: An Ecological History of the Australasian Lands and People*. Braziller, New York.

Flannery, T. F. 1999. Debating extinction. *Science* 283:182–183.

Fleagle, J. G. 1994. Anthropoid origins. Pp. 17–35 in R. S. Corruccini and R. L. Ciochon, eds., *Integrative Paths to the Past: Paleoanthropological Advances in Honor of F. Clark Howell*. Prentice Hall, Englewood Cliffs, NJ.

Fleagle, J. G. 1999. *Primate Adaptation and Evolution*. 2nd ed. Academic Press, San Diego.

Fletcher, D. J. C., and C. D. Michener. 1987. *Kin Recognition in Animals*. Wiley, New York.

Flinn, M. V. 1986. Correlates of reproductive success in a Caribbean village. *Human Ecology* 14:225–243.

Flint, J., R. M. Harding, A. J. Boyce, and J. B. Clegg. 1993. The population genetics of the haemoglobinopathies. *Ballière's Clinical Haematology* 6:215–261.

Flynn, J. R. 2000. IQ trends over time: Intelligence, race, and meritocracy. Pp. 35–60 in K. Arrow, S. Bowles, and S. Durlauf, eds., *Meritocracy and Economic Inequality*. Princeton Univ. Press, Princeton.

Fodor, J. A. 1983. *The Modularity of Mind*. MIT Press, Bradford Books, Cambridge.

Fodor, J. A. 1985. Précis of *The Modularity of Mind*. *Behavioral and Brain Sciences* 8:1–42.

Fodor, J. A. 1998. When is a dog a DOG? *Nature* 396:325–327.

Foley, R. A. 1987. *Another Unique Species*. Longman, Harlow.

Foley, R. A. 1988. Hominids, humans, and hunter-gatherers: An evolutionary perspective. Pp. 207–221 in T. Ingold, D. Riches, and J. Woodburn, eds., *Hunters and Gatherers*. Vol. 1, *History, Evolution, and Social Change*. Berg, Oxford.

Foley, R. A. 1991a. How many species of hominid should there be? *Journal of Human Evolution* 20:413–427.

Foley, R. A. 1991b. Language origins: The silence of the past. *Nature* 353:114–115.

Foley, R. A. 1995. *Humans before Humanity*. Blackwell, Oxford.

Foley, R. A. 1996. Measuring cognition in extinct hominids. Pp. 57–65 in P. Mellars and K. Gibson, eds., *Modelling the Early Human Mind*. McDonald Institute for Archaeological Research (Oxbow Books), Oxford.

Foley, R. A. 1999. Hunting down the hunter-gatherers. *Evolutionary Anthropology* 8:115–117.

Foley, W. A. 1997. *Anthropological Linguistics: An Introduction*. Blackwell, Malden, MA.

Folk, G. E., Jr., and H. A. Semken Jr. 1991. The evolution of sweat glands. *International Journal of Biometeorology* 35:180–186.

Ford, C. S., and F. A. Beach. 1951. *Patterns of Sexual Behavior*. Harper and Brothers, New York.

Forster, C. A., S. D. Sampson, L. M. Chiappe, and D. W. Krause. 1998. The theropod ancestry of birds: New evidence from the Late Cretaceous of Madagascar. *Science* 279:1915–1919.

Fortey, R. 1997. *Life: A Natural History of the First Four Billion Years of Life on Earth*. Knopf, New York.

Fossey, D. 1983. *Gorillas in the Mist*. Houghton Mifflin, Boston.

Foucault, M. 1972. *The Archaeology of Language and the Discourse on Language*. Pantheon Books, New York.

Foucault, M. 1988. *Madness and Civilization: A History of Insanity in the Age of Reason*. Vintage Books, New York.

Foucault, M. 1994. *The Birth of the Clinic: An Archaeology of Medical Perception*. Vintage Books, New York.

Foucault, M. 1995. *Discipline and Punish: The Birth of Prison*. Vintage Books, New York.

Fox, C. A., H. S. Wolff, and J. A. Baker. 1970. Measurement of intra-vaginal and intra-uterine pressures during human coitus by radio-telemetry. *Journal of Reproductive Fertility* 22:243–251.

Fox, R. 1983. *The Red Lamp of Incest: An Enquiry into the Origins of Mind and Society.* Univ. of Notre Dame Press, Notre Dame.

Fox, R. 1983 (1967). *Kinship and Marriage: An Anthropological Perspective.* Cambridge Univ. Press, Cambridge.

Fox, R. 1991. Aggression: Then and now. Pp. 81–93 in M. Robinson and L. Tiger, eds., *Man and Beast Revisited.* Smithsonian Institution Press, Washington, DC.

Fox, R. 1994. *The Challenge of Anthropology: Old Encounters and New Excursions.* Transaction, New Brunswick, NJ.

Fox, R. 1997. *Conjectures and Confrontations: Science, Evolution, Social Concern.* Transaction, New Brunswick, NJ.

Fox, S. 1981. *John Muir and His Legacy.* Little, Brown, Boston.

Fox, W. 1984. Deep ecology: A new philosophy for our time? *Ecologist* 14:194–204.

Fox, W. 1989. *Toward a Transpersonal Ecology: The Context, Influence, Meanings, and Distinctiveness of the Deep Ecology Approach to Ecophilosophy.* Shambhala, Boston.

Fox, W. 1990. *Toward a Transpersonal Ecology: Developing New Foundations for Environmentalists.* Shambhala, Boston.

Frängsmyr, T. 1999. The new academies and the scientific climate of the eighteenth century. *Proceedings of the American Philosophical Society* 143:109–115.

Frank, R. H. 1999. *Luxury Fever: Why Money Fails to Satisfy in an Era of Excess.* Free Press, New York.

Frayer, D. W., M. H. Wolpoff, A. G. Thorne, F. H. Smith, et al. 1993. Theories of modern human origins: The paleontological test. *American Anthropologist* 95:14–50.

Frazer, S. J. 1910. *Totemism and Exogamy.* Vol. 4. Macmillan, London.

Freedman, D. 1979. *Human Sociobiology: A Holistic Approach.* Free Press, New York.

Freedman, J. L., S. Klevansky, and P. R. Ehrlich. 1971. The effects of crowding on human task performance. *Journal of Applied Social Psychology* 1:17–25.

Freeman, S., and J. C. Herron. 1998. *Evolutionary Analysis.* Prentice Hall, Upper Saddle River, NJ.

Freud, S. 1961 (1930). *Civilization and Its Discontents.* Norton, New York.

Freud, S. 1977 (1909). *Five Lectures on Psycho-Analysis.* Norton, New York.

Freund, P., and G. Martin. 1993. *The Ecology of the Automobile.* Black Rose Books, New York.

Fridlund, A. J. 1991. Evolution and facial action in reflex, social motive, and paralanguage. *Biological Psychology* 32:3–100.

Friedrich, L. K., and A. H. Stein. 1975. Prosocial television and young children: The effects of verbal labeling and role playing on learning and behavior. *Child Development* 46:27–38.

Friedrich, P. 1975. The lexical symbol and its relative non-arbitrariness. Pp. 199–247 in M. D. Kinkade et al., eds., *Linguistics and Anthropology.* Peter De Ridder Press, Lisse.

Frison, G. C. 1986. Prehistoric, plains-mountain, large-mammal, communal hunting strategies. Pp. 177–223 in M. H. Nitecki and D. V. Nitecki, eds., *The Evolution of Human Hunting.* Plenum Press, New York.

Frith, C. D., and U. Frith. 1999. Interacting minds: A biological basis. *Science* 286:1692–1695.

Fruth, B., and G. Hohmann. 1994. Comparative analyses of nest building behavior in bonobos and chimpanzees. Pp. 109–128 in R. W. Wrangham, W. C. McGrew, F. B. M. de Waal, and P. G. Heltne, eds., *Chimpanzee Cultures.* Harvard Univ. Press, Cambridge.

Fryer, G., and T. D. Iles. 1972. *The Cichlid Fishes of the Great Lakes of Africa: Their Biology and Evolution.* TFH, Neptune, NJ.

Fullagar, R. L. K., D. M. Prince, and L. M. Head. 1996. Early human occupation of northern Australia: Archaeology and thermoluminescence dating of Jinmium rock-shelter, Northern Territory. *Antiquity* 70:751–773.

Fulton, R., and S. W. Anderson. 1992. The Amerindian "man-woman": gender, liminality, and cultural continuity. *Current Anthropology* 33:603–610.

Furuichi, T., and H. Ihobe. 1994. Variation in male relationships in bonobos and chimpanzees. *Behavior* 130:211–228.

Futuyma, D. J. 1983. Evolutionary interactions among herbivorous insects and plants. Pp. 207–231 in D. J. Futuyma, ed., *Coevolution.* Sinauer, Sunderland, MA.

Futuyma, D. J. 1998. *Evolutionary Biology.* 3rd ed. Sinauer, Sunderland, MA.

Futuyma, D. J., and M. Slatkin, eds. 1983. *Coevolution.* Sinauer, Sunderland, MA.

Gabunia, L., and A. Vekua. 1995. A Plio-Pleistocene hominid from Dmanisi, East Georgia, Caucasus. *Nature* 373:509–512.

Gabunia, L., A. Vekua, D. Lordkipanidze, C. Swisher, et al. 2000. Earliest Pleistocene hominid cranial remains from Dmanisi, Republic of Georgia: Taxonomy, geological setting, and age. *Science* 288:1019–1025.

Gadgil, M. 1991. Conserving India's biodiversity: The societal context. *Evolutionary Trends in Plants* 5:3–8.

Gadgil, M., F. Berkes, and C. Folke. 1993. Indigenous knowledge for biodiversity conservation. *Ambio* 22:151–156.

Gagneux, P., D. S. Woodruff, and C. Boesch. 1997. Furtive mating in female chimpanzees. *Nature* 387:358–359.

Galis, F., and J. A. J. Metz. 1998. Why are there so many cichlid species? *Trends in Ecology and Evolution* 13:1–41.

Gallardo, M. H., J. W. Bickham, R. L. Honeycutt, R. A. Ojeda, et al. 1999. Discovery of tetraploidy in a mammal. *Nature* 401:341.

Gallup, G. G., Jr. 1970. Chimpanzees: Self-recognition. *Science* 167:86–87.

Gallup, G. G., Jr. 1979. Self-awareness in primates. *American Scientist* 67:417–421.

Gallup, G. G., Jr. 1983. Toward a comparative psychology of mind. Pp. 473–510 in R. Mellgren, ed., *Animal Cognition and Behavior.* North Holland, Amsterdam.

Gallup, G. G., Jr. 1985. Do minds exist in species other than our own? *Neuroscience and Biobehavioral Reviews* 9:631–641.

Gallup, G. H., Jr., and J. Castelli. 1987. *The People's Religion: American Faith in the Nineties.* Macmillan, New York.

Gallup, G. H., Jr., and F. Newport. 1991. Belief in paranormal phenomena among adult Americans. *Skeptical Inquirer* 15:137–146.

Gamlin, L. 1990. The big sneeze. *New Scientist* 2:37–41.

Gannon, P. J., R. L. Holloway, D. C. Broadfield, and A. R. Braun. 1998. Asymmetry of chimpanzee planum temporale: Humanlike pattern of Wernicke's brain language area homolog. *Science* 279:220–222.

Gao, F., E. Bailes, D. L. Robertson, Y. Chen, C. M. Rodenburg, et al. 1999. Origin of the HIV-1 in the chimpanzee *Pan troglodytes troglodytes. Nature* 397:436–440.

Garber, P. A. 1997. One for all and breeding for one: Cooperation and competition as a tamarin reproductive strategy. *Evolutionary Anthropology* 5:187–199.

Garcia, J., and F. R. Ervin. 1968. Gustatory-visceral and telereceptor-cutaneous conditioning: Adaptation in internal and external milieus. *Communications in Behavioral Biology*, ser. A, 1:389–415.

Gardner, B. T., and R. A. Gardner. 1971. Two-way communication with an infant chimpanzee. Pp. 117–183 in A. Schrier and F. Stollnitz, eds., *Behavior of Nonhuman Primates.* Academic Press, New York.

Gardner, G. 1999. Why share? *World Watch* (July–August): 10–20.

Gardner, H. 1975. *The Shattered Mind.* Knopf, New York.

Gardner, H. 1998. Do parents count? *New York Review of Books* (5 November): 19–22.

Gardner, H. 1999. Who owns intelligence? *Atlantic Monthly* (February): 67–76.

Gardner, R. A., and B. T. Gardner. 1984. A vocabulary test for chimpanzees *(Pan troglodytes). Journal of Comparative Psychology* 98:381–404.

Gardner, R. A., B. T. Gardner, and T. E. Van Cantfort. 1989. *Teaching Sign Language to Chimpanzees.* State Univ. of New York Press, Albany.

Gargett, R. H. 1989. Grave shortcomings: The evidence for Neandertal burial. *Current Anthropology* 30:157–190.

Garnick, M. B., and W. R. Fair. 1998. Combating prostate cancer. *Scientific American* 279:75–83.

Gay, G. H. 1980. *Sole Survivor.* Midway, Naples, FL.

Gazzaniga, M. S. 1989. Organization of the human brain. *Science* 245:947–951.

Gazzaniga, M. S. 1998. The split brain revisited. *Scientific American* (July): 50–55.

Geary, D. C. 2000. Evolution and proximate expression of human paternal investment. *Psychological Bulletin* 126:55–77.

Gebo, D. L., M. Dagosto, K. C. Beard, T. Qi, et al. 2000. The oldest known anthropoid postcranial fossils and the early evolution of higher primates. *Nature* 404:276–278.

Geertz, C. 1966. Religion as a cultural system. Pp. 1–46 in M. Banton, ed., *Anthropological Approaches to the Study of Religion.* Tavistock, London.

Geertz, C. 1973. *The Interpretation of Cultures.* Basic Books, New York.

Gehring, W. J. 1986. Homeotic genes. *Annual Review of Genetics* 20:147–173.

Gehring, W. J. 1998. *Master Control Genes in Development and Evolution: The Homeobox Story.* Yale Univ. Press, New Haven.

Gell, A. 1980. The gods at play: Vertigo and possession in Muria religion. *Man* 15:219–248.

Gellner, E. 1983. *Nations and Nationalism.* Basil Blackwell, Oxford.

Gellner, E. 1994. *Conditions of Liberty: Civil Society and Its Rivals.* Penguin Books, New York.

Gelman, S. 1988. The development of induction within natural kind and artifact categories. *Cognitive Psychology* 20:65–95.

George, W. C. 1962. Biology and the race problem. Report for the National Putnam Letters Committee; reprint prepared by the Commission of the Governor of Alabama.

Georghiou, G. P. 1986. The magnitude of the resistance problem. Pp. 14–43 in National Research Council, ed., *Pesticide Resistance: Strategies and Tactics for Management.* National Academy Press, Washington, DC.

Georghiou, G. P., and A. Lagunes-Tejeda. 1991. *The Occurrence of Resistance to Pesticides in Arthropods.* Food and Agriculture Organization of the United Nations, Rome.

Getty, T. 1999. What do experimental studies tell us about group selection in nature? *American Naturalist* 154:596–598.

Gewertz, D. 1981. A historical reconsideration of female dominance among the Chambri of Papua New Guinea. *American Ethnologist* 8:94–106.

Ghiglieri, M. P. 1987. Sociobiology of the great apes and the hominid ancestor. *Journal of Human Evolution* 16:319–357.

Ghiglieri, M. P. 1999. *The Dark Side of Man: Tracing the Origins of Male Violence.* Perseus Books, Reading, MA.

Ghiselin, M. T. 1975. A radical solution to the species problem. *Systematic Zoology* 23:536–544.

Gibbons, A. 1996. The peopling of the Americas. *Science* 274:31–33.

Gibbons, A. 1997a. Archaeologists rediscover cannibals. *Science* 277:635–637.

Gibbons, A. 1997b. Ideas on human origins evolve at anthropology meeting. *Science* 276:535–536.

Gibbons, A. 1997c. Y chromosome shows that Adam was an African. *Science* 278:804–805.

Gibbons, A. 1998a. Ancient island tools suggest *Homo erectus* was a seafarer. *Science* 279:1635–1637.

Gibbons, A. 1998b. Mother tongues trace steps of earliest Americans. *Science* 279:1306–1307.

Gibbons, A. 1998c. Old, old skull has a new look. *Science* 280:1525.

Gibbons, A. 1998d. Solving the brain's energy crisis. *Science* 280:1345–1347.

Gibbons, A. 1998e. Which of our genes make us human? *Science* 281:1432–1434.

Gibbons, A. 1998f. Young ages for Australian rock art. *Science* 280:1351.

Gibbons, A. 2000. Chinese stone tools reveal high-tech *Homo erectus. Science* 287:1566.

Gibson, E. J., and R. D. Walk. 1960. The "visual cliff"? *Scientific American* (April): 64–71.

Gibson, K. R. 1990. New perspectives on instincts and intelligence: Brain size and the emergence of hierarchical mental constructional skills. Pp. 97–128 in S. T. Parker and K. R. Gibson, eds., *"Language" and Intelligence in Monkeys and Apes.* Cambridge Univ. Press, Cambridge.

Gibson, K. R. 1991. Tools, language, and intelligence: Evolutionary implications. *Man*, n.s., 26:255–264.

Gibson, K. R., and T. Ingold, eds. 1993. *Tools, Language, and Cognition in Human Evolution.* Cambridge Univ. Press, Cambridge.

Gilbert, L. E., and P. H. Raven, eds. 1975. *Coevolution of Animals and Plants.* Univ. of Texas Press, Austin.

Gilbert, M. 1985. *The Holocaust: A History of the Jews of Europe during the Second World War.* Holt, Rinehart and Winston, New York.

Gilbert, S. F. 1997. *Developmental Biology.* 5th ed. Sinauer, Sunderland, MA.

Gilderhus, M. T. 1996. *History and Historians: A Historiographical Introduction.* 3rd ed. Prentice Hall, Englewood Cliffs, NJ.

Gill, F. B., and L. L. Wolf. 1975. Economics of territoriality in the golden-winged sunbird. *Ecology* 56:333–345.

Gillham, N. W. 1965. Geographic variation and the subspecies concept in butterflies. *Systematic Zoology* 5:100–120.

Gillham, N. W. 1994. *Organelle Genes and Genomes.* Oxford Univ. Press, New York.

Gilman, A. 1981. The development of stratification in Bronze Age Europe. *Current Anthropology* 22:1–23.

Glaze, A. J. 1981. *Art and Death in a Senufo Village.* Indiana Univ. Press, Bloomington.

Godfrey-Smith, P. 1998. Maternal effects: On Dennett and *Darwin's Dangerous Idea. Philosophy of Science* 65:709–720.

Godfrey-Smith, P. 1999. Adaptationism and the power of selection. *Biology and Philosophy* 14:181–194.

Godfrey-Smith, P. 2000. Three kinds of adaptationism. In S. Orzack and E. Sober, eds., *Optimality and Adaptationism.* Cambridge Univ. Press, Cambridge. In press.

Goebel, T., A. Derevianko, and V. Petrin. 1993. Dating the Middle to Upper Paleolithic transition at Kara-Bom. *Current Anthropology* 34:452–458.

Goldberger, A. S., and L. J. Kamin. 1998. Behavior-genetic modeling of twins: A deconstruction. Social Systems Research Institute, Univ. of Wisconsin, Madison.

Goldberger, A. S., and C. F. Manski. 1995. Review article: *The Bell Curve* by Herrnstein and Murray. *Journal of Economic Literature* 33:762–776.

Goldfoot, D. A., H. Westerborg-van Loon, W. Groenveld, and A. Koos Slob. 1980. Behavioral and physiological evidence of sexual climax in the female stumptail macaque *(Macaca arctoides)*. *Science* 208:1477–1479.

Goldhagen, D. J. 1997. *Hitler's Willing Executioners: Ordinary Germans and the Holocaust.* Vintage Books, New York.

Goldin-Meadow, S. 1978. A study of human capacities. *Science* 200:649–651.

Goldin-Meadow, S. 1982. The resilience of recursion: A study of a communication system developed without a conventional language model. Pp. 51–77 in E. Wanner and L. R. Gleitman, eds., *Language Acquisition: The State of the Art.* Cambridge Univ. Press, Cambridge.

Goldin-Meadow, S. 1993. When does gesture become language? A study of gesture used as a primary communication system by deaf children of hearing parents. Pp. 63–85 in K. Gibson and T. Ingold, eds., *Tools, Language, and Cognition.* Cambridge Univ. Press, Cambridge.

Goldin-Meadow, S., C. Butcher, C. Mylander, and M. Dodge. 1994. Nouns and verbs in a self-styled gesture system: What's in a name? *Cognitive Psychology* 27:259–319.

Goldin-Meadow, S., and H. Feldman. 1977. The development of language-like communication without a language model. *Science* 197:401–403.

Goldin-Meadow, S., and C. Mylander. 1984. Gestural communication in deaf children: The effects and noneffects of parental input on early language development. *Monographs of the Society for Research in Child Development* 49:1–120.

Goldin-Meadow, S., and C. Mylander. 1990. Beyond the input given: The child's role in the acquisition of language. *Language* 66:323–355.

Goldman-Rakic, P. S. 1992. Working memory and the mind. *Scientific American* 267:73–79.

Goldschmidt, R. B. 1940. *The Material Basis of Evolution.* Yale Univ. Press, New Haven.

Goldschmidt, W. 1960. Culture and human behavior. Pp. 98–104 in A. Walker, ed., *Men and Cultures: Selected Papers of the Fifth International Congress.* Univ. of Pennsylvania Press, Philadelphia.

Goldsmith, T. H. 1990. Optimization, constraint, and history in the evolution of eyes. *Quarterly Review of Biology* 65:281–322.

Goldstein, D. B., A. R. Linares, L. L. Cavalli-Sforza, and M. W. Feldman. 1995. Genetic absolute dating based on microsatellites and the origin of modern humans. *Proceedings of the National Academy of Sciences USA* 92:6723–6727.

Goodall, J. 1986. *The Chimpanzees of Gombe: Patterns of Behavior.* Harvard Univ. Press, Cambridge.

Goodall, J. 1994. Postscript: Conservation and the future of chimpanzee and bonobo research in Africa. Pp. 397–404 in R. W. Wrangham, W. C. McGrew, F. de Waal, and P. G. Heltne, eds., *Chimpanzee Cultures.* Harvard Univ. Press, Cambridge.

Goodall, J., A. Bandora, E. Bergman, C. Busse, et al. 1979. Intercommunity interactions in the chimpanzee population of the Gombe National Park. Pp. 13–53 in D. A. Hamburg and E. R. McCown, eds., *The Great Apes.* Benjamin-Cummings, Menlo Park, CA.

Goodman, C. S., and C. J. Shatz. 1993. Developmental mechanisms that generate precise patterns of neuronal connectivity. *Cell* 72:77–98.

Goodman, M. 1999. The genomic record of humankind's evolutionary roots. *American Journal of Human Genetics* 64:31–39.

Goodman, M., D. A. Tagle, D. H. A. Fitch, W. Bailey, et al. 1990. Primate evolution at the DNA level and a classification of hominoids. *Journal of Molecular Evolution* 30:260–266.

Gopnik, M., and M. B. Crago. 1991. Familial aggregation of a developmental language disorder. *Cognition* 39:1–50.

Gordon, P. 1985. Level-ordering in lexical development. *Cognition* 21:73–93.

Gottesman, I. 1991. *Schizophrenia Genesis: The Origins of Madness.* Freeman, New York.

Gottesman, I., and J. Shields. 1972. *Schizophrenia: A Twin Study Vantage Point.* Academic Press, New York.

Gottesman, I. I., and J. Shields. 1982. *Schizophrenia: The Epigenetic Puzzle.* Cambridge Univ. Press, Cambridge.

Gotthardt, M. 1999. Fifty ways to survive the fat season. *Prevention* (November): 138–145.

Gould, J. L., and P. Marler. 1987. Learning by instinct. *Scientific American* 256:74–85.

Gould, R. 1982. To have and not to have: The ecology of sharing among hunter-gatherers. Pp. 69–92 in N. Williams and E. Hunn, eds., *Resource Managers: North American and Australian Hunter-Gatherers.* Westview Press, Boulder.

Gould, S. J. 1987. Freudian slip. *Natural History* 2:14–21.

Gould, S. J. 1989. *Wonderful Life: The Burgess Shale and the Nature of Natural History.* Norton, New York.

Gould, S. J. 1996. *Full House.* Harmony Books, New York.

Gould, S. J. 1997. The paradox of the visibly irrelevant. *Natural History* (December 1997, January 1998): 12–18, 60–64.

Gould, S. J., and R. C. Lewontin. 1979. The spandrels of San Marcos and the Panglossian paradigm. *Proceedings of the Royal Society of London,* ser. B, 205:581–598.

Gouldner, A. W. 1977. Stalinism: A study of internal colonialism. *Telos* 34:5–48.

Gourevitch, P. 1998. *We Wish to Inform You That Tomorrow We Will Be Killed with Our Families: Stories from Rwanda.* Farrar, Straus and Giroux, New York.

Gowlett, J. 1986. Culture and conceptualisation: The Oldowan-Acheulian gradient. Pp. 243–260 in G. Baily and P. Callow, eds., *Stone Age Prehistory.* Cambridge Univ. Press, Cambridge.

Goy, R. W., F. B. Bercovitch, and McBrair. 1988. Behavioral masculinization is independent of genital masculinization in prenatally androgenized female rhesus macaques. *Hormones and Behavior* 22:552–571.

Graber, R. B. 1989. A population-pressure alternative to a sociobiological theory of the rise of escalatory intergroup competition. *Politics and the Life Sciences* 7:203–206.

Graber, R. B. 1991. Population pressure, agricultural origins, and cultural evolution: Constrained mobility or inhibited expansion? *American Anthropologist* 93:692–695.

Graber, R. B. 1992. Population pressure, agricultural origins, and global theory: Comment on McCorriston and Hole. *American Anthropologist* 94:443–445.

Graber, R. B., and P. B. Roscoe. 1988. Introduction: Circumscription and the evolution of society. *American Behavioral Scientist* 31:405–414.

Graham, C. A., and W. C. McGrew. 1980. Menstrual synchrony in female undergraduates living on a coeducational campus. *Psychoneuroendocrinology* 5:245–252.

Graham, N. M. H., C. J. Burrell, R. M. Douglas, P. Debelle, et al. 1990. Adverse effects of aspirin, acetaminophen, and ibuprofen on immune function, viral shedding, and clinical status in rhinovirus-infected volunteers. *Journal of Infectious Diseases* 162:1277–1282.

Grant, B. R., and P. R. Grant. 1998. Hybridization and speciation in Darwin's finches. Pp. 404–422 in D. J. Howard and S. H. Berlocher, eds., *Endless Forms: Species and Speciation.* Oxford Univ. Press, Oxford.

Grant, B. S. 1996. Parallel rise and fall of melanic peppered moths in America and Britain. *Journal of Heredity* 87:351–357.

Grant, J. C., and A. H. Bittles. 1997. The comparative role of consanguinity in infant and childhood mortality in Pakistan. *Annals of Human Genetics* 61:143–149.

Grant, P. R. 1986. *Ecology and Evolution of Darwin's Finches.* Princeton Univ. Press, Princeton.

Grant, P. R. 1998. *Evolution on Islands.* Oxford Univ. Press, Oxford.

Graves, P. 1991. New models and metaphors for the Neanderthal debate. *Current Anthropology* 32:513–541.

Greenawalt, K. 1995. *Private Consciences and Public Reason.* Oxford Univ. Press, New York.

Greenawalt, K. 1998. Has religion any place in the politics and law of liberal democracy? *Proceedings of the American Philosophical Society* 142:378–387.

Greenberg, J. H. 1975. Research on language universals. *Annual Review of Anthropology* 4:75–94.

Greenberg, J. H. 1987. The present status of markedness theory: A reply to Scheffler. *Journal of Anthropological Research* 43:367–374.

Greene, J. C. 1999a. *Debating Darwin.* Regina Books, Claremont, CA.

Greene, J. C. 1999b. Reflections on Ernst Mayr's *This Is Biology. Biology and Philosophy* 14:103–116.

Greenfield, P. M. 1991. Language, tools, and brain: The ontogeny and phylogeny of hierarchically organized behavior. *Behavioral and Brain Sciences* 14:531–595.

Greenwood, P. H. 1974. The cichlid fishes of Lake Victoria, East Africa: The biology and evolution of a species flock. *British Museum (Natural History) Bulletin,* supp. 6.

Gregor, T. 1985. *Anxious Pleasures: The Sexual Lives of an Amazonian People.* Univ. of Chicago Press, Chicago.

Gregor, T. 1990. Uneasy peace: Intertribal relations in Brazil's upper Xingu. Pp. 105–123 in J. Haas, ed., *The Anthropology of War.* Cambridge Univ. Press, Cambridge.

Gregory, R. L. 1973. The confounded eye. Pp. 61–63 in R. L. Gregory and E. H. Gombrich, eds., *Illusion in Nature and Art.* Duckworth, London.

Gregory, R. L. 1980. Perceptions as hypotheses. *Philosophical Transactions of the Royal Society of London,* ser. B, 290:181–197.

Gregory, W. K., and M. Hellman. 1939. The South African fossil man-apes and the origin of the human dentition. *Journal of the American Dental Association* 26:558–564.

Griffin, D. 1976. *The Question of Animal Awareness.* Rockefeller Univ. Press, New York.

Griffin, D. 1992. *Animal Minds.* Univ. of Chicago Press, Chicago.

Griffin, J. 1996. *Value Judgement: Improving our Ethical Beliefs.* Clarendon Press, Oxford.

Griffiths, P. E. 1995. The Cronin controversy. *British Journal of the Philosophy of Science* 46:122–138.

Grine, F. E. 1981. Trophic differences between "gracile" and "robust" australopithicines: A scanning electron microscope analysis of occlusal events. *South African Journal of Science* 77:203–230.

Grine, F. E. 1985. Was interspecific competition a motive force in early hominid evolution? *Transvaal Museum Monograph* 4:143–152.

Grine, F. E. 1987. The diet of South African australopithecines based on a study of dental microwear. *L'Anthropologie* 91:467–482.

Grossman, D. 1995. *On Killing: The Psychological Cost of Learning to Kill in War and Society.* Little, Brown, Boston.

Groth, A. N., and A. W. Burgess. 1977a. Rape: A sexual deviation. *American Journal of Orthopsychiatry* 47:400–406.

Groth, A. N., and A. W. Burgess. 1977b. Sexual dysfunction during rape. *New England Journal of Medicine* 297:764–766.

Gruber, H. E. 1977. *The Essential Piaget.* Routledge and Kegan Paul, London.

Guenther, M. 1988. Animals in Bushman thought, myth, and art. Pp. 192–202 in T. Ingold, D. Riches, and J. Woodburn, eds., *Hunters and Gatherers.* Vol. 2, *Property, Power, and Ideology.* Berg, Oxford.

Guiller, A., and L. Madec. 1993. A contribution to the study of morphological and biochemical differentiation in French and Iberian populations of *Cepaea nemoralis. Biochemical Systematics and Ecology* 21:323–339.

Gullu, S., M. F. Erdogan, A. R. Uysal, N. Baskal, et al. 1998. A potential risk for osteomalacia due to sociocultural lifestyle in Turkish women. *Endocrine Journal* 45:675–678.

Gumperz, J. C., and S. C. Levinson. 1991. Rethinking linguistic relativity. *Current Anthropology* 32:613–623.

Habermas, J. 1979. *Communication and the Evolution of Society.* Beacon Press, Boston.

Hacking, I. 1999. *The Social Construction of What?* Harvard Univ. Press, Cambridge.

Haig, D. 1993. Genetic conflicts in human pregnancy. *Quarterly Review of Biology* 68:495–532.

Haldane, J. B. S. 1932. *The Causes of Evolution.* Longmans, Green, London.

Hall, B. G. 1988. Adaptive mutation that requires multiple spontaneous mutations. I. Mutations involving an insertion sequence. *Genetics* 120:887–897.

Hall, E. D. 1966. *The Hidden Dimension.* Doubleday, New York.

Hallowell, A. I. 1943. The nature and function of property as a social institution. *Journal of Legal and Political Sociology* 1:115–138.

Halverson, J. 1987. Art for art's sake in the Paleolithic. *Current Anthropology* 28:63–89.

Hamburg, D. 1992. *Today's Children.* Times Books, New York.

Hamer, D., and P. Copeland. 1998. *Living with Our Genes: Why They Matter More Than You Think.* Doubleday, New York.

Hamer, D. H., et al. 1993. A linkage between DNA markers on the X chromosome and male sexual orientation. *Science* 261:321–327.

Hames, R. 1991. Wildlife conservation in tribal societies. Pp. 172–199 in B. Oldfield and J. Alcorn, eds., *Biodiversity: Culture, Conservation, and Ecodevelopment.* Westview Press, Boulder.

Hamilton, M. 1982. Sexual dimorphism in skeletal samples. Pp. 107–163 in R. Hall, ed., *Sexual Dimorphism in Homo sapiens.* Praeger, New York.

Hamilton, W. D. 1964. The genetical evolution of social behavior. *Journal of Theoretical Biology* 7:1–52.

Hamilton, W. D. 1966. The moulding of senescence by natural selection. *Journal of Theoretical Biology* 12:12–45.

Hamilton, W. D. 1999. At the world's cross roads. Pp. 125–148 in S. P. Wasser, ed., *Evolutionary Theory and Processes: Modern Perspectives.* Kluwer, Dordrecht.

Hamilton, W. D., R. Axelrod, and R. Tanese. 1990. Sexual selection as an adaptation to resist parasites (a review). *Proceedings of the National Academy of Sciences USA* 87:3566–3573.

Hamilton, W. J. 1987. Omnivorous primate diets and human overconsumption of meat. Pp. 117–132 in M. Harris and E. Ross, eds., *Food and Evolution: Toward a Theory of Human Food Habits.* Temple Univ. Press, Philadelphia.

Hammer, M. F. 1995. A recent common ancestry for human Y chromosomes. *Nature* 378:376–378.

Harcourt, A. H. 1981. Intermale competition and the reproductive behavior of the great apes. Pp. 301–318 in C. E. Graham, ed., *Reproductive Biology of the Great Apes.* Academic Press, New York.

Harcourt, A. H. 1995. Sexual selection and sperm competition in primates: What are male genitalia good for? *Evolutionary Anthropology* 4:121–129.

Harcourt, A. H. 1997. Sperm competition in mammals. *American Naturalist* 149:189–194.

Harcourt, A. H., and J. Gardiner. 1994. Sexual selection and genital anatomy of male primates. *Proceedings of the Royal Society of London*, ser. B, 255:47–53.

Harcourt, A. H., and S. A. Harcourt. 1984. Insectivory by gorillas. *Folia Primatologica* 43:229–233.

Harcourt, A. H., P. H. Harvey, S. G. Larson, and R. V. Short. 1981. Testis weight, body weight, and breeding system in primates. *Nature* 293:55–57.

Harcourt, A. H., A. Purvis, and L. Liles. 1995. Sperm competition: Mating system, not breeding system, affects testes size of primates. *Functional Ecology* 9:468–476.

Harcourt, A. H., and K. J. Stewart. 1989. Functions of alliances in contests within wild gorilla groups. *Behaviour* 109:176–190.

Harcourt, A. H., and F. B. M. de Waal, eds. 1992. *Coalitions and Alliances in Humans and Other Animals.* Oxford Univ. Press, Oxford.

Hardin, G. 1968a. *Exploring New Ethics for Survival: The Voyage of the Spaceship Beagle.* Viking, New York.

Hardin, G. 1968b. The tragedy of the commons. *Science* 162:1243–1247.

Hardin, G. 1977. *The Limits of Altruism: An Ecologist's View of Survival.* Indiana Univ. Press, Bloomington.

Hardin, G. 1993. *Living within Limits.* Oxford Univ. Press, New York.

Hardin, G. 1996. *Stalking the Wild Taboo.* Social Contract Press, Petoskey, MI.

Harding, R. 1975. Meat eating and hunting in baboons. Pp. 245–257 in R. H. Tuttle, ed., *Socioecology and Psychology of Primates.* Mouton, The Hague.

Harding, R. M., S. M. Fullerton, R. C. Griffiths, J. Bond, et al. 1997. Archaic African *and* Asian lineages in the genetic ancestry of modern humans. *American Journal of Human Genetics* 60:772–789.

Harding, S. 1991. *Whose Science? Whose Knowledge? Thinking from Women's Lives.* Cornell Univ. Press, Ithaca.

Harlan, J. R. 1989. Wild-grass seed harvesting in the Sahara and Sub-Sahara of Africa. Pp. 79–98 in D. R. Harris and G. C. Hillman, eds., *Foraging and Farming: The Evolution of Plant Exploitation.* Unwin Hyman, London.

Harlan, J. R., J. M. J. de Wet, and E. G. Price. 1973. Comparative evolution of cereals. *Evolution* 27:311–325.

Harlow, H. F. 1958. The evolution of learning. Pp. 269–290 in A. Roe and G. G. Simpson, eds., *Behavior and Evolution.* Yale Univ. Press, New Haven.

Harlow, J. M. 1868. Recovery from the passage of an iron bar through the head. *Publications of the Massachusetts Medical Society* 2:327–347.

Harner, M. 1977. The ecological basis for Aztec sacrifice. *American Ethnologist* 4:117–135.

Harpending, H. C. 1994. Gene frequencies, DNA sequences, and human origins. *Perspectives in Biology and Medicine* 37:384–394.

Harpending, H. C., and J. Relethford. 1997. Population perspectives on human origins research. Pp. 361–368 in G. A. Clark and C. M. Willermet, eds., *Conceptual Issues in Modern Human Origins Research.* Aldine, New York.

Harrell, B. B. 1981. Lactation and menstruation in cultural perspective. *American Anthropologist* 83:796–823.

Harris, D. R. 1977a. Alternative pathways toward agriculture. Pp. 179–243 in C. A. Reed, ed., *Origins of Agriculture.* Mouton, The Hague.

Harris, D. R. 1977b. Settling down: An evolutionary model for the transformation of mobile bands into sedentary communities. Pp. 401–417 in J. Friedman and M. Rowlands, eds., *The Evolution of Social Systems.* Duckworth, London.

Harris, D. R., and G. C. Hillman. 1989. *Foraging and Farming: The Evolution of Plant Exploration.* Unwin Hyman, London.

Harris, J. R. 1995. Where is the child's environment? A group socialization theory of development. *Psychological Review* 102:458–489.

Harris, J. R. 1998. *The Nurture Assumption: Why Children Turn Out the Way They Do.* Free Press, New York.

Harris, M. 1968. *The Rise of Anthropological Theory.* Crowell, New York.

Harris, M. 1977. *Cannibals and Kings.* Random House, New York.

Harris, M. 1980. *Cultural Materialism: The Struggle for a Science of Culture.* Vintage Books, New York.

Harris, M. 1984. Animal capture and Yanomamö warfare: Retrospect and new evidence. *Journal of Anthropological Research* 40:183–201.

Harris, M. 1989. *Our Kind: Who We Are, Where We Came From, Where We Are Going.* Harper and Row, New York.

Harris, M. 1997. *Culture, People, and Nature: An Introduction to General Anthropology.* Longman, New York.

Harris, M., and E. B. Ross. 1987a. *Death, Sex, and Fertility: Population Regulation in Pre-industrial and Developing Societies.* Columbia Univ. Press, New York.

Harris, M., and E. B. Ross, eds. 1987b. *Food and Evolution: Toward a Theory of Human Food Habits.* Temple Univ. Press, Philadelphia.

Harris, R. 1986. *The Origin of Writing.* Duckworth, London.

Harrison, R. G. 1998. Linking evolutionary pattern and process: The relevance of species concepts for the study of speciation. Pp. 19–31 in D. J. Howard and S. H. Berlocher, eds., *Endless Forms: Species and Speciation.* Oxford Univ. Press, New York.

Harrison, T., and L. Rook. 1997. Enigmatic anthropoid or misunderstood ape? The phylogenetic status of *Oreopithecus bambolii* reconsidered. Pp. 327–362 in D. R. Begun, C. V. Ward, and M. D. Rose, eds., *Function, Phylogeny, and Fossils: Miocene Hominid Evolution and Adaptation.* Plenum Press, New York.

Hart, J., Jr., and B. Gordon. 1992. Neural subsystems for object knowledge. *Nature* 359:60–64.

Hartl, D. L., and A. G. Clark. 1997. *Principles of Population Genetics.* 3rd ed. Sinauer, Sunderland, MA.

Hartl, D. L., and E. W. Jones. 1998. *Genetics: Principles and Analysis.* 4th ed. Jones and Bartlett, London.

Hartung, J. 1985. Review of *Incest: A Biosocial View* by J. Shepher. *American Journal of Physical Anthropology* 67:169–171.

Hashimoto, C., and T. Furuichi. 1994. Social role and development of noncopulatory sexual behavior of wild bonobos. Pp. 155–168 in R. W. Wrangham, W. C. McGrew, F. B. M. de Waal, and P. G. Heltne, eds., *Chimpanzee Cultures.* Harvard Univ. Press, Cambridge.

Hata, Y., and M. P. Stryker. 1994. Control of thalamocortical afferent rearrangement by postsynaptic activity in developing visual cortex. *Science* 265:1732–1735.

Hatch, E. 1983. *Cultures and Morality: The Relativity of Values in Anthropology.* Columbia Univ. Press, New York.

Hauer, C. 1988. The rise of the Israelite monarchy. *American Behavioral Scientist* 31:428–437.

Haught, J. F. 1995. *Science and Religion: From Conflict to Conversation.* Paulist Press, Mahwah, NJ.

Hauser, R. M. 1998. Trends in black-white test-score differentials. I. Uses and misuses of NAEP/SAT data. Pp. 219–249 in U. Neisser, ed., *The Rising Curve: Long-Term Gains in IQ and Related Measures.* American Psychological Association, Washington, DC.

Hausfater, G., and S. B. Hrdy. 1984. *Infanticide: Comparative and Evolutionary Perspectives.* Aldine, New York.

Havel, V. 1999. Kosovo and the end of the nation state. *New York Review of Books* (10 June): 3.

Hawkes, K., J. F. O'Connell, and N. G. Blurton Jones. 1997. Hadza women's time allocation, offspring provisioning, and the evolution of long postmenopausal life spans. *Current Anthropology* 38:551–577.

Hawkes, K., J. F. O'Connell, N. G. Blurton Jones, H. Alvarez, et al. 1998. Grandmothering, menopause, and the evolution of human life histories. *Proceedings of the National Academy of Sciences USA* 95:1336–1339.

Hayden, B. 1979. *Lithic Use-Wear Analysis.* Academic Press, New York.

Hayden, B. 1987. Alliances and ritual ecstasy: Human responses to resource stress. *Journal for the Scientific Study of Religion* 26:81–91.

Hayden, B. 1993. The cultural capacities of Neandertals: A review and re-evaluation. *Journal of Human Evolution* 24:113–146.

Hayden, B., M. Deal, A. Cannon, and J. Casey. 1986. Ecological determinants of women's status among hunter/gatherers. *Human Evolution* 1:449–474.

Hayek, F. A. 1944. *The Road to Serfdom.* Univ. of Chicago Press, Chicago.

Hayek, F. A. 1979. *Law, Legislation, and Liberty: The Political Order of a Free People.* Univ. of Chicago Press, Chicago.

Hayes, R. O. 1975. Female genital mutilation, fertility control, women's roles, and the patrilineage in modern Sudan: A functional analysis. *American Ethnologist* 2:617–633.

Hebb, D. O. 1949. *The Organization of Behavior.* Wiley, New York.

Hebb, D. O. 1953. Heredity and environment in mammalian behaviour. *British Journal of Animal Behaviour* 1:43–47.

Hedges, S. B., P. H. Parker, C. G. Sibley, and S. Kumar. 1996. Continental breakup and the ordinal diversification of birds and mammals. *Nature (London)* 381:226–229.

Hedrick, P. W., and F. L. Black. 1997. HLA and mate selection: No evidence in South Amerindians. *American Journal of Human Genetics* 61:505–511.

Hegel, G. W. F. 1991 (1821). World history. Pp. 63–72 in R. Sältzer, ed., *German Essays on History.* Continuum, New York.

Hegi, U. 1997. *Stones from the River.* Simon and Schuster, New York.

Heino, M., J. A. J. Metz, and V. Kaitala. 1998. The enigma of frequency selection. *Trends in Ecology and Evolution* 13:367–370.

Heinzelin, J. de, J. D. Clark, T. White, W. Hart, et al. 1999. Environment and behavior of 2.5-million-year-old Bouri hominids. *Science* 284:625–636.

Held, D. 1989. *Political Theory and the Modern State: Essays on State, Power, and Democracy*. Stanford Univ. Press, Stanford.

Heller, T. C. 1997. Modernity, membership, and multiculturalism. *Stanford Humanities Review* 5:2–69.

Henderson, B. E., R. K. Ross, and M. C. Pike. 1993. Hormonal chemoprevention of cancer in women. *Science* 259:633–638.

Henderson, J. B., M. G. Dunnigan, W. B. McIntosh, A. A. Abdul-Motaal, et al. 1987. The importance of limited exposure to ultraviolet radiation and dietary factors in the aetiology of Asian rickets: A risk factor model. *Quarterly Journal of Medicine* 63:413–425.

Henderson, L. J. 1913. *The Fitness of the Environment*. Macmillan, New York.

Herdt, G. 1988. Cross-cultural forms of homosexuality and the concept of "gay." *Psychiatric Annals* 18:36–39.

Herdt, G. H., ed. 1982. *Rituals of Manhood: Male Initiation in Papua New Guinea*. Univ. of California Press, Berkeley.

Herdt, G. H., ed. 1984. *Ritualized Homosexuality in Melanesia*. Univ. of California Press, Berkeley.

Herrnstein, R. J., and C. Murray. 1994. *The Bell Curve: Intelligence and Class Structure in American Life*. Free Press, New York.

Heston, L. 1966. Psychiatric disorders in foster home reared children of schizophrenic mothers. *British Journal of Psychiatry* 112:819–825.

Hewes, G. W. 1961. Food transport and the origin of human bipedalism. *American Anthropologist* 63:687–710.

Hewes, G. W. 1973a. An explicit formulation of the relationship between tool-using, tool-making, and the emergence of language. *Visible Language* 7:101–127.

Hewes, G. W. 1973b. Primate communication and the gestural origins of language. *Current Anthropology* 14:5–24.

Hewes, G. W. 1994. The baseline for comparing human and nonhuman primate behavior. Pp. 59–93 in D. Quiatt and J. Itani, eds., *Hominid Culture in Primate Perspective*. Univ. Press of Colorado, Niwot.

Hewlett, B. S., and L. L. Cavalli-Sforza. 1986. Cultural transmission among Aka Pygmies. *American Anthropologist* 88:922–934.

Heyes, C. M. 1993. Anecdotes, training, trapping, and triangulating: Do animals attribute mental states? *Animal Behaviour* 46:177–188.

Heywood, V. H., ed. 1995. *Global Biodiversity Assessment*. Cambridge Univ. Press, Cambridge.

Higashi, M., G. Takimoto, and N. Yamamura. 1999. Sympatric speciation by sexual selection. *Nature* 402:523–526.

Higgins, K. 1996. Schopenhauer. Pp. 508–511 in T. Mautner, ed., *Penguin Dictionary of Philosophy*. Penguin Books, London.

Hill, J. H., and B. Mannheim. 1992. Language and world view. *Annual Review of Anthropology* 21:381–406.

Hill, K., and A. M. Hurtado. 1991. The evolution of premature reproductive senescence and menopause in human females: An evaluation of the "grandmother hypothesis." *Human Nature* 2:313–350.

Hill, K., and H. Kaplan. 1988. Trade-offs in male and female reproductive strategies among the Ache. Pp. 291–305 in L. Betzig, M. B. Mulder, and P. Turke, eds., *Human Reproductive Behavior: A Darwinian Perspective*. Cambridge Univ. Press, Cambridge.

Hill, R. A., and P. C. Lee. 1998. Predation risk as an influence on group size in cercopithecoid primates: Implications for social structure. *Journal of Zoology London* 245:447–456.

Hillman, G. C., and M. S. Davies. 1990. Domestication rates in wild-type wheats and barley under primitive cultivation. *Biological Journal of the Linnean Society* 39:39–78.

Hinde, R. A. 1966. *Animal Behaviour: A Synthesis of Ethology and Comparative Psychology*. McGraw-Hill, New York.

Hirschfeld, L. A. 1989. Rethinking the acquisition of kinship terms. *International Journal of Behavioral Development* 12:541–568.

Hirschfeld, L. A., and S. A. Gelman, eds. 1994. *Mapping the Mind: Domain Specificity in Cognition and Culture*. Cambridge Univ. Press, Cambridge.

Hirschman, A. O. 1977. *The Passions and the Interests: Political Arguments for Capitalism before Its Triumph*. Princeton Univ. Press, Princeton.

Hirsh-Pasek, K., and R. M. Golinkoff. 1991. Language comprehension: A new look at some old themes. Pp. 301–320 in N. Krasnegor, D. Rumbaugh, M. Studdert-Kennedy, and R. Schiefenbusch, eds., *Biological and Behavioral Aspects of Language Acquisition*. Erlbaum, Hillsdale, NJ.

Hirth, F., and H. Reichert. 1999. Conserved genetic programs in insect and mammalian brain development. *Bioessays* 21:677–684.

Hölldobler, B., and E. O. Wilson. 1990. *The Ants*. Harvard Univ. Press, Cambridge.

Hobbes, T. 1997 (1651). *Leviathan.* Norton, New York.

Hobsbawm, E. J. 1992. *Nations and Nationalism since 1780: Programme, Myth, Reality.* 2nd ed. Cambridge Univ. Press, Cambridge.

Hobson, J. A. 1994. *The Chemistry of Conscious States.* Little, Brown, Boston.

Hobson, J. A., and R. W. McCarley. 1977. The brain as a dream state generator: An activation-synthesis hypothesis of the dream process. *American Journal of Psychiatry* 134:1335–1348.

Hobson, S., and J. Lubchenco, eds. 1997. *Revelation and the Environment, A.D. 95–1995.* World Scientific, Singapore.

Hodell, D. A., J. H. Curtis, and M. Brenner. 1995. Possible role of climate in the collapse of classic Maya civilization. *Nature* 375:391–394.

Hoebel, E. A. 1949. *Man in the Primitive World.* McGraw-Hill, New York.

Hoebel, E. A. 1964. *The Law of Primitive Man: A Study of Comparative Legal Dynamics.* Harvard Univ. Press, Cambridge.

Hoffman, D. D. 1998. *Visual Intelligence: How We Create What We See.* Norton, New York.

Hohmann, G. W. 1966. Some effects of spinal cord lesions on experienced emotional feelings. *Psychophysiology* 3:143–156.

Hoijer, H. 1954. *Language in Culture.* Univ. of Chicago Press, Chicago.

Holdren, C., and A. Ehrlich. 1984. The Virunga volcanoes: Last redoubt of the mountain gorilla. *Not Man Apart* 8:8–9.

Holdren, J. P. 1991. Population and the energy problem. *Population and Environment* 12:231–255.

Holdren, J. P., and P. R. Ehrlich. 1974. Human population and the global environment. *American Scientist* 62:282–292.

Hole, F., and J. McCorriston. 1992. Reply to Graber. *American Anthropologist* 94:445–446.

Holliday, T. W. 1997. Postcranial evidence of cold adaptations in European Neanderthals. *American Journal of Physical Anthropology* 104:245–258.

Holling, C. S., and G. F. Meffe. 1996. Command and control and the pathology of natural resource management. *Conservation Biology* 10:328–337.

Holloway, R. L., Jr. 1967. Human aggression: The need for a species-specific framework. *Natural History* 76:44–48.

Holloway, R. L., Jr. 1969. Culture: A human domain. *Current Anthropology* 10:395–412.

Holloway, R. L. 1995. Toward a synthetic theory of human brain evolution. Pp. 42–54 in J. P. Changeux and J. Chavaillon, eds., *Origins of the Human Brain.* Oxford Univ. Press, Oxford.

Holmes, W. G., and P. W. Sherman. 1983. Kin recognition in animals. *American Scientist* 71:46–55.

Homer-Dixon, T., and V. Percival. 1996. *Environmental Scarcity and Violent Conflict: Briefing Book.* American Association for the Advancement of Science, Washington, DC.

Honderich, T., ed. 1995. *The Oxford Companion to Philosophy.* Oxford Univ. Press, Oxford.

Hookway, C. 1988. *Quine: Language, Experience, and Reality.* Stanford Univ. Press, Stanford.

Hopkins, K. 1980. Brother-sister marriage in Roman Egypt. *Comparative Studies in Society and History* 22:303–354.

Horai, S., K. Hayaska, R. Kondo, K. Tsugane, et al. 1995. Recent African origin of modern humans revealed by complete sequences of hominoid mitochondrial DNAs. *Proceedings of the National Academy of Sciences USA* 92:532–536.

Hori, M. 1993. Frequency-dependent natural selection in the handedness of scale-eating cichlid fish. *Science* 260:216–219.

Horowitz, I. L. 1997. *Taking Lives: Genocide and State Power.* 4th ed. Transaction, New Brunswick, NJ.

Hosoi, J., G. F. Murphy, C. L. Egan, E. A. Lerner, et al. 1993. Regulation of Langerhans cell function by nerves containing calcitonin gene-related peptide. *Nature* 363:159–163.

Hou, L., L. D. Martin, Z. Zhou, and A. Feduccia. 1996. Early adaptive radiation of birds: Evidence from fossils from northeastern China. *Science* 274:1164–1167.

Housman, A. E. 1951. *A Shropshire Lad.* Shakespeare House, New York.

Howard, D. J., and S. H. Berlocher, eds. 1998. *Endless Forms: Species and Speciation.* Oxford Univ. Press, Oxford.

Howard, M. 1984. *The Causes of Wars.* 2nd ed. Harvard Univ. Press, Cambridge.

Howell, F. C. 1996. Thoughts on the study and interpretation of the human fossil record. Pp. 1–45 in W. E. Meikle, F. C. Howell, and N. G. Jablonski, eds., *Contemporary Issues in Human Evolution.* California Academy of Sciences, San Francisco.

Howells, W. 1997. *Getting Here: The Story of Human Evolution.* New ed. Compass Press, Washington, DC.

Hrdy, S. B. 1977. *The Langurs of Abu: Male and Female Strategies of Reproduction.* Harvard Univ. Press, Cambridge.

Hrdy, S. B. 1981. *The Woman Who Never Evolved.* Harvard Univ. Press, Cambridge.

Hrdy, S. B. 1983. Heat loss. *Science '83* 1983:73–78.

Hrdy, S. B. 1999. *Mother Nature: A History of Mothers, Infants, and Natural Selection.* Pantheon Books, New York.

Hrdy, S. B., and G. C. Williams. 1983. Behavioral biology and the double standard. Pp. 3–17 in S. K. Wasser, ed., *Social Behavior of Female Vertebrates.* Academic Press, New York.

Hu, S., A. M. L. Pattatucci, C. Patterson, L. Li, et al. 1995. Linkage between sexual orientation and chromosome Xq28 in males but not in females. *Nature Genetics* 11:248–256.

Huang, W., Y. Ciochon, Y. Gu, R. Larick, et al. 1995. Early *Homo* and associated artifacts from Asia. *Nature* 378:275–278.

Hubel, D. H. 1988. *Eye, Brain, and Vision.* Scientific American Library, New York.

Hublin, J.-J. 1996. The first Europeans. *Archaeology* 49:36–44.

Hublin, J.-J., F. Spoor, M. Braun, F. Zonneveld, et al. 1996. A late Neanderthal associated with upper Palaeolithic artefacts. *Nature* 381:224–226.

Hudson, W. 1960. Pictorial depth perception in sub-cultural groups in Africa. *Journal of Social Psychology* 52:183–208.

Huesmann, L. R., and L. D. Eron. 1986. The development of aggression in American children as a consequence of television violence viewing. Pp. 44–80 in L. R. Huesmann and L. D. Eron, eds., *Television and the Aggressive Child: A Cross-National Comparison.* Erlbaum, Hillsdale, NJ.

Hughes, J. B., G. C. Daily, and P. R. Ehrlich. 1997. Population diversity: Its extent and extinction. *Science* 278:689–692.

Hughes, J. D. 1975. *Ecology in Ancient Civilizations.* Univ. of New Mexico Press, Albuquerque.

Hull, D. 1986. On human nature. Pp. 3–13 in A. Fine and P. Machamer, eds., *Proceedings of the Philosophy of Science Association.* Philosophy of Science Association, East Lansing, MI.

Hull, D. L. 1988. *Science as a Process.* Univ. of Chicago Press, Chicago.

Hume, D. 1977 (1777). *An Enquiry Concerning Human Understanding.* Hackett, Indianapolis.

Hume, D. 1978 (1739). *A Treatise of Human Nature.* 2nd ed. Oxford Univ. Press, Oxford.

Humphrey, N. K. 1976. The social function of intellect. Pp. 303–317 in P. P. G. Bateson and R. A. Hinde, eds., *Growing Points in Ethology.* Cambridge Univ. Press, Cambridge.

Humphrey, N. K. 1992. *A History of the Mind: Evolution and the Birth of Consciousness.* Chatto and Windus, London.

Humphrey, N. K. 1998. Cave art, autism, and the evolution of the human mind. *Cambridge Archaeological Journal* 8:165–191.

Hunt, K. D. 1994. The evolution of human bipedality: Ecology and functional morphology. *Journal of Human Evolution* 26:183–202.

Huntington, S. P. 1993. The clash of civilizations? *Foreign Affairs* 72:22–49.

Huntington, S. P. 1996. *The Clash of Civilizations and the Remaking of World Order.* Simon and Schuster, New York.

Hurford, J. 1991. The evolution of the critical period for language acquisition. *Cognition* 40:159–201.

Hurford, J. R., M. Studdert-Kennedy, and C. Knight, eds. 1998. *Approaches to the Evolution of Language.* Cambridge Univ. Press, Cambridge.

Hurtado, A. M., I. A. D. Hurtado, R. Sapien, and K. Hill. 1999. The evolutionary ecology of childhood asthma. Pp. 101–134 in W. R. Trevathan, E. O. Smith, and J. J. McKenna, eds., *Evolutionary Medicine.* Oxford Univ. Press, New York.

Hussain, R., and A. H. Bittles. 1998. The prevalence and demographic characteristics of consanguineous marriages in Pakistan. *Journal of Biosocial Science* 30:261–275.

Huxley, A. 1932. *Brave New World.* Harper and Row, New York.

Huxley, J. 1943. *Evolution: The Modern Synthesis.* Harper and Brothers, New York.

Huxley, T. H. 1863. *Evidence as to Man's Place in Nature.* Appleton, New York.

Huxley, T. H. 1910. *Lectures and Lay Sermons.* Dutton, New York.

Hyatt, C. W., and W. D. Hopkins. 1994. Self-awareness in bonobos and chimpanzees. Pp. 248–253 in S. T. Parker, R. W. Mitchell, and M. L. Boccia, eds., *Self-Awareness in Animals and Humans: Developmental Perspectives.* Cambridge Univ. Press, Cambridge.

Idani, G. 1991a. Cases of inter-unit group encounters in pygmy chimpanzees at Wamba, Zaire. Pp. 235–238 in A. Ehara, ed., *Primatology Today: Proceedings of the Thirteenth Congress of the International Primatological Society.* Elsevier, Amsterdam.

Idani, G. 1991b. Social relationships between immigrant and resident bonobo *(Pan paniscus)* females at Wamba. *Folia Primatologica* 57:83–95.

Iggers, G. G. 1997. *Historiography in the Twentieth Century: From Scientific Objectivity to the Postmodern Challenge.* Wesleyan Univ. Press and Univ. Press of New England, Hanover, NH.

Ihobe, H. 1997. Non-antagonistic relations between wild bonobos and two species of guenons. *Primates* 38:351–357.

Iltis, H. H., O. L. Loucks, and P. Andrews. 1970. Criteria for an optimum human environment. *Bulletin of the Atomic Scientists* 26:2–6.

Ingold, T. 1986. *The Appropriation of Nature: Essays on Human Ecology and Social Relations.* Manchester Univ. Press, Manchester.

Ingold, T., D. Riches, and J. Woodburn, eds. 1991 (1988). *Hunters and Gatherers.* Vol. 1, *History, Evolution, and Social Change.* Vol. 2, *Property, Power, and Ideology.* Berg, Oxford.

Ingram, V. M. 1956. A specific chemical difference between the globins of normal human and sickle-cell anemia hemoglobin. *Nature* 178:792–794.

Ingram, V. M. 1959. Abnormal human hemoglobins. III. The chemical difference between normal and sickle cell hemoglobins. *Biochemical and Biophysical Acta* 36:402–411.

Intergovernmental Panel on Climate Change (IPCC). 1996. *Climate Change, 1995.* Cambridge Univ. Press, Cambridge.

International Working Group on Indigenous Affairs. 1992. Declaration by the indigenous peoples. *IWGIA Yearbook 1991.* International Working Group on Indigenous Affairs, Copenhagen.

Irvine, W. 1955. *Apes, Angels, and Victorians.* McGraw-Hill, New York.

Isaac, G. 1978. The food-sharing behavior of protohuman hominids. *Scientific American* 238:90–108.

Iverson, J. M., and S. Goldin-Meadow. 1998. Why people gesture when they speak. *Nature* 396:228.

Iwamoto, T. 1993. Food digestion and energetic conditions in *Theropithecus gelada.* Pp. 453–463 in N. G. Jablonski, ed., *Theropithecus: The Rise and Fall of a Primate Genus.* Cambridge Univ. Press, Cambridge.

Iwamoto, T., and R. Dunbar. 1983. Thermoregulation, habitat quality, and the behavioral ecology of gelada baboons. *Journal of Animal Ecology* 52:357–366.

Jablonski, D. 1997. Body-size evolution in Cretaceous molluscs and the status of Cope's rule. *Nature* 385:250–252.

Jablonski, D. 1999. The future of the fossil record. *Science* 284:2114–2116.

Jablonski, N. G. 1992. Sun, skin color, and spina bifida: An exploration of the relationship between ultraviolet light and neural tube defects. *Proceedings of the Australasian Society of Human Biology* 5:455–462.

Jablonski, N. G., ed. 1993. *Theropithecus: The Rise and Fall of a Primate Genus.* Cambridge Univ. Press, Cambridge.

Jablonski, N. G., and G. Chaplin. 1993. Origin of habitual terrestrial bipedalism in the ancestor of the Hominidae. *Journal of Human Evolution* 24:259–280.

Jablonski, N. G., and G. Chaplin. 2000. The evolution of human skin pigmentation. *Journal of Human Evolution,* in press.

Jackendoff, R. 1987. *Consciousness and the Computational Mind.* MIT Press, Cambridge.

Jackendoff, R. 1994. *Patterns in the Mind: Language and Human Nature.* Basic Books, New York.

Jacob, F. 1977. Evolution and thinking. *Science* 196:1161–1166.

Jacob, F. 1982. *The Possible and the Actual.* Univ. of Washington Press, Seattle.

Jacobs, J. 1993 (1961). *The Death and Life of Great American Cities.* Random House, New York.

Jacobsen, T., and R. M. Adams. 1958. Salt and silt in ancient Mesopotamian agriculture. *Science* 128:1251–1258.

James, S. R. 1989. Hominid use of fire in the Lower and Middle Pleistocene. *Current Anthropology* 30:1–26.

James, W. 1880. Great men, great thoughts, and the environment. *Atlantic Monthly* 46:441–459.

James, W. 1975 (1907, 1909). *Pragmatism and the Meaning of Truth.* Harvard Univ. Press, Cambridge.

Jankowiak, W. R., and E. F. Fischer. 1992. A cross-cultural perspective on romantic love. *Ethnology* 31:149–155.

Jankowiak, W. R., E. M. Hill, and J. M. Donovan. 1992. The effects of sex and sexual orientation on attractiveness judgements: An evolutionary interpretation. *Ethology and Sociobiology* 13:73–85.

Jardin, C. 1967. *List of Foods Used in Africa.* Food and Agriculture Organization of the United Nations, Rome.

Järvi, T., and M. Bakken. 1984. The function of variation in the breast stripe of the great tit *(Parus major). Animal Behaviour* 32:590–596.

Jay, V., and L. E. Becker. 1995. Surgical pathology of epilepsy resections in childhood. *Seminars in Pediatric Neurology* 2:227–236.

Jaynes, J. 1976. *The Origin of Consciousness in the Breakdown of the Bicameral Mind.* Houghton Mifflin, Boston.

Jefferson, T. 1801. *Notes on the State of Virginia, with an Appendix.* 3rd American ed. Furman and Loudon, New York.

Jenkins, W. M., M. M. Merzenich, M. T. Ochs, T. Allard, et al. 1990. Functional reorganization of primary somatosensory cortex in adult owl monkeys after behaviorally controlled tactile stimulation. *Journal of Neurophysiology* 63:82–104.

Jensen, A. R. 1969. How much can we boost IQ and scholastic achievement? *Harvard Educational Review* 39:1–123.

Johanson, D., and B. Edgar. 1996. *From Lucy to Language.* Simon and Schuster, New York.

Johns, T. 1989. A chemical-ecological model of root and tuber domestication in the Andes. Pp. 504–519 in D. R. Harris and G. C. Hillman, eds., *Foraging and Farming: The Evolution of Plant Exploitation.* Unwin Hyman, London.

Johnson, A. 1983. Machiguenga gardens. Pp. 29–64 in R. Hames and W. Vickers, eds., *Adaptive Response of Native Amazonians.* Academic Press, New York.

Johnson, A. W., and T. Earle. 1987. *The Evolution of Human Societies: From Foraging Group to Agrarian State.* Stanford Univ. Press, Stanford.

Johnson, O., ed. 1997. *Information Please Almanac.* Houghton Mifflin, Boston.

Johnson, T. C., C. A. Scholz, M. R. Talbot, K. Kelts, et al. 1996. Late Pleistocene desiccation of Lake Victoria and rapid evolution of cichlid fishes. *Science* 273:1091–1093.

Johnson-Laird, P. N., and K. Oatley. 1992. Basic emotions, rationality, and folk theory. *Cognition and Emotion* 6:201–223.

Johnston, T. 1981. Contrasting approaches to a theory of learning. *Behavioral and Brain Sciences* 4:125–139.

Jolly, A. 1966. Lemur social behavior and primate intelligence. *Science* 153:501–506.

Jolly, A. 1972. *The Evolution of Primate Behavior.* Macmillan, New York.

Jolly, C. J. 1970. The seed-eaters: A new model of hominid differentiation based on a baboon analogy. *Man* 5:1–26.

Jonaitis, A. 1991. *Chiefly Feasts: The Enduring Kwakiutl Potlatch.* Univ. of Washington Press, Seattle.

Jones, A. 1999. Review: *Hitler's Pope. National Catholic Reporter* (19 November): 13.

Jones, J., and R. Sawhill. 1992. Just too good to be true: Another reason to beware of false eco-prophets. *Newsweek* (4 May): 68.

Jones, J. S., B. H. Leith, and P. Rawlings. 1977. Polymorphism in *Cepaea:* A problem with too many solutions. *Annual Review of Ecology and Systematics* 8:109–143.

Jones, S., R. Martin, and D. Pilbeam, eds. 1992. *The Cambridge Encyclopedia of Human Evolution.* Cambridge Univ. Press, Cambridge.

Joravsky, D. 1970. *The Lysenko Affair.* Univ. of Chicago Press, Chicago.

Jurmain, R. 1997. Skeletal evidence of trauma in African apes, with special reference to Gombe chimpanzees. *Primates* 38:1–14.

Kagan, D. 1995. *On the Origins of War and the Preservation of Peace.* Doubleday, New York.

Kagan, J. 1998. *Three Seductive Ideas.* Harvard Univ. Press, Cambridge.

Kahneman, D. 1980. Human engineering of decisions. Pp. 190–192 in M. Kranzberg, ed., *Ethics in an Age of Pervasive Technology.* Westview Press, Boulder.

Kaidbey, K. H., P. Poh Agin, R. M. Sayre, and A. M. Kligman. 1979. Photoprotection by melanin: A comparison of black and Caucasian skin. *Journal of the American Academy of Dermatology* 1:249–260.

Kandel, E. R., J. H. Schwartz, and T. M. Jessell. 1995. *Essentials of Neural Science and Behavior.* Appleton and Lange, Stamford, CT.

Kane, R. 1998. *The Significance of Free Will.* Oxford Univ. Press, New York.

Kane, R. H. 1999. *The Quest for Meaning: Values, Ethics, and the Modern Experience.* The Teaching Company, Springfield, VA.

Kanin, E. J. 1957. Male aggression in dating-courtship relations. *American Journal of Sociology* 63:197–204.

Kano, T. 1992. *The Last Ape: Pygmy Chimpanzee Behavior and Ecology.* Stanford Univ. Press, Stanford.

Kano, T. 1996. Male rank order and copulation rate in a unit-group of bonobos at Wamba, Zaire. Pp. 135–145 in W. C. McGrew, L. F. Marchant, and T. Nishida, eds., *Great Ape Societies.* Cambridge Univ. Press, Cambridge.

Kant, I. 1996 (1797). *The Metaphysics of Morals.* Cambridge Univ. Press, Cambridge.

Kaplan, H., and K. Hill. 1985a. Food sharing among Ache foragers: Tests of explanatory hypotheses. *Current Anthropology* 26:223–239.

Kaplan, H., and K. Hill. 1985b. Hunting ability and reproductive success among Ache foragers. *Current Anthropology* 26:131–133.

Kaplan, H., K. Hill, J. Lancaster, and A. M. Hurtado. 1999. A theory of human life history evolution: Brains, learning, and longevity. *Evolutionary Anthropology*, submitted.

Kaplan, M. A. 1998. *Between Dignity and Despair: Jewish Life in Nazi Germany*. Oxford Univ. Press, New York.

Kaplan, S. 1987. Aesthetics, affect, and cognition: Environmental preference from an evolutionary perspective. *Environment and Behavior* 19:3–32.

Kappeler, P. M., and C. P. van Schaik. 1992. Methodological and evolutionary aspects of reconciliation among primates. *Ethology* 92:51–69.

Karmiloff-Smith, A. 1992. *Beyond Modularity: A Developmental Perspective on Cognitive Issues*. MIT Press, Cambridge.

Katz, L. C., and C. J. Shatz. 1996. Synaptic activity and the construction of cortical circuits. *Science* 274:1133–1138.

Katz, N., E. Baker, and J. Macnamara. 1974. What's in a name? A study of how children learn common and proper names. *Child Development* 45:469–473.

Kaufman, L. S. 1991. Progress in the conservation of endemic fishes from Lake Victoria. Pp. 403–408 in *Proceedings of the 1992 Annual Conference of the American Association of Zoological Parks and Aquaria*. American Association of Zoological Parks and Aquaria, San Diego.

Kaufman, L. S. 1997. Asynchronous taxon cycles in haplochromine fishes of the greater Lake Victoria region. *South African Journal of Science* 93:601–606.

Kaufman, L. S., C. A. Chapman, and L. J. Chapman. 1997. Evolution in fast forward: Haplochromine fishes of the Lake Victoria region. *Endeavour (London)* 21:23–30.

Kaufman, L. S., and K. F. Liem. 1982. Fishes of the suborder Labroidei (Pisces: Perciformes): Phylogeny, ecology, and evolutionary significance. *Breviora* 217:1–19.

Kaufman, L. S., and P. Ochumba. 1993. Evolutionary and conservation biology of cichlid fishes as revealed by faunal remnants in Lake Victoria. *Conservation Biology* 7:719–730.

Kay, C. E. 1994. Aboriginal overkill. *Human Nature* 5:359–398.

Kay, C. E. 1995. Aboriginal overkill and native burning: Implications for modern ecosystem management. *Western Journal of Applied Forestry* 10:121–126.

Kay, P., B. Berlin, and W. Merrifield. 1991. Biocultural implications of systems of color naming. *Journal of Linguistic Anthropology* 1:12–25.

Kay, R. F., M. Cartmill, and M. Balow. 1998. The hypoglossal canal and the origin of human vocal behavior. *Proceedings of the National Academy of Sciences USA* 95:5417–5419.

Kay, R. F., C. Ross, B. A. Williams, and D. Johnson. 1997. Cladistic analysis and anthropoid origins. *Science* 278:2135–2136.

Kazazian, H. H., Jr., P. G. Waber, C. D. Boehm, J. I. Lee, et al. 1984. Hemoglobin E in Europeans: Further evidence for multiple origins of the XB^E-globin gene. *American Journal of Human Genetics* 36:212–217.

Keegan, J. 1976. *The Face of Battle*. Viking, New York.

Keegan, J. 1988. *The Price of Admiralty: The Evolution of Naval Warfare*. Penguin Books, London.

Keegan, J. 1990. *The Second World War*. Viking, New York.

Keegan, J. 1993. *A History of Warfare*. Random House, New York.

Keegan, J. 1998. *The First World War*. Knopf, New York.

Keeley, L. H. 1996. *War before Civilization: The Myth of the Peaceful Savage*. Oxford Univ. Press, New York.

Keller, L., ed. 1999. *Levels of Selection*. Princeton Univ. Press, Princeton.

Keller, L., and H. K. Reeve. 1998. Familiarity breeds cooperation. *Nature* 394:121–122.

Kellert, S. R., and E. O. Wilson, eds. 1993. *The Biophilia Hypothesis*. Island Press, Washington, DC.

Kelley, D. M. 1972. *Why Conservative Churches Are Growing*. Harper and Row, New York.

Kelly, R. C. 1976. *Etoro Social Structure: A Study in Structural Contradiction*. Univ. of Michigan Press, Ann Arbor.

Kelman, H. C. 1973. Violence without moral restraint: Reflections on the dehumanization of victims and victimizers. *Journal of Social Issues* 29:25–61.

Kendon, A. 1997. Gesture. *Annual Review of Anthropology* 26:109–128.

Kennedy, D. M. 1999a. *Freedom from Fear: The American People in Depression and War, 1929–1945*. Oxford Univ. Press, New York.

Kennedy, D. M. 1999b. Victory at sea. *Atlantic Monthly* 283:51–76.

Kennedy, P. 1987. *The Rise and Fall of the Great Powers*. Random House, New York.

Kerr, R. A. 1999. Early life thrived despite earthly travails. *Science* 284:2111–2113.

Kestenberg, J. S., and M. Kestenberg. 1987. Child killing and child rescuing. Pp. 139–154 in G. G. Neuman, ed., *Origins of Human Aggression: Dynamics and Etiology.* Human Sciences Press, New York.

Kettlewell, B. 1973. *The Evolution of Melanism: The Study of a Recurring Necessity.* Clarendon Press, Oxford.

Kidder, A. V. 1940. Looking backward. *Proceedings of the American Philosophical Society* 83:527–537.

Killworth, P. D., H. R. Bernard, and C. McCarty. 1984. Measuring patterns of acquaintanceship. *Current Anthropology* 25:391–397.

Kimbel, W. H. 1991. Species, species concepts, and hominid evolution. *Journal of Human Evolution* 20:335–371.

Kimbel, W. H. 1995. Hominid speciation and Pliocene climatic change. Pp. 425–437 in E. S. Vrba, G. H. Denton, T. C. Partridge, and L. H. Burckle, eds., *Paleoclimate and Evolution, with Emphasis on Human Origins.* Yale Univ. Press, New Haven.

Kimbel, W. H., D. C. Johanson, and Y. Rak. 1994. The first skull and other new discoveries of *Australopithecus afarensis* at Hadar, Ethiopia. *Nature* 368:449–451.

King, J. L., and T. H. Jukes. 1969. Non-Darwinian evolution. *Science* 164:788–798.

King, M.-C., and A. C. Wilson. 1975. Evolution at two levels in humans and chimpanzees. *Science* 188:107–116.

King, R. B. 1986. Population ecology of the Lake Erie water snake, *Nerodia sipedon insularum. Copeia* 1986:757–772.

King, R. B. 1992. Lake Erie water snakes revisited: Morph- and age-specific variation in relative crypsis. *Evolutionary Ecology* 6:115–124.

King, R. B. 1993. Color pattern variation in Lake Erie water snakes: Prediction and measurement of selection. *Evolution* 47:1819–1833.

King, R. B., and R. Lawson. 1995. Color pattern variation in Lake Erie water snakes: The role of gene flow. *Evolution* 49:885–896.

Kipling, R. 1902. *Just So Stories: For Little Children.* Macmillan, London.

Kipling, R. 1912. *Barrack-Room Ballads and Other Verses.* Macmillan, London.

Kirch, P. V. 1984. *The Evolution of the Polynesian Chiefdoms.* Cambridge Univ. Press, Cambridge.

Kirch, P. V. 1988. Circumscription theory and sociopolitical evolution in Polynesia. *American Behavioral Scientist* 31:416–427.

Kirch, P. V. 1990. The evolution of sociopolitical complexity in prehistoric Hawaii: An assessment of the archaeological evidence. *Journal of World Prehistory* 4:311–345.

Kirch, P. V. 1991a. Chiefship and competitive involution: The Marquesas Islands of eastern Polynesia. Pp. 119–145 in T. Earle, ed., *Chiefdoms: Power, Economy, and Ideology.* Cambridge Univ. Press, Cambridge.

Kirch, P. V. 1991b. Prehistoric exchange in western Melanesia. *Annual Review of Anthropology* 20:141–165.

Kirch, P. V. 1994. *The Wet and the Dry: Irrigation and Agricultural Intensification in Polynesia.* Univ. of Chicago Press, Chicago.

Kirch, P. V. 1996a. Lapita and its aftermath: The Austronesian settlement of Oceania. Pp. 57–70 in W. H. Goodenough, ed., *Prehistoric Settlement of the Pacific.* American Philosophical Society, Philadelphia.

Kirch, P. V. 1996b. Late Holocene human-induced modifications to a central Polynesian island ecosystem. *Proceedings of the National Academy of Sciences USA* 93:5296–5300.

Kirch, P. V. 1997. Microcosmic histories: Island perspectives on "global" change. *American Anthropologist* 99:30–42.

Kirch, P. V., and T. L. Hunt, eds. 1997. *Historical Ecology in the Pacific Islands.* Yale Univ. Press, New Haven.

Kirch, P. V., and M. I. Weisler. 1994. Archaeology in the Pacific islands: An appraisal of recent research. *Journal of Archaeological Research* 2:285–328.

Kitcher, P. 1985. *Vaulting Ambition: Sociobiology and the Quest for Human Nature.* MIT Press, Cambridge.

Kitcher, P. 1996. *The Lives to Come: The Genetic Revolution and Human Possibilities.* Penguin Books, London.

Klein, R. G. 1978. Stone Age predation on large African bovids. *Journal of Archaeological Science* 5:95–217.

Klein, R. G. 1982. Age (mortality) profiles as a means of distinguishing hunted species from scavenged ones. *Paleobiology* 8:151–158.

Klein, R. G. 1987. Reconstructing how early people exploited animals: Problems and prospects. Pp. 11–45 in M. H. Nitecki and D. V. Nitecki, eds., *The Evolution of Human Hunting.* Plenum Press, New York.

Klein, R. G. 1988. The causes of "robust" australopithecine extinction. Pp. 499–505 in F. E. Grine, ed., *Evolutionary History of the "Robust" Australopithecines.* Aldine, New York.

Klein, R. G. 1992. The archeology of modern human origins. *Evolutionary Anthropology* 1:5–14.

Klein, R. G. 1995. Anatomy, behavior, and modern human origins. *Journal of World Prehistory* 9:167–198.

Klein, R. G. 1998. Why anatomically modern people did not disperse from Africa 100,000 years ago. Pp. 509–521 in T. Akazawa, K. Aoki, and O. Bar-Yosef, eds., *Neanderthals and Modern Humans in Western Asia.* Plenum Press, New York.

Klein, R. G. 1999. *The Human Career: Human Biological and Cultural Origins.* 2nd ed. Univ. of Chicago Press, Chicago.

Klein, R. G. 2000a. Archeology and the evolution of human behavior. *Evolutionary Anthropology* 9:17–36.

Klein, R. G. 2000b. Fully modern humans. In T. D. Price and G. Feinman, eds., *Archaeology at the Millennium.* Plenum Press, New York. In press.

Klein, R. G., G. Avery, K. Cruz-Uribe, R. G. Milo, et al. 1999. Duinefontein 2: An Acheulean site in the western Cape Province of South Africa. *Journal of Human Evolution* 37:153–190.

Klemperer, V. 1998. *I Will Bear Witness: A Diary of the Nazi Years, 1933–1941.* Random House, New York.

Klemperer, V. 2000. *I Will Bear Witness: A Diary of the Nazi Years, 1941–1945.* Random House, New York.

Klicka, J., and R. M. Zink. 1997. The importance of recent ice ages in speciation: A failed paradigm. *Science* 277:1666–1669.

Kling, A. S. 1986. Neurological correlates of social behavior. *Ethology and Sociobiology* 7:175–186.

Kluger, M. J., W. Kozak, C. A. Conn, L. R. Leon, et al. 1997. The adaptive value of fever. Pp. 255–266 in P. A. Mackowiak, ed., *Fever: Basic Mechanisms and Management.* Lippincott-Raven, Philadelphia.

Knauft, B. 1987. Reconsidering violence in simple human societies: Homicide among the Gebusi of New Guinea. *Current Anthropology* 28:457–500.

Knauft, B. 1991. Violence and sociality in human evolution. *Current Anthropology* 32:391–428.

Knauft, B. 1994. Reply to Erdal and Whiten. *Current Anthropology* 35:181–182.

Knoll, A. H., and S. B. Carroll. 1999. Early animal evolution: Emerging views from comparative biology and geology. *Science* 284:2129–2137.

Koblinsky, M., J. Timyan, and J. Gay, eds. 1993. *The Health of Women: A Global Perspective.* Westview Press, Boulder.

Koenig, W. D., and R. Mumme. 1987. *Population Ecology of the Cooperatively Breeding Acorn Woodpecker.* Princeton Univ. Press, Princeton.

Kohlberg, L. 1984. *The Psychology of Moral Development.* Harper and Row, San Francisco.

Köhler, M., and S. Moyà-Solà. 1997a. Ape-like or hominid-like? The positional behavior of *Oreopithecus bambolii* reconsidered. *Proceedings of the National Academy of Sciences USA* 94:11747–11750.

Köhler, M., and S. Moyà-Solà. 1997b. Fossil muzzles and other puzzles. *Nature* 388:327–328.

Kohn, B., and M. Dennis. 1974. Selective impairments of visual-spatial abilities in infantile hemiplegics after right hemidecortation. *Neuropsychologia* 12:505–512.

Kondrashov, A. S., and F. Y. Kondrashov. 1999. Interactions among quantitative traits in the course of sympatric speciation. *Nature* 400:351–354.

Konner, M. 1990. *Why the Reckless Survive.* Viking, New York.

Kooijmans, L. P. L., Y. Smirnov, R. S. Solecki, P. Villa, et al. 1989. On the evidence for Neanderthal burial. *Current Anthropology* 30:322–330.

Koopman, K. F. 1950. Natural selection for reproductive isolation between *Drosophila pseudoobscura* and *Drosophila persimilis. Evolution* 4:135–148.

Korten, D. C. 1995. *When Corporations Rule the World.* Kumarian Press and Berrett-Koehler, West Hartford and San Francisco.

Kortlandt, A. 1980. How might early hominids have defended themselves against large predators and food competitors? *Journal of Human Evolution* 9:79–112.

Kraut, R., ed. 1992. *The Cambridge Companion to Plato.* Cambridge Univ. Press, Cambridge.

Krebs, J. R., and N. B. Davies, eds. 1997. *Behavioral Ecology: An Evolutionary Approach.* 4th ed. Blackwell, Oxford.

Kringlen, E. 1967. *Heredity and Environment in the Functional Psychosis: An Epidemiological-Clinical Twin Study.* Universitets-forlaget, Oslo.

Krings, M., A. Stone, R. W. Schmitz, H. Krainitzki, et al. 1997. Neandertal DNA sequences and the origin of modern humans. *Cell* 90:19–30.

Kristof, N. D. 1999. Fourteen Ninety-Two: The prequel. *New York Times Magazine* (6 June): 80–86.

Kruuk, H. 1972. *The Spotted Hyena: A Study of Predation and Social Behavior.* Univ. of Chicago Press, Chicago.

Kuhl, P. K. 1981. Discrimination of speech by nonhuman animals: Basic auditory sensitivities conducive to the perception of speech-sound categories. *Acoustic Society of America* 70:340–349.

Kuhl, P. K., et al. 1992. Linguistic experience alters phonetic perception in infants by six months of age. *Science* 255:606–608.

Kulozik, A. E., J. S. Wainscoat, G. R. Serjeant, B. C. Kar, et al. 1986. Geographic survey of XB^s-globin gene haplotypes: Evidence for an independent Asian origin of the sickle-cell mutation. *American Journal of Human Genetics* 39:239–244.

Kumm, J., and M. W. Feldman. 1997. Gene-culture coevolution and sex ratios: Sex chromosome distorters and cultural preferences for offspring sex. *Theoretical Population Biology* 52:1–15.

Kumm, J., K. N. Laland, and M. W. Feldman. 1994. Gene-culture coevolution and sex ratios: The effects of infanticide, sex-selective abortion, sex selection, and sex-biased parental investment on the evolution of sex ratios. *Theoretical Population Biology* 46:249–278.

Kundera, M. 1991. *Immortality.* Faber and Faber, London.

Kuper, L. 1981. *Genocide: Its Political Use in the Twentieth Century.* Penguin Books, London.

Kuroda, S., S. Suzuki, and T. Nishihara. 1996. Preliminary report on predatory behavior and meat sharing in Tschego chimpanzees *(Pan troglodytes troglodytes)* in the Ndoki Forest, northern Congo. *Primates* 37:253–259.

Kuttruff, J. T., S. G. DeHart, and M. J. O'Brien. 1998. Seven thousand, five hundred years of prehistoric footwear from Arnold Research Cave, Missouri. *Science* 281:72–75.

Kuznar, L. A. 1997. *Reclaiming a Scientific Anthropology.* Altamira Press, Walnut Creek, CA.

La Barre, W. 1954. *The Human Animal.* Univ. of Chicago Press, Chicago.

Labine, P. A. 1964. Population biology of the butterfly *Euphydryas editha.* I. Barriers to multiple inseminations. *Evolution* 18:335–336.

Labine, P. A. 1966. Population biology of the butterfly *Euphydryas editha.* IV. Sperm precedence: A preliminary report. *Evolution* 20:580–586.

Labov, W. 1969. The logic of non-standard English. *Georgetown Monographs on Language and Linguistics* 22:1–31.

Lack, D. 1947. *Darwin's Finches.* Cambridge Univ. Press, Cambridge.

Ladurie, E. L. 1979. *Montaillou: The Promised Land of Error.* Random House, New York.

Laitinen, L. V., and K. E. Livingston. 1973. *Surgical Approaches in Psychiatry.* Univ. Park Press, Baltimore.

Lakoff, G. 1987. *Women, Fire, and Dangerous Things: What Categories Reveal about the Mind.* Univ. of Chicago Press, Chicago.

Laland, K. N. 1993. The mathematical modelling of human culture and its implications for psychology and the human sciences. *British Journal of Psychology* 84:145–169.

Laland, K. N., J. Kumm, and M. W. Feldman. 1995. Gene-culture coevolutionary theory: A test case. *Current Anthropology* 36:131–156.

Laland, K. N., F. J. Odling-Smee, and M. W. Feldman. 1996. The evolutionary consequences of niche construction: A theoretical investigation using two-locus theory. *Journal of Evolutionary Biology* 9:293–316.

La Mettrie, J. O. de. 1996 (1747). *Machine Man.* Cambridge Univ. Press, Cambridge.

Lamotte, M. 1952. Le rôle des fluctuations furtuites dans la diversité des populations naturelles de *Cepaea nemoralis. Heredity* 6:333–343.

Lamotte, M. 1959. Polymorphism of natural populations of *Cepaea nemoralis.* Pp. 65–86 in M. Demerec, ed., *Genetics and Twentieth Century Darwinism.* Waverly Press, Baltimore.

Landes, D. S. 1998. *The Wealth and Poverty of Nations: Why Some Are So Rich and Some So Poor.* Norton, New York.

Lane, C. 1995. Tainted sources. Pp. 125–139 in R. Jacoby and N. Glauberman, eds., *The Bell Curve Debate: History, Documents, Opinions.* Times Books, New York.

Lane, R., E. Reiman, B. Axelrod, L. Yun, et al. 1998. Neural correlates of levels of emotional awareness: Evidence of an interaction between emotion and attention in the anterior cingulate cortex. *Journal of Cognitive Neuroscience* 10:525–535.

Langlois, J. H., and L. A. Roggman. 1990. Attractive faces are only average. *Psychological Science* 1:115–121.

Langlois, J. H., L. A. Roggman, R. J. Casey, J. M. Ritter, et al. 1987. Infant preferences for attractive faces: Rudiments of a stereotype? *Developmental Psychology* 23:363–369.

Langness, L. L. 1974. Ritual, power, and male dominance in the New Guinea highlands. *Ethos* 2:189–212.

Larson, E. J. 1997. *Summer for the Gods: The Scopes Trial and America's Continuing Debate over Science and Religion.* Harvard Univ. Press, Cambridge.

Larson, E. J., and L. Witham. 1999. Inherit an ill wind. *Nation* (4 October): 25–29.

Lassman, P., and R. Speirs. 1994. *Weber: Political Writings.* Cambridge Univ. Press, Cambridge.

Laughlin, C., Jr., and E. G. d'Aquili. 1974. *Biogenetic Structuralism.* Columbia Univ. Press, New York.

Laughlin, W. S. 1968. The demography of hunters: An Eskimo example. Pp. 241–243 in R. B. Lee and I. Devore, eds., *Man the Hunter.* Aldine, Chicago.

Laumann, E. O., J. H. Gagnon, R. T. Michael, and S. Michaels. 1994. *The Social Organization of Sexuality.* Univ. of Chicago Press, Chicago.

Lawson, R., and G. W. Humphreys. 1996. View specificity in object processing: Evidence from picture matching. *Journal of Experimental Psychology* 22:395–416.

Layton, R. 1991. *The Anthropology of Art.* 2nd ed. Cambridge Univ. Press, Cambridge.

Layton, R. 1992. *Australian Rock Art: A New Synthesis.* Cambridge Univ. Press, Cambridge.

Lazarus, R. S. 1982. Thoughts on the relations between emotion and cognition. *American Psychologist* 37:1019–1024.

Leach, E. 1967. Virgin birth. *Proceedings of the Royal Anthropological Institute of Great Britain and Ireland for 1966,* 34–49.

Leacock, E. 1978. Women's status in egalitarian society: Implications for social evolution. *Current Anthropology* 19:247–275.

Leahy, T. H., and R. J. Harris. 1985. *Human Learning.* Prentice Hall, Englewood Cliffs, NJ.

Leakey, L. S. B., P. V. Tobias, and J. R. Napier. 1964. A new species from the genus *Homo* from Olduvai Gorge. *Nature* 202:7–9.

Leakey, M. G., C. S. Feibel, I. McDougall, and A. Walker. 1995. New four-million-year-old hominid species from Kanapoi and Allia Bay, Kenya. *Nature* 376:565–571.

Leakey, M. G., C. S. Feibel, I. McDougall, C. Ward, et al. 1998. New specimens and confirmation of an early age for *Australopithecus anamensis. Nature* 393:62–66.

Leakey, R., and R. Lewin. 1992. *Origins Reconsidered: In Search of What Makes Us Human.* Little, Brown, London.

Leavitt, G. C. 1977. The frequency of warfare: An evolutionary perspective. *Sociological Inquiry* 47:49–58.

Leavitt, G. C. 1989. Disappearance of the incest taboo: A cross-cultural test of general evolutionary hypotheses. *American Anthropologist* 91:116–131.

Lecours, A. R. 1995. The origins and evolution of writing. Pp. 213–235 in J. P. Changeux and J. Chavaillon, eds., *Origins of the Human Brain.* Oxford Univ. Press, Oxford.

Ledoux, J. E. 1994. Emotional processing, but not emotions, can occur unconsciously. Pp. 291–292 in P. Ekman and R. J. Davidson, eds., *The Nature of Emotions: Fundamental Questions.* Oxford Univ. Press, New York.

Lee, R. B. 1968. What hunters do for a living, or, how to make out on scarce resources. Pp. 30–48 in R. B. Lee and I. DeVore, eds., *Man the Hunter.* Aldine, Chicago.

Lee, R. B. 1991 (1988). Reflections on primitive communism. Pp. 252–268 in T. Ingold, D. Riches, and J. Woodburn, eds., *Hunters and Gatherers.* Vol. 1, *History, Evolution, and Social Change.* Berg, Oxford.

Lee, R. B., and I. DeVore. 1968a. Problems in the study of hunters and gatherers. Pp. 3–12 in R. B. Lee and I. DeVore, eds., *Man the Hunter.* Aldine, Chicago.

Lee, R. B., and I. DeVore, eds. 1968b. *Man the Hunter.* Aldine, Chicago.

Lee, R. D. 1987. Population dynamics of human and other animals. *Demography* 24:443–465.

Leeds, A. 1963. The functions of war. Pp. 69–83 in J. H. Masserman, ed., *Violence and War: With Clinical Studies.* Grune and Stratton, New York.

Le Gros Clark, W. E. 1964. *The Fossil Evidence for Human Evolution.* 2nd ed. Univ. of Chicago Press, Chicago.

Lehrman, D. S. 1953. Critique of Konrad Lorenz's theory of instinctive behavior. *Quarterly Review of Biology* 28:337–363.

Lemonick, M. D. 1999. Smart genes? A new study sheds light on how memory works and raises questions about whether we should use genetics to make people brainier. *Time* (13 September): 54–58.

Lenneberg, E. H. 1953. Cognition in ethnolinguistics. *Language* 29:463–471.

Lenormand, T., T. Guillemund, D. Bourguet, M. Raymond. 1998. Appearance and sweep of a gene duplication: Adaptive response and potential for new functions in the mosquito *Culex pipiens. Evolution* 52:1705–1712.

Lenormand, T., D. Bourguet, T. Guillemaud, and M. Raymond. 1999. Tracking the evolution of insecticide resistance in the mosquito *Culex pipiens. Nature (London)* 400:861–864.

Lenormand, T., and M. Raymond. 1998. Resistance management: The stable zone strategy. *Proceedings of the Royal Society of London,* ser. B, 26:1985–1990.

Lenski, R. E., M. R. Rose, S. C. Simpson, and S. C. Tadler. 1991. Long-term experimental evolution in *Escherichia coli.* I. Adaptation and divergence during 2,000 generations. *American Naturalist* 138:1315–1341.

Lenski, R. E., and M. Travisano. 1994. Dynamics of adaptation and diversification: A 10,000 generation experiment with bacterial populations. *Proceedings of the National Academy of Sciences USA* 91:6808–6814.

Leopold, A. 1966. *A Sand County Almanac, with Essays from Round River.* Ballantine Books, New York.

Lepofsky, D., P. V. Kirch, and K. P. Lertzman. 1996. Stratigraphic and paleobotanical evidence for prehistoric human-induced environmental disturbance on Mo'orea, French Polynesia. *Pacific Science* 50:253–273.

Leslie, A. 1982. The perception of causality in infants. *Perception* 11:173–186.

Leslie, A. M. 1994. ToMM, ToBy, and agency: Core architecture and domain specificity. Pp. 119–148 in L. A. Hirschfeld and S. A. Gelman, eds., *Mapping the Mind: Domain Specificity in Cognition and Culture.* Cambridge Univ. Press, Cambridge.

Leslie, A. M., and S. Keeble. 1987. Do six-month-old infants perceive causality? *Cognition* 25:265–288.

Lessard, N., M. Paré, F. Lepore, and M. Lassonde. 1998. Early blind human subjects localize sound sources better than sighted subjects. *Nature* 395:278–280.

Lesthaeghe, R. 1980. On the social control of human reproduction. *Population and Development Review* 6:527–548.

Lethmate, J., and G. Dücker. 1973. Untersuchungen zum Selbsterkennen im Spiegel bei Orang-Utans und einigen anderen Affenarten. *Zeitschrift für Tierpsychologie* 33:248–269.

Leuteneggar, W. 1987. Origin of hominid bipedalism. *Nature* 325:305.

LeVay, S. 1991. A difference in hypothalamic structure between heterosexual and homosexual men. *Science* 253:1034–1037.

LeVay, S. 1996. *Queer Science.* MIT Press, Cambridge.

Levenson, R. W. 1994. Human emotion: A functional view. Pp. 123–126 in P. Ekman and R. J. Davidson, eds., *The Nature of Emotion: Fundamental Questions.* Oxford Univ. Press, Oxford.

Levin, B. R., and R. M. Anderson. 1999. The population biology of anti-infective chemotherapy and the evolution of drug resistance: More questions than answers. Pp. 125–137 in S. C. Stearns, ed., *Evolution in Health and Disease.* Oxford Univ. Press, Oxford.

Levin, B. R., and R. E. Lenski. 1983. Coevolution in bacteria and their viruses and plasmids. Pp. 99–127 in D. J. Futuyma and M. Slatkin, eds., *Coevolution.* Sinauer, Sunderland, MA.

Levin, S. 1999. *Fragile Dominion.* Perseus Books, Reading, MA.

Levy, S. 1999. Death by fire. *New Scientist* (1 May): 38–43.

Levy, S. B. 1992. *The Antibiotic Paradox: How Miracle Drugs Are Destroying the Miracle.* Plenum Press, New York.

Lewin, R. 1998. *Principles of Human Evolution.* Blackwell, Malden, MA.

Lewis, B. 1995. *The Middle East: A Brief History of the Last Two Thousand Years.* Scribner, New York.

Lewis, W. H., ed. 1980. *Polyploidy: Biological Relevance.* Plenum Press, New York.

Lewontin, R. C. 1958. A general method for investigating the equilibrium of gene frequencies in a population. *Genetics* 43:419–434.

Lewontin, R. C. 1970. The units of selection. *Annual Review of Ecology and Systematics* 1:1–18.

Lewontin, R. C. 1978. Adaptation. *Scientific American* 239:212–230.

Lewontin, R. C. 1982a. *Human Diversity.* Scientific American Library, New York.

Lewontin, R. C. 1982b. Organism and environment. Pp. 151–170 in H. C. Plotkin, ed., *Learning, Development, and Culture.* Wiley, New York.

Lewontin, R. C. 1985 (1983). The organism as the subject and object of evolution. Pp. 85–106 in R. Levins and R. C. Lewontin, eds., *The Dialectical Biologist.* Harvard Univ. Press, Cambridge.

Lewontin, R. C. 1998. Survival of the nicest? *New York Review of Books* (22 October): 59–62.

Lewontin, R. C. 1999. The problem with an evolutionary answer. *Nature* 400:728–729.

Lewontin, R. C., S. Rose, and L. J. Kamin. 1984. *Not in Our Genes: Biology, Ideology, and Human Nature.* Pantheon Books, New York.

Lieberman, D., and D. Symons. 1998. Sibling incest avoidance: From Westermarck to Wolf. *Quarterly Review of Biology* 73:463–466.

Lieberman, D. E. 1993. Variability in hunter-gatherer seasonal mobility in the southern Levant: From the Mousterian to the Natufian. Pp. 207–219 in G. L. Peterkin, H. M. Bricker, and P. Mellars, eds., *Hunting and Animal Exploitation in the Later Palaeolithic and Mesolithic of Eurasia.* American Anthropological Association, Washington, DC.

Lieberman, D. E. 1995. Testing hypotheses about recent human evolution from skulls: Integrating morphology, function, development, and phylogeny. *Current Anthropology* 36:159–178.

Lieberman, D. E. 1998. Sphenoid shortening and the evolution of modern human cranial shape. *Nature* 393:158–162.

Lieberman, D. E., and J. J. Shea. 1994. Behavioral differences between archaic and modern humans in the Levantine Mousterian. *American Anthropologist* 96:300–332.

Lieberman, L. S. 1987. Biocultural consequences of animals versus plants as sources of fats, proteins, and other nutrients. Pp. 225–260 in M. Harris and E. B. Ross, eds., *Food and Evolution: Toward a Theory of Human Food Habits.* Temple Univ. Press, Philadelphia.

Lieberman, P. 1993. On the evolution of language. *Current Anthropology* 34:166.

Lieberman, P. 1998. *Eve Spoke: Human Language and Human Evolution.* Norton, New York.

Lieberman, P., and E. Crelin. 1971. On the speech of Neanderthal man. *Linguistic Inquiry* 2:203–222.

Lieberman, P., E. S. Crelin, and D. H. Klatt. 1972. Phonetic ability and related anatomy of the newborn and adult human, Neanderthal man, and the chimpanzee. *American Anthropologist* 74:287–307.

Liem, K. F. 1974. Evolutionary strategies and morphological innovations: Cichlid pharyngeal jaws. *Systematic Zoology* 22:425–441.

Liem, K. F., and L. S. Kaufman. 1984. Intraspecific macroevolution: Functional biology of the polymorphic cichlid species *Cichlasoma minckleyi.* Pp. 203–215 in A. A. Echelle and I. Kornfeld, eds., *Evolution of Fish Species Flocks.* Univ. of Maine Press, Orono.

Lillie, R. S. 1913. The fitness of the environment. *Science* 38:337–342.

Limongelli, L., S. T. Boysen, and E. Visalberghi. 1995. Comprehension of cause-effect relations in a tool-using task by chimpanzees *(Pan troglodytes). Journal of Comparative Psychology* 109:18–26.

Lindahl, T. 1997. Facts and artifacts of ancient DNA. *Cell* 90:1–3.

Lindly, J. M., and G. A. Clark. 1990. Symbolism and modern human origins. *Current Anthropology* 31:233–262.

Lindner, E. W. 1998. *Yearbook of American and Canadian Churches, 1998.* Abingdon Press, Nashville.

Lindner, E. W. 1999. *Yearbook of American and Canadian Churches, 1999.* Abingdon Press, Nashville.

Lindqvist, S. 1992. *Exterminate All the Brutes.* New Press, New York.

Linnaeus, C. 1758. *Systema Naturae: Regnum Animale.* 10th ed. Engelmann, Leipzig.

Lipstadt, D. 1993. *Denying the Holocaust: The Growing Assault on Truth and Memory.* Free Press, New York.

Liska, J. 1994. The foundation of symbolic communication. Pp. 233–251 in D. Quiatt and J. Itani, eds., *Hominid Culture.* Univ. Press of Colorado, Niwot.

Livingstone, F. B. 1962. Reconstructing man's Pliocene pongid ancestor. *American Anthropologist* 64:301–305.

Livingstone, F. B. 1963. Blood groups and ancestry: A test case from the New Guinea highlands. *Current Anthropology* 4:541–542.

Lizot, J. 1977. Population, resources, and warfare among the Yanomami. *Man* 12:497–517.

Locke, J. 1975 (1690). *Essay concerning Human Understanding.* Oxford Univ. Press, Oxford.

Locke, J. 1978 (1689). Of property. Pp. 18–20 in C. B. Macpherson, ed., *Property: Mainstream and Critical Positions.* Blackwell, Oxford.

Lockett, S. F., A. Alonso, R. Wyld, M. P. Martin, et al. 1999. Effect of chemokine receptor mutations on heterosexual human immunodeficiency virus transmission. *Journal of Infectious Diseases* 180:614–620.

Loehlin, J. C. 1998. Whither dysgenics? Comments on Lynn and Preston. Pp. 389–398 in U. Neisser, ed., *The Rising Curve: Long-Term Gains in IQ and Related Measures.* American Psychological Association, Washington, DC.

Loewe, L., and S. Scherer. 1997. Mitochondrial Eve: The plot thickens. *Trends in Ecology and Evolution* 12:422–423.

Long, J. C. 1993. Human molecular phylogenetics. *Annual Review of Anthropology* 22:251–272.

Loomis, W. F. 1967. Skin-pigment regulation of vitamin-D biosynthesis in man. *Science* 157:501–506.

Lorenz, K. 1937. Über den Begriff der Instinkthandlung. *Folia Biotheoretica* 2:17–50.

Lorenz, K. 1950. The comparative method in studying innate behavior patterns. *Symposia of the Society for Experimental Biology* 4:221–268.

Lorenz, K. 1966. *On Aggression.* Harcourt, Brace, New York.

Losos, J. B., K. I. Warheit, and T. W. Schoener. 1997. Adaptive differentiation following experimental island colonization in *Anolis* lizards. *Nature* 387:70–73.

Lovejoy, C. O. 1981. The origin of man. *Science* 211:341–350.

Low, B. S. 1989. Cross-cultural patterns in the training of children: An evolutionary perspective. *Journal of Comparative Psychology* 103:311–319.

Low, B. S. 1996. Behavioral ecology of conservation in traditional societies. *Human Nature* 7:353–379.

Lowenthal, D. 1987. Montesquieu. Pp. 513–534 in L. Strauss and J. Cropsey, eds., *History of Political Philosophy.* 3rd ed. Univ. of Chicago Press, Chicago.

Lubbock, J. 1870. *The Origin of Civilization and the Primitive Condition of Man; Mental and Social Condition of Savages.* Longmans, Green, London.

Lubchenco, J. 1998. Entering the century of the environment: A new social contract for science. *Science* 279:491–497.

Lucy, J. A. 1985. Whorf's view of the linguistic mediation of thought. Pp. 73–97 in E. Mertz and R. Parmentier, eds., *Semiotic Mediation: Sociocultural and Psychological Perspectives.* Academic Press, New York.

Lucy, J. A. 1992a. *Grammatical Categories and Cognition: A Case Study of the Linguistic Relativity Hypothesis.* Cambridge Univ. Press, Cambridge.

Lucy, J. A. 1992b. *Language Diversity and Thought: A Reformulation of the Linguistic Relativity Hypothesis.* Cambridge Univ. Press, Cambridge.

Lucy, J. A. 1997. Linguistic relativity. *Annual Review of Anthropology* 26:291–312.

Lucy, J. A., and R. A. Shweder. 1979. Whorf and his critics: Linguistic and nonlinguistic influences on color memory. *American Anthropologist* 81:581–615.

Lumsden, C. J., and E. O. Wilson. 1981. *Genes, Mind, and Culture.* Harvard Univ. Press, Cambridge.

Lumsden, C. J., and E. O. Wilson. 1983. *Promethean Fire.* Harvard Univ. Press, Cambridge.

Luria, A. R. 1980. *Higher Cortical Functions in Man.* Rev. ed. Basic Books, New York.

Lynn, R. 1998. The decline of genotypic intelligence. Pp. 335–364 in U. Neisser, ed., *The Rising Curve: Long-Term Gains in IQ and Related Measures.* American Psychological Association, Washington, DC.

MacArthur, R. H. 1962. Some generalized theorems of natural selection. *Proceedings of the National Academy of Sciences USA* 232:123–138.

Mace, G. M., and P. H. Harvey. 1983. Energetic contraints on home range size. *American Naturalist* 121:120–132.

Machiavelli, N. 1979. *The Portable Machiavelli.* Viking, New York.

Machiavelli, N. 1981 (1513). *The Prince.* Bantam Books, New York.

MacIntyre, A. 1984. *After Virtue.* 2nd ed. Univ. of Notre Dame Press, Notre Dame.

MacIntyre, F., and K. W. Estep. 1993. Sperm competition and the persistence of genes for male homosexuality. *BioSystems* 31:223–233.

Macionis, J. J. 1997. *Sociology.* 6th ed. Prentice Hall, Upper Saddle River, NJ.

Mackintosh, N. J. 1983. *Conditioning and Associative Learning.* Clarendon Press, Oxford.

Mackintosh, N. J. 1987. Animal minds. Pp. 113–120 in C. Blakemore and S. Greenfield, eds., *Mindwaves.* Basil Blackwell, Oxford.

Macnamara, J. 1982. *Names for Things: A Study of Human Learning.* MIT Press, Cambridge.

MacNeilage, P. F. 1998. Evolution of mechanisms of language output: Comparative neurobiology of vocal and manual communication. Pp. 222–241 in J. R. Hurford, M. Studdert-Kennedy, and C. Knight, eds., *Approaches to the Evolution of Language.* Cambridge Univ. Press, Cambridge.

MacNeilage, P. F., M. G. Studdert-Kennedy, and B. Lindblom. 1984. Functional precursors to language and its lateralization. *American Journal of Physiology* 246:R912–R914.

MacNeilage, P. F., M. G. Studdert-Kennedy, and B. Lindblom. 1987. Primate handedness reconsidered. *Behavioral and Brain Sciences* 10:247–303.

MacNeish, R. S. 1992. *The Origins of Agriculture and Settled Life.* Univ. of Oklahoma Press, Norman.

Mahowald, M. B. 2000. *Genes, Women, Equality.* Oxford Univ. Press, Oxford.

Majerus, M. E. N. 1998. *Melanism: Evolution in Action.* Oxford Univ. Press, Oxford.

Malamuth, N. M. 1986. Prediction of naturalistic sexual aggression. *Journal of Personality and Social Psychology* 50:953–962.

Malenka, R. C., and R. A. Nicoll. 1997. Never fear, LTP is hear. *Nature* 390:552–553.

Malenka, R. C., and R. A. Nicoll. 1999. Long-term potentiation: A decade of progress? *Science* 285:1870–1874.

Mallet, J. 1995. A species definition for the modern synthesis. *Trends in Ecology and Evolution* 10:294–299.

Mallet, J., and N. Barton. 1989. Strong natural selection in a warning color hybrid zone. *Evolution* 43:421–431.

Malmberg, T. 1980. *Human Territoriality.* Mouton, The Hague.

Malotki, E. 1983. *Hopi Time: A Linguistic Analysis of the Temporal Concepts in the Hopi Language.* Mouton, Berlin.

Malthus, T. R. 1970 (1798). *An Essay on the Principle of Population.* Pelican Books, Baltimore.

Mander, J., and E. Goldsmith, eds. 1996. *The Case against the Global Economy—and for a Turn toward the Local.* Sierra Club Books, San Francisco.

Manderscheid, E. J., and A. R. Rogers. 1996. Genetic admixture in the Late Pleistocene. *American Journal of Physical Anthropology* 100:1–5.

Manderson, L. 1995. The pursuit of pleasure and the sale of sex. Pp. 305–329 in P. R. Abramson and S. D. Pinkerton, eds., *Sexual Nature/Sexual Culture.* Univ. of Chicago Press, Chicago.

Manson, W. C. 1986. Sexual cyclicity and concealed ovulation. *Journal of Human Evolution* 15:21–30.

Marcus, G. F., S. Vijayan, S. Bandi Rao, and P. M. Vishton. 1999. Rule learning by seven-month-old infants. *Science* 283:77–80.

Marean, C. W., and S. Y. Kim. 1998. Mousterian large-mammal remains from Kobeh Cave. *Current Anthropology* 39:S79–S113.

Marr, D., and H. K. Nishihara. 1978. Representation and recognition of the spatial organization of three-dimensional shapes. *Proceedings of the Royal Society of London*, ser. B, 200:269–294.

Marsh, G. P. 1874. *The Earth as Modified by Human Action.* Scribner, New York.

Marshack, A. 1972. *The Roots of Civilization.* McGraw-Hill, New York.

Marshack, A. 1996. A Middle Paleolithic symbolic composition from the Golan Heights: The earliest known depictive image. *Current Anthropology* 37:357–365.

Marshack, A. 1997. Paleolithic image making and symboling in Europe and the Middle East: A comparative review. Pp. 53–91 in M. W. Conkey, O. Soffer, D. Stratmann, and N. G. Jablonski, eds., *Beyond Art: Pleistocene Image and Symbol.* California Academy of Sciences, San Francisco.

Marshall, C. R., and P. D. Ward. 1996. Sudden and gradual molluscan extinctions in the latest Cretaceous of western European Tethys. *Science* 274:1360–1363.

Marshall, E. 1998. DNA studies challenge the meaning of race. *Science* 282:654–655.

Marshall, S. L. A. 1978. *Men against Fire.* Smith, Gloucester, MA.

Martan, J., and B. A. Shepard. 1976. The role of the copulatory plug in reproduction of the guinea pig. *Journal of Experimental Zoology* 196:79–84.

Martin, A. P., and S. R. Palumbi. 1993. Body size, metabolic rate, generation time, and the molecular clock. *Proceedings of the National Academy of Sciences USA* 90:4087–4091.

Martin, C. 1978. *Keepers of the Game.* Univ. of California Press, Berkeley.

Martin, L. 1986. "Eskimo words for snow": A case study in the genesis and decay of an anthropological example. *American Anthropologist* 88:418–423.

Martin, P. S. 1967. Prehistoric overkill. Pp. 75–120 in P. S. Martin and H. E. Wright Jr., eds., *Pleistocene Extinctions.* Yale Univ. Press, New Haven.

Martin, P. S. 1984. Prehistoric overkill: The global model. Pp. 354–403 in P. S. Martin and R. G. Klein, eds., *Quaternary Extinctions: A Prehistoric Revolution.* Univ. of Arizona Press, Tucson.

Martin, P. S., and R. G. Klein, eds. 1984. *Quaternary Extinctions: A Prehistoric Revolution.* Univ. of Arizona Press, Tucson.

Martin, R. 1990. *Primate Origins and Evolution: A Phylogenetic Reconstruction.* Princeton Univ. Press, Princeton.

Martin, R. D., L. A. Willner, and A. Dettling. 1994. The evolution of sexual size dimorphism in primates. Pp. 159–200 in R. V. Short and Balaban, eds., *The Differences between the Sexes.* Cambridge Univ. Press, Cambridge.

Martinson, J. J., L. Excoffier, C. Swinburn, A. J. Boyce, et al. 1995. High diversity of alpha-globin haplotypes in a Senegalese population, including many previously unreported variants. *American Journal of Human Genetics* 57:1186–1198.

Marx, K. 1967 (1867). *Capital: A Critique of Political Economy.* 3 vols. International, New York.

Marx, K. 1970 (1859). *A Contribution to the Critique of Political Economy.* International, New York.

Matlock, J. F., Jr. 1999. Can civilizations clash? *Proceedings of the American Philosophical Society* 143:428–439.

Maton, K. 1989. The stress-buffering role of spiritual support: Cross-sectional and prospective investigations. *Journal for the Scientific Study of Religion* 28:310–323.

Matson, P., P. Vitousek, J. Ewel, M. Mazzarino, et al. 1987. Nitrogen transformations following tropical forest felling and burning on a volcanic soil. *Ecology* 68:491–502.

Matsuzawa, T. 1985a. Color naming and classification in a chimpanzee. *Journal of Human Evolution* 14:283–291.

Matsuzawa, T. 1985b. The use of numbers by a chimpanzee. *Nature* 315:57–59.

Matsuzawa, T. 1990. Form perception and visual acuity in a chimpanzee. *Folia Primatologica* 55:24–32.

Matsuzawa, T. 1991. Nesting cups and metatools in chimpanzees. *Behavioral and Brain Sciences* 14:570–571.

Matsuzawa, T. 1996. Chimpanzee intelligence in nature and in captivity: Isomorphism of symbol use and tool use. Pp. 196–209 in W. C. McGrew, L. F. Marchant, and T. Nishida, eds., *Great Ape Societies.* Cambridge Univ. Press, Cambridge.

Maurus, M., D. Barclay, and K. M. Streit. 1988. Acoustic patterns common to human communication and communication between monkeys. *Language and Communication* 8:87–94.

May, R. M. 1979. When to be incestuous. *Nature* 279:192–195.

May, R. M. 1993. Ecology and evolution of host-virus associations. Pp. 58–68 in S. S. Morse, ed., *Emerging Viruses.* Oxford Univ. Press, New York.

Maynard Smith, J. 1978. *The Evolution of Sex.* Cambridge Univ. Press, Cambridge.

Maynard Smith, J. 1982. *Evolution and the Theory of Games.* Cambridge Univ. Press, Cambridge.

Maynard Smith, J. 1983. The genetics of stasis and punctuation. *Annual Review of Genetics* 17:11–25.

Maynard Smith, J. 1998. *Evolutionary Genetics.* 2nd ed. Oxford Univ. Press, Oxford.

Maynard Smith, J., D. J. P. Barker, C. E. Finch, S. L. R. Kardia, et al. 1999. The evolution of non-infectious and degenerative disease. Pp. 267–272 in S. C. Stearns, ed., *Evolution in Health and Disease.* Oxford Univ. Press, Oxford.

Maynard Smith, J., and P. W. Price. 1973. The logic of animal conflict. *Nature* 248:15–18.

Mayr, E. 1940. Speciation phenomena in birds. *American Naturalist* 74:249–278.

Mayr, E. 1942. *Systematics and the Origin of Species.* Columbia Univ. Press, New York.

Mayr, E. 1944. On the concepts and terminology of vertical subspecies and species. *National Research Council Bulletin* 2:11–16.

Mayr, E. 1950. Taxonomic categories in fossil hominids. *Cold Spring Harbor Symposia in Quantitative Biology* 15:109–117.

Mayr, E. 1963. *Animal Species and Evolution.* Harvard Univ. Press, Cambridge.

Mayr, E. 1982. *The Growth of Biological Thought.* Harvard Univ. Press, Cambridge.

Mayr, E. 1988. *Toward a New Philosophy of Biology.* Harvard Univ. Press, Belknap Press, Cambridge.

Mayr, E. 1997. *This Is Biology: The Science of the Living World.* Harvard Univ. Press, Cambridge.

Mayr, E., E. G. Linsley, and R. L. Usinger. 1953. *Methods and Principles of Systematic Zoology.* McGraw-Hill, New York.

Mazon, L. I., A. Vicario, M. A. Martinez de Pancorbo, and C. M. Lostao. 1990. Polymorphism in *Cepaea hortensis* in marginal populations in Spain. *Genetica* 81:109–115.

McCall, G. 1980. *Rapanui: Tradition and Survival on Easter Island.* Univ. of Hawaii Press, Honolulu.

McCarthy, J. 1983. *Muslims and Minorities: The Population of Ottoman Anatolia and the End of the Empire.* New York Univ. Press, New York.

McCarthy, T. 1998. The butcher of Cambodia. *Time* 151:40.

McClelland, J. L., and M. S. Seidenberg. 2000. Why do kids say *goed* and *brang*? *Science* 287:47–48.

McClintock, B. 1948. Mutable loci in maize. *Carnegie Institute of Washington Yearbook* 47:155–169.

McClintock, M. K. 1971. Menstrual synchrony and suppression. *Nature* 229:224–245.

McCollum, M. A. 1999. The robust australopithecine face: A morphogenetic perspective. *Science* 284:301–305.

McCorriston, J., and F. Hole. 1991. The ecology of seasonal stress and the origins of agriculture in the Near East. *American Anthropologist* 93:46–69.

McDaniel, C. N., and J. M. Gowdy. 2000. *Paradise for Sale: A Parable of Nature.* Univ. of California Press, Berkeley.

McElvaine, R. S. 2000. *Eve's Seed: Biology, the Sexes, and the Course of History.* McGraw-Hill, New York.

McEwen, B. S. 1991. Non-genomic and genomic effects of steroids on neural activity. *Trends in Pharmacological Sciences* 12:141–147.

McGinn, C. 1999. *The Mysterious Flame: Conscious Minds in a Material World.* Basic Books, New York.

McGinnis, W., and R. Krumlauf. 1992. Homeobox genes and axial patterning. *Cell* 68:283–302.

McGrew, W. C. 1989. Why is ape tool use so confusing? Pp. 457–472 in V. Standen and R. A. Foley, eds., *Comparative Socioecology: The Behavioral Ecology of Humans and Other Mammals.* Blackwell Scientific, Oxford.

McGrew, W. C. 1992. *Chimpanzee Material Culture: Implications for Human Evolution.* Cambridge Univ. Press, Cambridge.

McGrew, W. C. 1993. The intelligent use of tools: Twenty propositions. Pp. 151–170 in K. Gibson and T. Ingold, eds., *Tools, Language, and Cognition in Human Evolution.* Cambridge Univ. Press, Cambridge.

McGrew, W. C. 1999. Comments: The raw and the stolen. *Current Anthropology* 40:582–583.

McGrew, W. C., and L. F. Marchant. 1996. On which side of the apes? Ethological study of laterality of hand use. Pp. 255–272 in W. C. McGrew, L. F. Marchant, and T. Nishida, eds., *Great Ape Societies.* Cambridge Univ. Press, Cambridge.

McHenry, H. M. 1992a. Body size and proportions in early hominids. *American Journal of Physical Anthropology* 87:407–431.

McHenry, H. M. 1992b. How big were early hominids? *Evolutionary Anthropology* 1:15–20.

McHenry, H. M. 1996. Homoplasy, clades, and hominid phylogeny. Pp. 77–92 in W. E. Meikle, F. C. Howell, and N. G. Jablonski, eds., *Contemporary Issues in Human Evolution.* California Academy of Sciences, San Francisco.

McHenry, R., and C. Van Doren, eds. 1972. *A Documentary History of Conservation in America.* Praeger, New York.

McKenzie, J. A. 1996. *Ecological and Evolutionary Aspects of Insecticide Resistance.* Landes, Austin.

McMurtry, J. A., C. B. Huffaker, and M. van de Vrie. 1970. Ecology of tetranychid mites and their natural ene-

mies: A review. I. Tetranychid enemies: Their biological characters and the impact of spray practices. *Hilgardia* 40:331–390.
McNeill, D. 1985. So you think gestures are nonverbal? *Psychological Review* 92:350–371.
McNeill, W. H. 1976. *Plagues and Peoples.* Doubleday, New York.
McNeill, W. H. 1992. *The Global Condition: Conquerors, Catastrophes, and Community.* Princeton Univ. Press, Princeton.
McPherson, J. M. 1999. Was blood thicker than water? Ethnic and civic nationalism in the American Civil War. *Proceedings of the American Philosophical Society* 143:102–108.
Medawar, P. B. 1952. *An Unsolved Problem in Biology.* Lewis, London.
Medin, D., and A. Ortony. 1989. Psychological essentialism. Pp. 179–195 in S. Vosniadou and A. Ortony, eds., *Similarity and Analogical Reasoning.* Cambridge Univ. Press, Cambridge.
Mehler, J., and R. Fox, eds. 1985. *Neonate Cognition: Beyond the Blooming Buzzing Confusion.* Erlbaum, Hillsdale, NJ.
Meier, R. P., and E. Newport. 1990. Out of the hands of babes: On a possible sign advantage in language acquisition. *Language* 66:1–23.
Meister, M., R. O. L. Wong, D. A. Baylor, and C. J. Shatz. 1991. Synchronous bursts of action potentials in ganglion cells of the developing mammalian retina. *Science* 252:939–943.
Melko, M. 1994. World-systems theory: A Faustian delusion? *Comparative Civilizations Review* 30:8–12.
Mellars, P. 1989. Major issues in the emergence of modern humans. *Current Anthropology* 30:349–385.
Mellars, P. 1991. Cognitive changes and the emergence of modern humans in Europe. *Cambridge Archeological Journal* 1:63–76.
Mellars, P. 1996. *The Neanderthal Legacy: An Archaeological Perspective from Western Europe.* Princeton Univ. Press, Princeton.
Mellars, P. 1998. The fate of the Neanderthals. *Nature* 395:539–540.
Mellor, M. 1997. *Feminism and Ecology.* New York Univ. Press, New York.
Mencken, H. L. 1922. *Prejudices: Third Series.* Knopf, New York.
Merchant, C. 1992. *Radical Ecology: The Search for a Livable World.* Routledge, Chapman and Hall, New York.
Merton, R. K. 1938. Social structure and anomie. *American Sociological Review* 8:672–682.
Merton, R. K. 1968. *Social Theory and Social Structure.* Free Press, New York.
Merzenich, M. M., W. M. Jenkins, P. Johnston, C. Schreiner, et al. 1996. Temporal processing deficits of language-learning impaired children ameliorated by training. *Science* 271:77–81.
Metter, E., D. Kempler, C. Jackson, W. Hanson, et al. 1987. Cerebral glucose metabolism in chronic aphasia. *Neurology* 37:1599–1606.
Meyer, A., T. D. Kocher, P. Basasibwaki, and A. C. Wildson. 1990. Monophyletic origin of Lake Victoria cichlid fishes suggested by mitochondrial DNA sequences. *Nature* 347:550–553.
Michener, C. D. 1974. *The Social Behavior of the Bees: A Comparative Study.* Harvard Univ. Press, Cambridge.
Michener, C. D., and R. R. Sokal. 1957. A quantitative approach to a problem in classification. *Evolution* 11:130–162.
Michod, R. E. 1981. Positive heuristics in evolutionary biology. *British Journal of the Philosophy of Science* 32:1–36.
Michod, R. E. 1982. The theory of kin selection. *Annual Review of Ecology and Systematics* 13:23–55.
Michotte, A. E. 1965. *The Perception of Causality.* Basic Books, New York.
Midgley, M. 1979. Gene-juggling. *Philosophy* 54:439–458.
Midgley, M. 1985. *Evolution as a Religion: Strange Hopes and Stranger Fears.* Methuen, London.
Midgley, M. 1994. *The Ethical Primate: Humans, Freedom, and Morality.* Routledge, London.
Mies, M., and V. Shiva. 1993. *Ecofeminism.* Zed Books, London.
Miles, H. L. 1991. The development of symbolic communication in apes and early hominids. Pp. 9–20 in W. von Raffler-Engel, J. Wind, and A. Jonker, eds., *Studies in Language Origins.* Benjamins, Amsterdam.
Miles, H. L. W. 1994. Me Chantek: The development of self-awareness in a signing orangutan. Pp. 254–272 in S. T. Parker, R. W. Mitchell, and M. L. Boccia, eds., *Self-Awareness in Animals and Humans: Developmental Perspectives.* Cambridge Univ. Press, Cambridge.
Milgram, S. 1974. *Obedience to Authority: An Experimental View.* Harper and Row, New York.
Milius, S. 1999. Should we junk Linnaeus? *Science News* 156:268–270.
Mill, J. S. 1998 (1863). *Utilitarianism.* Oxford Univ. Press, Oxford.
Miller, G. H., J. W. Magee, B. J. Johnson, M. L. Fogel, et al. 1999. Pleistocene extinction of *Genyornis newtoni:* Human impact of Australian megafauna. *Science* 283:205–208.

Miller, P. H., and P. A. Aloise. 1989. Young children's understanding of the psychological causes of behavior: A review. *Child Development* 60:257–285.

Mills, C. W. 1959. *The Sociological Imagination.* Oxford Univ. Press, London.

Milner, G. R., D. A. Humpf, and H. C. Harpending. 1989. Pattern matching of age-at-death distributions in paleodemographic analysis. *American Journal of Physical Anthropology* 80:49–58.

Milner, P. M. 1993. The mind and Donald O. Hebb. *Scientific American* 268:124–129.

Milo, R. D., and D. Quiatt. 1994. Language in the Middle and Late Stone Ages: Glottogenesis in anatomically modern *Homo sapiens.* Pp. 321–339 in D. Quiatt and J. Itani, eds., *Hominid Culture in Primate Perspective.* Univ. Press of Colorado, Niwot.

Milton, K. 1993. Diet and primate evolution. *Scientific American* (August): 86–93.

Milton, K. 1999. The hypothesis to explain the role of meat-eating in human evolution. *Evolutionary Anthropology* 8:11–21.

Mineka, S., R. Keir, and V. Price. 1981. Fear of snakes in wild- and laboratory-reared rhesus monkeys *(Macaca mulatta). Animal Learning and Behavior* 8:653–663.

Minsky, M. 1985. *The Society of Mind.* Simon and Schuster, New York.

Mischler, B. D., and M. J. Donoghue. 1982. Species concepts: A case for pluralism. *Systematic Zoology* 31:491–503.

Mitani, J. C. 1996. Comparative studies of African ape vocal behavior. Pp. 241–254 in W. C. McGrew, L. F. Marchant, and T. Nishida, eds., *Great Ape Societies.* Cambridge Univ. Press, Cambridge.

Mitchell, R. W., and H. L. Miles. 1993. Apes have mimetic culture. *Behavioral and Brain Sciences* 16:768.

Mitchell, W. P. 1973. The hydraulic hypothesis: A reappraisal. *Current Anthropology* 14:532–534.

Mithen, S. 1993. Simulating mammoth hunting and extinction: Implications for the Late Pleistocene of the central Russian plain. Pp. 25–31 in G. L. Peterkin, H. M. Bricker, and P. Mellars, eds., *Hunting and Animal Exploitation in the Later Palaeolithic and Mesolithic of Eurasia.* American Anthropological Association, Washington, DC.

Mithen, S. 1996. *The Prehistory of the Mind.* Thames and Hudson, London.

Miyamoto, M. M., J. L. Slighton, and M. Goodman. 1987. Phylogenetic relations of humans and African apes from DNA sequences in the __-globin region. *Science* 238:369–373.

Mlot, C. 1998. Probing the biology of emotion. *Science* 280:1005–1007.

Modelski, G. 1978. The long cycle of global politics and the nation state. *Comparative Studies in Society and History* 20:214–235.

Molitor, F., and K. W. Hirsch. 1994. Children's tolerance of real-life aggression after exposure to media violence: A replication of the Drabman and Thomas studies. *Child Study Journal* 24:191–207.

Money, J. 1981. The development of sexuality and eroticism in humankind. *Quarterly Review of Biology* 56:379–404.

Money, J., and A. A. Ehrhardt. 1972. *Man and Woman/Boy and Girl.* Johns Hopkins Univ. Press, Baltimore.

Montagu, A. 1964. *The Concept of Race in the Human Species in the Light of Genetics.* Collier-Macmillan, London.

Montesquieu, C. S., Baron de. 1989 (1748). *The Spirit of the Laws.* Cambridge Univ. Press, Cambridge.

Moore, G. E. 1903. *Principia Ethica.* Cambridge Univ. Press, Cambridge.

Moorehead, A. 1963. *Cooper's Creek.* Harper and Row, New York.

Morata, G. 1993. Homeotic genes of *Drosophila. Current Opinion in Genetics and Development* 3:606–614.

Morell, V. 1997. The origin of dogs: Running with wolves. *Science* 276:1647–1648.

Morell, V. 1998. Genes may link ancient Eurasians, Native Americans. *Science* 280:520.

Morison, S. E. 1947–1962. *History of the United States Naval Operations in World War II.* Little, Brown, New York.

Morphy, H. 1988. Maintaining cosmic unity: Ideology and the reproduction of Yolngu clans. Pp. 249–271 in T. Ingold, D. Riches, and J. Woodburn, eds., *Hunters and Gatherers.* Vol. 2, *Property, Power, and Ideology.* Berg, Oxford.

Morphy, H. 1992. From dull to brilliant: The aesthetics of spiritual power among the Yolngu. Pp. 181–208 in J. Coote and A. Shelton, eds., *Anthropology, Art, and Aesthetics.* Clarendon Press, Oxford.

Morris, D. 1967. *The Naked Ape.* McGraw-Hill, New York.

Morris, J. 1860. Darwin's *On the Origin of Species. Dublin Review* 48:50–81.

Morwood, M. J., P. B. O'Sullivan, F. Aziz, and A. Raza. 1998. Fission-track ages of stone tools and fossils on the east Indonesian island of Flores. *Nature* 392:173–176.

Motluk, A. 1997. Wise mothers fake it. *New Scientist* (13 December): 9

Moxon, E. R., and C. Wills. 1999. DNA microsatellites: Agents of evolution? *Scientific American* 280:94–99.

Moyà-Solà, S., M. Köhler, and L. Rook. 1999. Evidence of hominid-like precision grip capability in the hand of the Miocene ape *Oreopithecus. Proceedings of the National Academy of Sciences USA* 96:313–317.

Muchmore, E. A., S. Diaz, and A. Varki. 1998. A structural difference between the cell surfaces of humans and the great apes. *American Journal of Physical Anthropology* 107:187–198.

Mukerjee, M. 1999. Out of Africa, into Asia. *Scientific American* 280:24.

Müller, W. A. 1997. *Developmental Biology.* Springer, New York.

Murdoch, W. W. 1980. *The Poverty of Nations: The Political Economy of Hunger and Population.* Johns Hopkins Univ. Press, Baltimore.

Murdock, G. P. 1967. *Ethnographic Atlas.* Macmillan, New York.

Muroyama, Y., and Y. Sugiyama. 1994. Grooming relationships in two species of chimpanzees. Pp. 169–180 in R. W. Wrangham, W. C. McGrew, F. B. de Waal, and P. G. Heltne, eds., *Chimpanzee Cultures.* Harvard Univ. Press, Cambridge.

Myers, D., and E. Diener. 1995. Who is happy? *Psychological Science* 6:10–18.

Myers, N. 1979. *The Sinking Ark.* Pergamon Press, New York.

Myers, N. 1993. *Ultimate Security: The Environmental Basis of Political Stability.* Norton, New York.

Myers, N. 1996a. The biodiversity crisis and the future of evolution. *Environmentalist* 16:37–47.

Myers, N. 1996b. The world's forests: Problems and potentials. *Environmental Conservation* 23:156–168.

Myers, R. E. 1975. Neurology of social behavior and affect in primates: A study of prefrontal and anterior temporal cortex. Pp. 161–170 in K. J. Zuelch, ed., *Cerebral Localization.* Springer-Verlag, New York.

Nadel, S. F. 1954. *Nupe Religion.* Routledge, London.

Naess, A. 1973. The shallow and the deep, long-range ecology movement: A summary. *Inquiry* 16:95–100.

Naess, A. 1989. *Ecology, Community, and Lifestyle: Outline of an Ecophilosophy.* Cambridge Univ. Press, Cambridge.

Nag, M. 1972. Sex, culture, and human fertility: India and the United States. *Current Anthropology* 13:231–267.

Nagel, R. L. 1984. The origin of the hemoglobin S gene: Clinical, genetic, and anthropological consequences. *Einstein Quarterly Journal of Biology and Medicine* 2:53–62.

Nagel, T. 1974. What it's like to be a bat. *Philosophical Review* 83:435–450.

Napier, J. 1964. The evolution of the hand. *Scientific American* 207:308–312.

Nash, R. F. 1989. *The Rights of Nature: A History of Environmental Ethics.* Univ. of Wisconsin Press, Madison.

National Academy of Sciences USA. 1989. *On Being a Scientist.* National Academy Press, Washington, DC.

National Academy of Sciences USA. 1993. A joint statement by fifty-eight of the world's scientific academies. *Population Summit of the World's Scientific Academies.* National Academy Press, New Delhi.

National Academy of Sciences USA. 1998. *Teaching about Evolution and the Nature of Science.* National Academy Press, Washington, DC.

Naylor, R., R. J. Goldburg, H. Mooney, M. Beveridge, et al. 1998. Nature's subsidies to shrimp and salmon farming. *Science* 282:883–884.

Needham, J. 1956. *Science and Civilization in China.* Cambridge Univ. Press, Cambridge.

Neer, R. M. 1975. The evolutionary significance of vitamin D, skin pigment, and ultraviolet light. *American Journal of Physical Anthropology* 43:409–416.

Nei, M. 1995. Genetic support for the out-of-Africa theory of human evolution. *Proceedings of the National Academy of Sciences USA* 92:6720–6722.

Nei, M., and A. K. Roychoudhury. 1993. Evolutionary relationships of human populations on a global scale. *Molecular Biology and Evolution* 10:927–943.

Neisser, U., ed. 1998. *The Rising Curve: Long-Term Gains in IQ and Related Measures.* American Psychological Association, Washington, DC.

Nelson, J., M. Smith, and A. H. Bittles. 1997. Consanguineous marriage and its clinical consequences in migrants to Australia. *Clinical Genetics* 52:142–146.

Nesse, R. M., and A. T. Lloyd. 1992. The evolution of psychodynamic mechanisms. Pp. 601–624 in J. H. Barkow, L. Cosmides, and J. Tooby, eds., *The Adapted Mind.* Oxford Univ. Press, New York.

Nesse, R. M., and G. C. Williams. 1995. *Why We Get Sick: The New Science of Darwinian Medicine.* Times Books, New York.

Nettleship, M. A., R. Dalegivens, and A. Nettleship, eds. 1975. *War: Its Causes and Correlates.* Mouton, The Hague.

Neville, H. 1977. EEG testing of cerebral specialization in normal and congenitally deaf children: A preliminary report. Pp. 121–131 in S. J. Segalowitz and F. A. Gruber, eds., *Language Development and Neurological Theory.* Academic Press, New York.

Nevo, E. 1999. *Mosaic Evolution of Subterranean Mammals: Regression, Progression, and Global Convergence.* Oxford Univ. Press, Oxford.

Newmeyer, F. J. 1991. Functional explanation in linguistics and the origins of language. *Language and Communication* 11:3–28.

Newport, E. 1990. Maturational constraints on language learning. *Cognitive Science* 14:11–28.

Nichols, J. 1998. The origin and dispersal of languages: Linguistic evidence. Pp. 127–170 in N. G. Jablonski and L. C. Aiello, eds., *The Origin and Diversification of Language.* California Academy of Sciences, San Francisco.

Nichols, M., and J. Goodall. 1999. *Brutal Kinship.* Aperture, New York.

Nicholson, M. 1987. *The New Environmental Age.* Cambridge Univ. Press, Cambridge.

Niebuhr, G. 1998. American religion at the millennium's end. *Word and World* 18:5–13.

Nietzsche, F. W. 1974 (1887). *The Gay Science.* Random House, New York.

Nietzsche, F. W. 1996 (1878). *Human, All Too Human: A Book for Free Spirits.* Univ. of Nebraska Press, Lincoln.

Nilsson, D. E., and S. Pelger. 1994. A pessimistic estimate of the time required for an eye to evolve. *Proceedings of the Royal Society of London,* ser. B, 256:53–58.

Nimchinsky, E., E. Gilissen, J. Allman, D. Perl, et al. 1999. A neuronal morphologic type unique to humans and great apes. *Proceedings of the National Academy of Sciences USA* 96:5268–5273.

Nishida, T. 1983. Alpha status and agnostic alliance in wild chimpanzees. *Primates* 24:318–336.

Nishida, T. 1997. Sexual behavior of adult male chimpanzees of the Mahale Mountains National Park, Tanzania. *Primates* 38:379–398.

Nitecki, M. H., ed. 1983. *Coevolution.* Univ. of Chicago Press, Chicago.

Nitecki, M. H., and D. V. Nitecki, eds. 1993. *Evolutionary Ethics.* State Univ. of New York Press, Albany.

Nobile, P. 1974. Review of *Hefner. New York Times Magazine* (15 December): 4–5.

Noble, W., and I. Davidson. 1991. The evolutionary emergence of modern human behavior: Language and its archeology. *Man* 26:222–253.

Nolan, P., and G. Lenski. 1999. *Human Societies: An Introduction to Macrosociology.* 8th ed. McGraw-Hill, New York.

Noll, R. 1983. Shamanism and schizophrenia: A state-specific approach to the "schizophrenia metaphor" of shamanic states. *American Ethnologist* 10:443–459.

Nordborg, M. 1992. Female infanticide and human sex ratio evolution. *Journal of Theoretical Biology* 158:195–198.

Norris, S. 1999. Family secrets. *New Scientist* (19 June): 42–46.

North, D. C., and R. P. Thomas. 1977. The first economic revolution. *Economic History Review* 30:229–241.

Notestein, F. W. 1945. Population: The long view. Pp. 36–57 in T. W. Schultz, ed., *Food for the World.* Univ. of Chicago Press, Chicago.

Nottebohm, F. 1993. Neural lateralization of vocal control in a passerine bird. I. Song. *Journal of Experimental Zoology* 177:229–262.

Novick, P. 1988. *That Noble Dream.* Cambridge Univ. Press, Cambridge.

Nowak, R. M. 1991. *Walker's Mammals of the World.* 2 vols. Johns Hopkins Univ. Press, Baltimore.

Nozick, R. 1974. *Anarchy, State, and Utopia.* Basic Books, New York.

Ober, C., L. R. Weitkamp, N. Cox, H. Dytch, et al. 1997. HLA and mate choice in humans. *American Journal of Human Genetics* 61:497–504.

Oberman, H. A. 1997. The devil and the devious historian: Reaching for the roots of modernity. *KNAW Heineken Lectures—1996* 1996:33–44.

O'Connell, J. F., and J. Allen. 1998. When did humans first arrive in greater Australia, and why is it important to know? *Evolutionary Anthropology* 6:132–146.

O'Connell, J. F., K. Hawkes, and N. G. Blurton Jones. 1999. Grandmothering and the evolution of *Homo erectus. Journal of Human Evolution* 36:461–485.

O'Dea, K. 1991. Traditional diet and food preferences of Australian Aboriginal hunter-gatherers. *Philosophical Transactions of the Royal Society of London,* 334:233–241.

Oldroyd, H., and K. G. V. Smith. 1973. Eggs and larvae of flies. Pp. 289–323 in K. G. V. Smith, ed., *Insects and Arthropods of Medical Importance.* British Museum (Natural History), London.

O'Leary, D. D. M., P. A. Yates, and T. McLaughlin. 1999. Molecular development of sensory maps: Representing sights and smells in the brain. *Cell* 96:255–269.

Oliner, S. P., and P. M. Oliner. 1988. *The Altruistic Personality: Rescuers of Jews in Nazi Europe.* Free Press, New York.

Olson, D. R. 1986. The cognitive consequences of literacy. *Canadian Psychology* 27:109–121.

Ong, W. J. 1982. *Orality and Literacy.* Methuen, London.

Oosten, J. 1992. Representing the spirits: The masks of the Alaskan Inuit. Pp. 113–134 in J. Coote and A. Shelton, eds., *Anthropology, Art, and Aesthetics.* Clarendon Press, Oxford.

Opdyke, I. G., and J. Armstrong. 1999. *In My Hands: Memories of a Holocaust Rescuer.* Knopf, New York.

Oppenheim, J. S., J. E. Skerry, M. J. Tramo, and M. S. Gazzaniga. 1989. Magnetic resonance imaging morphology of the corpus callosum in monozygotic twins. *Annals of Neurology* 26:100–104.

Orians, G. H., and J. H. Heerwagen. 1992. Evolved responses to landscapes. Pp. 555–579 in J. H. Barkow, L. Cosmides, and J. Tooby, eds., *The Adapted Mind: Evolutionary Psychology and the Generation of Culture.* Oxford Univ. Press, Oxford.

Ornstein, R. 1986a. *Multimind: A New Way of Looking at Human Behavior.* Houghton Mifflin, Boston.

Ornstein, R. 1986b. *The Psychology of Consciousness.* Penguin Books, New York.

Ornstein, R. 1988. *Psychology: The Study of Human Experience.* 2nd ed. Harcourt Brace Jovanovich, Orlando, FL.

Ornstein, R. 1997. *The Right Mind.* Harcourt, Brace, New York.

Ornstein, R., and P. Ehrlich. 1989. *New World/New Mind: Moving toward Conscious Evolution.* Doubleday, New York.

Ornstein, R., and R. F. Thompson. 1984. *The Amazing Brain.* Houghton Mifflin, Boston.

Orr, H. A., R. Dawkins, D. C. Dennett, and A. M. Shapiro. 1996. Denying Darwin: David Berlinski and critics. *Commentary* (September): 4–39.

Orr, M. R., and T. B. Smith. 1998. Ecology and speciation. *Trends in Ecology and Evolution* 13:502–506.

Ortiz de Montellano, B. 1978. Aztec cannibalism: An ecological necessity. *Science* 200:611–617.

Ortner, S. B. 1974. Is female to male as nature is to culture? Pp. 67–87 in M. Rosaldo and L. Lamphere, eds., *Women, Culture, and Society.* Stanford Univ. Press, Stanford.

Otte, M. 1995. Testing hypotheses about recent human evolution from skulls: Integrating morphology, function, development, and phylogeny: Comments. *Current Anthropology* 36:181–182.

Otto, S. P. 1997. Unravelling gene interactions. *Nature* 390:343.

Otto, S. P., and Y. Michalakis. 1998. The evolution of recombination in changing environments. *Trends in Ecology and Evolution* 13:145–151.

Owen, D. F. 1997. Natural selection and evolution in moths: Homage to J. W. Tutt. *Oikos* 78:177–181.

Owen, R. B., R. Crossley, T. C. Johnson, D. Tweddle, et al. 1990. Major low levels of Lake Malawi and their implications for speciation rates in cichlid fishes. *Proceedings of the Royal Society of London*, ser. B, 240:519–553.

Owens, K., and M.-C. King. 1999. Genomic views of human history. *Science* 286:451–453.

Ozguc, T. 1973. Ancient Ararat. *Cities: Their Origin, Growth, and Human Impact.* Freeman, San Francisco.

Pääbo, S. 1995. The Y chromosome and the origin of all of us (men). *Science* 268:1141–1142.

Packer, C., M. Tatar, and A. Collins. 1998. Reproductive cessation in female mammals. *Nature* 392:807–811.

Padian, K., and L. M. Chiappe. 1998. The origin of birds and their flight. *Scientific American* (February): 38–47.

Paik, H., and G. Comstock. 1994. The effects of television violence on antisocial behavior: A meta-analysis. *Communication Research* 21:516–546.

Paley, W. 1803. *Natural Theology; or, Evidences of the Existence and Attributes of the Deity, Collected from the Appearances of Nature.* 5th ed. Faulder, London.

Palmer, C. 1989. Is rape a cultural universal? A re-examination of the ethnographic data. *Ethnology* 28:1–16.

Palombit, R. A. 1994. Dynamic pair bonds in hylobatids: Implications regarding monogamous social systems. *Behaviour* 128:65–101.

Palombit, R. A. 1996. Pair bonds in monogamous apes: A comparison of the siamang *Hylobates syndactylus* and the white-handed gibbon *Hylobates lar. Behaviour* 133:321–356.

Pälsson, G. 1988. Hunters and gatherers of the sea. Pp. 189–204 in T. Ingold, D. Riches, and J. Woodburn, eds., *Hunters and Gatherers.* Vol. 1, *History, Evolution, and Social Change.* Berg, Oxford.

Panksepp, J. 1994. Basic emotions ramify widely in the brain, yielding many concepts that cannot be distinguished unambiguously . . . yet. Pp. 86–88 in P. Ekman and R. J. Davidson, eds., *The Nature of Emotions: Fundamental Questions.* Oxford Univ. Press, New York.

Paradis, J., and G. C. Williams. 1989. *Evolution and Ethics.* Princeton Univ. Press, Princeton.

Parker, G. A. 1989. Hamilton's rule and conditionality. *Ethology, Ecology, and Evolution* 1:195–211.

Parker, S. T., and M. L. McKinney. 1999. *Origins of Intelligence: The Evolution of Cognitive Development in Monkeys, Apes, and Humans.* Johns Hopkins Univ. Press, Baltimore.

Parmentier, R. J. 1994. Comments. *Current Anthropology* 35:388–389.

Parr, L. A., and F. B. M. de Waal. 1999. Visual kin recognition in chimpanzees. *Nature* 399:647–648.

Parsons, T. 1951. *The Social System.* Free Press, Glencoe, IL.

Parsons, T. 1964. Evolutionary universals in society. *American Sociological Review* 29:339–357.

Partridge, L., and N. H. Barton. 1993. Optimality, mutation, and the evolution of aging. *Nature* 362:305–311.

Partridge, L., and K. Fowler. 1992. Direct and correlated responses to selection on age at reproduction in *Drosophila melanogaster. Evolution* 46:76–91.

Partridge, T. C., G. C. Bond, J. H. Hartnady, P. B. deMenocal, et al. 1995. Climatic effects of Late Neogene tectonism and volcanism. Pp. 8–23 in E. S. Vrba, G. H. Denton, T. C. Partridge, and L. H. Burckle, eds., *Paleoclimate and Evolution, with Emphasis on Human Origins.* Yale Univ. Press, New Haven.

Partridge, T. C., B. A. Wood, and P. B. deMenocal. 1995. The influence of global climate change and regional uplift on large-mammalian evolution in East and southern Africa. Pp. 331–355 in E. S. Vrba, G. H. Denton, T. C. Partridge, and L. H. Burckle, eds., *Paleoclimate and Evolution, with Emphasis on Human Origins.* Yale Univ. Press, New Haven.

Pascal, B. 1995 (1670). *Pensées.* Penguin Books, London.

Pascual-Leone, J. 1987. Organismic processes for neo-Piagetian theories: A dialectical causal account of cognitive development. *International Journal of Psychology* 22:531–570.

Passingham, R. E. 1973. Anatomical differences between the neocortex of man and other primates. *Brain, Behaviour, and Evolution* 7:337–359.

Pattatucci, A. M., and D. M. Hamer. 1995. The genetics of sexual orientation: From fruit flies to humans. Pp. 154–174 in P. R. Abrahamson and S. D. Pinkerton, eds., *Sexual Nature/Sexual Culture.* Univ. of Chicago Press, Chicago.

Patterson, F., and R. Cohn. 1994. Self-recognition and self-awareness in lowland gorillas. Pp. 273–290 in S. T. Parker, R. W. Mitchell, and M. L. Boccia, eds., *Self-Awareness in Animals and Humans.* Cambridge Univ. Press, New York.

Patterson, H. E. H. 1985. The recognition concept of species. Pp. 21–29 in E. S. Vrba, ed., *Species and Speciation.* Transvaal Museum, Pretoria.

Paul, R. 1978. Instinctive aggression in man: The Semai case. *Journal of Psychological Anthropology* 1:65–79.

Pavelka, M. S. M. 1995. Sexual nature: What can we learn from a cross-species perspective? Pp. 17–36 in P. R. Abramson and S. D. Pinkerton, eds., *Sexual Nature/Sexual Culture.* Univ. of Chicago Press, Chicago.

Pavlov, I. P. 1927. *Conditioned Reflexes.* Oxford Univ. Press, New York.

Pawlowski, B. 1999. Loss of oestrus and concealed ovulation in human evolution: The case against the sexual-selection hypothesis. *Current Anthropology* 40:257–275.

Pearsall, D. 1989. *Paleoethnobotany: A Handbook of Procedures.* Academic Press, San Diego.

Peirce, C. S. 1955. *Philosophical Writings of Peirce.* Dover, New York.

Pennisi, E. 1998. A genomic battle of the sexes. *Science* 281:1984–1985.

Pennisi, E. 1999a. Are our primate cousins "conscious"? *Science* 284:2073–2076.

Pennisi, E. 1999b. Did cooked tubers spur the evolution of big brains? *Science* 283:2004–2005.

Penny, D., M. Steel, P. J. Waddell, and M. D. Hendy. 1995. Improved analyses of human mtDNA sequences support a recent African origin for *Homo sapiens. Molecular Biology and Evolution* 12:863–882.

Penton-Voak, I. S., D. I. Perrett, D. L. Castles, T. Kobayashi, et al. 1999. Menstrual cycle alters face preference. *Nature* 399:741–742.

Perlmutter, A. D., and C. Reitelman. 1992. Surgical management of intersexuality. In P. C. Walsh, A. B. Retik, T. A. Stamey, and J. R. Vaughan, eds., *Campbell's Urology.* 6th ed. Saunders, Philadelphia.

Perls, T. T., L. Alpert, and R. C. Fretts. 1997. Middle-aged mothers live longer. *Nature* 389:133.

Perrett, D. I., K. J. Lee, I. Penton-Voak, D. Rowland, et al. 1998. Effects of sexual dimorphism on facial attractiveness. *Nature* 394:884–886.

Perrett, D. I., K. A. May, and S. Yoshikawa. 1994. Facial shape and judgements of female attractiveness. *Nature* 368:239–242.

Perrin, N. 1979. *Giving Up the Gun: Japan's Reversion to the Sword.* Hall, Boston.

Peter, K. A. 1987. *The Dynamics of Hutterite Society: An Analytical Approach.* Univ. of Alberta Press, Edmonton.

Peterkin, G. L., H. M. Bricker, and P. Mellars, eds. 1993. *Hunting and Animal Exploitation in the Later Palaeolithic and Mesolithic of Eurasia.* American Anthropological Association, Washington, DC.

Peters, R. H. 1976. Tautology in evolution and ecology. *American Naturalist* 110:1–12.

Peters, R. H. 1978. Predictable problems with tautology in evolution and ecology. *American Naturalist* 112:759–762.

Peters, R. H. 1991. *A Critique for Ecology.* Cambridge Univ. Press, Cambridge.

Peterson, A. T., J. Soberón, and V. Sánchez-Cordero. 1999. Conservatism of ecological niches in evolutionary time. *Science* 285:1265–1267.

Peterson, N. 1993. Demand sharing: Reciprocity and the pressure for generosity among foragers. *American Anthropologist* 95:860–874.

Petitto, L., and P. Marentette. 1991. Babbling in the manual mode: Evidence for the ontogeny of language. *Science* 22:1493–1496.

Petrie, M., and B. Kempenaers. 1998. Extra-pair paternity in birds: Explaining variation between species and populations. *Trends in Ecology and Evolution* 13:52–58.

Petulla, J. M. 1980. *American Environmentalism: Values, Tactics, Priorities.* Texas A&M Univ. Press, College Station.

Phoenix, C. H., R. W. Goy, A. A. Gerall, and W. C. Young. 1959. Organizing action of prenatally administered testosterone propionate on the tissues mediating mating behavior in the female guinea pig. *Endocrinology* 65:369–382.

Piaget, J. 1952. *The Origins of Intelligence in Children.* International Univ. Press, New York.

Piattelli-Palmarini, M. 1989. Evolution, selection, and cognition: From "learning" to parameter setting in biology and the study of language. *Cognition* 31:1–44.

Pickford, M. 1990. Uplift of the roof of Africa and its bearing on the evolution of mankind. *Human Evolution* 5:1–20.

Pickford, M. 1993. Climatic change, biogeography, and *Theropithecus*. Pp. 227–243 in N. G. Jablonski, ed., *Theropithecus: The Rise and Fall of a Primate Genus.* Cambridge Univ. Press, Cambridge.

Pimentel, D., and H. Lehman, eds. 1993. *The Pesticide Question: Environment, Economics, and Ethics.* Chapman and Hall, New York.

Pimm, S. L. 1991. *Balance of Nature? Ecological Issues in the Conservation of Species and Communities.* Univ. of Chicago Press, Chicago.

Pinker, S. 1991. Rules of language. *Science* 253:530–535.

Pinker, S. 1994. *The Language Instinct: How the Mind Creates Language.* Morrow, New York.

Pinker, S. 1997. *How the Mind Works.* Norton, New York.

Pinker, S. 1999. *Words and Rules: The Ingredients of Language.* Basic Books, New York.

Pinker, S., and P. Bloom. 1990. Natural language and natural selection. *Behavior and Brain Sciences* 13:707–784.

Pinker, S., and A. Prince. 1988. On language and connectionism. *Cognition* 28:73–193.

Pirages, D. C. 1996. *Building Sustainable Societies.* Sharpe, Armonk, NY.

Pirages, D. C., ed. 1977. *The Sustainable Society: Social and Political Implications.* Praeger, New York.

Pirages, D. C., and P. R. Ehrlich. 1974. *Ark II: Social Response to Environmental Imperatives.* Viking, New York.

Plato. 1991 (ca. 387). *The Republic: Translated with Notes, an Interpretive Essay, and a New Introduction by Allan Bloom.* 2nd ed. Basic Books, New York.

Platt, J. 1973. Social traps. *American Psychologist* 28:642–651.

Plattner, S. 1989a. Markets and marketplaces. Pp. 171–208 in S. Plattner, ed., *Economic Anthropology.* Stanford Univ. Press, Stanford.

Plattner, S. 1989b. Marxism. Pp. 379–396 in S. Plattner, ed., *Economic Anthropology.* Stanford Univ. Press, Stanford.

Pleijel, F. 1999. Phylogenetic taxonomy, a farewell to species, and a revision of *Heteropodarke (Hesionidae, Polychaeta, Annelida). Systematic Biology* 48:755–798.

Plomin, R. 1989. Environment and genes: Determinants of behavior. *American Psychologist* 44:105–111.

Plomin, R., M. J. Owen, and P. McGuffin. 1994. The genetic basis of complex human behaviors. *Science* 264:1733–1739.

Podolefsky, A. 1984. Contemporary warfare in the New Guinea highlands. *Ethnology* 23:73–87.

Polgar, S. 1972. Population history and population policies from an anthropological perspective. *Current Anthropology* 13:202–215.

Polunin, N. 1949. *Arctic Unfolding.* Hutchinson, London.

Pomerantz, S. M., R. W. Goy, and M. M. Roy. 1986. Expression of male-typical behavior in adult female pseudohermaphroditic rhesus: Comparisons with normal males and neonatally gonadectomized males and females. *Hormones and Behavior* 20:483–500.

Pope, G. G. 1995. The influence of climate and geography on the biocultural evolution of the Far Eastern hominids. Pp. 493–506 in E. S. Vrba, G. H. Denton, T. C. Partridge, and L. H. Burckle, eds., *Paleoclimate and Evolution, with Emphasis on Human Origins.* Yale Univ. Press, New Haven.

Popper, K. 1968. *Conjectures and Refutations: The Growth of Scientific Knowledge.* Harper and Row, New York.

Popper, K. 1968 (1935). *The Logic of Scientific Discovery.* Routledge, London.

Population Reference Bureau. 1998. *World Population Data Sheet.* Population Reference Bureau, Washington, DC.

Porter, R. 1997. *The Greatest Benefit to Mankind: A Medical History of Humanity.* Norton, New York.

Posner, M. I., and M. E. Raichle. 1994. *Images of Mind.* Scientific American Library, New York.

Posner, M. I., and M. K. Rothbart. 1998. Attention, self-regulation, and consciousness. *Philosophical Transactions of the Royal Society of London,* ser. B, 353:1915–1927.

Postel, S. L., G. C. Daily, and P. R. Ehrlich. 1996. Human appropriation of renewable fresh water. *Science* 271:785–788.

Potts, R. 1984. Home bases and early hominids. *American Scientist* 72:338–347.

Potts, R. 1996a. Evolution and climate variability. *Science* 273:922–923.

Potts, R. 1996b. *Humanity's Descent: The Consequences of Ecological Instability.* Morrow, New York.

Potts, R. 1998. Environmental hypotheses of hominid evolution. *Yearbook of Physical Anthropology* 41:93–136.

Povinelli, D. J. 1987. Monkeys, apes, mirrors, and minds: The evolution of self-awareness in primates. *Human Evolution* 2:493–507.

Povinelli, D. J., and L. R. Godfrey. 1993. The chimpanzee's mind: How noble in reason? How absent of ethics? Pp. 277–324 in M. H. Nitecki and D. V. Nitecki, eds., *Evolutionary Ethics.* State Univ. of New York Press, Albany.

Povinelli, D. J., K. E. Nelson, and S. T. Boysen. 1990. Inferences about guessing and knowing by chimpanzees *(Pan troglodytes). Journal of Comparative Psychology* 104:203–210.

Povinelli, D. J., K. E. Nelson, and S. T. Boysen. 1992. Comprehension of role reversal in chimpanzees: Evidence of empathy? *Animal Behaviour* 43:633–640.

Povinelli, D. J., K. A. Parks, and M. A. Novak. 1991. Do rhesus monkeys *(Macaca mulatta)* attribute knowledge and ignorance to others? *Journal of Comparative Psychology* 105:318–325.

Povinelli, D. J., K. A. Parks, and M. A. Novak. 1992. Role reversal by rhesus monkeys, but no evidence of empathy. *Animal Behaviour* 44:269–281.

Powell, J. R. 1997. *Progress and Prospects in Evolutionary Biology.* Oxford Univ. Press, New York.

Prange, G. W. 1982. *Miracle at Midway.* McGraw-Hill, New York.

Premack, D. 1984. Pedagogy and aesthetics as sources of culture. Pp. 15–35 in M. S. Gazzaniga, ed., *Handbook of Cognitive Neuroscience.* Plenum Press, New York.

Premack, D. 1985. "Gavagai!" or the future history of the animal language controversy. *Cognition* 19:207–296.

Premack, D. 1986. Pangloss to Cyrano de Bergerac: "Nonsense, it's perfect!" A reply to Bickerton. *Cognition* 23:81–88.

Premack, D., and G. Woodruff. 1978. Does the chimpanzee have a theory of mind? *Behavioral and Brain Sciences* 4:515–526.

Press, F., and R. Siever. 1972. *Earth.* Freeman, San Francisco.

Preston, S. H. 1999. Differential fertility by IQ and the IQ distribution of a population. Pp. 377–387 in U. Neisser, ed., *The Rising Curve: Long-Term Gains in IQ and Related Measures.* American Psychological Association, Washington, DC.

Price, P. W. 1996. *Biological Evolution.* Saunders, Fort Worth.

Pringle, H. 1998a. Eight millennia of footwear fashion. *Science* 281:23–24.

Pringle, H. 1998b. North America's wars. *Science* 279:2038–2040.

Pringle, H. 1998c. The slow birth of agriculture. *Science* 282:1446–1450.

Pringle, H. 1999. Temples of doom. *Discover* (March): 78–85.

Pritchard, J. K., M. T. Seielstad, A. Perez-Lezaun, and M. W. Feldman. 1999. Population growth of human Y chromosomes: A study of Y chromosome microsatellites. *Molecular Biology and Evolution* 16:1791–1798.

Profet, M. 1988. The evolution of pregnancy sickness as protection to the embryo against Pleistocene teratogens. *Evolutionary Theory* 8:177–190.

Profet, M. 1991. The function of allergy: Immunological defense against toxins. *Quarterly Review of Biology* 66:1991.

Profet, M. 1993. Menstruation as a defense against pathogens transported by sperm. *Quarterly Review of Biology* 68:335–381.

Pugh, G. E. 1977. *The Biological Origin of Human Values.* Basic Books, New York.

Pullum, G. K. 1991. *The Great Eskimo Vocabulary Hoax, and Other Irreverent Essays on the Study of Language.* Univ. of Chicago Press, Chicago.

Purves, W. K., G. H. Orians, H. C. Heller, and D. Sadava. 1997. *Life: The Science of Biology.* 5th ed. Sinauer, Sunderland, MA.

Pusey, A. E. 1980. Inbreeding avoidance in chimpanzees. *Animal Behaviour* 28:543–552.

Pyke, G. H. 1986. Human diets: A biological perspective. Pp. 273–281 in L. Manderson, ed., *Shared Wealth and Symbol: Food, Culture, and Society in Oceania and Southeast Asia.* Cambridge Univ. Press, Cambridge.

Quigley, C. 1961. *The Evolution of Civilizations.* Macmillan, New York.

Quine, W. V. 1977. *Ontological Relativity and Other Essays.* Columbia Univ. Press, New York.

Quine, W. V. 1992. *Pursuit of Truth.* Rev. ed. Harvard Univ. Press, Cambridge.

Ragir, S. 1993. On the evolution of language. *Current Anthropology* 34:167–168.

Raikow, R. J. 1977. The origin and evolution of the Hawaiian honeycreepers (Drepanididae). *Living Bird* 15:95–117.

Rainey, P. B., and M. Travisano. 1998. Adaptive radiation in a heterogeneous environment. *Nature (London)* 394:69–72.

Rakove, J. N. 1996. *Original Meanings: Politics and Ideas in the Making of the Constitution.* Random House, New York.

Ralls, K. 1977. Sexual dimorphism in mammals: Avian models and unanswered questions. *American Naturalist* 111:917–938.

Ralls, K., J. D. Ballou, and A. Templeton. 1987. Estimates of lethal equivalents and the cost of inbreeding in mammals. *Conservation Biology* 2:185–193.

Ralls, K., K. Brugger, and J. Ballou. 1979. Inbreeding and juvenile mortality in small populations of ungulates. *Science* 206:1101–1103.

Raloff, J. 1998. Very hot grills may inflame cancer risks. *Science News* 154:341.

Rapoport, A. 1994. Spatial organization and the built environment. Pp. 460–502 in T. Ingold, ed., *Companion Encyclopedia of Anthropology: Humanity, Culture, and Social Life.* Routledge, London.

Rapoport, J. L. 1989. The biology of obsessions and compulsions. *Scientific American* (March): 83–89.

Rappaport, R. A. 1971. The sacred in human evolution. *Annual Review of Ecology and Systematics* 2:23–44.

Rappaport, R. A. 1979. *Ecology, Meaning, and Religion.* North Atlantic Books, Richmond, CA.

Raup, D. M. 1991. *Extinction: Bad Genes or Bad Luck?* Norton, New York.

Rauschecker, J. P., and U. Kniepert. 1987. Auditory localization behavior in cats deprived of vision. *Society of Neuroscience Abstracts* 13:871.

Rauschecker, J. P., and M. Korte. 1993. Auditory compensation for early blindness in cat cerebral cortex. *Journal of Neuroscience* 13:4538–4548.

Raven, P. H. 1990. The politics of preserving biodiversity. *BioScience* 40:769–774.

Raven, P. H. 1993. *A Biological Survey for the Nation.* National Academy Press, Washington, DC.

Raven, P. H., ed. 1980. *Research Priorities in Tropical Biology.* National Academy of Sciences, Washington, DC.

Raven, P. H., and G. B. Johnson. 1999. *Biology.* 5th ed. WCB/McGraw-Hill, St. Louis.

Rawls, J. 1971. *A Theory of Justice.* Harvard Univ. Press, Cambridge.

Rawls, J. 1996. *Political Liberalism.* Columbia Univ. Press, New York.

Ray, D. J. 1967. *Eskimo Masks: Art and Ceremony.* Univ. of Washington Press, Seattle.

Read, C., and P. Schreiber. 1982. Why short subjects are harder to find than long ones. Pp. 78–101 in E. Wanner and L. R. Gleitman, eds., *Language Acquisition: The State of the Art.* Cambridge Univ. Press, Cambridge.

Real, L. A. 1991. Animal choice behavior and the evolution of cognitive architecture. *Science* 253:980–986.

Recher, H. F., D. Lunney, and I. Dunn. 1986. *A Natural Legacy: Ecology in Australia.* 2nd ed. Pergamon Press, Sydney.

Redford, K. H., and A. M. Stearman. 1993a. Forest-dwelling native Amazonians and the conservation of biodiversity: Interests in common or in collision? *Conservation Biology* 7:248–255.

Redford, K. H., and A. M. Stearman. 1993b. On common ground? Response to Alcorn. *Conservation Biology* 7:427–428.

Reed, C. A. 1977a. Origins of agriculture: Discussion and some conclusions. Pp. 881–953 in C. A. Reed, ed., *Origins of Agriculture.* Mouton, The Hague.

Reed, C. A., ed. 1977b. *Origins of Agriculture.* Mouton, The Hague.

Reff, D. T. 1991. *Disease, Depopulation, and Culture Change in Northwestern New Spain, 1518–1764.* Univ. of Utah Press, Salt Lake City.

Regan, T. 1983. *The Case for Animal Rights.* Univ. of California Press, Berkeley.

Regan, T., and P. Singer, eds. 1989. *Animal Rights and Human Obligations.* 2nd ed. Prentice Hall, Englewood Cliffs, NJ.

Reich, D. E., and D. B. Goldstein. 1998. Genetic evidence for a Paleolithic human population expansion in Africa. *Proceedings of the National Academy of Sciences USA* 95:8119–8123.

Reid, T. 1785. *Essays on the Intellectual Powers of Man.* MIT Press, Cambridge.

Reid, T. 1970 (1764). *An Inquiry into the Human Mind, on the Principles of Common Sense.* Univ. of Chicago Press, Chicago.

Relethford, J. H. 1995. Genetics and modern human origins. *Evolutionary Anthropology* 4:53–63.

Rendel, J. M. 1970. The time scale of genetic change. Pp. 27–47 in S. V. Boyden, ed., *The Impact of Civilization on the Biology of Man.* National Univ. Press, Canberra.

Renfrew, C. 1998. The origins of world linguistic diversity: An archaeological perspective. Pp. 171–192 in N. G. Jablonski and L. C. Aiello, eds., *The Origin and Diversification of Language.* California Academy of Sciences, San Francisco.

Rensch, B. 1959 (1954). *Evolution above the Species Level.* Columbia Univ. Press, New York.

Restak, R. M. 1995. *Receptors.* Bantam Books, New York.

Reynolds, J. D., and P. H. Harvey. 1994. Sexual selection and the evolution of sex differences. Pp. 53–70 in R. V. Short and E. Balaban, eds., *The Differences between the Sexes.* Cambridge Univ. Press, Cambridge.

Reynolds, P. C. 1981. *On the Evolution of Human Behavior: The Argument from Animals to Man.* Univ. of California Press, Berkeley.

Reynolds, P. C. 1993. The complementation theory of language and tool use. Pp. 407–428 in K. R. Gibson and T. Ingold, eds., *Tools, Language, and Cognition in Human Evolution.* Cambridge Univ. Press, Cambridge.

Reynolds, V. 1994. Kinship in nonhuman and human primates. Pp. 137–165 in D. Quiatt and J. Itani, eds., *Hominid Culture in Primate Perspective.* Univ. Press of Colorado, Niwot.

Reznik, D. N., F. Shaw, H., F. H. Rodd, and R. G. Shaw. 1997. Evaluation of the rate of evolution in natural populations of guppies *(Poecilia reticulata). Science* 275:1934–1936.

Rice, G., C. Anderson, N. Rische, and G. Ebers. 1999. Male homosexuality: Absence of linkage to microsatellite markers at Xq28. *Science* 284:665–667.

Rice, W. R., and E. E. Hostert. 1993. Laboratory experiments on speciation: What have we learned in forty years? *Evolution* 47:1637–1653.

Rice, W. R., and G. W. Salt. 1988. Speciation via disruptive selection on habitat preference: Experimental evidence. *American Naturalist* 131:911–917.

Rice, W. R., and G. W. Salt. 1990. The evolution of reproductive isolation as a correlated character under sympatric conditions: Experimental evidence. *Evolution* 44:1140–1152.

Richards, D. D., and R. S. Siegler. 1986. Children's understanding of the attributes of life. *Journal of Experimental Child Psychology* 42:1–22.

Richards, R. J. 1986a. A defense of evolutionary ethics. *Biology and Philosophy* 1:265–293.

Richards, R. J. 1986b. Justification through biological faith: A rejoinder. *Biology and Philosophy* 1:337–354.

Richardson, L. F. 1960. *The Statistics of Deadly Quarrels.* Boxwood Press, Pittsburgh.

Richerson, P. J., and R. Boyd. 1978. A dual inheritance model of human evolutionary process. I. basic postulates and a simple model. *Journal of Social and Biological Structures* 1:148–153.

Richerson, P. J., and R. B. Boyd. 1984. Natural selection and culture. *BioScience* 34:430–434.

Richerson, P. J., and R. Boyd. 1989. The role of evolved predispositions in cultural evolution. *Ethology and Sociobiology* 10:195–219.

Ricketts, T. H., E. Dinerstein, D. M. Olson, C. J. Loucks, et al. 1999. *Terrestrial Ecoregions of North America: A Conservation Assessment.* Island Press, Washington, DC.

Ridley, M. 1996a. *Evolution.* Blackwell Scientific, Cambridge, MA.

Ridley, M. 1996b. *The Origins of Virtue: Human Instincts and the Evolution of Cooperation.* Penguin Books, London.

Ridley, M. 1999. *Genome: The Autobiography of a Species in Twenty-Three Chapters.* HarperCollins, New York.

Rieff, D. 1999. The false dawn of civil society. *Nation* (22 February): 11–16.

Rightmire, G. P. 1992. *Homo erectus:* Ancestor or evolutionary side-branch? *Evolutionary Anthropology* 1:43–49.

Rightmire, G. P. 1995. Diversity within the genus *Homo.* Pp. 483–492 in E. S. Vrba, G. H. Denton, T. C. Partridge, and L. H. Burckle, eds., *Paleoclimate and Evolution, with Emphasis on Human Origins.* Yale Univ. Press, New Haven.

Rindos, D. 1984. *The Origins of Agriculture: An Evolutionary Perspective.* Academic Press, Orlando.

Rivers, W. H. R. 1924. *Social Organization.* Kegan Paul, London.

Roan, S. 1989. *Ozone Crisis: The Fifteen-Year Evolution of a Sudden Global Emergency.* Wiley, New York.

Robarchek, C. A. 1989. Primitive warfare and the ratomorphic image of mankind. *American Anthropologist* 91:903–920.

Robarchek, C. A., and R. K. Dentan. 1987. Blood drunkenness and the bloodthirsty Semai: Unmaking another anthropological myth. *American Anthropologist* 89:356–365.

Roberts, D. F., and D. P. S. Kahlon. 1976. Environmental correlations of skin color. *Annals of Human Biology* 3:11–22.

Roberts, G. 1998. Development of cooperative relationships through increasing investment. *Nature* 394:175–178.

Roberts, N. 1992. Climatic change in the past. Pp. 174–178 in S. Jones, R. Martin, and D. Pilbeam, eds., *The Cambridge Encyclopedia of Human Evolution.* Cambridge Univ. Press, Cambridge.

Roberts, R. G., R. Jones, N. A. Spooner, and M. J. Head. 1994. The human colonization of Australia: Optical dates of 53,000 and 60,000 years bracket human arrival at Deaf Adder Gorge, Northern Territory. *Quaternary Science Reviews* 13:575–586.

Roberts, S. J. 1998. The role of diffusion in the genesis of Hawaiian creole. *Language* 74:1–39.

Robins, A. 1991. *Biological Perspectives on Human Pigmentation.* Cambridge Univ. Press, Cambridge.

Robinson, D. N., ed. 1977. *Darwinism.* University Publications of America, Washington, DC.

Robinson, J. G. 1996. Hunting wildlife in forest patches: An ephemeral resource. Pp. 111–130 in J. Schelhas and R. Greenberg, eds., *Forest Patches in Tropical Landscapes.* Island Press, Washington, DC.

Robinson, J. T. 1963. Adaptive radiation in the australopithecines and the origin of man. Pp. 385–416 in F. C. Howell and F. Bourlière, eds., *African Ecology and Human Evolution.* Aldine, Chicago.

Roche, H., A. Delagnes, J.-P. Brugal, C. Feibel, et al. 1999. Early hominid stone tool production and technical skill 2.34 million years ago in West Turkana, Kenya. *Nature* 399:57–60.

Rodman, P. S., and H. M. McHenry. 1980. Bioenergetics and the origin of hominid bipedalism. *American Journal of Physical Anthropology* 52:103–106.

Rodseth, L., R. W. Wrangham, A. M. Harrigan, and B. B. Smuts. 1991. The human community as a primate society. *Current Anthropology* 32:221–241.

Roe, A. 1953. *The Making of a Scientist.* Dodd, Mead, New York.

Roe, F. G. 1970. *The North American Buffalo: A Critical Study of the Species in Its Wild State.* 2nd ed. Univ. of Toronto Press, Toronto.

Roff, D. A. 1992. *The Evolution of Life Histories: Theory and Analysis.* Chapman and Hall, New York.

Roff, D. A. 1997. *Evolutionary Quantitative Genetics.* Chapman and Hall, New York.

Rogers, A. R., and H. C. Harpending. 1992. Population growth makes waves in the distribution of pairwise differences. *Molecular Biology and Evolution* 9:552–569.

Rogers, A. R., and L. B. Jorde. 1995. Genetic evidence on modern human origins. *Human Biology* 67:1–36.

Rogers, E. M. 1995. *Diffusion of Innovations.* 4th ed. Free Press, New York.

Rolston, H. 1988. *Environmental Ethics: Duties to and Values in the Natural World.* Temple Univ. Press, Philadelphia.

Roosevelt, A. C. 1984. Population, health, and the evolution of subsistence: Conclusions from the conference. Pp. 559–583 in M. N. Cohen and G. J. Armelagos, eds., *Paleopathology at the Origins of Agriculture.* Academic Press, Orlando.

Root, T. L., and S. H. Schneider. 1995. Ecology and climate: Research strategy and implications. *Science* 269:334–341.

Roper, C., P. Pignatelli, and L. Partridge. 1993. Evolutionary effects of selection on age at reproduction in larval and adult *Drosophila melanogaster. Evolution* 47:445–455.

Rorty, R. 1982. *Consequences of Pragmatism (Essays: 1972–1980).* Univ. of Minnesota Press, Minneapolis.

Rosch, E. 1973. Natural categories. *Cognitive Psychology* 4:328–350.

Rose, M. D. 1976. Bipedal behavior of olive baboons *(Papio anubis)* and its relevance to an understanding of the evolution of human bipedalism. *American Journal of Physical Anthropology* 44:247–261.

Rose, M. D. 1991. The process of bipedalization in hominids. Pp. 37–48 in Y. Coppens and B. Senut, eds., *Origine(s) de la Bipédie chez les Hominidés.* Éditions du CNRS, Paris.

Rose, M. R. 1984. Laboratory evolution of postponed senescence in *Drosophila melanogaster. Evolution* 38:1004–1010.

Rose, M. R. 1991. *The Evolutionary Biology of Aging.* Oxford Univ. Press, Oxford.

Rose, M. R., and B. Charlesworth. 1981. Genetics of life history in *Drosophila melanogaster.* II. Exploratory selection experiments. *Genetics* 97:187–196.

Rose, M. R., and G. V. Lauder, eds. 1996. *Adaptation.* Academic Press, San Diego.

Rosen, S. P. 1991. *Winning the Next War: Innovation and the Modern Military.* Cornell Univ. Press, Ithaca.

Rosenberg, M. 1990. The mother of invention: Evolutionary theory, territoriality, and the origins of agriculture. *American Anthropologist* 92:399–415.

Rosenberg, M. 1991. Population pressure, locational constraints, and the evolution of culture: A reply to Graber. *American Anthropologist* 93:695–697.

Rosenblatt, R., ed. 1999. *Consuming Desires: Consumption, Culture, and the Pursuit of Happiness.* Island Press, Washington, DC.

Rosin, H. 1999. Creationism evolves as Kansans dispute evolution. *Fresno (California) Bee* (9 August): A4.

Rosten, L., ed. 1975. *Religions of America: Ferment and Faith in an Age of Crisis.* Simon and Schuster, New York.

Roughgarden, J. 1971. Density-dependent natural selection. *Ecology* 52:453–468.

Roughgarden, J. 1975. Evolution of marine symbiosis: A simple cost-benefit model. *Ecology* 56:1201–1208.

Roughgarden, J. 1976. Resource partitioning among competing species: A coevolutionary approach. *Theoretical Population Biology* 9:388–424.

Roughgarden, J. 1979. *Theory of Population Genetics and Evolutionary Ecology: An Introduction.* Macmillan, New York.

Roughgarden, J. 1999. *Evolution's Rainbow: Gender and Sexuality in Nature.* Manuscript.

Roughgarden, J. D., D. Heckel, and E. Fuentes. 1983. Coevolutionary theory and the biogeography and community structure of *Anolis.* Pp. 371–410 in R. Huey, E. Pianka, and T. Schoener, eds., *Lizard Ecology: Studies of a Model Organism.* Harvard Univ. Press, Cambridge.

Roush, R. T., and J. C. Daly. 1990. The role of population genetics in resistance research and management. Pp. 97–152 in R. T. Roush and B. E. Tabashnik, eds., *Pesticide Resistance in Arthropods.* Chapman and Hall, New York.

Roush, R. T., and J. A. McKenzie. 1987. Ecological genetics of insecticide and acaracide resistance. *Annual Review of Entomology* 32:361–380.

Roush, W. 1997. Herbert Benson: Mind-body maverick pushes the envelope. *Science* 276:357–359.

Rousseau, J.-J. 1762. *The Social Contract.* Dent, London.

Rubinstein, W. D. 2000. Review: *Hitler's Pope. First Things: A Monthly Journal of Religion and Public Life* (January): 39.

Ruddle, F., and C. Kappen. 1995. Mammalian homeobox genes: Evolutionary and regulatory aspects of a network gene system. Pp. 137–150 in J. P. Changeux and J. Chavaillon, eds., *Origins of the Human Brain.* Oxford Univ. Press, Oxford.

Ruff, C. B., E. Trinkhaus, and T. W. Holliday. 1997. Body mass and encephalization in Pleistocene *Homo. Nature* 387:173–176.

Ruff, C. B., and A. Walker. 1993. Body size and body shape. Pp. 234–263 in A. Walker and R. Leakey, eds., *The Nariokotome Homo erectus Skeleton.* Harvard Univ. Press, Cambridge.

Rugg, M. D. 1998. Memories are made of this. *Science* 281:111–112.

Ruhlen, M. 1994. *The Origin of Language.* Wiley, New York.

Rumelhart, D. E., and J. L. McClelland. 1986. On learning the past tenses of English verbs. Pp. 216–271 in J. L. McClelland and D. E. Rumelhart, eds., *Psychological and Biological Models.* Vol. 2, *Parallel Distributed Processing: Explorations in the Microstructure of Cognition.* MIT Press, Cambridge.

Rumsey, A. 1990. Wording, meaning, and linguistic ideology. *American Anthropologist* 92:346–361.

Rundle, H. D., L. Nagel, J. W. Boughman, and D. Schluter. 2000. Natural selection and parallel speciation in sympatric sticklebacks. *Science* 287:306–308.

Ruse, M. 1984. The morality of the gene. *Monist* 67:167–199.

Ruse, M. 1986. Evolutionary ethics: A phoenix arisen. *Zygon* 21:95–112.

Ruse, M. 1987. Darwinism and determinism. *Zygon* 22:419–442.

Ruse, M. 1989. Is the theory of punctuated equilibria a new paradigm? *Journal of Biological and Social Structures* 12:195–212.

Ruse, M. 1993. The new evolutionary ethics. Pp. 133–162 in M. H. Nitecki and D. V. Nitecki, eds., *Evolutionary Ethics.* State Univ. of New York Press, Albany.

Ruse, M. 1998. Darwinism and atheism: Different sides of the same coin? *Endeavor* 22:17–20.

Ruse, M., and E. O. Wilson. 1985. The evolution of ethics. *New Scientist* 108:50–52.

Ruse, M., and E. O. Wilson. 1986. Moral philosophy as applied science. *Philosophy* 61:173–192.

Russell, B. 1945. *A History of Western Philosophy.* Simon and Schuster, New York.

Russell, B. 1957. *Why I Am Not a Christian and Other Essays on Religion and Related Subjects.* Simon and Schuster, New York.

Russell, C., and W. M. S. Russell. 1973. The natural history of violence. Pp. 240–273 in C. M. Otten, ed., *Aggression and Evolution.* Xerox College Publishing, Lexington.

Russell, M. J., G. M. Switz, and K. Thompson. 1980. Olfactory influences on the human menstrual cycle. *Pharmacology, Biochemistry, and Behavior* 13:737–738.

Ruwenda, C., S. C. Khoo, R. W. Snow, S. N. R. Yates, et al. 1995. Natural selection of hemi- and heterozygotes for G6PD deficiency in Africa by resistance to severe malaria. *Nature* 376:246–249.

Ryle, G. 1949. *The Concept of Mind.* Hutchinson, London.

Ryner, L. C., S. F. Goodwin, D. H. Castrillon, A. Anand, et al. 1996. Control of male sexual behavior and sexual orientation in *Drosophila* by the fruitless gene. *Cell* 87:1079–1089.

Sabloff, J. A. 1971. The collapse of the classic Maya civilization. Pp. 16–27 in J. Harte and R. Socolow, eds., *Patient Earth.* Holt, Rinehart and Winston, New York.

Sabloff, J. A. 1995. Drought and decline. *Nature* 375:357.

Sachs, J., and J. Devin. 1976. Young children's use of age-appropriate speech styles in social interaction and role-playing. *Journal of Child Language* 3:81–98.

Sack, D. 1987. *Human Territoriality.* Cambridge Univ. Press, Cambridge.

Sade, D. S. 1968. Inhibition of son-mother mating among free-ranging *Rhesus* monkeys. *Science and Psychoanalysis* 12:18–38.

Saffran, J. R., R. N. Aslin, and N. E. L. 1996. Statistical learning by eight-month-old infants. *Science* 274:1926–1928.

Safina, C. 1997. *Song for the Blue Ocean.* Holt, New York.

Sagan, C. 1995. *Demon-Haunted World: Science as a Candle in the Dark.* Random House, New York.

Sahlins, M. 1960. The origin of society. *Scientific American* (September): 76–89.

Sahlins, M. 1968. Notes on the original affluent society. Pp. 85–89 in R. B. Lee and I. Devore, eds., *Man the Hunter.* Aldine, Chicago.

Sahlins, M. 1972. *Stone Age Economics.* Aldine, Chicago.

Sahlins, M. 1976. Colors and cultures. *Semiotica* 16:1–22.

Samelson, F. 1981. Struggle for scientific authority: The reception of Watson's behaviorism, 1913–1920. *Journal of the History of the Behavioral Sciences* 17:399–425.

Sandel, M. J. 1988. *Liberalism and the Limits of Justice.* 2nd ed. Cambridge Univ. Press, Cambridge.

Sandel, M. J. 1996. *Democracy's Discontent: America in Search of a Public Philosophy.* Harvard Univ. Press, Cambridge.

Sanderson, S. K. 1998. *Macrosociology: An Introduction to Human Societies.* Longman, New York.

Sanderson, S. K., ed. 1995. *Civilizations and World Systems: Studying World-Historical Change.* Sage, Walnut Creek, CA.

Sapolsky, R. 1997. *The Trouble with Testosterone and Other Essays on the Biology of the Human Predicament.* Scribner, New York.

Sapolsky, R. 1998. *Why Zebras Don't Get Ulcers: An Updated Guide to Stress, Stress-Related Diseases, and Coping.* Freeman, New York.

Sapolsky, R. 2000. *Cultural Desert.* Manuscript.

Sargent, L. T. 1993. *Contemporary Political Ideologies: A Comparative Analysis.* 9th ed. Wadsworth, Belmont, CA.

Sartre, J.-P. 1960 (1956). *Being and Nothingness.* Washington Square Press, New York.

Saunders, D. A., R. J. Hobbs, and P. R. Ehrlich, eds. 1993. *Reconstruction of Fragmented Ecosystems.* Surrey Beatty, Chipping Norton, New South Wales.

Savage-Rumbaugh, E. S. 1991. Language learning in the bonobo: How and why they learn. Pp. 209–223 in N. A. Krasnegor, D. M. Rumbaugh, R. L. Schiefelbusch, and M. Studdert-Kennedy, eds., *Biological and Behavioral Determinants of Language Development.* Erlbaum, Hillsdale, NJ.

Savage-Rumbaugh, E. S., and D. M. Rumbaugh. 1993. The emergence of language. Pp. 86–108 in K. R. Gibson and T. Ingold, eds., *Tools, Language, and Cognition in Human Evolution.* Cambridge Univ. Press, Cambridge.

Savage-Rumbaugh, E. S., S. L. Williams, T. Furuichi, and T. Kano. 1996. Language perceived: *Paniscus* branches out. Pp. 173–184 in W. C. McGrew, L. F. Marchant, and T. Nishida, eds., *Great Ape Societies.* Cambridge Univ. Press, Cambridge.

Savage-Rumbaugh, S., D. M. Rumbaugh, and S. T. Boysen. 1978. Sarah's problems in comprehension. *Behavioral and Brain Sciences* 1:555–557.

Sawaguchi, T., and H. Kudo. 1990. Neocortical development and social structure in primates. *Primates* 31:283–289.

Schacht, R. M. 1988. Circumscription theory: A critical review. *American Behavioral Scientist* 31:438–448.

Schaik, C. P. van, and A. Paul. 1996. Male care in primates: Does it ever reflect paternity? *Evolutionary Anthropology* 5:152–156.

Schaller, G. B. 1963. *The Mountain Gorilla: Ecology and Behavior.* Univ. of Chicago Press, Chicago.

Scharloo, W. 1987. Constraints in selection response. Pp. 125–149 in V. Loeschcke, ed., *Genetic Constraints on Adaptive Evolution.* Springer-Verlag, Berlin.

Schelling, T. 1978. *Micromotives and Macrobehavior.* Norton, New York.

Schick, K. D., and N. Toth. 1993. *Making Silent Stones Speak.* Simon and Schuster, New York.

Schjelderup-Ebbe, T. 1935. Social behavior of birds. Pp. 947–972 in C. Murchison, ed., *Handbook of Social Psychology.* Clark Univ. Press, Worcester.

Schlichting, C. D., and P. Massimo. 1998. *Phenotypic Evolution: A Reaction Norm Perspective.* Sinauer, Sunderland, MA.

Schlottmann, A., and N. H. Anderson. 1993. An information integration approach to phenomenal causality. *Memory and Cognition* 21:785–801.

Schlottmann, A., and D. R. Shanks. 1992. Evidence for a distinction between judged and perceived causality. *Quarterly Journal of Experimental Psychology* 44A:321–342.

Schmalhausen, J. J. 1949. *Factors of Evolution.* Blakiston, Philadelphia.

Schmandt-Besserat, D. 1978. The earliest precursor of writing. *Scientific American* 238:38–47.

Schneider, S. H. 1989. *Global Warming.* Sierra Club Books, San Francisco.

Schneider, S. H. 1995. The future of climate: Potential for interaction and surprises. Pp. 77–113 in T. E. Downing, ed., *Climate Change and World Food Security.* Springer-Verlag, Heidelberg.

Schneider, S. H. 1997. *Laboratory Earth: The Planetary Gamble We Can't Afford to Lose.* Basic Books, New York.

Schneider, S. H., and R. Londer. 1984. *The Coevolution of Climate and Life.* Sierra Club Books, San Francisco.

Schneirla, T. C. 1952. A consideration of some conceptual trends in comparative psychology. *Psychological Bulletin* 49:559–597.

Schor, J. B. 1998. *The Overspent American: Upscaling, Downshifting, and the New Consumer.* Basic Books, New York.

Schröder, I. 1992. Human sexual behavior, social organization, and fossil evidence: A reconsideration of human evolution. *Homo* 43:263–277.

Schulze, E.-D., and H. A. Mooney, eds. 1993. *Biodiversity and Ecosystem Function.* Springer-Verlag, Berlin.

Schustack, M. W., and R. J. Sternberg. 1981. Evaluation of evidence in causal inference. *Journal of Experimental Psychology: General* 110:101–120.

Schwarz, J. C. 1998. *Global Population from a Catholic Perspective.* Twenty-Third, Mystic, CT.

Scott, G. M. 1997. *Political Science: Foundations for a Fifth Millennium.* Prentice Hall, Upper Saddle River, NJ.

Scott, M. P., and A. J. Weiner. 1984. Structural relationships among genes that control development: Sequence homology between *Antennapedia, Ultrabithorax,* and *fushi tarazu* loci of *Drosophila. Proceedings of the National Academy of Sciences USA* 81:4115–4119.

Scrimshaw, N. E., C. E. Taylor, and J. E. Gordon. 1968. *Interactions of Nutrition and Infection.* World Health Organization, Geneva.

Scrimshaw, S. C. M. 1984. Infanticide in human populations: Societal and individual concerns. Pp. 439–462 in G. Hausfater and S. B. Hrdy, eds., *Infanticide: Comparative Evolutionary Perspectives.* Aldine, New York.

Scully, D., and J. Marolla. 1985. "Riding the bull at Gilley's": Convicted rapists describe the rewards of rape. *Social Problems* 32:251–263.

Searle, J. R. 1980. Minds, brains, and programs. *Behavioral and Brain Sciences* 3:417–425.

Searle, J. R. 1992. *The Rediscovery of the Mind.* MIT Press, Cambridge.

Searle, J. R. 1993. The problem of consciousness. *Social Research* 60:3–16.

Searle, J. R. 1997. *The Mystery of Consciousness.* New York Review of Books, New York.

Seeger, A. 1994. Music and dance. Pp. 686–705 in T. Ingold, ed., *Companion Encyclopedia of Anthropology: Humanity, Culture, and Social Life.* Routledge, London.

Seehausen, O. 1996. *Lake Victoria Rock Cichlids: Taxonomy, Ecology, and Distribution.* Verduyn Cichlids, Zevenhuizen, Netherlands.

Seehausen, O. 1999. Speciation and species richness in African cichlids: Effects of sexual selection by mate choice. Ph.D. diss., Leiden Univ.

Seehausen, O., J. J. M. van Alphen, and F. Witte. 1999. Can ancient colour polymorphisms explain why some cichlid lineages speciate rapidly under disruptive sexual selection? *Belgium Journal of Zoology* 129:279–294.

Seehausen, O., J. J. M. van Alphen, and F. Witte. 1997. Cichlid fish diversity threatened by eutrophication that curbs sexual selection. *Science* 277:1808–1811.

Seehausen, O., F. Witte, E. F. Katunzi, J. Smits, et al. 1997. Patterns of the remnant cichlid fauna in southern Lake Victoria. *Conservation Biology* 11:890–904.

Seeley, R. H. 1986. Intense natural selection caused a rapid morphological transition in a living marine snail. *Proceedings of the National Academy of Sciences USA* 83:6897–6901.

Segall, M. H. 1963. Acquiescence and "identification with the aggressor" among acculturating Africans. *Journal of Social Psychology* 61:247–262.

Segall, M. H., D. T. Campbell, and M. J. Herskovits. 1963. Cultural differences in the perception of geometric illusions. *Science* 139:769–771.

Segall, M. H., D. T. Campbell, and M. J. Herskovits. 1966. *The Influence of Culture on Visual Perception*. Bobbs-Merrill, Indianapolis.

Segerstråle, U. 2000. *Defenders of the Truth: The Battle for Science in the Sociobiology Debate and Beyond*. Oxford Univ. Press, Oxford.

Seligman, M. E. P. 1970. On the generality of the laws of learning. *Psychological Review* 77:406–418.

Sen, A. 1992. Missing women. *British Medical Journal* 304:916–917.

Sereno, P. C. 1999. The evolution of dinosaurs. *Science* 284:2137–2147.

Sereno, P. C., and C. Rao. 1992. Early evolution of avian flight and perching: New evidence from the Lower Cretaceous of China. *Science* 255:845–848.

Service, E. R. 1978. Classical and modern theories of the origins of government. Pp. 21–34 in R. Cohen and E. Service, eds., *Origins of the State*. Institute for the Study of Human Issues, Philadelphia.

Sessions, G. 1981. Shallow and deep ecology: A review of the philosophical literature. Pp. 391–462 in R. C. Schultz and J. D. Hughes, eds., *Ecological Consciousness: Essays from the Earth Day X Colloquium*. Univ. Press of America, Washington, DC.

Seyfarth, R. M., D. L. Cheney, and P. Marler. 1980. Monkey responses to three different alarm calls: Evidence of predator classification and semantic communication. *Science* 210:801–803.

Shabecoff, P. 1993. *A Fierce Green Fire: The American Environmental Movement*. Hill and Wang, New York.

Sharpton, V. L., L. M. Marin, J. L. Carney, S. Lee, et al. 1996. A model of the Chicxulub impact basin based on evaluation of geophysical data, well logs, and drill core samples. *Geological Society of America Special Papers* 307:55–74.

Shatin, R. 1968. Evolution and lactase deficiency. *Gastroenterology* 54:992–993.

Shatz, C. J. 1992. The developing brain. *Scientific American* 267:35–41.

Shea, J. J. 1989. A functional study of the lithic industries associated with hominid fossils in the Kebara and Qafzeh Caves, Israel. Pp. 611–625 in P. A. Mellars and C. B. Stringer, eds., *The Human Revolution: Behavioural and Biological Perspectives in the Origins of Modern Humans*. Edinburgh Univ. Press, Edinburgh.

Sheets-Johnstone, M. 1989. Hominid bipedality and sexual-selection theory. *Evolutionary Theory* 9:57–70.

Shennan, S. 1989. Cultural transmission and cultural change. Pp. 330–346 in S. E. van der Leeuw and R. Torrence, eds., *What's New? A Closer Look at the Process of Innovation*. Unwin Hyman, London.

Shepher, J. 1971. Mate selection among second-generation kibbutz adolescents and adults: Incest avoidance and negative imprinting. *Archives of Sexual Behavior* 1:293–307.

Shepher, J. 1983. *Incest: A Biosocial View*. Academic Press, New York.

Sheppard, P. M. 1951. Fluctuations in the selective value of certain phenotypes in the polymorphic land snail *Cepaea nemoralis* (L.). *Heredity* 5:125–134.

Sheppard, P. M. 1952. Natural selection in two colonies of the polymorphic land snail *Cepaea nemoralis*. *Heredity* 6:233–238.

Sherman, P. W. 1991. Multiple mating and kin recognition by self-inspection. *Ethology and Sociobiology* 12:377–386.

Sherman, P. W. 1998. The evolution of menopause. *Nature* 392:759–761.

Sherry, D. F., and D. L. Schacter. 1987. The evolution of multiple memory systems. *Psychological Review* 94:439–454.

Sherry, S. T., A. R. Rogers, H. Harpending, H. Soodyall, et al. 1994. Mismatch distributions of mtDNA reveal recent human population expansions. *Human Biology* 66:761–775.

Shipman, P. 1984. The earliest tools: Re-assessing the evidence from Olduvai Gorge. *Anthroquest* 29:9–10.

Shipman, P. 1986. Scavenging or hunting in early hominids: Theoretical framework and tests. *American Anthropologist* 88:27–43.

Shipman, P. 1998. *Taking Wing: Archaeopteryx and the Evolution of Bird Flight*. Simon and Schuster, New York.

Shiva, V. 1989. *Staying Alive: Women, Ecology, and Development*. Zed Books, London.

Shockley, W. 1972. Dysgenics, geneticity, raceology: A challenge to the intellectual responsibility of educators. *Phi Delta Kappan* (January): 297–307.

Short, R. V. 1979. Sexual selection and its component parts, somatic and genital selection, as illustrated by man and the great apes. *Advances in the Study of Behavior* 9:131–158.

Short, R. V. 1994. Why sex? Pp. 3–22 in R. V. Short and E. Balaban, eds., *The Differences between the Sexes*. Cambridge Univ. Press, Cambridge.

Siegel, J. M. 1986. The Multidimensional Anger Inventory. *Journal of Personal and Social Psychology* 51:191–200.

Silberpfennig, J. 1945. Psychological aspects of current Japanese and German paradoxa. *Psychoanalytic Review* 32:73–85.

Sillén-Tullberg, B., and A. Møller. 1993. The relationship between concealed ovulation and mating systems in anthropoid primates. *American Naturalist* 141:1–25.

Silverman, J. 1967. Shamans and acute schizophrenia. *Journal of Anthropology* 69:21–31.

Simberloff, D., and J. Cox. 1987. Consequences and costs of conservation corridors. *Conservation Biology* 1:63–71.

Simmel, G. 1902. The number of members as determining the sociological form of the group. *American Journal of Sociology* 8:1–46.

Simon, H. A. 1990. A mechanism for social selection and successful altruism. *Science* 250:1665–1668.

Simons, E. L. 1989. Human origins. *Science* 245:1343–1350.

Simpson, G. G. 1944. *Tempo and Mode in Evolution.* Columbia Univ. Press, New York.

Simpson, G. G. 1953. *The Major Features of Evolution.* Columbia Univ. Press, New York.

Simpson, G. G. 1969. Biology and ethics. Pp. 130–148 in G. G. Simpson, ed., *Biology and Man.* Harcourt, Brace, New York.

Singer, D. G., and J. L. Singer. 1990. *The House of Make-Believe.* Harvard Univ. Press, Cambridge.

Singer, I. 1973. Fertility and the female orgasm. Pp. 159–197 in I. Singer, ed., *The Goals of Human Sexuality.* Wildwood, London.

Singer, P. 1972. Famine, affluence, and morality. *Philosophy and Public Affairs* 1:229–243.

Singer, P. 1975. *Animal Liberation.* New York Review of Books, New York.

Singer, P. 1981. *The Expanding Circle.* Farrar, Straus and Giroux, New York.

Singer, P. 1986. Life, the universe, and ethics. *Biology and Philosophy* 1:367–371.

Singer, P. 1993. *How Are We to Live? Ethics in an Age of Self-Interest.* Text, Melbourne.

Singh, D. 1993a. Adaptive significance of female physical attractiveness: Role of waist-to-hip ratio. *Journal of Personality and Social Psychology* 65:293–307.

Singh, D. 1993b. Body shape and women's attractiveness. *Human Nature* 4:297–321.

Singh, D. 1994. Is thin really beautiful and good? Relationship between waist-to-hip ratio (WHR) and female attractiveness. *Personality and Individual Differences* 16:123–132.

Sisk, T. D., A. E. Launer, K. R. Switky, and P. R. Ehrlich. 1994. Identifying extinction threats. *BioScience* 44:592–604.

Skinner, B. F. 1938. *The Behavior of Organisms.* Appleton, Century, Crofts, New York.

Skocpol, T. 1979. *States and Social Revolutions: A Comparative Analysis of France, Russia, and China.* Cambridge Univ. Press, Cambridge.

Skybreak, A. 1984. *Of Primal Steps and Future Leaps: An Essay on the Emergence of Human Beings, the Source of Women's Oppression, and the Road to Emancipation.* Banner, Chicago.

Slater, M. K. 1959. Ecological factors in the origin of incest. *American Anthropologist* 61:1042–1059.

Slobin, D. 1990. The development from child speaker to native speaker. Pp. 233–256 in J. W. Stigler, R. A. Shweder, and G. Herdt, eds., *Cultural Psychology: Essays on Comparative Human Development.* Cambridge Univ. Press, Cambridge.

Slobin, D. 1991. Learning to think for speaking: Native language, cognition, and rhetorical style. *Pragmatics* 1:7–26.

Slobodkin, L. 1993. The complex questions relating evolution to ethics. Pp. 337–347 in M. H. Nitecki and D. V. Nitecki, eds., *Evolutionary Ethics.* State Univ. of New York Press, Albany.

Small, M. F. 1988. Female primate sexual behavior and conception. *Current Anthropology* 29:81–88.

Small, M. F. 1992a. The evolution of human sexuality and mate selection in humans. *Journal of Human Nature* 3:133–156.

Small, M. F. 1992b. Female choice in mating. *American Scientist* 80:142–151.

Small, M. F. 1993. *Female Choices: Sexual Behavior in Female Primates.* Cornell Univ. Press, Ithaca.

Small, M. F. 1996. "Revealed" ovulation in humans? *Journal of Human Evolution* 30:483–488.

Small, M. F. 1999. A woman's curse? *Sciences* (January–February): 24–29.

Smart, N. 1994. *The Dimensions of the Sacred: An Anatomy of the World's Beliefs.* HarperCollins, London.

Smith, A. 1974 (1759). *The Theory of Moral Sentiments.* Clarendon Press, Oxford.

Smith, A. 1976 (1776). *An Inquiry into the Nature and Causes of the Wealth of Nations.* Univ. of Chicago Press, Chicago.

Smith, A. D. 1986. *The Ethnic Origins of Nations.* Blackwell, Oxford.

Smith, B. D. 1995. *The Emergence of Agriculture.* Scientific American Library, New York.

Smith, D. D. 1976. The social content of pornography. *Journal of Communication* 26:16–24.

Smith, E. A. 1991 (1988). Risk and uncertainty in the "original affluent society": Evolutionary ecology of resource-sharing and land tenure. Pp. 222–251 in T. Ingold, D. Riches, and J. Woodburn, eds., *Hunters and Gatherers.* Vol. 1, *History, Evolution, and Social Change.* Berg, Oxford.

Smith, E. O. 1999. Evolution, substance abuse, and addiction. Pp. 375–405 in W. R. Trevathan, E. O. Smith, and J. J. McKenna, eds., *Evolutionary Medicine.* Oxford Univ. Press, New York.

Smith, F. H. 1992. Models and realities in modern human origins: The African fossil evidence. *Philosophical Transactions of the Royal Society of London,* ser. B, 337:243–250.

Smith, F. H., A. B. Falsetti, and S. M. Donnelly. 1989. Modern human origins. *Yearbook of Physical Anthropology* 32:35–68.

Smith, F. H., E. Trinkaus, P. B. Pettitt, and I. Karavanic. 1999. Direct radiocarbon dates for Vindija G_1 and Velika Pecina Late Pleistocene hominid remains. *Proceedings of the National Academy of Sciences USA* 96:12281–12286.

Smith, K. 1987. *Biofuels, Air Pollution, and Health: A Global Review.* Plenum Press, New York.

Smith, R. L. 1984. Human sperm competition. Pp. 601–659 in R. L. Smith, ed., *Sperm Competition and the Evolution of Animal Mating Systems.* Academic Press, Orlando.

Smouse, P. E. 1976. The implications of density-dependent population growth for frequency- and density-dependent selection. *American Naturalist* 110:849–860.

Smuts, B., and R. Smuts. 1993. Male aggression and sexual coercion of females in non-human primates and other mammals: Evidence and theoretical implications. Pp. 1–63 in P. J. B. Slater, M. Milinski, J. S. Rosenblatt, and C. T. Snowdon, eds., *Advances in the Study of Behavior.* Academic Press, New York.

Sniegowski, P. D., and R. E. Lenski. 1995. Mutation and adaptation: The directed mutation controversy in evolutionary perspective. *Annual Review of Ecology and Systematics* 26:553–578.

Snow, C. P. 1998 (1959, 1964). *The Two Cultures.* Cambridge Univ. Press, Cambridge.

Snow, D. 1995. Microchronology and demographic evidence relating to the size of the pre-Columbian North American Indian population. *Science* 268:1601–1604.

Snyderman, M. 1994. How to think about race. *National Review* (12 September): 78–80.

Sober, E. 1988. What is evolutionary altruism? Pp. 75–99 in M. Matthen and B. Linsky, eds., *Philosophy and Biology.* Univ. of Calgary Press, Calgary.

Sober, E. 1993. Evolutionary altruism, psychological egoism, and morality: Disentangling the phenotypes. Pp. 199–216 in M. H. Nitecki and D. V. Nitecki, eds., *Evolutionary Ethics.* State Univ. of New York Press, Albany.

Sober, E., and D. S. Wilson. 1998. *Unto Others: The Evolution and Psychology of Unselfish Behavior.* Harvard Univ. Press, Cambridge.

Socolow, R., C. Andrews, F. Berkhout, and V. Thomas. 1994. *Industrial Ecology and Global Change.* Cambridge Univ. Press, Cambridge.

Sohnle, P. G., and S. R. Gambert. 1982. Thermoneutrality: An evolutionary advantage against aging? *Lancet* 1982:1099–1101.

Sokal, R. R. 1966. Pupation site differences in *Drosophila melanogaster. University of Kansas Science Bulletin* 46:697–715.

Sokal, R. R., and T. J. Crovello. 1970. The biological species concept: A critical evaluation. *American Naturalist* 104:127–153.

Sokal, R. R., P. R. Ehrlich, P. E. Hunter, and G. Schlager. 1960. Some factors affecting pupation site of *Drosophila. Annals of the Entomological Society of America* 53:174–182.

Sokal, R. R., and P. E. Hunter. 1954. Reciprocal selection for correlated quantitative characters in *Drosphila. Science* 119:649–651.

Sokal, R. R., N. L. Oden, and B. A. Thomson. 1999. A problem with synthetic maps. *Human Ecology* 71:1–13.

Sokal, R. R., N. L. Oden, and C. Wilson. 1991. Genetic evidence for the spread of agriculture in Europe by demic diffusion. *Nature* 351:143–145.

Solbrig, O. 1971. The population biology of dandelions. *American Scientist* 59:686–694.

Sontag, S., and C. Drew. 1998. *Blind Man's Bluff: The Untold Story of American Submarine Espionage.* Public Affairs, New York.

Soulé, M. E. 1995a. The social siege of nature. 137–170 in M. E. Soulé and G. Lease, eds., *Reinventing Nature? Responses to Postmodern Deconstruction.* Island Press, Washington, DC.

Soulé, M. E. 1995b. An unflinching vision: Networks of people defending networks of lands. Pp. 1–8 in D. A. Saunders, J. L. Craig, and E. M. Mattiske, eds., *The Role of Networks.* Surrey Beatty, Chipping Norton, New South Wales.

Soulé, M. E., and G. Lease, eds. 1995. *Reinventing Nature? Responses to Postmodern Deconstruction.* Island Press, Washington, DC.

Soulé, M. E., and B. A. Wilcox, eds. 1980. *Conservation Biology: An Evolutionary-Ecological Perspective.* Sinauer, Sunderland, MA.

Soulé, M. E., B. A. Wilcox, and C. Holtby. 1979. Benign neglect: A model of faunal collapse in the game reserves of East Africa. *Biological Conservation* 15:259–271.

Southwick, C. H., M. A. Beg, and M. R. Siddiqi. 1965. Rhesus monkeys in north India. Pp. 111–159 in I. DeVore, ed., *Primate Behavior: Field Studies of Monkeys and Apes.* Holt, Rinehart and Winston, New York.

Spelke, E. S., K. Breinlinger, J. Macomber, and K. Jacobson. 1992. Origins of knowledge. *Psychological Review* 99:605–632.

Spencer, H. 1891 (1860). *Essays: Scientific, Political, and Speculative.* Vol. 1. Appleton, New York.

Spengler, O. 1996 (1923). *The Decline of the West.* Random House, New York.

Sperber, D. 1985. Anthropology and psychology: Towards an epidemiology of representations. *Man*, n.s., 20:73–89.

Sperry, R. W. 1972. Science and the problem of values. *Perspectives in Biology and Medicine* 16:115–130.

Spinage, C. A. 1998. Social change and conservation misrepresentation in Africa. *Oryx* 32:265–276.

Spiro, M. E. 1958. *Children of the Kibbutz.* Harvard Univ. Press, Cambridge.

Squire, L. R., and E. R. Kandel. 1999. *Memory: From Mind to Molecules.* Scientific American Library, New York.

Stanford, C. B. 1999. *The Hunting Apes.* Princeton Univ. Press, Princeton.

Stanley, S. M. 1975. A theory of evolution above the species level. *Proceedings of the National Academy of Sciences USA* 72:757–760.

Staub, E. 1986. A conception of the determinants and development of altruism and aggression: Motives, the self, and the environment. Pp. 135–164 in C. Zahn-Waxler et al., eds., *Altruism and Aggression: Biological and Social Origins.* Cambridge Univ. Press, Cambridge.

Steadman, D. W. 1977. Extinction of Polynesian birds: Reciprocal impacts of birds and people. Pp. 51–79 in P. V. Kirch and T. L. Hunt, eds., *Historical Ecology in the Pacific Islands: Prehistoric Environmental and Landscape Change.* Yale Univ. Press, New Haven.

Stearns, S. C. 1992. *The Evolution of Life Histories.* Oxford Univ. Press, Oxford.

Stearns, S. C. 1999. *Evolution in Health and Disease.* Oxford Univ. Press, Oxford.

Stebbins, G. L. 1950. *Variation and Evolution in Plants.* Columbia Univ. Press, New York.

Stebbins, G. L. 1959. The role of hybridization in evolution. *Proceedings of the American Philosophical Society* 103:231–251.

Stebbins, G. L. 1977. In defense of evolution: Tautology or theory? *American Naturalist* 111:386–390.

Steele, J. 1999. Stone legacy of skilled hands. *Nature* 399:24–25.

Steenbeek, R., R. C. Piek, M. van Buul, and J. A. R. A. M. van Hooff. 1999. Vigilance in wild Thomas's langurs *(Presbytis thomasi):* The importance of infanticide risk. *Behavioral Ecology and Sociobiology* 45:137–150.

Steenhoven, G. van den. 1959. *Legal Concepts among the Netsilik Eskimos of Pelly Bay, N.W.T.* Northern Co-ordination and Research Centre, Department of Northern Affairs and National Resources, Ottawa.

Stefanini, F. M., and M. W. Feldman. 1999. Microsatellite loci and the origin of modern humans: A Bayesian analysis. Pp. 249–272 in S. P. Wasser, ed., *Evolutionary Theory and Processes: Modern Perspectives.* Kluwer, Dordrecht.

Stein, N. L., and L. Levine. 1987. Thinking about feelings: The development and organization of emotional knowledge. Pp. 165–198 in R. Snow and M. Farr, eds., *Aptitude, Learning, and Instruction: Cognition, Conation, and Affect.* Vol. 3. Erlbaum, Hillsdale, NJ.

Steinhauer, K., K. Alter, and A. D. Friederici. 1999. Brain potentials indicate immediate use of prosodic cues in natural speech processing. *Nature Neuroscience* 2:191–196.

Stent, G. S. 1998. Epistemic dualism of mind and body. *Proceedings of the American Philosophical Society* 142:578–588.

Stephan, H., H. Frahm, and G. Baron. 1984. Comparison of brain structure volumes in insectivora and primates. IV. Non-cortical visual structures. *Journal für Hirnforschung* 25:385–403.

Stephan, H., H. Frahm, and G. Baron. 1987. Comparison of brain structure volumes in insectivora and primates. V. Amygdaloid components. *Journal für Hirnforschung* 28:571–584.

Sterelny, K., and P. E. Griffiths. 1999. *Sex and Death: An Introduction to the Philosophy of Biology.* Univ. of Chicago Press, Chicago.

Stern, K., and M. K. McClintock. 1998. Regulation of ovulation by human pheromones. *Nature* 392:177–179.

Steudel, K. L. 1994. Locomotor energetics and hominid evolution. *Evolutionary Anthropology* 3:42–48.

Stevenson, L., and D. L. Haberman. 1998. *Ten Theories of Human Nature.* Oxford Univ. Press, New York.

Steward, J. H. 1949. Cultural causality and law. *American Anthropologist* 51:1–27.

Stiassny, M. L. J., and A. Meyer. 1999. Cichlids of the rift lakes. *Scientific American* (February): 64–69.

Stiner, M. C., N. D. Munro, T. A. Surovell, E. Tchernov, et al. 1999. Paleolithic population growth pulses evidenced by small animal exploitation. *Science* 283:190–193.

Stoddart, D. M. 1990. *The Scented Ape: The Biology and Culture of Human Odour.* Cambridge Univ. Press, Cambridge.

Stolberg, S. G. 1999. After four deaths, scientists fear germ's threat. *New York Times* (20 August): A1, A17.

Stone, A. R., and D. L. Hawksworth, eds. 1986. *Coevolution and Systematics.* Clarendon Press, Oxford.

Stone, C. D. 1974. *Should Trees Have Standing? Towards Legal Rights for Natural Objects.* Kaufmann, Los Altos, CA.

Stone, C. D. 1987. *Earth and Other Ethics: The Case for Moral Pluralism.* Harper and Row, New York.

Stoneking, M. 1993. DNA and recent human evolution. *Evolutionary Anthropology* 2:60–73.

Stoneking, M., K. Bhatia, and A. C. Wilson. 1986. Rate of sequence divergence estimated from restricted maps of mitochondrial DNAs from Papua New Guinea. *Cold Spring Harbor Symposia on Quantitative Biology* 51:433–439.

Storfer, A., J. Cross, V. Rush, and J. Caruso. 1999. Adaptive coloration and gene flow as a constraint to local adaptation in the streamside salamander *Ambystoma barbouri. Evolution* 53:889–898.

Storr, A. 1989. *Freud.* Oxford Univ. Press, Oxford.

Strait, D. S., and F. E. Grine. 1999. Cladistics and early hominid phylogeny. *Science* 285:1210–1211.

Strait, D. S., F. E. Grine, and M. A. Moniz. 1997. A reappraisal of early hominid phylogeny. *Journal of Human Evolution* 32:17–82.

Strassmann, B. I. 1991. The function of menstrual taboos among the Dogon: Defense against cuckoldry? *Human Nature* 3:89–131.

Strassmann, B. I. 1996a. Energy economy in the evolution of menstruation. *Evolutionary Anthropology* 5:157–164.

Strassmann, B. I. 1996b. The evolution of endometrial cycles and menstruation. *Quarterly Review of Biology* 71:181–220.

Strassmann, B. I. 1996c. Menstrual hut visits by Dogon women: A hormonal test distinguishes deceit from honest signaling. *Behavioral Ecology* 7:304–315.

Strassmann, B. I. 1997. The biology of menstruation in *Homo sapiens:* Total lifetime menses, fecundity, and nonsynchrony in a natural-fertility population. *Current Anthropology* 38:123–129.

Strassmann, B. I., and R. I. M. Dunbar. 1999. Human evolution and disease: Putting the Stone Age in perspective. Pp. 91–101 in S. C. Stearns, ed., *Evolution in Health and Disease.* Oxford Univ. Press, Oxford.

Straus, L. G. 1989. On early hominid use of fire. *Current Anthropology* 30:488–491.

Straus, L. G., J. L. Bischoff, and E. Carbonell. 1993. A review of the Middle to Upper Paleolithic transition in Iberia. *Prehistoire Européenne* 3:11–27.

Strauss, E. 1998. Writing, speech separated in split brain. *Science* 280:827.

Strauss, L., and J. Cropsey, eds. 1987. *History of Political Philosophy.* 3rd ed. Univ. of Chicago Press, Chicago.

Strawson, P. 1962. Freedom and resentment. *Proceedings of the British Academy* 48:1–25.

Strayer, J. 1987. Affective and cognitive perspectives on empathy. Pp. 218–244 in N. Eisenberg and J. Strayer, eds., *Empathy and Its Development.* Cambridge Univ. Press, Cambridge.

Street, B. V., and N. Besnier. 1994. Aspects of literacy. Pp. 527–562 in T. Ingold, ed., *Companion Encyclopedia of Anthropology.* Routledge, London.

Streuver, S. 1968. Flotation techniques for the recovery of small-scale archaeological remains. *American Antiquity* 33:353–362.

Stringer, C. B. 1992. Reconstructing recent human evolution. *Philosophical Transactions of the Royal Society of London,* ser. B, 337:217–224.

Stringer, C. B. 1995. The evolution and distribution of later Pleistocene human populations. Pp. 524–531 in E. S. Vrba, G. H. Denton, T. C. Partridge, and L. H. Burckle, eds., *Paleoclimate and Evolution, with Emphasis on Human Origins.* Yale Univ. Press, New Haven.

Stringer, C. B. 1996. Current issues in modern human origins. Pp. 115–134 in W. E. Meikle, F. C. Howell, and N. G. Jablonski, eds., *Contemporary Issues in Human Evolution.* California Academy of Sciences, San Francisco.

Stringer, C. B., and P. Andrews. 1988. Genetic and fossil evidence for the origin of modern humans. *Science* 239:1263–1268.

Stringer, C. B., and C. Gamble. 1993. *In Search of the Neanderthals: Solving the Puzzle of Human Origins.* Thames and Hudson, New York.

Stringer, C. B., and R. McKie. 1996. *African Exodus: The Origins of Modern Humanity.* Holt, New York.

Strum, S. C. 1987. Baboon models and muddles. Pp. 87–104 in W. G. Kinzey, ed., *The Evolution of Human Behavior: Primate Models.* State Univ. of New York Press, Albany.

Sturgeon, N. 1997. *Ecofeminist Natures: Race, Gender, Feminist Theory, and Political Action.* Routledge, New York.

Suarez, S. D., and G. G. Gallup Jr. 1981. Self-recognition in chimpanzees and orangutans, but not gorillas. *Journal of Human Evolution* 10:175–188.

Sugiyama, S. 1965. On the social change of Hanuman langurs *(Presbytis entellus)* in their natural condition. *Primates* 6:381–418.

Sugiyama, Y. 1994. Tool use by wild chimpanzees. *Nature* 367:327.

Sugiyama, Y. 1997. Social tradition and the use of tool-composites by wild chimpanzees. *Evolutionary Anthropology* 6:23–27.

Sugiyama, Y., and J. Koman. 1979. Tool-using and making behavior in wild chimpanzees at Bossou, Guinea. *Primates* 20:513–524.

Surtees, G. 1970. Effects of irrigation on mosquito populations and mosquito-borne disease in man, with particular attention to rice field extension. *International Journal of Environmental Studies* 1:35–42.

Susman, R. L. 1987. Pygmy chimpanzees and common chimpanzees: Models for the behavioral ecology of the earliest hominids. Pp. 72–86 in W. G. Kinzey, ed., *The Evolution of Human Behavior: Primate Models.* State Univ. of New York Press, Albany.

Susman, R. L. 1994. Fossil evidence for early hominid tool use. *Science* 265:1570–1573.

Sutton, G. M. 1932. The exploration of Southampton Island, Hudson Bay. *Memoirs of the Carnegie Museum* 12:1–75.

Suwa, G., B. Asfaw, Y. Beyene, T. D. White, et al. 1997. The first skull of *Australopithecus boisei. Nature* 389:489–492.

Swartz, K., and S. Evans. 1994. Social and cognitive factors in chimpanzee and gorilla mirror behavior and self-recognition. Pp. 189–205 in S. T. Parker, R. W. Mitchell, and M. L. Boccia, eds., *Self-Awareness in Animals and Humans.* Cambridge Univ. Press, New York.

Swartz, K. B., D. Sarauw, and S. Evans. 1999. Comparative aspects of mirror self-recognition in great apes. Pp. 283–294 in S. T. Parker, R. W. Mitchell, and H. L. Miles, eds., *The Mentalities of Gorillas and Orangutans: Comparative Perspectives.* Cambridge Univ. Press, Cambridge.

Swisher, C. C., III. 1994. Dating hominid sites in Indonesia. *Science* 266:1727.

Swisher, C. C., III, G. H. Curtis, T. Jacob, A. G. Getty, et al. 1994. The age of the earliest known hominids in Java, Indonesia. *Science* 263:1118–1121.

Swisher, C. C., III, W. J. Rink, S. C. Anton, H. P. Schwarcz, et al. 1996. Latest *Homo erectus* of Java: Potential contemporaneity with *Homo sapiens* in Southeast Asia. *Science* 274:1870–1874.

Symons, D. 1979. *The Evolution of Human Sexuality.* Oxford Univ. Press, New York.

Symons, D. 1982. Another woman that never existed. *Quarterly Review of Biology* 57:297–300.

Tabashnik, B. E. 1990. Modeling and evaluation of resistance management tactics. Pp. 153–182 in R. T. Roush and B. E. Tabashnik, eds., *Pesticide Resistance in Arthropods.* Chapman and Hall, New York.

Tabashnik, B. E., Y.-B. Liu, N. Finson, L. Masson, et al. 1997. One gene in diamondback moth confers resistance to four *Bacillus thuringiensis* toxins. *Proceedings of the National Academy of Sciences USA* 94:1640–1644.

Tabashnik, B. E., Y.-B. Liu, T. Malvar, D. G. Heckel, et al. 1997. Global variation in the genetic and biochemical basis of diamondback moth resistance to *Bacillus thuringiensis. Proceedings of the National Academy of Sciences USA* 94:12780–12785.

Tague, R. G. 1992. Sexual dimorphism in the human bony pelvis, with a consideration of the Neanderthal pelvis from Kebara Cave, Israel. *American Journal of Physical Anthropology* 88:1–21.

Tainter, J. A. 1988. *The Collapse of Complex Societies.* Cambridge Univ. Press, Cambridge.

Takahata, Y., H. Ihobe, and G. Idani. 1996. Comparing copulations of chimpanzees and bonobos: Do females exhibit proceptivity or receptivity? Pp. 146–155 in W. C. McGrew, L. F. Marchant, and T. Nishida, eds., *Great Ape Societies.* Cambridge Univ. Press, Cambridge.

Takeshita, H., and V. Walraven. 1996. A comparative study of the variety and complexity of object manipulation in captive chimpanzees *(Pan troglodytes)* and bonobos *(Pan paniscus). Primates* 37:423–441.

Talmon, Y. 1964. Mate selection on collective settlements. *American Sociological Review* 29:491–508.

Tang, Y.-P., E. Shimizu, G. R. Dube, C. Rampon, et al. 1999. Genetic enhancement of learning and memory in mice. *Nature* 401:63–69.

Tarr, M. J. 1995. Rotating objects to recognize them: A case study on the role of viewpoint dependency in the recognition of three-dimensional objects. *Psychonomic Bulletin and Review* 2:55–82.

Tarr, M. J., and H. H. Bülthoff. 1995. Is human object recognition better described by geon structural descriptions or by multiple views? Comment on Biederman and Gerhardstein (1993). *Journal of Experimental Psychology* 21:1494–1505.

Tattersall, I. 1992. Species concepts and species identification in human evolution. *Journal of Human Evolution* 22:341–349.

Tattersall, I. 1995. *The Fossil Trail.* Oxford Univ. Press, Oxford.

Tattersall, I. 1998a. *Becoming Human: Evolution and Human Uniqueness.* Harcourt, Brace, New York.

Tattersall, I. 1998b. Neanderthal genes: What do they mean? *Evolutionary Ecology* 6:157–158.

Taylor, C. 1992. *The Ethics of Authenticity.* Harvard Univ. Press, Cambridge.

Taylor, C. 1992 (1989). *Sources of the Self: The Making of the Modern Identity.* Harvard Univ. Press, Cambridge.

Taylor, C. R., and E. R. Weibel. 1981. Design of the mammalian respiratory system. I. Problem and strategy. *Respiratory Physiology* 44:1–10.

Taylor, E. B., and P. Bentzen. 1993. Evidence for multiple origins and sympatric divergence of trophic ecotypes of smelt *(Osmerus)* in northeastern North America. *Evolution* 47:813–832.

Taylor, R. B. 1988. *Human Territorial Functioning: An Empirical, Evolutionary Perspective on Individual and Small Group Territorial Cognitions, Behaviors, and Consequences.* Cambridge Univ. Press, Cambridge.

Teitelbaum, M. S. 1975. Relevance of demographic transition theory for developing countries. *Science* 188:420–425.

Templeton, A. R. 1992. Human origins and analysis of mitochondrial DNA sequences. *Science* 255:737.

Templeton, A. R. 1993. The "Eve" hypotheses: A genetic critique and reanalysis. *American Anthropologist* 95:51–72.

Terborgh, J. 1988. The big things that run the world: A sequel to E. O. Wilson. *Conservation Biology* 2:402–403.

Terrace, H. S., L. A. Petitto, R. J. Sanders, and T. G. Bever. 1979. Can an ape create a sentence? *Science* 206:891–902.

Tessier-Lavigne, M., and C. S. Goodman. 1996. The molecular biology of axon guidance. *Science* 274:1123–1133.

Textor, R. B. 1967. *A Cross-Cultural Summary.* HRAF Press, New Haven.

Thomas, K. 1983. *Man and the Natural World: A History of the Modern Sensibility.* Pantheon Books, New York.

Thomas, L. 1986. Biological moralism. *Biology and Philosophy* 1:316–325.

Thomas, M. K., D. M. Lloyd-Jones, R. I. Thadani, A. C. Shaw, et al. 1998. Hypovitaminosis D in medical patients. *New England Journal of Medicine* 338:777–783.

Thomason, S. G., and T. Kaufman. 1988. *Language Contact, Creolization, and Genetic Linguistics.* Univ. of California Press, Berkeley.

Thompson, J. B. 1990. *Ideology and Modern Culture.* Stanford Univ. Press, Stanford.

Thompson, J. N. 1989. Concepts of coevolution. *Trends in Ecology and Evolution* 4:179–183.

Thompson, J. N. 1993. Preference hierarchies and the origin of geographic specialization in host use in swallowtail butterflies. *Evolution* 47:1585–1594.

Thompson, J. N. 1998. Rapid evolution as an ecological process. *Trends in Ecology and Evolution* 13:329–332.

Thompson, J. N. 1999. The evolution of species interactions. *Science* 284:2116–2118.

Thomson, R., J. K. Pritchard, P. Shen, P. J. Oefner, et al. 2000. Recent common ancestry of human Y chromosomes: Evidence from DNA sequence data. In preparation.

Thoreau, H. D. 1986 (1849, 1854). *Walden and Civil Disobedience.* Penguin Books, New York.

Thornhill, R., and S. W. Gangestad. 1996. The evolution of human sexuality. *Trends in Ecology and Evolution* 11:98–102.

Thornton, B. S. 1997. *Eros: The Myth of Ancient Greek Sexuality.* Westview Press, Boulder.

Thwaites, T. 1998. Ancient mariners: Early humans were much smarter than we suspected. *New Scientist* (11 March): 6.

Tierney, A. J. 1986. The evolution of learned and innate behavior: Contributions from genetics and neurobiology to a theory of behavioral evolution. *Animal Learning and Behavior* 14:339–348.

Tiger, L., and R. Fox. 1971. *The Imperial Animal.* Holt, Rinehart and Winston, New York.

Tilman, D., D. Wedin, and J. Knops. 1996. Productivity and sustainability influenced by biodiversity in grassland ecosystems. *Nature* 379:718–720.

Tinbergen, L. 1960. The natural control of insects in pinewoods. I. Factors influencing the intensity of predation by song-birds. *Archives Néerlandaises de Zoologie* 13:265–336.

Tinbergen, N. 1951. *The Study of Instinct.* Oxford Univ. Press, Oxford.

Tisdell, C. A. 1989. Environmental conservation: Economics, ecology, and ethics. *Environmental Conservation* 16:107–112.

Tishkoff, S. A., E. Dietzsch, W. Speed, A. J. Pakstis, et al. 1996. Global patterns of linkage disequilibrium at the CD4 locus and modern human origins. *Science* 271:1380–1387.

Tobias, M., ed. 1985. *Deep Ecology.* Avant Books, San Diego.

Toennies, F. 1971. *Ferdinand Toennies on Sociology: Pure, Applied, and Empirical.* Univ. of Chicago Press, Chicago.

Tomasello, M., and J. Call. 1997. *Primate Cognition.* Oxford Univ. Press, New York.

Tomonaga, M., S. Itakura, and T. Matsuzawa. 1993. Superiority of conspecific faces and reduced inversion effect in face perception by a chimpanzee. *Folia Primatologica* 61:110–114.

Tooby, J., and L. Cosmides. 1992. The psychological foundations of culture. Pp. 19–136 in J. H. Barkow, L. Cosmides, and J. Tooby, eds., *The Adapted Mind: Evolutionary Psychology and the Generation of Culture.* Oxford Univ. Press, New York.

Tooby, J., and I. Devore. 1987. The reconstruction of hominid evolution through strategic modeling. Pp. 183–237 in W. G. Kinzey, ed., *The Evolution of Human Behavior: Primate Models.* State Univ. of New York Press, Albany.

Torrance, R. M. 1994. *The Spiritual Quest: Transcendence in Myth, Religion, and Science.* Univ. of California Press, Berkeley.

Toth, N. 1985. Archaeological evidence for preferential right-handedness in the Lower and Middle Pleistocene, and its possible implications. *Journal of Human Evolution* 14:607–614.

Toth, N., K. Schick, E. S. Savage-Rumbaugh, R. Sevcik, et al. 1993. *Pan* the tool-maker: Investigations into the stone tool-making and tool-using capabilities of a bonobo *(Pan paniscus). Journal of Archaeological Science* 20:81–91.

Townsend, P. 1987. Deprivation. *Journal of Social Policy* 16:125–146.

Toynbee, A. 1934–1961. *A Study of History.* 12 vols. Oxford Univ. Press, London.

Tramo, M. J., W. C. Loftus, C. E. Thomas, R. L. Green, et al. 1995. Surface area of human cerebral cortex and its gross morphological subdivisions: *In vivo* measurements in monozygotic twins suggest differential hemisphere effects of genetic factors. *Journal of Cognitive Neuroscience* 7:292–301.

Tratz, E. P., and H. Heck. 1954. Der afrikanische Anthropoide "Bonobo": Eine neue Menschenaffengattung. *Säugetierkundliche Mitteilungen* 2:97–101.

Travis, J. 1999. Two genes for the price of one. *Science News* 156:239.

Travisano, M., J. A. Mongold, A. F. Bennett, and R. E. Lenski. 1995. Experimental tests of the roles of adaptation, chance, and history in evolution. *Science* 267:87–90.

Treat, A. E. 1957. Unilaterality in infestations of the moth ear mite. *Journal of the New York Entomological Society* 65:41–50.

Treat, A. E. 1975. *Mites of Moths and Butterflies.* Cornell Univ. Press, Ithaca.

Tregenza, T., and R. K. Butlin. 1999. Speciation without isolation. *Nature* 400:311–312.

Trevathan, W. R., E. O. Smith, and J. J. McKenna, eds. 1999. *Evolutionary Medicine.* Oxford Univ. Press, New York.

Trigg, R. 1986. Evolutionary ethics. *Biology and Philosophy* 1:325–335.

Trinkaus, E. 1986. The Neanderthals and modern human origins. *Annual Review of Anthropology* 15:193–217.

Trinkaus, E. 1989. The Upper Pleistocene transition. Pp. 42–66 in E. Trinkaus, ed., *The Emergence of Modern Humans: Biocultural Adaptations in the Later Pleistocene.* Cambridge Univ. Press, Cambridge.

Trinkaus, E. 1992. Evolution of human manipulation. Pp. 346–349 in S. Jones, R. Martin, and D. Pilbeam, eds., *The Cambridge Encyclopedia of Human Evolution.* Cambridge Univ. Press, Cambridge.

Trinkaus, E. 1995. Comments: Testing hypotheses about recent human evolution from skulls: Integrating morphology, function, development, and phylogeny. *Current Anthropology* 36:185–186.

Trinkaus, E., and P. Shipman. 1993. *The Neandertals: Changing the Image of Mankind.* Knopf, New York.

Trivers, R. L. 1971. The evolution of reciprocal altruism. *Quarterly Review of Biology* 46:35–57.

Trivers, R. L. 1972. Parental investment and sexual selection. Pp. 136–179 in B. Campbell, ed., *Sexual Selection and the Descent of Man, 1871–1971.* Aldine, Chicago.

Trotter, M. M., and B. McCulloch. 1984. Moas, men, and middens. Pp. 708–727 in P. S. Martin and R. G. Klein, eds., *Quaternary Extinctions: A Prehistoric Revolution.* Univ. of Arizona Press, Tucson.

Tsutsumi, H., S. Nakamura, K. Sato, K. Tamaki, et al. 1989. Comparative studies on an antigenicity of plasma proteins from humans and apes by ELISA: A close relationship of chimpanzee and human. *Comparative Biochemistry and Physiology* 94B:647–649.

Tucker, G. M. 1991. Apostatic selection by song thrushes *(Turdus philomelos)* feeding on the snail *Cepaea hortensis. Biological Journal of the Linnean Society* 43:149–156.

Tudge, C. 1997. Relative danger. *Natural History* 1997:28–31.

Turco, R., O. Toon, T. Ackerman, J. Pollack, et al. 1983. Nuclear winter: Global consequences of multiple nuclear weapons explosions. *Science* 222:1283–1292.

Turing, A. 1950. Computing machinery and intelligence. *Mind: A Quarterly Review of Psychology and Philosophy* 59:433–460.

Turkewitz, G., and P. Kenny. 1982. Limitations on input as a basis for neural organization and perceptual development: A preliminary theoretical statement. *Developmental Psychobiology* 15:357–368.

Turnbull, C. M. 1961. Some observations regarding the experiences and behavior of BaMbuti Pygmies. *American Journal of Psychology* 74:304–308.

Turnbull, C. M. 1982. The ritualization of potential conflict between the sexes among the Mbuti. Pp. 133–155 in E. Leacock and R. Lee, eds., *Politics and History in Band Societies.* Cambridge Univ. Press, Cambridge.

Turner, A. 1992. Large carnivores and earliest European hominids: Changing determinants of resource availability during the Lower and Middle Pleistocene. *Journal of Human Evolution* 22:109–126.

Turner, A. 1994. Evolution and dispersion of larger mammals in Europe during the time span of *Homo erectus. Courier Forschungsinstitut Senckenberg* 171:241–247.

Turner, B. L., II, ed. 1990. *The Earth as Transformed by Human Action: Global and Regional Changes in the Biosphere over the Past Three Hundred Years.* Cambridge Univ. Press, Cambridge.

Tutin, C. E. G. 1979. Mating patterns and reproductive strategies in a community of wild chimpanzees *(Pan troglodytes schweinfurthii). Behavioral Ecology and Sociobiology* 6:29–38.

Tutin, C. E. G. 1996. Ranging and social structure of lowland gorillas in the Lopé Reserve, Gabon. Pp. 58–70 in W. C. McGrew, L. F. Marchant, and N. Toshisada, eds., *Great Ape Societies.* Cambridge Univ. Press, Cambridge.

Tutt, J. W. 1896. *British Moths.* Routledge, London.

Tuttle, R. H. 1994. Up from electromyography: Primate energetics and the evolution of bipedalism. Pp. 269–312 in R. S. Corruccini and R. L. Ciochon, eds., *Integrative Paths to the Past: Paleoanthropological Advances in Honor of F. Clark Howell.* Prentice Hall, Englewood Cliffs, NJ.

Tuttle, R. H., D. M. Webb, and N. I. Tuttle. 1991. Laetoli footprint trails and the evolution of hominid bipedalism. Pp. 187–198 in Y. Coppens and B. Senut, eds., *Origine(s) de la Bipédie chez les Hominidés.* Éditions du CNRS, Paris.

Tylor, E. B. 1888. On a method of investigating the development of institutions; applied to laws of marriage and descent. *Journal of the Royal Anthropological Institute* 18:245–269.

Tylor, E. B. 1920 (1871). *Primitive Culture: Researches into the Development of Mythology, Philosophy, Religion, Language, Art, and Custom.* 6th ed. Murray, London.

Union of Concerned Scientists. 1993. *World Scientists' Warning to Humanity.* Union of Concerned Scientists, Cambridge, MA.

United Nations. 1991a. *Women: Challenges to the Year 2000.* United Nations Department of Public Education, New York.

United Nations. 1991b. *The World's Women, 1970–1990: Trends and Statistics.* United Nations, New York.

U.S. General Accounting Office. 1989. *Traffic Congestion: Trends, Measures, and Effects.* U.S. Government Printing Office, Washington, DC.

Uyenoyama, M. K. 1979. Evolution of altruism under group selection in large and small populations in fluctuating environments. *Theoretical Population Biology* 17:380–414.

Uyenoyama, M. K., and M. W. Feldman. 1980. Theories of kin and group selection. *Theoretical Population Biology* 17:380–414.

Valladas, H., I. L. Joron, B. Valladas, P. Arensburg, et al. 1987. Thermoluminescence dates for the Neanderthal burial site at Kebara (Mount Carmel), Israel. *Nature* 330:159–160.

Valladas, H., J. L. Reyss, J. L. Joron, G. Valladas, et al. 1988. Thermoluminescence dating of Mousterian "Proto-Cro-Magnon" remains from Israel and the origin of modern man. *Nature* 331:614–616.

Valladas, H., et al. 1992. Direct radiocarbon dates for prehistoric paintings at the Altamira, El Castillo, and Niaux Caves. *Nature* 357:68–70.

Vamosi, S. M., and D. Schluter. 1999. Sexual selection against hybrids between sympatric stickleback species: Evidence from a field experiment. *Evolution* 53:874–879.

Van den Berghe, P. L. 1983. Human inbreeding avoidance: Culture in nature. *Behavioral and Brain Sciences* 6:91–123.

Van den Berghe, P. L., and P. Frost. 1986. Skin color preference, sexual dimorphism, and sexual selection: A case of gene–culture co-evolution? *Ethnic and Racial Studies* 9:87–113.

Van der Leeuw, S. E., and R. Torrence, eds. 1989. *What's New? A Closer Look at the Process of Innovation.* Unwin Hyman, London.

Vandermeersch, B. 1970. Une sépulture moustérienne avec offrandes découverte dans la grotte de Qafzeh. *Comptes Rendus de l'Académie des Sciences de Paris* 270:298–301.

Van Gelder, R. G. 1977. Mammalian hybrids and generic limits. *American Museum Novitates* 2635:1–25.

Van Lawick, H., and J. van Lawick-Goodall. 1971. *Innocent Killers.* Houghton Mifflin, Boston.

Van Lennep, E. 1990. Arctic to Amazonia: An alliance for the earth. *Cultural Survival Quarterly* 14:46–47.

Van Petten, C., and P. Bloom. 1999. Speech boundaries, syntax, and the brain. *Nature Neuroscience* 2:103–104.

Van Tilburg, J. A. 1994. *Easter Island: Archaeology, Ecology, and Culture.* Smithsonian Institution Press, Washington, DC.

Van Zeist, W. 1986. Some aspects of early Neolithic plant husbandry in the Near East. *Anatolica* 15:29–67.

Vayda, A. P. 1961. Expansion and warfare among swidden agriculturalists. *American Anthropologist* 63:346–358.

Veblen, T. 1967 (1899). *The Theory of the Leisure Class.* Penguin Books, New York.

Vermeij, G. J. 1987. *Evolution and Escalation: An Ecological History of Life.* Princeton Univ. Press, Princeton.

Via, S., and A. J. Shaw. 1996. Short-term evolution in the size and shape of pea aphids. *Evolution* 50:163–173.

Vicario, A., L. I. Mazón, A. Aguirre, A. Estomba, et al. 1989. Relationships between environmental factors and morph polymorphism in *Cepaea nemoralis,* using canonical correlation analysis. *Genome* 32:908–912.

Vilà, C., P. Savolainen, J. E. Maldonado, I. R. Amorim, et al. 1997. Multiple and ancient origins of the domestic dog. *Science* 276:1687–1689.

Vines, G. 1993. The hidden cost of sex selection. *New Scientist* (1 May): 12–13.

Vines, G. 1999. Queer creatures. *New Scientist* (7 August): 32–35.

Visalberghi, E. 1993. Capuchin monkeys: A window into tool use in apes and humans. Pp. 138–150 in K. R. Gibson and T. Ingold, eds., *Tools, Language, and Cognition in Human Evolution.* Cambridge Univ. Press, Cambridge.

Visalberghi, E., and L. Limongelli. 1994. Lack of comprehension of cause-effect relations in tool-using capuchin monkeys *(Cebus apella). Journal of Comparative Psychology* 108:15–22.

Vitousek, P. M., J. D. Aber, R. W. Howard, G. E. Likens, et al. 1997. Human alteration of the global nitrogen cycle: Sources and consequences. *Ecological Applications* 7:737–750.

Vitousek, P. M., J. D. Aber, R. Howarth, G. Likens, et al. 1997. *Human Alteration of the Nitrogen Cycle.* Ecological Society of America, Washington, DC.

Vitousek, P. M., C. M. D'Antonio, L. L. Loope, and R. Westbrooks. 1996. Biological invasions as global environmental change. *American Scientist* 84:468–478.

Vitousek, P. M., P. R. Ehrlich, A. H. Ehrlich, and P. A. Matson. 1986. Human appropriation of the products of photosynthesis. *BioScience* 36:368–373.

Vitousek, P. M., H. A. Mooney, J. Lubchenco, and J. M. Melillo. 1997. Human domination of Earth's ecosystems. *Science* 277:494–499.

Vogel, F., and A. G. Motulsky. 1996. *Human Genetics: Problems and Approaches.* 3rd ed. Springer-Verlag, New York.

Von Koenigswald, G. H. R., and F. Weidenreich. 1939. The relationship between *Pithecanthropus* and *Sinanthropus. Nature* 144:926–927.

Vrba, E. S. 1988. Late Pliocene events and hominid evolution. Pp. 405–426 in F. E. Grine, ed., *Evolutionary History of the "Robust" Australopithecines.* Aldine, New York.

Vrba, E. S. 1995a. The fossil record of African antelopes (Mammalia, Bovidae) in relation to human evolution and paleoclimate. Pp. 385–424 in E. S. Vrba, G. H. Denton, T. C. Partridge, and L. H. Burckle, eds., *Paleoclimate and Evolution, with Emphasis on Human Origins.* Yale Univ. Press, New Haven.

Vrba, E. S. 1995b. On the connections between paleoclimate and evolution. Pp. 24–45 in E. S. Vrba, G. H. Denton, T. C. Partridge, and L. H. Burckle, eds., *Paleoclimate and Evolution, with Emphasis on Human Origins.* Yale Univ. Press, New Haven.

Vrba, E. S., G. H. Denton, T. C. Partridge, and L. H. Burckle. 1995. *Paleoclimate and Evolution, with Emphasis on Human Origins.* Yale Univ. Press, New Haven.

Waal, F. de. 1982. *Chimpanzee Politics: Power and Sex among Apes.* Harper and Row, New York.

Waal, F. de. 1989. *Peacemaking among Primates.* Harvard Univ. Press, Cambridge.

Waal, F. de. 1995. Sex as an alternative to aggression in the bonobo. Pp. 37–56 in P. R. Abrahamson and S. D. Pinkerton, eds., *Sexual Nature/Sexual Culture.* Univ. of Chicago Press, Chicago.

Waal, F. de. 1996a. Conflict as negotiation. Pp. 159–172 in W. C. McGrew, L. F. Marchant, and N. Toshisada, eds., *Great Ape Societies*. Cambridge Univ. Press, Cambridge.

Waal, F. de. 1996b. *Good Natured: The Origins of Right and Wrong in Humans and Other Animals*. Harvard Univ. Press, Cambridge.

Waal, F. de. 1997. *Bonobo: The Forgotten Ape*. Univ. of California Press, Berkeley.

Waal, F. de. 1999. Cultural primatology comes of age. *Nature* 399:635–636.

Waal, F. de., and A. van Roosmalen. 1979. Reconciliation and consolation among chimpanzees. *Behavioral Ecology and Sociobiology* 5:55–66.

Wackernagel, M., and W. Rees. 1996. *Our Ecological Footprint: Reducing Human Impact on the Earth*. New Society, Gabriola Island, British Columbia.

Waddington, C. H. 1960. *The Ethical Animal*. Allen and Unwin, London.

Wade, M. J. 1977. An experimental study of group selection. *Evolution* 31:134–153.

Wade, M. J. 1979. The primary characteristics of *Tribolium* populations group selected for increased and decreased population size. *Evolution* 33:749–764.

Wade, M. J. 1985. Soft selection, hard selection, kin selection, and group selection. *American Naturalist* 125:61–73.

Wade, M. J., C. J. Goodnight, and L. Stevens. 1999. Design and interpretation of experimental studies of interdemic selection: A reply to Getty. *American Naturalist* 154:599–603.

Wagner, A. D., D. L. Schacter, M. Rotte, W. Koutstaal, et al. 1998. Building memories: Remembering and forgetting of verbal experiences as predicted by brain activity. *Science* 281:1188–1191.

Waldman, I. D. 1999. Problems in inferring dysgenic trends for intelligence. Pp. 365–376 in U. Neisser, ed., *The Rising Curve: Long-Term Gains in IQ and Related Measures*. American Psychological Association, Washington, DC.

Walker, A. C., R. E. F. Leakey, J. M. Harris, and F. H. Brown. 1986. 2.5 Myr *Australopithecus boisei* from west of Lake Turkana, Kenya. *Nature* 322:517–522.

Wallace, A. F. C. 1956. Revitalization movements. *American Anthropologist* 58:264–281.

Wallace, A. F. C. 1961. The psychic unity of human groups. Pp. 129–163 in B. Kaplan, ed., *Studying Personality Cross-Culturally*. Harper and Row, New York.

Wallace, A. R. 1870a. *Contributions to the Theory of Natural Selection: A Series of Essays*. Macmillan, London.

Wallace, A. R. 1870b. *Darwinism: An Exposition on the Theory of Natural Selection and Some of Its Applications*. Macmillan, London.

Wallace, I., and A. Wallace. 1978. *The Two*. Simon and Schuster, New York.

Wallace, J. W., and R. L. Mansell, eds. 1976. *Biochemical Interaction between Plants and Insects*. Plenum Press, New York.

Wallerstein, I., C. Juma, E. F. Keller, J. Kocka, et al. 1996. *Open the Social Sciences: Report of the Gulbenkian Commission on the Restructuring of the Social Sciences*. Stanford Univ. Press, Stanford.

Walzer, M. 1984. *Spheres of Justice: A Defense of Pluralism and Equality*. Basic Books, New York.

Wang, R.-L., A. Stec, J. Hey, L. Lukens, et al. 1999. The limits of selection during maize domestication. *Nature* 398:236–238.

Wang, X., M. M. Merzenich, K. Sameshima, and W. M. Jenkins. 1995. Remodelling of hand representation in adult cortex determined by timing of tactile stimulation. *Nature* 378:71–75.

Washburn, S. L. 1960. Tools and human evolution. *Scientific American* 203:63–75.

Washburn, S. L., and C. S. Lancaster. 1968. The evolution of hunting. Pp. 293–304 in R. B. Lee and I. DeVore, eds., *Man the Hunter*. Aldine, Chicago.

Waterlow, J. C., D. G. Armstrong, L. Fowden, and R. Riley, eds. 1998. *Feeding a World Population of More Than Eight Billion People: A Challenge to Science*. Oxford Univ. Press, New York.

Watson, J. B. 1970 (1924). *Behaviorism*. Norton, New York.

Watt, W. B. 1977. Adaptation at specific loci. I. Natural selection on phosphoglucose isomerase of *Colias* butterflies: Biochemical and population aspects. *Genetics* 87:177–184.

Watt, W. B., R. C. Cassin, and M. B. Swann. 1983. Adaptation at specific loci. III. Field behavior and survivorship differences among *Colias* PGI genotypes are predictable from *in vitro* biochemistry. *Genetics* 103:725–739.

Watts, D. P. 1996. Comparative socio-ecology of gorillas. Pp. 16–28 in W. C. McGrew, L. F. Marchant, and T. Nishida, eds., *Great Ape Societies*. Cambridge Univ. Press, Cambridge.

Wax, M. L. 1984. Religion as universal: Tribulations of an anthropological enterprise. *Zygon* 19:5–20.

Weatherall, D. J. 1987. Common genetic disorders of the red cell and the "malaria hypothesis." *Annals of Tropical Medicine and Parasitology* 81:539–548.

Weatherall, D. J. 1996. Disorders of the synthesis or function of haemoglobin. Pp. 3500–3520 in D. J. Weatherall, J. G. G. Ledingham, and D. A. Warrell, eds., *Oxford Textbook of Medicine.* 3rd ed. Oxford Univ. Press, Oxford.

Webb, M. C. 1988. The first states: How—or in what sense—did "circumscription" circumscribe? *American Behavioral Scientist* 31:449–458.

Webster, D. 1975. Warfare and the evolution of the state. *American Antiquity* 40:464–470.

Webster, G. S. 1990. Labor control and emergent stratification in prehistoric Europe. *Current Anthropology* 31:337–347.

Weiner, S., Q. Xu, P. Goldberg, J. Liu, et al. 1998. Evidence for the use of fire at Zhoukoudian, China. *Science* 281:251–253.

Weiskrantz, L. 1986. *Blindsight: A Case Study and Implications.* Oxford Univ. Press, Oxford.

Weiskrantz, L. 1990. Outlooks for blindsight: Explicit methodologies for implicit processes. The Ferrier lecture. *Proceedings of the Royal Society of London,* ser. B, 239:247–278.

Weiskrantz, L. 1995. The origins of consciousness. Pp. 239–248 in J. P. Changeux and J. Chavaillon, eds., *Origins of the Human Brain.* Oxford Univ. Press, Oxford.

Weisler, M., and P. V. Kirch. 1996. Interisland and interarchipelago transfer of stone tools in prehistoric Polynesia. *Proceedings of the National Academy of Sciences USA* 93:1381–1385.

Weiss, K. M., R. E. Ferrell, and C. L. Hanis. 1984. A new world syndrome of metabolic disease with a genetic and evolutionary basis. *Yearbook of Physical Anthropology* 27:153–178.

Weiss, S. B., D. D. Murphy, C. Metzler, and P. Ehrlich. 1994. Adult emergence phenology in checkerspot butterflies: The effects of macroclimate, topoclimate, and population history. *Oecologia* 96:261–270.

Weller, A. 1998. Communication through body odor. *Nature* 392:126–127.

Weller, L., A. Weller, and O. Avinir. 1995. Menstrual synchrony: Only in roommates who are close friends? *Physiology and Behavior* 58:883–889.

Wells, H. G. 1946. *A Short History of the World.* Penguin Books, Harmondsworth.

Wells, P. A. 1987. Kin recognition in humans. Pp. 395–415 in D. J. C. Fletcher and C. D. Michener, eds., *Kin Recognition in Animals.* Wiley, New York.

Werner, D. 1979. A cross-cultural perspective on theory and research on male homosexuality. *Journal of Homosexuality* 4:345–362.

Westermarck, E. 1891. *The History of Human Marriage.* Macmillan, London.

Wheeler, P. E. 1984. The evolution of bipedalism and the loss of functional body hair in hominids. *Journal of Human Evolution* 13:91–98.

Wheeler, P. E. 1991. The thermoregulatory advantages of hominid bipedalism in open equatorial environments: The contribution of increased convective heat loss and cutaneous evaporative cooling. *Journal of Human Evolution* 21:107–115.

Wheeler, P. E. 1992. The influence of the loss of functional body hair on the water budgets of early hominids. *Journal of Human Ecology* 23:379–388.

Wheeler, P. E. 1993. The influence of stature and body form on hominid energy and water budgets: A comparison of *Australopithecus* and early *Homo* physiques. *Journal of Human Evolution* 24:13–28.

Whitam, F. L., and R. M. Mathy. 1986. *Male Homosexuality in Four Societies: Brazil, Guatemala, the Philippines, and the United States.* Praeger, New York.

White, F. J. 1996. Comparative socio-ecology of *Pan paniscus.* Pp. 29–41 in W. C. McGrew, L. F. Marchant, and T. Nishida, eds., *Great Ape Societies.* Cambridge Univ. Press, Cambridge.

White, L., Jr. 1967. The historical roots of our ecologic crisis. *Science* 155:1203–1207.

White, L. A. 1959. *The Evolution of Culture: The Development of Civilization to the Fall of Rome.* McGraw-Hill, New York.

White, M. J. D. 1945. *Animal Cytology and Evolution.* Cambridge Univ. Press, Cambridge.

White, R. 1989. Visual thinking in the ice age. *Scientific American* (July): 92–99.

White, T. D. 1995. African omnivores: Global climate change and Plio-Pleistocene hominids and suids. Pp. 369–384 in E. S. Vrba, G. H. Denton, T. C. Partridge, and L. H. Burckle, eds., *Paleoclimate and Evolution, with Emphasis on Human Origins.* Yale Univ. Press, New Haven.

White, T. D., G. Suwa, and B. Asfaw. 1994. *Australopithecus ramidus,* a new species of early hominid from Aramis, Ethiopia. *Nature* 371:306–312.

White, T. D., G. Suwa, and B. Asfaw. 1995. *Australopithecus ramidus,* a new species of early hominid from Aramis, Ethiopia (corrigendum). *Nature* 375:88.

White, T. D., G. Suwa, W. K. Hart, R. C. Walter, et al. 1993. New discoveries of *Australopithecus* at Maka in Ethiopia. *Nature* 366:261–265.

Whitehead, A. N. 1978 (1929). *Process and Reality.* Free Press, New York.

Whitehead, J., ed. 1995. *The Barrack-Room Ballads of Rudyard Kipling.* Centenary ed. Hearthstone, Munslow.

Whiten, A., and R. W. Byrne, eds. 1997. *Machiavellian Intelligence II.* Cambridge Univ. Press, Cambridge.

Whiten, A., J. Goodall, W. C. McGrew, T. Nishida, et al. 1999. Cultures in chimpanzees. *Nature* 399:682–685.

Whitley, B. E., Jr. 1990. The relationship of heterosexuals' attributions for the causes of homosexuality to attitudes toward lesbians and gay men. *Personality and Social Psychology Bulletin* 16:369–377.

Whitman, W. B., D. C. Coleman, and W. J. Wiebe. 1998. Prokaryotes: The unseen majority. *Proceedings of the National Academy of Sciences USA* 95:6578–6583.

Whitten, W. 1999. Pheromones and regulation of ovulation. *Nature* 40:232.

Whorf, B. L. 1941. The relation of habitual thought and behavior to language. Pp. 75–93 in L. Spier, A. I. Hallowell, and S. S. Newman, eds., *Language, Culture, and Personality: Essays in Memory of Edward Sapir.* Sapir Memorial Publication Fund, Menasha, WI.

Whorf, B. L. 1956. *Language, Thought, and Reality: Selected Writings of Benjamin Lee Whorf.* MIT Press, Cambridge.

Whyte, M. K. 1978. Cross-cultural codes dealing with the relative status of women. *Ethnology* 17:211–237.

Wichman, H. A., M. R. Badgett, L. A. Scott, C. M. Boulianne, et al. 1999. Different trajectories of parallel evolution during viral adaptation. *Science* 285:422–424.

Wickelgren, I. 1999. Discovery of "gay gene" questioned. *Science* 284:571.

Wickler, S., and M. Spriggs. 1988. Pleistocene human occupation of the Solomon Islands, Melanesia. *Antiquity* 62:703–706.

Wiesenfeld, S. L. 1967. Sickle-cell trait in human biological and cultural evolution. *Science* 157:1134–1140.

Wilcove, D. S. 1994. Turning conservation goals into tangible results: The case of the spotted owl in old-growth forests. Pp. 313–329 in P. J. Edwards, R. M. May, and N. R. Webb, eds., *Large-Scale Ecology and Conservation Biology.* Blackwell Scientific, Oxford.

Wilford, J. N. 1998. Neanderthal or cretin? A debate over iodine. *New York Times* (1 December).

Wilkinson, G. S. 1984. Reciprocal food sharing in the vampire bat, *Desmodus rotundus. Nature* 308:181–184.

Wilkinson, G. S. 1986. Social grooming in the common vampire bat, *Desmodus rotundus. Animal Behaviour* 34:1880–1889.

Williams, G. C. 1957. Pleiotropy, natural selection, and the evolution of senescence. *Evolution* 11:398–411.

Williams, G. C. 1966. *Adaptation and Natural Selection: A Critique of Some Current Evolutionary Thought.* Princeton Univ. Press, Princeton.

Williams, G. C. 1992. *Natural Selection: Domains, Levels, and Challenges.* Oxford Univ. Press, New York.

Williams, G. C. 1993. Mother Nature is a wicked old witch. Pp. 217–231 in M. H. Nitecki and D. V. Nitecki, eds., *Evolutionary Ethics.* State Univ. of New York Press, Albany.

Williams, J. 1837. *A Narrative of Missionary Enterprises in the South Pacific.* Snow, London.

Williams, M. B. 1973. Falsifiable predictions of evolutionary theory. *Philosophy of Science* 40:518–537.

Williams, N. 1997. Biologists cut reductionist approach down to size. *Science* 277:476–477.

Williams, P. A. 1993. Can beings whose ethics evolved be ethical beings? Pp. 233–239 in M. H. Nitecki and D. V. Nitecki, eds., *Evolutionary Ethics.* State Univ. of New York Press, Albany.

Williams, W. R. 1896. Remarks on the mortality rate from cancer. *Lancet* 2:481–482.

Williamson, S., and R. Nowak. 1998. The truth about women. *New Scientist* (1 August): 34–35.

Wills, C. 1993. Stages versus continuity. *Behavioral and Brain Sciences* 16:773.

Wills, C. 1995. When did Eve live? An evolutionary detective story. *Evolution* 49:593–607.

Wilson, C. 1999. Evolutionary theory and historical fertility change. *Population and Development Review* 25:531–541.

Wilson, C., and P. Airey. 1999. How can a homeostatic perspective enhance demographic transition theory? *Population Studies* 53:117–128.

Wilson, D. S. 1975. A general theory of group selection. *Proceedings of the National Academy of Sciences USA* 72:143–146.

Wilson, D. S. 1983. The group selection controversy: History and current status. *Annual Review of Ecology and Systematics* 14:159–187.

Wilson, D. S., and E. Sober. 1994. Reintroducing group selection to the human behavioral sciences. *Behavioral and Brain Sciences* 17:585–654.

Wilson, E. O. 1975. *Sociobiology: The New Synthesis.* Harvard Univ. Press, Cambridge.

Wilson, E. O. 1976. The central problems of sociobiology. Pp. 205–217 in R. M. May, ed., *Theoretical Ecology: Principles and Applications.* Saunders, Philadelphia.

Wilson, E. O. 1978. *On Human Nature.* Harvard Univ. Press, Cambridge.

Wilson, E. O. 1984. *Biophilia.* Harvard Univ. Press, Cambridge.

Wilson, E. O., ed. 1988. *Biodiversity.* National Academy Press, Washington, DC.

Wilson, E. O. 1992. *The Diversity of Life.* Harvard Univ. Press, Cambridge.

Wilson, E. O. 1994. *Naturalist.* Island Press, Washington, DC.

Wilson, E. O. 1998. *Consilience: The Unity of Knowledge.* Knopf, New York.

Wilson, E. O., and W. L. Brown. 1953. The subspecies concept and its taxonomic application. *Systematic Zoology* 2:97–111.

Wilson, I. F. 1995. Application of ecological genetics techniques to test for selection by habitat on allozymes in *Cepaea nemoralis. Heredity* 77:324–335.

Wilson, M., and M. Daly. 1985. Competitiveness, risk taking, and violence: The young male syndrome. *Ethology and Sociobiology* 6:59–73.

Wind, J. 1983. Primate evolution and the emergence of speech. Pp. 15–35 in E. De Grolier, ed., *Glossogenetics: The Origins and Evolution of Language.* Harwood Academic, Paris.

Wind, J., B. Chiarelli, B. Bichakjian, A. Nocentini, et al., eds. 1992. *Language Origin: A Multidisciplinary Approach.* Kluwer, Dordrecht.

Winterhalder, B., and F. Lu. 1997. A forager-resource population ecology model and implications for indigenous conservation. *Conservation Biology* 11:1354–1364.

Witkowski, S. R., and C. H. Brown. 1982. Whorf and universals of color nomenclature. *Journal of Anthropological Research* 38:411–420.

Witte, F., T. Goldschmidt, J. H. Wanink, M. J. P. van Oijen, et al. 1992. The destruction of an endemic species flock: Quantitative data on the decline of the haplochromine species from the Mwanza Gulf of Lake Victoria. *Environmental Biology* 34:1–28.

Wittfogel, K. 1957. *Oriental Despotism.* Yale Univ. Press, New Haven.

Wittgenstein, L. 1921. *Tractatus Logico-Philosophicus.* Kegan Paul, London.

Wittgenstein, L. 1953. *Philosophical Investigations.* Basil Blackwell, Oxford.

WoldeGabriel, G., T. D. White, G. Suwa, and P. Renne. 1994. Ecological and temporal placement of early Pliocene hominids at Aramis, Ethiopia. *Nature* 371:330–333.

Wolf, A. P. 1970. Childhood association and sexual attraction: A further test of the Westermarck hypothesis. *American Anthropologist* 72:503–515.

Wolf, A. P. 1995. *Sexual Attraction and Childhood Association: A Chinese Brief for Edward Westermarck.* Stanford Univ. Press, Stanford.

Wolpoff, M. H. 1984. Modern *Homo sapiens* origins: A general theory of hominid evolution involving the fossil evidence from East Asia. Pp. 411–483 in F. H. Smith and F. Spencer, eds., *The Origins of Modern Humans: A World Survey of the Fossil Evidence.* Liss, New York.

Wolpoff, M. H. 1989. Multiregional evolution: The fossil alternative to Eden. Pp. 62–108 in P. Mellars and C. B. Stringer, eds., *The Human Revolution: Behavioural and Biological Perspectives on the Origins of Modern Humans.* Vol. 1. Edinburgh Univ. Press, Edinburgh.

Wolpoff, M. H. 1999. The systematics of *Homo. Science* 284:1774–1775.

Wolpoff, M. H., and R. Caspari. 1997. *Race and Human Evolution.* Simon and Schuster, New York.

Wood, B. A. 1991. *Koobi Fora Research Project.* Vol. 4, *The Hominid Cranial Remains.* Clarendon Press, Oxford.

Wood, B. A. 1992. Origin and evolution of the genus *Homo. Nature* 355:783–790.

Wood, B. A. 1994. The oldest hominid yet. *Nature* 371:280–281.

Wood, B. A. 1996. Origin and evolution of the genus *Homo.* Pp. 105–114 in W. E. Meikle, F. C. Howell, and N. G. Jablonski, eds., *Contemporary Issues in Human Evolution.* California Academy of Sciences, San Francisco.

Wood, B. A. 1997. Ecce *Homo*—behold mankind. *Nature* 390:120–121.

Wood, B. A., and M. Collard. 1999. The human genus. *Science* 284:65–71.

Wood, J. W., G. R. Milner, H. C. Harpending, and K. M. Weiss. 1992. The osteological paradox. *Current Anthropology* 33:343–370.

Woodburn, J. 1982. Egalitarian societies. *Man* 17:431–451.

Woolfenden, G. E., and J. W. Fitzpatrick. 1984. *The Florida Scrub Jay: Demography of a Cooperative-Breeding Bird.* Princeton Univ. Press, Princeton.

World Bank. 1991. *Gender and Poverty in India.* World Bank, Washington, DC.

World Council of Churches. 1980. *Faith and Science in an Unjust World.* Vols. 1, 2. World Council of Churches, Geneva.

Worster, D. 1994. *Nature's Economy: A History of Ecological Ideas.* 2nd ed. Cambridge Univ. Press, New York.

Worster, D. 1995. Nature and the disorder of history. Pp. 65–85 in M. E. Soulé and G. Lease, eds., *Reinventing Nature? Responses to Postmodern Deconstruction.* Island Press, Washington, DC.

Wrangham, R. W. 1979. On the evolution of ape social systems. *Social Science Information* 18:335–368.

Wrangham, R. W. 1980. Bipedal locomotion as a feeding adaptation in gelada baboons, and its implications for hominid evolution. *Journal of Human Evolution* 9:329–331.

Wrangham, R. W. 1987. The significance of African apes for reconstructing human social evolution. Pp. 51–71 in W. G. Kinzey, ed., *The Evolution of Human Behavior: Primate Models.* State Univ. of New York Press, Albany.

Wrangham, R. W., J. H. Jones, G. Laden, D. Pilbeam, et al. 1997. The raw and the stolen: Cooking and the ecology of human origins. *Current Anthropology* 40:567–594.

Wrangham, R. W., and D. Peterson. 1996. *Demonic Males: Apes and the Origins of Human Violence.* Houghton Mifflin, New York.

Wrangham, R. W., F. de Waal, and W. C. McGrew. 1996. The challenge of behavioral diversity. Pp. 1–18 in W. C. McGrew, L. F. Marchant, and N. Toshisada, eds., *Chimpanzee Cultures.* Harvard Univ. Press, Cambridge.

Wright, G. 1997. Decline and fall. *New Scientist* (4 October): 33–36.

Wright, R. 1994. *The Moral Animal: The New Science of Evolutionary Psychology.* Random House, New York.

Wright, R. V. S. 1972. Imitative learning of a flaked tool technology: The case of an orangutan. *Mankind* 8:296–306.

Wright, S. 1931. Evolution in Mendelian populations. *Genetics* 16:97–159.

Wright, S. 1968–1978. *Evolution and the Genetics of Populations.* 4 vols. Univ. of Chicago Press, Chicago.

Wright, S. 1969. *Evolution and the Genetics of Populations.* Vol. 2, *The Theory of Gene Frequencies.* Univ. of Chicago Press, Chicago.

Wrigley, E. A. 1978. Fertility strategy for the individual and the group. Pp. 135–154 in C. Tilly, ed., *Historical Studies of Changing Fertility.* Princeton Univ. Press, Princeton.

Wuethrich, B. 1998. Why sex? Putting theory to the test. *Science* 281:1980–1982.

Wuethrich, B. 1999. Mexican pairs show geography's role. *Science* 285:1190.

Wuthnow, R. 1988. *The Restructuring of American Religion: Society and Faith since World War II.* Princeton Univ. Press, Princeton.

Wynn, T. 1991. Archaeological evidence for modern intelligence. Pp. 52–66 in R. A. Foley, ed., *The Origins of Human Behavior.* Unwin Hyman, London.

Wynn, T. 1993. Two developments in the mind of early *Homo. Journal of Anthropological Archaeology* 12:299–322.

Wynne-Edwards, V. C. 1962. *Animal Dispersion in Relation to Social Behavior.* Oliver and Boyd, Edinburgh.

Yamei, H., R. Potts, Y. Baoyin, G. Zhengtang, et al. 2000. Mid-Pleistocene Acheulean-like stone technology of the Bose Basin, South China. *Science* 287:1622–1626.

Yanofsky, C. 1967. Gene structure and protein structure. *Scientific American* (May): 80–94.

Yanofsky, C., B. C. Carlton, J. R. Guest, D. R. Helinski, et al. 1964. On the colinearity of gene structure and protein structure. *Proceedings of the National Academy of Sciences USA* 51:266–272.

Yellen, J. E., A. S. Brooks, E. Cornelissen, M. J. Mehlman, et al. 1995. A Middle Stone Age worked bone industry from Katanda, Upper Semliki Valley, Zaire. *Science* 268:553–556.

Yesner, D. R. 1980. Maritime hunter-gatherers: Ecology and prehistory. *Current Anthropology* 21:727–750.

Yoffee, N. 1995. Political economy in early Mesopotamian states. *Annual Review of Anthropology* 24:281–311.

Yoffee, N., and G. L. Cowgill, eds. 1988. *The Collapse of Ancient States and Civilizations.* Univ. of Arizona Press, Tucson.

Yoo, B. H. 1980. Long-term selection for a quantitative character in large replicate populations of *Drosophila melanogaster.* I. Response to selection. *Genetical Research* 35:1–17.

Yu, D. W., and G. H. Shepard Jr. 1998. Is beauty in the eye of the beholder? *Nature* 396:321–322.

Zajonc, R. B. 1984. On the primacy of affect. *American Psychologist* 39:117–123.

Zhou, J.-N., M. A. Hofman, L. J. G. Gooren, and D. F. Swaab. 1995. A sex difference in the human brain and its relation to transsexuality. *Nature* 378:68–70.

Zihlman, A. 1996. Reconstructions reconsidered: Chimpanzee models and human evolution. Pp. 293–304 in W. C. McGrew, L. F. Marchant, and T. Nishida, eds., *Great Ape Societies.* Cambridge Univ. Press, Cambridge.

Zillmann, D. 1984. *Connections between Sex and Aggression.* Erlbaum, Hillsdale, NJ.

Zimbardo, P. G., and R. J. Gerrig. 1999. *Psychology and Life.* Longman, New York.

Zimmerman, M. E. 1986. Implications of Heidegger's thought for deep ecology. *Modern Schoolman* 64:19–43.

Zimmet, P. 1991. The epidemiology of diabetes mellitus and associated disorders. *Diabetes Annual* 6:1–19.

Zischler, H., H. Geisert, A. von Haeseler, and S. Pääbo. 1995. A nuclear "fossil" of the mitochondrial D-loop and the origin of modern humans. *Nature* 378:489–492.

Zohary, D., and M. Hopf. 1988. *Domestication of Plants in the Old World.* Oxford Univ. Press, Oxford.

Zoli, M., C. Torri, R. Ferrari, A. Jansson, et al. 1998. The emergence of the volume transmission concept. *Brain Research Reviews* 26:136–147.

Zubrow, E. 1989. The demographic modelling of Neanderthal extinction. Pp. 212–231 in P. Mellars and C. Stringer, eds., *The Human Revolution: Behavioural and Biological Perspectives on the Origins of Modern Humans.* Princeton Univ. Press, Princeton.

ACKNOWLEDGMENTS

The following people were kind enough to read and comment on part or all of the manuscript: David Ackerly, Aviv Bergman, Carol Boggs, Brendan Bohannan, Gretchen Daily, Anne Ehrlich, Marcus Feldman, Elizabeth Hadly, Jessica Hellmann, Aaron Hirsh, Jennifer Hughes, Susan McConnell, Taylor Ricketts, Joan Roughgarden, Robert Sapolsky, Stephen Schneider, and Charles Yanofsky, all of the Department of Biological Sciences, Stanford University; John Allman, Division of Biology, California Institute of Technology; Virginia Barber, Virginia Barber Agency; Derek Bickerton, Department of Linguistics, University of Hawaii; D. L. Bilderback, Department of History, California State University, Fresno; Iris Brest, attorney, Stanford, California; Ellyn Bush, Neurology Service, Department of Veteran's Affairs, Medical Center, Palo Alto, California; Jared Diamond, Department of Physiology, University of California, Los Angeles; Walter Falcon, Donald Kennedy, and Rosamond Naylor, Institute for International Studies, Stanford University; Sylvia Fallon, Department of Biology, University of Missouri, St. Louis; Corey S. Goodman, Department of Molecular and Cell Biology, University of California, Berkeley; Geoffrey Heal, Graduate School of Business, Columbia University; Thomas Heller, Law School, Stanford University; Ann Holdren, anthropologist, Monterey, California; Jean-Jacques Hublin, CNRS IRESCO, Paris; Nina Jablonski, Department of Anthropology, California Academy of Sciences; Allen W. Johnson, Department of Anthropology, University of California, Los Angeles; Leslie Kaufman, Department of Biology, Boston University; Patrick Kirch, Department of Anthropology, University of California, Berkeley; Richard Klein, Department of Anthropological Sciences, Stanford University; Simon Levin, Department of Biology, Princeton University; Charles D. Michener, Department of Entomology, University of Kansas; Robert Ornstein, Institute for the Study of Human Knowledge; Sarah P. Otto, Department of Zoology, University of British Columbia; Merritt Ruhlen, linguist, Palo Alto, California; Jim Salzman, professor of law, American University; Robert R. Sokal, Department of Ecology and Evolution, State University of New York, Stony Brook; Donald Symons, Department of Anthropology, University of California, Santa Barbara; and Wren Wirth, Winslow Foundation. Their help has been invaluable, and many of their suggestions have been incorporated into the book.

I am especially indebted to Loy Bilderback, Gretchen Daily, Marc Feldman, Annie Holdren, and Richard Klein, all of whom struggled with the manuscript more than once and often provided much-improved wording and ideas, and to Anne Ehrlich, who, besides carefully criticizing and editing the manuscript numerous times, helped me to organize the literature.

Discussions with many colleagues have contributed to this book. I can't begin to mention them all, but a few stand out especially in my memory. The late Richard Holm and I worked closely together on evolutionary issues for more than three decades, and my views on these issues have been much influenced by long associations with Marcus Feldman, Charles Michener, Peter Raven, Joan Roughgarden, Robert Sokal, and Charles Yanofsky and by conversations with Ernst Mayr and the late Theodosius Dobzhansky (who was kind enough to read and criticize the first edition of the evolution textbook that Dick Holm and I wrote long ago). I've learned from each person on that list and have had disagreements with every one of them—I've found the ability to differ and remain friends one of the pleasant hallmarks of the scientific enterprise. Bob Ornstein, Sally Mallam, Anne, and I have met regularly for years to talk about issues in psychology and at the interface of science and society. Terry Root and Steve Schneider have joined us in places as strange as Sulawesi, Bhutan, and Chez Sophie to hash over issues related to this book, and how scientists should deal with them, more times than I can remember. Charles Birch started debating philosophical issues with me when I spent my first sabbatical with him at the University of Sydney in 1965–1966. We did so most recently in October 1999; Charles is always fun to argue with. Loy Bilderback and I have enjoyed discussing history and biology since we roomed together at the University of Kansas in 1953. Friends at Stanford University's Center for Conservation Biology and Department of Biological Sciences and at the Beijer International Institute of Ecological Economics of the Royal Swedish Academy of Sciences have often helped me with ideas and critiques. Finally, the immediate impetus for writing this book was co-teaching a course in human evolution and ecology in Stanford's Science, Mathematics, and Engineering Program with Gretchen Daily; she encouraged me to undertake the project and gave me critical feedback on my lectures (in addition to her work on the manuscript).

Colleen Mitchell Bilderback of the Henry Madden Library, California State University, Fresno, used her great skill to run down some very obscure bits of information for me. Other people who helped by supplying information or suggestions include John Brauman, Department of Chemistry, Stanford University; Robert Foley, University of Cambridge; Ellen Gruenbaum, professor of anthropology, California State University, Fresno; Philip Hanawalt, Department of Biological Sciences, Stanford University; Victor Hanson, professor of classics and humanities, California State University, Fresno; John Harte, Energy and Resources Group, University of California, Berkeley; J. P. Myers, Alton Jones Foundation; Michael Price, Department of Anthropology, University of California, Santa Barbara; Barton H. Thompson Jr., Law School, Stanford University; Bruce S. Thornton, professor of foreign languages and humanities, California State University, Fresno; and Gideon Yoffe, Novalux, Inc.

Needless to say, I would be delighted to have any error of fact or interpretation blamed on any one of the many colleagues just mentioned, but in the end I suppose I'll have to take the responsibility.

Scott Daily, Sylvia Fallon, Wendy Fox, Margie Mayfield, and Margaret Vas Dias of the Center for Conservation Biology, Stanford University, were of enormous aid in tracking down references, carting books from the library and back, and generally making themselves helpful with promptness, accuracy, and cheer. The staff of the Falconer Biology Library, especially Jill Otto, once again was extremely helpful in dealing with the literature, and, as always, Pat Browne and Steve Masley did reams of photocopying with dispatch and without complaint. Julia Harte was a big help in reworking the index.

I am extremely grateful for the good advice, detailed, on-the-mark comments, and fine suggestions I received from my editor at Island Press, Jonathan Cobb. His hard work, broad knowledge, keen appreciation of the issues, great patience, and steady good humor made interacting with him on the manuscript a genuine pleasure. Pat Harris did a wonderful job of copyediting, working with enormous skill and diligence, often far into the night. I will miss post-midnight e-mail exchanges with both of my editors—certainly more than they will.

Working with Island Press once again has proven to me that there are highly competent publishing houses.

I also want to acknowledge the late Tom Freeman of the Canadian Department of Agriculture, who gave me the opportunity in 1951 and 1952 to work on the Department/Defense Research Board Northern Insect Survey. In 1952, I was on Southampton Island with the Inuit in what for me was a life-transforming experience. That's where this book really began—I can remember hunting walrus and dog-sledding across the sea ice as if it were yesterday instead of almost a half century ago.

Once again, I acknowledge the great debt I owe to my departed friend LuEsther Mertz, who made so much possible and, through the wonderful support of the LuEsther T. Mertz Charitable Trust, still makes so much possible. LuEsther was a wonderful example of human nature at its very best.

About the Author

Paul R. Ehrlich is Bing Professor of Population Studies, Department of Biological Sciences, Stanford University. An expert in the fields of evolution, ecology, taxonomy, and population biology, Ehrlich has devoted his career to studying such topics as human ecology, the dynamics and genetics of insect populations, the coevolution of plants and herbivores, and the effects of crowding on human beings. His fieldwork has carried him to all continents, from the Arctic and the Antarctic to the tropics, and from high mountains to the ocean floor. Professor Ehrlich has written more than 800 scientific papers and popular articles as well as many books (some coauthored with his wife, Anne), including *The Population Bomb*, *The Process of Evolution*, *The Machinery of Nature*, *Extinction*, *Science of Ecology*, *The Birder's Handbook*, *New World/New Mind*, *The Population Explosion*, *Healing the Planet*, *Birds in Jeopardy*, *The Stork and the Plow*, *A World of Wounds*, and *Betrayal of Science and Reason*.

Among his many scientific honors, Ehrlich is a Fellow of the American Association for the Advancement of Science and the American Academy of Arts and Sciences, an honorary member of the British Ecological Society, and a member of the United States National Academy of Sciences and the American Philosophical Society. He was awarded the first AAAS/Scientific American Prize for Science in the Service of Humanity, and the Crafoord Prize in Population Biology and the Conservation of Biological Diversity, an explicit substitute for a Nobel Prize in fields of science where the latter is not given. Ehrlich has also received a MacArthur Prize Fellowship, the Volvo Environment Prize, the International Center for Tropical Ecology's World Ecology Medal, the International Ecology Institute Prize, the Dr. A. H. Heineken Prize for Environmental Science, the Blue Planet Prize, and the First Peterson Memorial Medal.

Together with his wife, Paul Ehrlich has also received the United Nations Environment Programme (UNEP) Sasakawa Environment Prize, the Heinz Award for the Environment, the American Humanists Association Distinguished Service Award, the Nuclear Age Peace Foundation Distinguished Peace Leadership Award, and the Tyler Prize for Environmental Achievement.

About the Center for Conservation Biology

In 1984, Paul R. Ehrlich founded Stanford University's Center for Conservation Biology to develop the science of conservation biology and to help devise ways and means for protecting Earth's life-support systems.

In pursuit of its mission, the Center for Conservation Biology designs experiments to address specific and general questions in conservation biology; conducts research on broad-scale policy issues, including human population growth, resource use, and environmental deterioration; communicates the results of this scientific and policy research to conservation biologists, reserve managers, planners, nongovernmental organizations, decision makers, and the public; and educates students and professionals as well as fostering collaboration with other scientists and conservation groups around the world.

The Center for Conservation Biology is part of the Stanford University Department of Biological Sciences and is supported by donations and grants from individuals, private foundations, and corporations.

INDEX

Definitions and discussions of technical terms are found on pages indicated by boldface.